XXIInd National Conference on Thermodynamics with International Participation (NACOT2019)

IOP Conference Series: Materials Science and Engineering
Volume 595

Galati, Romania
23 – 24 May 2019

ISBN: 978-1-5108-9816-5
ISSN: 1757-8981

Printed from e-media with permission by:

Curran Associates, Inc.
57 Morehouse Lane
Red Hook, NY 12571

Some format issues inherent in the e-media version may also appear in this print version.

Additional copies of this publication are available from:

Curran Associates, Inc.
57 Morehouse Lane
Red Hook, NY 12571 USA
Phone: 845-758-0400
Fax: 845-758-2633
Email: curran@proceedings.com
Web: www.proceedings.com

Table of contents

Volume 595

The XXIInd National Conference on Thermodynamics with International Participation

23–24 May 2019, Galati, Romania

Accepted papers received: 11 July 2019
Published online: 20 September 2019

Preface

Papers

Applied thermodynamics, Heat and mass transfer, Fluid dynamics and CFD

Refrigeration and heat pumps, Power plants and CHP, Internal combustion engines

IOP Conf. Series: Materials Science and Engineering **595** (2019) 011001 doi:10.1088/1757-899X/595/1/011001

Preface

The National Conference on Thermodynamics with International Participation (NACOT2019) held in Galaţi is the 22nd meeting of a series of conferences initiated in 1990 by the Romanian Society of Thermotechnics (SRT).

NACOT2019 was organized by the Department of Thermal Systems and Environmental Engineering, Faculty of Engineering from "Dunărea de Jos" University of Galati in collaboration with the Romanian Association of Frigotehnists and Cryogenics (AFCR).

More than 80 academicians, researchers, industry professionals, experts in thermodynamics from 14 countries participated and presented papers at the event. The conference topics included: applied thermodynamics, heat and mass transfer, fluid mechanics and CFD, refrigeration and heat pumps, power plants and CHP, internal combustion engines, renewable energy, energy use and storage, environmental impact of energy conversion.

The conference offered participants the opportunity to present new research, discuss the results, exchange ideas and establish links between different research directions in the field.

The format of the conference was based on plenary presentations, oral presentations, poster presentations, and presentations of equipment and technologies.

The NACOT2019 organizing committee thanks the reviewers, plenary speakers and sponsors.

NACOT2019 organizing committee

IOP Conf. Series: Materials Science and Engineering **595** (2019) 011002 doi:10.1088/1757-899X/595/1/011002

Peer review statement

All papers published in this volume of *IOP Conference Series: Materials Science and Engineering* have been peer reviewed through processes administered by the proceedings Editors. Reviews were conducted by expert referees to the professional and scientific standards expected of a proceedings journal published by IOP Publishing.

Application areas and calculation method for capillary-porous heat exchangers of the new heat-removing class

A Genbach[1], N Jamankulova[1] and I Iliev[2]

[1]Almaty University of Power Engineering & Telecommunications, Almaty, Kazakhstan
[2]"Angel Kanchev" University of Ruse, Ruse, Bulgaria

E-mail: dnellya@mail.ru

Abstract. The paper is centered on research focused on the fundamental studies of capillary-porous systems working in the mass forces field. Examined are the effects of pressure, excess fluid, the methods for heating and for supplying coolant, as well as the type of material used, in addition to the intensifier type amongst others, on the integral and thermo-hydraulic characteristics of the heat transfer process. The information obtained from the research can be used in the design, engineering calculations and operation of different energy installations of power plants. Experimentally obtained approximation for the law of the growth of a bubble in a cell of a porous structure, taking into account the influence of underheating, velocity and thermal properties of liquid and a heating surface, was used for determination of the radius of a "dry" spot. The heat transfer control permits to form a new class in the system for heat removing. Devices for heat exchange were built to raise their efficiency, reliability and to be loyal to environment. Capillary-porous heat exchanger shaped as a box provides safety from an explosion. The engineering method for calculation is done for limiting and operational characteristics of capillary-porous heat exchanger. The heat transfer control permits the development of a new class in the system for heat removal. Devices for heat exchange were built with the purpose of increasing both efficiency and dependability in an environmentally conscious manner. A capillary-porous heat exchanger shaped as a box provides protection against an explosion. Obtaining Its limiting and operational characteristics is done by means of an engineering method for calculation.

1. Introduction

The development and study of capillary-porous cooling systems acting under both mass and capillary forces, which create underheating and flow rate in the cross section and on the surface of the structure, made it possible to expand the removal of heat fluxes, increasing the heat transfer rate [1-3] and intensify heat transfer processes [4-8]. Intensification of heat exchange processes [4] is achieved by controlling the heat wave energies, the characteristics of boiling [5] and the heat transfer parameters [6]. Experimental data were summarized by means of the theory of similarity, modelling and analogy of heat exchange phenomena [7], which allowed receiving the calculated dependences on definition of heat flows and temperature pressures depending on regime and design parameters of structure and a heat exchange surface [8-14].

The most important factor in the intensification of energy production is the strengthening of the economy with regard to environmental requirements. It is necessary to satisfy the growing needs for fuel, energy and raw materials mainly due to economy of material and raw resources, improvement of using of secondary resources. It can be promoted by porous systems.

Energy savings in the industry are particular importance now, since primary energy resources will be depleted in the foreseeable future if they are used in the same way as at the present stage. The solution for the problem of pollution of the biosphere can't be postponed anymore and this problem should be solved not locally, but on a global scale. A considerable number of equipment in thermal power installations needs the utilization of heat recovery systems. Even a small list of equipment and processes reveals significant reserves of energy saving, but requires the development of new technical solutions.

2. Application areas of capillary-porous heat exchangers

Porous systems made in the form of heat pipes are being introduced in power plants. They allow to take away and transport the high temperature heat fluxes, with high intensity and reliability, solving a number of the environmental problems which are put forward as a result of anthropogenous impact on surrounding environment: to promote economy of natural resources, water, oxygen, to reduce the amount of harmful emissions.

Porous cooling has proven effective for gas turbine blades. Air was introduced into the internal cavities of the blade and forced through the porous wall. Compared to convective or convective-film cooling, the air flow was reduced, but it was necessary to thoroughly clean the air of dust and prevent pores from clogging with particles contained in combustion.

The heat transferring device found practical application in electrical machines. Rotor cooling systems by means of a centrifugal thermosiphon and the stator cooling by heat pipes and vaporizing contours were considered. Heat pipes create areas with the uniform temperature field in radio-electronic devices, thermally stabilize the heating surfaces, and structurally solve the heat removal problem by moving their surface out of limits of blocks and knots.

The heat exchanger with ridge heat pipes is used in utilization of heat ventilating emissions. The wall of a pipe can be considered as additional thermal resistance; however, efficiency of a heat transfer in the pipe is so considerable that about 70% of waste heat was utilized in the device. The main advantage of heat exchangers is their reliability in comparison with the rotating regenerator and the device with the intermediate heat carrier. Failure of several pipes only slightly reduces efficiency of work; there are no rotating parts and the related noise and vibration.

The field of application of porous mesh materials was cooling systems, dust-gas cleaning and foam generation devices, filtering elements for cleaning station wastewater with a rather wide range of filter material cleaning fineness – from 5 to 100×100^{-6} m, mixers, fire prevention devices, aerators, noise mufflers, chemical towers, evaporators, distillation columns.

The mesh materials from knitted grids possessing high porosity were applied in sound-absorbing designs of power stations. Acoustic efficiency of the noise- absorbing panels installed with a clearance of 10×10^{-3} m in the range of frequencies (200 … 10000) Hz, was (5 … 10) dB. The development of the proposed porous systems operating in the field of mass forces in the future should be aimed at creating compact, reliable and explosion-proof devices for cooling heat-loaded boilers and gas turbines, utilization heat exchangers that can drastically reduce low-temperature corrosion of heating surfaces and utilize the heat of exhaust gases of boiler rooms, metallurgical furnaces compressor station. Porous systems have to increase reliability, efficiency and manoeuvrability of work of the heat-loaded elements of boilers, turbines, recovery boilers, burners, drills, to reduce concentration of nitrogen oxides in furnace chambers and to improve the ecological environmental conditions. Solving the problems of preventing water and soil pollution from oil and oil products, utilizing low-grade heat in cooling towers, cooling ponds, for heating buildings, structures, greenhouses and air conditioning are required. New technical devices of porous filters for gas cleaning are required. Porous systems can be useful in solving the food program and a number of tasks to ensure electrical and fire safety when heating studs and flanges of turbines and production and domestic premises [2-4, 6-9].

Approximate values of heat exchange coefficients α, W/m^{2}K and heat transfer coefficients k, W/m^{2}K in heat-exchange devices of power plants can be presented in a general view in table 1.

Table 1. Approximate values of heat exchange and heat transfer coefficients.

No	Heat exchange process	Heat exchange coefficient α, Wm^{-2}K^{-1}
1	When heating and cooling air	α = 1 … 50
2	When heating and cooling superheated steam	α = 20 … 100
3	When heating and cooling oils	α = 50 … 1500
4	When heating and cooling water	α = 200 … 10000
5	When boiling	α = 500 … 45000
6	At film condensation of water vapour	α = 4000 … 15000
7	At drop condensation of water vapour	α = 40000 … 120000
	Heat transfer process	Heat transfer coefficients *k*, Wm^{-2}K^{-1}
8	At a heat transfer from gas to gas	25
9	At a heat transfer from gas to water	50
10	At a heat transfer from water to water	1000
11	At a heat transfer from the condensing vapours to water	3500

To increase of reliability and efficiency of stationary heat exchangers taking into account ecology we developed the following devices [2, 4, 8, 9]:

1) Condensers of turbines on porous structures;
2) Intensifiers of deaeration in condensate collectors;
3) Recyclers of waste heat through the use of the "triad": heat pipes, vortex tubes, heat pumps, the establishment of power plants without cooling towers and without chimneys;
4) Cooling tower with porous elements;
5) The porous inserts to prevent cavitation in the condensate and feed pumps;
6) Heat exchange intensifiers in heaters with porous elements;
7) The porous structures in deaerators increasing efficiency of decontamination;
8) Porous structures for increase of efficiency of separation of steam and heat exchange of superheaters;
9) Oil coolers on heat pipes, excluding the contact of oil with the water and vice versa;
10) Porous oil coolers;
11) Bubblers in porous deaerators;
12) Porous evaporators;
13) Heat pipes in greenhouse facility (management of blossoming phenophases, porous watering, storage of fruits);
14) Wavy porous two-phase heat exchangers;
15) Porous network heaters;
16) Porous heaters;
17) Porous heat exchangers on Coanda effect;
18) Porous heat exchangers in the form of a penetrator;
19) The porous heat exchangers using effect of separation, concentration, transport, drain and energy control of a wave and gases;
20) Heat exchangers on the basis of the operated flexible porous structures.

3. Capillary-porous heat exchangers

We will consider as an illustration the use of capillary and porous systems in the cooling of melting units for the purpose of their explosion safety.

The design of caissons (Figure 1) represents a box-shaped form. They consist of a housing 1 and cover 2 which allow for detachment and are tightly closed around the perimeter using bolts 3. The interior side of the caisson's 4 is covered by the capillary-porous structure 5. Perforated plates 6 are intended to press this structure to the wall's internal surface. The top ends of the structure are connected with feed liquid channels 7. Cooling liquid is supplied to the cooled surface through the upper end of the structure by mass and capillary forces. The lower ends of the structure are generally loose and submerged in tray 8 where the liquid accumulates due to leakage, dripping or excess. On the surfaces of the plate, which are smoothed with holes 9, there is provided exit steam of structure in groove 10, as well as traps which are formed by the structure of the glass and the outer surface of the outer surface. The liquid channel is connected to a branch pipe 11, with the conducting tubes 12 and the collector 13. The excess refrigerant accumulates at the caisson's base, where it is removed using siphon 14 to the lower collector 15, which circulates it back to the supply system.

To facilitate construction and to maintain sufficient rigidity, the spacers 16 fitted in the caissons are composed of either reinforcing ribs or perforated plates.

The ribs may be located either internally or externally of the caisson's housing and its cover. On the cover, at its upper part, the junction tubes 17 are welded with flanges for the connection to the steam pipe. The structure may be enlarged vertically or horizontally, with the upper or lower ends (or both) of the structure connected to feed liquid channels. The perforated plates are made in shape and size commensurate with the structure. Stamped perforated grooves in the plates can take the shape of a truncated cone, or longitudinal grooves with upwards facing holes.

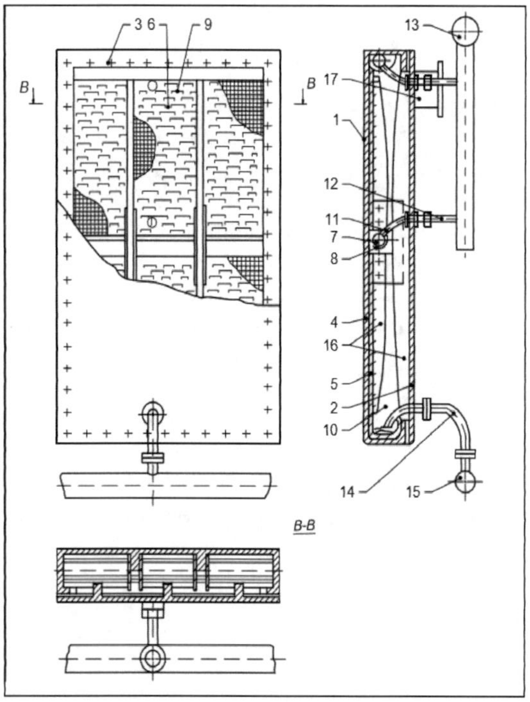

Figure. 1 The caisson of capillary-porous system with internal ribbing: 1 – housing; 2 – cover; 3 – bolt; 4 – wall; 5 - capillary-porous structure; 6 – plate; 7 – feed liquid channel; 8 – tank; 9 – hole; 10 – channel; 11, 17 – branch pipe; 12 – tube; 13, 15 – collector; 14 – siphon; 16 – internal ribbing.

4. Calculation method

Engineering calculation of operating and limit characteristics of the capillary-porous cooling system can be executed with use of nomogram's method.

For creation of characteristics the geometrical parameters of the cooling system and the porous structure are previously determined.

Calculation is made on the basis of equation for the thermal power:

$$Q = \frac{T_W^E - T_W^C}{\dfrac{1}{\alpha_E F_E} + \dfrac{1}{\alpha_C F_C}}, W \tag{1}$$

The coefficient of heat exchange of the evaporator α_E is determined by the criteria equation [7] received by us, or by the calculated dependences [3, 6, 8].

Assuming the wall temperature of the condenser T_W^C = constant for a number of steam temperature values T_S, the required physical parameters of the liquid in the condenser are defined and α_C = f (T_S) is plotted. From the equation:

$$Q = \alpha_C F_C (T_S - T_L^C), \text{ W} \tag{2}$$

where $T_L^C = 0.5(T_S + T_W^L)$, for the number of values T_S the corresponding values of Q are defined.

Assuming the several values for T_W^E, the parameters of liquid in an evaporator at the chosen temperature of steam are determined and schedules for α_E = f(T_W^E) are built on mentioned formulas and on the formula:

$$\alpha_E = \frac{Q}{(T_W^E - T_S)F_E}, \text{ W/m}^2\text{K} \tag{3}$$

The cross point of curves gives the required temperature T_W^E. Thus, the grid of equidistant lines in the plane Q = f(T_W^E) for various values T_W^C can be put, and it is necessary to take into account the heat-transferring opportunities limited by crisis phenomena [2, 3, 9].

To construct a nomogram it is also necessary to know the heat exchange laws of the cooling system with a surrounding medium (α_C^{SM} = constant). For this purpose, for example, the coefficients of heat exchange with external environment are assumed, or the conditions of environment temperature (T_{env} = constant) are accepted.

The Q value is determined by the formula characterizing the heat exchange of an external wall of the condenser with environment

$$Q = \alpha_C^{SM} F_C (T_W^C - T_{env}), \text{ W} \tag{4}$$

Then to each value of Q (or T_W^C) corresponds a certain value of T_W^C (or Q). Therefore, by changing of external conditions of heat exchange with environment it is possible to regulate T_W^E for a given heat emission.

In case the wall temperature of the cooling element is given, it is necessary to ask a number of Reynolds (Re) values to calculate the number of Stanton (St) values and, having determined the value of α_E clarify the temperature T_W^E.

If the wall temperature exceeds a given value, it is necessary to reduce Re value and, hence withdrawn heat load.

5. The mechanism of the heat transfer process

In the process of boiling, the radius of the "dry" spot at the base of the steam bubble is given by:

$$R_{ds} = \frac{\delta_0}{tg\alpha} = \frac{\sqrt{3}}{2tg\alpha} \frac{\lambda'}{\sqrt{\rho c \lambda}} \frac{P\sqrt{\tau_0}}{\sqrt{f'(P)}}, \tag{5}$$

where: α - the angle between the heating surface and a tapered microlayer located under the bubble with a radius R_{ds}; τ_0 - time which determines of complete evaporation of microlayer with thickness δ_0, when under the bubble the "dry" spot is settled; λ, λ' - thermal conductivity of the wall and liquid; c, ρ - heat capacity and density of the wall.

For capillary-porous system which working in the field of mass forces, the law of steam bubble growth [10] is defined as:

$$R_d = 2\sqrt{54.1a'Ja\,\tau_0}\left[1+\left(\frac{m_{liq}}{m_v}\right)^{0.1}\right]^{-1} \tag{6}$$

Then the equation for R_{ds} (5) is converted to the form:

$$R_{ds} = \frac{\sqrt{3}}{2tg\alpha}\frac{\lambda'}{\sqrt{\rho c\lambda}}\frac{P}{\sqrt{f'(P)}}\frac{R_d\left[1+(m_{liq}/m_V)\right]^{0.1}}{2\sqrt{54.1\,a'Ja}} \tag{7}$$

where: $Ja = \frac{c'_p\Delta T}{r}\frac{\rho'}{\rho''}$ - Jakobi number; m_{liq} - liquid flow rate; m_v - vapor flow rate; m_{liq}/m_v – the influence of the liquid flow rate expressed in terms of liquid excess; ρ',ρ'' - density of liquid and steam; r - evaporation heat; $K = \frac{R_{ds}}{R_d}$ represents a coefficient of "dry" spot.

Generally speaking, in the studied capillary-porous cooling system the influence of subcooling, liquid velocity, liquid thermal properties and the characteristics of heating surface can be approximated by the averaged expression of the form:

$$\overline{R} = \frac{\overline{R}_d}{\overline{R}_0} = 2.42\left(k_{liq}k_W\right)^{-1} \tag{8}$$

where $k_{liq} = 1+\tilde{m}^{0.1}$; $k_W = 1+\left[\frac{(\rho c\lambda)'}{(\rho c\lambda)}\right]^{0,5}$ - coefficients taking into account the excess of fluid and heat storage capacity of the wall; $\tilde{m}$ =1-14; $\tilde{m} = \frac{m_{liq}}{m_V}$ - option, taking into account the excess of fluid.

Excess of fluid $\tilde{m}$ in the cross-section of the porous structure makes a flow with low subcooling and low speed W_0, which reduces the averaged value of the detachable bubble radius $\overline{R}_0$ to the value of $\overline{R}_d$. The values of W_0 = (1.1×10 - 0.1) m/s; $W_0 = \frac{m_{liq}lq}{m_V\varepsilon\delta_\varphi r\rho'}$ was found from experimental parameters, where l - height of the heat exchange surface; ε - porosity of structure; δ_φ - thickness of porous structure. This is due to a decrease in the average weight temperature, which leads to overheating falling of liquid film surrounding the bubble, and may cause partial condensation [11].

The low of steam bubble growth R_d takes into account deformation and outline of bubbles due to excess of fluid m_{liq} with in relation to the flow of generated steam m_v.

6. Conclusion

Thus, on the basis of the fundamental research of the models and the nature of the thermal hydraulic properties of the heat transfer, the necessary information was obtained for the development, design, engineering and operation of capillary-porous systems in different thermal power plants of power plants. Examined are the effects of pressure, excess fluid, the methods for heating and for supplying coolant, as well as the type of material used, in addition to the intensifier type amongst others, on the integral and thermo-hydraulic characteristics of the heat transfer process obtained. The information obtained from the research can be used in the design, engineering calculations and operation of various energy installations within power plants. Heat transfer regulation enables the formation of a new class in the system for heat removal. Heat exchangers were built with the purpose of increasing both efficiency and dependability in an environmentally conscious manner. A capillary-porous heat exchanger shaped as a box provides protection against explosions. Obtaining its limiting and operational characteristics is done by means of an engineering method for calculation.

Further development of high-performance apparatuses with porous structures makes it possible to turn energy production into environmentally friendly process, improve occupational safety conditions, significantly save natural resources, intensify processes in alternative sources, protect air, water and soil from pollution, including "thermal", solve methodological problems and long-term problems of generation, transport and energy storage.

References

[1] Polyaev V M, Genbach A N and Genbach A A 1992 An experimental study of thermal stress in porous materials by methods of holography and photoelasticity *Experimental thermal and fluid science* **5**(6) pp 697-702

[2] Genbach A A and Burmistrov A V 2011 Investigation of the thermal condition of the cylinders of steam turbines (Promyshlennost Kazakhstana No **2** (65)) (in Russian) pp 91-93

[3] Polyaev V M, Genbach A N and Genbach A A 1991 redel'noye sostoyaniye poverkhnosti pri teplovom vozdeystvii High Temperature (TVT) **29**(5) (in Russian) pp 923-934

[4] Popescu F and Ion V I 2012 The study of a separating and reattaching flow topology (*Proceedings of the 4th International Conference "Advanced Composite Materials Engineering" COMAT Braşov, 18-20 October*) pp 189-192

[5] Polyaev V M and Genbach A A 1991 Upravlenie vnutrennimi kharakteristikami kipeniya poristoy sistemy (in Russian) (*Sbornik trudov MGTU*) pp 224-237

[6] Polyaev V M and Genbach A A 1992 Upravleniye teploobmenom v poristyh strukturah (in Russian) (*Isvestiya Rossiyskoy akademii nauk. Energetika i transport*) **38**(6) pp 105-110

[7] Polyaev V M and Genbach A A 1993 Teploobmen v poristoy sisteme, rabotayushey pri sovmestnom deystvii kapillyarnyh i gravitatsionnyh sil (in Russian) (*Teploenergetika* No 7) pp 55-58

[8] Polyaev V M and Genbach A A 1993 Intensivnost teploobmena poristoy sistemy v gravitatsionnom pole (in Russian) (*Isvestiya vuzov Energetika.* **1-2** pp 91-95

[9] Polyaev V M and Genbach A A 1991 Predelnye teplovye nagruzki v poristyh strukturah (in Russian) (*Isvestiya vuzov. Aviatzionnaya tekhnika.* **1** pp 33-37

[10] Polyaev V M and Genbach A A 1990 The speed of growth of steam bubbles in porous structures (in Russian) (*Isvestiya vuzov. Mashinostroyenie.* **10** pp 61-65

[11] Polyaev V M and Genbach A A 1990 Detachable diameter and frequency separation of vapor bubbles in porous structures (*Bulletin MSTU ser. Mashinostroenie* No 1) pp 69-72

[12] Ion V I et al. 2009 Analysis of the energetic-environmental performances of gas turbine plant - effect of thermal barrier coatings and mass of cooling air *Thermal Science Journal.* **13**(1) pp 147-164

[13] Genbach A A, Olzhabaeva K S and Iliev I K 2016 Boiling process in oil coolers on porous elements *Thermal Science.* **20**(5) pp 1777 - 89

[14] Genbach A A, Jamankulova N O and Bakic V V 2017 The processes of vaporization in the porous structures working with the excess of liquid *Thermal Science.* **21**(1A) pp 363 – 373

IOP Conf. Series: Materials Science and Engineering **595** (2019) 012002 doi:10.1088/1757-899X/595/1/012002

Experimental validation data for CFD of heat transfer processes in a heat exchanger thermosyphon type with complex geometry

A K Terziev[1], C I Iliev[2], I K Iliev[3], I H Beloev[3] and E V Velkov[3]

[1] Department of Power Engineering and Power Machines, Technical University of Sofia, Bulgaria
[2] College of Engineering and Physical Science, University of Birmingham, United Kingdom
[3] Department of Thermal Engineering, Hydraulics and Ecology, University of Ruse, Bulgaria

E-mail: aterziev@tu-sofia.bg

Abstract. The paper explores the validation of experimental studies by means of numerical modelling of heat transfer processes occurring in a two-phase thermosyphon air heater. The object analysed is an air heater, used for the utilisation of the heat from flue gases emitted by a steam generator running on solid fuel with nominal load of 125 t/h, and temperature of the flue gases at nominal load 220C. The experimental study performed shows heat transfer coefficients of the evaporation and condensation zones of h_{fgas}=104.9 and h_{fgas} = 84.9 W(m^2·K) respectively, and a reduction of the exhaust gas temperature down to 184°C. The numerical study pertains to a thermosyphon tube with complex geometry. Three zones can be distinguished for the air heater in question. In the evaporation zone the tube is inclined at 30 degrees in the direction of the flue gasses, in addition to the inclines of the adiabatic zone – 30 degrees, and the evaporation zone of the thermosyphon – 90 degrees to the horizon. The commercial software Ansys Fluent was used to perform the numerical calculations.

1. Introduction

Waste heat utilisation from the exit flue gases of steam generators is the most direct way of increasing their energy efficiency. In most cases, however, the lowering of the gases' temperature leads to the creation of conditions whereby corrosive processes in the exit heating surfaces of steam boilers take place. The application of a technology of heat pipes (the term "heat exchanger with intermediate heat carrier" is used in practice, however it does not correspond to what is used in contemporary literature on the topic - "heat pipe" or more precisely "thermosyphon") or thermosyphons is suitable for the 'cold' part of the first step of air heaters; this has the added benefit of protecting the primary air heater from said corrosion. This paper focuses on the achieved improvement of steam generators №1 and №2 of "Republica" TPP (Thermal Power Plant), where air heaters thermosyphons type have been mounted.

A literature review performed shows that interest in two-phase thermosyphons is great. Different studies are conducted both experimentally and from an analytical point of view. In [1- 4] experimental studies with the performance of two phase thermosyphons are presented. The focus of the research is the behavior and efficiency of thermosyphons in different modes of operation. The main conclusion is that the geometry and the composition of the fluid in the two-phase thermosiphon have a significant effect on its efficiency.

Validation of results between analytical and experimental studies on the heat transfer coefficients is presented in [5]. Good correlations are observed, giving recommendations on model studies of two-phase thermosyphons.

The processes of heat transfer through the surrounding surface of a two-phase thermosyphon are described in [6]. It is shown both numerically and experimentally how the diameter, evaporation and condenser lengths, in addition to the vapour temperature all affect the operation envelope. Given the huge area of application of thermosyphons, in recent years there has also been an increase in numerical modeling. In [7] numerical study of two-phase thermosyphon is performed where mass flow rate, saturation pressure and also the temperatures have been evaluated. A similar numerical study is presented in [8] where the difference in modeling of single-phase and two-phase thermosyphons are discussed.

2. General information

The operational process of the thermosyphon comes down to the following: hermetically sealed tubes are filled 30% with chemically treated water. This part of the tubes, also called the zone of evaporation is under the effect of flue gas flow. The steam resulting from the gases' heat condenses at the other end of the tube – the zone of condensation, which is affected by the air stream. The heat from the condensation heats the air, with the condensate trickling down into the evaporation zone owing to the gravitational forces.

In this type of air heaters, the heat taken off from the gases is released to the air by means of an "intermediate heat carrier" (in our case water-steam), due to which the section of the tubes located in the gas duct is characterized by a constant temperature along its length. The main advantages of using air heaters with thermosyphons are:

- their application makes the use of steam calorifiers for preheating the cold air in order to prevent the low temperature corrosion of the heated surfaces redundant;
- they have very high surface temperatures in the gas part, making their exploitation safer with regard to corrosive activity;
- they ensure the heating of outdoor air to a certain temperature, thereby ensuring a non-corrosive regime of the air heater of first degree in all exploitation regimes of the steam boiler;
- they have low aerodynamic resistance in flue gas and air media and do not require the replacement of existing ventilators;
- air leakages are practically absent in them, the eventual malfunction of separate thermosyphons does not lead to an increase in leakages.

All aforementioned benefits serve to make air heaters with thermosyphons highly competitive (and sometimes without alternative) to the air heaters used as exit heating surfaces, in particular when the price of the former is reasonable.

The performed on-site experimental tests show that the outlet flue gas temperature is high and there is a possibility of utilizing certain amount of heat. The plant's management has taken the decision to install additional thermosyphon type heat exchangers to the steam generators to increase their efficiency.

The steam generators object of the study are very old and run on solid fuel "a mix from the local mine" with the following characteristics:

- Moisture content: $W^r = 16\%$;
- Ash content: $A^r = 43\%$;
- Volatile matter: $V^r = 51.5\%$;
- Lower Heating Value: $Q_r^i = 10500$ kJ/kg.

During the years the calorific value of the fuel decreasing as at the end of 2001 its value reaches 8164kJ/kg or 22.6% lower than to the first coal used. The proposed thermosyphons serving the operation of the steam generators were implemented at the end of 2010. A schematic representation of the realized and put into operation air heater with thermosyphons is shown in Figure 1. It is evident from the scheme that the implementation of the thermosyphons differs from the standard one. Their form has been altered in order to reduce the aerodynamic resistance of the heat exchanger and respectively to increase the surface temperature of the thermosyphon, in turn making the heat exchanger compact and safe from corrosion. Using this technical solution does not require any displacements/alterations nor reconstructions of the existing ventilators, and thus reduces the ultimate cost of the overall reconstruction. The technical and economic parameters for which the additional Air heater thermosyphon (AH-TS) was designed together with the nominal parameters of №1 and №2 can be found in table 1 below.

Table 1. Operational parameters of the steam boiler and thermosyphons.

Name	Nom	Dimension	Options. Custom sizing of AH-TS
Elementary composition of the design fuel for a work mass for AH-TS	-	%	C^r 21.4; H^r 2.1; S^r 1.3; O^r 7.0; N^r 0.5
Moisture and ash content of the design fuel of AH-TS	W^r, A^{dry}	%	until 2000 year: 15.1, 59; after 2000 year: 14.5, 53.2
LHV by operational data	Q_i^r	kJ/kg	until 2000 year 8 540; after 2000 year 8 164
Combustion air per kg fuel	V_{ca}^0	Nm^3/kg	2.269
Fuel gas volume per kg fuel	V_g	Nm^3/kg	3.508
Excess of air coefficient in the combustion chamber	α_c	-	1.2
Volume ratio of air per kg fuel to AH-1	V_b^0	Nm^3/kg	2.831
Dew point temperature of the flue gases	t^{dp}	^oC	80
Nominal operating parameters of steam generator (SG) №1			
Nominal load of the steam generator	D_{ps}	t/h	125
Pressure of preheated steam	p_{ps}	MPa	7.7
Temperature of preheated steam	t_{ps}	^oC	500
Pressure of feed water	p_{fw}	MPa	13.2
Temperature of feed water	t_{fw}	^oC	130
Enthalpy of preheated steam	I_{ps}	kJ/kg	813.5
Enthalpy of feed water	I_{fw}	kJ/kg	132.5
Heat losses			Based on the average load of the boiler
Heat loss due to flue gas	q_2	%	$q_2=\left[\left(I''_{fgas}-\alpha'' I^0_{air}\right)/Q_i^r\right].\left[\left(100-q_4\right)/100\right]$ (balance equation) where: I^0_{air} - theoretical enthalpy of combustion air, kJ/kg; α'' - excess air rate, -;
Heat loss due to carbon remained	q_3	%	0 (according to data from the operation)
Heat loss due to unburnt fuel in residual	q_4	%	1.5 (according to data from the operation)
Heat loss due to radiation	q_5	%	0.67 (according to data from the operation)
Heat loss due to fly ash and slag	q_6	%	0.18
Enthalpy of outgoing gases for the designed fuel of AH-TS	I''_{fgas}	kJ/kg	$I''_{fgas}=I_g^0+(a-1).I^0_{air}+I_{ash}$ where: I_g^0 - theoretical enthalpy of flue

			gas, kJ/kg ; I^0_{ash} - enthalpy of ash, kJ/kg ;
Temperature of gases at nominal load of the boiler	t'_{fgas}	^oC	220 (data from boiler workshop)
Temperature of gases after modernization	t''_{fgas}	^oC	184 (data from boiler workshop)
Air temperature at the inlet of AH-TS	t'^{AH-TS}_{air}	^oC	25 (data from boiler workshop)
Air temperature at the outlet of AH-TS	t''^{AH-TS}_{air}	^oC	84 (data from boiler workshop)
Enthalpy of outgoing gases At $t'_{fgas}=220^oC$	$I''^{220^oC}_{fgas}$	kJ/kg	975.2
Enthalpy of outgoing gases At $t''_{fgas}=184^oC$	$I''^{184^oC}_{fgas}$	kJ/kg	179.7
Enthalpy of cold air at the entrance of the AH-TS	I'_{air}	kJ/kg	74.9
Excess of air in the flue gases	α''_{fgas}	-	1.37 (data from boiler workshop)
Flue gas losses q''_2 at $t''_{fgas}=184^oC$ and $Q^r_i=8164kJ/kg$	q''_2	-	$q_2=\left[\left(I''_{fgas}-I_{air}\right)/Q^r_i\right].\left[\left(100-q_4\right)/100\right]=10.48$
Flue gas losses q'_2 at $t'_{fgas}=220^oC$ and $Q^r_i=8164kJ/kg$	q'_2	-	$q_2=\left[\left(I'_{fgas}-I_{air}\right)/Q^r_i\right].\left[\left(100-q_4\right)/100\right]=12.86$ where: I'_{fgas} - enthalpy of ingoing gases, kJ/kg ; I_{air} - enthalpy of air, kJ/kg .
Boiler efficiency at $t'_{fgas}=220^oC$; $\alpha''_{fgas}=1.37$; $t_{air}=25^oC$ and $Q^r_i=8164kJ/kg$	η'	%	$\eta'=100-\left(q'_2+q_4+q_5+q_6\right)=84.79$
Boiler efficiency at $t''_{fgas}=184^oC$; $\alpha''_{fgas}=1.37$; $t_{air}=25^oC$ and $Q^r_i=8164kJ/kg$	η''	%	$\eta''=100-\left(q''_2+q_4+q_5+q_6\right)=87.17$
Difference in boiler efficiency at $t''_{fgas}=184^oC$ and $t'_{fgas}=220^oC$	$\Delta\eta$	%	$\Delta\eta=\left(\eta''-\eta'\right)/\eta''=2.72\%$
Amendment in losses q_2 at change of t'_{fgas} by 10^oC	Δq_2	%	0.774
Determination of fuel consumption at $t''_{fgas}=184^oC$	B'	t/h	$B'=\left[D_{ps}-\left(i_{ps}-i_{fw}\right)\right]/Q^r_i.\eta'=48.972$
Determination of fuel consumption at $t'_{fgas}=220^oC$	B'	t/h	$B''=\left[D_{ps}\left(i_{ps}-i_{fw}\right)\right]/Q^r_i.\eta''=50.343$
Fuel savings at nominal load	ΔB	t/h	$\Delta B=B'-B''=1.371$
Annual usage of the boiler #1	τ	h/yr	2000 year 4030; Current year 4818
Annual fuel savings	B	t/yr	$B=\Delta B.\tau=6604$
Fuel price	$\Pr_{fuel}$	$EUR/tcoal$	12.61
Price of electrical energy	$\Pr_{el}$	EUR/kWh	0.09
Annual cash savings	L_{fuel}	EUR/yr	$L_{fuel}=\Pr_{fuel}.B=83276$
Amount of electricity for coal milling	E	kWh/t	25
Energy savings achieved for coal milling	L_{el}	kWh/yr	$L_{el}=B.e=165100$
Annual net incomes from electricity savings.	L_{el}	EUR/yr	14 859
Total annual cash savings	L_a	EUR/yr	$L_a=L_{fuel}+L_{el}=98135$

Total heating surface of AH-TS	A_{Total}	m^2	860
Total mass of AH-TS	G	kg	26 144
Investment costs for installation of AH-TS	L_{total}	EUR	160 350
Payback period	PB	$years$	1.63

3. Experimental study

A number of experimental tests at different operating modes of boilers have been carried as the impact of the air-heaters on the boiler efficiency is established.

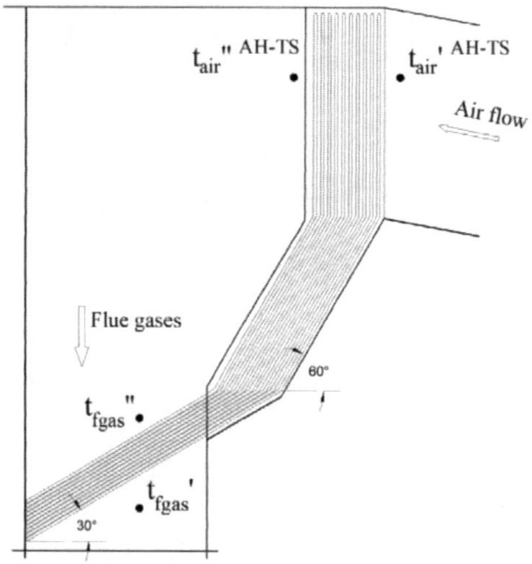

Figure 1. Location of the parameters measured.

Experimental studies include measurement of airflow and flue gas temperatures before and after thermosyphons. On fig. 1 the measurement points are displayed. The results from the on-site measurements performed are presented in Table 2.

There are two zones in which heat transfer processes occur in the thermosyphone heat exchanger - the evaporation and condensation zone. The heat transfer coefficient is determined using the normative method [9], based on the temperatures measured for both flows (air and flue gases).

Based on the methodology applied the following values for the heat transfer coefficients of for the evaporation and condensation zones respectively are obtained: $h_{fgas} = 104.9$ W(m^2·K) and $h_{air} = 84.9$ W(m^2·K).

Table 2. Experimental data.

Parameter	Label	Trial 1	2..	..23	24	Average
Air temperature in front of AH-TS, ^oC	t'^{AH-TS}_{air}	25	23.6	24.8	26.2	**25.0**
Air temperature behind AH-TS, ^oC	t''^{AH-TS}_{air}	84.8	81.7	73.0	80.2	**81.2**
Flue gas temperature in front of AH-TS, ^oC	t'_{fgas}	219.5	214	218	217.5	**218.4**
Flue gas temperature behind AH-TS, ^oC	t''_{fgas}	184.3	179	189	185	**184.9**
Steam generator load, t/h	D_S	122.5	118	121	123	**121.0**

From the data in Table 1 it is evident that after the installation of air-heater the heat losses due to wasted heat in flue gases are 2.72% lower. It must also be emphasised that lowering the exit gases' temperature by 10°C (given $\alpha''_{fgas} = 1.37$ and $Q^r_i = 8164$ kJ/kg) leads to a reduction in the losses q_2 by 0.77%. When burning fuel with lower heating value equal to the project one ($Q^r_i = 10550$ kJ/kg), heat losses due to wasted heat in flue gases are 9.96%, and at typical lower heating value (LHV) for the burnt fuel after the year 2000 $Q_i^r = 8164$ kJ/kg, $q_2 = 12.86\%$, that is, there is a noticeable increase in the losses as a result of the worsened fuel quality.

Steam generator efficiency (η_{SG}) which is the primary criterion for thermodynamic excellency of the steam generator increased by 2.72% as a result of the reconstruction realizing the air heater with thermosyphons. This does not take into account the fact that in operational conditions such an optimal coefficient of air excess for the exit gases ($\alpha''_{fgas} = 1.37$) by lower heating value of the fuel ($Q_i^r = 8164$ kJ/kg) is hard to obtain; this in turn means that the increase in boiler efficiency is higher than 2.72% in reality.

When determining the economic effect of the realisation of the AH-TS (air heater with thermosyphons) in addition to the direct fuel economy of 1.37 t/h the economy of electricity consumption amounting to 165 MWh/year, necessary for the fuel milling savings, was also taken into consideration.

4. Numerical study

A detailed numerical study of the heat exchange in the air heater has been carried out. Given that the existing air heater is composed of a few assembled parts, only a single tube bundle was thoroughly analysed in order to reduce the time taken for the calculation procedures. Two different processes can be established – the heat transfer between air flow and exhaust flue gases, and the two-phase flow behaviour in the thermosyphon. The focus of the current work is the heat transfer between flue gases and fresh air via a thermosyphon.

4.1. Geometric model of the waste heat recovery system

Numerical procedures are performed on a real size 3D model of the air-heater (Figure 2). The thermosyphon pipes are arrange in a square tube layout. The air-heater consist of 1710 pipes. The distances between pipes in horizontal and vertical directions are 42 mm and 45 mm respectively. Three working zones of the thermosyphon are identifiable:

- evaporation zone – 3 m;
- adiabatic zone 2 m;
- condensation zone 2 m.

The middle zone connects evaporation and condensation zones and it is used to prevent the heat losses as in terms of the above is well isolated. The total head loss is calculated for both tracts: flue gases section 303 Pa; fresh air section 219 Pa.

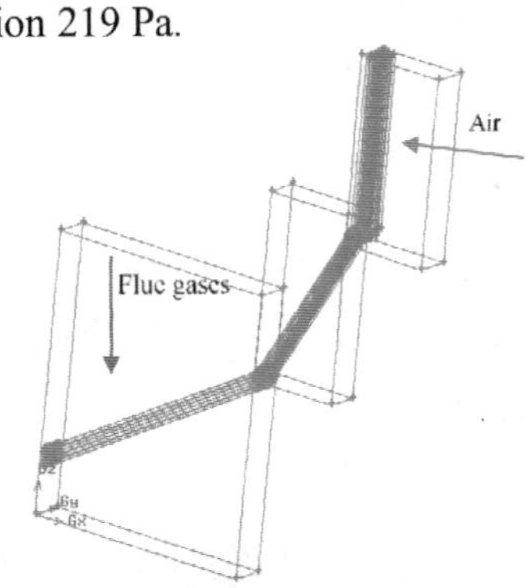

Figure 2. 3D Model of the proposed unit.

The experimental studies show that the average velocity of the flue gases in the space between the tubes is $w_{fg} = 14.4$ m/s, compared to that of the air $w_{fg} = 10.1$ m/s. Taking into account that the object analyzed is a single section of the suggested air heater, then the heat transfer processes concern the flue gases' temperatures before and after the first air heater, as well as the air temperature before and after the second air heater (bearing in mind the opposite direction of movement of both flows).

The numerical procedures are performed on a CFD software. The model used for the study is three dimensional. Because of the size only part of the air-heater (30 pipes) is modeled. Triangle type of meshing elements have been used during the meshing procedure as the total number of cells amounted to 1250000 pcs.

4.2. Mathematical modeling

It was noted above that heat transfer processes are running only at evaporation and condensation zones. The middle zone is adiabatic. Also, it is accepted the steady state process into the two phase thermosyphon.

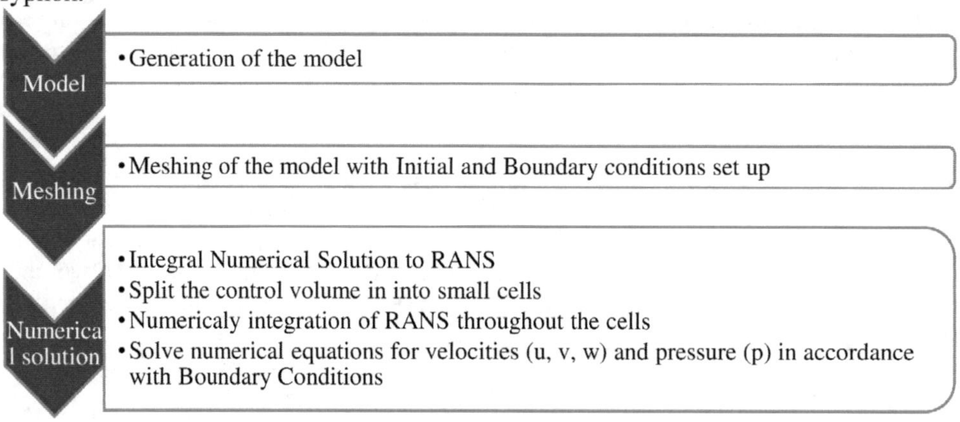

Figure 3. Numerical procedure block-scheme.

The CFD study concerns the heat transfer processes taking place both between the air heater and flu gases and air-heater and fresh air. The continuity, momentum and energy equations are applied during the numerical procedure. Reynold Averaged Navier-Stokes Equations (RANS) is accepted as an appropriate turbulent model. Different methods are available for discretization of RANS as here is perceived Finite Volume Method (FVM) The fundamental equations were derived in FVM using integral approach. The solution procedure is according Figure 3. The accepted mathematical model for the numerical procedures is according [10].

4.3. Initial conditions

The results from the numerical procedure will show the amount of energy transferred between air heater and air and flue gases. During the numerical study the impact of the middle zone (adiabatic) is also considered. Input parameters for the study are velocities, and temperatures of both media. Also, different turbulent intensities are considered. Based on the on-site measurements the velocity of the flue gases (w_{fgas}) before the thermosyphon is 7.8 m/s, and the fresh air velocity - 5.6 m/s. The analysis is also geared to study the impact of turbulent intensity (TI) and heat transfer coefficient (h) on the improvement of heat transfer processes. The numerical study is performed with the following input data (Table 3).

Table 3. Input data.

Parameter	**Label**	**Mode**		
		1	**2**	**3**
Air temperature before AH-TS, ^oC	$t'_{airAH-TS}$	25.0	25.0	25.0
Air temperature after AH-TS, ^oC	$t''_{airAH-TS}$	81.2	81.2	81.2
Flue gas temperature before AH-TS, ^oC	t'_{fgas}	218.4	218.4	218.4

Flue gas temperature after AH-TS, ^oC	t''_{fgas}	184.9	184.9	184.9
Turbulent Intensity, %	TI	5	10	20
Heat transfer coefficient (evaporation zone), W(m^2·K)	h_{ev}	92	105	116
Heat transfer coefficient (condenser zone), W(m^2·K)	h_{con}	74	85	104

5. Results

The results obtained during the numerical study are presented below. On Figure 4a the temperature field of flue gases passing through the thermosyphon air heater is presented (Mode 2, Table 3). It is quite clear that flue gases passing through the air heater were cooled down to 185°C (assuming initial turbulent intensity of 10% for flue gases). Velocity distribution for the current case is presented in Figure 4b. The average velocity in the vicinity of the air heater is 7.8 m/s while the average velocity between pipes reaches up to 14.1m/s.

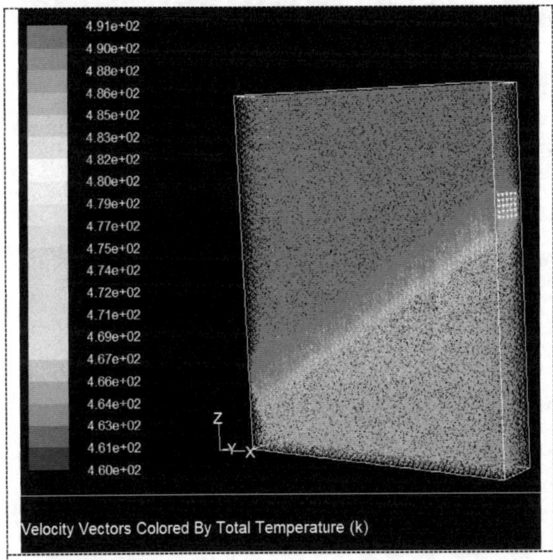

Figure 4a. Flue gases temperature filed distribution.

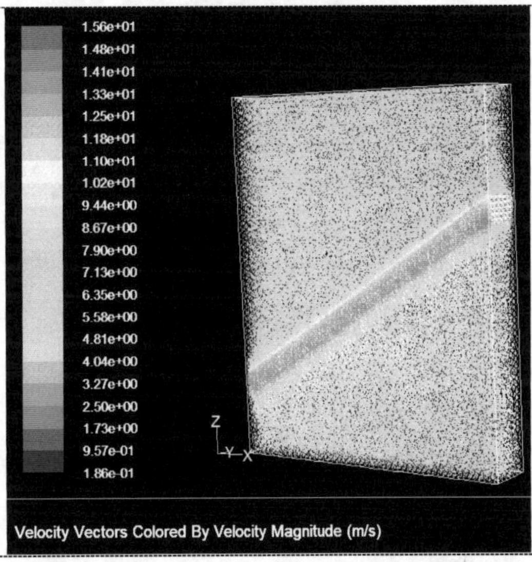

Figure 4b. Flue gases velocity filed distribution.

Perpendicular and parallel temperature contours of the flue gases in the vicinity of thermosyphon heat exchanger are presented on Figure 5a. The flue gas specific zones (cooling zones around air heater) are clearly identifiable. The initial turbulent intensity (TI) is important for heat transfer exchange. Figure 5b presents the turbulence distribution via the path of flue gases. TI in the inter-tubular space reaches values of 220%, which is one of the reasons for the intensive heat exchange in the proposed type of inclined thermosyphon air heater.

The results of the numerical solution give a clear picture for the distribution of the gas phase and flue gas parameters before, after and in the vicinity of the thermosyphon. Summarized results for main parameters of the flows are presented in Table 4. It is important to note that the heat exchange is enhanced by increasing the turbulent intensity or at higher velocity of the two flows.

IOP Conf. Series: Materials Science and Engineering **595** (2019) 012002 doi:10.1088/1757-899X/595/1/012002

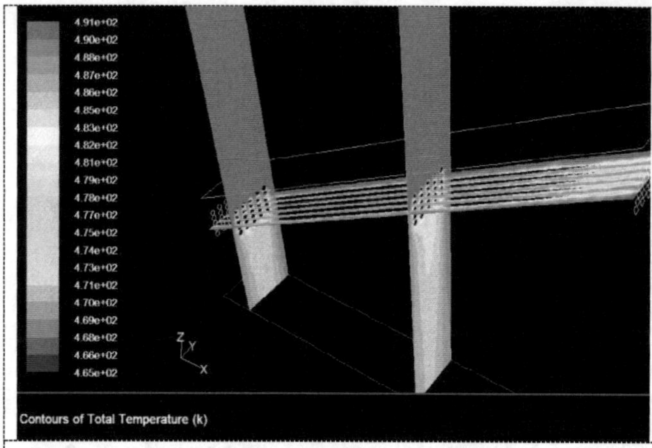

Figure 5a. Flue gases temperature contours.

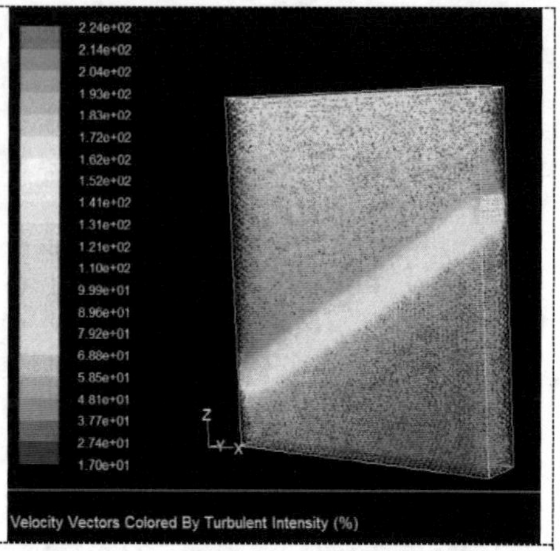

Figure 5b. Flue gases turbulent intensity.

Table 4 summarizes the results of the numerical procedure.

Table 4. Results obtained during the numerical study.

Numerical procedure case number	t_{flugas} ^{o}C	w_{flugas} m/s	$TI_{flue-gas}$ %	t_{air} ^{o}C	w_{air} m/s	TI_{air} %
TI - 5%, $\alpha_{gas}=92$, $\alpha_{air}=74$ W(m^2·K)	177	13.9	220	79.0	9.5	295
TI - 10%, $\alpha_{gas}=92$, $\alpha_{air}=74$ W(m^2·K)	179	14.1	225	79.5	9.56	304
TI - 20%, $\alpha_{gas}=116$, $\alpha_{air}=104$ W(m^2·K)	186	14.9	234	82.0	9.8	315
TI - 10%, $\alpha_{gas}=105$, $\alpha_{air}=85$ W(m^2·K)	183	14.2	224	82	10.0	301
TI - 10%, $\alpha_{gas}=116$, $\alpha_{air}=104$ W(m^2·K)	192	14.1	225	84.0	10.5	303

6. Conclusion

The paper has studied the possibility of increasing the efficiency of a steam boiler by installing a two-phase thermosiphon. As a result of the measure, a reduction of the flue gas temperature 218 to 185°C was achieved, resulting in an increasing the efficiency with 2.72%. With the amount of heat transfered, fresh air is heated to 81°C, which eliminate the corrosion of the air path.

The performed numerical study gives a detailed picture for the distribution of the both flows parameters. The impact of turbulent intensity (TI) and heat transfer coefficient (h) on the heat transfer processes is clearly presented. Increasing the TI or velocities of two flows improve the heat transfer processes.

The middle (adiabatic) zone certainly deteriorates the heat transfer processes and also increase the total head loss of the system. Despite the above, due to the constructional features, its avoidance is impossible. In addition, the 15 years of exploitation of the thermosyphon is expressed through leakages (mixing between flue gases and fresh air) between the perlitoconcrete and pipe because of the significant thermal expansion and contraction.

The numerical results obtained by using the initial data presented in row 4 of Table 4 show the best correlation with experimental data. This is the case where the initial turbulence of the two flows does not exceed 10%, as the heat transfer coefficients in both zones (evaporation and condensation) are 105 and 85 W(m^2·K) respectively.

Validation of the numerical data with one obtained during the experimental studies will allow an express evaluation of the efficiency of a two-phase thermosyphon air heater with complex geometry, all whilst being in accordance with the proposed initial and boundary conditions.

IOP Conf. Series: Materials Science and Engineering **595** (2019) 012002 doi:10.1088/1757-899X/595/1/012002

References

[1] Luikov A V 2005 Heat pipes in modern heat exchangers, *Applied Thermal Engineering.* **25** 1-19
[2] Chehade A A, Louahlia-Gualous H, Le Masson S, Victor I and Abouzahab-Damaj N 2014 Experimental investigation of thermosyphon loop thermal performance, *Energy Conversion and Management.* **84** pp 671-680
[3] Bai Y, Wang L, Zhang S, Xie N and Chen H 2018 Heat transfer characteristics of a natural circulation separate heat pipe under various operating conditions, *International-Journal-of-Heat-and-Mass-Transfer.* **126** pp 191-200
[4] Cataldo F and Thome J 2017 Experimental evaluation of the thermal performance of a thermosyphon cooling system rejecting heat by natural and forced convection, *Applied Thermal Engineering.* **127** pp 1404-1415
[5] Mercado M, Wong N and Hartwig J 2019 Assessment of two-phase heat transfer coefficient and critical heat flux correlations for cryogenic boiling pipe heating experiments, *International Journal of Heat and Mass Transfer.* **133** pp 295-315
[6] El-Genk M and Saber H 1999 Determination of operation envelope for closed, two-phase thermosyphons, *International Journal of Heat and Mass Transfer.* **42(5)** pp 889-903
[7] Joudi K and Al-tabbakh A 1999 Computer simulation of a two-phase thermosyphon solar domestic hot water heating system, *Energy Conversion & Management.* **40** pp 775-793
[8] Bay Y et al. 2019 Numerical analysis of a closed loop two-phase thermosyphon under states of single-phase, two-phase and supercritical, *International Journal of Heat and Mass Transfer.* **135** pp 354-367
[9] Thermal calculation of boilers (normative method), third edition, revised and expanded, St. Petersburg, 1998 (in Russian);
[10] *Fluent 6.3 User's Guide,* September, 2006

Acknowledgments

The authors would like to thank the Research and Development Sector at the Technical University of Sofia for the financial support.

Finite elements analysis of mass transfer and mechanical processes in ceramic ware at convective drying

N Y Penkova[1], B M Mladenov[1] and K S Krumov[1]

[1] University of Chemical Technology and Metallurgy, Sofia, Bulgaria

E-mail: boian.m@gmail.com

Abstract. Mathematical models of the coupled moisture diffusion and mechanical processes in ceramic ware at industrial drying are composed. They allow taking into account the impact of the drying regime on the plastic deformation and shrinkage of the articles. The models are validated and applied for numerical investigations of the transient fields of the moisture content, stresses and strains in building bricks in continuous dryers. Ways for their application for improving of the energy efficiency of drying installations are discussed.

1. Introduction

The reduction of energy consumption and subsequent carbon footprint at the industrial systems is an actual problem that provokes many research efforts [1]. The computational modelling and simulation are preferable approaches for analysis of technologies, aggregates and their performance [2]. They allow obtaining detailed information about different multi-physics problems at all stages of the manufacturing. The drying is a widespread technological process in the mining and processing industries. At the same time it is difficult for modelling due to the complex conjugate heat and mass transfer in the dryers and thermo-structural-diffusion processes in the dried materials. The non-uniform shrinkage and plastic deformation of the ceramic mass at the first period of drying can cause defects in the dried body not only during the drying, but also at the next firing [3]. These coupled processes are subject of a number of theoretical and experimental studies. Their prediction and proper organization are particularly important for faultless manufacturing in the ceramic industry. Investigations about the heat, mass transfer and mechanical behaviour in dried samples at laboratory conditions are known from the literature [4, 5 and 6]. But complex studies of coupled thermo-structural- mechanical behaviour in ceramic articles at drying in industrial conditions have not published to now. The investigations, presented in this paper aim developing of an algorithm for modelling and finite element analysis of the coupled mass transfer and mechanical processes in ceramic ware at industrial convective drying, suitable for analysis and improvement of the efficiency of the process.

2. Mathematical modelling and conceptions for numerical simulations

2.1. Kinetics of convective drying

The convective drying of the ceramic material can be examined in three main periods [3]: preheating period, constant drying rate period (CDRP) and falling drying rate period (FDRP). During the preheating period the material with an initial temperature is heated up/down to the wet bulb temperature of the drying media. The change of the moisture content is insignificant and that relatively short period can be examined as a part of the next one.

The CDRP begins at surface temperature of the material, equal to the wet bulb temperature, which remains constant during that period. A film of free water is available at the evaporating surface. The moisture migrates from the inside of the porous material to the surface by diffusion. The drying rate remains constant as long as the moisture transport rate from the interior of the material to the exchanging surface is equal to the evaporation rate. The moisture content at the end of the CDRP is known as critical moisture content (W_c). The evaporation rate starts to decrease at that point because of the disconnection of the liquid meniscus from the surface. This decreasing trend continues at FDRP until all liquid meniscuses are disconnected. The internal moisture transport rate specifies the drying rate of the FDRP. The temperature of material increases to the drying media temperature. The drying process stops when the equilibrium moisture content is reached (W_{eq}).

Several types of strains are expected to exist during the drying of wet bodies: shrinkage, provoked by the mass transfer, plastic deformation due to the gravity and temperature strains. The shrinkage is one of the most important factors effecting on drying behaviour of highly shrinkable porous media such as clay-like materials. It occurs during CDRP period and stops at the critical moisture content. The non-uniform shrinkage at that process results in stresses that may cause cracking in the dried products or even make them useless. If drying is performed sufficiently slowly, the shrinkage would be closer to uniform and the drying induced stresses are not dangerous for the material. By other hand the increasing of the drying time results in higher embodied energy and lower dryer productivity. For economic reasons the industrial drying processes are usually performed faster - that results in non-uniform shrinkage and subsequent high tensile stresses in the dried bodies.

It is important to choose suitable conditions of drying for the specific products to prevent failures of the dried articles and to save energy. The time durations of the CDRP and FDRP are between the most important factors determining the efficiency of drying. They depend on the potential of drying, expressed by the difference between the dry and wet bulb temperatures of the drying media. As smaller is it as longer is the drying time. To prevent failure of the production due to non-uniform shrinkage, the CDRP is maintain at a relatively small drying potential via keeping of high relative humidity and velocity, and relatively low temperature of the drying gas during that period. The boundary conditions of the models below allow taking into account the influence of these fluid flow parameters on the coupled mass transfer and mechanical processes in the dried solid domain.

2.2. Modeling of mass transfer and mechanical processes in the dried bodies

System of equations

The transient fields of the moisture content and subsequent mechanical behaviour of the materials are obtained by coupled numerical solution of diffusion equation (1) and stress-strain relationship (2) for 3-dimensional finite element mesh, approximating the geometry of the ceramic body [7]:

$$\frac{\partial C}{\partial \tau} = [D_{ef}]\nabla^2 C \tag{1}$$

where: C = water concentration, kgm^{-3}; D_{ef} - effective diffusion coefficient, m^{2}s^{-1}; τ - time, s.

$$\{\sigma\} = [D]\{\varepsilon\} \tag{2}$$

where: $\{\sigma\}$ - stress vector; $\{\varepsilon\}$- elastic strain vector; $[D]$ - elastic stiffness matrix, formed by module of elasticity E and Poisson ratio ν.

In a coupled structural –diffusion analysis the total strain is formed of elastic $\{\varepsilon^{el}\}$ and diffusion $\{\varepsilon^{d}\}$ parts, respectively:

$$\{\varepsilon\} = \{\varepsilon^{el}\} + \{\varepsilon^{d}\} = [D]^{-1}\{\sigma\} + \{\beta\}\Delta C \tag{3}$$

where: $\{\beta\}$= vector of coefficient of diffusion expansion, m^{3}kg^{-1}; ΔC= concentration change according reference value C_{ref}:

$$\Delta C = C - C_{ref} \tag{4}$$

The modulus of elasticity E, Poisson ratio, coefficient of diffusion expansion β and D_{ef} can be obtained experimentally and used in the models as function of moisture content in the dried material [7].

Boundary conditions

The mass transfer from the boundaries S of the ceramic body to the drying gas can be computed by mass flux, depending on the thermodynamically and fluid flow conditions in the dryer [3 and 8]:

$$D_{ef}\left.\frac{\partial C}{\partial n}\right|_S = \dot{q}_m \quad (5)$$

The intensity of drying in CDRP is expressed by:

$$\dot{q}_{mI} = \frac{h_{mI}}{R_w \mathrm{T}}(\mathrm{p}_s - \mathrm{p}_w) \quad (6)$$

where: $\dot{q}_{mI}$ - mass flux of evaporated water by unit surface, kg m^{-2}s^{-1}; h_{mI} - mass transfer coefficient in CDRP, ms^{-1}; T - temperature of the drying media, K; R_w - specific gas constant for water vapour: R_w=462 J kg^{-1}K^{-1}; p_w and p_s are partial pressures of unsaturated and saturated water vapor respectively, Pa.

The coefficient h_{mI} depend on the fluid velocity, relative humidity and temperature field in the drier and can be determined by Nusselt number at the mass transfer:

$$Nu_{mI} = \frac{h_{mI}.l}{D_{d.m}} \quad (7)$$

where: l - length of the ceramic body on the drying gas way; $D_{d.g}$ - diffusion coefficient of the wet drying gas, m^{2}s^{-1}.

At isobaric process it is function of the temperature of the drying media t, °C:

$$D_{d.m} = 22.63.10^{-6}\left(\frac{t+273}{273}\right)^{1.81} \quad (8)$$

The Nusselt number is given in the common form in the literature [3]:

$$Nu_{mI} = 2 + A_1.\mathrm{Pr}_m^{0.33}.\mathrm{Re}^{n1}.Gu^{m1} \quad (9)$$

where: A_1, m_1 and n_1 depend on Reynolds number (Re). They are known for large limits of Re and can be calibrated on the base of operating conditions in the dryer. Re, Prandtl number at mass transfer and Guhman number, used as non-dimensional drying potential, correspondently are:

$$\mathrm{Re} = \frac{wl}{\nu};\ \mathrm{Pr}_m = \frac{\nu}{D_{d.m}};\ Gu = \frac{T-T_w}{\mathrm{T}} \quad (10)$$

where: w - average velocity, ms^{-1}; ν - kinematic viscosity of the drying gas, m^{2}s^{-1}; T_w - wet bulb temperature, K.

The mass flux, evaporated from ceramic ware at FDRP can be obtained according a hypothesis for linear decreasing of the drying rate [3]:

$$\dot{q}_{mII} = \frac{\dot{q}_{mI}}{C_c - C_{eq}}(C - C_{eq}) = h_{mII}(C - C_{eq}) \quad (11)$$

where: C_c - water concentration at the critical moisture content, kgm^{-3}; C_c - equilibrium water concentration, kgm^{-3}; h_{mII} - mass transfer coefficient in FDRP, ms^{-1}.

The boundary conditions for the structural analysis include zero normal displacements at the contact surface between the transport and the dried body and the symmetry boundaries (if there exists any).The gravity is taken into account.

Analogy between the heat and mass transfer
The advanced software for numerical simulation [9] allows structural-diffusion analysis at limited opportunities for definitions of boundary conditions and material properties for solution of equations (1) and (2). The modelling of the mass flux $\dot{q}_{mII}$, mechanical properties and coefficient of mass diffusion as function of the concentration is not possible. An alternative approach is to implement the structural-diffusion simulation using the analogies between the mass and heat transfer and between the diffusion and thermal stresses.

The effective diffusion coefficient in equation (1) can be expressed by:

$$D_{ef} = \frac{K_{art}}{\rho_d c} \tag{12}$$

where: K_{art}=artificial thermal conductivity, $Wm^{-1}K^{-1}$; ρ - density of the dry mass, kdm^{-3}; c - specific heat capacity, J $kg^{-1}K^{-1}$.

The concentration can be referred to the dry mass of the material:

$$C = u\rho_d = \frac{W}{100}\rho_d \tag{13}$$

where: u - moisture content of the wet mass, kg water/kg dry material; W - moisture content, %.

Then equation (1) is transformed to:

$$\rho_d c \frac{\partial W}{\partial \tau} = [K_{art}]\nabla^2 W \tag{14}$$

and the boundary conditions (5) for its solution can be expressed by artificial heat flux q_{art}:

$$\frac{K_{art}}{100c}\frac{\partial W}{\partial n}\bigg|_S = \dot{q}_m \quad \text{or} \quad K_{art}\frac{\partial W}{\partial n}\bigg|_S = 100c\dot{q}_m = \dot{q}_{art} \tag{15}$$

If W is replaced by T in equations (14) and (15), they are transformed respectively to Fourier equation for heat transfer and the second kind of boundary conditions for its solution, respectively. Taking into account equations (11) and (15) the mass flux, used as boundary condition to model the mass transfer between the ceramic ware and the dried media can be written in common form:

$$\dot{q}_{art} = \begin{cases} -100c\dot{q}_{mI} = constant \ \ (average\ value)\ at\ W \geq W_c \ \ (CDRP) \\ \dfrac{100c\dot{q}_{m1}}{W_c - W_{eq}}\left(W_{eq} - W\right) \ \ at\ W < W_c \ \ (FDRP) \end{cases} \tag{16}$$

The diffusion strain can be expressed as thermal strain in equation (2):

$$\varepsilon^d = \beta\left(C - C_{ref}\right) \rightarrow \varepsilon^{th} = \alpha\left(T - T_{ref}\right) \tag{17}$$

where α= instantaneous coefficient of thermal expansion, K^{-1}. It is accepted equal to β:

$$\beta = \frac{\Delta l}{l(W_{in} - W_c)} \tag{18}$$

where W_{in} is the initial moisture content, %.

3. Finite element analysis of structural-diffusion processes in row building bricks in an industrial dryer

The proposed models are used to investigate the structural-diffusion behaviour of building bricks in industrial continuously working dryer. Detail information about the operating conditions in the dryer is not possible due to confidential rules.

The drying gas (air-fluc gases mixture with predominant air fraction) and the wet ceramic articles are moved in counter flow through the drying tunnel. The change of the fluid flow temperature, relative humidity and velocity are obtained by in situ measurements, mass and thermal balances. The figures below illustrate the change of the Re number and the drying potential ε of the drying media with the time:

$$\varepsilon = T - T_w \tag{19}$$

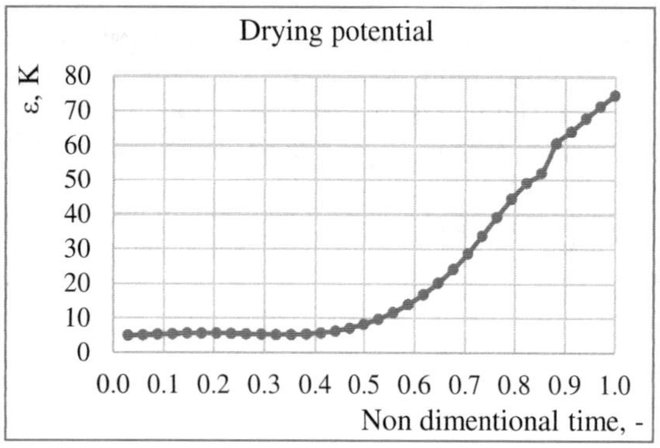

Figure 1. Drying potential variation through the channel (with the time).

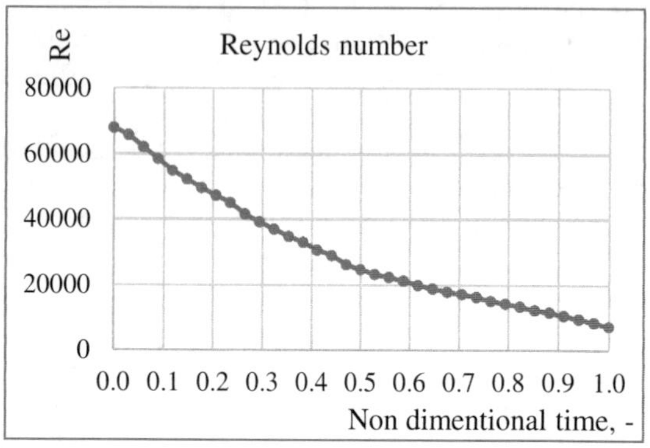

Figure 2. Reynolds number variation through the channel (with the time).

The wet raw mass is a mixture of clay and combustible supplements. Standard geometry samples made by the wet material after the extruder in the plant are dried to different moisture content in a laboratory installation. They are tested at series of experiments to obtain the critical moisture, relative deformations at the shrinkage, modulus of elasticity, Poisson ratio and effective diffusion coefficients as function of the water content [7]. These physical properties variations with the moisture content are used in the numerical analyses below. The change of the compressive and flexure strength (modulus of rupture) of the material with W are also determined experimentally to be used as criteria for failure.

Three dimensional geometrical model, representing the real shape of a wet article after the forming by the extruder, is created. It is discretised by a finite element mesh in ANSYS Mechanical APDL using elements, suitable for coupled thermal –structural analysis (figures 4 and 5).

Series of numerical solutions are implemented in order to achieve agreement between the numerical results and the real behaviour of the body during the drying process. Final average moisture content, shrinkage mode during the process (measured by retractometers in the plant) and final dimensions are accepted as criteria to be reached at the simulations.

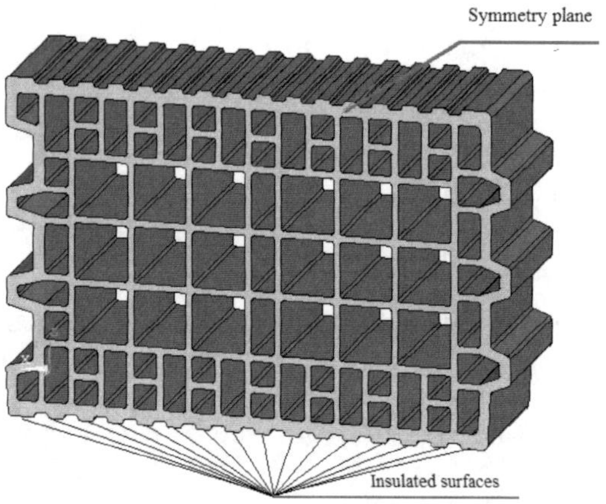

Figure 3. Geometrical model.

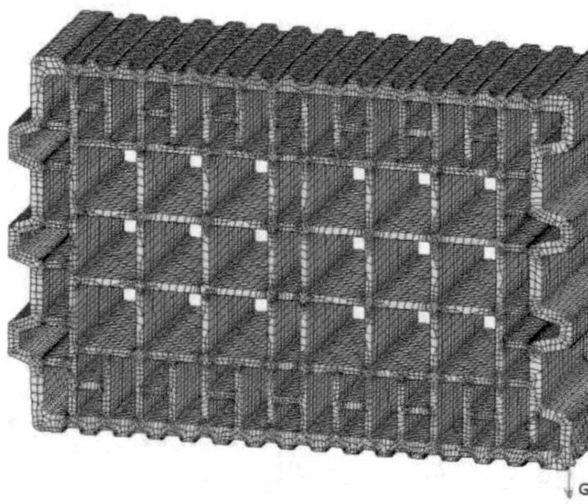

Figure 4. Finite element mesh at elements Solid 226.

The first numerical simulation was done at the assumption that the time duration of the CDRP is 50% of the full time duration, corresponding to the maintained small drying potential in the dryer (Figure 1). The exponent of Re and Gu in equation (8) were calibrated to obtain the mass transfer coefficient and mass flux in equation (5) correspondent to the evaporated amount of water per article at the change of the moisture content from the initial to the critical value according to factory data. So obtained mass flux was used in (15) to compute the boundary conditions.

The simulation was implemented at discretised model using elements Solid 226 – hexaeder with intermediate nodes on each side (total 20 nodes). That allows performing the analyses at a relatively coarse mesh (figure 2) with enough nodes to estimate the gradients in the body. The computation procedure was performed at a relatively long CPU time (19 days at workstation with 128 RAM Memory and processor Intel Xeon E5 2603).

Transient 3D fields of the moisture content, stresses and strains are obtained [8]. Validations of the models were implemented comparing the computed and the maintained in the factory final sizes and moisture content at the dried body. The differences between the computed and real final sizes of the deformed body are less than 2 %. That is a proof for the adequacy of the models of mechanical processes. But the computed final average moisture content was three time higher than the real one. That results in difference of 10 % between the computed and the maintained evaporated amount of water.

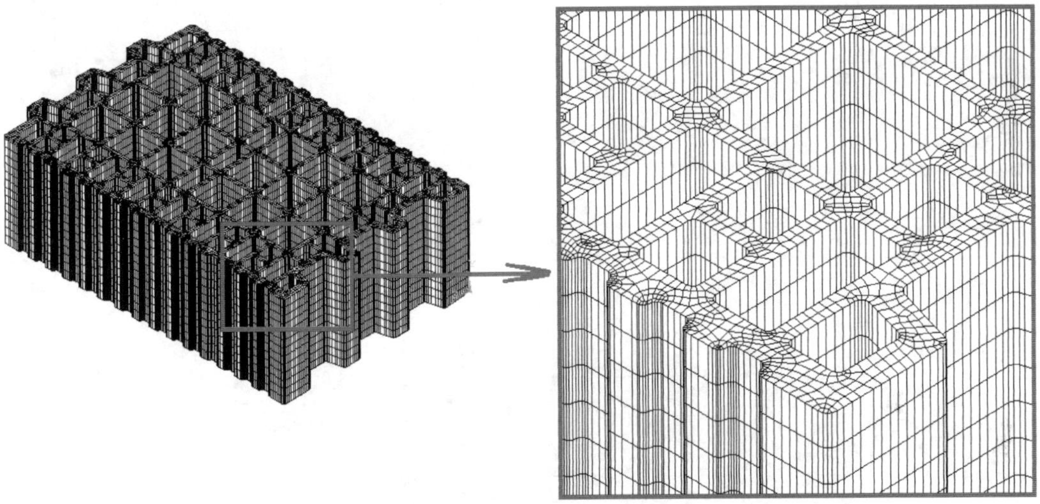

Figure 5. Finite element mesh at elements Solid 5.

Studies were undertaken in several directions to improve adequacy and the efficiency of the models:

1) Finite element mesh with elements Solid 5 (hexaeders with 8 modes) is used. That allows faster computation procedures due to the lower nodes number in comparison to finite element mesh with Solid 226.
2) Numerical analyses of the pressure and velocity fields of the drying media, surrounding the dried bodies were implemented in order to estimate possibilities for model inadequacy due to the assumption of equal mass flux of the mass exchanging surfaces. Such CFD analyses were done at the beginning, middle and end of the drying process (figures 6 and 7). Maximal difference of 10 % in the local velocities around the mass transferring surfaces and subsequent change of the mass flux of 4% according to equation (8) were obtained. That gives reasons to continue the investigations with equal mass fluxes for all surfaces, contacting with the drying media.
3) Correction of the mass flux taking into account the reduction of the sizes of the mass exchanging surfaces. That computed mass flux, used in the boundary conditions is increased dividing the initial value by the square of the average relative linear deformation. As results closer average final moisture content to the real one is obtained. The difference between the computed and reached in the factory amount of evaporated water is reduced to 7 %, which is also unsatisfactory.

IOP Conf. Series: Materials Science and Engineering **595** (2019) 012003 doi:10.1088/1757-899X/595/1/012003

4) Subsequent increasing of the mass fluxes was done at the assumption that the real time duration of CDRP is shorter than the half of the drying process. Actually the moisture content field is non-uniform and the nodal critical moisture content is achieved at different moments of the drying. After several numerical simulations final average moisture content, enough close to the real one was obtained at a difference between the computed and real amount of evaporated water of 3%. Additional reduction of this difference may be sought through a non-linear dependence of the mass flux of the moisture in equations (11) and (16) at FDRP [3].

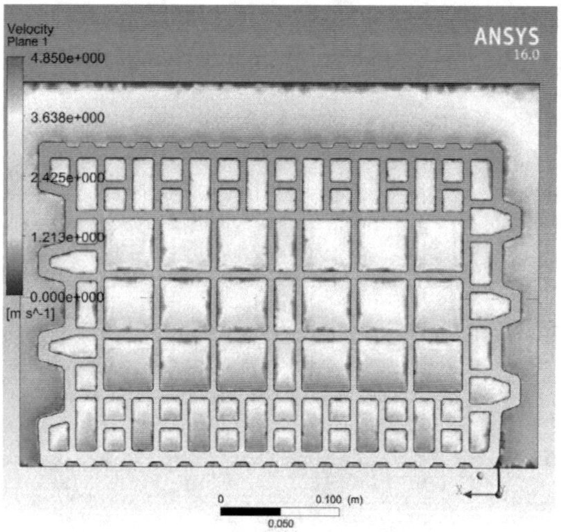

Figure 6. Velocity field in the start of the drying process, m s^{-1} (cross section).

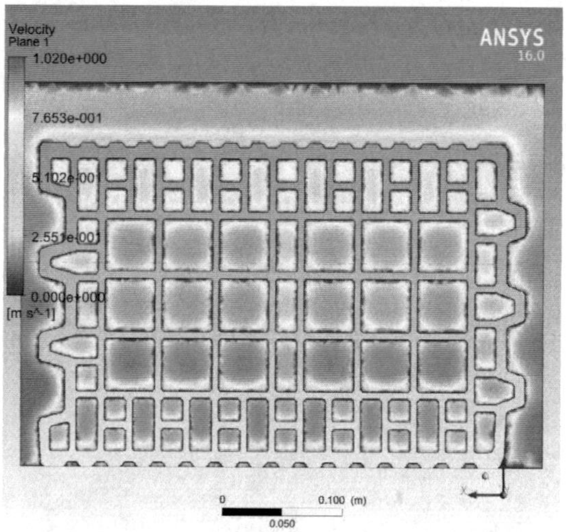

Figure 7. Velocity field in the end of the drying process, m s^{-1} (cross section).

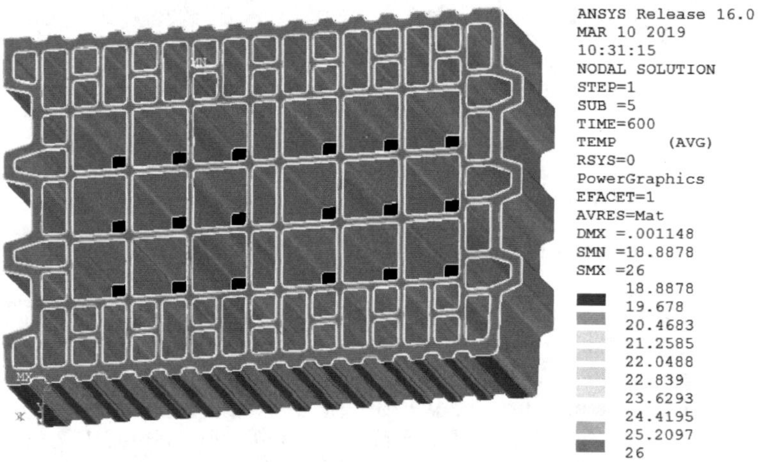

Figure 8. Moisture content at the beginning of the process, % (symmetry plane and insulated surfaces view).

Results about the investigated fields are shown on the Figures 8-12. The non-uniform moisture fields are obvious from the figures bellow. The domains around the insulated boundaries are dried with a smaller rate. That leads to non-uniform deformations, resulting in stresses. The maximal first principal tensile stresses are several times smaller than the modulus of rupture of the material at the correspondent moisture content. Also the maximal absolute values of the first principal compressive stresses don't exceed the maximal permissible compressive stress of the material. So defects in the articles due to non-uniform shrinkage and plastic deformations are not expected. That corresponds to the faultless drying in the factory at the maintained regime.

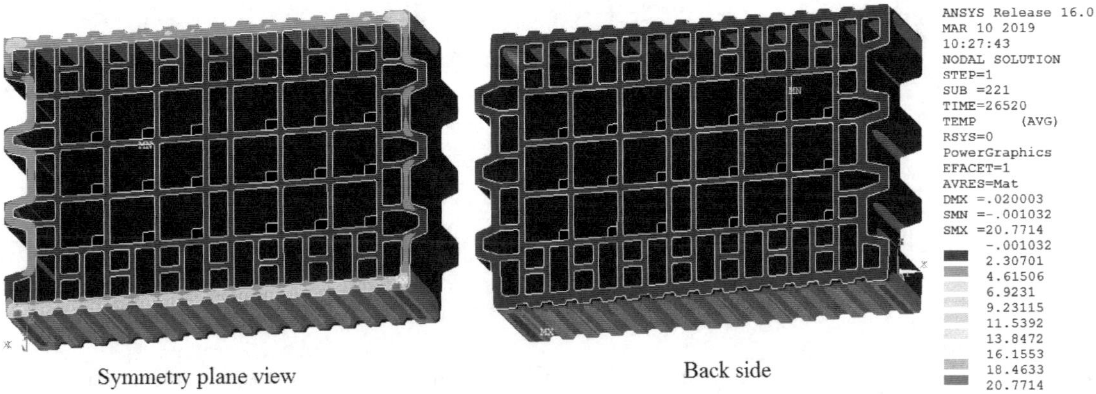

Figure 9. Moisture content at the final of the process, %.

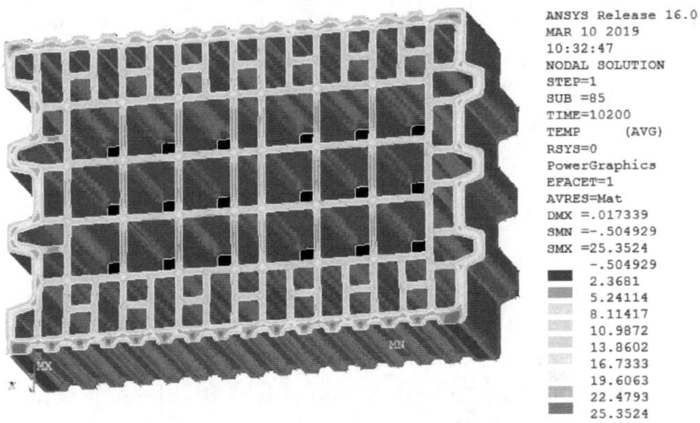

Figure 10. Moisture content at the moment with the maximal stresses in the body, % (symmetry plane and insulated surfaces view).

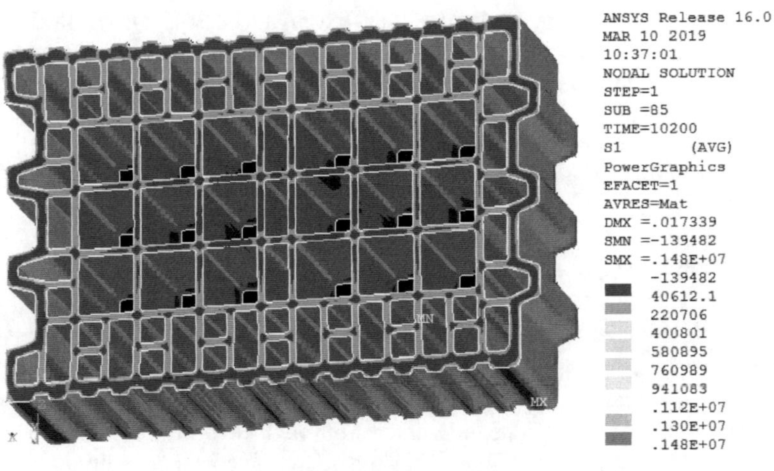

Figure 11. First principal stress at the moment with the maximal stresses, Pa (symmetry plane and insulated surfaces view).

IOP Conf. Series: Materials Science and Engineering **595** (2019) 012003 doi:10.1088/1757-899X/595/1/012003

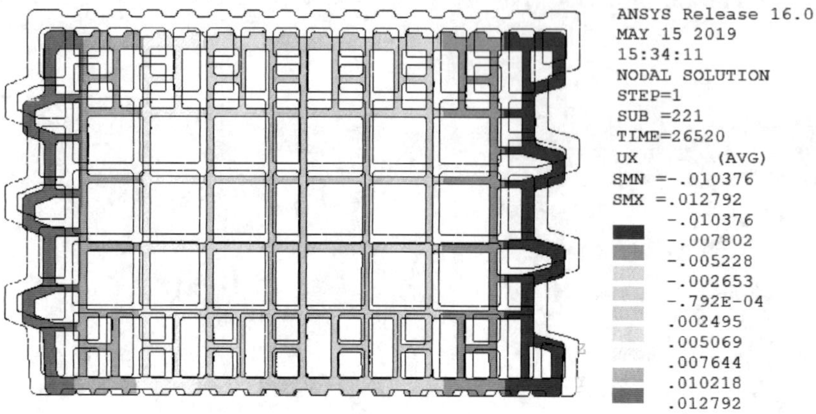

Figure 12. Longitudinal displacements at the end of the process, m (deformed and initial shape).

4. Conclusions

An algorithm for numerical investigation of the structural-diffusion processes in ceramic ware at drying in industrial conditions is composed, tested on a real object and validated. It will be used for investigation of the possibilities for improvements of the efficiency of the dryer installations by reduction of the time duration of the process. That is possible if the moment local stresses, obtained at the numerical simulations at different drying regimes don't exceed the strength at the correspondent moisture content, reduced by a safety coefficient.

References

[1] Valtchev G, Rasheva V and Georgieva M 1998 Possibilities for reduction of fuel consumption in bread ovens *Proceedings of University of Food Technology* Vol XLIII – 3 pp 65-69

[2] Yordanov K, Mechkarova T, Stoyanova A and Zlateva P 2018 Determination of the temperature of cathode unit of indirect plasma burner through a computer simulation model *Proceedings of the Second International Scientific Conference IITI 2017 - Advances in Intelligent Systems and Computing* Vol 680 (Springer Cham) pp 403-409

[3] Lykov A V 1968 *Theory of Drying* (Moscow, in Russian)

[4] Sander A, Skansi D and Bolf N 2003 Heat and mass transfer models in convection drying of clay slabs *Ceramics International* **29(6)** pp 641-653

[5] Heydari M, Khalili K and Ahmadi-Brooghani S Y 2017 Simulation of stresses induced by heat and mass transfer in drying process of clay-like material *JCAMECH* **48(2)** pp 171-184

[6] Chandramohan V P 2016 Experimental analysis and simultaneous heat and moisture transfer with coupled CFD model for convective drying of moist objects *International Journal for Computational Methods in Engineering Science and Mechanics* **17(1)** pp 1550-2295

[7] Penkova N, Chervenliev P, Mladenov B and Krumov K 2018 Modelling and numerical simulation of mass transfer and mechanical processes at ceramic ware in industrial dryers Proceedings *of 6th European Conference on Computational Mechanics and the 7th European Conference on Computational Fluid Dynamics* pp 4368-4379

[8] Crank J. 1975 *The Mathematics of Diffusion* (Oxford University Press, Second edition)

[9] *ANSYS Release 16* 2016 (© SAS IP, Inc.)

Acknowledgments

This study has been financially supported by the Operational Programme "Science and education for smart growth" 2014-2020 of the European Union cofounded by the European Social Fund through the project BG05M2OP001-2.009-0015 “Support for the development of capacity of doctoral students and young researchers in the field of engineering, natural and mathematical sciences”.

The computational procedures are implemented in Centre of mathematical modelling and numerical simulation at University of Chemical Technology and Metallurgy, Sofia.

IOP Conf. Series: Materials Science and Engineering **595** (2019) 012004 doi:10.1088/1757-899X/595/1/012004

Analytical and numerical investigation of heat exchange in fin-tube refrigerating heat exchanger

M Georgieva, P Raynov, D Atanasov and I Tashev

University of Food Technologies – Plovdiv, Bulgaria

E-mail: georgieva_mariya@abv.bg

Abstract. Intensification by tube fin is intended to increase the heat exchange surface between the heat exchange surface and surrounding fluid. This method of intensification is applied to the heat exchangers type air coolers (refrigeration heat exchanger with indirect cooling). The interaction of cooled air in refrigeration equipment is most often a natural convection (quiet cooling). For the purposes of this work, a finned tube heat exchanger with a coolant a brine solution of $CaCl_2$ was viewed. The numerical studies for ribbed cross sections at three different heights were done. On the basis of the results obtained for heat flow in a rib and throughout the exchanger, the effectiveness of the apparatus was determined. The model studies in Flow Simulation were carried out. For the purpose of the simulations a geometric models of two elements from cooling fin - tubes at three different heights (h) of the ribs (10 mm, 20 mm and 30 mm) were made. Numerical studies to investigate the velocity and temperature distribution at different rib height were carried out.

1. Introduction

The intensification of heat transfer without phase transition has been directed to the main thermal resistance along the path of the heat flow - the boundary layer. Methods for intensifying convective heat exchange have been divided into two groups: intensive and extensive. Intensive methods refer to the possibilities for influence on the thickness and character of the flow in the boundary layer. The second group of methods have been applied to these cases and constructive solutions associated with the increase of the heat exchange surface by ribbing. Ribbing to intensify heat exchange on the side with higher thermal resistance (with a lower heat transfer coefficient) has been intended. Accurate demarcation between the two groups of intensification methods cannot be done. Therefore, the application of a specific principle of intensification has been required knowledge of heat transfer and hydraulic or aerodynamic technical features of the interaction of the fluid flows with the heat exchange surface.

In many cases, fluids streamlined the heat exchange surface have very different heat-physical properties, have under different pressure or have a different physical condition.

This is the reason to observe very different values of the two coefficients of convective heat exchange.

Thus, if one surface of the heat exchange wall has streamlined from a gas stream, the coefficient of heat transfer has usually not more than 100 W/(m^2·K), while on the other hand, the liquid refrigerant having a low viscosity or in the condensation of water vapor coefficient of convection can to have one or even two orders of dimension higher [7, 8].

It has been proven that the ribbing has to be done on the side of the low heat transfer coefficient. That's why has been developed of a wide variety of ribbed surfaces - pipes with different cross-section of the pipe, flat or with other geometry.

The method of ribbing has been also different in terms of the geometry of the ribs and in the manufacturing methods.

Cooling fin - tubes may be by direct cooling (via evaporating refrigerant) and with indirect cooling (using a salt solution, etc.). They have been made of smoothness or ribbed pipes with a horizontal or vertical position. Smoothness pipes for cooling fin - tubes have been hardly used, as the thermal load of one linear meter of the pipe has been very low. This led to an increase in the length of the tubes and the amount of refrigerant required for their filling. Application has been mainly for cooling fin-tubes, in which the cost of the metal (tubes) has been decreased about three times and their dimensions have been considerably - small [3].

Ribs of heat exchange surfaces can be of a different profile (rectangular, round, triangular, etc.). Depending on their purpose, their design has been subject to different conditions: effective use of the material, maximum heat transfer, maximum mass, minimum dimensions, etc.

The main characteristics that determine the efficiency of the rib and ribbed wall have been the distribution of the temperature at height and the heat flow. They have been depending on the profile, the shape of the cross section, the material, the dimensions and the boundary conditions of the heat exchange [3, 7, 8].

In the present work have been used analytical and model studies of heat exchanger with applied intensification method. As a particular type of the heat exchanger, cooling fin - tubes have been used to cool the air in cold-storage rooms (chambers) with free convection. They have been used mainly in the chambers for the storage of frozen foods at temperatures $t_f \leq -20$ °C as auxiliary equipment. They can also be combined with air coolers [3].

2. Materials and methods

The investigated cooling fin-tubes are with following geometrical performance: ∅ 38x2; rib thickness– $\delta = 2$ mm; step between ribs – $s = 10$ mm; heights of the ribs (h) – 10 mm; 20 mm; 30 mm (figure 1). The pipe and the ribs have been made of the same material 304L(1.4307), with thermal conductivity $\lambda = 17$ W/(m·K) [10].

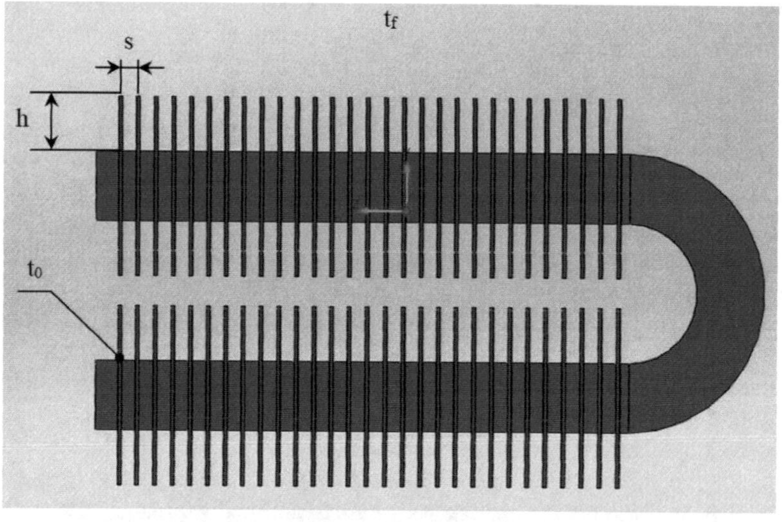

Figure 1. 3D model in h=30mm.

Water solution of $CaCl_2$ has been used for liquid refrigerant at temperature (t_{f1} = -25 °C) [5]. Numerical investigation were conducted at chamber temperature t_{f2} = -20 °C. The fin-tube refrigerating heat exchanger has been used as an auxiliary element in a cooling chamber and therefore a small temperature difference was used. The temperature at the base of the ribs t_0 via applying a heat balance has been determined from:

$$t_0 = \frac{\alpha_2 . t_{f2} + \alpha_1 . t_{f1}}{\alpha_1 + \alpha_2} \tag{1}$$

where α_1 has been the heat transfer coefficient between liquid refrigerant and the inner side of the pipe, $\alpha_1 = 1100$ W/(m^2·K) [3]; α_2 has been the heat transfer coefficient between the ribs and the air in the refrigeration chamber, $\alpha_2 = 3.6$ W/(m^2·K) [3]. Solving equation (1) has been received $t_0 = -24.983$ °C.

The efficiency coefficient E has been used for comparison of the ribs. The same has been the ratio of the heat flow has been taken away from real rib $\dot{Q}_R$ and the heat flow $\dot{Q}_0$ has been taken away from the sample rib under the same conditions and dimension.

$$E = \frac{\dot{Q}_R}{\dot{Q}_0} \tag{2}$$

The heat flow of a real round rib $\dot{Q}_R$ has been determined by the following method:

- An existing rib (conventional rib on a plane surface) has been considered with length $l = 1$ m, thickness δ, adjusted height h'.

$$h' = (r_2 - r_1) + 0.5\delta \tag{3}$$

where: r_2 has been an outer radius of the rib, mm; r_1 - inner radius of the rib, mm.

The heat flow $\dot{Q}_{\text{adj}}$ of this conventional rib has been calculated according to the following equation:

$$\dot{q}_{\text{adj}} = \frac{\dot{Q}_{\text{adj}}}{F_{\text{adj}}} \approx \frac{\lambda f \vartheta_0 m . th(mh')}{2h'} \tag{4}$$

where: f has been the fore-part surface of the rib, m^2; F_{adj} - the around surface of the conventional rib, m^2; ϑ_0 - the temperature difference between the temperature of the fluid and the temperature in the base of the rib, K; m has been the parameter of the rib, m^{-1}. The same have been calculated by following equations:

$$\vartheta_0 = t_0 - t_{f2} \tag{5}$$

$$m = \sqrt{\frac{2\alpha_2}{\lambda\delta}} \tag{6}$$

The heat flow has been calculated for real round rib with the dependence:

$$\dot{Q}_R = \varepsilon_r F_r \dot{q}_{adj} \tag{7}$$

where ε_r has been the correction coefficient accounting the influence of the curvedness of the rib; F_r, m^2 has been calculated the around surface for the real round rib with the dependence:

$$F_r = 2\pi\left[\left(r_2 + \frac{\delta}{2}\right)^2 - r_1^2\right] \tag{8}$$

The correction factor has been accounting by graphic dependencies via figure 2 [7], as a function of relations $\frac{\vartheta_h}{\vartheta_0}$ and $\frac{r_2}{r_1}$, ϑ_h - the temperature difference between the temperature at the edge of the rib and the surrounding fluid

$$\vartheta_h = t_h - t_{f2} \tag{9}$$

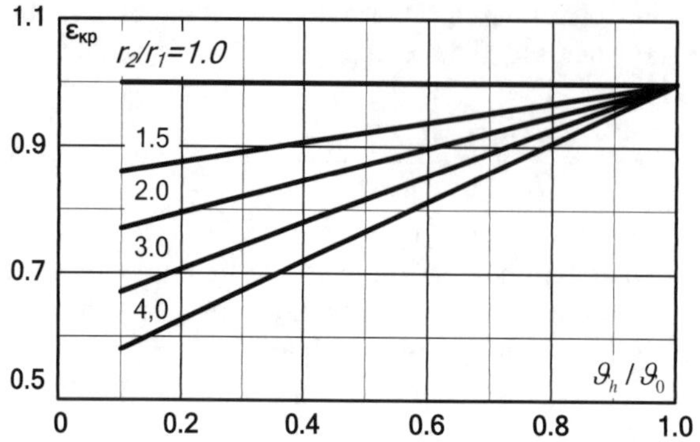

Figure 2. Correction coefficient accounting the influence of the curvedness of the rib.

The heat flow of the conventional rib $\dot{Q}_0$ has been determined according to the Newton equation:

$$\dot{Q}_0 = \alpha_2 F_r \vartheta_0 \tag{10}$$

The rib is conventional, where the temperature on its entire surface is equal to and equal to that at the base.

Numerical studies to investigate the velocity and temperature distribution at different rib height were carried out. For the purpose of the simulations a geometric models of two elements from cooling fin - tubes at three different heights of the ribs (10 mm, 20 mm and 30 mm) were made (figure 1). To create the 3D models a software product SolidWorks was used. It has two CFD tools - FloXpress and FlowSimulation. There are studies of heat exchangers in FloXpress were carry out [4], but due to a number of limitations regarding the initial and boundary conditions of this tool, the numerical studies in the current work in FlowSimulation were carry out. For working fluid, a water solution of $CaCl_2$ was used. $CaCl_2$ is missing in the program library and therefore additionally was added (manually all thermophysical characteristics required for calculation were imported). A square mesh was used, as the specific locations a local mesh with a lesser density was created (Figure 3).

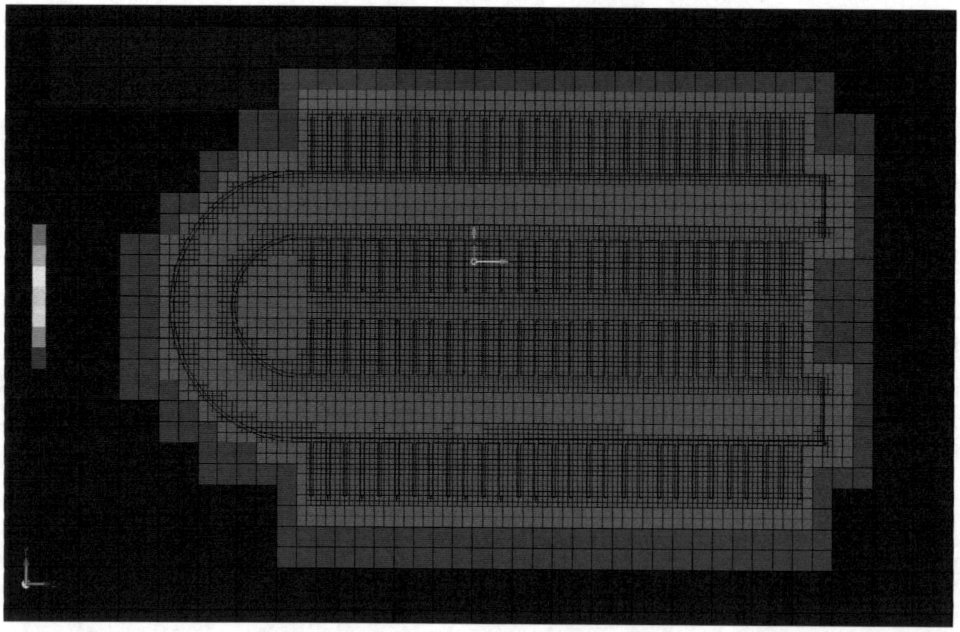

Figure 3. Mesh preview.

Studies of the solution's independence by the density of the mesh were done. Two mesh densities (233173 and 451413 cell number) were studied, and the results obtained concluded that the optimal model for the study was that of 451413 cells. For the verification, the analytically calculated temperature at the base of the rib and that calculated by the program was adopted:

$$d\dot{Q} = \alpha_2(t - t_f)udx = \alpha_2 \vartheta udx \tag{11}$$

The resulting difference with the optimal mesh was 0.03 %. To close the system of partial differential equations for motion [1] the *k-ε* turbulence model was used [2, 6, 9]. The simulations on a computer with processor Intel Core I7 and 16 GB RAM were carried out. Performing a simulation with shown mesh takes approximately 1 hour.

3. Results and discussion

Analytical results were obtained for the thermal flow of the real rib $\dot{Q}_R$ and the heat flow $\dot{Q}_0$ take away from the conventional rib and the efficiency of the round rib.

Table 1. Efficiency of a round rib at different rib heights.

λ	α_1	α_2	r_2	h'	mh	ϑ_h/ϑ_0	$\dot{Q}_{adj}$	$\dot{Q}_0$	$\dot{Q}_R$	E
W/m·K	W/m²K	W/m²K	mm	mm			W	W	W	%
			29	11	0.160	0.987	0.391	0.610	0.051	97.2
17	1100	3.6	39	21	0.305	0.955	0.731	0.134	0.133	95.1
			49	31	0.451	0.906	1.042	0.241	0.225	91.9

On the basis of analytical studies were obtained graphic dependences of efficiency and heat flow from the heights of the ribs (Figure 4).

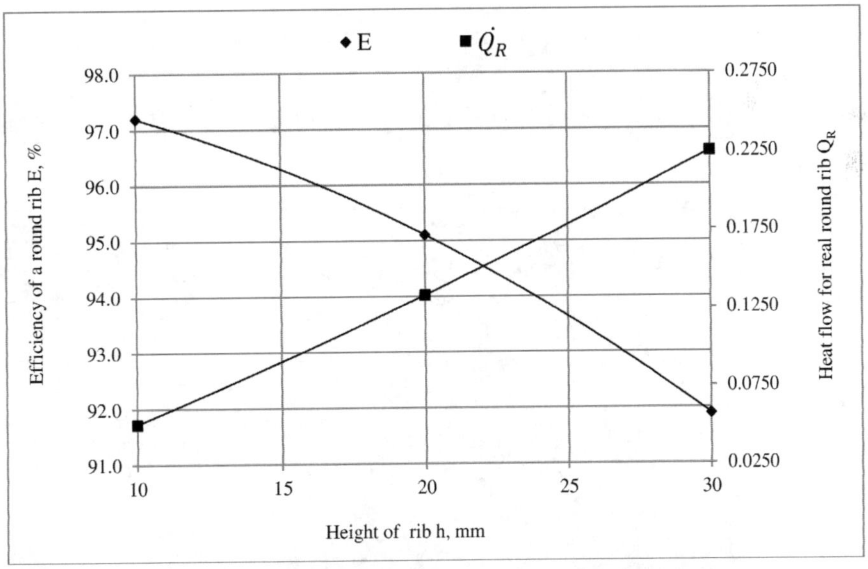

Figure 4. Dependences of efficiency and heat flow from the heights of the ribs.

The results were present that increasing the height of the rib increases the heat flow but reduces its efficiency.

Figures (5, 6, 7) shows the distribution of velocity at the three rib heights.

IOP Conf. Series: Materials Science and Engineering **595** (2019) 012004 doi:10.1088/1757-899X/595/1/012004

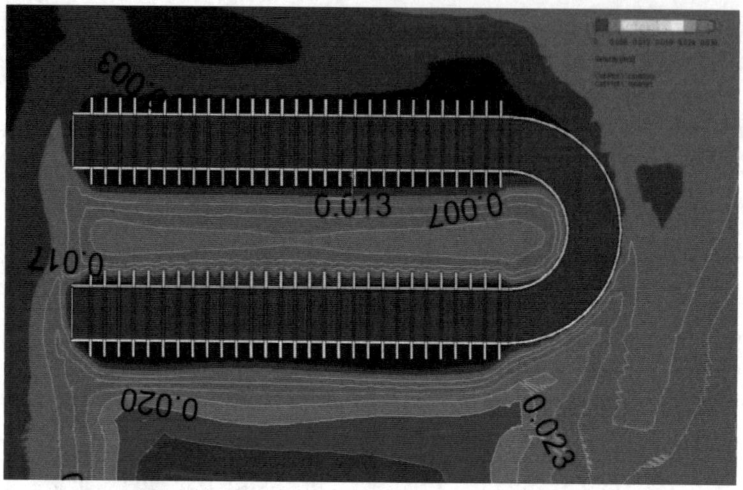

Figure 5. Distribution of velocity in $h = 10$ mm.

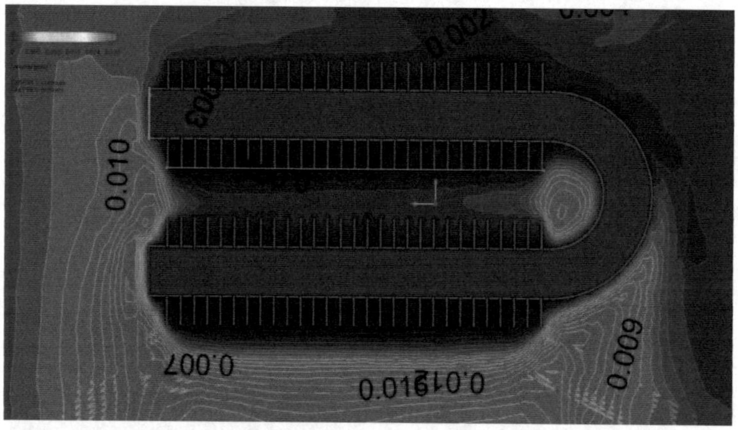

Figure 6. Distribution of velocity in $h = 20$ mm.

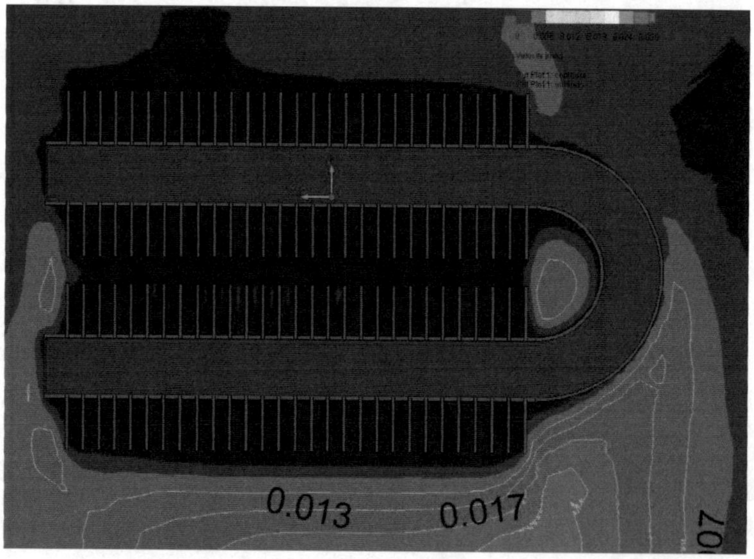

Figure 7. Distribution of velocity in $h = 30$ mm.

It is seen that with increasing height of the ribs, airflow velocity around the ribs between the two elements decreases, it can be argued that almost no natural circulation of air.

Figures 8, 9 and 10 shows the temperature distribution of the first and last rib.

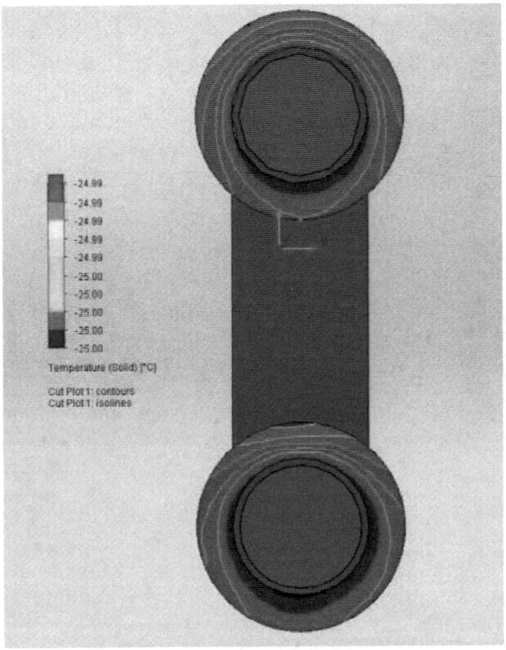

Figure 8. Distribution of temperature in $h = 10$ mm.

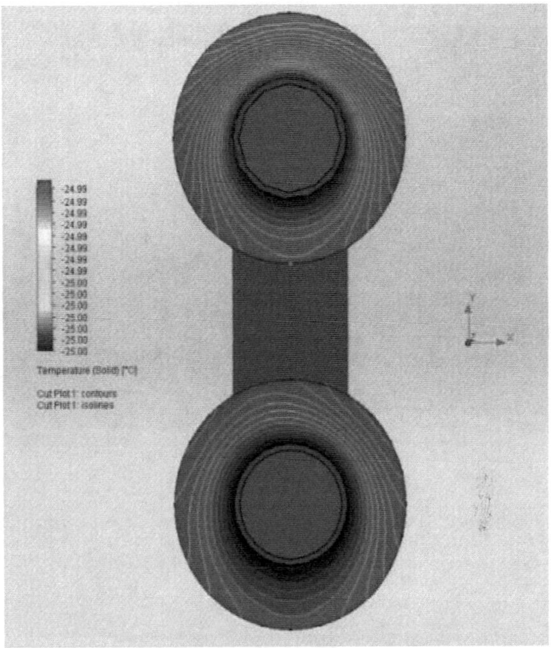

Figure 9. Distribution of temperature in $h = 20$ mm.

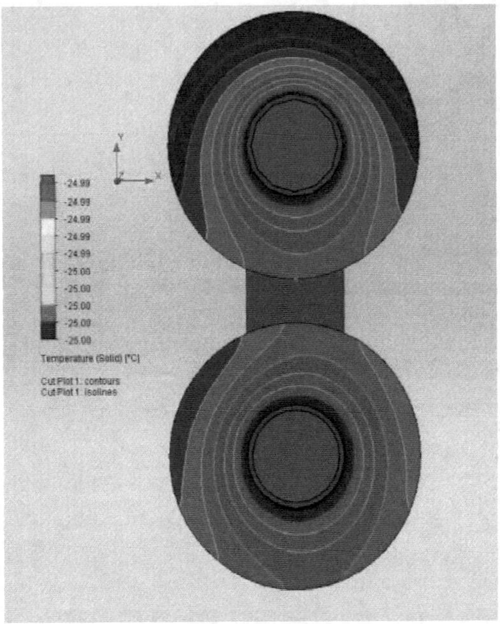

Figure 10. Distribution of temperature in $h = 30$ mm.

At $h = 30$ mm it is noticed that the temperature of a part of the surface of the ribs is close to that of the environment.

IOP Conf. Series: Materials Science and Engineering **595** (2019) 012004 doi:10.1088/1757-899X/595/1/012004

Figures 12, 13 and 14 shows the temperature distribution in length - L1 (Figure 11) for one element of the cooling fin - tubes.

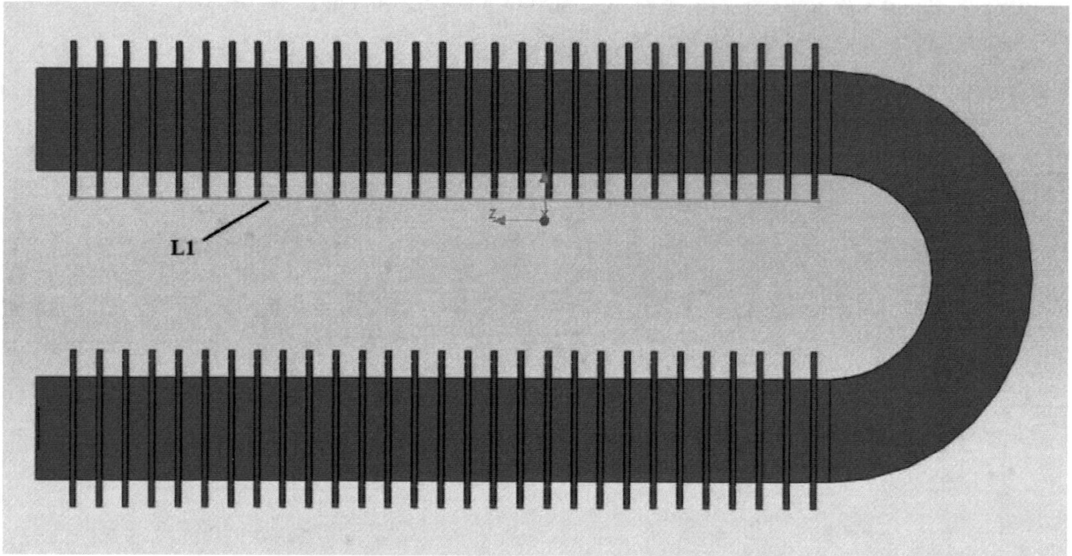

Figure 11. Length (L1) preview.

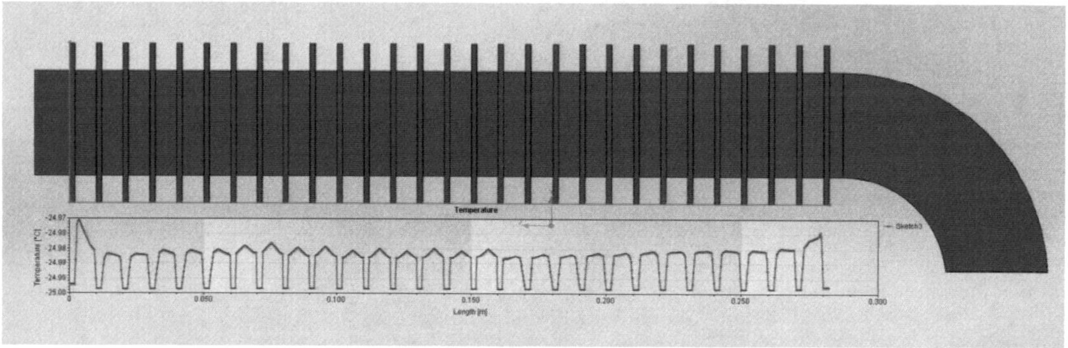

Figure 12. Distribution of temperature for length (L1) in h = 10 mm.

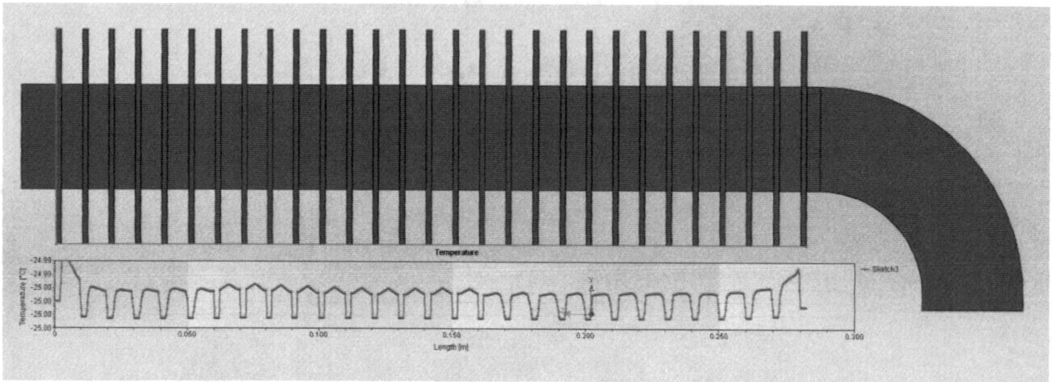

Figure 13. Distribution of temperature for length (L1) in h = 20 mm.

IOP Conf. Series: Materials Science and Engineering **595** (2019) 012004 doi:10.1088/1757-899X/595/1/012004

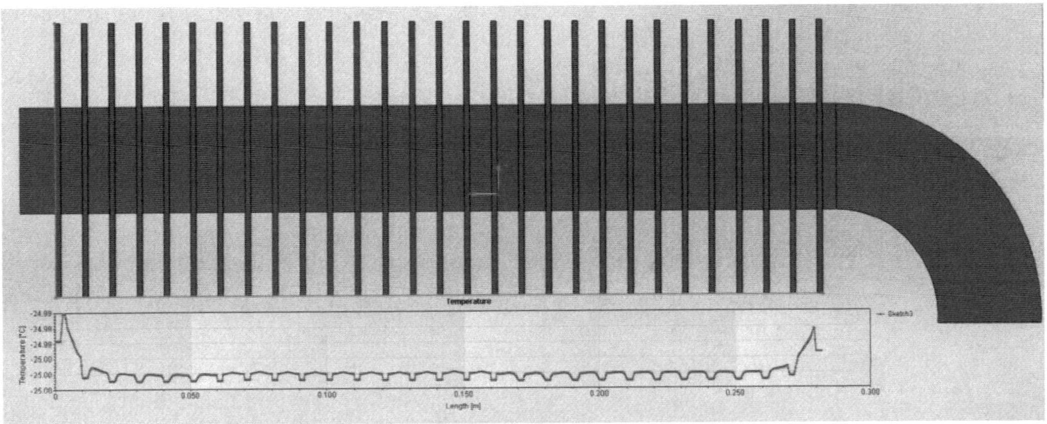

Figure 14. Distribution of temperature for length (L1) in $h = 30$ mm.

Well known that at h = 30 mm the temperature of the air between the ribs is almost equal to their surface temperature, which is related to the lack of good natural convection also is associated with (figure 7).

4. Conclusion

Under the given conditions, the optimal height of the rib is $h = 20$ mm;

On the basis of the simulations it can be argued that the use of ribs with height h = 30 is inappropriate, which is also confirmed by the analytical results.

Modeling the transmission processes for this type of cooling fin - tubes can be successfully done using FLOW SIMULATION from the SOLID WORKS software.

The obtained results can be successfully used in the design, optimization and constructing of this type cooling fin - tubes, as well as in the educational process.

References

[1] Angelov M and Raynov P 2013 Model studies of the hydrodynamics of flow in corrugated tubes, *Journal of food and packaging Science Technique and Technologies*. vol 2

[2] Angelov M, Raynov P and Stoeva D 2012 Possibilities for numerical modeling of the exchanged processes in a double-pass heat, (in Bulg.), *Science conference: Food science, engineering and technologies*. Plovdiv, 19-20 October, Scientific works, vol LIX, pp 768-773

[3] Dichev S 1993 Refrigeration machines, (in Bulg.). Sofia, *Technics*

[4] Hodzheva Z and Rasheva V 2014 Analysis of fluid dynamics in shell and tube heat exchanger with floating head, (in Bulg.), *Scientific research of the Union of Scientists in Bulgaria-Plovdiv, series C. Technics and Technologies*, vol XII, Union of Scientists, Session 31 October – 1 November 2014

[5] Kimenov G 1995 Thermodynamic and Thermophysical Properties of Substances - Guidebook, (in Bulg.), *Technics*

[6] Launder B E and Spalding D B 1974 The numerical computation of turbulent flow. *Comp. Mech. In Appl. Mech. Engng*. **3**, pp 269-289

[7] Raichkov G 2011 Heat and masstransfer, (in Bulg.), *Academic Publishing House of UFT*, Plovdiv

[8] Sendov S and Yordanov P 1981 Intensification of heat exchange, (in Bulg.), Sofia, *Technics*

[9] Stoeva D 2014 Intensification of the heat transfer through corrugated wall, *Journal of Faculty of Food Engineering*. vol XIII, Issue 1, pp 14 – 22

[10] http://www.matweb.com.

IOP Conf. Series: Materials Science and Engineering **595** (2019) 012005 doi:10.1088/1757-899X/595/1/012005

Numerical analysis of the transient heat transfer in high temperature chamber furnaces

K S Krumov and N Y Penkova

University of Chemical Technology and Metallurgy, Sofia, Bulgaria

E-mail: kkrumov@uctm.edu

Abstract. An algorithm for numerical analysis of the transient thermal processes in high temperature gas chamber furnaces for ceramic firing is developed. It is based on mathematical models of the conjugate heat transfer, including combustion, turbulent flows of the torch and gas mixtures, heat transfer through radiation, conduction and convection. The algorithm is used for analysis and improvement of the efficiency of a furnace for firing of technical ceramics.

1. Introduction

The intermittent chamber furnaces are widely used in the ceramic industry because of the possibilities for flexible thermal regimes for firing of different ceramic ware. The firing process includes heating and cooling of the production by increasing, keeping and decreasing of the temperature in the furnace space with the time (so-called temperature curve). That variation depends on the ceramic material, geometry of the fired articles and their arrangement in the space. The non-stationary temperature field in the furnace usually is maintained by means of an adjustable combustion process at unsteady fuel and airflows through the burners. The efficiency of such thermal aggregates is relatively small due to the high thermal loses to the environment by the exhausted gases, heat exchange with the environment and heat accumulation by the furnace envelopes [1]. It also depends on the temperature and velocity fields in the furnace space – they have to be uniform to ensure faultless firing of the ceramics. Computational fluid dynamics and heat transfer, based on finite volume method are useful tools for investigation of the possibilities for improvement of the efficiency of chamber furnaces [2, 3, 4 and 5]. The complete finite volume analysis of the conjugate heat transfer can be performed at a geometrical model, including the solid and gaseous domains of the furnace. It requires large computer resources and computational time at fine finite element meshes and minor steps of the unsteady firing process. Such complex study of the coupled combustion and heat transfer in chamber furnaces at industrial operating conditions was not published so far. An algorithm for investigation of the transient thermal processes in periodically working chamber furnaces via different approaches is presented in this paper. It is validated and applied for improvement of the construction of a gas furnace for firing of technical ceramics.

2. Conceptions for modelling and numerical analysis of the transient heat transfer in gas chamber furnaces

The conjugate heat transfer in the gas chamber furnaces covers coupled transient processes:

- combustion at unsteady turbulent fuel and air flows;
- heat transfer by convection, radiation and conduction from the hot gas mixture to the furnace envelopes and fired ware, arranged on auxiliary refractory construction;
- heat transfer by conduction in the solid media. Usually it is accompanied by endothermic and

IOP Conf. Series: Materials Science and Engineering **595** (2019) 012005 doi:10.1088/1757-899X/595/1/012005

exothermic chemical and phase change processes in the fired materials;

- surface to surface radiation heat exchange by the solid surfaces, seeing each other.

For a complex analysis of the processes above the geometrical model has to include:

- gas domain of the internal furnace space and the burners where fuel and airflows are fed and combustion proceeds (Part 1);
- solid domain of the ware to be fired, arranged on auxiliary refractory construction (Part 2);
- solid envelopes of the furnace: multilayer walls, floor and arch, made by refractory bricks and insulation (Part 3).

The geometrical model is discretized by a finite element mesh – finer in the fluid regions and coarse in the solid regions. The different domains have to share common faces in order to simplify the mesh and the connections between the nodes at the interfaces. The equations below are solved to model the transient conjugate heat transfer in the furnace.

- **Fluid domain (Part 1)**: continuity equation, momentum equations, energy equation, standard κ-ε model turbulence model, boundary layer model [6, 7], ideal gas relations for the air, fuel components and products of combustion, eddy dissipation combustion model and radiation model P1 [8].
- **Solid domains (Part 2 and 3)**: energy equation with non-zero heat generation rate in the ceramic bodies to be fired to model the thermal effects of endo- and exothermic processes in the material (if there exist any). The surface to surface radiation exchange cam be neglected at relatively near temperatures of the solid surfaces.

The temperature in the furnace changes several times during the firing regime: the physical properties of the gases, surrounding elements, furnace masonry and thermal treated products have to be accepted as functions of the temperature.

The initial values of the variables to be computed can be accepted as follows:

- zero velocity components and pressure (according a reference pressure) for each node in the gas space;
- zero mass fractions of fuels components and products of combustion in the gas space: it contains air only (nitrogen and oxygen);
- initial temperatures of the gas and solid space according to the technology stages and the temperature curve.

The models can be solved numerically by CFX of FLUENT at ANSYS Workbench [8] at the boundary conditions in Table 1. The fields of the pressure, gas fractions, velocity components, turbulence parameters and temperatures in the fluid and solid space are obtained as results. The heat fluxes and heat transfer coefficients at all fluid/solid interfaces are computed as output derived variables. That approach (named Variant 1) requires large computer resources and computational time.

Another way to solve the problem (Variant 2) is to generate separate geometrical models of the internal space and furnace envelops and to simulate consequently the heat transfer in them to calibrate the boundary conditions at the interface between the gas space and the walls. Variant 2 is implemented at the steps below.

Generation of geometrical model of Part 1 and 2 (internal furnace space), and a distinct model of Part 3 (solid furnace envelopes).

Numerical simulating of the transient heat transfer in the furnace envelopes (Part 3) according to the boundary conditions in Table 2. The heating and cooling rates are computed according to the temperature curve, accepted at the technology. Two subsequent simulations have to be done. The final nodal temperatures after the first simulation are used as initial ones for a second numerical simulation of the heat transfer. The variation of the heat fluxes to the internal surfaces with the time is determined as results of the second simulation. Their integration over the time gives the accumulated heat by the furnace envelopes ant thermal losses to the environment.

Computation of the moment values of artificial heat transfers coefficients U^* for the furnace envelopes by equation (1):

$$U^* = \frac{\dot{q}_{si}}{\Delta T} \tag{1}$$

IOP Conf. Series: Materials Science and Engineering **595** (2019) 012005 doi:10.1088/1757-899X/595/1/012005

where: $\dot{q}_{si}$- heat flux, Wm^{-2}; ΔT - temperature deference between the internal temperature according the boundary condition T_{si} in Table 2 and the ambient temperature: $\Delta T = T_{si} - T_0$

Numerical simulation of the transient processes in the furnace space (combustion and conjugate heat transfer in Part 1 + Part 2) at the boundary conditions in Table 3.

Comparison of measured temperatures in fixed points in the gaseous domain to the computed nodal temperatures in the same points. In the case of differences exceeding accepted one, ΔT, used at the temperature boundary conditions in Table 2 are corrected. The steps 2, 3 and 4 are repeated until the measured and calculated temperatures are close enough - it is considered that the model of heat transfer with the environment is calibrated and validated.

Both approaches for numerical simulation are demonstrated on Figure 1.

Table 1. Boundary conditions for complex CFD analysis of the furnace (Part 1+Part 2+Part 3).

Boundaries	Boundary conditions
1. External surfaces of vertical walls, floor and arch	$-\lambda \frac{\partial T}{\partial n} + h_{se}(T - T_o) = 0$ $h_{se} = A\sqrt[4]{T - T_o} + \sigma_s \varepsilon \frac{(T^4 - T_o^4)}{T - T_o}$ where: h_{se} - heat transfer coefficient by convection and radiation from the external wall surfaces to the environment, $Wm^{-2}K^{-1}$; T_0 = 293K; λ - thermal conductivity of the external layer, $Wm^{-1}K^{-1}$; ε - emissivity of the surfaces; σ - Stefan-Boltzmann constant; A is a constant that can be accepted as follows [5]: $A = \begin{cases} 3.3 \ at\ arch \\ 2.6 \ at\ vertical\ walls \\ 1.6 \ at\ floor \end{cases}$;
2. Inlets	
2.1 – Air inlets	Air temperature: $T_{air} = T_{air}(\tau)$ Velocity, normal to the boundary: $V_{n,\,air} = V_{n,\,air}(\tau)$* Oxygen and nitrogen mass fractions: g_{O2}=0.22; g_{N2}=0.78
2.2 – Fuel inlets	$T_f = T_o = const.$ Velocity, normal to the boundary: $V_{n,fuel} = V_{n,\,fuel}(\tau)$ * Mass fractions of the i-th fuel components: g_i
3. Exhaust gases outlet	p = 0 Pa according a reference pressure

Table 2. Boundary conditions for thermal analyses of the furnace envelopes (Part 3).

Boundaries	Boundary conditions
1. Internal surfaces of arch, vertical walls and floor	Temperature change mode: $T_{si} = \begin{cases} T_{in} + C_h \tau & at \ \ \tau \le \tau_{heating} \\ T_{max} - \Delta T & at \ \ \tau_{heating} \le \tau \le \tau_{firing} \\ T_{max} - \Delta T - C_c \tau & at \ \ \tau_{firing} \le \tau \le \tau_{cooling} \end{cases}$ C_h= heating rate, Ks^{-1}; C_c= cooling rate, Ks^{-1}; $\tau_{heating}$ = time duration of the heating period, s; τ_{firing}=time duration of the period with constant maximal temperature T_{max} in the furnace, s; $\tau_{cooling}$ = time duration of cooling period, s; T_{in}=initial temperature, K. If C_h=const: $C_h = \frac{T_{max} - \Delta T - T_{in}}{\tau_{heating}}$; Temperature differences: $\Delta T = \begin{cases} 0K & at \ \ arch \\ 50K & at \ \ vertical \ \ walls \\ 100K & at \ \ floor \end{cases}$
2. Burner walls	$T = T_{max}$
3. External surfaces	$-\lambda \frac{\partial T}{\partial n} + h_{se}(T - T_o) = 0$

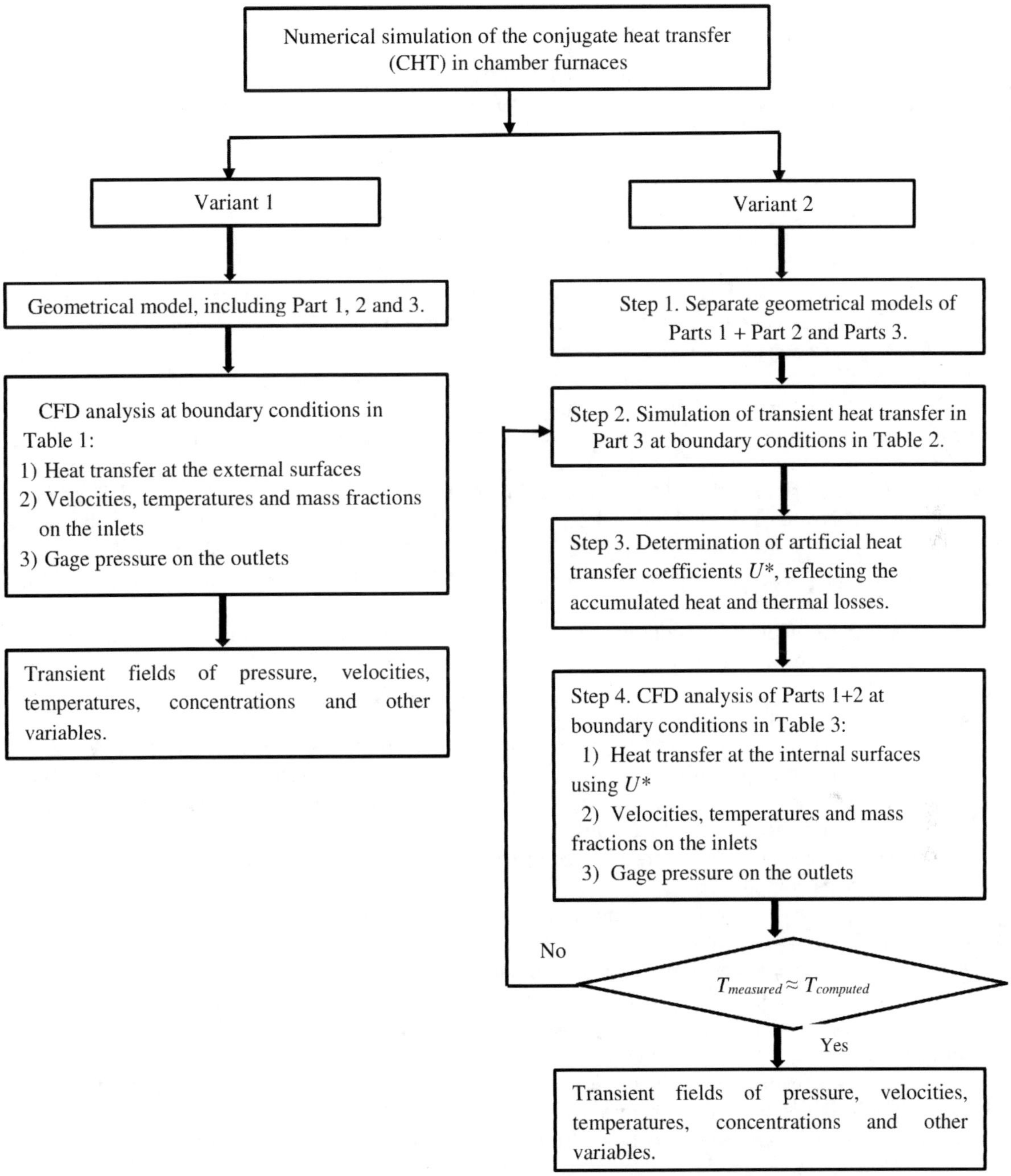

Figure 1. Algorithm for numerical simulation of the transient heat transfer in gas chamber furnaces.

Table 3. Boundary conditions for CFD analysis of the furnace space (Part 1 and 2).

Boundarics	Boundary conditions
1. Internal surfaces of vertical walls, floor and arch	$-\lambda \frac{\partial T}{\partial n} + U^*(T - T_o) = 0$ where U^* is different for arch, floor and walls
2. Inlets	
2.1 – Air inlets	Air temperature: $T_{air} = T_{air}(\tau)$ Velocity, normal to the boundary: $V_{n,\,air} = V_{n,\,air}(\tau)$ *

IOP Conf. Series: Materials Science and Engineering **595** (2019) 012005 doi:10.1088/1757-899X/595/1/012005

	Oxygen and nitrogen mass fractions: g_{O2}=0.22; g_{N2}=0.78
2.2 – Fuel inlets	$T_f = T_o = const.$
	Velocity, normal to the boundary: $V_{n,fuel} = V_{n,fuel}(\tau)$ *
	Mass fractions of the fuel components i: g_i
3. Exhaust gases outlet	p = 0 Pa according a reference pressure

* The moment values of the air and fuel flows are determined by in situ measurements at working aggregates or can be computed by a thermal balance of the furnace and mass balance of the combustion process.

3. Numerical investigations

Variant 2 of the algorithm above is used to analyse and improve the performance of a high temperature chamber furnace for firing of technical ceramics. The heating periods includes increasing of gas temperature to approximately 1600°C for 15 hours and keeping that temperature for an hour. These processes are realized by subsequent rising of the fuel (natural gas) and air flows. The fuel and the combustion air are conducted in the furnace by six burners (Figure 3). The ceramic articles (chocks with relatively small sizes) are arranged on a refractory construction (checker work). During the operations of the furnace non-uniform temperature and velocity fields in the furnace space are established, resulting in unsatisfactory firing of the production and subsequent wastes. Detail information about the furnace and the production is not possible due to confidential rules.

Initially a geometrical model of the ¼ of furnace envelopes, divided by symmetry planes, was generated. The arch, floor and walls are included in the geometry, the burners are neglected. Transient temperature fields are simulated numerically at the overall firing regime to determine the final temperature field after the cooling period that is the initial one for the next heating process (Figure 2). The instantaneous values of artificial overall heat transfer coefficients U^* are calculated for the established heating period.

The geometrical model of the internal furnace space (Parts 1 and 2) is obvious from Figure 3. The articles to be fired are not present in the geometrical models: they are modelled by increasing of the density and the roughness of the horizontal checker work elements. The geometry of the burners is generated according their technical data.

A numerical simulation of the transient combustion, fluid flow and heat transfer processes were implemented at boundary conditions, corresponding to Table 3. The variations of fuel and air flows and temperatures with the time at the inlets of the burners are determined at in situ measurements. They are used to compute the inlet velocities.

The models are calibrated and validated comparing the controlled (measured) temperatures in the furnace space to the computed ones.

The conducted studies proved the inappropriate topology and capacity of the burners and the inability to maintain a uniform temperature field in the furnace chamber by adjusting the combustion process (Figure 3). The power of the burners is higher than the necessary one, computed by a thermal balance. That results in difficulty at the flow control and non-fully combustion in the top row burners (Figure 5). Their positions on the relatively small furnace volume lead to non-uniform thermal loads to the furnace envelopes and the ceramic ware.

A variant for reconstruction of the furnace, changing the burner installation and auxiliary refractory checker work is accepted for further investigations (Figure 4). It is accepted to change the number of the burners, their positions and their power in order to ensure possibilities for smooth flow control, uniform forced convection in the furnace space and full combustion of the fuel.

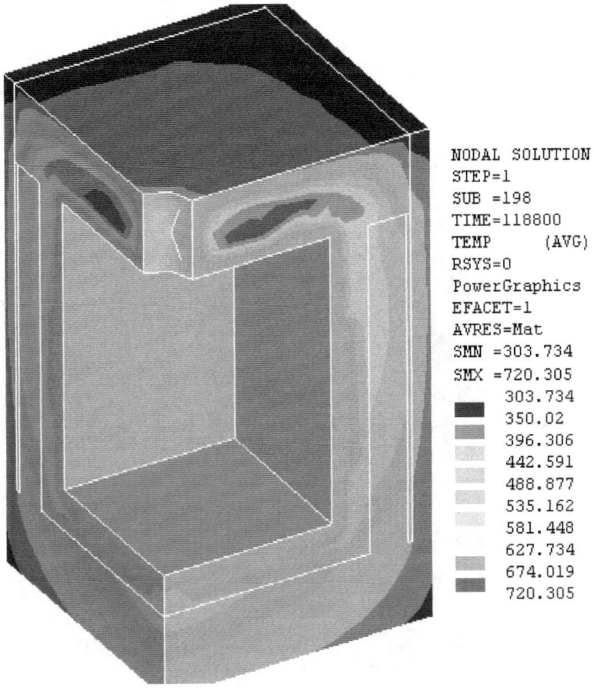

Figure 2. Initial temperature field in the furnace envelope (at the start of the heating process).

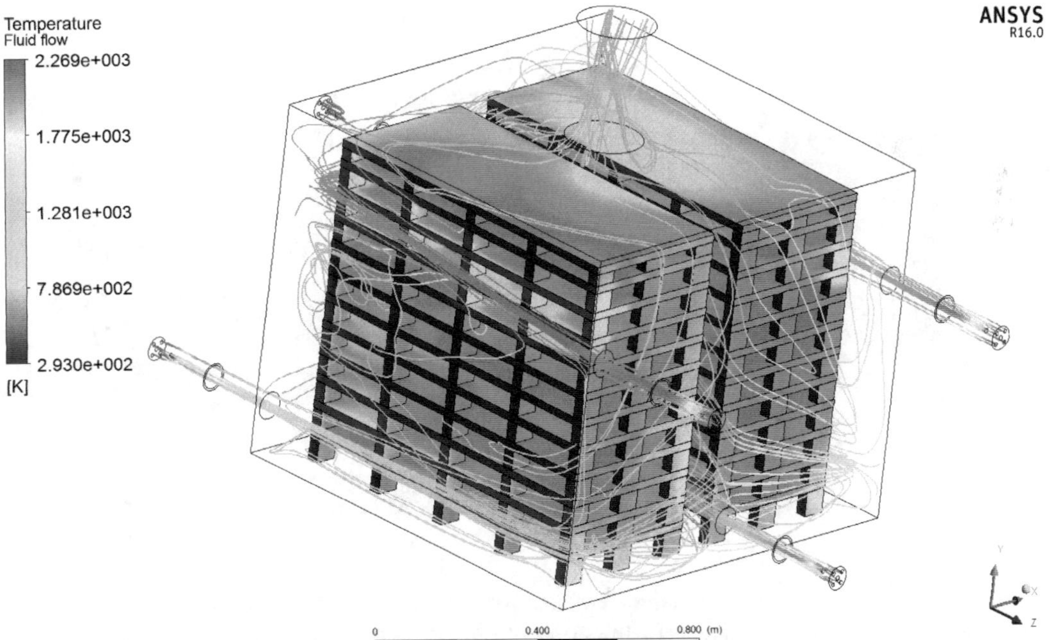

Figure 3. Gas streamlines and temperature fields at the moment of the firing process at the existing construction.

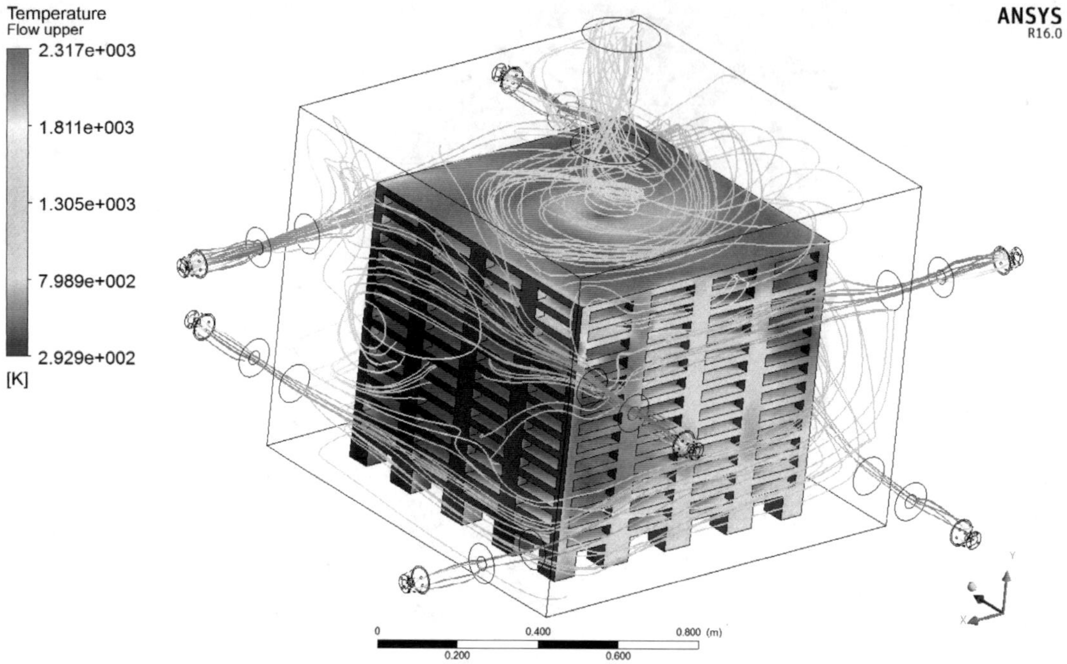

Figure 4. Gas streamlines and temperature fields at the moment of the firing process at the proposed changes.

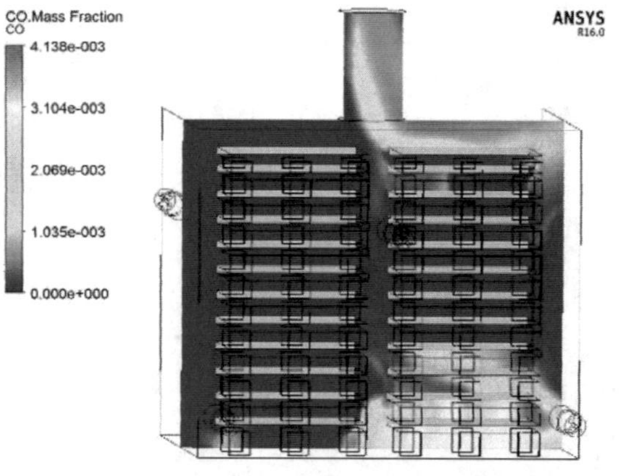

Figure 5. Mass fraction of CO in cross section of the existing variant.

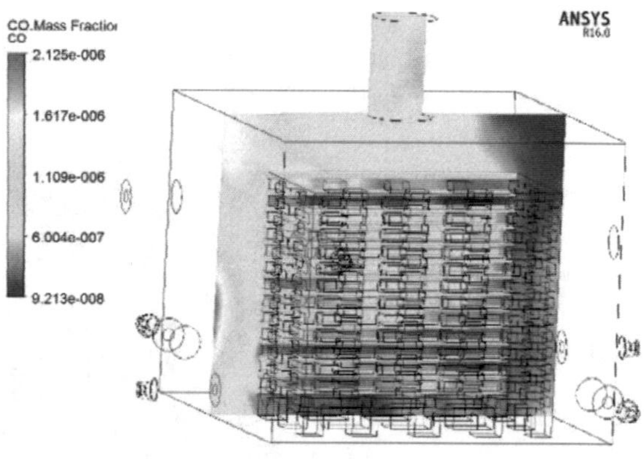

Figure 6. Mass fraction of CO in cross section of the proposed variant for reconstruction.

Numerical simulations of the transient combustion process, temperature, velocity and pressure fields are implemented in accordance to the conception for reconstruction. An increase of time for circulation of hot gases in chamber space at the new burner topology in comparison to the existing one is established. It is due to the extended flow path provoked by the counter current action of the lower and upper burners. As results, complete combustion of the fuel is observed (Figures 5 and 6).

A reduction of the fuel consumption per firing cycle of 48 % is expected after the reconstruction of the furnace aggregate as a result of wastes decreasing, possibility for smooth flow control and a full combustion.

IOP Conf. Series: Materials Science and Engineering **595** (2019) 012005 doi:10.1088/1757-899X/595/1/012005

4. Conclusions

The proposed algorithm for numerical investigation of the transient conjugate heat transfer allows obtaining of detail information of the thermal and fluid flow processes in periodically working gas chamber furnaces. The rapid development of the computer equipment's is a precondition for its complex use and improvement. The algorithm can be applied for analysis of the efficiency of furnaces and connected with them burner installations at operation conditions and project stages.

References

[1] Valtchev G, Rasheva V and Georgieva M 1998 Possibilities for reduction of fuel consumption in bread ovens *Proceedings of University of Food Technology.* **XLIII (3)** p 65-69

[2] Yordanov K, Mechkarova T, Stoyanova A and Zlateva P 2018 Determination of the temperature of cathode unit of indirect plasma burner through a computer simulation model *Proceedings of the Second International Scientific Conference IITI 2017 - Advances in Intelligent Systems and Computing* vol 680 (Springer Cham) p 403-409

[3] Kaushal P, Mohammadazhar A and Arjunsinh M 2018 Development of mathematical model for heat transfer analysis of cylindrical induction furnace wall *International Journal of Advance Research and Innovative Ideas in Education.* **4(6)** p 451-461

[4] Chougule, V N, Daud O F, Gupta G R, Shelke V H and Ugale R V 2013 Numerical analysis of combustion furnace performance by CFD *In Second National Conference on Recent Developments in Mechanical Engineering* p 26-27

[5] Zashkova L 2008 Mathematical modelling of the heat behaviour in the ceramic chamber furnaces at different temperature baking curves *Simulation Modelling Practice and Theory.* **16** p 1640–1658

[6] Launder B and Spalding D 1974 The numerical computation of turbulent flows *Computer Methods in Applied Mechanics and Engineering* **3** p 269-289

[7] Spalding D 1958 Heat transfer from surfaces of non-uniform temperature *Journal of Fluid Mechanics.* **4** p 22-32

[8] ANSYS Release 16 2016 © SAS IP Inc

Acknowledgments

The study has been financially supported by the Operational Programme "Science and education for smart growth" 2014-2020 of the European Union cofounded by the European Social Fund through the project BG05M2OP001-2.009-0015 "Support for the development of capacity of doctoral students and young researchers in the field of engineering, natural and mathematical sciences".

The authors acknowledge the research ideas of National Program Young scientist and postdoctoral students, funded by Bulgarian Ministry of Education.

IOP Conf. Series: Materials Science and Engineering **595** (2019) 012006 doi:10.1088/1757-899X/595/1/012006

Numerical Modelling of Heat Transfer in Convector's Pipes by ABAQUS

Zh D Kolev[1] and S Y Kadirova[2]

[1]Department of Heat, Hydraulics and Environmental Engineering, University of Ruse, Bulgaria
[2]Department of Electronics, University of Ruse, Bulgaria

Email: zkolev@uni-ruse.bg

Abstract. The paper presents method for implementation of numerical modelling of heat transfer process between the water flow and the internal surface of a "water-air" convector's pipes. The convector is a heat consumer in a reversible heat pump installation. To implement the CFD simulations, which are part of the numerical modelling of the processes, the ABAQUS software has been used. The main features and results from the simulations have been analysed. The method includes experimental investigations aiming determination of the boundary conditions for CFD simulation. The distribution of temperature, pressure and velocity in the fluid flowing through the pipe have been investigated. The results of the investigation of the temperature distribution present reduced heat transfer intensity in pipe's sections, characterized by higher local hydraulic resistances.

1. Introduction

One of the most common heat transfer processes in the heat exchangers is the heat transfer process between flowing in a pipe fluid and the pipe's internal surface. These processes are characterized by complexity of influencing factors, which lead to difficulties in their numerical modeling. On the other hand, the investigation of the processes of heat transfer making accurate prediction for design of the heat exchangers [1, 2]. The numerical simulations verify the accuracy and reliability of experimental results [3].

In order to accomplish numerical modeling of such processes and to obtain results regarding the distribution of certain parameters (temperature, pressure, velocity, etc.), experimental investigation is needed to define the input data for CFD simulation [1, 2, 3, 4, 5].

When solving numerical heat transfer problems, use of correct boundary conditions is compulsory. The problem and its numerical modeling will lead to establishment of adequate model with high precision [6].

The three-dimensional CFD simulation of heat transfer processes allow obtaining of information about parameters, which can't be observed experimentally [5, 7, 8, 9].

One of the softwares suitable for computer simulations of heat transfer processes is ABAQUS (ABAQUS FEA engineering software). It is based on finite elements method (FEM) and can be applyed for differents engineering fileds [3, 4, 9, 10].

The "water-water" heat pump installations are characterized with water temperature variation in the heat consumers. Therefore, the boundary conditions are defined for specific operation parameters of the installation [11].

In many studies heat transfer processes have been considered generally. Average values of the heat transfer parameters have been calculated and no detailed analysis has been implemented in respect of certain zones characterized by reduced heat transfer intensity. In this study a quantitative investigation of zones characterized by reduced intensity of the heat transfer process has been implemented.

The aim of the paper is to define heat transfer processes parameters in the convector's pipes, based on CFD modelling.

2. Methods for determination of the input data for CFD simulation

The experimental reversible "water-water" heat pump installation can operate in two regimes - "heating" and "cooling". The installation consists of internal and external circles. The heat consumer is a "water-air" convector, located in the internal circle of the installation [11]. This research focuses on investigation of the heat transfer processes by forced convection in the convector's pipes.

The CFD modelling of the processes can be divided into a few parts. The first one refers to preparing the geometry [1, 3, 5, 9]. Three dimensional model (finite element 3D model) for the water flow has been developed in the current research. CFD is fulfilled by splitting a fluid domain into small cells for creating the mesh [9].

Schematic diagram of the considered simulation model is presented in Figure 1.

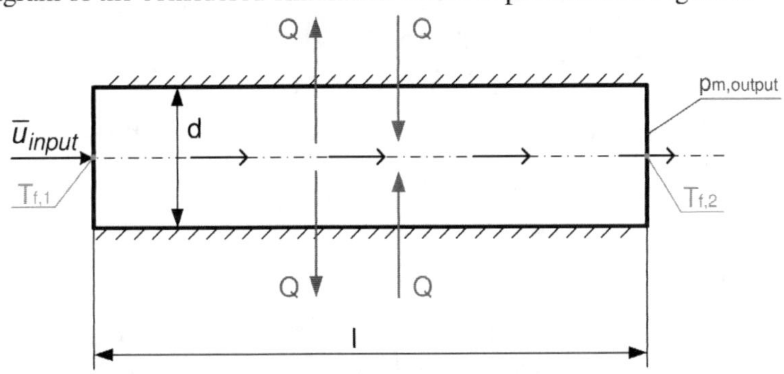

Figure 1. Schematic diagram of the simulation model.

- Measurement of the temperature in the center of the pipe's inlet cross-section (inlet bulk temperature) $T_{f,1}$, °C;
- Measurement of the temperature in the center of the pipe's outlet cross-section (outlet bulk temperature) $T_{f,2}$, °C;
- Calculation of the heat flow Q, W (Eq.1);

$$Q = \dot{m}.c_p.\Delta T_f, W \tag{1}$$

where: $\dot{m}$ is the mass flow rate, kg/s; c_p - mass specific heat capacity (at constant pressure), J/(kg.K); ΔT_f – fluid temperature difference between pipe's inlet and outlet cross-sections, K.

- Calculation of the input average fluid velocity $\overline{U}_{input}$, m/s, based on the measured water volumetric flow rate, measured temperature $T_{f,1}$, the number of the pipes, and calculated cross-sectional area of the pipe.
- Measurement of the outlet fluid manometric pressure $p_{m,output}$, Pa.

3. Results and discussion

3.1. Initial conditions for CFD simulation

The "Tet" shape of the computing cells has been set. In order to receive good regular mesh, an approximate size of 0.005 m of the computing cells is set. The computational domain is discretized by using 69293 cells (calculation model's elements). The result for computing nodes is 18384.

3.2. Input data for CFD simulation

Numerical modelling of two heat transfer processes in the convector's pipes has been implemented. The input data is determined by the developed methods. Dimensions of fluid flow are assumed to: d = 8 mm, and l = 4087 mm. These are the geometric dimensions of the internal surface of the convector's pipes [11]. The input data of the simulated processes are shown in Table 1.

Table 1. Input data of the simulated heat transfer processes.

Operation regime of the installation	$\overline{U}_{input}$, m/s	$p_{m,output}$, Pa	$T_{f,1}$, °C	$T_{f,2}$, °C	q, W/m^2
"heating"	1.472	150000	50.8	42.0	26220.2
"cooling"	2.536	150000	7.5	12.4	25569.2

3.3. Validation of CFD model

Validation aims to verify the accuracy of CFD model [1, 3]. In study [1] the numerical results on heat transfer and pressure drop are validated using theoritical correlations. In [2] the used parameter for validation of CFD model is the average temperature of the investigated process.

In the present study, the validation is performed by comparing the simulation and experimental values of the average heat convection coefficient, between the fluid, flowing in the convector's pipes, and their internal heat exchange surface. For this purpose, the data of the „cooling" regime of the installation is used (Table 1).

The simulation value of the heat convection coefficient is calculated using the average cross-section temperature difference, regarding to the pipe's length. For this purpose results of the temperature field were taken from the CFD post.

The experimental value of the heat convection coefficient is calculated according to a criterion equation, using the measured inlet and outlet pipe's temperatures.

The calculated error obtained from the validation of the model is 4.92 %.

3.4. Simulation results for temperature field

The simulation results, regarding the data in Table 1 are presented in Figure 2.

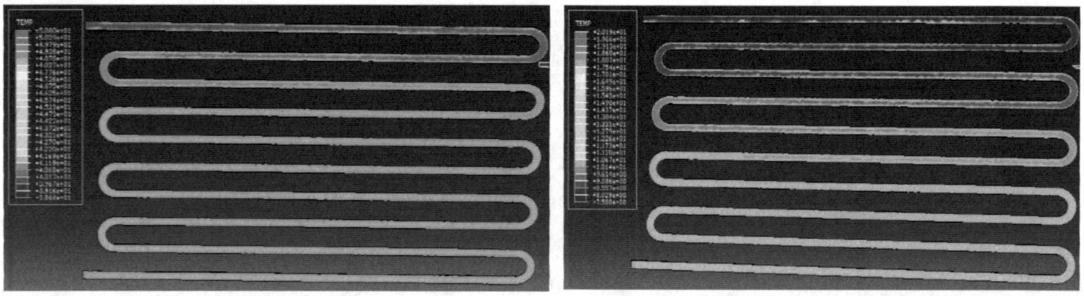

a) "heating" regime b) "cooling" regime

Figure 2. Temperature field in longitudinal cross-section.

The variation of the fluid temperature depends on the heat flow direction. At "heating"regime the temperature decreases in direction of flow, forasmuch as the heat flow direction is to the pipe's internal surface. In "cooling" regime is seen the opposite, respectively.

3.5. Investigation of zones with reduced heat transfer intensity

Figure 3 presents zones with relatively decreased amount of temperature of the external cylindrical surface of the fluid, which is the pipe's internal heat exchange surface, in "heating" regime. Low temperatures indicate relatively low intensity of the heat transfer. This is an indicator of high temperature difference in the pipe's cross-section. There is a necessity of quantitative investigation of the influence of hydraulic resistances on the heat transfer intensity.

IOP Conf. Series: Materials Science and Engineering **595** (2019) 012006 doi:10.1088/1757-899X/595/1/012006

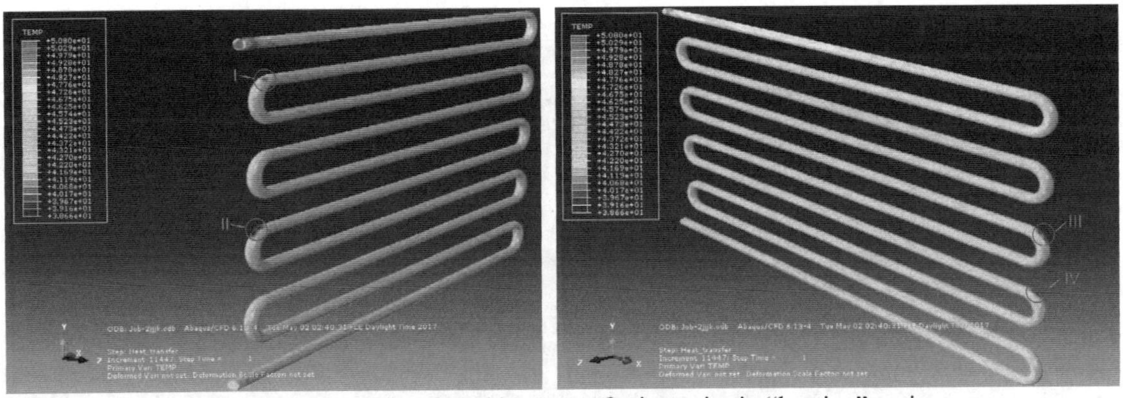

Figure 3. Zones with reduced heat transfer intensity in “heating” regime

Similarly, the zones with visible reduced intensity of the heat transfer process in "cooling" regime are presented in Figure 4. The location of these zones is analogous to the location of the zones, formed during the operation of the installation in "heating" regime. Relatively high surface temperatures are observed, which are indicators of a relatively large temperature difference in the pipe’s cross-sections and for reduced heat transfer intensity.

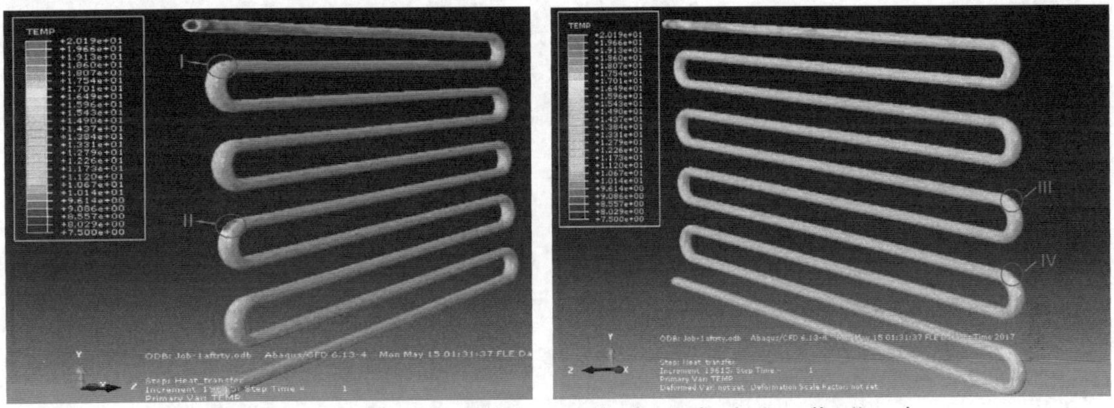

Figure 4. Zones with reduced heat transfer intensity in “cooling” regime.

The investigated cross-sections (Fig. 3 and Fig. 4) are presented in Table 2.

Table 2. Location of the cross-sections.

Cross-section	A	B	C	D
Zone	I	II	III	IV
Distance from the pipe’s inlet, m	0.021267	0.023827	0.405538	0.403126

Indicator of the reduced heat transfer intensity in the region of the respective cross-section is the ratio $\Delta T_H/\Delta T_N$. The parameters ΔT_H and ΔT_N are respectively relatively high temperature difference and normal temperature difference, specific for relevant the cross-section.

The nodes for temperature differences ΔT_H and ΔT_N are defined in “heating“ regime (Figure 5). A node located closed to of the cross-section’s center has been chosen for determination of the temperature differences. For each investigated cross-section it has been established in advance that the difference between the temperatures of the nodes, located near the center of the cross-section, is negligible. For determination of ΔT_H, a node located on the external cylindrical surface of the fluid, in the region with relatively low temperature, has been selected. For ΔT_N, the selected node from the external surface is in a region with a normal surface temperature for the cross-section.

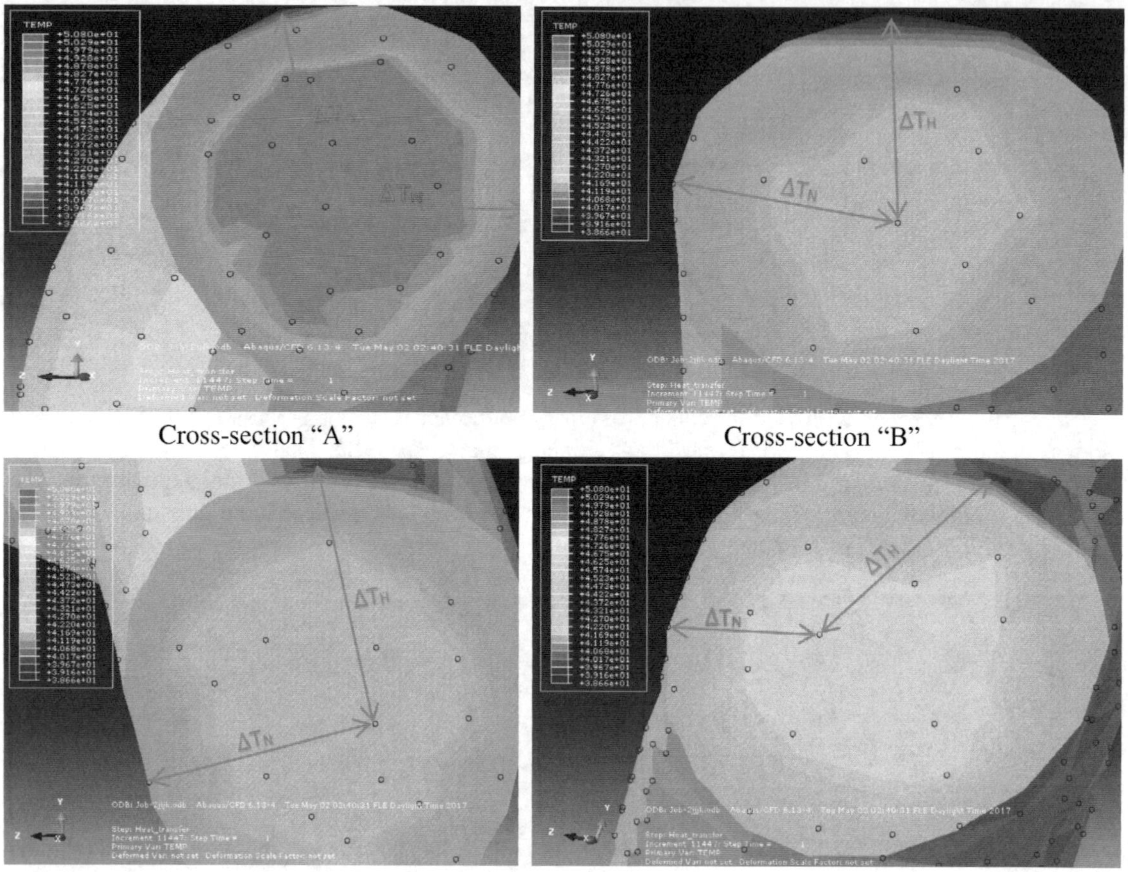

Cross-section "A" Cross-section "B"

Cross-section "C" Cross-section "D"

Figure 5. Cross-sections in "heating" regime.

Similarly, the CFD results for the cross-sections in "cooling" regime have been ploted in Figure 6. The node selected from the external fluid surface, used for calculation of ΔT_H , is with relatively high temperature.

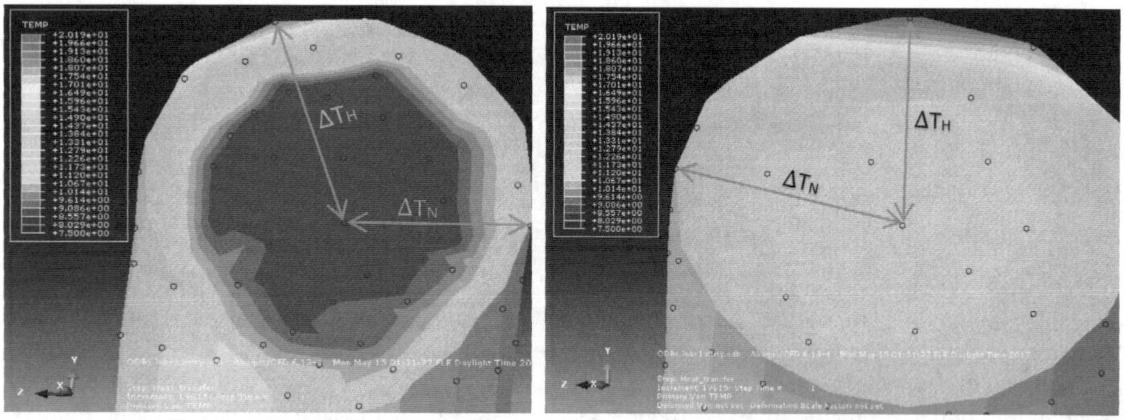

Cross-section "A" Cross-section "B"

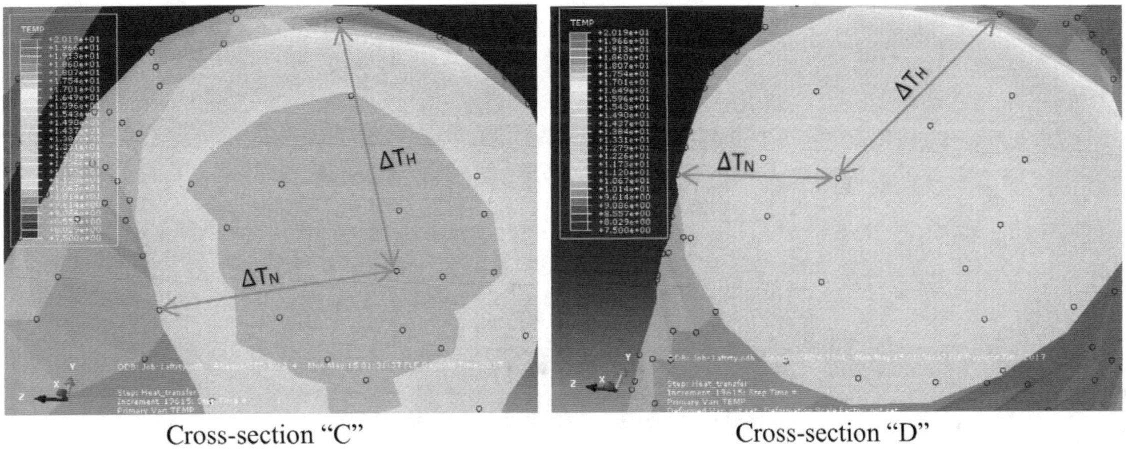

Cross-section “C” Cross-section “D”

Figure 6. Cross-sections in “cooling” regime.

Table 3 contains results obtained from the CFD post, for the investigated in "heating" and "cooling" regimes.

Table 3. Results for "heating" and "cooling" regimes.

Regime	Cross-section	ΔT_H, °C	ΔT_N, °C	$\Delta T_H/\Delta T_N$, [-]
“heating”	“A”	8.752	3.415	2.563
	“B”	8.676	3.919	2.214
	“C”	7.587	3.133	2.423
	“D”	7.342	3.500	2.098
“cooling”	“A”	9.781	3.634	2.691
	“B”	9.259	3.959	2.338
	“C”	8.181	3.207	2.551
	“D”	7.729	3.531	2.189

The ratio $\Delta T_H/\Delta T_N$ is with higher values for the cross-sections, closer to the knees's centers, comparing to those closer to the linear pipe's sections ("A" with "B"; and "C" with "D", respectively). Therefore, the local hydraulic resistances in the cross-sections, closer to the knees' centers, have higher impact on the heat transfer intensity.

3.6. Investigation of the influence of the fluid velocity on the heat transfer intensity

The investigation has been realized at various heat transfer process parameters, obtained by control of the average fluid velocity $\overline{U}_{input}$ in the pipe. Based on the uniform quality change of the temperature at the different fluid velocity studied, the selected cross-section is "D". The study is in “heating” regime.

The parameters of the heat transfer process and the results obtained from CFD modelling, for the temperature differences ΔT_H and ΔT_N, are presented in Table 4.

Table 4. Results for the temperature differences.

$\overline{U}_{input}$, m/s	$T_{f,1}$, °C	$T_{f,2}$, °C	q,W/m^2	ΔT_H, °C	ΔT_N, °C
1.389	50.8	42.4	23624.9	6.857	3.289
1.592	50.8	43.4	23836.2	6.327	2.982
1.860	50.8	44.5	23709.9	5.658	2.618
2.267	50.8	45.6	23838.6	4.852	2.229
2.470	50.8	45.8	24974.3	4.768	2.179

Figure 7 presents the graphical relation between the ratio of temperature differences and the average fluid velocity in the pipe.

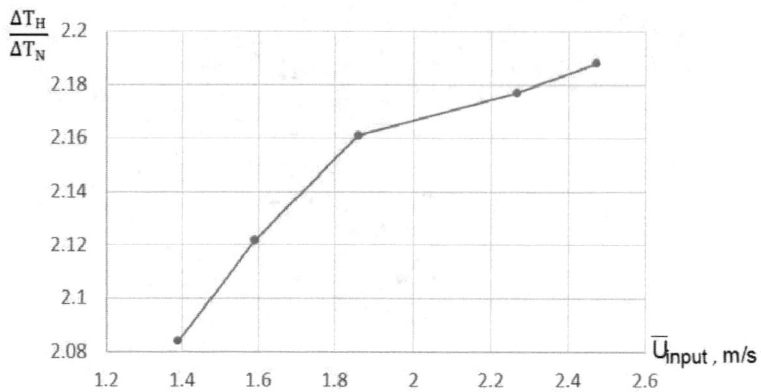

Figure 7. Graphical representation of the function $\frac{\Delta T_H}{\Delta T_N} = f(\overline{U}_{input})$.

From the investigation can be concluded that at higher velocity values, the curve is less sloping, i.e. the difference in ratio values decrease. Hence the increase in fluid velocity results in decrease in the rate of decrease of the heat transfer intensity.

4. Conclusion

This investigation based on CFD modeling of the heat transfer process in the convector's pipes, obtains results for the temperature distribution. This information is used for determination of the heat convection coefficient in the pipes of the heat exchanger.

Results for quantitative temperature variation between defined computing nodes have been obtained.

The accuracy of the results obtained from the CFD simulation depends on the set number of computing cells (the grid density). The calculated model's error is 4.92 %. The built finite element computing mesh provides the necessary quality of results simultanesly with not very long duration of the simulations.

From the three-dimensional numerical analysis is found that the local hydraulic resistances in the pipe influence on the intensity of the heat transfer process. The increase of the fluid velocity results in a decrease in the heat transfer intensity in the investigated cross-sections' regions of the pipe's knees. The variation of the fluid velocity in the range of 1.39 to 2.47 m/s corresponds to increase of the temperature differences ratio from 2.084 to 2.188 (by 4.75 %).

References

[1] Cebrucean D and Cebrucean V 2017 Numerical study of a shell-and-tube heat exchanger for heating rich monoethalonamine using hot flue gases – Part II. Tube-side heat transfer enhancement, *Revista Termotehnica*. XXII (2) pp 37-41

[2] Osley W, Droegemueller P and Ellerby P 2013 CFD investigation of heat transfer and flow patterns in tube side laminar flow and the potential for enhancement, *Chemical Engineering Transactions* **35** pp 997–1002

[3] Oon C, Togun H, Kazi S, Badarudin A, Zubir M and Sadeghinezhad E 2012 Numerical simulation of heat transfer to separation air flow in an annular pipe, *International Communications in Heat and Mass Transfer* **39** pp 1176–1180

[4] Abdelkader K, Youcef K and Sabiha T 2016 Numerical simulation of free vibrations of multilayer composite beams, Proceedings of 2nd Conference on Advances in Mechanical Engineering ICAME2016, Istanbul, Turkey, p. 105–109

[5] Yordanov K, Mechkarova T, Stoyanova A and Zlateva P 2018 Determination of the temperature of cathode unit of indirect plasma burner through a computer simulation model, Proceedings of the Second International Scientific Conference "Intelligent Information Technologies for Industry" (IITI'17), pp 403-409

[6] Yordanov K, Zlateva P, Hadzhidimov I and Stoyanova A 2018 Testing and clearing the high temperature module error from 0 to 1250°C for measurement with 16 K-type thermocouples, Proceedings of the 20th International Symposium on Electrical Apparatus and Technologies (SIELA), pp 480-483

[7] Han-Taw C, Yu-Jie C, Chein-Shan L and Jiang-Ren C 2017 Numerical and experimental study of natural convection heat transfer characteristics for vertical annular finned tube heat exchanger, *International Journal of Heat and Mass Transfer* **109** pp 378–392

[8] Majewski K and Grądziel S 2016 CFD simulations of heat transfer in internally helically ribbed tubes, *Chemical and Process Engineering* **37** **(2)** pp 251-260

[9] Saternus Z and Piekarska W 2017 Numerical analysis of thermomechanical phenomena in laser welded pipe-to-flat, *Procedia Engineering* **17** pp 196-203

[10] Steinheimer R and Engel B 2014 Thermal influences during rotary draw bending of tubes from stainless steel, *Procedia Engineering* **81** pp 2165–2170

[11] Kolev Zh, Kadirova S and Nenov T 2017 Research of reversible heat pump installation for greenhouse heating, *INMATEH - Agricultural Engineering* **2** pp 77-84

IOP Conf. Series: Materials Science and Engineering **595** (2019) 012007 doi:10.1088/1757-899X/595/1/012007

Research of the heat transfer crisis and limit state of the heat exchange surface covered by capillary porous medium

A A Genbach[1], K K Shokolakov[1], D Y Bondartsev[1] and I Iliev[2]

[1]Almaty University of Power Engineering & Telecommunications, Almaty, Kazakhstan
[2]"Angel Kanchev" University of Ruse, Ruse, Bulgaria

E-mail: iiliev@enconservices.com

Abstract. Model of the vapour bubbles formed at the solid surface in porous structures and steam generating wall (base) was developed. The model is based on filming and photography by high-speed camcorder SKS-1M. Heat flow extraction (up to 2.10^6 Wm^{-2}) is ensured by the joint action of capillary and mass forces with application of intensifier. The analytical model is based on the theory of thermoelectricity. Limit state of the poor conductive porous coating and metal base are determined. Heat flows were calculated from the time of spontaneous birth of vapour bubble (10-8) to the material destruction time ($10^2 \div 10^3$s), i.e. the time interval was described from the relaxation process to macro process (destruction). The size of pulled particles at the moment of porous coating destruction in model gave a good match with trial at the optic stand.

1. Introduction

The internal characteristics of the liquid boiling process in capillary porous structures were assessed with high speed filming and photography from birth of steam phase [1] to its destruction. That helped develop models and mechanisms of the heat transfer, and obtain single dependencies for different modes of boiling up to the critical state [2]. Heat exchanges were controlled by the joint action of capillary and mass forces. Visualization of heat influence was also performed by holography, which helped research the limit state of good and poor conductive materials in form of porous structures and steam generating surface. Control of heat exchange in porous structures was ensured by influence upon the internal boiling characteristics and the integral values [1, 8-9].

It is required to give a comparative analysis to the newly proposed capillary porous cooling system that applies a joint action of capillary and mass forces with the traditional systems such as boiling in large volume with the help of thin-filmed evaporators and heat pipes [3-7]. Authors [3] perform a comparative analysis of the calculation methods of heat transfer as per the water boiling along with subcooling in vertical channels, calculating corrosion cases as analogy of the capillary porous structure [4, 5]. However, research of heat exchange on the regular structured surface was not conducted; also, the role of speed and subcooling of liquid towards the boiling crisis in porous surfaces is yet not clear.

In the authors' opinion [6, 7] the surface boiling on porous surfaces could cause or intensify corrosion, and possible erosion, as well as the surface strength when vapour bubbles collapse in the subcooled liquid. Therefore, heat exchange shall be examined in field of mass and capillary forces, taking into account the excessive liquid.

There is a series of experimental devices that allow to research the integral heat transfer characteristics: specific heat flows q, consumption of liquid and steam m_l, m_s, distribution of the temperature field as per height and length of the heat exchanged surface. Methods of holography [8], summarizing similar [9] and identical cases [1, 2] are applied to research the heat exchange mechanism.

It is interesting to provide comparison to intensity of heat transfer and the surface limit state (boiling crisis) with heat pipes [10, 11], as well as make assessment of the fracture mechanics of heated (cooled) surfaces, coated by capillary porous structures applicable to strength, resource and justification of safe operation of the heat mechanical equipment of power plants [12, 13]. This problem arose a long time ago due to the upgrade and extension of the service life of the steam turbine power plant.

2. Trial units for research of the heat exchange crisis and the surface limit state

We developed and studied the trial units to determine the integral (average) heat exchange characteristics of the capillary porous cooling system; the operational diagram and methodology of measurement were established, as well as the cooling element device with pipe arteries, perforated spring and micro arteries. The different factors were studied, such as height of heat exchange surface, pressure in the cooling system up to the wall burning-off (Figures 1 and 2).

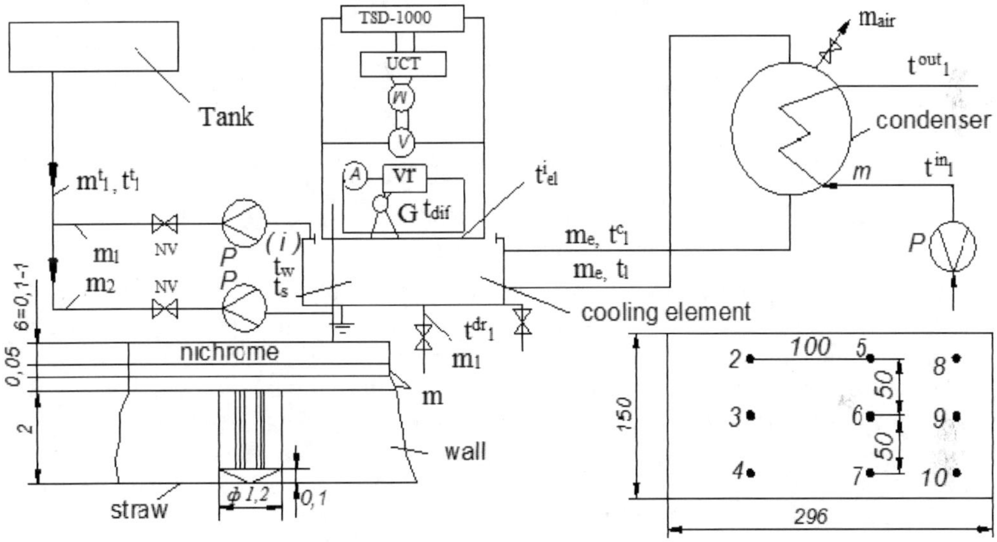

Figure 1. Operation scheme of a porous system and measurement technique: TSD-1000 - welding transformer; UCT - universal current transformer; W - power meter; V-voltmeter; A - ammeter; VR - voltage regulator; G - galvanometer; R - rotameter; NV - needle valve.

Power supply to the main heater is maintained by the welding transformer TSD-1000, output voltage values are as follows: 2.5; 5; 7.5 and 10. Current that feeds the heater is measured as per diagram with universal transformer UTT-6M2 cl. 0,2. Secondary current is up to 5 A, primary is 100-2000 A. Voltage drop at heater is measured by voltmeter D523 cl. 0,5. The maximum admissible tolerance when measuring current is ± 0.6%, voltage drop is ± 1%, capacity is ± 1.6%. Power at the guarding heater is supplied from the voltage controller RNO.

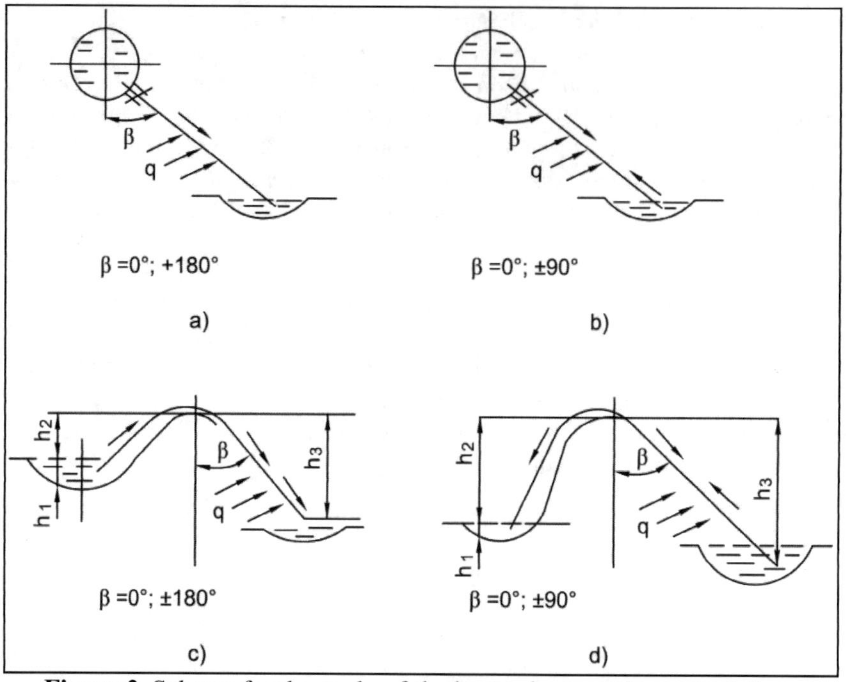

Figure 2. Scheme for the study of the heat-release surface orientation:
a, b – liquid is supplied by the artery; c, d – the "siphon" liquid supply: β is the angle between the cooling surface and the gravitational force.

Current transformer TSD-1000 is used when studying the beginning of liquid boiling and critical loads with output no-load voltage 71 V. Current rate varies within (200÷1200)A. Temperature of liquid and environment are measured by mercury thermometers TL – 4 with scale of $(0\ldots50)^0$C and $(50\ldots100)^0$C and pressure 0.1^0C. Temperature of the drained liquid and steam are measured by chromel-alumel thermocouples made of wire of diameter $0.1\cdot10^{-3}$ m. Diameter of the thermocouple head is $0.4\cdot10^{-3}$ m. Thermocouple electrodes are isolated by dual straws of diameter $1\cdot10^{-3}$ m, which are fixed by glue BF-2 inside injection needles of diameter $1{,}2.10^{-3}$ m. To measure temperature, the electrode, walls of diameter $0.2\cdot10^{-3}$ m, are welded with arc welding formed during discharge of condensers. For this purpose, surface of wall with thickness $2\cdot10^{-3}$ m is drilled at a depth of $1.9\cdot10^{-3}$ m of diameter $1.2\cdot10^{-3}$ with precision $\pm\ 0.05\cdot10^{-3}$ m. The thermocouple electrodes are isolated by porcelain straw of diameter $1.2\cdot10^{-3}$ m and are placed at the wall surface between two layers of mica with thickness $0.05\cdot10^{-3}$ m, glued to the heater surface. Cold ends of thermocouple are thermostated in the melting ice. Electrodes of thermocouple are jointed with two dihexagonal switches PP-63 cl. 0.05. In order to avoid the induced ground current upon readings of thermocouples, the device and tools are grounded. Consumption of the cooling and circulating liquid is measured by the electric flow meters RED with secondary electric tool KSDZ 43 cl.1, calibrated volumetric method. Consumption of the drained liquid and condensate are measured by tank with pressure rate $0.5\cdot10^{-3}$, and time of filling is measured by stopwatch S-P-1b with graduation 0,1 second. The maximum admissible tolerance when measuring liquid consumption by flow meters is ± 3%, and by volumetric method is ± 2%. Imbalance of heat supplied by power and heat discharged by circulating and excessive water with Q shall not exceed ± 12%, and for heat supplied by steam in condenser and heat discharged by circulating water is ± 11%. The discrepancy of balance between consumption of cooling liquid, discharge and condensate is not more than ± 10%.

3. Model of the capillary porous structure of cooling system

Growth of the vapour bubble of radius R was reviewed as separate cell structure. The assumption was that the heat flow q, that determines growth of vapour bubble supplied from the heat surface q1 with "dry" spot through the micro layer of liquid located underneath the vapour bubble, and part of heat q_2 is extracted from the overheated liquid that surrounds a growing bubble since value of the overheated liquid in porous structure could reach larger values, and increases reserve of enthalpy of the adjacent layers of liquid. The cooled liquid is transported due to the joint action of capillary and mass forces ΔP_{g+c}. The "dry" spot in the bubble centre is described by radius r, which at the breakaway torque is proportional to $R_{d.s.} = k{\cdot}R$, whereas micro layer of liquid underneath bubble forms an angle α' with sides δ'_o and δ_o.

Vapour bubble was introduced in volume of the sphere segment, where truncated cone should be deducted, formed by micro layer. Thickness of micro layer δ_o, feeding a vapour bubble due to its evaporation during the growth of bubble will be a constant value as the capillary and gravity forces ensure leaking of fresh portions of cooling liquid to the bubble centre. In model of the vapour bubble growth there is direct transfer from the developed bubble boiling to the possible crisis occurrence, when the imbalance of forces occurs and thickness of micro layer goes to zero ($\delta_o \to 0$), which is very important to study the limit state of system.

The interphase surface and steam generating wall form an average dynamic angle θ during growth of vapour bubble. Since task refers to not so low pressures, the dynamic processes at the initial stage of growth of vapour bubble are not under review. Then viscosity forces and superficial tension will also be comparable to the inertia forces and may not be reviewed. The boiling crisis is related to the active growth of "dry" spot size in the bubble centre.

4. Calculation of the boiling crisis

Calculation of value q_{cr} applicable to the described model could be performed in terms of subcooling and flow speed as per equations [1], which means that subcooling of liquid help extend the heat transfer possibilities in the porous system of cooling. Because the heat transfer processes occur in the thin porous structures, then even insignificant excess of the free flowing film along the outer side of structure determined with parameter $\tilde{m}$, at this hydrostatic pressure ΔP_g and relative permeability factor K_f, forms a liquid core as a source of the permanent supply to under-heated cooler because of difference of temperature and capillary forces.

Excessive liquid in the porous system creates directional movement of flow that leads to deformation of vapour bubbles in structure, reduction of its diameter, rising frequency of bubble birth. The increased flow speed results in the increased energy that is allocated for liquid displacement from the wall boundary layer, and therefore leads to the increased speed of steam generation V_{cr} and value q_{cr}. Although, at some value of the liquid flow speed which is determined by parameter $\tilde{m}_{cr}$, of energy, allocated for liquid displacement from the two-phase wall layer is not sufficient that leads to the heat transfer crisis. Certainly, increase of q_{cr} will be reached at major liquid rates which results in the increased energy rate for the drive of supply machines. When a certain value of the moisture content $\bar{\varphi}'_{cr}$ is reached, the flow speed will not increase value q_{cr}, and in some cases could even decrease value q_{cr}, because of the complicated steam extraction from the wall layer. The increased speed of liquid film adjacent to wall due to the parameter $\tilde{m}$, begins to give way to the dominant effect of decreased moisture content $\bar{\varphi}'$ in that zone, which primarily impact upon value q_{cr}, even minimizing it. Therefore, in each separate case it is required to set a favourable rate of the excessive liquid $\tilde{m}$ due to the type of porous structure. For porous systems of cooling mostly for all mode and geometric parameters under the bubble boiling of water the depth of permeability for temperature wave is $h_{av} < \delta_w$, so thickness of wall δ_w is not input into the design ratio for q_{cr} in works [1, 8-9].

The design equation for qcr in case when $P \geq 0.1$ MPa, a $b_g > 0.28{\cdot}10^{3}$ m is as follows:

$$q_{cr} = 0.0347\bar{r}[g(\rho_l - \rho_s)\rho_s\bar{D}_{br.bub}]0.5(b_g / b_i)^{0.3}(\delta_f / \delta_i)^{0.5}(1+cos\beta)^{0.6} \quad (1)$$

where: $\bar{r}$ – heat of vaporization; $\bar{D}_{br.bub}$ – average size of a steam conglomerate that meets the condition $\Delta T = \Delta T_{cr}$, determined by the formula of work [2]; b_o=0,14·10^{-3} m; δ_f - structure thickness; δ_o=0.18·10^{-3}m.

It follows from equation (1) that $q_{cr} = \bar{D}^{0.5}_{br.bub}$ ($p \geq 0.1$ MPa) and $q_{cr} = \bar{f}^{0.5}$ (p < 0.1 MPa).

Quantity $\bar{D}_{br.bub}$ and $\bar{f}$ depend on the thermophysical properties of the heat-dissipating surface: $\bar{D}_{br.bub} = K_w^{-1}$, $\bar{f}_{cr}^{-1} = K_w^2$, where $K_w=1+[(\rho c\lambda)_l/(\rho c\lambda)_w]^{0.5}$. Then for surfaces made of copper and stainless steel and covered with mesh structures, we have the following: q_{cr} =1.07 ($p \geq$ 0.1MPa), $\tilde{q}_{cr} = 1.15$ (p < 0.1 MPa).

The wall material affects value q_{cr} by set $(\rho c\lambda)_w$, where c, λ – the wall heat capacity and heat conductivity. However, it is not appropriate to make a firm statement because practically it is impossible to retain equal conditions for clarity of processing and micro structure.

When designing a combustion chamber and a nozzle in particular, it is required to take into consideration some reserve for thickness of the heated surface. The boiling crisis previously occurs at "thin" heaters since at the pre-crisis boiling area a size of "dry" spot will increase in the bubble centre, the heat exchange process will abruptly worsen, and wall temperature will increase. Surfaces, that have larger thickness will require more time for their heating. For surfaces with porous coating it is very critical as inside them a time of bubble growth is ten times less, hydrodynamic conditions of liquid supply are sharply changed, and therefore time of steam at wall may extend, which avoids contact of liquid with the heat exchanged surface despite the excessive liquid $\tilde{m}$.

The described process is a pre-history of the boiling crisis development. Further "destiny" of the process at other equal conditions is determined by the heat accumulating capacity $(\rho c\lambda)_w$. When large value is selected, there is possibility of delaying the boiling crisis, the heat extension along the heat surface increases, and again the favourable conditions are established for contact of liquid phase and wall. Only wall thickness increased ten times increases value q_{cr} a few times whereas this situation is more distinctive for highly conductive materials and at greater atmospheric pressure.

As shown by calculations, during time $\tau \leq 5$ s heat flows reach values ~ 8·10^7 Wm^{-2} for copper and 1.3·10^8 Wm^{-2} – for stainless steel. Although, they will be screened by melting curves through ~ 0.01 s. High heat tension stresses occur as a result of abrupt increase of gradients in wall. The impact of different materials and wall thickness at time of beginning destruction at the boiling crisis was studied. With methods of holography and photo elasticity the most hazardous place was determined at the moment of destruction for porous surface [1, 9].

Cases of the liquid discharge from cell of porous structure [1, 2] deteriorate intensity of heat exchange at movement of some boundary heat flow. By selecting an appropriate type of structure this case could be minimized. The smallest discharge is obtained for the single layer meshes with cells more than 0.28·10^{-3}m. The rising deteriorated modes as per their mechanism are identical to processes that exist under movement of the steam-water mixture in pipes that have no porous coating. These modes are characterized by the resistance crisis when friction resistance is getting reduced at the heated area. This is related to the fact that due to the major discharge of drops the liquid consumption is reduced. At the initial stage of the discharge process liquid drops tubulise process, then at the critical discharge quantity of liquid becomes insufficient for watering the heat exchange wall (Fig. 3).

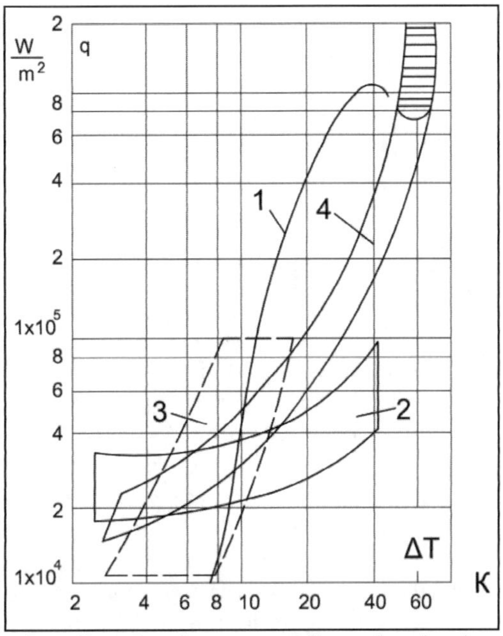

Figure 3. Influence of the heat flow density upon wall overheating in relation to water steam temperature (p = 0.1 MPa): 1 – boiling in large volume at surface w/o coating [3-6]; 2 – Operational area of heat pipes [7, 10, 11]; 3 – operational area of thin filmed evaporators [3]; 4 – researched porous cooling system [1, 2].

Shaded area refers to application of intensifier in porous system. As shown in Figure 3, it is a comparative analysis of the system research (4) with curve in large volume (area 1), by thin-filmed evaporators (area 3) and area of the heat pipes (2). The system (4) extends a limit for discharge of heat loads approaching the liquid boiling in large volume, and in case of applying intensifiers could also discharge large heat flows (shaded part of area 4). As the heat exchange intensifiers the following was studied: wave porous elements with gas-and-liquid dispersoids and the upgraded capillary porous structure, vibrated high heat conductive t-offs and flexible turbulizer.

5. Conclusions

Model of the steam generating bubbles at solid surface with porous coating contains filming and photography observations over the internal characteristics of liquid boiling. High boosting of heat transfer is maintained by the joint action of capillary and mass forces. For such model a thermoelasticity problem is solved and the limit state of system was determined: good and poor conductive materials (porous coating at metal base). Heat flows were calculated from explosive birth of vapor bubble (10^{-8}s), i.e. from relaxation time to the time of macro process. There is interrelation during destruction process only by compression stress, either by melting or tension stress. Sizes of pulled particles are confirmed by high-speed filming and photography. Comparative analysis was provided for the studied capillary porous system along with boiling in large volume by thin-filmed evaporators and heat pipes applying the heat transfer intensifiers. In future, it is required to expand researches for the other mineral medium.

References

[1] Genbach A A, Bondartsev D Yu and Iliev I K 2018 Investigation of a high-forced cooling system for the elements of heat power installations, *Journal of machine Engineering*. **18(2)** pp 106-117

IOP Conf. Series: Materials Science and Engineering **595** (2019) 012007 doi:10.1088/1757-899X/595/1/012007

[2] Genbach A A, Bondartsev D Yu and Iliev I K. 2018 Modelling of capillary coatings and heat exchange surfaces of elements of thermal power plants. *Bulgarian Chemical Communications*, **50**, Special Issue G pp 133 - 139

[3] Jamialahmadi M et al. 2008 Experimental and Theoretical Studies on Subcooled Flow Boiling of Pure Liquids and Multicomponent Mixtures, *Intern. J Heat Mass Transfer*. **51** pp 2482-2493

[4] Ose Y, Kunugi T 2011 Numerical Study on Subcooled Pool Boiling, *Progr. In Nucl. Sci. and Technology*. **2** pp 125-129

[5] Krepper E, et al. 2007 CFD Modeling Subcooled Boiling-Concept, Validation and Application to Fuel Assembly Design, *Nucl. Eng. and Design*, **237(7)** pp 716-731

[6] Ovsyanik A V 2012 Modelling of Processes of Heat Exchange at Boiling Liquids (in Russian), Gomel State Technical University named after P.O., Sukhoy, Gomel, Belarus

[7] Alekseik, O S and Kravets V Yu 2013 Physical Model of Boiling on Porous Structure in the Limited Space, *Eastern-European Journal of Enterprise Technologies*. **(64)** [4/8] pp 26-31

[8] Genbach A A and Bondartsev D Yu 2018 Experimental method of investigation of the heat transfer crisis in a capillary-porous cooling system. *News of the National Academy of Sciences of the Republic of Kazakhstan*, Series of Geology and Technical Sciences. **2**, No. 4 [28] pp 81-88

[9] Genbach A A and Shokolakov K K 2018 Development of an experimental plant of a non-nozzle porous foam generator for producing with foam generating and defoaming structures. *News of the National Academy of Sciences of the Republic of Kazakhstan,* Series of Geology and Technical Sciences. **3(429)** pp 300 – 308

[10] Polyaev V M, Maiorov V A and Vasiliev L L 1998 Hydrodynamics and heat exchange in porous elements of flying devices. *M.: Machine engineering*, pp 168

[11] Kovalyov S A and Soloviyov S L 1980 Evaporation and condensation in heat pipes. *M.: Science* pp 112

[12] Kupetz M, Jeni Heiew E and Hiss F 2014 Upgrade and extension of operations for steam turbine power plants in the Eastern Europe and Russia. *Heat & Power*. **6** pp 35-43

[13] Grin E A 2013 Opportunities of the destruction mechanics applicable to problems of strength, resource and substantiation of safe operations for heat mechanical power equipment. *Heat & Power*. **1**, pp 25-32

IOP Conf. Series: Materials Science and Engineering **595** (2019) 012008 doi:10.1088/1757-899X/595/1/012008

The E4 brickhouse of Romania – a local proved nZEB

E C Mladin

Department of Engineering Thermodynamics, University POLITEHNICA of Bucharest, Romania

E-mail: cerna_mladin@yahoo.fr

Abstract. The nZEB concept, intensively required by the European directives, is intended to include buildings with very low energy consumptions from fossil fuels and it will be mandatory for all new buildings starting with 2021. This paper explores the design stages of a process that meets the requirements of the "E4 program" initiated in Europe by the Wienerberger company for buildings intensively using ceramic elements. The 4 Es stand for Energy, Economy, Ecology and Emotion. Therefore, the paper presents a house in terms of architecture and installations for a healthy, energy efficient, environmentally friendly and affordable home, where ceramic building elements simultaneously stand for increased thermal and mechanical resistance and thermal mass. Energy performance assessment is accompanied by economic analysis and primary energy calculations. Once the building has been completed and occupied by the owner family of 2 adults and two children, the energy use and indoor comfort were monitored for two years. The real energy performance could thus be assessed and compared to the design stage estimate. Results indicated an over-compliance with the nZEB criteria set for Romania. The study reflects the value of simulations at the building design stage and the differences that occur between statistical approach and real building performance.

1. Introduction

Buildings are responsible for 40% of energy consumption and more than 50% of carbon dioxide emissions in the European Union countries. The number of residential buildings in Romania is constantly increasing in recent years, although at a lower rate due to the economic crisis. The reduction of energy needs and the increased use of energy from renewable sources are very important, not only in terms of costs or reduction of entropic greenhouse gas emissions, but also to ensure a sustainable development. Directive 2010/31/EU (EPBD2) for energy performance of buildings specifies that the Member States shall ensure that, by 31st of December 2020, all new buildings are "nearly zero energy buildings" (NZEB) [1]. The NZEB concept refers to a building that has a very high energy performance, in terms of energy needs and use of fossil fuel energy as well. As a consequence, the low amount of energy required to ensure proper indoor comfort all year round has to be covered to a very significant extent from renewable energy sources, wherever possible with equipment placed on-site or nearby, to reduce transportation and storage losses. The calculation and performance evaluation methodology of NZEB is provided by each Member State with guidelines from the EPB2 and associated European documents [2].

Noteworthy as well, most people in urban areas spend about 90% of their time indoor of buildings (homes, offices, stores, etc.). The proper comfort, not only from the thermal point of view, but also from the many other aspects, including emotion, is crucial when a type of building is intended to have success on the market.

The present paper illustrates a house designed and built to be at nZEB level, using ceramic elements, as specified in the "E4" program defined by the Wienerberger company [3]. The 4 Es mean *Energy*, *Economy*, *Ecology* and *Emotion*. "Energy efficiency" is achieved through the minimization of energy consumption by an optimum design of a highly thermally insulating building envelope built with clay blocks. The design also seeks to use innovative heating and cooling technologies that would lead to a referenced nZEB. For "Ecology", it is based on intensive integration of equipment based on renewable energy sources: solar radiation, geothermal heat, and/or biomass. For "Emotion and health", this pilot project is intended to guarantee widely accepted indoor climate quality for many generations, using mostly natural building material such as clay products. "Economy" for the concept means affordable construction and maintenance costs for the users' comfort, otherwise said, it means economic efficiency.

2. The E4 concept applied in climate conditions of Romania

The E4 House concept uses the principles of the major energy efficient building design approaches, paying attention to the cultural differences and the local climate adaptation. In architectural terms, the building elements designed for energy efficiency are also most visible and therefore must connect to the local tradition and culture. Therefore, the design included the pitched roof covered with ceramic tiles, the massive brick walls, terraces covered with ceramic elements, large windows for the day time area, and balanced textures and colors for outside and inside finishing.

The house is designed as a two-storey building for a family of four, and provided with an attic that can be partially used as living area. It is located in the South of Romania, in a village surrounded by forest and with a climate having the statistical minimum temperature of -15°C. The structure is made of reinforced concrete frames and ceramic blocks and ceiling beams. A pitched roof with a variable shape is positioned on top of the house. PVC-framed windows and roof windows with triple glazing were used. The windows mounted on South and West façades are equipped with exterior shutters. Total built floor area of the house is of 228.15 square meters (attic not included). Figure 1 presents a view from the courtyard of the Romanian E4 house, as built, showing all the main spaces facing South (bedrooms, living and dining room). In the right side, a vertical section shows the extended use of clay-based products.

Energy efficiency passive strategies in the building design include *the placement* of building into the available site, its orientation relative to the cardinal points and, as well as to the predominant wind, placement and properties of the transparent and opaque parts of the envelope, as well as the utility of spaces facing different cardinal points. As partially illustrated in figure 2, the design considered:

- Orientation of the main façade towards South, to facilitate solar gains for the living room area and the bedrooms, which represent the main living area of the house. All secondary spaces (technical room, bathrooms, stair and circulation halls) that have limited daily use are facing North, lacking significant use of solar radiation, but protecting the main area from the outdoor conditions.
- Use of large windows facing south for direct passive solar gain and small windows on the north façade in order to minimize the thermal losses. Few windows were positioned on the East and West façades in order to control overheating during summer period.
- Protection against overheating using consoles above the transparent doors of the dinning area and a recessed external solar shading to eliminate excessive heat gains inside the bedrooms.
- Use of the attic as a buffer space against overheating in summer and against cold in winter.
- Opening windows both in external walls and roof for natural ventilation wherever possible. Night-time cross-natural ventilation of the building, to cool the structure in hot summer conditions.
- Unobstructed air paths across the house because of the void above living room and the stair case that alongside with the use of heavy thermal mass with a high surface area (i.e. clay products and concrete structure) exposed to the internal airstream, stabilise internal temperatures.
- Enhanced gain of natural light using roof windows and transom dormer windows for rooms on the first floor. Also the use of relatively narrow circulation zones to ensure that daylight levels are high.

- Greenhouse effects from glass panels mounted in front of the French windows of bedrooms; the gap between panels and the wall allows for hot air circulation during summer, when the greenhouse effect is not desirable.

Figure 1. Southern view of the Romanian E4 house, as built.

The energy efficient architectural design of a NZEB also focuses on the building materials. The considerations for the materials used included as a key element recyclable or recycled, as well aslocally available materials. The main materials (bricks, concrete, glass, wood) have high recycled content and have to be regionally available, thereby reducing transportation/manufacturing energy. The design team made a good selection of materials, as exemplified below:

- High levels of thermal insulation for foundations and ground floor slab, for façades and under the consoles, for the attic slab and roof.
- Additional insulation for thermal bridges.
- Use of metal consoles and insulation to minimize the thermal bridges usually generated by concrete ground slabs (for terraces).
- Terrace slab from the first floor is recessed in order to gain additional insulation space and keep a comfortable walking level from indoor to outdoor.
- Intelligent adaptive membranes with active moisture management at the attic surfaces.
- Energy performant triple-glazing windows and doors, with frame made of PVC.

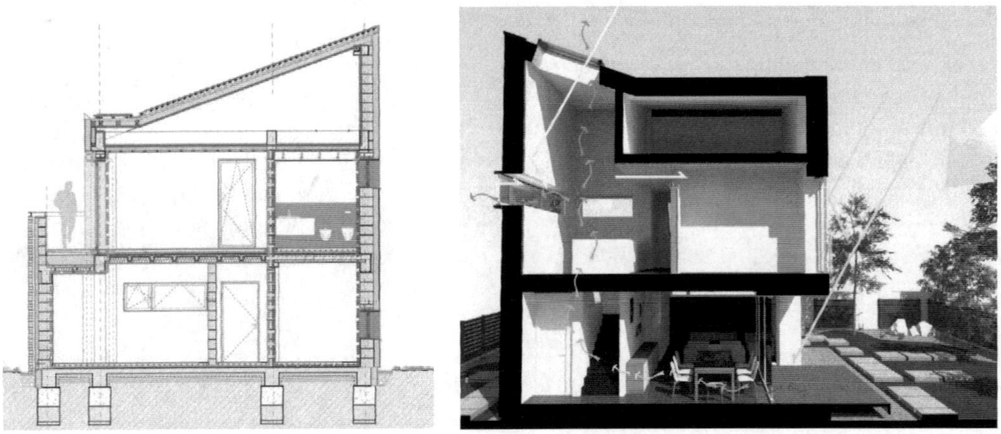

Figure 2. South-North Sections showing the orientation of spaces and passive strategies.

Energy efficiency active strategies in the building design added more value in the overall energy performance of the building. The systems implemented are described below and partly illustrated in Figure 3:

- a heat pump, which uses the underground as an energy source or sink, as it may work reversibly to produce heat in the winter or cooling in the summer. If indoor conditions allow, the heat pump may provide passive cooling by simply running the underground water of 7-12 °C through indoor or built-in-walls tubes. If a heat pump cannot be used for technical, environmental or economic reasons, an alternative option may consist of using a biomass boiler on pellets.
- solar collectors that supply heated water to a buffer storage unit to partially support the preparation of the domestic hot water and/or of the hot water used for space heating. The innovation of the buffer tank is that it stores thermally stratified water that transfers heat to the coils with domestic water flows. This solution avoids the risk of developing germs and micro-organisms in the domestic water and thus various possible diseases.
- low-temperature heating pipes that are placed underfloor and walls with built-in tubes used for both heating and cooling along various seasons.

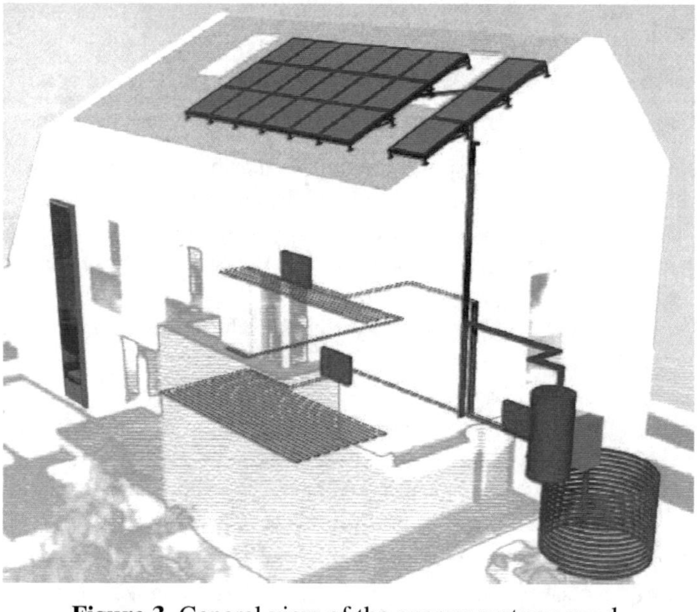

Figure 3. General view of the energy systems used.

- a central home automation technology that allows users to control heating/cooling levels, window shading devices , on/off lighting, as well as security elements and video cameras. This system may register indoor temperature and humiditiy levels in real time or various energy consumptions over predefined time intervals, educating thus the users into what experts usually call „energy behaviour".

Innovative strategies in the building consists here in the use of a Trombe wall, optimized to provide additional heating, cooling, or just fresh air by a combination of greenhouse and thermal mass effects. The Trombe wall is placed on the Southern surface of the living room, where the solar radiation has highest levels and the heat stored in the ceramic elements may be exchanged with the indoor air during day time and night time as well.

3. Theoretical performance analysis

Design team used modular systems during design phase, running several optimization scenarios in order to choose the best solution for a fast building rate, with reduced pollution, and with at least 50% less energy waste than buildings made according to the current minimal standards in Romania. The building was not intended to be a high-tech exercise, but, on the contrary, to be friendly, i.e., traditionally operational, and affordable. The cost of nearly zero energy buildings is usually compared with that of regular buildings, and it is expected to be higher due to the better quality of envelope elements, more energy savings and/or unconventional HVAC installations. For the purpose of illustrating the advantages of the proposed design, it is presented here a comparison between a standard house and the designed one. Both have same geometry and orientation, but different levels of energy performance, as described below. Construction prices were estimated and a prognosis for energy price increase was used to evaluate the economic efficiency for each case.

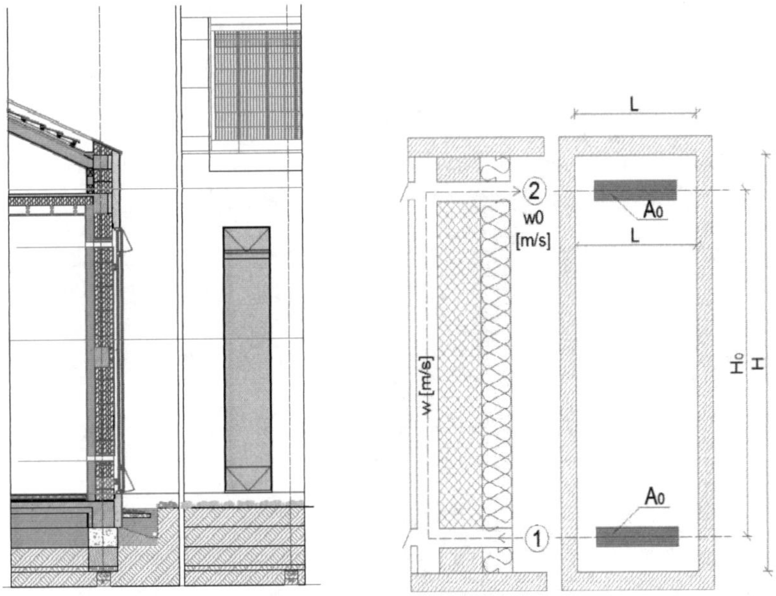

Figure 4. Trombe wall.

- **Study Case 01: Standard new house** – built with the current national standards [4], with no renewable energy sources. It uses a gas boiler for heating and DHW and a commercial split system for cooling. This case is used as baseline for the next ones, both in terms of costs and energy savings.
- **Study Case 02: E4 design** – masonry 38 cm + 14 cm mineral wool, ground slab insulated with 10 cm XPS, consoles with 15 cm mineral wool, roof with 20 cm cellulose and adaptive humidity membranes, 20 cm XPS for terrace, PVC windows with triple glazing, solar panels for partial DHW, PV panels, Trombe wall, and geothermal heat pump for heating and cooling.

Heat transfer coefficient U-values [W/(m^{2}K)] for each building envelope construction are shown in table 1. For comparison, standardized value for regular Romanian buildings [4] are also included in the table. U-values for 2020 are not yet regulated.

The solution that induces energy savings for lighting and also brings a lot of comfort and emotion to the users remains the natural light. This means large window area. but also shading devices and low-emissivity glazing to reduce heating over the summer or thermal losses during the winter. Roof windows are used at the first floor to ensure enough light to the stairs and hall areas. as well as in the bedrooms / offices. The goal was to obtain a daylight factor in every single room a few times higher than the usual standard.

Table 1. Table with the main U-values [W/(m^2·K)] corrected to account for thermal bridges.

Structural element	Romanian standard in 2013	E4 – case 02
External walls	0.55	0.22
Attic slab	0.20	0.15
Slab over the ground	0.22	0.18
Window (PVC frame)	1.30	0.85
Roof window	1.30	0.94
Slab over exterior (e.g. arcade/bowindow)	0.22	0.22
Flat roof	0.33	0.19
Slab between first floor and attic	0.33	0.22
Roof	0.20	0.17

Table 2. Energy and environment performance of the two study-cases.

	Heating kWh/(m^{2}y)	DHW kWh/(m^{2}y)	Lighting kWh/(m^{2}y)	Cooling kWh/(m^{2}y)	TOTAL kWh/(m^{2}y)	difference %	CO2 kg/(m^{2}y)	difference %
SC 01	106.8	23.5	4.4	6.3	**141.0**	0%	**35.4**	0%
	100% from natural gas boiler	100% from natural gas boiler	Normal use, simultaneity coefficient. =0.7	100% indvidual split with COP = 3.0			Emission coeff. = 0.21 kg/kWh – nat. gas	
SC 02	10.5	2.2	2.5	2.4	**17.7**	-87%	**8.6**	-76%
	5%/yr from Trombe wall and the rest from heat pump of COP= 4.5	60%/yr from solar panels and the rest from heat pump of COP = 4.5	1/2 less time used during the day and simultaneity coefficient =0.5	Cooling with heat pump of COP = 4.0			Emission coeff. = 0.485 kg/kWh - electricity from the grid (2013)	

The energy analysis of the two cases led to the values summarized in Table 2. The calculation was performed according to the Romanian methodology for energy auditing of buildings [5]. It is a monthly-averaged approach over a statistical year in terms of outdoor temperature solar radiation. Energy need for heating or cooling considers energy transfer through building envelope (thermal resistances including thermal bridges) and air exchange, as well as internal and solar gains. Corrected degree-days for the particular house analysed could thus be calculated and used for heating and cooling demands, as illustrated in Figure 5. In this figure, Tint is the averaged indoor temperature based on standard values [5], Text is the monthly averaged outdoor temperature based on statistical data, TiR is the indoor temperature corrected for internal gains, and TeR is the outdoor temperature corrected for solar gains. The beginning and end of the heating/cooling seasons are marked by the arrows and the area closed by TiR and TeR curves represents the degree-days for heating (a) and cooling (b). Noteworthy, the cooling energy demand calculation included also the latent heat of water vapour condensed during the cooling process [5]. Noteworthy. energy needs for heating considered in details the internal load and solar loads for a South-orientation of the building in the vicinity of Bucharest. Corrected degree-days for the particular house analysed could thus be calculated and used for heating and cooling demands. as illustrated in figure 5.

The energy values in table 2 are final non-renewable energies used in the building. meaning that the renewable energies have been subtracted from the theoretical values. The emissions of CO_2 are calculated using values specified in the Romanian methodology for environmental loads [5]. Below values. some brief explanations are given in order to facilitate the understanding of what changes from case to case. All differences are then expressed in percentages relative to study case 01. for both total final energy and total emissions. It must be said that. in this table. thermal and electrical energies for the four uses were added up. as it is done in some legal documents. although it is not a correct approach. For this reason, the energy and environment performance must be judged together.

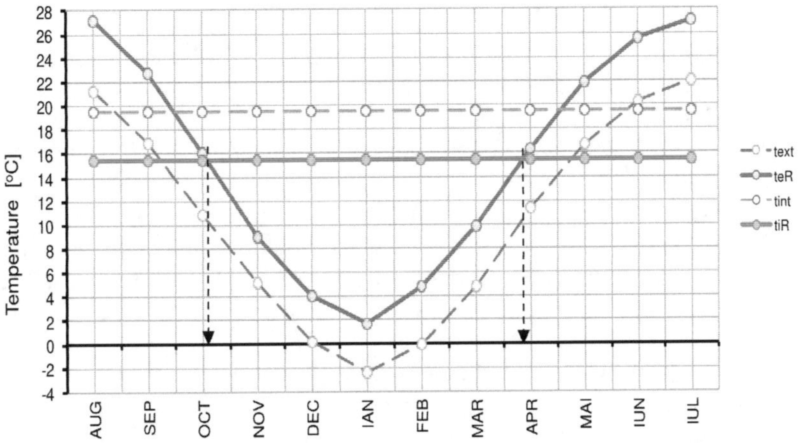

(a)

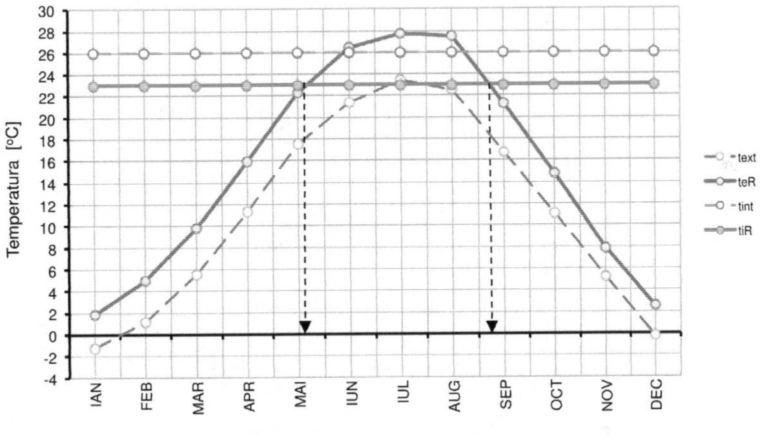

(b)

Figure 5. Corrected degree-days for heating (a) and cooling (b).

Noteworthy, additional energy savings may be achieved with the design approach, as it aimed for reduction of embodied carbon-dioxide emissions, through reducing waste by modular design based on standard dimensions of each building element, prefabrication and site management.

The engineering approach is always completed with an economical evaluation of the solutions being promoted, such as to ensure marketable products. Given the long life of buildings, the instabilities of prices and fuels availability, as well as the legal increasing demands for energy and environmental performances of new buildings, careful calculations must be performed in order to show to the client the advantages of obviously more expenses designs. The client must be educated to appreciate in long term the savings that he/she can make.

In order to emphasize only costs associated with energy efficiency measures, from the total costs of the real building was subtracted the total cost of the standard case SC-01. These extra-costs were then compared to money savings from energy savings relative to the same case SC-01 during the operation of the house over a certain period of time. In the economic efficiency evaluation presented in this paper, the annual inflation of euros and increase of energy tariffs are considered at rates of 3% and 10%, respectively. The extra-investment payback time period for the design case SC-02 resulted from calculation 17.9 years, compared to the estimated life-time of energy saving solutions of 30 years.

4. Energy performance analysis from measurements

The real performance results clearly from measurements on comfort and energy use in the built environment. For this purpose, the house was monitored for full two years. Temperature, relative humidity and illuminance levels were measured in all major rooms (living, bedrooms, kitchen), simultaneously with similar outdoor conditions, with a time step of 30 minutes. Inlet and outlet air temperatures at the Trombe wall, electricity meters for the general use and for the PV panels were also continuously registered.

For proper understanding the building energy response, Figure 6 presents the outdoor temperature and humidity levels measured in a sunny spot near the South façade of the building. This placement makes the measured temperature to play the role of TeR in Figure 5. Panels (a) and (b) are for the first trimester of 2017 (January-March) and panels (c) and (d) are for the third trimester of 2017 (July-September).

The building has a BMS implemented that ensures pre-set indoor temperatures, such as, no major problems were indicated by monitoring devices, regardless the outdoor conditions. However, for relative humidity it was noticed that high discomfort was created by high levels in the living room and dehumidifiers were installed. This was caused by the outdoor high relative humidity associated with a lot of greenery in the garden and in the nearby forest. Figure 7 indicates such measurements before and after installing the dehumidifier device.

All indoor parameters were used to compute the Fanger comfort indicators PMV (Predicted Mean Vote) and PPD (Predicted Percentage of Dissatisfied) and family members were inquired about their level of satisfaction, indicating very small and temporary discomfort levels in the kitchen and living room.

The daily energy use and production are represented in the figure below (Figure 8), indicating periods when production was higher than consumption (summers) and periods when production was lower (winters). In summer, cooling is of passive type, so the heat pump (PQ) does not consume much energy, with the exception of circulating water pumps.

The overall energy balanced is illustrated by the cumulative representation of energy use/production in Figure 9. It results that after two years of use, the difference between heat pump consumption and PV production is 1700 kWh, which, with a tariff of 0.1 Euro/kWh, means 170 Euro net cost with electricity for the thermal comfort conditions in the house.

When the total energy use and production is considered, a representation of terms as a function of outdoor mean monthly temperatures is significant, as illustrated in Figure 10. It is shown that at high outdoor temperatures (as measured by the weather station placed on the house external walls), the heat pump does not consume much and the electricity production in the PV panels is superior to the total electricity consumption from the national network (SEN). Other electricity users (i.e., computers, appliances, lighting and electrical correction for hot water) besides the heat pump result from the differences, assuming that all PV energy has been used.

Besides the inherent errors stemmed from various theoretical assumptions, the differences occurred also due to the following:

- The estimated cooling indicator included night cooling by additional ventilation, which was not operated by the users for safety purposes.

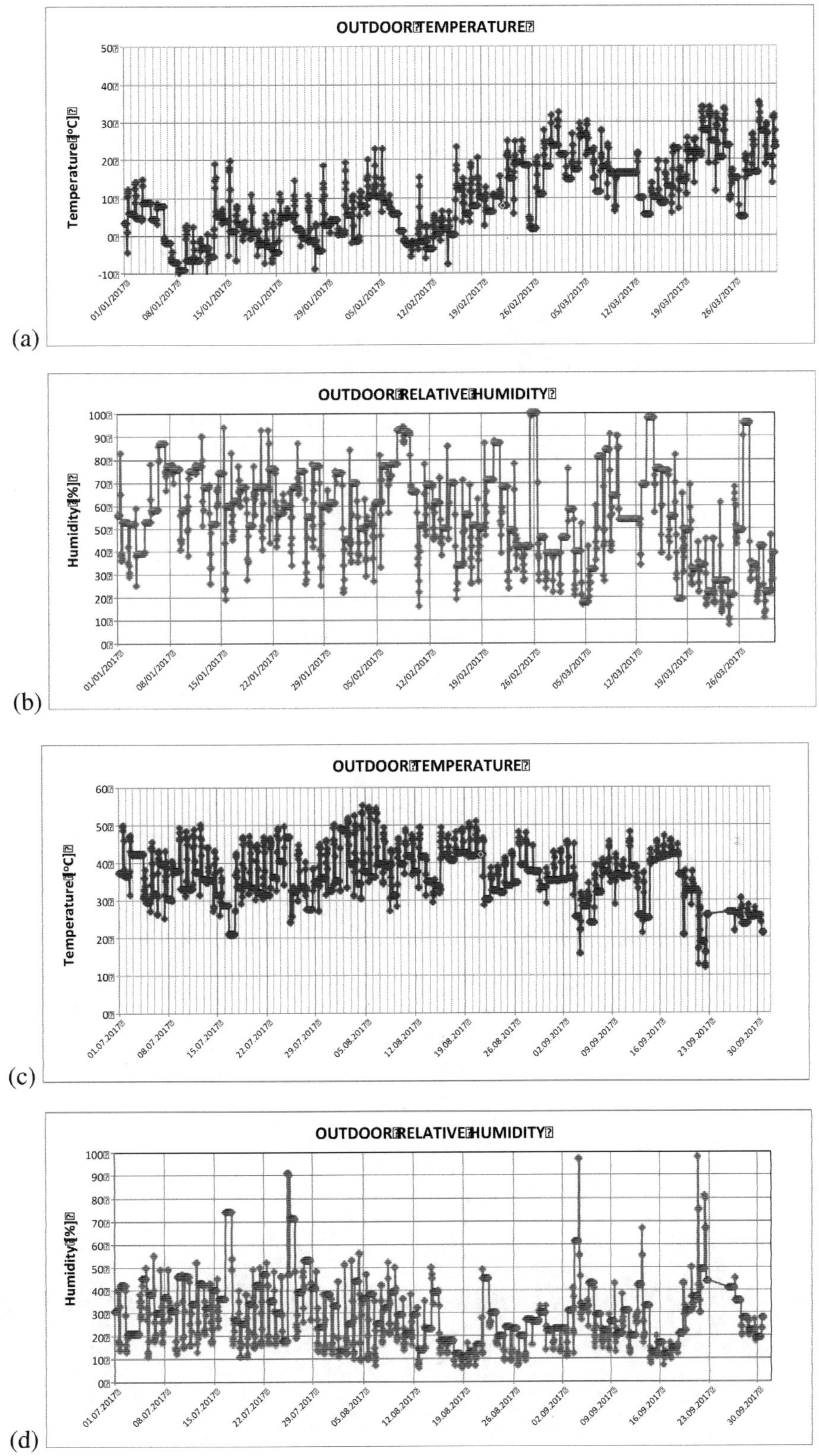

Figure 6. Outdoor temperature and relative humidity.

- The PV-panels did not produce any electricity while being covered with snow during winter times. The longer periods with cloudy days than assumed are another reason for less PV long-term performance.
- The Trombe wall did not function properly, neither for heating, nor for cooling.
- The behaviour pattern of people may significantly influence energy use in any house.

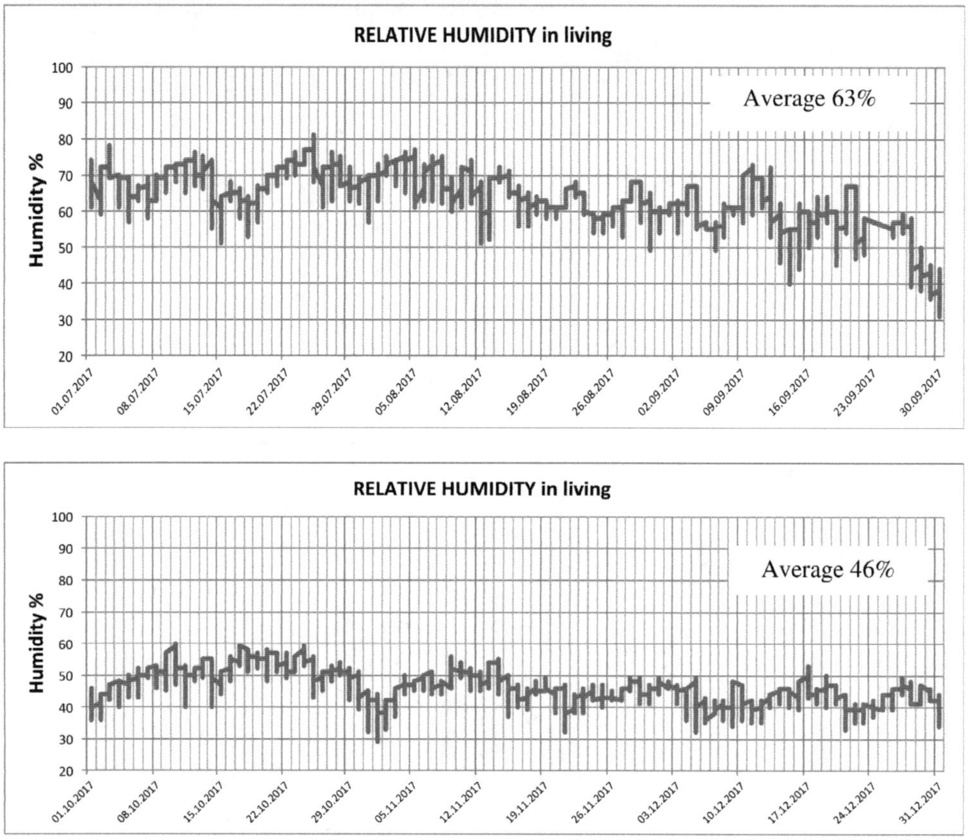

Figure 7. Indoor relative humidity before and after dehumidification.

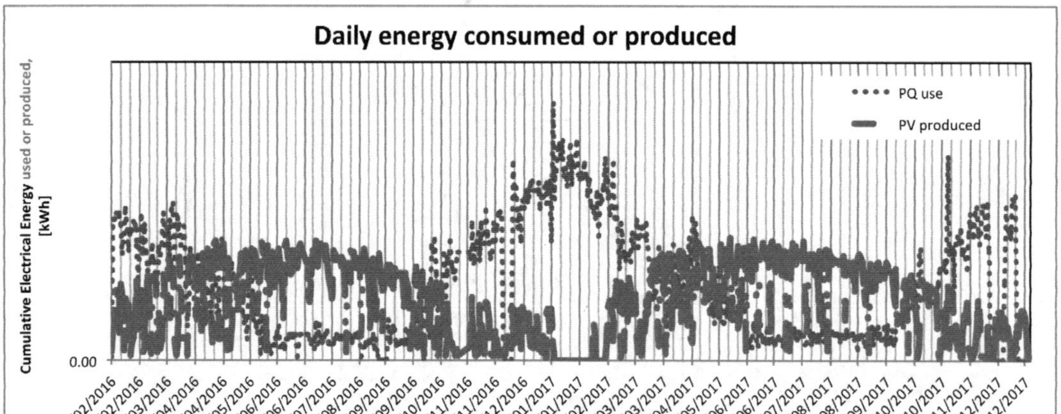

Figure 8. Daily energy terms over two years' period.

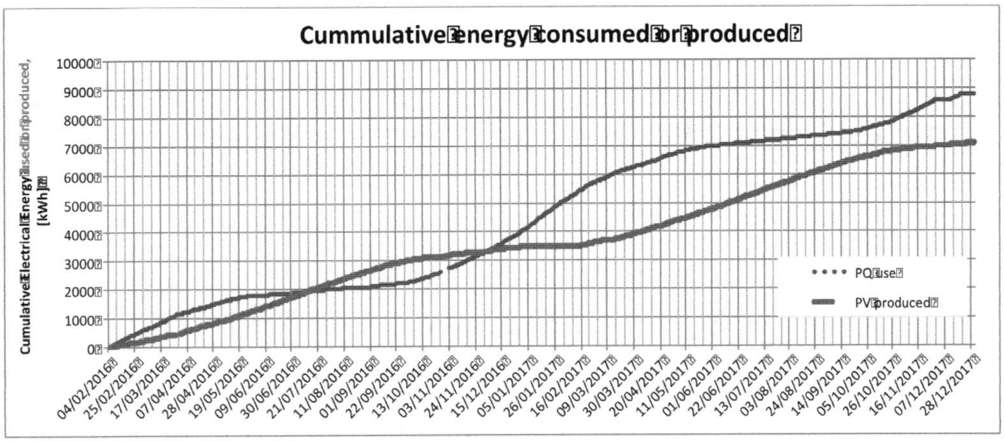

Figure 9. Energy terms along two years.

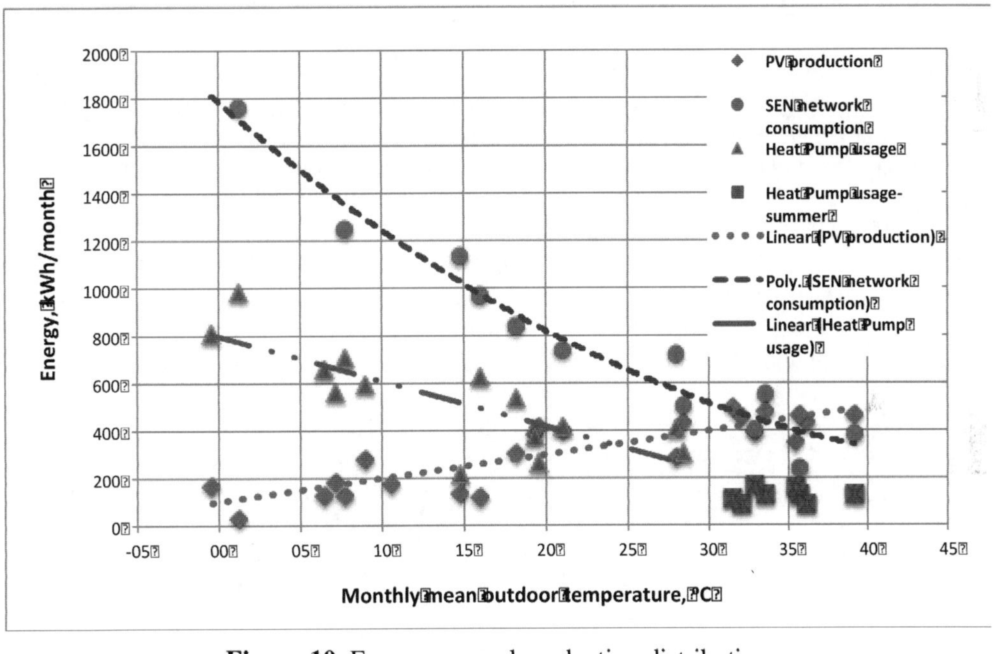

Figure 10. Energy use and production distribution.

For the Trombe wall, measurements indicated large air infiltration through the inlet throttles and, as a result, a minor contribution to the house heating. Future repairing is expected to fix this drawback and increase the Trombe efficiency.

One last result of the analysis is a comparison between the estimated energy performance and the actual energy performance (Table 3), after scaling the real measurements to the indoor and outdoor conditions considered in the theoretical calculation.

Table 3. Comparison of the estimated and actual energy performance.

	Heating kWh/(m^2y)	Domestic hot water kWh/(m^2y)	Lighting kWh/(m^2y)	Cooling kWh/(m^2y)	PV- production kWh/(m^2y)
Estimated	10.50	2.20	2.50	2.40	31.00
Measured	12.84	2.39	2.19	4.03	18.69
Difference, %	**22.3%**	**-8.6%**	**-3.1%**	**6.8%**	**-39.7%**

5. Conclusion

The paper presents the importance of three stages in the proper analysis of the energy performance of any building.

I. Architectural design based on energy performance concepts
II. Theoretical analysis of multiple solutions for associated energy indicators
III. Monitoring of real building under use conditions.

In the particular case of the house analysed here, all stages led to valuable conclusions and decisions for improvements. No matter the differences between estimations and measurements, the real energy performance of the building is outstanding: a total of annual 40.14 kWh/m^2 final non-renewable energy. As primary non-renewable energy, the indicator becomes 105 kWh/m^2, which is below the value of 111 kWh/m^2 stated in the Romanian legal norms for the nZEB level for an individual house.

The project aims to prove that it is possible to have a healthy living in energy-efficient and sustainable buildings made of clay products, that are attractive and affordable at the same time. A low-energy house design according to E4 house concept proved to be at nZEB level with already widely disseminated design concepts and building and materials. It remains the role of the designer to educate the beneficiary, no matter who is, about the advantages in medium and long terms of having a NZEB type of house.

6. References

[1] EPDB Energy Performance of Building Directive European Union 2010 http://europa.eu/legislation summaries/other/l27042_en.htm
[2] The EU climate and energy package, http://ec.europa.eu/clima/policies/package/
[3] https://wienerberger-building-solutions.com/expertise/the-e4-brick-house-concept-1366259717625
[4] C107/2010 – Romanian Standard for thermal properties of construction elements
[5] Mc001/1-3 2007 and 2013 6 - Romanian Methodology for assessing energy performance of buildings *Official Journal* **126 bis** and *Official Journal* **561bis**
[6] Mladin E C et al 2001 *Building Energy Auditing Technique* (Bucharest: MatrixRom Publishing)

IOP Conf. Series: Materials Science and Engineering **595** (2019) 012009 doi:10.1088/1757-899X/595/1/012009

Two-dimensions finite-difference heat transfer study for a single borehole heat storage system during a seasonal charge-discharge cycle

B Horbaniuc, G Dumitraşcu and A Dumencu

Automotive and Mechanical Engineering Department, "Gheorghe Asachi" Technical University of Iasi, Iasi, Romania

E-mail: bogdan_horbaniuc@yahoo.com

Abstract. Borehole heat exchangers are used to remove and/or add heat from/into the ground in different applications involving low-temperature heat usage, such as residential or social buildings heating by means of heat pumps. In such types of applications, seasonal heat storage is necessary since during the warm season an additional solar system is used to charge heat into the ground via the borehole system in order to regenerate the heat source for usage during the cold season. Heat transfer in the ground by means of the borehole results in both radial and axial propagations of the perturbation (ground heating/cooling). The magnitude of this perturbation and the added/ removed heat are important parameters of the process that must be evaluated. The theoretical model presented in the paper uses the decomposition finite difference technique to determine the temperature distribution in the ground at different moments (1, 2, 4, and 6 months) during a charge/discharge cycle of the heat storage. The numerical results are necessary in order to evaluate the penetration depth of the thermal perturbation along the radial and longitudinal axes and the amount of stored/extracted heat.

1. Introduction

The continuous growth of the energy demand of humankind in order to satisfy its mounting necessities requires burning enormous amounts of fossil fuels. The consequence is the environment's degradation which is about to reach the threshold of irreversibility. Phenomena such mountain glaciers retraction and polar caps melting, increasingly unpredictable weather patterns, etc. rise serious concerns regarding the alteration of living conditions at a global scale.

One solution to overcome this drawback is to use alternate forms of energy, such as renewables. Since the ground contains huge amounts of quasi-isothermal, low-temperature heat at depths less than a few tens of meters, a feasible idea might be to use the ground as a heat source for heat pumps - called Ground-Source Heat Pumps (GSHPs) - in residential heating applications. This solution only applies to countries with no geothermal potential; privileged countries such as Iceland or New Zealand [1] benefit from high-temperature ground heat due to volcanism and therefore heat pumps are not necessary. Most of the countries with volcanic activity on their territory use geothermal energy to generate power: 24 of them report that 15-22% of the power is obtained from geothermal energy [1], whereas 82 countries use directly this energy for residential heating [2, 3].

Direct utilization of geothermal heat represented at the end of 2014 a total 70,329 MWt of installed thermal power covering a consumption of 163,287 GWh/yr, distributed as follows: 55.3% ground-source heat pumps, 20.3% bathing and swimming, 15.9% space heating, and the rest for other purposes (greenhouse heating, industrial process heating, snow melting, etc.) [2].

Although the technology dates from the 1940's [4], the interest in this solution was very weak until the increase of the greenhouse gases effects on the global climate made necessary a radically different approach in obtaining energy (for power and heating).

GSHPs are increasingly used in residential heating and domestic hot water production applications in regions or countries with low geothermal potential where only the heat naturally contained in the ground is available, at temperatures in the range 8 … 20°C. Presently, 50,000 units are installed annually in the USA, with a capacity of around 9,600 MWht [1]. One of the iconic examples is a 19.6 MW heating/15.8 MW cooling unit for 100 apartments and 89,000 m^2 of office surface area (161,650 m^2 in total) which supplies heat for Galt House Hotel (Louisville, Kentucky), thus saving 47 percent of the necessary energy, representing $ 25,000 per month [2]. In Europe, one must mention Mining Engineering Building of the National Technical University of Athens, which is heated and cooled by a GSHP, as a result from a pilot project from 1993, accomplished with Swiss support [4].

There are three configurations for the underground heat exchangers [5, 6]: horizontal loop, spiral loop, and vertical loop (borehole). When the available surface area is small, the third solution is recommended. A specific advantage of this configuration is the fact that below a depth of 15-20 meters, the ground temperature remains practically constant all year long, which provides a quasi-isothermal heat reservoir. The depth of the borehole is usually in the range 23 … 100 m [6].

The paper deals with the investigation of the conduction heat transfer in the ground during injection followed by extraction of the heat by a single borehole working as a heat exchanger for a ground heat storage unit. A six-month operation is supposed for each of the two modes (injection or charging, respectively removal or discharging), roughly representing the warm and the cold seasons respectively. Heat is stored in the ground during the warm season, and is removed when there is a heat demand in order to operate a heat pump for residential heating purposes.

Analytical methods are very difficult to use to study the unsteady conduction heat transfer within the ground (and only in very simple situations - one-dimension heat propagation in bodies with simple geometric shapes, constant thermo-physical properties of the heat transfer medium, and Dirichlet boundary conditions) [7, 8]. For instance, one-dimension unsteady heat conduction in long cylinders involves the use of Bessel functions, which are very difficult to manipulate especially when it comes to write the boundary conditions [8].

With respect to the considered mathematical model for a single borehole, Bandos *et al.* [9], Lamarche and Beauchamp [10], use the finite-source model to determine the effect of vertical temperature variations, and Philippe *et al.* [11], compare the results provided by three analytical methods: the infinite line source, the infinite cylindrical source and the finite line source respectively.

The actual phenomenon of heat propagation across the ground involve two- or three- dimension unsteady conduction heat transfer with Neumann boundary conditions, making the analytical approach almost impossible to use, due to the complications associated with finding solutions for the differential equations that describe the theoretical model. A different, more pragmatic, less accurate, approach involves the use of numerical techniques, among which one of the most popular is the finite difference method [12], [13], [14]. Due to the difficulties arisen by the analytical model involving two-dimension unsteady conduction heat transfer in the studied case, the approach presented in the paper makes use of the finite differences technique.

When modelling a two-dimension heat conduction, two variations of the finite difference technique can be used: the explicit finite difference method (EFD), which allows the calculation of the present time temperature in the current node, respectively the implicit scheme (IFD), where three nodal temperatures are simultaneously considered at the present time step. The EFD determines the unknown temperature in a sequential manner, but the magnitude of the time step is drastically limited in order to secure the stability of the scheme. In the EFD such a constraint does not operate the scheme being unconditionally stable, but the method involves the necessity to solve a set of linear algebraic equations corresponding to the total number of nodes [14, 15].

The paper deals with the use of the decomposition technique (DT) of the IFD scheme in two dimensions, in cylindrical coordinates, for a cycle of heat injection in the ground followed by heat removal from the ground by a single borehole, with an intermediate "standby" time gap when no heat transfer from the borehole occurs.

2. The physical model

A long hollow cylinder of length *L*, inner radius R_0, and outer radius R_d represents the ground domain, initially isothermal at temperature T_∞. The ground material is supposed to be homogenous and its physical properties (density – ρ, thermal conductivity – λ, specific heat – *c*, and thermal diffusivity – *a)* are not temperature-dependent. Figure 1 shows the schematic of half the domain, due to the symmetrical configuration of the cylindrical geometry.

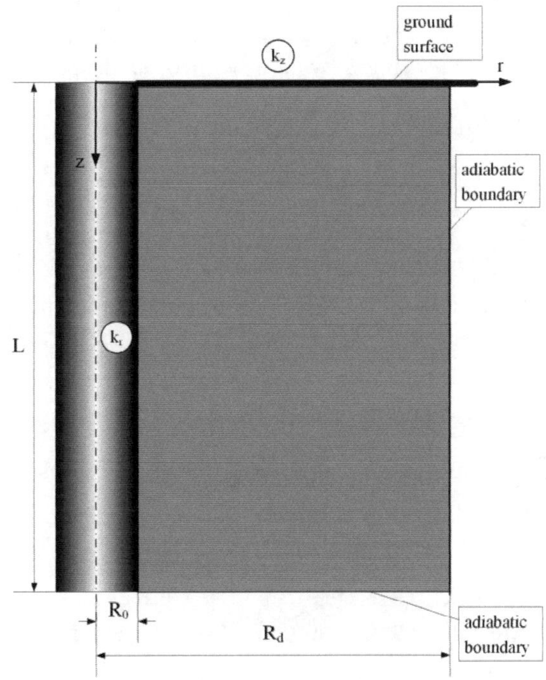

Figure 1. The domain.

During the charging stage of the cycle, a fluid of temperature $T_H > T_\infty$ flows along the vertical tube (representing the borehole), transferring heat by convection (convection heat transfer coefficient k_r) to the cylindrical domain that represents the ground. The latter is supposed to have adiabatic boundaries on its outer and bottom surfaces, whereas the frontal surface (in contact with air) allows the convection heat transfer between the ground and the atmospheric air to occur (convection heat transfer coefficient: k_z).

During the heat removal stage (discharge), a cold fluid of temperature T_L, lower than T_∞ flows along the central channel (tube) that represents the borehole, cooling the cylindrical domain (the ground). Heat is removed by convection with the heat transfer coefficient k_r.

The essence of the decomposition approach is described in [15]. During a time step of the procedure, the two-dimension conduction heat transfer is split into two one-dimension heat transfer processes that occur sequentially: first in radial direction (intermediate stage), and then axially (final stage). The order of the two one-dimension processes for a time step is arbitrary. The final temperatures at the end of the previous (superscript $p - 1$) time step $\Delta\tau$ are the start temperatures (known) of the intermediate time step (superscript *i*). The intermediate temperatures resulting from the application of the decomposition procedure in its first (intermediate) stage are the input temperatures for the second (final) one the output of which is represented by the final temperature (superscript p).

The "trick" of splitting a two-dimension heat conduction process into two two-dimension ones results in a significant simplification, because one only must deal with very simple finite difference equations, which are much easier to manipulate.

The heat conduction equation written corresponding to the direction of propagation is [7], [8]:

– radial propagation of the conduction heat transfer (in cylindrical coordinates):

$$\frac{\partial \theta}{\partial \tau} = \frac{\lambda}{\rho c}\left(\frac{\partial^2 \theta}{\partial r^2} + \frac{1}{r}\frac{\partial \theta}{\partial r} + \frac{\partial^2 \theta}{\partial z^2}\right) \tag{1}$$

– axial propagation (in Cartesian coordinates):

$$\frac{\partial \theta}{\partial \tau} = \frac{\lambda}{\rho c}\frac{\partial^2 \theta}{\partial z^2} \tag{2}$$

Since the model deals with a two stages-cycle governed by different values of the heat transfer fluid temperature, it is necessary to define a unique criterion when considering both heat removal from the ground (discharging stage) and heat injection into the ground (charging stage). Therefore, a dimensionless temperature θ is defined with respect to the hot fluid temperature T_H and the temperature T_∞ of the initially isothermal ground:

$$\theta = \frac{T - T_\infty}{T_H - T_\infty} \tag{3}$$

As a result of the above description of the domain, the boundary conditions can be written as:

– at $r = R_0$:

$$k_r \theta_0 = -\lambda \left(\frac{\partial \theta}{\partial r}\right)\bigg|_{r=R_0} \tag{4}$$

where θ_0 represents the dimensionless temperature of the central channel's surface.

– at $z = 0$:

$$k_z(\theta_a - \theta_0) = -\lambda \left(\frac{\partial \theta}{\partial z}\right)\bigg|_{z=0} \tag{5}$$

where θ_a represents the dimensionless temperature of atmospheric air above the frontal surface of the domain.

– at $r = R_d$:

$$\left(\frac{\partial \theta}{\partial r}\right)\bigg|_{r=R_d} = 0 \tag{6}$$

– at $z = L$:

$$\left(\frac{\partial \theta}{\partial z}\right)\bigg|_{z=L} = 0 \tag{7}$$

3. Finite difference approach

A rectangular mesh has been attached to the cylindrical domain, consisting of N_r nodes in the radial direction (radial step: h_r) and N_z nodes in the axial direction (axial step: h_z) – see figure 2.

The heat transfer equation (1) can be rewritten in finite difference form for a time step $\Delta\tau$, by using the finite difference operator D^2 as follows [13], [14]:

$$\frac{\theta^i_{m,n} - \theta^{p-1}_{m,n}}{\Delta\tau} = aD^2_r \theta^i_{m,n} \tag{8}$$

where $a = \dfrac{\lambda}{\rho c}$ stands for the thermal diffusivity of the ground, $\Delta\tau$ for the time step, and p for the current time step number.

The operator D_r^2 can be expressed as [16]:

$$D_r^2\theta_{m,n}^i = \frac{1}{h_r^2}\left(\frac{c_m - 1}{c_m}\theta_{m-1,n}^i - 2\theta_{m,n}^i + \frac{c_m + 1}{c_m}\theta_{m+1,n}^i\right) \tag{9}$$

where:

$$c_m = 2\left(N_r \frac{R_0}{R_d - R_0} + m\right) \tag{10}$$

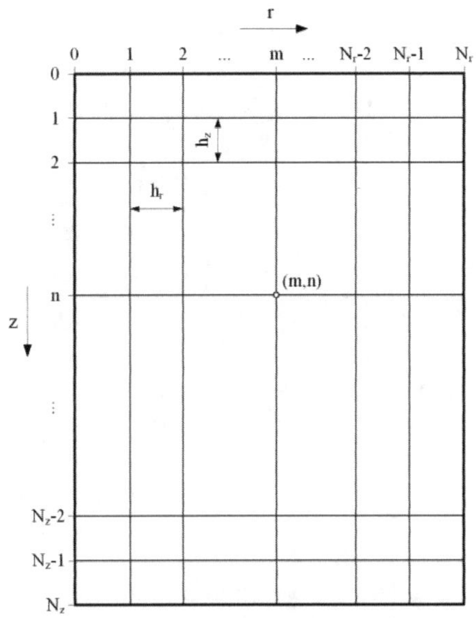

Figure 2. The rectangular mesh attached to the domain.

Equation (2) can be written in finite difference form as:

$$\frac{\theta_{m,n}^p - \theta_{m,n}^i}{\Delta\tau} = aD_z^2\theta_{m,n}^i \tag{11}$$

where

$$D_z^2\theta_{m,n}^p = \frac{1}{h_z^2}\left(\theta_{m,n-1}^p - 2\theta_{m,n}^p + \theta_{m,n+1}^p\right) \tag{12}$$

The convergence and stability criteria for the two directions of propagation of the decomposition technique are:

$$\alpha_r = \frac{a\Delta\tau}{h_r^2} \text{ (radial direction)} \tag{13}$$

$$\alpha_z = \frac{a\Delta\tau}{h_z^2} \text{ (axial direction)} \tag{14}$$

After a series of manipulations [16], equation (9) becomes:

$$-\theta_{m-1,n}^i + \sigma_{r,m}\theta_{m,n}^i - \gamma_{r,m}\theta_{m+1,n}^i = \beta_{r,m} \tag{15}$$

where:

$$\sigma_{r,m} = \frac{c_m}{c_m - 1}\left(2 + \frac{1}{\alpha_r}\right) \tag{16}$$

$$\gamma_{r,m} = \frac{c_m + 1}{c_m - 1} \tag{17}$$

$$\beta_{r,m} = \frac{c_m}{c_m - 1}\frac{1}{\alpha_r}\theta_{m,n}^{p-1} \tag{18}$$

In a similar manner [16], one obtains the finite difference form of equation (12):

$$-\theta_{m,n-1}^{p} + \sigma_z \theta_{m,n}^{p} - \theta_{m,n+1}^{p} = \beta_{z,n} \tag{19}$$

where:

$$\sigma_z = 2 + \frac{1}{\alpha_z} \tag{20}$$

$$\beta_{z,n} = \frac{1}{\alpha_z}\theta_{m,n}^{i} \tag{21}$$

The temperatures on the boundaries of the domain result from the boundary conditions written in the finite difference form.

The decomposition technique, as described in [15], involves the use, during a time step $\Delta\tau$, in a successive manner, of two one-dimension implicit finite difference schemes corresponding to the two directions of propagation (radial and axial respectively - the order is arbitrary). From the physical standpoint, the two-dimension heat propagation is firstly "decomposed" in a radial propagation by inserting thermal insulating barriers (separated by a space step) between the nodes, parallel to the direction of propagation thus preventing heat from diffusing in the normal direction. This stage is followed by the axial heat propagation stage by removing the insulating barriers and placing them parallel to the axial propagation and thus forcing heat to propagate in this direction.

4. . Results and discussion

The decomposition method described above applied to the finite difference technique has been used to study a single borehole ground thermal storage system consisting of a 20 m long tube 0.25 m across and a surrounding ground region of radius R_d = 5.125 m. The temperatures were: T_H = 100°C, T_∞ = 10°C, and T_L = 5°C. The thermo-physical properties of the ground were: λ = 0.25 $Wm^{-1}K^{-1}$, a = $0.18 \cdot 10^{-6}$ m^2s^{-1}. The convection heat transfer coefficients were k_r = 1000 $Wm^{-2}K^{-1}$ and k_r = 10 $Wm^{-2}K^{-1}$ respectively. The mesh attached to the cylindrical domain had N_r = 50 and N_z = 50 nodes respectively.

A computer code was written for the algorithm of the decomposition technique and was run in order to obtain the nodal temperatures at 30, 60, 120, and 180 days respectively for the two stages of the heat storage cycle. The computer code had also the option to list the stored heat's evolution during each stage. The final nodal temperatures supplied by running the computer code written for the charging (heat injection) stage as well as the other necessary data have been stored in a data file, which has been used to initiate the computer code for the discharge (heat removal) stage.

The output data supplied by the computer codes for the charging, respectively the discharge stages of the cycle have been used to generate two families of 3-D plots: wireframe and colour contour.

The plots corresponding to the charging stage are displayed in figures 3 through 10.

One can notice that the heat diffusion is very slow: the penetration radius is about 10 nodes after 60 days and about 15 nodes after 180 days. This can be explained by the low conductivity of the ground material.

The numerical results confirm that the choice of the cylindrical domain's radius was good. This choice is justified by the fact that a semi-infinite domain in the radial direction is a hypothesis that cannot be taken into account for the numerical model, since the finite difference approach cannot operate with infinite values. For this reason, the only choice left is to consider a boundary of finite radius at which an adiabatic boundary condition must be imposed. If this radius is long enough, the heat perturbation induced by the presence of the central channel (tube) which adds or removes heat cannot reach the outer boundary.

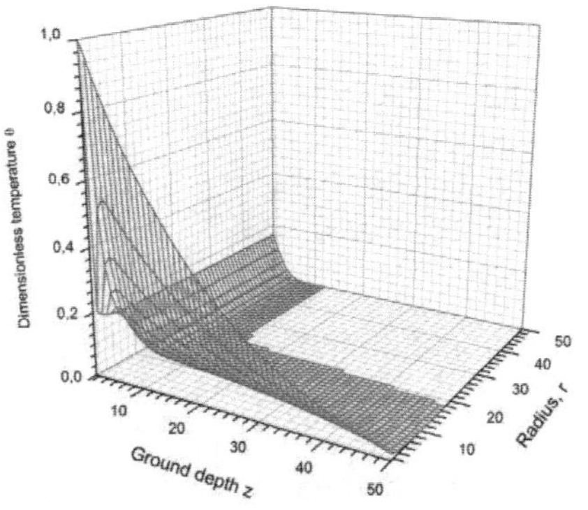

Figure 3. Wireframe plot of the temperatures at charging after 30 days.

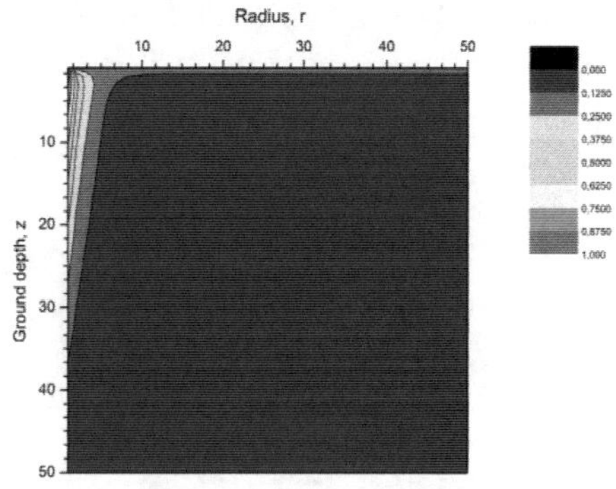

Figure 4. Contour plot of the temperatures at charging after 30 days.

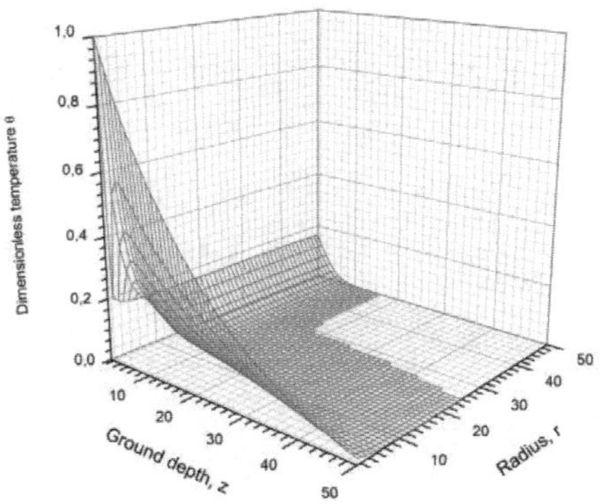

Figure 5. Wireframe plot of the temperatures at charging after 60 days.

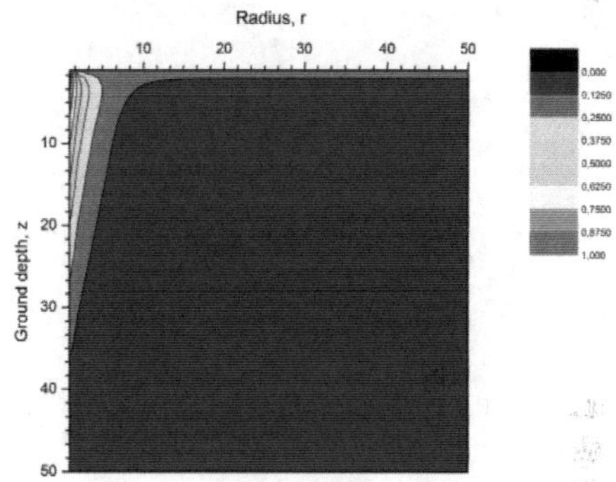

Figure 6. Colour contour plot of the temperatures at charging after 60 days.

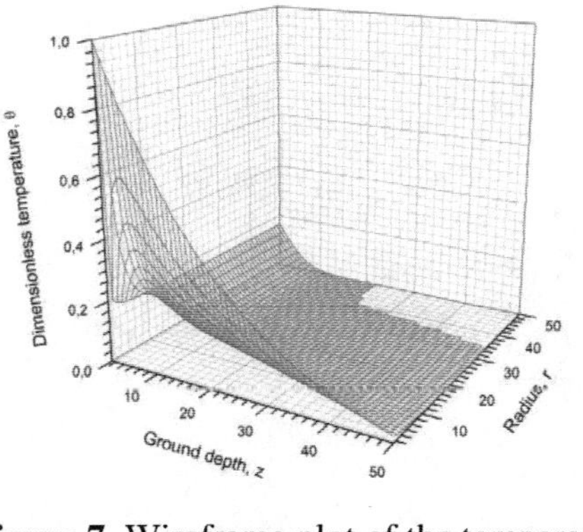

Figure 7. Wireframe plot of the temperatures at charging after 120 days.

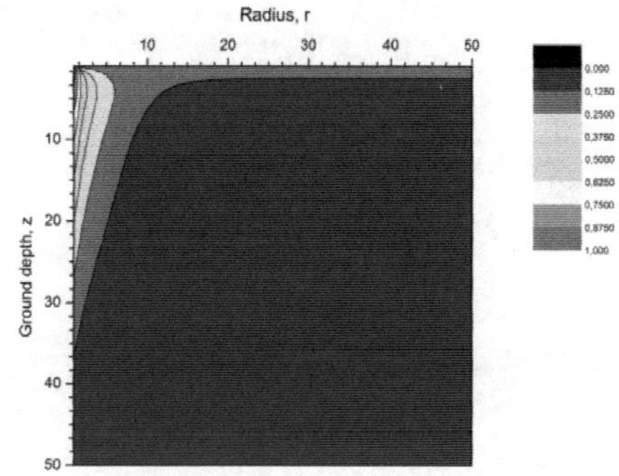

Figure 8. Colour contour plot of the temperatures at charging after 120 days.

IOP Conf. Series: Materials Science and Engineering **595** (2019) 012009 doi:10.1088/1757-899X/595/1/012009

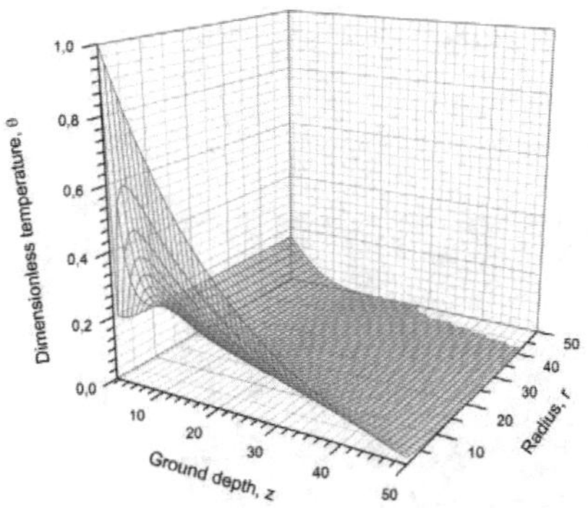

Figure 9. Wireframe plot of the temperatures at charging after 180 days.

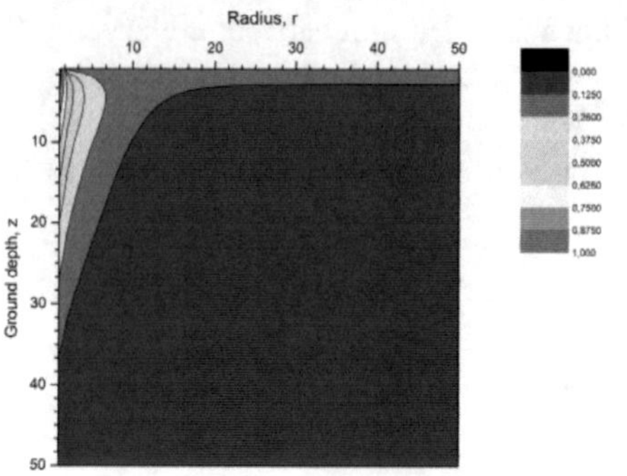

Figure 10. Colour contour plot of the temperatures at charging after 180 days.

As for the bottom surface of the cylindrical domain, the adiabatic boundary hypothesis is sufficiently accurate, because the end effects of the central tube are negligible.

The discharge stage of the yearly cycle represents 180 days. Figures 11 through 18 display the temperature plots in 3-D (wireframe) and colour contour respectively for elapsed periods of 30, 60, 120, and 180 days.

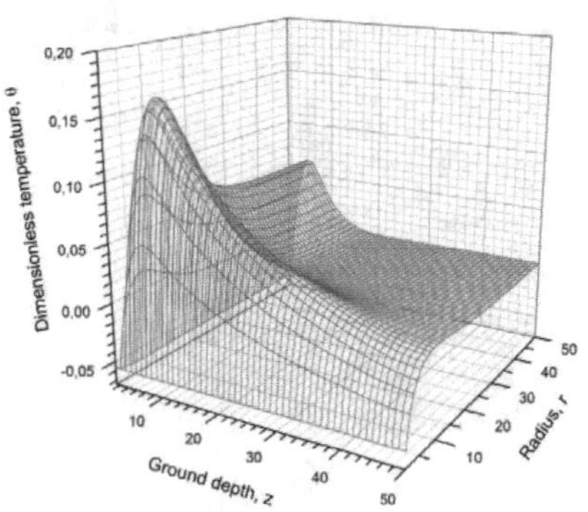

Figure 11. Wireframe plot of the temperatures during discharging after 30 days.

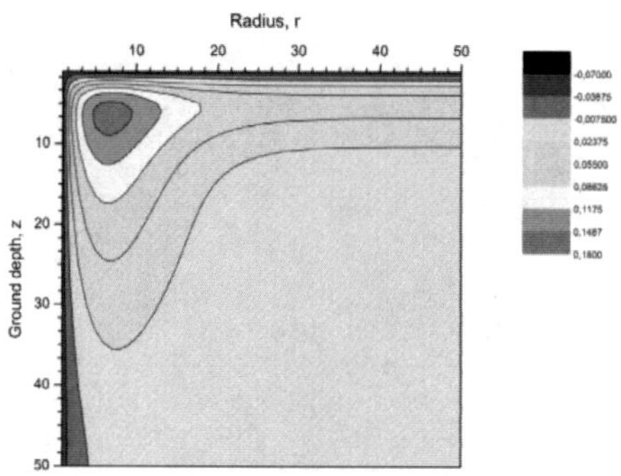

Figure 12. Colour contour plot of the temperatures during discharging after 30 days.

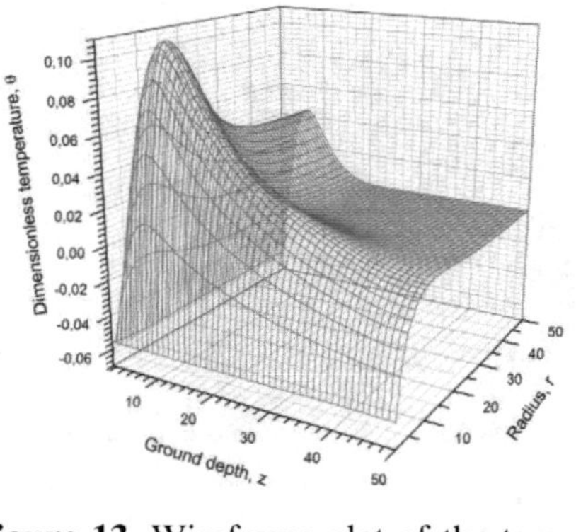

Figure 13. Wireframe plot of the temperatures during discharging after 60 days.

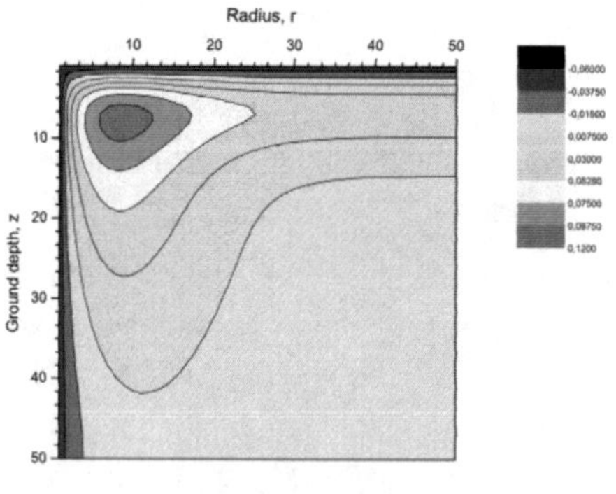

Figure 14. Colour contour plot of the temperatures during discharging after 60 days.

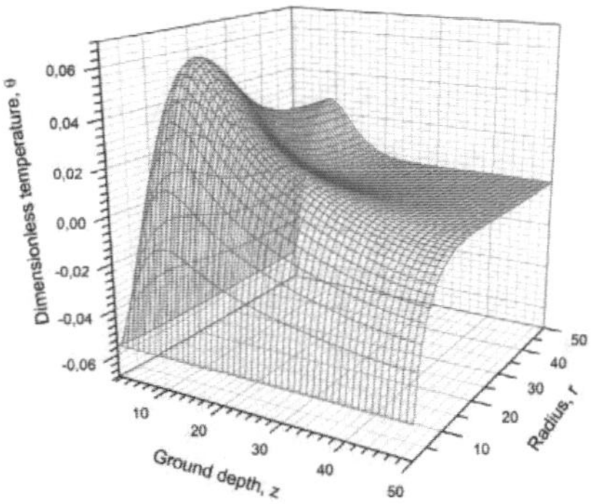

Figure 15. Wireframe plot of the temperatures during discharging after 120 days.

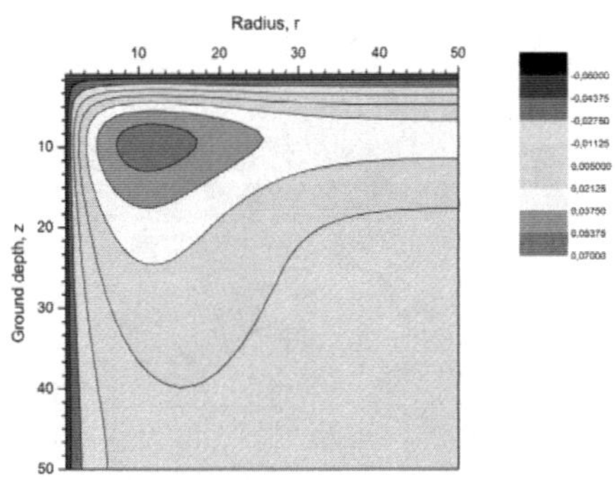

Figure 16. Colour contour plot of the temperatures during discharging after 120 days.

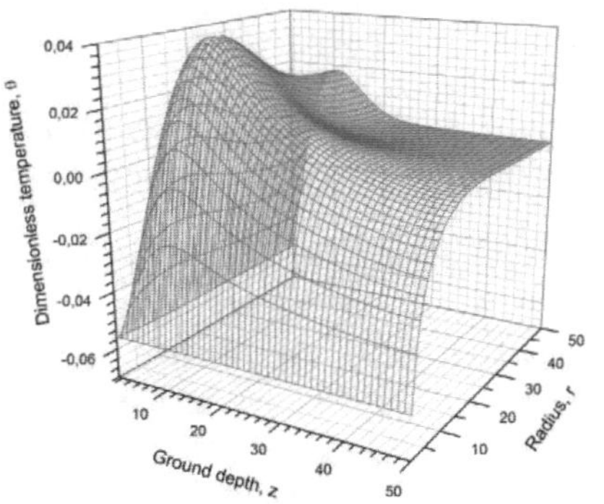

Figure 17. Wireframe plot of the temperatures during discharging after 180 days.

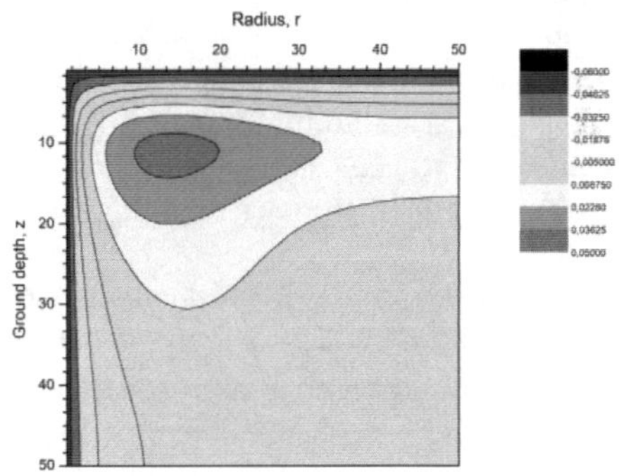

Figure 18. Colour contour plot of the temperatures during discharging after 180 days.

The change of the temperature distribution across the domain is very noticeable during discharging, along with the temperature decrease due to the heat removal. The colour contour plots (figures 12, 14, 16, and 18) highlight a warmer core (in red and orange) as a result of the concurrent heat removal by both the cold fluid flowing through the central channel and by the atmospheric air. The evolution of this central core is seemingly not in accordance with an expected shrinking as the result of the cooling of the ground. However, this discrepancy can be explained by the different temperature ranges allocated to every conventional colour at different elapsed times by the data analysis and graphing software used to generate the graphs displayed in the paper.

Negative dimensionless temperatures of the z-axis of the 3-D wireframe plots are explained by the fact that the ground is cooled in some regions (rendered more visible on the colour contour plots) to temperatures that are less than the initial temperature T_∞ of the ground.

Between the two stages of the storage cycle, we have considered a five days "standby" period, during which there is no heat transfer whatsoever except the heat exchange with atmospheric air.

The evolution of the stored heat during the charging stage of the heat storage is shown on figure 19 and on figure 20 for the discharging stage respectively. The end value for the charging period differs from the start value for the discharging one because of the time gap of five days between the two stages, when some heat is lost to the atmospheric air across the frontal surface of the domain.

After the discharging stage ends, a small amount of the stored heat remains in the ground.

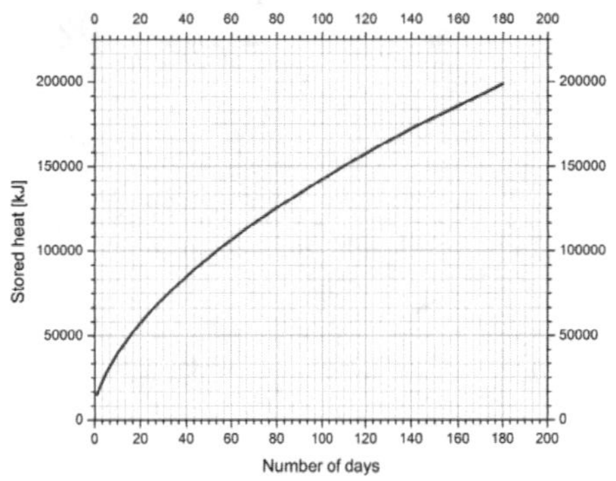

Figure 19. Evolution of the stored heat stored during the charging stage of the cycle.

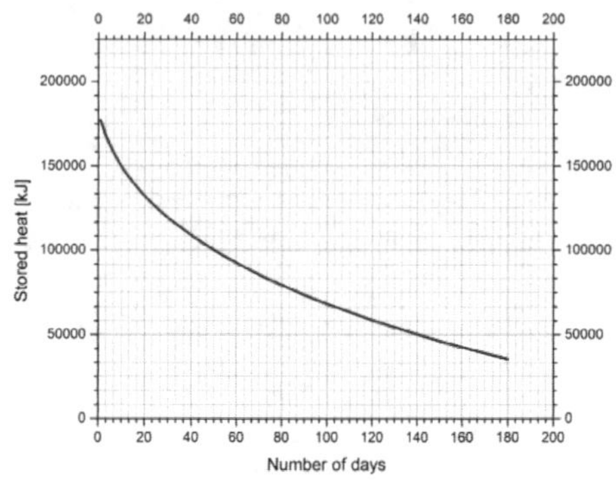

Figure 20. Evolution of the remaining heat during the discharging stage of the cycle.

5. Conclusions

Borehole heat storage systems represent a promising solution to store heat in the ground and to remove it in order to make it useful for ground-source heat pumps (GSHPs). This approach improves the performance of heat pumps used to provide heat and domestic hot water for residential applications.

Such a system adds heat into the ground during the warm season when this is available via solar collector heat conversion systems and extract it in order to operate the heat pump during the cold season.

In order to determine the amount of heat that can be accumulated in the ground with a borehole ground heat storage system, a study must be performed to evaluate the temperature distribution at every moment. A practical approach is to use the finite-difference technique for the two-dimension unsteady conduction heat transfer in the cylindrical setting representing the central tube and the ground.

The decomposition technique has been applied to a two-dimension implicit finite-difference scheme used to calculate the temperature distribution across a cylindrical domain that represents the ground around a central tube along which alternately flow hot and cold fluids respectively that transfer heat to/from the ground.

The numerical results obtained by running a computer software developed by the authors show that the penetration radius of the thermal perturbation during the charging stage is small because of the low thermal conductivity of the ground.

The computer software also allowed to determine the amount of heat stored in the ground at every moment and thus to provide data for future developments involving the calculation of the ground heat pump performance.

References

[1] Fridleifsson I B, Bertani R, Huenges E, Lund J W, Ragnarsson A, and Rybach L 2008 *Proc. IPCC Scoping Meeting on Renewable Energy Sources* ed O Hohmeyer and T Trittin (Luebeck, Germany) p 59

[2] Lund J, Sanner B, Rybach L, Curtis R, and Hellström G 2004 *GHC Bulletin* (Geo-Heat Center, Oregon Institute of Technology) September p 1

[3] Lund J W and Boyd T L 2015 *Proc. World Geothermal Congress* (Melbourne, Australia) p 1

[4] Sanner B, Karytsas C, Mendrinos D, and Rybach L 2003 *Geothermics* **32** p 579

[5] Yang H, Cui P, and Fang Z 2003 *Applied Energy* **87** p 16

[6] Omer A M 2008 *Renewable and Sustainable Energy Review* **12** p 344

[7] Carslaw H S and Jaeger J C 1959 *Conduction of Heat in Solids* (second Edition) (Oxford: Clarendon Press)

[8] Wang L Zhou X and Wei X 2008 *Heat Conduction. Mathematical Models and Analytical*

Solutions (Berlin Heidelberg: Springer-Verlag)

[9] Bandos T, Monter A, Fernandez E, Santander J L G, Isidro J M, Perez J, Fernadez de Cordoba P J, and Urchueguia J F 2009 *Geothermics* **38** p 263

[10] Lamarche L, and Beauchamp B 2007 *Energy and Buildings* **39** p 188

[11] Philippe M, Bernier M, and Marchio D 2009 *Geothermics* **38** p 407

[12] Causon D M, and Mingham C G 2010 *Introductory Finite Difference Methods for PDEs*, (Frederiksberg: Ventus Publishing Aps)

[13] LeVeque R 2007 *Finite Difference Methods for Ordinary and Partial Differential Equations* (Philadelphia: Society for Industrial and Applied Mathematics)

[14] Recktenwald G W 2004 *Mechanical Engineering* **10** p 1

[15] Godunov S K, and Reabenki V S 1977 *Finite Difference Computing Schemes* – Romanian translation – (Bucharest: Editura Tehnică)

[16] Dumencu A, Horbaniuc B and Dumitraşcu G, 2016 Study of a two-dimension transient heat propagation in cylindrical coordinates by means of two finite difference methods, *IOP Conf. Series: Materials Science and Engineering* **147** 012139

IOP Conf. Series: Materials Science and Engineering **595** (2019) 012010 doi:10.1088/1757-899X/595/1/012010

Behaviour of transmission and distribution networks with big consumption, the stress test

C N Eparu[1], S Neacsu[1], A P Prundurel[1], R Radulescu[1] and A Neacșa[2]

[1] Petroleum-Gas University of Ploiesti, Faculty of Oil and Gas Engineering, Drilling, Extraction and Transportation of Hydrocarbons Department, Romania
[2] Petroleum-Gas University of Ploiesti, Faculty of Mechanical Engineering, Mechanical Engineering Department, Romania

e-mail: cristian.eparu@gmail.com

Abstract. Customers of gas transmission and distribution networks consume gas as they needed. Such consumptions vary continuously. It was found that within one day, in most of the cases, there are two major peaks. Because the gas is compressible, the load peaks can be covered from the network's linepack, with the condition that their dynamics are known and the network is prepared in advance, that is, "today" for "tomorrow". The paper presents the tools and the analysis of the natural gas networks and the limits of their safe operation. The paper also proposes a network stress test to verify the future behaviour of the network at high loads as well as to determine the necessary parameters for the next period of maintenance.

1. Introduction

EU legislation puts the consumers in the spotlight. Thus, the clients connected to the natural gas networks needs to receive this product in the parameters written in the contract. Because consumers use the gas depending on necessities, optimum management of the transmission and distribution network must be maid taking into consideration client satisfaction and safety condition fulfilment.

Natural gas volumes are traded by the providers and consumers on the gas market. These transactions are the base of the transport loads definition. Gas has to be delivered by the providers to each client respecting the quantities and qualities from the contract.

Preparing the transport schedule for tomorrow and/or the next days need to take into consideration the quantities required by the big consumers and the gas volumes needed by the small consumers. For this, the nominations are made based on consumption profiles.

Security of natural gas supply is affected by several factors: the amount of gas in the transmission network (linepack), the quantities of gas entering the system, the quantities of gas coming out of the system, the time of year, time of the day, imbalances introduced in the electrical system, problems with transport infrastructure etc.

In order to know the amount of gas needed for tomorrow forecasts algorithms are used. The importance of a good forecast on the natural gas market is given by the difference between nomination and allocation. If the forecasts are closer to real consumption then the cost of supply, gas stocks and cost penalties are diminished.

When the transport program for the next day [1] is prepared by the transmission system operator, the precision of the forecast is essential to balance the network, check if the nominated volumes can be delivered and refuse the capacity when certain values are exceeded.

In general, daily consumption values are forecasted for the small clients and loaded on the exit points from the network. Where this is a big problem arise because with daily average values for gas consumption a static simulation for the current day is made and the operational pressures are calculated.

With SCADA system entering the market a very big difference was observed between the calculated pressure with static simulation and real values from the gas network. The static simulation cannot reflect correctly the dynamic situations from the gas transmission or distribution network because the client's consumption has hourly variations. This can only be simulated using defined hourly profiles.

In figure 1 a comparison between nomination obtained using a forecast with hourly profiles and real consumption is presented along with the corresponding errors [2].

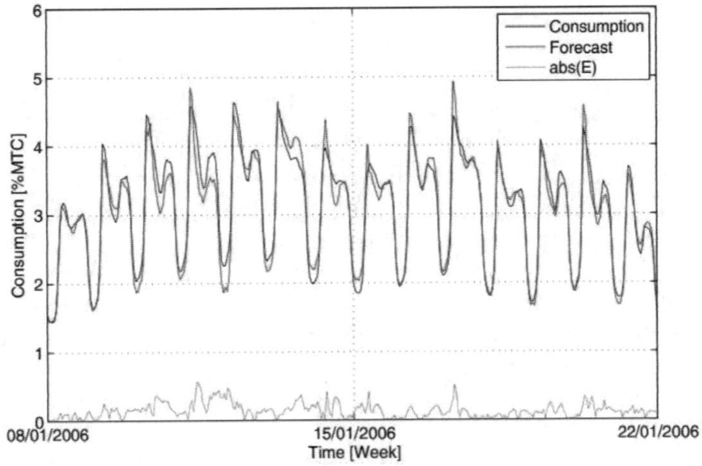

Complete forecast (daily + hourly), January 2006, training data

Figure 1. Comparison between nomination based on hourly profiles and real consumption [2].

In Figure 2 is presented the hourly variation of pressure due to flow rates used by a big consumer (a town). One can notice that the gas consumption has 2 consumption peaks. The first one determines a local pressure drop and simultaneously the Linepack decreasing. After that, the pressure is building again, but it drops very much after the second peak.

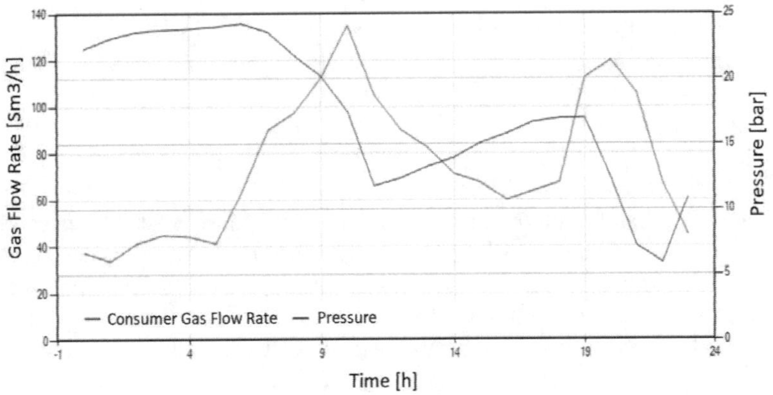

Figure 2. Variation of pressure in the RMS area.

If the Measurement Regulation Station (MRS) supply pressure is reduced with only 1 bar (see figure 3) in the middle of the second gas consumption peak a blackout appears. To overcome this kind of problems the pressure must be increase in that point. This will also influence the pressure regime in the entire network. The final pressures must be determined based on dynamic simulations.

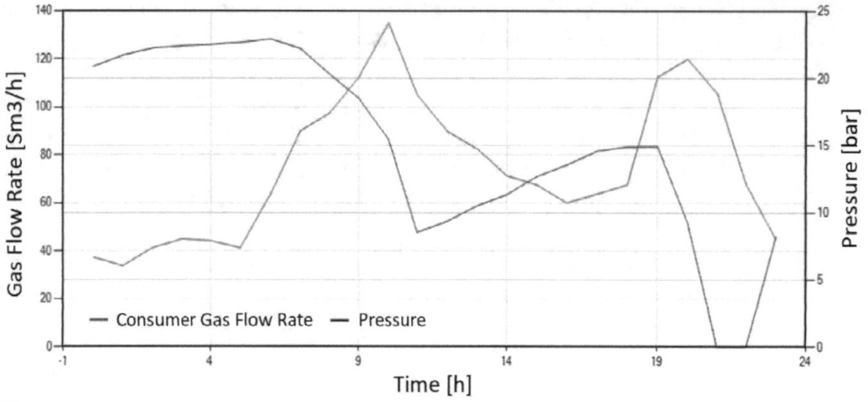

Figure 3. Variation of pressure in the RMS area - alimentation problem.

Form the above we can have a partial conclusion: Operating pressure in transmission and distribution gas network must be generated by dynamic simulations. The loads for each exit point from the network must be modelled with hourly consumption profiles.

From our experience we find out that if a daily load of the transport network is used a pressure regime will result. Using this regime in dynamic simulations blackouts appear in more than 80% of the cases. In order to obtain an optimum pressure regime, that will ensure a safety functioning of the equipment's in the transport networks, dynamic simulations must be carried out. We named this ″network stress test″.

2. Model simulation

In Romania there are a lot of big consumers directly connected to the National Transmission System (NTS). They have a consumption of between 720,000 and 6,000,000 Nm^3/day, but hourly consumption may vary depending on demand [3]. Superposition of such high hourly consumption can lead to a crisis or emergency situation that may culminate with the gas supply interruption in the area.

For a safe operation of a big consumer, gas network must be prepared to deliver gas under the most critical situation [4]. There result critical parameters that must be observed so that there may not occur crisis or emergency situations [5]. Close observance of these parameters lead to decisions on the system operation.

Given the size of the Romanian transport system and the implications of such decisions, simulations of real situations are used. For the data on which decisions are taken to be as accurate as possible, the numerical simulator that process them should be based on a correct and calibrated model of the transportation network [6].

The process of gas transport is continuous and in order to obtain a snapshot of the network parameters at any moment in time it is necessary to have a dynamic simulator [1], because the SCADA systems can offer just a limited number of parameters and the cost are very high.

The simulator can complete the picture of parameters based on the real data received from some points in the network. The simulator can return as results also parameters that cannot be measured, like line pack for pipes [7]. The simulator can present the dynamic in space and time of transport parameters (pressure, flow rate, temperature, composition, calorific value) distribution [8]. A sample of how this simulator looks like is presented in figure 4. One can see how the network looks, what equipment are available, what parameter are used in the simulations and can be seen on graphs, how results can be presented and thematic colouring for parameters.

As one can see the simulator use different composition for all the entry points and can track this composition all over the system.

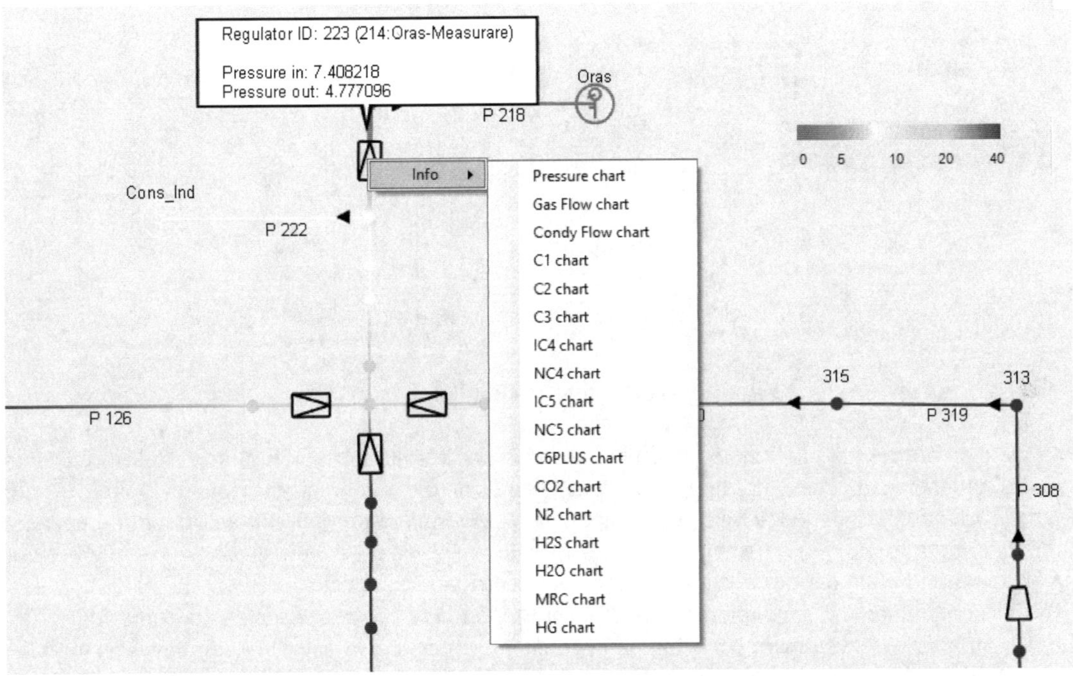

Figure 4. Simulation network and parameters taken into consideration.

Output flows of the transmission system are established on the basis of daily nominations of gas volumes required by customers [9]. These flows are examined on the basis of consumer profiles for each output point, in order to see the level of solicitation [1, 4]. Without these checks, there is a risk that customers may use quantities of gas daily supplied from the NTS without paying them, as payment will be made on a monthly basis.

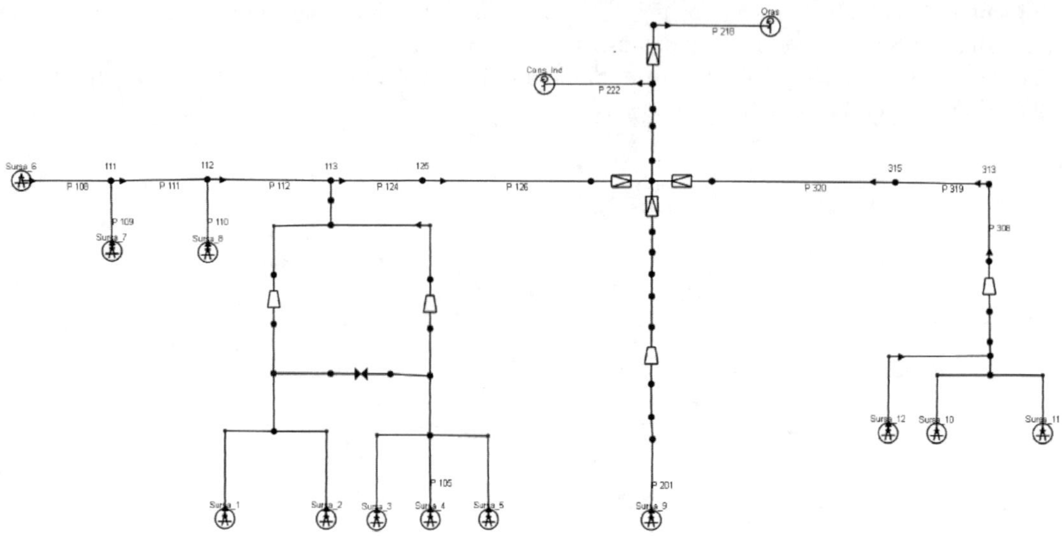

Figure 5. Transmission simulation network.

Consumer profiles for an exit point (measurement control station - MRS) are established taking into account the main consumption factor determined for the previous gas year to which a daily, variable correction is applied, a variable component with temperature [8] and a correction depending on the day of the week [1]. Consumption forecast for the next 3 days is adjusted based on the previous days' consumption that define consumer behaviour specific to each client.

In the next figure the simulation network is presented. We have a complex transmission system with 12 suppliers, 2 customers (one big industrial consumer and a town), 4 compressor station, 3 regulators and 1 Measurement control Station - MRS.

The network has 3 major branches and a total of more than 800 km of pipes. The town is similar with Bucharest in the terms of gas consumption volumes, who has also 3 major delivery lines. Each of the 3 major branches has at least 150 km, so the travel time of the gas will be very important in simulations. All the supply nodes have different qualities of gas. We used gas composition until C6+. There are platforms with 90% C1 and others with just 70 %. The quality of the gas is tracked all over the network.

In the paper we present 2 simulations made for a summer gas day that means from 6:00 today until 6:00 tomorrow morning. The difference between this 2 is that one is with daily loads for the consumers (50 and 992 MMscfd) and the other one is with hourly profiled consumptions

The small clients' (non-monitored) consumptions behaviour is analysed with a performant profiling software (Figure 6). The consumption profile is created based on daily consumption data, temperature, day of the week and season. This software is used to analyse the consumption behaviour of an area in order to create good nomination forecasts for the gas volumes needed.

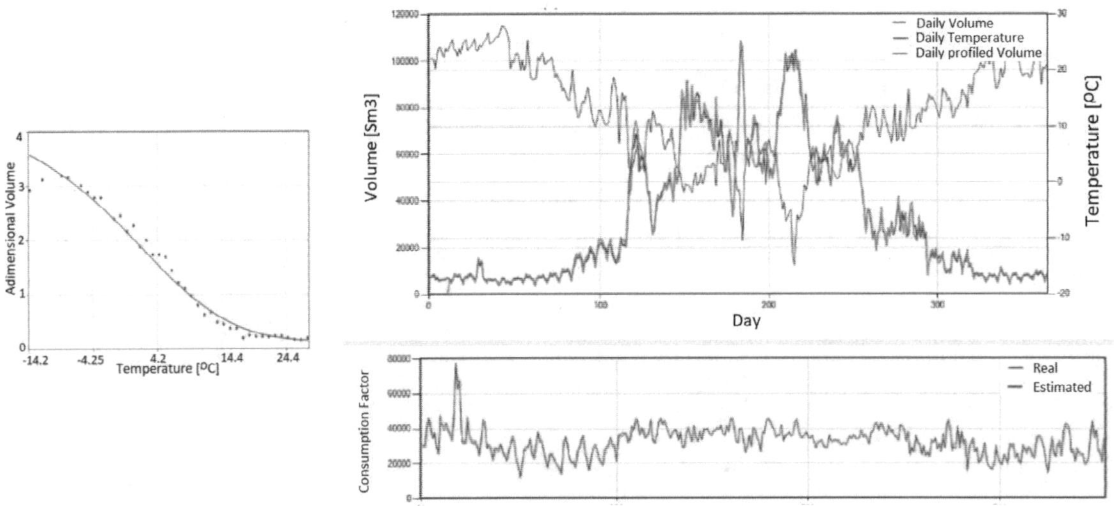

Figure 6. Profile use and gas volume nomination.

In Romania and perhaps in many other South-Eastern European countries the introduction of Smart Metering involves high costs. Because of this there has been introduced an AMR automatic measuring system which is designed for large consumers classified under B4-B6 categories. As for the small gas consumers, classified under B1-B3 categories, in order to reduce costs, readings are performed once every three months.

3. Result analysis

For a daily (every 24 hours) balancing of the network, dynamic scenarios are carried out with an hour-long pace, as shown in Figure 2. The input data employed comprise the gas supplies in the network measured at Source, the gas exits at the large consumers, the initial linepack of the network and the operating manoeuvres specific to the simulated time span.

The gas volumes consumed by the B1-B3 small consumers are initially estimated based on both the consumption profiles, illustrated in Figure 3, and the customer's consumption factor. These values are grouped for the whole town in a single point of the network.

The increased accuracy of the forecasted nomination volumes leads to a minimization of penalties for gas companies. This accuracy increases if the forecast is based on variable consumption factor for each consumer. Once the transport system load (gas values at the entry and exit points) is defined, the scenario creation for tomorrow transport programme can be made.

3.1. The first scenario - with static loads

In the next figure is presented the starting state of the network as a pressure coloured map. The colouring is based on 4 colours: red - if $p<5$ bar, yellow - if pressure is between 5 and 10 bar, green - if pressure is between 10 and 20 bar and bleu - if $p>20$ bar. As one can see only the town is at lower pressure, because is a distribution network.

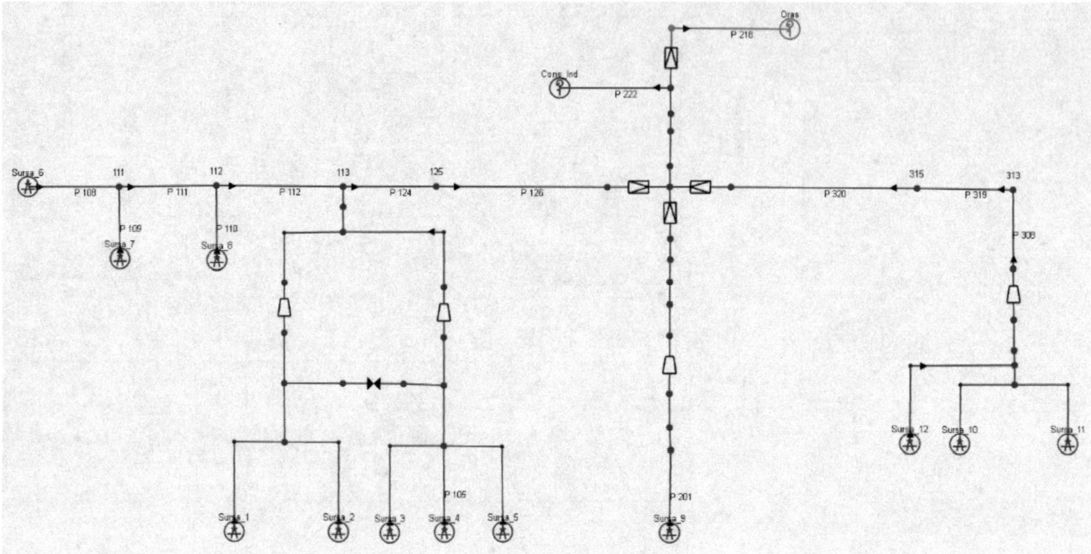

Figure 7. Pressure variation in the simulation network - hour 6.

Because the consumer loads are bigger than the supply the pressure in the system is dropping until the last hour, when the pressure graph looks like in the next picture.

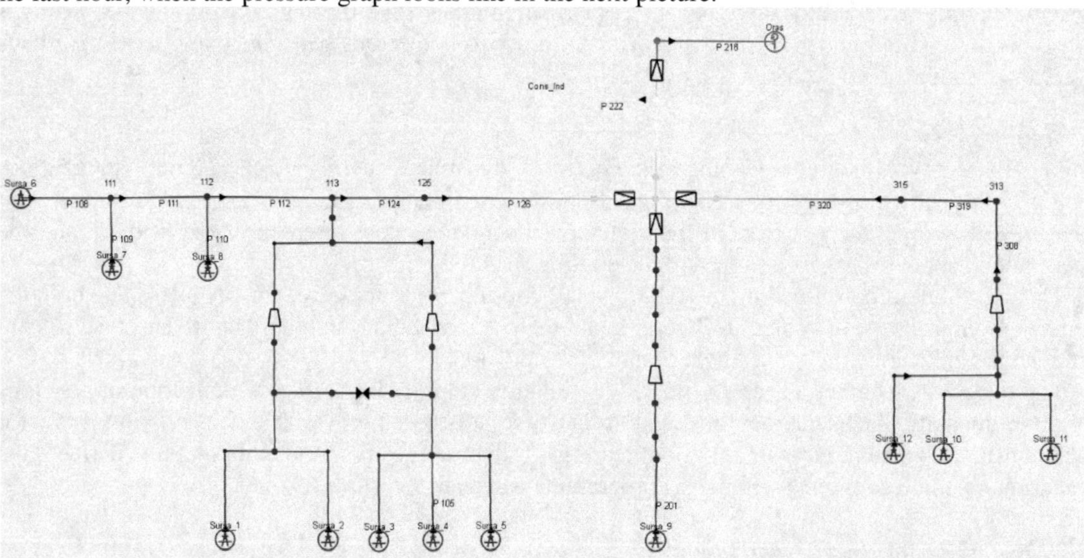

Figure 8. Pressure variation in the simulation network - hour 5 next day.

For a better understanding in figures 9 and 10 are presented the pressure and flow rates variation at city Gates (MRS). The pressure has a 20 bar drop during this day, but still the system can function. This is a day when the nomination is very low compared with the allocation.

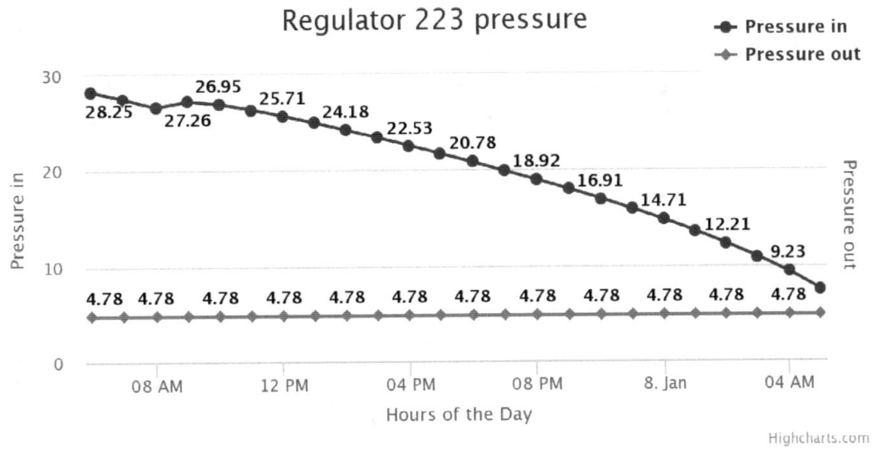

Figure 9. Pressure variation at the town RMS.

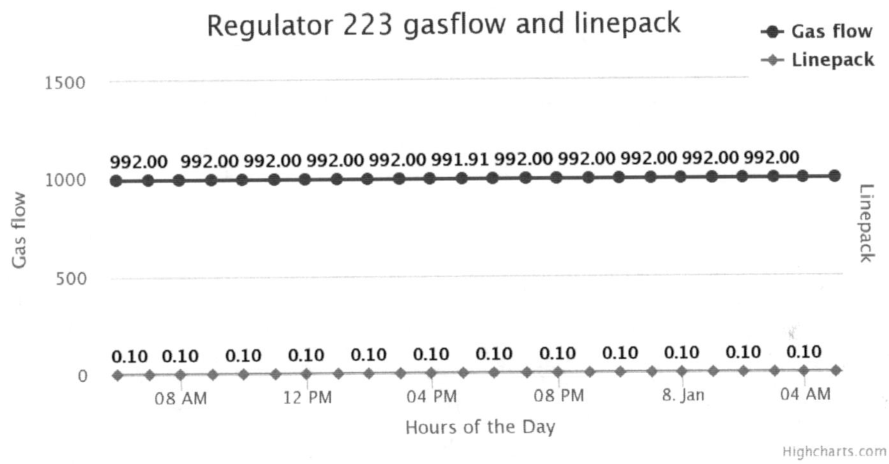

Figure 10. Gas flow (in MMscfd) at the town RMS.

3.2. The second scenario - with dynamic loads

The consumption volumes used as loads for the entry and exit points from the network derived from the gas market. Each registry for each consumption point must be divided using the hourly profiles defined from historical data [5] in order to be used in dynamic simulations (Figure 11). The town profile is generated with 3 peaks due to the (30-70) % of gas volume consumed by the industry and small consumer. The Industrial consumer situation is common because of the daily program of the companies, i.e. from 6 to 18 hrs.

With these new loads, the simulation is made starting from the same starting state as before, thus means that the first hour pressure map will be similar with the one in Figure 7.

In the case of profiled consumption for the Town and Industrial consumer the simulation encounters a negative pressure at hour 21 (see the red part of the network from Figure 12), thus meaning a blackout in gas delivery to the clients.

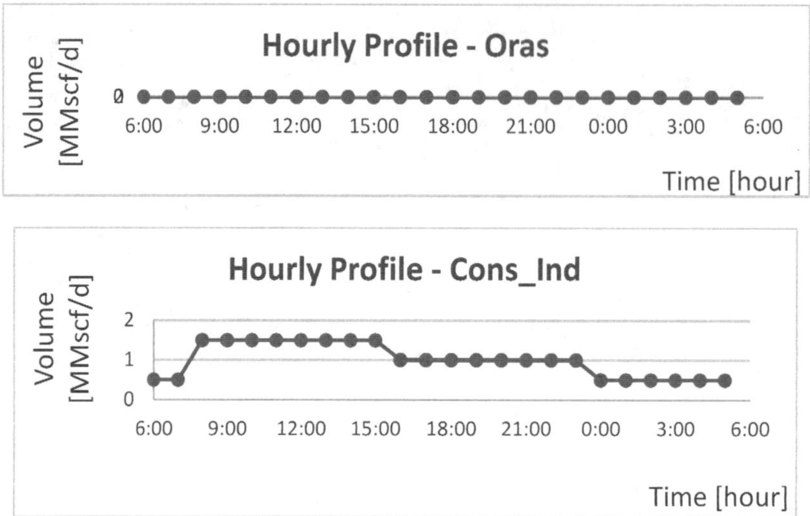

Figure 11. Hourly profile for the town and the Industrial consumer.

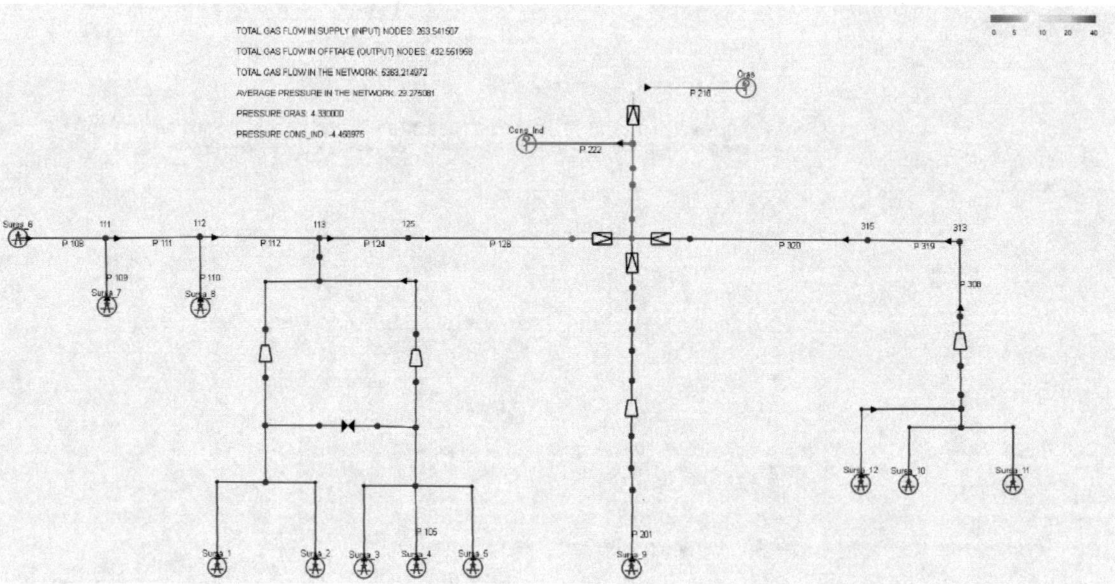

Figure 12. Pressure variation in the simulation network - hour 5 next day.

In order to explain why the pressure has such a big drop let us analyse the profiled flow rates of the customers. In Figure 13 one can see the hourly gas flow rates for the Industrial consumer (values in kg/s) and the Town (values in MMscfd). The main problem occurs because of the high volume of gas consumed by the town in the afternoon. Form 17 hour to 23 the consumption is bigger than the average due to many home activities (cooking, washing, etc.).

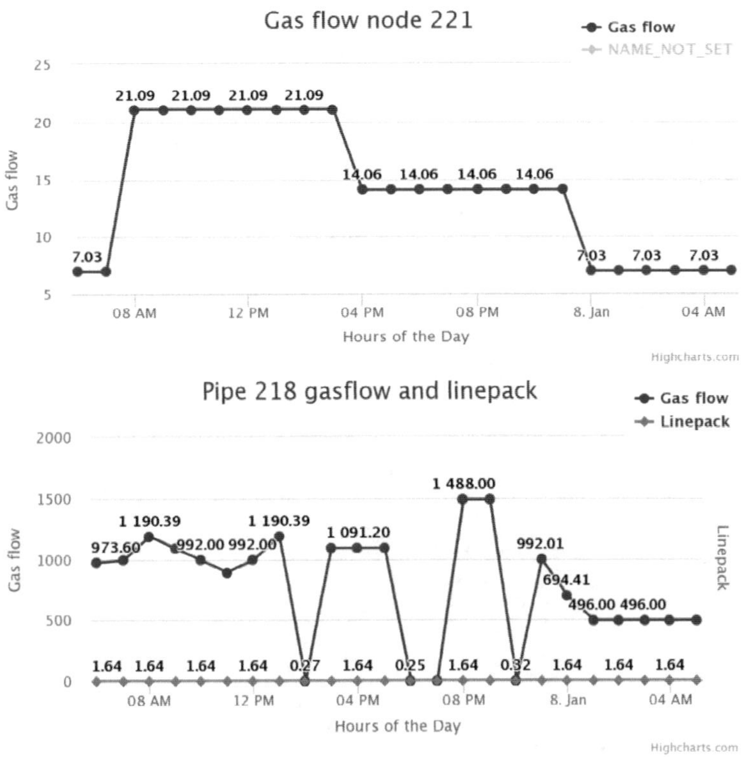

Figure 13. Hourly gas flow rates for the Industrial consumer and the Town.

Because the gas entering the system is less than the exit gas the linepack is consumed. Even if the linepack is big, because of the long distances it cannot be used quickly, as is needed here. Thus, the pressure before the regulator of MRS is variating a lot, with a spike at 15 hours at 7.84 bar and going under 4 bar after hour 20. Before 21 the system is crushing as the pressure goes to 0, thus means no gas will be delivered.

In this simulation we keep even the negative pressure in order to show that after some hours, when the town consumption is reduced the system is going back to the initial state.

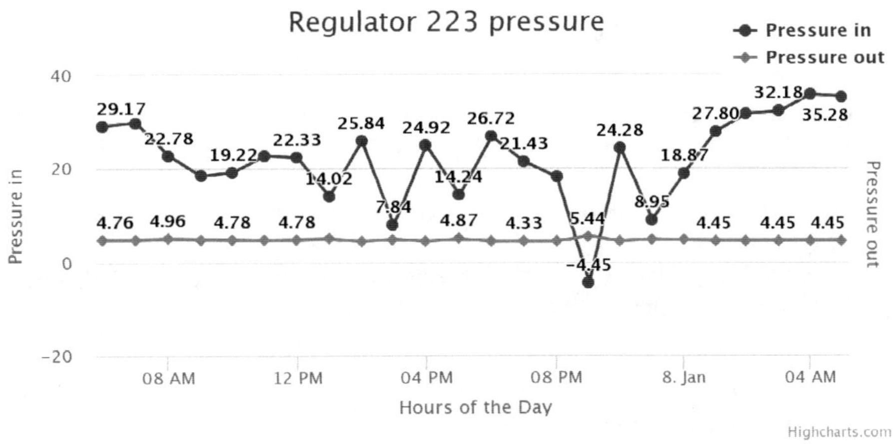

Figure 14. Simulated Hourly Pressure variation in the town RMS.

This simulation shows that the system is not prepared to cope with the hourly consumption for a day, so there should be taken measures in order to ensure a higher gas pressure in the system at the start-up time. Therefore, the energy security of the town, and by similitude the Bucharest area, depends very much on the way the gas transmission network is operated and on gas parameters.

In order to find the security conditions in which the town can work, there are made several changes in the network operation in order to have a running system, but the availabilities of some sources are insufficient to meet the demands and the pressure in the town MRS area is low.

Another measure that can be used is that the Industrial consumer to change his working hours to night shift when the town consumptions are reduced. In this way the hourly consumption are better balanced and the network can be functional for the entire day.

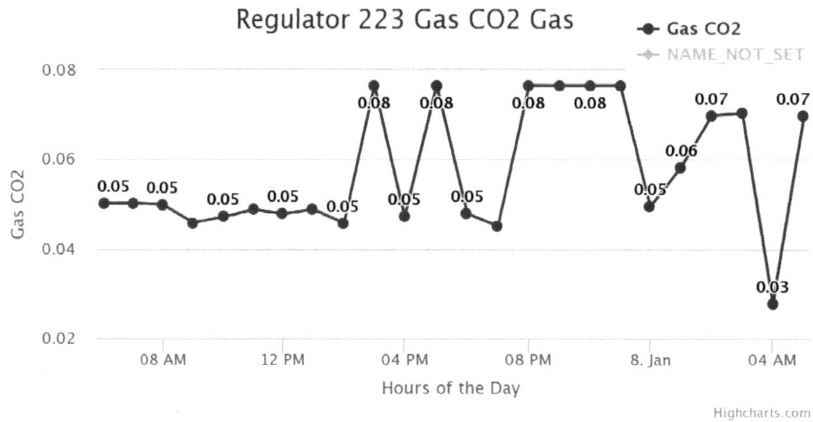

Figure 15. Simulated Hourly Composition variation in the town RMS - CO_2.

If the gas delivery assurance is the first priority and from that we obtain the pressure variation in the system, the second priority is the gas quality. Depending from which part of the network the gas is coming and the volume and composition of these Sources we can have big variation of quality at Town MRS. For exemplification purposes we present in Figure 15 the variation of CO_2 during a gas day. As the CO_2 is highly corrosive, the equipment's must handle this big CO_2 composition. In this case, there is necessary to have a CO_2 removal unit.

4. Conclusion

To comply with safety requirements in the operation of the gas transport network, simulations to prepare next day's schedule are dynamic simulations. In order to obtain the dynamics of the gas flows, alongside daily consumption profiles there are also used hourly consumption profiles specific to each exit point.

The influence of the hourly consumption profile can be seen when there are compared the situation in which power consumption is constant throughout the day and the situation in which consumption relates to a specific profile.

The problem of overcoming peak demands is a very important one to consider whenever establishing the transport program for the next day, in order to avoid any problems in ensuring the security of gas supply to consumers. Safe operation of the network is more difficult as these consumption peaks coincide in point of time.

References

[1] Neacşu S and Eparu C 2013 System for Real Time Managing the Natural Gas Transport Network, *21st IBIMA Conference on Vision 2020: Innovation, Development Sustainability and Economic Growth, 27-28 June Viena,* Austria, 9 p

[2] https://www.intechopen.com/books/natural-gas/practical-results-of-forecasting-for-the-natural-gas-market
[3] Neacşu S, Trifan C and Albulescu M 2008 Considerations on the Errors Associated to the Measuring of the Amounts of Natural Gas Delivered to Household Consumers *Chemistry Magazine* **59** **(7)** 796
[4] Albulescu M A et al. 2011 Consideraţii referitoare la capacitatea maximă de transport a conductelor/reţelelor de gaze naturale *Revista Termotehnica* **1** 16
[5] * * * ERGEG principles: Capacity allocation and congestion management in natural gas transmission networks 15 Jan 2008
[6] Neacşu S and Eparu C 2012 Presentation of some aspects related to the use of SIMONE in modelling the gas transport network in Romania *11th SIMONE Congress, 13-15 June 2012,* Bratislava, Slovakia, **05**
[7] Eparu C N, Radulescu R and Stoica D 2013 The dynamic simulation of the natural gas transportation, *Analele Universităţii Ovidius din Constanţa, Seria Chimie* **24** **(2)** 83
[8] Neacsu S, Suditu S and Popescu C 2010 Educational Software for the Calculation of Thermodynamic Values of Real Gases for the Training of Engineering Students in Romania *9th WSEAS International Conference on Education and Educational Technology (EDU '10), Iwate Prefectural University,* Japan
[9] Albulescu M, Neacşu S, Trifan C and Ionescu E M 2008 Theoretical Considerations and Experimental Measurements Concerning the Defining of Correction Coefficients in Case of Measuring Natural Gas Volumes for Household Consumers *Materiale Plastice* **45** **(1)** 38-41

Development of the highly boosted capillary porous foam generator for gas (steam) mechanical foam

A A Genbach[1], K K Shokoloakov[1], I K Iliev[2] and A K Terziev[3]

[1] Deptartment of Heat Engineering Installations, Almaty University of Power Engineering and Telecommunications, Republic of Kazakhstan
[2] Department of Thermotechnics, Hydraulics and Ecology, Universiy of Ruse, Bulgaria
[3] Department of Power Engineering and Power Machines, Technical University of Sofia, Bulgaria

E-mail: aterziev@tu-sofia.bg

Abstract. A highly boosted capillary porous foam generator for gas (steam) mechanical foam is developed. It contains the inlet and outlet nozzles, nozzles for supply of the foam generating solution and foam discharge, package of the foam generating meshes with increased sizes and sprayer. In order to increase capacity and improve the foam characteristics, the generator is equipped with a capillary patch and forming channel. The capillary patch is located at the external surface of perforated pipe that supplies the foam generating solution, and the forming channel is installed outside of the capillary patch with feeder. Additionally, the foam generator has control plates located at both sides of the feeding device, where the plates at both sides pass into a spring guiding viewfinder, also the packet of foamed grids with increasing size in the direction of gas flow is made with dimensions 0.4x0.55.

1. Introduction

A foam generator of a new class is proposed - a nozzle-shaped one, having a capillary-porous structure specially optimized for foam generation instead of the usual mesh package. To do this, it was necessary to study the internal characteristics of the process of boiling liquid in capillary-porous structures: firstly clean [3-6], then with the addition of a additives [4, 7-13]. Experimental data were summarized using dimensional analysis with obtaining a criterial equation [7]. Nozzle-free foam generators have some peculiarities as follows: they are able to manage power processes [1, 4, 13], have defoaming structures [8], able to generate foam with the help of bubbling processes [9] and supply of power instead of gas [11], as well as have special materials with controlled geometry of micro channels [12].

The invention refers to the industrial and station heat and power, and designed for generation of two flows at the heat and power units, that could be used in fire engines for firefighting inside buildings and facilities, as well as be useful in the different fields of national economy for dust suppression [6, 10].

It is known that the similar invention was introduced in USSR 1202598 A62C 5/04, 1985, which contains inlet and outlet nozzles, package of meshes, sprayer in the form of tubular ring.

One disadvantage of the device is its low foam efficiency characterized by the limited penetration of the foam generating solution through the transverse section of meshes located in the tubular ring. In addition, the characteristics of the foam turned out to be low, since the package of standard metal grids does not allow turbulization and intensify the processes of foam generation. Besides, the boosting action of turbulator looks insignificant because the fluid flow in the meshes section is uneven and unstable. It is also impossible to control a fluid rate. Thus, the foam features come short of the expected ones. However, it is possible to increase efficiency of the foam generator and enhance its operational characteristics.

The similar invention USSR 1498511 A 62 C 5/04, 1989, which contains inlet and outlet nozzles, nozzles for supply of the foam generating solution and foam extraction, sprayer, package of the foam generating meshes in the form of turbulator with the expanded mesh cells along the foam flow.

2. Installation description and specifics

Foam generator is equipped with the capillary patch and forming channel in order to increase efficiency of the foam generator and improve the foam features. The capillary patch is located at the external surface of perforated pipe of foam generating solution, and the forming channel is located outside of capillary patch forming a feeder. In addition, foam generator contains the adjusting plates located at both sides of the feeding device whereas the plates at both sides transfer to the spring deflector, and package of foam generating meshes with extended sizing along the gas flows of 0.4x0.55.

Compared with the known technical solution in the proposed device due to the use of a perforated pipe for supplying a foam solution, the flow of fluid is significantly increased, rather than through the end of a packet of grids located in a tubular ring. Since there is a capillary patch at the external surface of the pipe that has capillary size dozens of times less than the size of metal cells, then a quite balanced transpiration of the increased fluid consumption is ensured into the forming channel, which serves as a feeder for the foam generating mesh package [7, 11]. That helps significantly increase efficiency and stability of film of the foam generating solution, in the mesh package 0.4x0.55, and therefore increases such features as frequency and dispersion.

Additionally, installed adjusting plates help control a fluid consumption rate and therefore the important internal foam characteristics (critical and detachable size of bubbles, their lifetime and silence, frequency of generation, density of the forming centers) [2, 3, 5]. The above values determine the device characteristics: specific foam efficiency, stability and steadiness of foam, frequency and dispersion. Upgrade of the mesh package 0.4x0.55 help reduce a gas dynamic resistance in comparison with prototype and retain a hydraulic resistance.

This is to emphasize that the significant enhancement and boosting of the foam generating process is ensured by the available spring deflector that creates spontaneous vibrational forces because of the gas flow. The last ones jointly with the capillary, gravity and inertial forces of pressure create a quite stable two-phase multicomponent pulse heat-and-hydrodynamic boundary layer that generates foam with high efficiency and enhanced characteristics.

Consequently, by increasing the specific performance and improving the foam performance, the material consumption and dimensions of the device are reduced, energy is saved by pumping the solution and gas due to the possibility of controlling the process of foam generation, especially with forced and variable modes of operation of the foam generator, the reliability and service life of the device increases.

Figure 1 describes a foam generator in plain view. Foam generator consists of inlet 1 and outlet 2 nozzles, package of the foam generating meshes 3, made of two metal standard meshes of capillary porous structure, each consequent mesh has an increased size of cells along the gas flow direction $0.4\text{x}10^{-3}m$; $0.55\text{x}10^{-3}$ m accordingly. The sprayer body 4 includes capillary patch 5, located at the external surface of perforated pipe 6 and forming channel 7, located outside of capillary patch 5 forming a feeder 8 with plates 9 at both sides, transferring to the spring deflector 10.

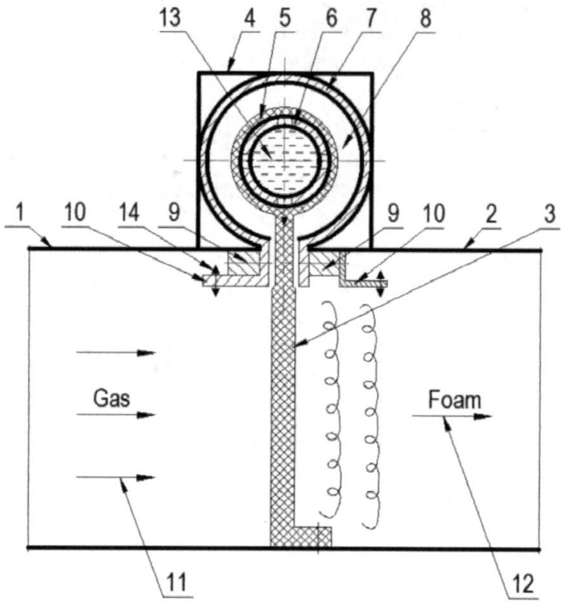

Figure 1. Highly-boosted nozzle-free capillary porous foam generator of gas (steam) mechanical foam.

Foam generator runs in the following sequence as follow: The foam generating solution 13 is injected by sprayer 4 to the package of foam generating meshes with increased sizes along the gas flow $0.4x10^{-3}$m и $0.55x10^{-3}$ m accordingly.

Solution is transported through the holes of perforated pipe 6 via capillary patch 5 and approaches feeder 8, formed by patch 5 and forming channel 7. It helps pump large volume of the foam generating solution 13 and quite evenly re-distributes it, feeding the mesh package 3. Fluid 13 in the mesh package begins distribution in a form of stable and steady films because of the joint action of the gravity and capillary forces.

The vibrational potential occurs as a result of spontaneous vibrational (wave) action of the spring deflector 10, streamlined by the gas flow 11 in nozzle 1 and foam flow 12 in nozzle 2. Sound waves 14 actively promote an even re-distribution of fluid 13 when it is transferred from capillary patch 5 in the mesh package 3, whereas the adjusting plates 9 help actively manage the process. Under oncoming gas flow 11 in volume and at the mesh surface 3 the gas bubbles arise and that leads to generation of two-phase flow in the form of gas mechanical foam 12, which is formed in spray with the help of attachment (not shown) and moves forward as per the process either in the heat and power unit or to the fire and dust source.

Availability of the capillary patch 5 ensures an even distribution of the foam generating solution 13 inside feeder 8, whereas an increased consumption of solution 13, pumped through pipe 6 forms a large specific capacity of foam generator. Evenly distributed thin liquid films in mesh package 3 have stable and steady features, which allows to obtain a foam of high technical characteristics such as frequency, dispersion and stability.

In comparison with prototype due to the joint action of capillary, gravity, inertial and vibrational potentials, the heat exchange factors are boosted by 1.5 times and specific capacity is increased. This helps accordingly reduce material consumption rate 1.5 times and the device dimensions, weight of device 2.5 times by saving capital expenses. Although, it leads to saving power and solution under various conditions (due to the capillary patch 5) and improved conditions due to the controlled process with the help of adjusting plates 9 and upgraded meshes 0.4 x 0.55. Therein, the operational conditions are getting simplified and the operational costs are reduced.

3. Comparative analysis of different dust collectors. Hydraulic losses specifics

In Table 1 a and b package of meshes 0.4×0.55 in size means that it is combined from two layers of wire meshes installed consequently along the incoming gas flow with cell sizes 0.4×10^{-3} m; 0.55×10^{-3} m accordingly.

The advantage of meshes with larger cells is that in smaller cells like in prototype (0.08×10^{-3} m) a.e. USSR 1498511, 1989, the foam generation process gets worse as bubbles block their cells, which leads to hydrodynamic crisis. Hence the production of fresh foam at the inlet is blocked. This picture was observed by the authors using optic research methods [1, 2, 4] (holographic interferometry and high-speed movie capturing). Package of meshes 0.4×0.55 forms quite stable two-phase multi-component hydrodynamic shear layer pulsed by heat, which is a prerequisite for high quality foam production.

Large cells like in prototype (1×10^{-3} m), lead to the abrupt reduction of value of capillary potential, liquid film becomes unstable and crisis of foam generation happens much earlier than in meshes with smaller cells 0.08×10^{-3} m.

Thus, due to boosting of the foam generation process in package of meshes 0.4×0.55, 1.5 times (see table 1) the efficiency is increased, the process is enhanced, and package of meshes is a kind of turbulizer that helps approximately 1.5 times reducing materials consumption and 2.5 times weight reduction.

In fact, enhancement of the foam generation process will be significantly higher due to the spring directing shield, which influences upon the adjusting plates installed at both sides of the feeding device and lead to more active and stable supply from foam generating package of meshes via liquid solution. Mutual impact on the process of the gravity, capillary, inertial and vibration potentials is observed. In case of vibration movements of surface cause by vertical cylinders, enhancement of processes may increase two times, and in some cases up to ten times [14].

Table 1. Comparative analysis of different dust collectors.

a)

No.	Indicator	Chamber of gravitational dust collector	Inertial (centrifugal) dust collector	Cyclone (centrifugal medium-pressure) Large diameter (1-3m)	Small diameter and battery type
1	2	3	4	5	6
1	Hydraulic resistance, Pa	100-200	200-300	700-800	800-1250
2	Minimum size of collected particles, 10^{-6}m and effectiveness (η%)	50-100 85	40-50 85	30-40 85	10-25 85
3	Maximum admissible temperature, °C	400	400 (as per steel grade)	400	400
4	Gas low limit temperature	Over dew point		Any	
5	Corrosion resistance		Too resistant		
6	Explosion and Fire Hazard		Minor		
7	Relative cost of cleanup of 1000m^{3}gas	1	1.5	2	3
8	Main settling mechanism	G	U	U	U
9	Dependency of effectiveness from: a) particle size b) temperature c) concentration (g/m^3) d) humidity upon condensate of water vapors	 No impact (2500…15) No impact	$\sim d^2_r\, C_k \sim d^2_r\, C_k$ $\sim C_k/\mu_r \sim C_k/\mu_r$	As per device diameter and dust adhesion Discharge might be complicated	

IOP Conf. Series: Materials Science and Engineering **595** (2019) 012011 doi:10.1088/1757-899X/595/1/012011

b)

No.	Indicator	Hose (fiber) filter	Granular filter	Wet filter	Wet Dust Collector		Electric filter	Wet filter of Author
					Low pressure	High pressure		
1	2	7	8	9	10	11	12	13
1	Hydraulic resistance, Pa	750-1500	1200-1600	1000-1800	750-1500	5000-12500	100-400	500-1000
2	Minimum size of collected particles, 10^{-6}m and effectiveness (η%)	0.5 99	0.1-1 97	1-2 97	2-5 97	0.1-1 97	0.25-1 99	0.1-0.25 99
3	Maximum admissible temperature, °C	<80 и <250 for glass fibers	Not limited				<425 as per gas content and dust properties	Not limited
4	Gas low limit temperature	Any					Over dew point	any
5	Corrosion resistance	Resistant at temperature exceeding dew point			Required corrosion protection if acids available in gases		Resistant at temperature exceeding dew point	
6	Explosion and Fire Hazard	Major (hazardous)	Minor	Minimum			Major	Minimum
7	Relative cost of cleanup of $1000m^3$gas	3-3.75	2.5-4	7-10	2.5-5	5-15	7-15	1-2.5
8	Main settling mechanism	at $d_r \leq 3x10^{-7}$m-D at $d_r > 3x10^{-7}$m-Stk			Stk	Stk	K_E	D; K; Stk
9	Dependancy of effectiveness from: a) particle size	at $d_r \leq 3x10^{-7}m \sim C_k/d_r$ at $d_r > 3x10^{-7}m \sim d^2_r/C_k$				$\sim d^2_r C_k$	$\sim d_r C_k$	n, 7…9
	b) temperature	at $d_r \leq 3x10^{-7}m \sim C_k T_r/\mu_r$ at $d_r > 3x10^{-7}m \sim C_k/\mu_r$						n, 7…9
	c) concentration (g/m³)		Not over 2g/m³	As per water supply system and power consumption (50…0.5)			Limited 50	n, 8…11
	d) humidity upon condensate of water vapors	Fiber is clogged	Improves settling	No impact	No impact		Complicated dust removal	Improves settling

In particular, enhancement of hydrodynamics processes reaches up to 10 times for vibrations with low frequency (high amplitude), and with high frequency (low amplitude) [6].

Below is presented some calculation related with hydraulic resistance of the package of meshes.

The pressure loss is presented with the following:

$$\Delta p = \mu_l m_l h \rho_l^{-1} F_e^{-1} k_y^{-1}, \text{ Pa} \tag{1}$$

where: $\mu_l = 282.10^{-6}$ Pa·s - dynamic viscosity of the liquid; $m_l = 2.4\left(10^{-3}\right)$ kg·s^{-1} liquid consumption for 1m^2 of package of meshes; h - length of layer, m; F_e - cross section area of package of meshes, m^2; k_y - relatively permeability factor, m^2; ρ_l - density of the liquid, kg/m^3.

Cross section area of the package is calculated with the following:

$$F_e = \varepsilon . F_C = 0.7\left(10^{-3}\right),\ \mathrm{m}^2 \tag{2}$$

where: $\varepsilon = 0.7$ is mesh porosity, $F_c = \delta L = 1.02\left(10^{-3}\right)1$. Here δ is thickness of the package. For stainless steel meshes 12X18H9T (4 M TC – 4-7-66) thickness of package equals to 0.4 × 0.55.

Relatively permeability factor is defined experimentally [1, 2]:

$$k_y = 5.5\left(10^{-7}\right)\left(\frac{b_r}{d}\right)^{-1.29} = 2.1\left(10^{-7}\right),\ \mathrm{m}^2 \tag{3}$$

where b_r is hydraulic diameter; d average diameter of the mesh wire.

Considering equations 1, 2 and 3 for the pressure loss is obtained:

$$\Delta p = 282\left(10^{-6}\right)2.4\left(10^{-3}\right)1\left(958\right)^{-1}\left(0.7\left(10^{-3}\right)\right)^{-1}\left(2.1\left(10^{-7}\right)\right)^{-1} = 4.8\ \mathrm{Pa}, \tag{4}$$

At the maximum efficiency (boosting) (see table 1):

$$\Delta p = 4.8\frac{8\left(10^5\right)}{0.5\left(10^4\right)} = 768\ \mathrm{Pa}, \tag{5}$$

The calculated pressure loss (Eq. 4) is significantly lower that than pressure loss of the nozzle foam generators where foam generating solution is injected directly to the nozzles under pressure $2 \div 5$ (10^5) Pa and nearly the same as in prototype since ratio (b_r / d) will be slightly changed.

The pressure loss of the package of meshes for the gas flow is as follow:

$$\Delta p = \sum_1^2 K \rho_g \frac{v^2}{2}\ \mathrm{Pa}, \tag{6}$$

where ρ is density of gas, kg/m^3; v - velocity of gas phase, m/s; K - minor loss coefficient ([15, 16, 17]).

The flow behavior can be established with the Reynold number:

$$\mathrm{Re} = \frac{v_0 . d}{\upsilon} = 62{,}9 \tag{7}$$

where: v_0 - velocity of gas phase in cellular meshes, m/s; d - average diameter of mesh wire, m; υ kinematic viscosity, m^2/s.

Since the Re number is higher than 50, the total minor loss coefficient is:

$$\sum_1^2 K = 2.6 \tag{8}$$

Based on the value of the minor loss coefficient, the pressure loss is:

$$\Delta p = 2.6\frac{1.2\left(3^2\right)}{2} = 14.04\text{, Pa} \quad (9)$$

Comparing with the prototype:

$$\text{Re} = \frac{v_0.d}{\upsilon} = 62.9 \quad (10)$$

Based on the diagram [15] minor loss coefficient is 7, and pressure loss is 37.8 Pa.

4. Conclusion

The advantage from implementation of the proposed foam generator in comparison with prototype is reduction in material consumption and dimensions by 1.5 times while the specific productivity growth achieved due to the intensification of the two-phase flow generation processes, 2.5 times the mass. At nominal mode of operation, the hydraulic resistance will be as in the prototype, but hundreds of times lower than in the injector generators, and the gas-dynamic resistance is 2.7 times less than in the prototype. Simplification of installation (no heater) leads to a reduction in operating costs, simplifies the structural design of the foam generator, increases its reliability and service life. A stable two-phase multicomponent pulse immersion enhances the internal and technical characteristics of device. Under various and emergency operational conditions a proper control of the foam generation process is ensured, and results in power saving when supplying gas and solution, and also saves the foam generator and reduces operational expenses.

Overall, the proposed foam generator reduces power and material consumption, increases the foam quality, simplifies operational conditions and ensures the process safety and protects environment (social effect).

References

[1] Polyaev V, Genbach A N and Genbach A A 1995 Methods of Monitoring Energy Processes. Experimental thermal and fluid Science, International of Thermodynamics, Experimental Heat Transfer and Fluid Mechanics. *Avenue of the Americas. – New York. 1995. V.10*. April. – pp 273-286

[2] Polyayev V M and Genbach A A 1990 Plotnost tsentrov paroobrazovaniya i vybros kapel iz poristoy struktury. *Izvestiya vuzov Mashinostroyeniye*. **9** pp 50-55

[3] Polyayev V M and Genbach A A 1990 Otryvnoy diametr i chastota otryva parovykh puzyrey v poristykh strukturakh, *Vestnik MGTU, seriya Mashinostroyeniye*. **1** pp 69-72

[4] Polyayev V M, Genbach A A and Minashkin D V 1991 Vizualizatsiya protsessov v poristom ellipticheskom teploobmennike, *Izvestiya vuzov, Mashinostroyeniye*, M. **10-12** pp 75-80

[5] Polyayev V M and Genbach A A 1990 Skorost rosta parovykh puzyrey v poristykh strukturakh, *Izvestiya vuzov, Mashinostroyeniye.* **10** pp 56-61

[6] Polyayev V M and Genbach A A 1991 Oblasti primeneniya poristoy sistemy, *Izvestiya vuzov, Energetika.* **12** pp 97-101

[7] Polyaev V and Genbach A A 1993 Heat Transfer in a Porous System in the Presence of Both Capillary and Gravity Forces, *Thermal Engineering*, M. **40(7)** pp 551-554

[8] Genbach A A and Shokolakov K 2011 Poristyy pennyy pyleulovitel, Poisk, MON RK, Almaty. **2** pp 266-271

[9] Genbach A A and Pylepodavleniye K N V 2011 Penogenerator s pronitsayemoy peregorodkoy, *Energetika i toplivnyye resursy Kazakhstana*. **5** pp 85-87

[10] Genbach A A and Genbach N A 2013 Puti polucheniya trebuyemoy informatsii pri razrabotke kapillyarno-poristykh sistem energoustanovok, *Vestnik AUES*, Almaty. **2(21)** pp 12-18

[11] Genbach A A and Genbach A N 2009 Issledovaniye penogeneratora s obogrevayemoy poverkhnostyu, *Vestnik AIES*, Almaty. **4** pp 24-27

IOP Conf. Series: Materials Science and Engineering **595** (2019) 012011 doi:10.1088/1757-899X/595/1/012011

[12] Genbach A A and Piontskovskiy M S 2010 Poristyy pyleulovitel s upravlyayemoy geometriyey mikrokanalov, *Energetika i toplivnyye resursy Kazakhstana.* **4** pp 59-61
[13] Polyayev V M, Genbach A A and Minashkin D V 1991 Protsessy v poristom ellipticheskom teploobmennike, *Izvestiya vuzov, Mashinostroyeniye*. **4-6** pp 73-77
[14] Spravochnik po bor'be s pyl'ju v gornodobyvajushhej promyshlennosti/Pod red. A.S.Kuz'micha. – M.:Nedra, 1982, S.103
[15] Idelchik I E 1975 Spravochnik po gidravlicheskim soprotivlenijam. – M.: Mashinostroenie. 559
[16] Litvinenko V V 2018 Intensifikatsiya protsessov teploobmena pri kipenii zhidkostej na kapillyarno-poristykh strukturakh // Molodoj uchenyj. — 2018. — №43. — S. 19-21. — URL https://moluch.ru/archive/229/53349/ (data obrashheniya: 24.04.2019).
[17] Sangionganova I K and Markin V B 2019 The organizations of the tasks implementation in the distributed automatic control systems of heat supply stations. News of NAS RK. Series of geology and technical sciences. https://doi.org/10.32014/2019.2518-170X.7 ISSN 2518-170X (Online), ISSN 2224-5278 (Print).

Acknowledgments
The authors would like to thank the Research and Development Sector at the Technical University of Sofia for the financial support.

Investigation of the energy consumption in regulating the flow rate of fan systems

G S Popov, K V Klimentov and B Z Kostov

Department of Heat, hydraulics and environmental engineering, University of Ruse, Bulgaria

E-mail: gspopov@uni-ruse.bg

Abstract. The ensuring of the required parameters of the air, used in air-conditioning systems, is directly related to the change of the flow rate of each fan that is installed in these systems. This change can be achieved by using different methods of flow rate regulation. However, it is well-known that the selected method determines the energy efficiency of the given fan system. This work represents a comparative theoretical analysis, concerning the energy efficiency of three different methods of regulating a fan's flow rate: by throttling, by using inlet guide vanes and by changing the voltage frequency - the motor's speed of rotation has been changed with the help of a variable frequency drive (VFD). For the aims of this research, the change of the coefficient of efficiency for some of the key system elements – fan, motor and VFD, at different work regimes, have also been indicated. To fulfil the objectives of this research a given number of fans, having different specific speeds of rotations: $n_q = 3.31\,; 7.11 \text{ and } 8.82 \text{ min}^{-1}$, have been selected. As a criteria of effectiveness, concerning the different methods of flow rate regulation, the specific fan power P_{SFP} - the invested energy in transporting a unit volume of fluid (gas), has been used: $P_{SFP} = \frac{p}{\eta_{tot}}\left[\frac{kW}{m^3/s}\right]$, where p is the total pressure of the fan; η_{tot} is the overall efficiency of the fan. As a result of this theoretical investigation some graphics representing the change of the specific energy consumption, when the three different methods of flow rate regulation are used and indicating the fan's initial work regime and its specific speed of rotation, have been given.

1. Introduction

Ventilation systems are widely used in practice. They are involved in different technological processes related to the transportation of gas fluids – most often air flows. The achieving of steady-state operating (work) regimes of these systems, providing the corresponding technological requirements, is a present-day applied engineering problem. This requires the using of specialized devices, such as throttle, inlet guide vane, VFD, etc., enabling the continuous changing of a fan's aerodynamic parameters. This change is known as "regulation".

There is a good number of contemporary (new) research works concerning the investigation of the effectiveness of the different methods of flow rate regulation, in terms of energy efficiency, of trubopump systems. In his PhD Thesis [1], Bakman has presented a method of high-efficiency frequency (VFD) regulation of pumps, working in parallel. A similar investigation has been presented by Kostov in his PhD Thesis. In the following literature sources [4], [7], [8] and [11], the authors have proven the advatnages of using VFD, instead of other methods of regulating the flow rate of a pump system. In [8], the authors have presented a number of system operating conditions, whose implementation leads to the decreasing of the effectiveness in using the frequency method of flow rate regulation. Source [3] compares the effectiveness, in terms of energy efficiency, where an inlet vane fan is used, instead of a flap-adjustment based centrifugal fan. This research indicates that where the inlet vane is used the fan operates less effectively in a wider range.

This work represents a comparative investigation concerning the effectiveness, in terms of energy efficiency, of three different methods of flow rate regulation of a centrifugal fan system: by using a VFD, by using an inlet guide vane (IVC) and by throttling. The main purpose is to determinate how much more effective, in terms of energy efficiency, it is to use VFD, instead of inlet guide vane or throtteling for the regulation of a fan system's flow rate. A comparison between the effectiveness in applying the last two methods has also been done. In addition to that, the impact of a fan's specific speed of rotation on the effectiveness of the selected methods for flow rate regulation is investigated. This research investigated three centrifugal fans named C8-18, C5-40 and C4-50, which are produced by Central Aerohydrodynamic Institute (TsAGI) of Moscow. Each fan has a different value for its specific speed of rotation, respectively: C8-18 - $n_q = 3.31\ min^{-1}$; C5-40 – $n_q = 7.11\ min^{-1}$; C4-50 – $n_q = 8.82\ min^{-1}$, where n_q has been estimated by the following equation:

$$n_q = \frac{nQ^{0,5}}{p^{0,75}}, [min^{-1}] \tag{1}$$

where: n_q is the fan's specific speed of rotation; $n, [min^{-1}]$ – the fan's speed of rotation; $Q, [m^3/s]$ – the fan's volumetric flow rate at best efficiency point (BEP); $p, [Pa]$ – the fan's total pressure at BEP. The fan's operating curves are given in [13].

2. Methodology of the investigation

In [10] it has been presented a methodology to investigate the effectiveness of three different methods, used in regulating the flow rate of pump systems. As a criteria of effectiveness the specific energy consumption e_V has been used:

$$e_V = \frac{g}{3600}\frac{H}{\eta}. [kWh/m^3] \tag{2}$$

where: $g, [m/s^2]$ is the gravity acceleration; $H, [m]$ – pump's head; η – the total pump system's coefficient of efficiency [10].

According to the standard prEN 13779:2006 (E) „Ventilation for non-residential buildings — Performance requirements for ventilation and room-conditioning systems", as a criteria of effectiveness of a fan system the "specific fan power" has been given [12]. For its estimation the foolowing equation can be used:

$$P_{SFP} = \frac{p}{\eta_{tot}}\left[\frac{kW}{m^3/s}\right] \tag{3}$$

where p is the total fan's pressure; η_{tot} – tptal coefficient of efficiency of the fan system, including the coefficients of efficiency of the fan and system's drive (mechanical transmission, motor and VFD).

The physical importance (meaning) of the parameters e_V and P_{SFP} is the same – the electrical energy, used for the transportantion of a cubic meter of fluid in a pump or fan system. In this research the "specific fan power" [12], is used.

The methodology, presented in [10], determining the specific energy consumption in case of flow rate regulation, is accomplished by using three different methods, requiring information about the equations describing the pump's operating curves: $H = f(Q)$ and $\eta = f(Q)$. The operating curves of the fans, which are objects of this investigation, are presented in a dimensionless form in [13]. It has to be indicated that the methodology, given in [10], takes into account the including of a static head. At the same time, it is a well-known fact that the fan systems usually work without static head, which requires the accomplishing of a simplification of the equations before they can be used in this work. As it is mentioned above, source [13] provides detailed information about the fans' dimensionless operating curves (characteristics) - $\psi = f(\varphi)$ and $\eta_f = f(\varphi)$, where:

$$\psi = \frac{2p}{\rho u_2^2} \tag{4}$$

is the coefficienct of fan's pressure; $\rho = 1.2\ kg/m^3$ – the air's density at given atmosphere pressure - $p_{atm} = 101325\ Pa$; air's temperature T = 293.15 K and relative humidity $hu = 0.5$ (these are the conditions at wich the operating curves have been determined); $u_2 = \frac{\pi n D_2}{60}, [m/s]$ – transmitted velocity at the fan impeller's outlet; $D_2, [m]$ – outer diameter of the fan's impeller.

The flow rate coefficient - φ, for each fan, can be estimated by using the following equation:

$$\varphi = \frac{4Q}{\pi D_2^2 u_2}, \tag{5}$$

where $Q, \left[\frac{m^3}{s}\right]$ comes from (5) and is the fan's volumetric flow rate. The fan's coefficient of efficiency is given with η_f.

The functions: $\psi_i = f(\varphi_i)$ and $\eta_{f,i} = f(\varphi_i)$, can be approximated by using the equations (6) and (7):

$$\psi_i = a_i + b_i\varphi_i + c_i\varphi_i^2 \tag{6}$$

and

$$\eta_{f,i} = d_i + e_i\varphi_i + f_i\varphi_i^2. \tag{7}$$

The coefficients a_i, b_i, c_i, d_i, e_i and f_i, used in equations (6) and (7), have been preliminary detrmined in case that each of the investigated fans works with an inlet vane whose blades are fixed at different angulars: $\alpha_i = 0, 20, 40$ and 60°. The corresponding values of these coefficients are given in Table 1. This theoretically found data has been used to accomplish the aims of this research.

Table 1. Values of the coefficients that are used in the equations (6) and (7).

Fan $C8-18$; $n_q = 3.31\ min^{-1}$; $\varphi_0 = 0.038$						
α_i, deg	a	b	c	d	e	f
0	0.8788	41.708	-536.78	0.2903	18.561	-240.19
20	1.0113	35.346	-570.84	0.1537	26.023	-357.37
40	0.8687	44.299	-925.58	0.1083	31.93	-538.99
60	0.9837	39.163	-1751.9	0.0644	40.238	-1099.7
Fan $C5-40$; $n_q = 7.11\ min^{-1}$; $\varphi_0 = 0.1$						
α_i, deg	a	b	c	d	e	f
0	1.2317	-0.8062	-28.158	0.3788	7.8764	-41.376
20	1.0874	2.7172	-63.287	0.1639	13.08	-73.322
40	1.0946	2.2013	-80.172	0.134	16.125	-108.89
60	1.0584	3.1195	-167.35	0.029	21.957	-215.96
Fan $C4-50$; $n_q = 8.82\ min^{-1}$; $\varphi_0 = 0.072$						
α_i, deg	a	b	c	d	e	f
0	1.1741	0.4603	-54.674	0.3154	12.03	-84.364
20	0.9875	4.5191	-103.71	0.1576	16.113	-117.2
40	0.9767	5.153	-160.17	0.1028	20.78	-193.06
60	0.9998	0.6827	-250.77	0.0597	26.244	-374.99

To compare the effectiveness of the selected different methods of a fan system's flow rate regulation, the following equation is used:

$$\frac{P_{SFP,i}}{P_{SFP,0}} = f\left(\frac{Q_i}{Q_0}\right), \tag{8}$$

where $P_{SFP,i}$ is the current value of SFP at given flow rate Q; $P_{SFP,0}$ is the value of SFP at the initial operating regime concerning the relevant method of flow rate regulation, corresponding to the provided flow rate Q_0. In defining the equation (8), the ratio $\frac{Q_i}{Q_0}$ for a given fan, can be replaced by the ratio $\frac{\varphi_i}{\varphi_0}$. For an initial operating regime in this research the nominal regime of each fan, given in [13], has been selected. To provide the graphically presented results, concerning a comparative analysis of the energy effectiveness for the three selected methods, equation (8) is used.

3. Throttle method of flow rate regulation

When the throttle method of flow rate regulation is used, the ratio $\frac{P_{SFP,i}}{P_{SFP,0}}$ can be determined by using the following equation:

$$\frac{P_{SFP,i}}{P_{SFP,0}} = \frac{\psi_i\, \eta_{tot,0}}{\psi_0\, \eta_{tot,i}}, \tag{9}$$

where the relevant coefficient of efficiency η_{tot} represents the multiplication of the coefficients of efficiency respectively of the fan η_f and motor η_m:

$$\eta_{tot} = \eta_f \eta_m. \tag{10}$$

It is assumed that the fan's impeller has been installed (mounted) directly on the motor's shaft, i.e. this excludes the existence of mechanical transmission between the fan and motor. In case that the flow rate has been regulated by throtteling the work points belong to the fan's operating characteristic when it works with inlet guide vane and its blades are fixed at an angle $\alpha_0 = 0^o$. In this regard, by using the equations (6) and (7), the following equation - (9), is found:

$$\frac{P_{SFP,i}}{P_{SFP,0}} = \frac{a_0+b_0\varphi_i+c_0\varphi_i^2}{a_0+b_0\varphi_0+c_0\varphi_0^2}\,\frac{d_0+e_0\varphi_0+f_0\varphi_0^2}{d_0+e_0\varphi_i+f_0\varphi_i^2}\,\frac{1}{\eta_{m,i}/\eta_{m,0}}. \tag{11}$$

The coefficients a_0, b_0, c_0, d_0, e_0 and f_0, belonging to equation (11), are valid only in case that the investigated fans work with inlet guide vanes, whose blades are fixed at a given angular - $\alpha_0 = 0^o$.

The literature sources [2] and [9] provides the necessity information, concerning the change of a motor's coefficient of efficiency as a function of load. In the accomplishing of this research, some data found for a 10 kW motor, given in [9], is used. The literature source [9] provides an equation, used for the estimation of a motor's nominal speed of rotation:

$$\frac{\eta_{m,i}}{\eta_{m,0}} = f\left(\frac{\mu_{,i}}{\mu_0}\right), \tag{12}$$

where $\mu_i = \frac{\psi_i \varphi_i}{\eta_{f,i}}$ is the fan's power coefficient.

By using the equations (11) and (12), it can be estimated the ratio $\frac{P_{SFP,i}}{P_{SFP,0}}$ at different values of the flow rate's relative coefficient - $\frac{\varphi_i}{\varphi_0}$ (or respectively $\frac{Q_i}{Q_0}$).

4. Flow rate regulation by using an inlet guide vane

A well-know fact is that ususally the fans are used in systems without static head. The resistance characterictic of such a system, containing a point whose coordinates respectively are: $[\varphi_0; \psi_0]$, can be presented in the following way:

$$\psi = \kappa \varphi^2, \tag{13}$$

where $\kappa = \psi_0/\varphi_0^2$ (a coefficient). The work points, corresponding to the achieved after flow rate regulation operating regimes, belong to a curve that has been described by using equation (13). The values of the flow rate's coefficient φ_i, when the inlet guide vane's blades are installed at different angle α_i, can be found by using the following equation:

$$\left|\begin{array}{l}\psi_i = a_i + b_i\varphi_i + c_i\varphi_i^2 \\ \psi_i = \kappa\varphi_i^2\end{array}\right.,$$

and its final solution is:

$$\varphi_i = \frac{-b_i - \sqrt{b_i^2 - 4a_i\left(c_i - \frac{\psi_0}{\varphi_0^2}\right)}}{2\left(c_i - \frac{\psi_0}{\varphi_0^2}\right)} \tag{14}$$

After the determination of the flow rate coefficient φ_i for each of the achieved operating regimes is accomplished, the ratio $\frac{P_{SFP,i}}{P_{SFP,0}}$ can also be determined:

$$\frac{P_{SFP,i}}{P_{SFP,0}} = \frac{\psi_i\, \eta_{tot,0}}{\psi_0\, \eta_{tot,i}} = \frac{a_i+b_i\varphi_i+c_i\varphi_i^2}{a_0+b_0\varphi_0+c_0\varphi_0^2}\,\frac{d_0+e_0\varphi_0+f_0\varphi_0^2}{d_i+e_i\varphi_i+f_i\varphi_i^2}\,\frac{1}{\eta_{m,i}/\eta_{m,0}}. \tag{15}$$

5. Flow rate regulation by using VFD (frequency method)

When the flow rate regulation is accomplished by using VFD the work points, corresponding to the achieved operating regimes, belong to the system's curve described by equation (13). In this case, the mentioned curve coincides with the parabola of similarity. According to the "Theory of similarity" when the gas' density has a constant value, the following relation is valid:

$$\frac{\psi_i}{\psi_0} = \frac{\varphi_i^2}{\varphi_0^2}.$$

Indicating the previous statement, equation (9) can be presented the following way:

$$\frac{P_{SFP,i}}{P_{SFP,0}} = \frac{\varphi_i^2}{\varphi_0^2} \cdot \frac{\eta_{tot,0}}{\eta_{tot,i}}. \tag{16}$$

The fan system's total coefficient of efficiency is the multiplication between the coefficients of efficiency respectively of the fan (η_f), motor (η_m) and VFD (η_{VFD}). For the determination of a fan's coefficient of efficiency operating at different Reynolds numbers (Re) the methodology given in [13] can be used:

$$\eta_{f,i} = 1 - \frac{2(1-\eta_{f,0})}{1+\left(\frac{Re_i}{Re_0}\right)^{0.2}}. \tag{17}$$

Taking into account that: $\frac{Re_i}{Re_0} = \frac{n_i}{n_0} = \frac{\varphi_i}{\varphi_0}$, then equation (17) can be given the following way:

$$\eta_{f,i} = 1 - \frac{2(1-\eta_{f,0})}{1+\left(\frac{\varphi_i}{\varphi_0}\right)^{0,2}}. \tag{18}$$

For the determination of the motor's coefficient of efficiency, equation (12) was used:

In [5] and [6] the authors are given equations that can be used to estimate a VFD's coeffcieint of efficiency. For the aims of this work, an equation given in [5] is used.

$$\eta_{VFD,i} = \frac{aX_{VFD}}{b+X_{VFD}} + cX_{VFD}, \tag{19}$$

where in case the VFD works at output power $P = 14.91\ kW$ (20 hP) the values of the coefficients are: $a = 0.9850$, $b = 0.0175$, $c = -0.0005$ and the parameter X_{VFD} represents the ratio between the current and maximal output power. If the values of $\frac{\eta_{f,i}}{\eta_{f,0}}$ and $\frac{\eta_{m,i}}{\eta_{m,0}}$ are preliminary known, the parameter X_{VFD} can be determined by using the following equation:

$$X_{VFD,i} = \frac{\psi_i\varphi_i}{\eta_{f,i}\eta_{m,i}} \frac{\eta_{f,0}\eta_{m,0}}{\psi_0\varphi_0} = \frac{\varphi_i^3}{\varphi_0^3} \frac{1}{\frac{\eta_{f,i}\eta_{m,i}}{\eta_{f,0}\eta_{m,0}}}. \tag{20}$$

6. Results and discussion

Figure 1 shows the change of the specific fan power as a function of the relative flow rate's change when the flow rate of a fan (C8-18) system, has been regulated by the three selected methods. The curves describing these methods are named as it follows: TH – throttle, IVC – Inlet (Guide) Vane Control and VFD – Variable Frequcny Drive.

The results concerning the investigated fan systems are similar with the results (related to pump systems) given in [11]. Therefore, the using of VFD regulation is most effective in terms of energy efficiency. In this case, the decreasing of the relative flow rate $\frac{Q}{Q_0}$ (i.e. increasing of the range of regulation: $\frac{Q_0-Q}{Q_0} = 1 - \frac{Q}{Q_0}$) leads to decreasing of the relative energy consumption, used for the transportation of a cubic meter of air. The reason is that the system's pressure decreases with the decreasing of the flow rate squared.

In applying the flow rate regulation by throttling the relative coefficient SFP increases with the decreasing of the relative flow rate and has highest values compared with the coefficients achieved by applying the other two methods. In case of applying the IVC regulation, it is indicated that the relative energy consumption decreases and this continues until the relative flow rate has a value of $\frac{Q}{Q_0} \approx 0.9$.

At the same time the relative coefficient SFP has also been decreased to $\frac{P_{SFP}}{P_{SFP,0}} \approx 0.9$. If the decreasing of the relative flow rate continues, the relative coefficient SFP stays constant.

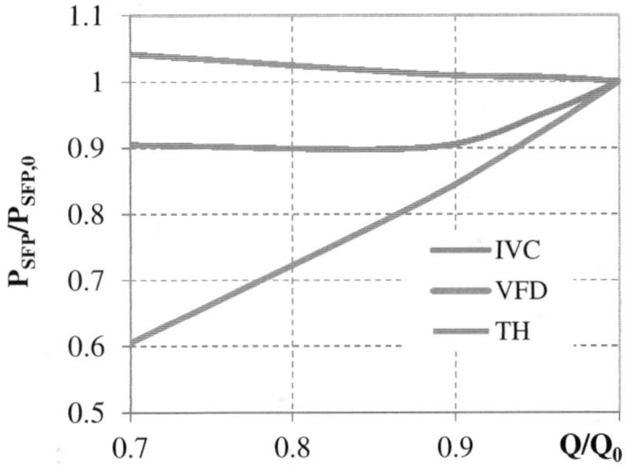

Figure 1. The relation $\frac{P_{SFP,i}}{P_{SFP,0}} = f(\frac{Q_i}{Q_0})$ found for the three methods of a fan (C8-18) system's flow rate regulation.

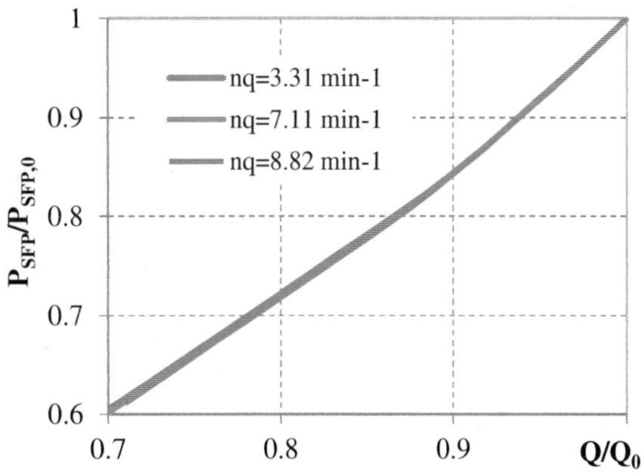

Figure 2. The relation $\frac{P_{SFP,i}}{P_{SFP,0}} = f(\frac{Q_i}{Q_0})$ found for the three investigated fans in case the flow rate has been regulated by VFD.

The trends describing the change of the relative energy consumption concerning the three selected methods of flow rate regulation are identical for the other two fans that are object of this research. The quantitative deviations depending on the fan's type and its specific speed of rotation are graphically presented in the figures 2, 3 and 4.

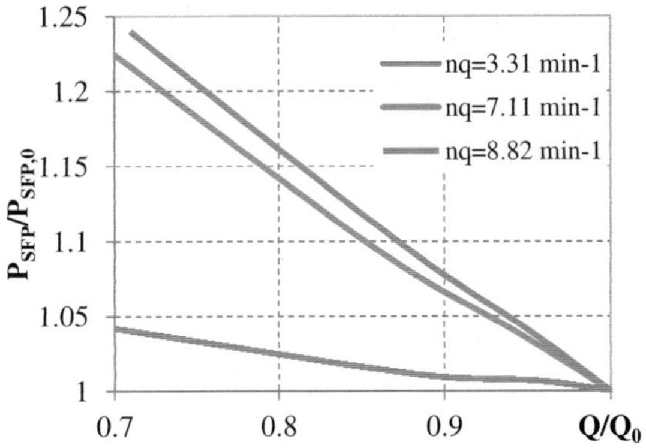

Figure 3. The relation $\frac{P_{SFP,i}}{P_{SFP,0}} = f(\frac{Q_i}{Q_0})$ found for the three investigated fans in case the flow rate has been regulated by throttling.

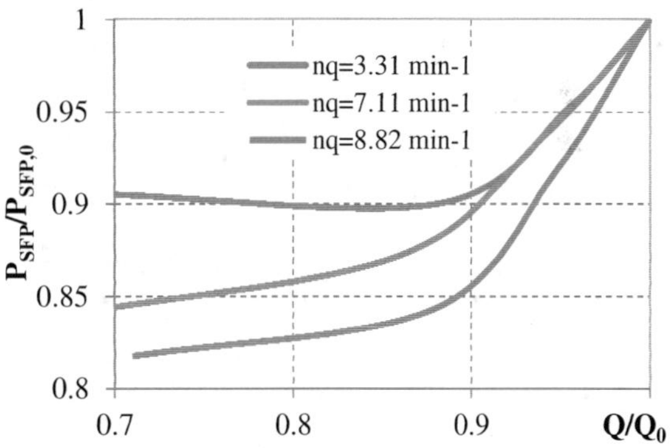

Figure 4. The relation $\frac{P_{SFP,i}}{P_{SFP,0}} = f(\frac{Q_i}{Q_0})$ found for the three investigated fans in case the flow rate has been regulated by IVC.

Figure 2 ilustrates the results found for the relation: $\frac{P_{SFP,i}}{P_{SFP,0}} = f(\frac{Q_i}{Q_0})$, in case that the three investigated fans are regulated by using VFD. Actually, the three curves have to coincide with each other because in using this method of regulation the operating point "moves" along the fan's curve, which is the same in the three cases and the change of the relative coefficient SFP is described by the same equation – (16). In this case, the fan's type do not affect the change of $\frac{P_{SFP,i}}{P_{SFP,0}}$ in the process of regulation.

Figure 3 ilustrates the results found the relation: $\frac{P_{SFP,i}}{P_{SFP,0}} = f(\frac{Q_i}{Q_0})$, in case that the three investigated fans are regulated by throttling. In this case, it can be seen that the specific speed of rotation n_q has a significant impact on the change of the relative energy consumption in regulating the fan's flow rate. For the fan having the lowest specific speed of rotation ($n_q = 3.31\ min^{-1}$) the value of the relative coefficient increases with about 5% at $\frac{Q_i}{Q_0} = 0.7$, as for the fans whose speed of rotations respectively are c $n_q = 7.11\ min^{-1}$ and $n_q = 8.82\ min^{-1}$, this increase is about 25%.

According to the results given in fig.4, it can be seen that the relative coefficient SFP decreases most intensively (about 20%, at $\frac{Q_i}{Q_0} = 0.7$) in case of the fan with highest specific speed of rotation operates, as for the fan with the lowest specific speed of rotation this decrease is about 10%. As a general conclusion concerning the three selected fans, it can be indicated that the fan's relative flow rate decreases intensively until it comes to $\frac{Q_i}{Q_0} = 0.9$, and then its value stays almost constant.

Summarizing the above-mentioened conclusions, it is clear that the most effective method of flow rate regulation is by using VFD and as most inefficient it can be indicated the throtteling. In order to assess the superiority, in terms of effectiveness, of a given method of flow rate regulation compared with the other selected methods by indicating the impact of the fan's type the graphical relations given in the figures 5, 6 and 7, is used. The abscissa axis represents the relative flow rate - $\frac{Q_i}{Q_0}$, and the ordinate axis represents the relative difference (RD), which can be estimated by using the following equation:

$$RD = \frac{\left(\frac{P_{SFP,i}}{P_{SFP,0}}\right)_{IVC} - \left(\frac{P_{SFP,i}}{P_{SFP,0}}\right)_{VFD}}{\left(\frac{P_{SFP,i}}{P_{SFP,0}}\right)_{VFD}} .100\% \tag{21}$$

$$RD = \frac{\left(\frac{P_{SFP,i}}{P_{SFP,0}}\right)_{TH} - \left(\frac{P_{SFP,i}}{P_{SFP,0}}\right)_{VFD}}{\left(\frac{P_{SFP,i}}{P_{SFP,0}}\right)_{VFD}} .100\% \tag{22}$$

$$RD = \frac{\left(\frac{P_{SFP,i}}{P_{SFP,0}}\right)_{TH} - \left(\frac{P_{SFP,i}}{P_{SFP,0}}\right)_{IVC}}{\left(\frac{P_{SFP,i}}{P_{SFP,0}}\right)_{IVC}} .100\% \tag{23}$$

Figures 5, 6 and 7 ilustrate the results found for the relation: $RD = f(\frac{Q_i}{Q_0})$, concerning the three investigated fans.

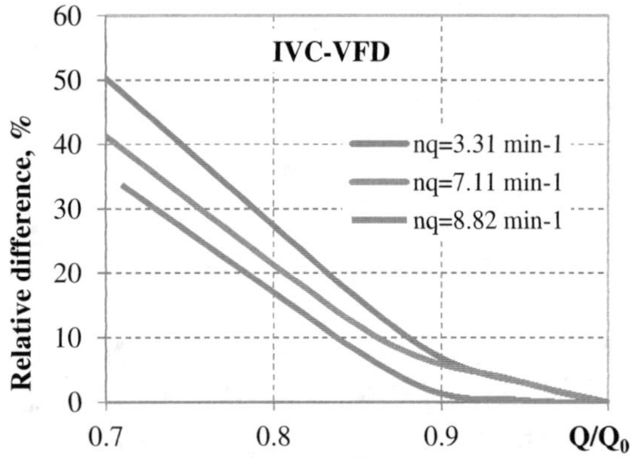

Fig. 5. The relation $RD = f(\frac{Q_i}{Q_0})$, found for the three investigated fans by using equation (21).

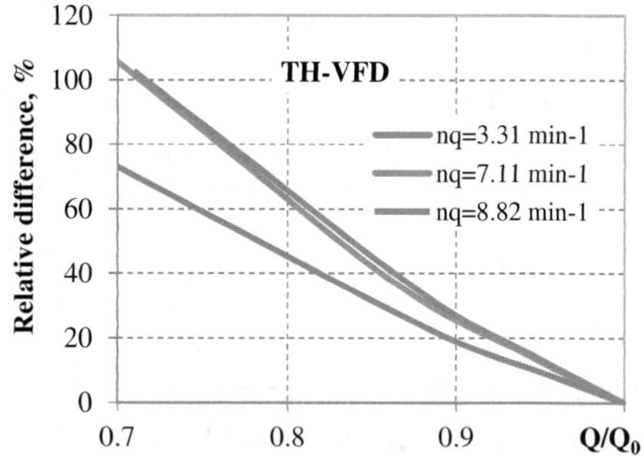

Fig. 6. The relation $RD = f(\frac{Q_i}{Q_0})$, found for the three investigated fans by using equation (22).

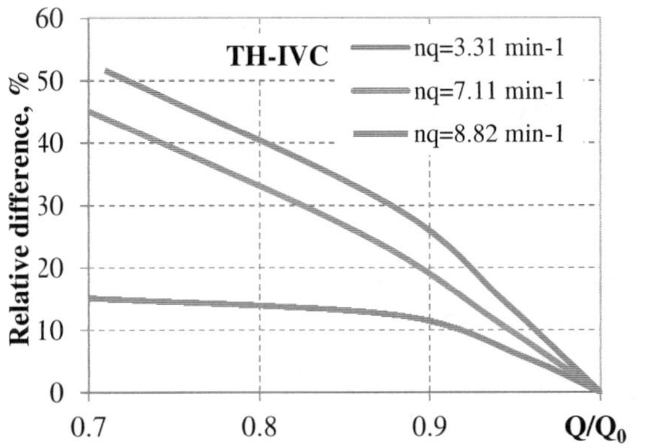

Figure 7. The relation $RD = f(\frac{Q_i}{Q_0})$, found for the three investigated fans by using equation (23)

Figure 5 illustrates the change of the relative difference RD for the three investigated fans, estimated by using the equation (21), given as a function of the fan's relative flow rate. A comparison between the effectiveness of regulation by using VFD or IVC shows similarity when the relative flow rate $\frac{Q_i}{Q_0}$ is less than 0.9. For the same range ($\frac{Q_i}{Q_0}$=0…0.9) the fan whose $n_q = 8.82\ min^{-1}$ has almost no relative difference $RD \approx 0$, as for the other two fans its value reaches 6-7%. In the range of: $0.7 \leq \frac{Q_i}{Q_0} \leq 0.9$, the relative difference RD of the fan with $n_q = 8.82\ min^{-1}$ increases to 35% and for the fan with $n_q = 3.31\ min^{-1}$ this value increases up to 50%.

Figure 6 illustrates a comparison between the using of throttle and VFD flow rate regulation and it can be clearly seen that relative difference RD increases almost linear and this is valid for the investigated full range of changing the relative flow rate. For the fan with the lowest specific speed of rotation n_q the value of RD reaches about 70%, as for the other fans with higher values of n_q the RD is over 100%.

The comparison between the throttle and IVC methods, given in figure 7, gives a clear indications that in case a fan has a higher value of n_q the value of RD increases. The indicated maximal value of RD for the investigated fans with $n_q = 7.11\ min^{-1}$ and $n_q = 8.82\ min^{-1}$ is about 50% at $\frac{Q_i}{Q_0} = 0.7$. For the fan with the lowest value of n_q the value of RD reaches about 15%. In this case, the increasing of RD is more intensive in the range of $0.9 \leq \frac{Q_i}{Q_0} \leq 1$ and the change is much less in the range of $0.7 \leq \frac{Q_i}{Q_0} \leq 0.9$. The last statement can be explained by paying attention to the type of the same fan's curve, given in figure 4, that is described by the relation $\frac{P_{SFP,i}}{P_{SFP,0}} = f(\frac{Q_i}{Q_0})$.

7. Conclusion

The accomplishment of this research work leads to the following significant conclusions:

- The using of VFD method of flow rate regulation is more efficient, in terms of energy efficiency, than using the other two methods - throttle and IVC, independently of the fan's specific speed of rotation;
- The regulation of a fan system's flow rate by using IVC is more efficient, in terms of energy efficiency, than throtteling, independently of the fan's specific speed of rotation;
- In case of using VFD to regulate the flow rate the specific speed of rotation does not affect the effectivess of this selected method;
- The using of throtteling for regulating a fan system's flow rate is more efficient, in terms of energy efficiency, in case that the fan has a lower value of specific speed of rotation;
- The using of IVC for regulating a fan system's flow rate is more efficient, in terms of energy efficiency, in case that the fan has a higher value of specific speed of rotation;
- In case the applying of a fan system's flow rate regulation is in the range of $\frac{Q_0 - Q}{Q_0} \leq 0{,}1$, the energy efficiency of using the two selected methods – IVC and VFD, is similar.

References

[1] Bakman I 2016 *High-Efficiency Predictive Control of Centrifugal Multi-Pump Stations with Variable-Speed Drives* (Dissertation of the degree of Doctor of Philosophy, Tallinn University of Technology, Faculty of Power Engineering)

[2] Burt C, Piao X, Gaudi F, Busch B and Taufik N 2008 Electric motor efficiency under variable frequencies and loads *Journal of Irrigation and Drainage Engineering.* **134(2)**

[3] Chen G, Xu W, Zhao J and Zhang H 2018 Energy-Saving Performance of Flap-Adjustment-Based Centrifugal Fan *Energies.* **11(162)** doi:10.3390/en11010162

[4] Kostov B 2014 *Investigation and analysis of methods concerning the effective use of energy in pump systems* (Dissertation of the degree of Doctor of Philosophy, "Angel Kanchev" University of Ruse. Department of Thermotechnics, Hydraulics and Ecology)

[5] Krukowski A and Wray C 2013 Standardizing Data for VFD Efficiency" *ASHRAE Journal.* **55(6)** pp 16-26

[6] Ma Z and Wang S 2009 Energy efficient control of variable speed pumps in complex building central air-conditioning systems *Energy and Buildings.* **41** pp 197–205

[7] Marchi A, Simpson A R and Ertugrul N 2012 Assessing variable speed pump efficiency in water distribution systems *Drink. Water Eng. Sci.* **5** pp 15–21

[8] Maxwell J. 2005 *How to Avoid Overestimating Variable Speed Drive Savings* (Proceedings of the Twenty-Seventh Industrial Energy Technology Conference, New Orleans, LA, May 10-13, 2005)

[9] Novakaov N and Popov G 2018 Some aspects of using variable frequency drive in pump systems *Thermal engineering.* **13** pp 73-77

[10] Popov G, Klimentov K and Kostov B 2011 Methods to estimate the energy consumption in regulating the flow rate of pump systems (10th Anniversary international conference an accomplishment in Electrical and Mechanical Engineering and Information Technology, DEMI'2011, Banja Luka) pp 495-500

[11] Popov G., Klimentov K and Kostov B 2011 *Investigation of the energy consumption in regulating the flow rate of pump systems* (10th Anniversary international conference an accomplishment in Electrical and Mechanical Engineering and Information Technology, DEMI'2011, Banja Luka) pp 481-487

[12] prEN 13779:2006 (E) Ventilation for non-residential buildings — Performance requirements for ventilation and room-conditioning systems

[13] Solomahova T and Chebaysheva K 1980 *Centrobezhnaye ventiljataray* Aerodinamicheskie shemay i harakteristiki. Mashinostroenie, Moskva

IOP Conf. Series: Materials Science and Engineering **595** (2019) 012013 doi:10.1088/1757-899X/595/1/012013

Numerical study of the heat transfer and flow characteristics of heat pipes

A Huminic[1] and G Huminic[1]

[1] Mechanical Engineering Department, Faculty of Mechanical Engineering, Transilvania University of Brasov, Romania

E-mails: angel.h@unitbv.ro

Abstract. In this work, a three-dimensional (CFD) analysis is used to investigate the heat transfer and fluid flow of a heat pipe using water and nanofluids as working fluids. The study focused mainly on the effects of volume concentrations of nanoparticles and the operating temperature on the heat transfer performance of the heat pipe using the nanofluids, but there is revealed also aspects concerning flow characteristics of fluids. The analysis was performed for water and two types of nanofluids, two volume concentrations of nanoparticles (1.0% and 4 vol.%) and three operating temperatures (70 °C, 80 °C and 90 °C). Nanofluids were composed of water and SiC and TiC nanoparticles with the diameter of 10nm. The numerical results indicate that the volume concentration of nanoparticles has an important effect in reducing of the temperature difference between evaporator and condenser. The present study shows also that the thermal performance of a heat pipe can be significantly improved by using of both wick structure and nanofluids.

1. Introduction

The use of heat pipes is becoming widespread in a large number of applications, from the space programs to electronic cooling, as these devices are generally passive in their operation and provide effective heat transport with minimal losses. A heat pipe operates on a closed two-phase cycle and utilizes the latent heat of vaporization to transfer heat with very small temperature gradients. Circulation of a working fluid inside the heat pipe is accompanied by phase changes at both the evaporator and the condenser [1], as shown in figure 1.

As a new research frontier, the nanofluid two-phase flows have the potential to improve heat transfer and energy efficiency in thermal management systems. It has been demonstrated that nanofluids can have remarkably higher thermal conductivities than those of conventional pure fluids and significantly better heat transfer characteristics than the base fluids. Up to now, studies concerning application of nanofluids in heat pipes are experimental and only a few reports the modelling of the heat pipe characteristics in the presence of a nanofluid [2-8].

Because, numerical works concerning the fluid flow and heat transfer processes in heat pipes using nanofluids are limited, the goals of the present study are the analysis of these processes coupled with liquid flow in the wick structure and finding of that nanofluid which could enhance significantly thermal performance of the heat pipes.

IOP Conf. Series: Materials Science and Engineering **595** (2019) 012013 doi:10.1088/1757-899X/595/1/012013

2. Thermo-physical properties of nanofluids

Thermo-physical properties of nanofluids are very important in the study of nanofluid two-phase flow. Temperature dependency of different physical properties (k, c_p, ρ, μ) of the working fluids has been considered to improve the accuracy of the calculations and results are tabulated in Table 1.

Table 1. Thermo-physical properties of the fluid and nanoparticles [9].

Fluid/ Nanoparticles	Density, ρ (kg/m^3)	Specific heat, c_p (J/kg·K)	Thermal conductivity, k (W/m K)	Dynamic viscosity, μ (kg/ms)
Water	997.0	4181.7	0.6069	0.0008899
SiC	3370	1340	350	-
TiC	4930	711	330	-

3. Simulation methodology

The CFD code used for this study was ANSYS CFX, which is fully integrated fluid dynamics analysis software of ANSYS Workbench platform [10].

3.1 Geometry and grid resolution

The geometry of the heat pipe studied in the present work, with a grooves shaped wick, is shown in Figure 1. The dimensions of the heat pipe are similar with those of Li et al. [11], as given in Table 2.

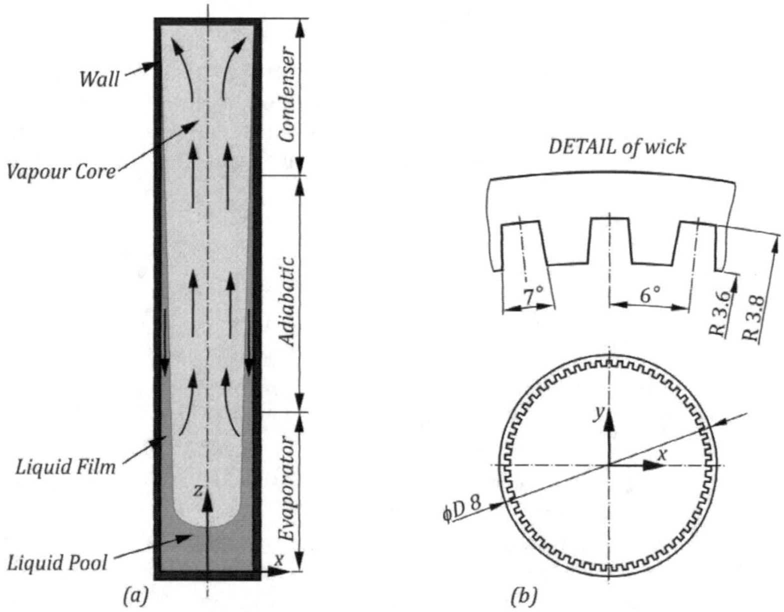

Figure 1. Sketch of heat pipes working principle (a) and cross section of current wick (b).

A multi-block structured grid has been used to generate the computational domain. It consists of two main regions with rectangular prism elements, in order to get a higher quality mesh. Transition between these two regions was made with some layers of triangular prism elements. This approach provides the possibility to concatenate the grid points nearest to surfaces of the walls, necessary to capture the flow of the liquid phase along the heat pipe, and also to use a reasonable grid in terms of number of elements. Several grid distributions have been tested to ensure the results are grid independent. Therefore, a mesh consisting of 4.896.000 elements has been selected for the present computations of the heat pipe.

Table 2. Dimensions of the heat pipe.

Dimensions of the heat pipe [mm]	Present study	Li et al. [11]
Length	300	300
Diameter (outer)	8	8
Evaporator	45	50
Adiabatic section	210	200
Condenser	45	50
Groove width	0.20	0.20
Groove height	0.20	0.25
Number of grooves	60	74

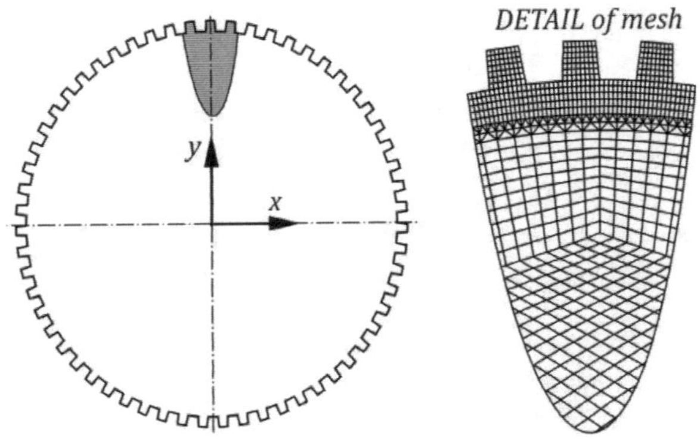

Figure 2. Multi-block structured grid used for the heat pipe in cross section.

3.2. Governing equations and conditions of simulations

The governing equations of flow and heat transfer inside of the heat pipes were solved in the Cartesian coordinate system for steady state conditions, assuming that both liquid and vapour phases are in equilibrium. Gravitational effect was considered, too. The set of equations solved were continuity equation, RANS (Reynolds Averaged Navier-Stokes) equations and thermal energy equation. For modelling of heat pipes, the phase change model [10] has been used. Vapour phase was modelled as continuous fluid and liquid phase as a dispersed fluid. As boundary conditions, constant wall temperatures on evaporator, T_e, and condenser, $T_c = 293\ K$, were imposed. An initial liquid water volume fraction of 15% (initial liquid filling ratio) was considered.

4. Results and discussion

The analysis was carried out by incorporating the effect of nanofluids in the geometrical model given by Li et al. [11]. Two different types of water based nanofluids, SiC and TiC, were used to perform the numerical simulations. A reasonable range of concentration was chosen for solid particles, 1.0%, and 4.0%, each of them studied for three different values of operating temperature of evaporator T_e = 343.353 and 363 K.

4.1. Comparison between heat pipe and wickless heat pipe

In order to reveal the influence of the wick on thermal performances of the heat pipe, a study was performed for both cases of heat pipe with and without grooves, for all three temperatures considered, using water as working fluid. The results, shown in Figure 3, reveal that in the case of the heat pipe (with groove) the temperature difference between evaporator and condenser sections is more higher compared to of the wickless heat pipe. In this figure, an increase with 70% of the temperature difference between evaporator and condenser is obtained for an operating temperature of 70°C, meaning that heat pipe works efficiently at low operating temperatures. It is also clearly shown that the wick structure of the heat pipe has a great effect on the reduction of the temperature along of the heat pipes.

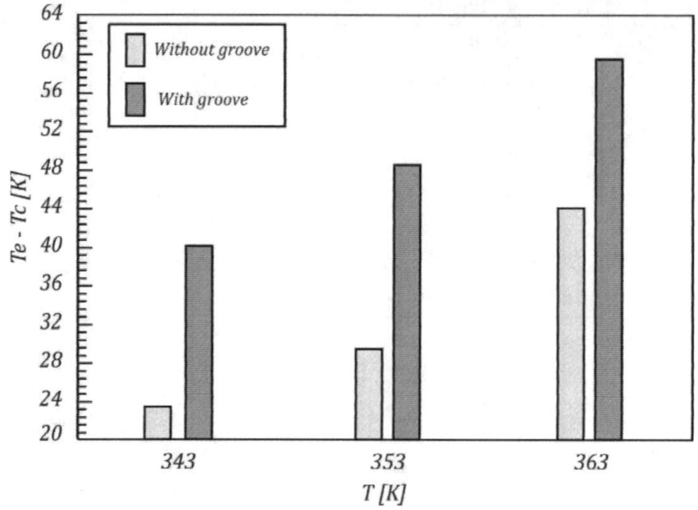

Figure 3. Comparison of the temperature difference between evaporator and condenser sections for heat pipe with groove and wickless heat pipe.

4.2. Heat transfer

In order to quantitatively evaluate the thermal performance of the heat pipes, the temperature profiles of the heat pipes using both water and the nanofluids are compared.

Figures 4 and 5 depict the effect of volume concentrations of nanoparticles on thermal performances of heat pipe under various the operating temperatures. As can be seen from these figures, increases the particle concentration decrease the temperature difference between the evaporator and condenser. For 4.0% SiC and TiC particles and an operating temperature of 80°C, the temperature difference is reduced up to 8.71% and 10.33% respectively. It is clearly shown that the volume fraction of nanoparticles has a great effect on the reduction of the temperature difference along of the heat pipes.

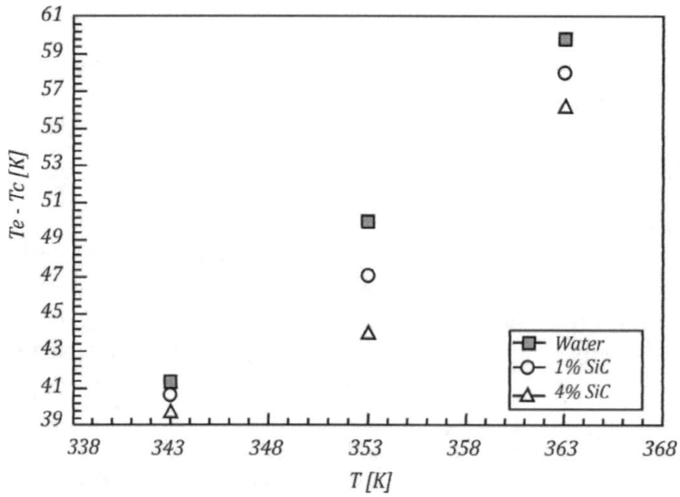

Figure 4. Effect of the SiC particle concentration on thermal performance of a heat pipe under various operating temperature.

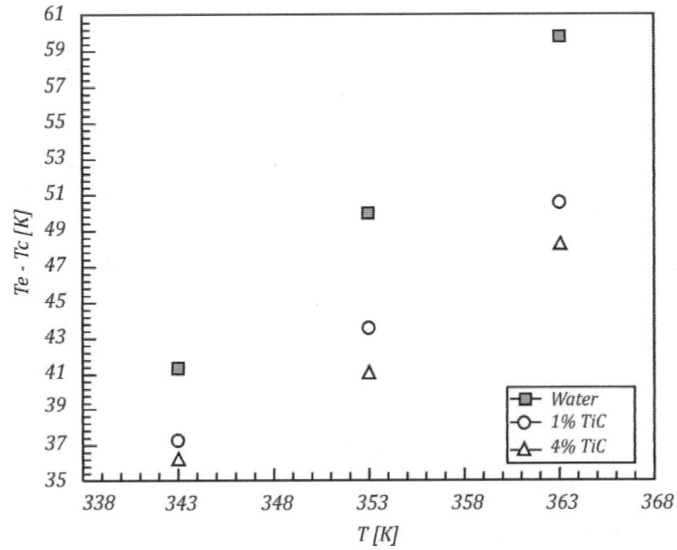

Figure 5. Effect of the TiC particle concentration on thermal performance of a heat pipe under various operating temperature.

4.3. Fluid flow

In order to characterize the flow inside the heat pipe were checked velocity fields and liquid volume fraction in the cross sections. Figures 6 and 7 depict averaged variations of the maximum velocity of vapours and liquids along the heat pipe under the operating temperatures.

Concerning the vapours velocity for cases studied, it decreases with decreasing of $T_e - T_c$, following a curve as in Figure 6. Thus, after a fast growing in the evaporator, followed by a maximum constant value until the middle of the heat pipe, the velocity of vapours decreases in the second part. The decreasing rate becomes smaller to the end of condenser.

In the opposite direction, the maximum velocity of liquid phase increases from condenser to last third of the adiabatic section then becomes constant until near the bottom wall of the evaporator. There are no significant variations in the liquid phase velocity with operating temperature range, but an increase of speed occurs for cases with nanofluids, higher for TiC, as in figure 7, showing that the mass forces dominate for the liquid phase.

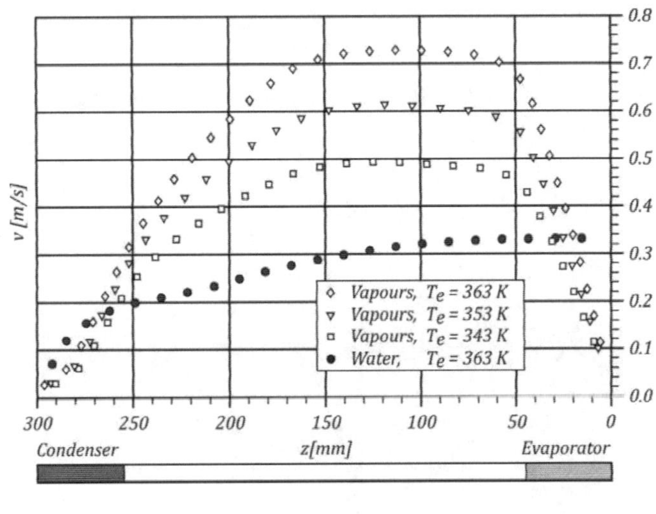

Figure 6. Variations of water velocity (r = 3.2 mm) and vapours velocity (r = 0.0 mm) along of the heat pipe.

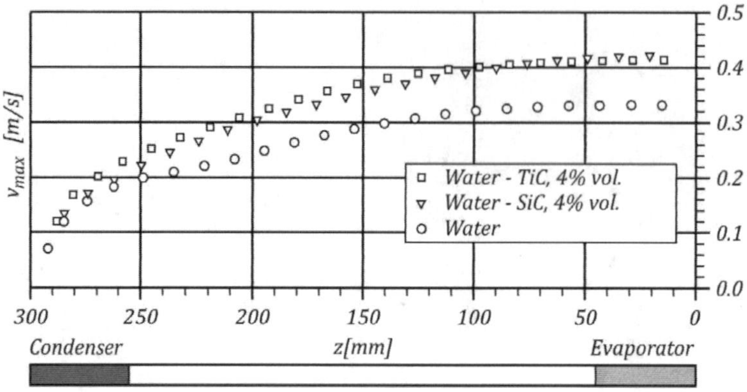

Figure 7. Variations of maximum liquid velocity for *Te* = 363 K.

To find values of the maximum liquid speed, the velocity profile was checked along the heat pipe, as shown in figure 8 for water. The velocity is constant inside grooves, where the adhesion forces dominate. Outside grooves, the influence of the mass forces increase from condenser to adiabatic section where are dominant, and therefore the velocity profile changes, revealing also the mechanism of flow due to phase change process.

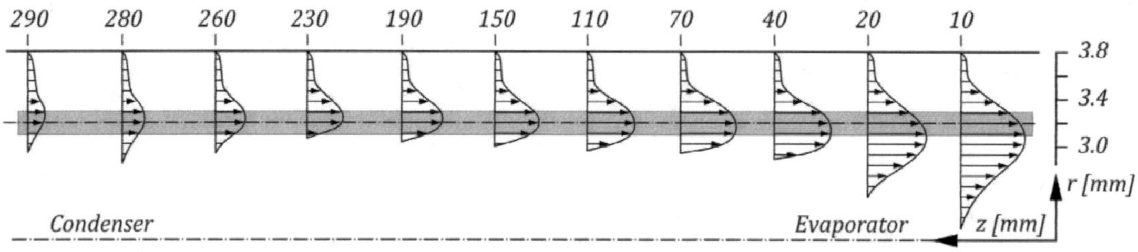

Figure 8. Water velocity profile along of the heat pipe, *Te* = 363 K.

Contours of water volume fraction and to an operating temperature of 90 °C are shown in Figure 9. Based on the volume fraction profile, the liquid phase returns on condenser walls due to capillary force in the wick. The vector plots show that the vapours flow upward and the liquid phase flows downward through the condenser walls.

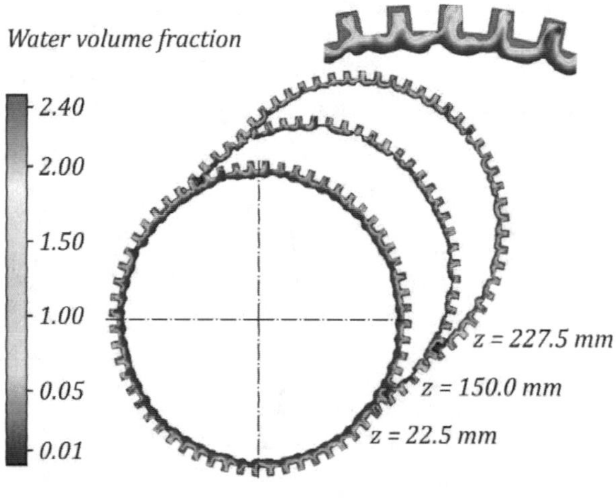

Figure 9. Contours of water volume fraction in cross section.

5. Conclusions

In this paper the thermal and hydrodynamic behaviors of water and nanofluids flowing inside a heat pipe were numerically investigated in stationary conditions and for laminar flow. Two types of nanofluids (i.e SiC/water and TiC/water) were analyzed. The heat pipe temperature and velocity are obtained for different volume concentrations of nanoparticles and different operating temperatures of the heat pipe. The results have showed that the heat transfer behaviours of the nanofluids were highly depended on the volume concentration of nanoparticles. Also, the numerical results confirm experimental results obtained of many researches so far as well as the thermal performances of the heat pipes are improved and the temperature difference between the evaporator and condenser is reduced when nanofluids are used as the working fluid.

References

[1] Suman B 2009 Microgrooved Heat Pipe, *Advances in Heat Transfer* **41** 1-5

[2] Kole M and Dey T K, 2013, Thermal performance of screen mesh wick heat pipes using water based copper nanofluids, *Applied Thermal Engineering* **50** 763–770

[3] Putra N, Septiadi W N, Rahman H and Irwansyah R, 2012, Thermal performance of screen mesh wick heat pipes with nanofluids, *Experimental Thermal and Fluid Science.* **40** 10–17

[4] Hajian R, Layeghi M and Sani K A 2012 Experimental study of nanofluid effects on the thermal performance with response time of heat pipe, *Energy Conversion and Management* **56** 63–68

[5] Solomon A B, Ramachandran K and Pillai B C, 2012, Thermal performance of a heat pipe with nanoparticles coated wick, *Applied Thermal Engineering* **36**,106-112

[6] Firouzfar E, Soltanieh M, Noie S H and Saidi M H 2012 Investigation of heat pipe heat exchanger effectiveness and energy saving in air conditioning systems using silver nanofluid, *Int. J. Environmental Science Technology* **9** 587-594

[7] Shafahi M, Bianco V, Vafai K and Manca O 2010 An investigation of the thermal performance of cylindrical heat pipes using nanofluids, *Int. J. Heat Mass Transfer* **53** 376-383

[8] Shafahi M, Bianco V, Vafai K and Manca O 2010 Thermal performance of flat-shaped heat pipes using nanofluids, *Int. J. Heat Mass Transfer* **53** 1438-1445

[9] Sarkar J 2011 A critical review on convective heat transfer correlations of nanofluids, *Renewable and Sustainable Energy Reviews* **15** 3271– 3277

[10] *** "ANSYS CFX 14.0, Theory Book", ANSYS Inc., 2011

[11] Li Y, He H F and Zeng Z X, 2013, Evaporation and condensation heat transfer in a heat pipe with a sintered-grooved composite wick, *Applied Thermal Engineering* **50** 342-351

Oxy-combustion simulation of ethylene

S Grigorean, G Dumitrascu and S Predoi

Faculty of Mechanical Engineering, „Gheorghe Asachi" Technical University of Iasi, Romania

E-mail: grigorean_stefan@yahoo.com

Abstract. The paper presents the simulation of the combustion of ethylene intended to be used in hybrid rocket propulsion. The simulation, was performed in ANSYS numerical codes, considering the following assumptions: fuel temperature 700 K and 800 K; pressure in the combustion chamber 1 to 3 bar; overall excess air from 1.5 to 5, in order to determine the case with lowest pollutant species. The simulation displayed the combustion temperature field, the flame OH concentration field, the exhaust gases composition fields, the gases velocity field, NO, CO, CO_2 concentration fields and the gases pressure field. The simulation showed the influence of fuel temperature upon the combustion spatial features.

1. Introduction

Fuels for hybrid rocket engines have some advantages in comparison to pure solid or liquid propellants, including increased safety regarding storage feeding system, propellant versatility, much simple design solution, low temperature sensitivity, and low relative cost. Although there are some disadvantages regarding low volumetric loading, potential fuel residuals, mixture ratio shift during motor firing, and mixing and combustion inefficiencies [1]. The most commonly cited drawback of their performance is actually low solid fuel regression rates. Research efforts devoted to overcoming these drawbacks have led to numerous fuel regression rate enhancement strategies including utilization of non-traditional fuels and oxidizers; manipulation of oxidizer flow to yield unique flows, such as swirl or vortex flows; inclusion of energetic additives such as metals, metal hydrides, and solid oxidizers; and augmentation of combustion port geometry to yield increased burn surface area. Compared to a liquid fuel system, the hybrid fuel system presents the advantage of a reduced explosion hazard and complexity due to the absence of regenerative cooling systems used for the combustion chamber and nozzle [2].

2. Combustion basic chemistry of ethylene

One hybrid rocket engines fuel used is polyethylene. Main combustible product obtained after ethylene pyrolysis process in hybrid rockets is ethylene. Thus, in this paper is presented a simplified simulation model of ethylene combustion process using pure O_2 as oxidizer. Few researches had been carried out in order to improve the knowledge of ethylene oxidation. In the paper [3], there were detailed some measurements of the laminar flame velocity of ethylene-air mixture. For some intermediate temperatures there were some studies using jet stirred reactors to derive analytical data [4, 5]. In the works [7-10] there were included studies regarding ethylene flame chemical composition at low and atmospheric pressure, using mass spectrometer.

At intermediate temperatures most reactions between C_2H_4 and chemical radical species like O and OH including reactions (1), (2), and (3), appear to involve the formation of an activated complex followed by a rearrangement and subsequent fragmentation [9].

The rates and product distributions for reactions between ethylene and O, OH and H radicals have been the subject of considerable study. At high temperatures abstraction reactions can be expected to be important, including reactions (4) and (5).

$$C_2H_4 + O \rightarrow CH_3 + HCO \quad (1)$$

$$C_2H_4 + O \rightarrow CH_2O + CH_2 \quad (2)$$

$$C_2H_4 + OH \rightarrow CH_3 + CH_2O \quad (3)$$

$$C_2H_4 + OH \rightarrow C_2H_3 + H_2O \quad (4)$$

$$C_2H_4 + OH \rightarrow C_2H_3 + H_2O \quad (5)$$

3. Combustion chamber design and ANSYS model

In order to analyse the oxidation process of ethylene was designed a simplified model of a combustion chamber with one oxidizer inlet located in the front, four fuel inlet ports located on the circumference of the combustion chamber and one outlet port for flue gases. Combustion chamber model have a length of 200 mm and diameter 52 mm.

For these simulations was used ANSYS software, in order to create the mesh and to define all the necessary input computational parameters of combustion process. To obtain a simplified model and shorter computational periods there were removed all components of the combustion chamber like walls, fixtures, spark plug, etc. The meshed model was developed on the inner volume of combustion chamber, and defined the inlet and outlet flow areas. The mesh has 335000 elements and 64682 nodes with finite element sizes of 1 mm. The mesh has a refinement in the oxidizer inlet area and fuel inlet area. They were assessed numerical simulations of reactive turbulent flow through the combustion chamber of ethylene burning, using pure O_2 as oxidizer. Using ANSYS CFX-RIF module they were defined the reactions for ethylene combustion using O_2. The carried out simulations used excess oxygen ratios of 1.5 and 3, fuel temperature of 700 K and 800 K. The presumed combustion pressures were set for 1 bar, 2 bar and 3 bar. Oxidizer inlet temperature was set at 300 K.

The adopted oxidizer inlet flow area has a round shape and 12.99 cm^2. The fuel, pure C_2H_4, enters the combustion volume using four circular inlets with a total area of 1.319 cm^2. The fuel mass flow rate is set to 1.169×10^{-4} kg/s.

The imposed selected chemical reactions and used in the combustion simulations are:

$$C_2H_4 + 3O_2 \rightarrow 2CO_2 + 2H_2O \quad (6)$$

$$2CO_2 \rightarrow 2CO + O_2 \quad (7)$$

$$CO_2 \rightarrow CO + O \quad (8)$$

$$2H_2O \rightarrow 2OH + H_2 \quad (9)$$

$$H_2O \rightarrow OH + H \quad (10)$$

$$CO_2 + H_2 \rightarrow CO + H_2O \quad (11)$$

All carried out simulations meet the convergence criteria after minimum 2500 computational steps.

4. Simulation results.

Below figures, 1 to 12, depict the temperature fields, the velocity fields, and spatial mass fractions of chemical species C_2H_4, CO, CO_2 and NO, for each imposed values of pressure, excess oxygen ratio, and fuel temperature for two demonstrative cases.

4.1. Fuel temperature 800 K, excess oxygen ratio 1.5, pressure 3 bar

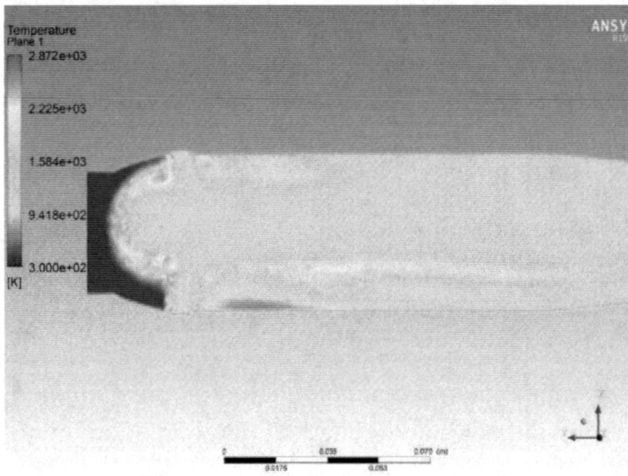

Figure 1. Temperature field.

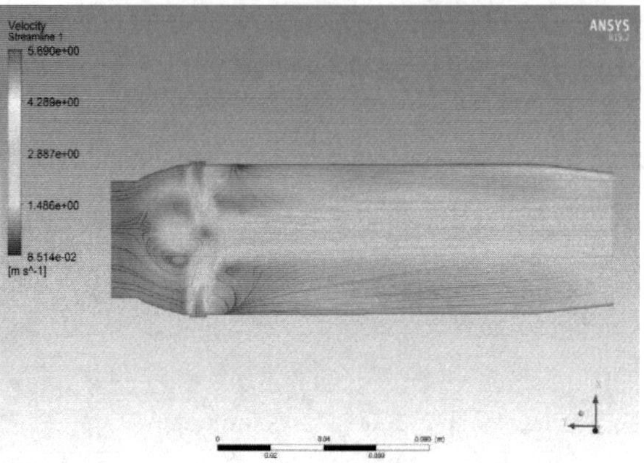

Figure 2. Velocity field.

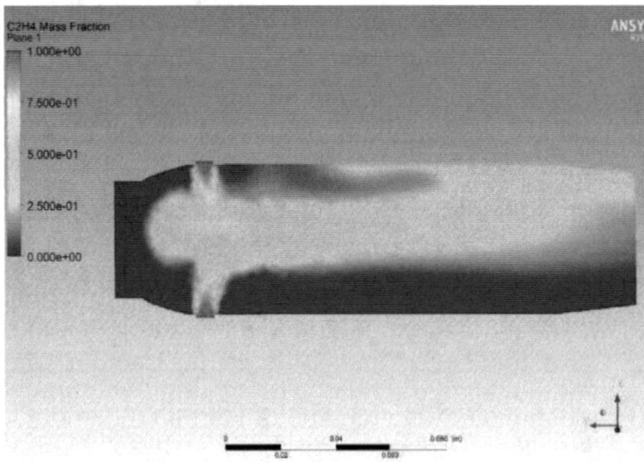

Figure 3. C_2H_4 mass fraction.

Figure 4. CO mass fraction.

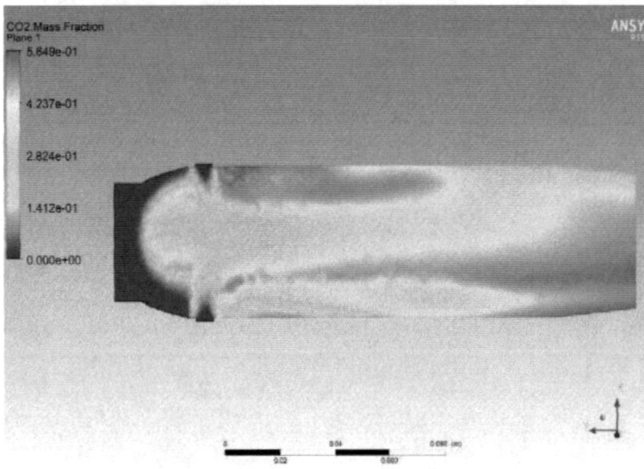

Figure 5. CO_2 mass fraction.

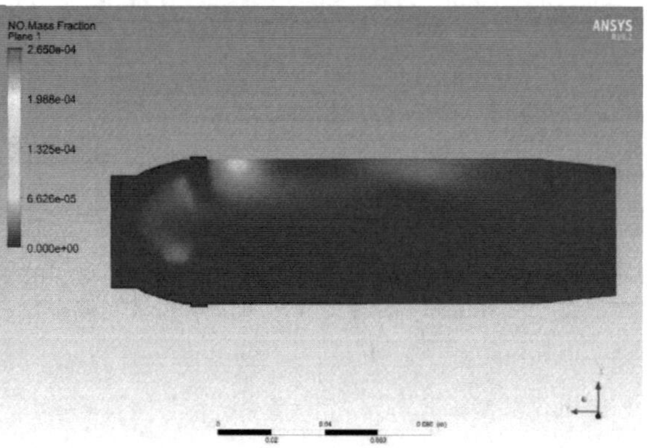

Figure 6. NO mass fraction.

Analysing the figure 2, it is visible that the flow reactive process has the following aspects:

- a laminar profile after the fuel injector ports toward flue gases outlet, and
- around the oxidizer inlet appears a backflow causing an intense recirculation of hot combustion products inside combustion chamber.

These particular aspects have strongly affected the other parameters fields inside combustion chamber.

4.2. Fuel temperature 800 K, excess oxygen ratio 3, pressure 3 bar

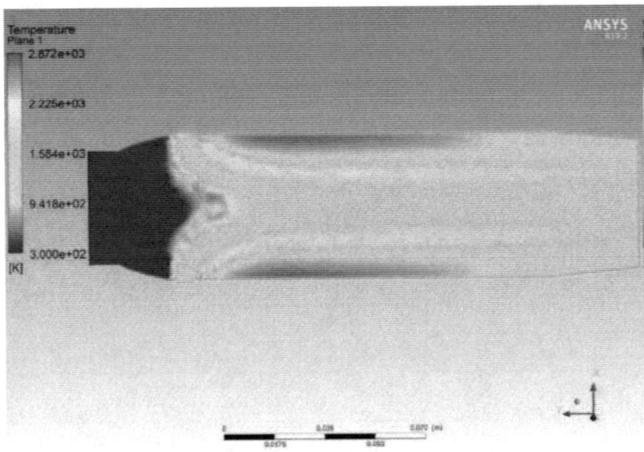

Figure 7. Temperature field.

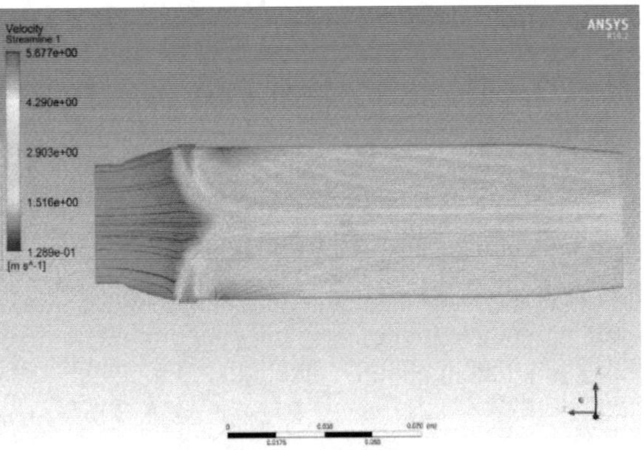

Figure 8. Velocity field.

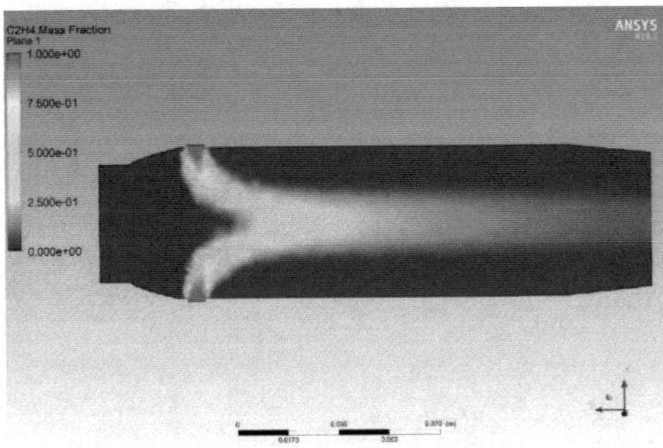

Figure 9. C_2H_4 mass fraction.

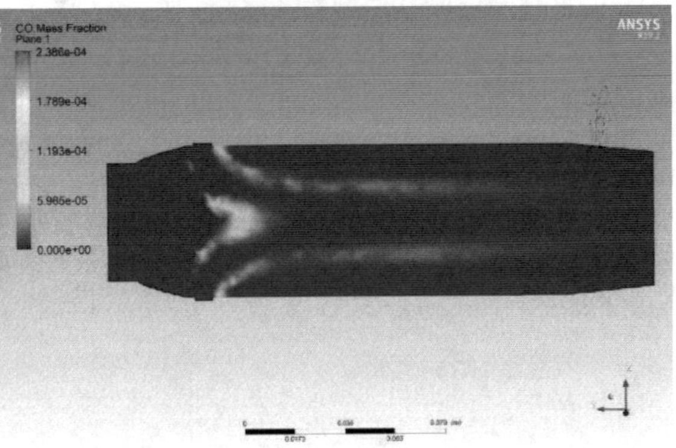

Figure 10. CO mass fraction.

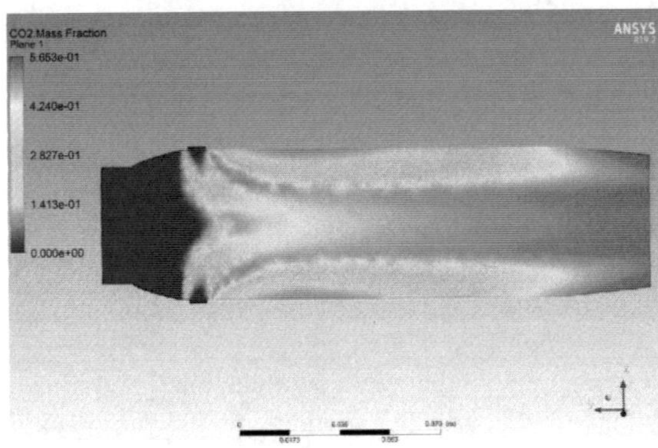

Figure 11. CO_2 mass fraction.

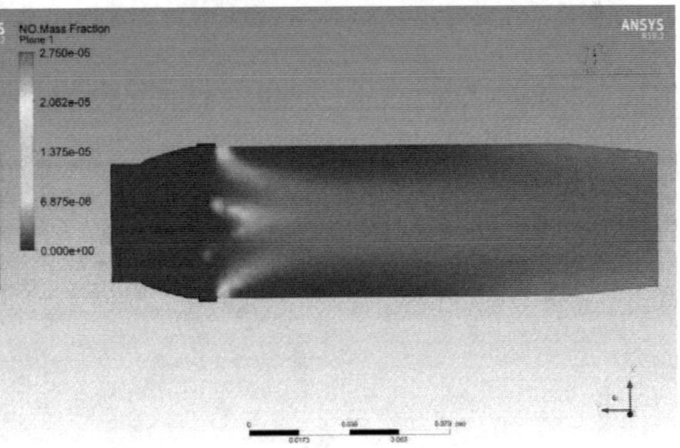

Figure 12. **NO** mass fraction.

The velocity field, figure 8, is characterised by a laminar flow, and therefore the temperature field has highest temperatures in the centre of combustion chamber. In order to obtain a more homogeneous combustion process it is necessary to use inlet flow geometries causing artificial turbulence.

The next below figures include the average values of NO, CO, CO_2 mass fractions, measured on the outlet section of combustion chamber. These averaged values were recorded in 25 points on this flow section. Figure 13 shows the presence of NO mass fraction, with highest value in the case of simulation with 800 K fuel temperature, excess oxygen ratio of 1.5 and pressure of 1 bar. The lowest values of NO mass fraction were obtained in the case of 700 K, excess oxygen ratio of 3, and pressure of 3 bar.

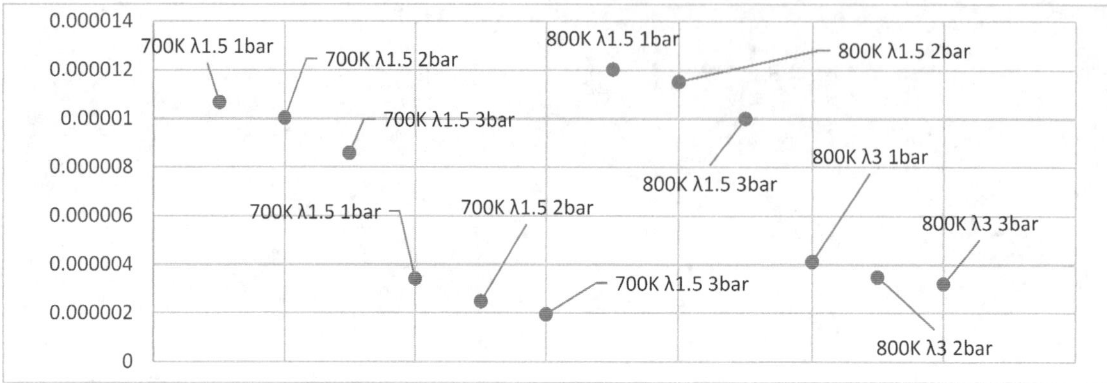

Figure 13. NO mass fraction for analyzed cases.

The Figure 14 includes the average mass fraction of CO in same points. Minimum value of CO mass fraction is obtained in the case of 700 K, excess oxygen ratio of 1.5, pressure of 1 bar, and maximum value is obtained in the case of 800 K, excess oxygen ratio of 1.5, pressure of 3 bar.

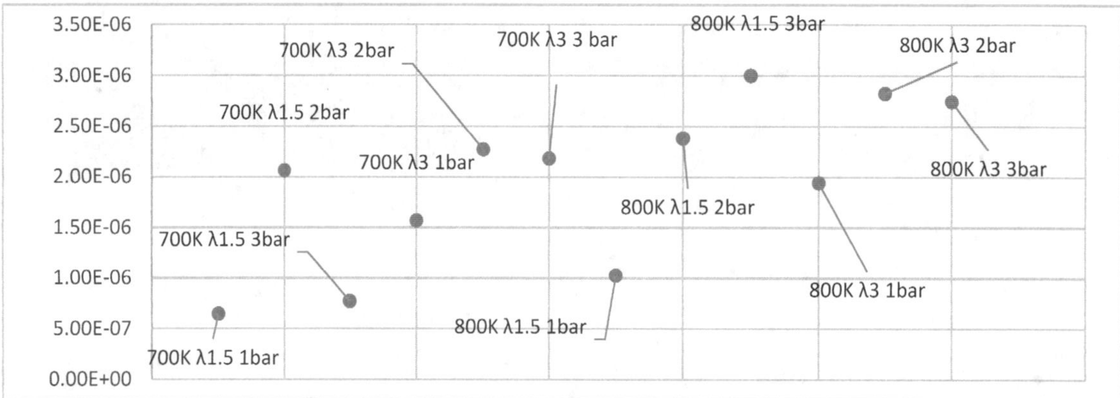

Figure 14. CO mass fraction for analyzed cases.

Figure 15 shows the average CO_2 mass fraction, with highest value in the case of simulation with 700 K fuel temperature, excess oxygen ratio of 3 and pressure of 3 bar. Lowest value for this mass fraction is obtained in the case of 700 K, excess oxygen ratio factor 1.5, and pressure 1 bar.

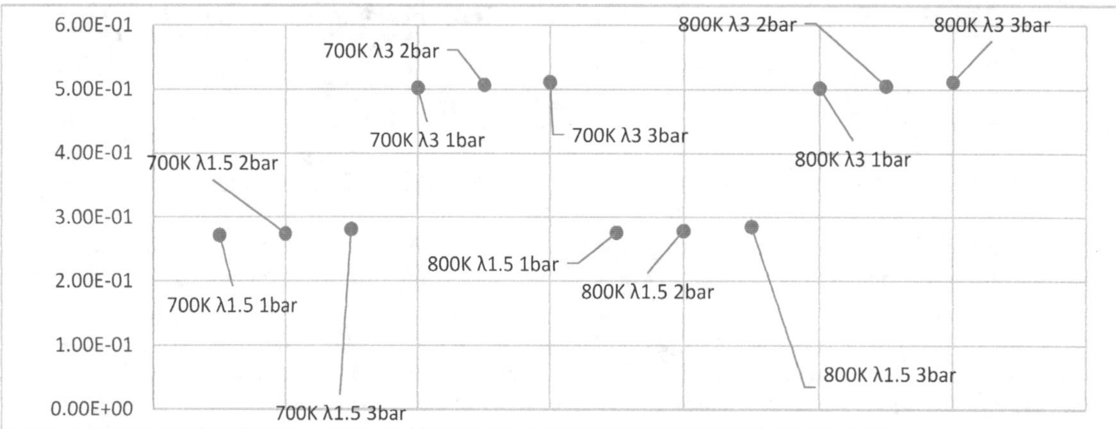

Figure 15. CO_2 mass fraction for analyzed cases.

5. Conclusions

After conducted simulations some conclusions are revealed. In order to obtain a more homogeneous combustion process it is necessary to use inlet flow geometries causing artificial turbulence. Minimum value of CO mass fraction is obtained in the case of 700 K, excess oxygen ratio of 1.5, pressure of 1 bar, and maximum value is obtained in the case of 800 K, excess oxygen ratio of 1.5, pressure of 3 bar. Minimum NO mass fraction concentrations are obtained in the case of 700 K, excess oxygen ratio of 3, and pressure of 3 bar, while highest value is obtained in the case of simulation with 800 K fuel temperature, excess oxygen ratio of 1.5 and pressure of 1 bar. For CO_2 mass fraction concentration, the highest value is obtained in the case of simulation with 700 K fuel temperature, of 3 and pressure of 3 bar while the lowest value is obtained in the case of 700 K, excess oxygen ratio 1.5, and pressure 1 bar.

References

[1] Thomas J C, Stahl J M, Tykol A J, Rodriguez F A and Petersen E L, Hybrid rocket studies using HTPB/Paraffin fuel blends in gaseous oxygen, *7th European Conference for Aeronautics and Space Science*

[2] Larson C W, Pfeil K L, DeRose M E and Carrick P G, 1996 High pressure combustion of cryogenic solid fuels for hybrid rockets, *32nd AIAA/ASME/SAE/ASEE Joint Propulsion Conference and Exibit*

[3] Westbrook C K, Dryer F L and Schug K P, 1983 A comprehensive mechanism for the pyrolysis and oxidation of ethylene, 19th Symposium on Combustion / The Combustion Institute

[4] Konopka T F, Figueira da Silva L F and Celis C, 2016 Comparative study of detailed chemical kinetics models of soot precursors for ethylene/air combustion, 16th Brazilian Congress of Thermal Sciences and Engineering

[5] Dagaut P, Cathonnet M, Boettner J C and Gaillard F 1988 Kinetic modelling of ethylene oxidation, *Combustion and Flame*. **71(3)**

[6] Shanshan C, Yong J, Rong Q and Jiangtao A, 2012 Numerical Study of laminar burning velocity and flame stability of premixed methane/ethylene/air flames, *Chinese Journal of Chemical Engineering*. *20(5)*

[7] Delfau J L, Biet J, Idir M, Pillier L and Vovelle C 2007 Experimental and numerical study of premixed, lean ethylene flames, *Proc. Combust. Inst.* **31(1)** p 357-365

[8] Varatharajan B and Williams F A 2002 Ethylene ignition and detonation chemistry, part 1: detailed modeling and experimental comparison. *Journal of Propulsion and Power*. **18(2)**

[9] Smith I and Zellner W M 1996 Journal of Chemical Society, Faraday Trans II, **92** 23

Acknowledgments

This work was supported by UEFISCDI Romania – research grant PCCDI 32/2018.

IOP Conf. Series: Materials Science and Engineering **595** (2019) 012015 doi:10.1088/1757-899X/595/1/012015

Research on the fire endurance of the silica sand membrane

A Kaya, S Çelen and U Akyol

Mechanical Engineering Department, "Tekirdag Namik Kemal University", Çorlu, Tekirdag, Turkey

E-mail: uakyol@nku.edu.tr

Abstract. In this study, the fire endurance, density, low-temperature deflection, penetration and ring-ball method tests of the membrane produced throughout the mixing of Bitumen, IPP (isotactic polypropylene), LDPE (Low-density polyethylene) calcite and silica sand at varying proportions were performed and the test results were compared with the membrane used today. In our study, TS 120 EN 1427, TS 118 EN 1426, TS EN 13501-1, KG-TL-029 testing methods were used. The density of the silica material we have obtained was measured at 1.3782 g/cm³, the endurance of the material was found out to be better, softening temperature was measured at 155°C, brittleness temperature at -7°C and the fire class was found out to be as E.

1. Introduction

With the energy resources coming closer to depletion, the need for isolating the buildings from the inside or outside with the purpose of using the resources efficiently has emerged. As a result of thermal insulation, it is both possible to save energy and keep the internal climatic conditions at the desired level. When it comes to buildings, sections such as roofing is among the places where thermal insulation should be applied [1].

The primary external factors that the buildings are exposed to are water and moisture. Various waterproofing and drainage applications are carried out in building constructions in order to eliminate or minimize the effect of water and moisture [2]. Considering the fact that we spend most of their time in closed environments; one of the ways to provide a safe structure and a comfortable indoor environment is to ensure that the building elements surrounding the place are sufficiently dry. In order to achieve this, it is necessary to take all measures against water. Measures that are taken to remove the water from a structure prolong its life while also increasing its endurance against earthquakes [3].

The fire must be taken under control before it spreads in order to minimize property damage during events of fire. Spreading is slowed down by setting up of fire-resistant parts inside the building. The resistant parts made of fireproofing material restrict the spreading of fire for certain periods and provide the firefighting teams with sufficient time to intervene before the fire expands. Thus, the risk of property loss is minimized by fireproofing. In addition, fireproofing measures should be also taken for the exterior façade to prevent a fire from spreading across the floors and adjacent buildings and to avoid life and property loss. The inability of the firefighters to intervene due to an increased temperature resulting from the rapid expansion of fire in buildings which lack fireproofing may result in total damage of the building [4].

As an alternative to expensive and imported insulation materials, producing a mineral based, naturally structured, artificial fibre and boron mineral enforced fireproof insulation material by using completely local resources (such as limestone and silica sand) is necessary [5].

IOP Conf. Series: Materials Science and Engineering **595** (2019) 012015 doi:10.1088/1757-899X/595/1/012015

In this work, we have conducted is to meet the increasing energy need of our country and to reduce dependence on foreign countries by increasing the fire endurance of the insulation materials that are used with the awareness of the fact that the energy use and loss (heat, sound, water, etc.) is at its most in residential areas. This way, both the energy conservation will be ensured by the prevention of unnecessary energy loss and comfort will be provided to people by ensuring property and life safety with the increase in fire endurance. As a result of this energy conservation, contributions to the country economy will be made with the decrease in energy imports and energy dependence on foreign countries. In addition to that, the fire endurance effect will be determined throughout the production of silica sand membrane.

2. Material and method

2.1. Bitumen

The bitumen is defined as a viscous liquid or a solid substance which contains mainly hydrocarbons and its derivatives; it is soluble in trichloroethylene, non-volatile and gradually softening when heated [6].

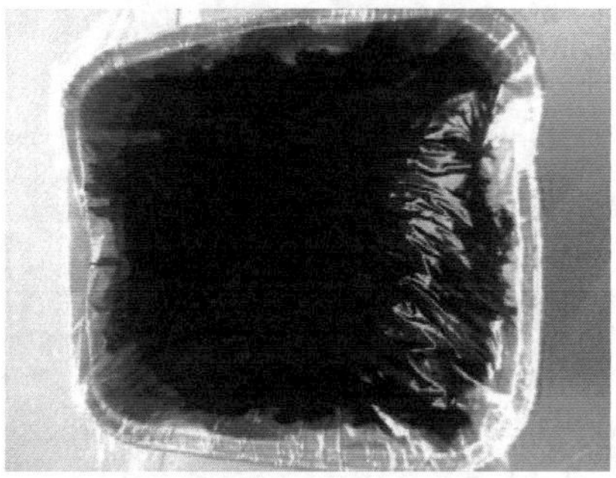

Figure 1. Bitumen.

2.2 Isotactic Polypropylene (IPP)

Polypropylene, specifically isotactic polypropylene (IPP), the most common commercial thermoplastic polymer, has a significant power: weight ratio due to its low density and high endurance. Polypropylene crystals have a melting point of about 160°C and are usually processed over 200°C [7].

Figure 2. Isotactic Polypropylene.

2.3 Low-density polyethylene (LDPE)

LDPE is produced by the high-pressure process (at 1600 atm and 200°C) of polyethylene. LDPE, which has many branching in its molecular structure compared to HDPE, therefore has lower crystallinity. For this reason, its hardness and durability are lower, and chemical endurance is limited. In contrast, its impact resistance is higher. The density of LDPE which can be used between -40-70°C is between 0.91-0.92 g/cm³ [8].

Figure 3. Low-density polyethylene.

2.4 Calcite

Calcite; chemical formula $CaCO_3$, is a mineral which is the constituent of limestone which has a crystal grain size between 1 mm-100 mm. According to the Mohs hardness scale, its hardness is 3, and its specific gravity is 2.7 g/cm^3 at 20°C [9].

Figure 4. Calcite.

2.5 Silica Sand

Silica sand belongs under the group named insulated lightweight building materials. Silica sand is composed of silica particles smaller than 2 mm that are formed as a result of the separation of granite-type rocks. Although the average surface area varies depending on the region that it is extracted, it is about between 100-200 m^2/g. It is used in the refractory industry, silica brick production, casting industry and glass industry [10]. The results of the analysis of silica sand are given in Table 1.

Figure 5. Silica Sand

Table 1. Sand Analysis Report.

Sieves	**Oversize**	**Multiply**	**Percentage**	**Experiments**	**Acceptance**	**Criteria**	**Experiment**
2.0		0	0.00		Max	Min	Result
1.6		0	0.00	AFS	+2	-2	87.35
1.0	0.00	0	0.00	Moisture %	3	8	
0.710	0.00	0	0.00	Clay %	0.1	0.5	0.35
0.500	0.03	0.75	0.06	SiO_2 %	98	99	98.10
0.355	0.28	9.8	0.56	Fe_2O_3 %	0.16	0.40	0.16
0.250	1.36	61.2	2.73	Al_2O_3 %	0.5	1.2	1.08
0.180	5.19	311.4	10.41	Sintering %			
0.125	31.35	2539.35	62.86	Temp (°C)	1500^{0}C	1750°C	>1500°C
0.090	10.69	1261.42	21.44	Temp (°C)	+4	-4	
0.063	0.85	139.4	1.70	Loss on ignition % (LOI)		0.02	
Pan	0.12	33	0.24	Acid demand value (ml)			
Total	49.87	4356.32	100	Sample amount		Vibration duration	
AFS	87.35			50 g	x	6 minutes	x
Avg. micron	151.29			100 g		10 minutes	

2.6 Method

In this study, the fire (heat) endurance, density, low-temperature deflection, penetration and ring-ball method tests of the membrane produced throughout the mixing of Bitumen, IPP (isotactic polypropylene), LDPE (Low-density polyethylene) calcite and silica sand at varying proportions were performed and the test results were compared. In this study, TS 120 EN 1427, TS 118 EN 1426, TS EN 13501-1, KG-TL-029 testing methods were used.

2.7 Production of the material (membrane)

The bitumen was heated and mixed in a steel pot to 150°C until it became liquid (Figure 6a). Then the recycling oil at a rate of 2.8% was melted and mixed into the liquid bitumen (Figure 6b). After that, IPP was added at a rate of 2.75% to give our mixture hardness (Figure 6c). The mixture was melted up to approximately 190°C-200°C by adding 3% LDPE for viscosity. After the polymers melted, we added our filler materials (45% calcite + 5% silica sand) (Figure 6d) and melted it by mixing it at 190°C-200°C for 1 hour.

We poured the mixture into the cylindrical moulds for penetration testing and we put it on the steel tray for other tests and let it cool down.

(a) (b) (c) (d)

Figure 6. Production of the membrane.

2.8 Penetration test

The penetration test was performed in two different ways at temperatures of 25°C and 60°C. The prepared material was poured into cylinder moulds with the diameters and heights of 4 cm. After 15 minutes of cooling, the mixture was placed in water baths (Figure 7b). It was kept for 35-40 minutes until the temperature is fixed. The penetration measurement of the needle was performed on the penetration device (Figure 7c).

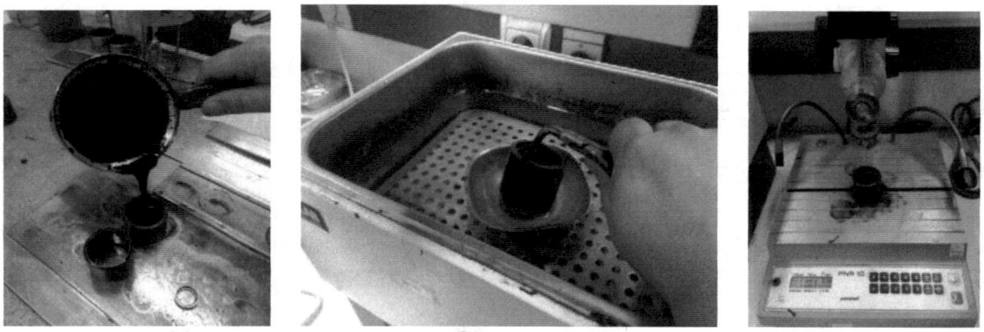

(a) **(b)** **(c)**

Figure 7. Penetration test.

2.9 Ring and ball test

In order to measure the degree of the softening, a Ring-ball mechanism was installed with 2 pieces of balls with the weight of 3.5 grams (Figure 8). It was subjected to a heating test in a glycerine bath. The temperature was noted at the exact time the balls fell down and its softening degree was recorded.

Figure 8. Ring-ball mechanism.

2.10 Low-temperature deflection test
Our 13 cm x 5 cm rectangular shaped and 3 mm thickness sample was put under deflection in cold test in a Cold-Flex device.

2.11 Density test
The density test of our material was done with a pycnometer. After the sample was placed in the slot of the pycnometer, the lid was closed and the gas in the sample was emptied. Then our density test was completed and density values were determined from the pycnometer.

2.12 Fire endurance test
The samples that were prepared at 90 mm x 250 mm x 3 mm were tested with an E-class fire testing device (Figure 9).

Figure 9. Fire testing device.

3. Results and discussion

3.1 Penetration Test Findings
As a result of the test conducted on the penetration device for 5 seconds at 100 g, the penetration measurements of the needle were found at 18.3 mm, 14.1 mm, 11.7 mm at 25°C, 97.6 mm and 99 mm at 60°C respectively.

3.2 Ring and Ball Test Findings
We noted the temperature at the moment the balls fell down and we recorded the softening by measuring it at 155°C.

3.3 Low-temperature Deflection Test Findings
After holding our sample at -5°C in the device for an hour, we have observed that our sample was not broken. Then, we held the sample at 0°C for an hour and we again didn't observe any breaking. Then after putting our sample to rest for 30 minutes, we again didn't observe any breaking at -5°C for an hour. Then, after letting the sample rest for another 30 minutes, we have tested the sample at -7°C and we have observed that breaking occurred in the sample.

3.4 Density Test Findings
We measured the density from the pycnometer at 1.3782 g/cm³.

3.5 Fire Endurance Test Findings
As a result of the test, no big difference was observed between the normal production (50% calcite) and the sample (5% silica + 45% calcite) and its fire class was determined as fire class E.

IOP Conf. Series: Materials Science and Engineering **595** (2019) 012015 doi:10.1088/1757-899X/595/1/012015

4. Conclusion

The concluding observations are as follows:

• The results of the penetration tests done for 5 seconds at 25°C and 100 g are measured at 18.3 mm, 14.1 mm and 11.7 mm respectively. As for 60°C, while the material that is used in the marked is measured at 125 mm and 143 mm, the measurement values of the material we have produced is at 97.6 mm and 99 mm. According to the results, it can be said that our material is harder.

• According to ring and ball test that we have conducted to determine the softening degree, the temperature at the moment the balls fell down was recorded as 155°C for the material we have produced. According to ring and ball test that we have conducted to determine the softening degree, the temperature at the moment the balls fell down was recorded as 155°C for the material we have produced.

• According to the low-temperature deflection test, we have produced was broken at -7°C.

• From the pycnometer, we have produced was measured at 1.3782 g/cm³.

• As a result of the test, no big difference was observed between the normal production (50% calcite) and the sample (5% silica + 45% calcite) and its fire class was found out as E.

• It is observed that fire class remained as E.

When all the results are taken into consideration, there was no change in class for the silica material we tested compared to the equivalent materials used in the market, however, improvements were made in terms of density and endurance.

References

[1] Uygunoğlu T Güneş İ Çalış M and Özgüven S 2015 EPS ve XPS malzemeleriyle yapılan mantolamaların yangın sırasındaki davranışlarının araştırılması *Journal of Polytechnic.* **18(1)** pp 21-28 (in Turkish)

[2] Bayraktar D and Bayraktar EA 2018 Unit cost of water insulation applications in school buildings *Journal of Engineering Sciences and Design.* **6(2)** pp 202–12

[3] Kartal S and Üstündağ S I 2016 The importance of waterproofing applications in buildings *DÜMF Mühendislik Dergisi.* **7(3)** pp 399-407

[4] http://www.ode.com.tr/fire-insulation 2019

[5] Yılmaz G 2016 Inorganic-based composite thermal insulation panel production *MSc Thesis Afyon Kocatepe University*

[6] Kuloğlu N Yılmaz M and Kök B V 2008 Investigation of resistance to permanent deformation and workability of the different penetration grades asphalt cements *Uludağ University Journal of the Faculty of Engineering.* **13(1)** pp 81-91

[7] Yıldız F and İlalan K 2013 Gıda ambalaj malzemeleri araştırma projesi *Orta Doğu Teknik Üniversitesi Gıda Mühendisliği Bölümü* (in Turkish)

[8] https://www.pluspolimer.com/urunler/polietilen/ldpe-alcak-yogunluklu-polietilen 2019

[9] Şahin N 2008 Kalsit hakkında bilgiler *Madencilik Bülteni* pp 48-51 (in Turkish)

[10] Güler D 2012 Synthesis of silica aerogel with silica sand feldspar and tetra ethylortosilicate by sol-gel method and characterization *MSc Thesis Gazi University*

Acknowledgements

As the authors, we thank ODE Yalıtım A.Ş. for letting us use their laboratories for this study.

IOP Conf. Series: Materials Science and Engineering **595** (2019) 012016 doi:10.1088/1757-899X/595/1/012016

Thermal - fluidics behaviour of water mist and gas fuel flame jets based on experimental results

D Pavel[1], A Chisacof[2], S Dimitriu[2] and P Cozma[3]

[1] Police Academy, Firefighter Faculty, Bucharest, Romania
[2] Thermal Engineering, Engines, Thermal and Refrigeration Equipment's Department, University Politehnica of Bucharest, Romania
[3] General Inspectorate for Emergency Situations, Bucharest, Romania

E-mail: pavel_dragos@yahoo.co.uk

Abstract. The main objective of the present paper is to provide a visual investigation of the fire and water mist interference during the extinguish process. This kind of the experimental study for these processes is imposed by the phenomena complexity. The experimental study provides us the information of real behaviour of jets assembly in operation. On this way the physical theoretical model may by better conceived. The main test was focused on the concave shape of ceiling, which gives the better suppression of the flame. From the heat balance, of the studied case, the extinguish efficiency is obtained, the values being around 20%. From these values result that a lot of working agent is lost. That loss is related to the finest of the droplets in the liquid jet. The experimental tests give us the optimal angles between the two jets in aim to realise an efficient extinguish operation.

1. Introduction

The actual systems used for the fire extinguish, must have a short time to suppress the fire and high efficiency of the agent used in this scope. The use of water mist is an opportunity in aim to have an economical method and an ecological way for fire suppression. The most available fluid used is the liquid water. This fluid is not pollutant or toxic, available and has an important heat absorption in comparison with the other agents [1-5]. Liquid dispersed in droplets and the transfer processes occurring in a spray surrounding are complicated and is difficult to model. Droplet dispersion in a jet depends of the atomizer system. The finesse of droplets must be high in aim to have small thermal inertia. Therefore, to accomplish these exigencies must have a droplets diameter less than 20 μm. The dimensions of the pulverized droplets are recommended to be in the range of 10 to 50 μm [2, 5]. Another problem consists in the time interval for evaporation of liquid, which must be a reduced one.

A way to realise this short time is to have an important pressure difference on the nozzle sides, even the magnitude of hundred bars [6]. The authors of this paper projected a heating of water before the entry in the nozzle. On this way a fast evaporation of liquid increases and the useful agent losses diminishes. Our experiences were made with the water heated up to 40°C, its pressure being in the range of (4-6) bar [5, 6]. In fact, the vapour generated by the liquid evaporation pushes the other gases from the affected zone. Consequently, the condition to maintain the flame is not satisfied. By this way the flame volume captured by jet is important and the agent is used efficiently. With these two effects, which take place simultaneously, an efficient cooling of the zone is realisable [5, 7]. Also, by the fast

IOP Conf. Series: Materials Science and Engineering **595** (2019) 012016 doi:10.1088/1757-899X/595/1/012016

evaporation of water the gaseous mixture created in the jet has a molar weight of vapour less than the molar weight of surrounding gas. By this way the mixture, rich in vapour, rises and protect the ceiling.

2. Experimental bench arrangement and particularities

The paper is based on the tests made on the experimental layout described below. Considering the above findings, an experimental bench producing the mist jet used for the flame suppression was conceived. The liquid having different temperature at the discharge head of the nozzle is modified in aim to put in evidence its influence on the mist spreading and its evaporation rate. The liquid feed water pressure is in the range of (2-6) bar and the temperature in the range of (20-50)°C. The experimental layout diagram is exposed on the Figure 1 [5, 6]. The tested nozzle shape and diameters are shown on the figures 2 and 3.

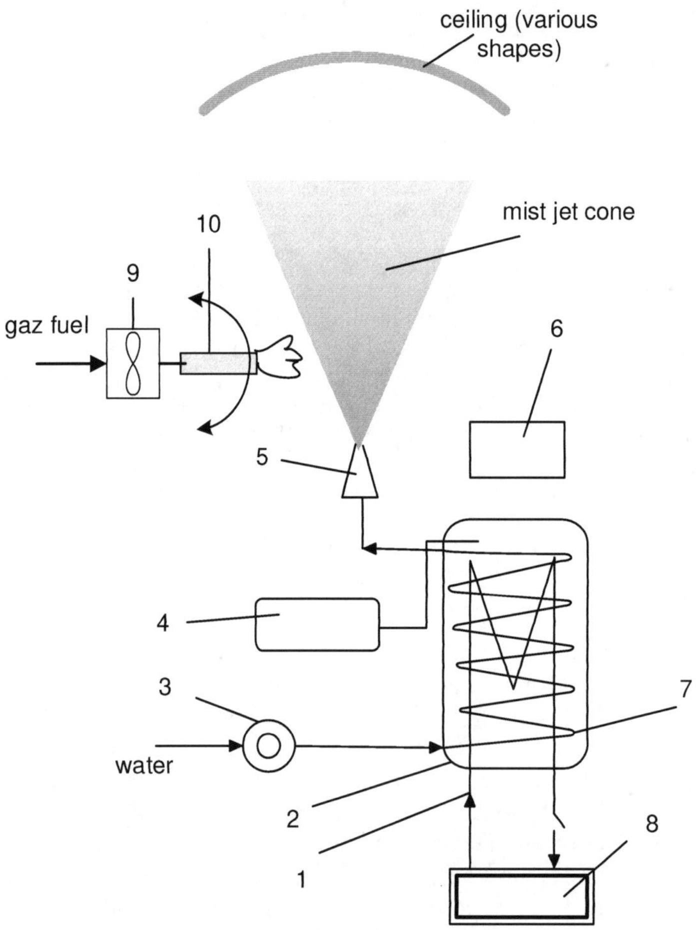

Figure 1. Experimental layout.
1 - electrical resistance; 2 - oil tank; 3 - water flowmeter; 4 - electronic thermostat; 5 - water nozzle; 6 - measurement bench; 7 - water heater; 8 - electrical supplier; 9 - gas flowmeter; 10 - gas burner.

The cold water passes the mass flowmeter type Coriolis, is driven to the water coil which is immersed in the oil tank. The oil is heated by an electrical resistance. The electronic thermostat is used to control the heating processes. The core of the liquid dispersion is represented by the nozzle assemblage with a swirl chamber. The swirl device has the function to create the vortex in liquid before that be pushed through the nozzle orifice. By an adequate architecture of the nozzle a short length of the liquid jet is obtained, and the breaking of the liquid is spread in small droplets.

A key element of the nozzle is the shape of the exit section and the angle of this. A qualitative design is presented on the figure 2, and the picture or various exit angles is displayed on the figure 3. These shapes were chosen taking into consideration the Coandă effect consequence which must be bring into consideration for the jet shape [8]. By reducing the angle α the jet angle envelope increases too. This enhancement of the cone jet volume is necessary in aim to encompasses a high space of flame and to use efficiently the water mist. The nozzle number I has a 2 mm orifice diameter and the angle α=40°. For the nozzle number II the diameter is of 1,7 mm and the angle α=60°, and for the device number III the diameter is 1,5 mm with the angle of 20°. After the tests we observe that the most efficient nozzle in the spray development, and the finest droplets are realised with the nozzle number III. As example on the jet mist dispersion of warm water followed by the liquid evaporation is shown on the figure 4.

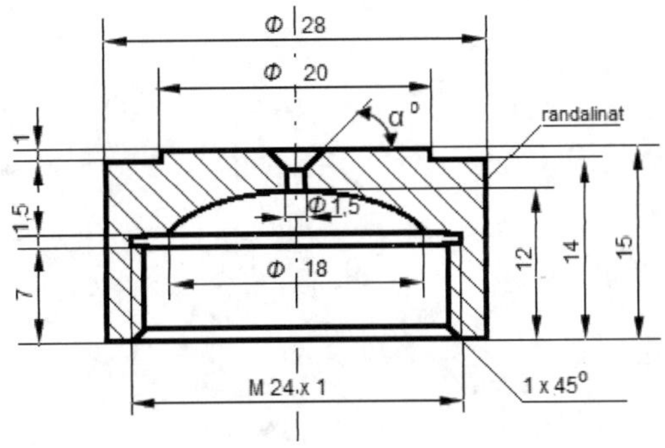

Figure 2. The nozzle shape and details (nozzle III).

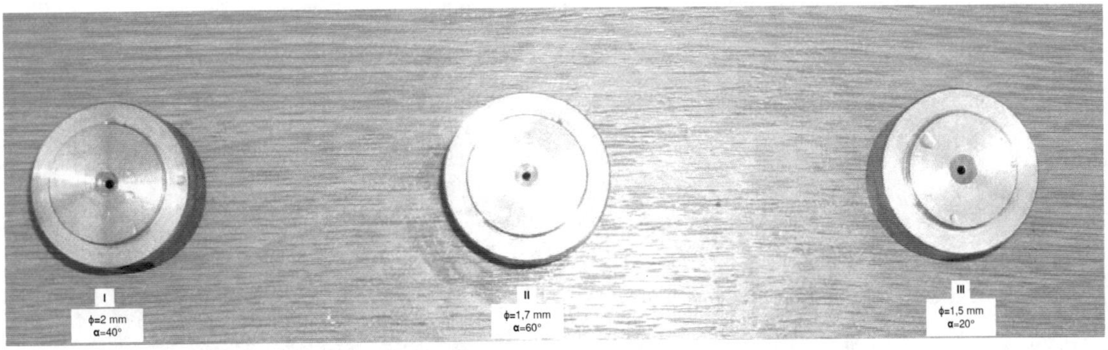

Figure 3. Nozzles shape.

An infrared image of the mist jet is displayed on the figure 5. The exit water temperature from the nozzle is around 34°C. Due of the reduced relative humidity of the adjacent environment the liquid evaporation takes place, and the oxygen concentration in this area decreases. Because of oxygen reduction may arrive at a condition with the flame instability followed by its suppression. From the figure 5 the temperature reduction is relatively obtained in the vertical distance of 70 cm from the nozzle plan, the values being of (19.8-21.5)°C.

For a premixed flame the burner cage is exposed to the mist, and a supplementary vaporisation is obtained. In this case the rate of evaporation is abundant and the local vapor concentration is larger than in the case of the diffusion flame. As a result of this high rate of vapour the suppression of the premixed flame is faster that in the case of diffusion flame. Our experiences were realised using a premixed burner.

Concerning the vertical orientation of the jet to the ceiling, it was chosen in aim to realise a more evaporation of the droplet and to avoid the loss of liquid not yet vaporised. On the other hand, the time passed by the droplet in contact with the hot medium is improved that in the case of the jet orientation to down. That allows also a good fulfilled of the zone with vapour. Using an artificial ceiling the vapour concentration of the zone of fire increase and the flame is suffocated. The figure 6 b) shows that for the concave ceiling a high concentration of vapour is obtainable.

Figure 4. Mist jet development using the nozzle II.

Measurements		
Sp1		30.1 °C
Sp2		26.9 °C
Sp3		25.0 °C
Sp4		24.7 °C
Sp5		22.6 °C
Sp6		23.3 °C
Sp7		24.5 °C
Sp8		21.9 °C
Sp9		23.8 °C
Sp10		22.0 °C
Sp11		21.5 °C
Sp12		21.1 °C
Sp13		20.1 °C
Sp14		19.8 °C
Li1	Max	32.3 °C
	Min	19.9 °C
	Average	22.4 °C

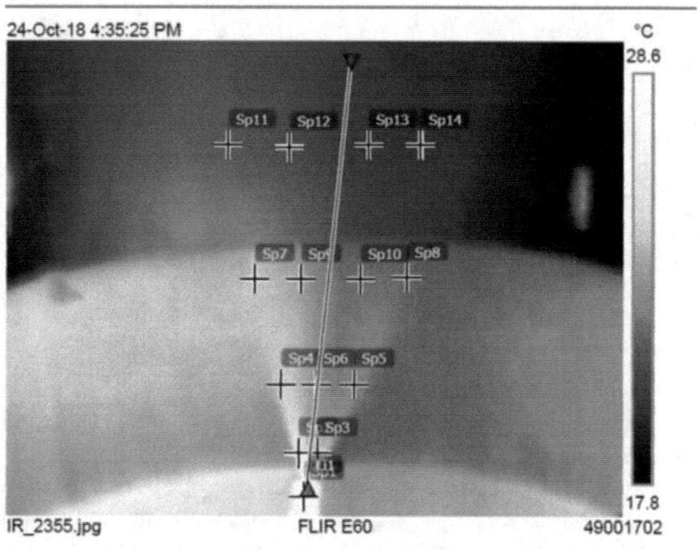

Figure 5. Infrared picture of the free mist jet with the nozzle II.

In aim to improve the experiment field on the flame, issued from burner, may have different angles with the mist jet to find the approached value of these between them, for the most probably extinguish process. The layout from the figure 1 was prepared for the various ceiling forms (plane, curve). The distance between the jet exit nozzle and ceiling was also modified to observe the extinguish behavior. From the first arrangement shown on the figure 6 a) we observe that the fluid line pattern is unfavorable for a mist concentration near the ceiling, a lot of unused extinguish agent being lost. From the figure 6 b) due of the ceiling concavity the fluid is concentrated, especially the vapor phase which remain near the internal surface of ceiling. By this method the increases in two-phase vapor concentration at the top makes impossible the flame stability, and its suppression is quickly. The third plane shape of the ceiling presented on the figure 6 c) has a certain vapor concentration on the below surface of plate, but not enough to ensure often the flame extinguish. In all the presented cases the Coandă effect is important and must be considered.

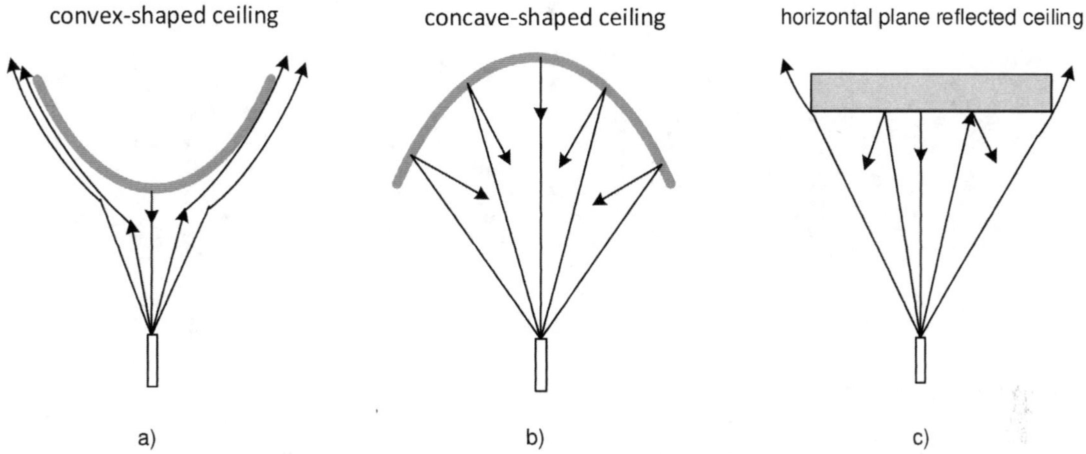

Figure 6. Different arrangement of jet-ceiling arrangement.

The figure 7 shown the sketch of the experimental assembly used in the tests with the relevant geometrical data. The nozzle is placed in the middle of the ceiling at 25 cm below the inferior ceiling plane. It is observable that the extinguish take place at about 45 cm in the interior of ceiling. Practically the flame suppression is at the cone limit of the mist jet. The nozzle used in the experiences was of 1.5 mm diameter (number III, figure 3)

Concerning the vertical position, the tests give a good process of extinguish at 45-50 cm high from the nozzle plane, or of (20-25) cm from the inferior plane of ceiling. These phenomena are the conjugated consequences of the high vapor concentration and the reflected droplet by the curvature of ceiling (figure6 b). On the figure 8 there are some images of the concave and plane ceilings.

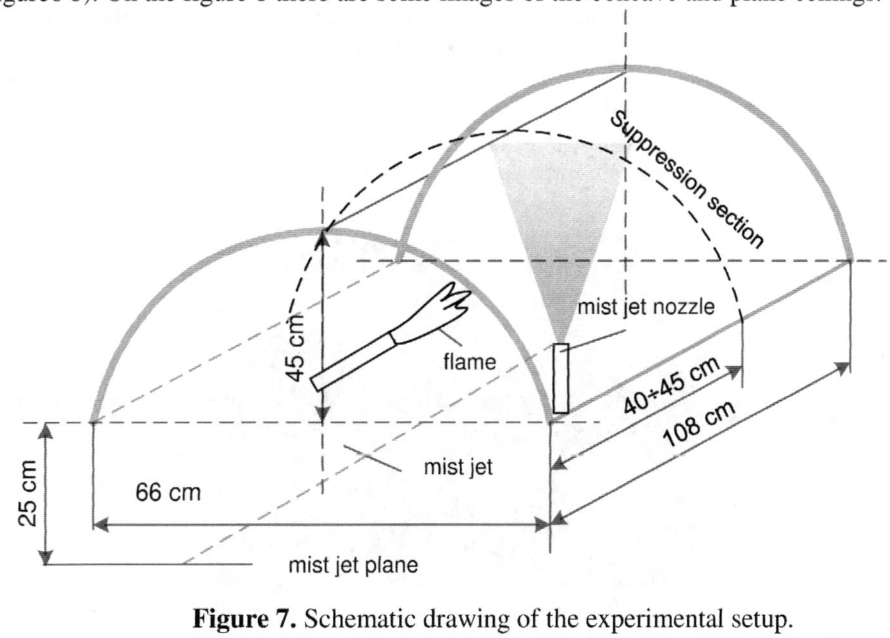

Figure 7. Schematic drawing of the experimental setup.

IOP Conf. Series: Materials Science and Engineering **595** (2019) 012016 doi:10.1088/1757-899X/595/1/012016

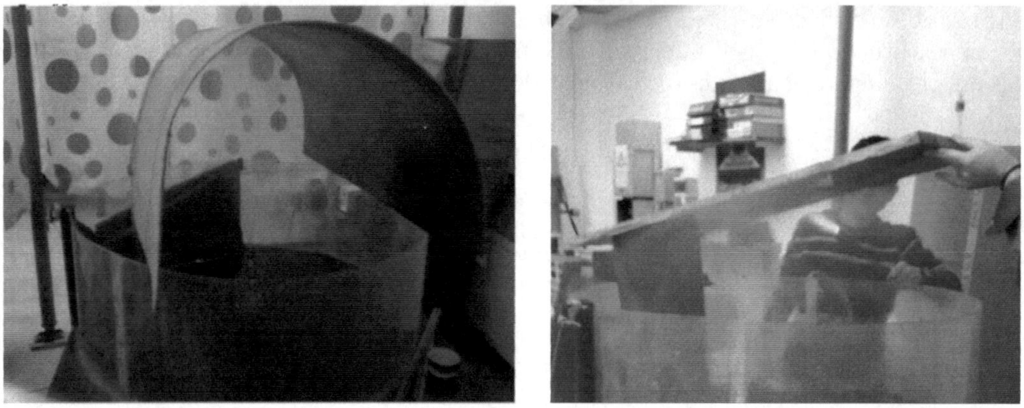

a) b)

Figure 8. Experimental modes a) concave ceiling; b) plane ceiling.

3. Results and discussion

Some elements of the jet behaviour issued from the experiments is observed from the figure 5. From other tests at higher temperature, we observe that by increasing the liquid temperature, the droplet dimension diminishes. Complementary, the formed cloudiness shows that due to the high evaporation rate the relative humidity around and, in the jet, increases and the saturation conditions are reached.

The consequence of the evaporation rate growing, the partial vapour pressure in the jet volume rises, and therefore, the oxygen concentration reduces. We found that on the vertical axes the temperature is greater that on the horizontal axes. On the other hand, from our tests, the decreasing in temperature for the temperature around 50°C at the nozzle exit, for a certain height, is more than for the jet outgoing at 35°C. That means that the temperature decrease rate is more important for the higher exit temperature. Therefore, the saturation of the concerned volume in vapor increases faster for higher exit temperature.

Concerning the extinguish efficiency the tests were made using the system with the concave ceiling and the flame issued from a burner supplied with the butane gas. The flowrate was measured with a volume flowmeter as was shown on the figure 1. The flame was introduced at different distance from the liquid nozzle and from the ceiling, implicitly. From the test the more efficient flame extinguish occurs at around (25-35) cm from the nozzle head, and at 40-45 cm deepness in the ceiling (Figure 7). The necessary time for the extinguish was of 4 to 6 s. Other tests were made with the flame at 5 to 10 cm from the ceiling, and the necessary time for the flame extinguish may arrives to 10 s and more, often the extinguish was problematic. On the figure 9 a sample of the flame suppression captured by of the infrared camera is displayed.

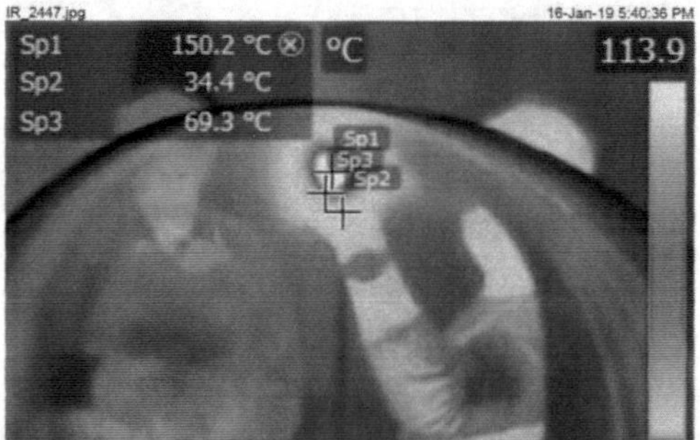

Figure 9. Infrared capture during the flame extinguish.

4. Heat balance of for the flame suppression

The experimental layout allows us to realise a heat balance of the system. The model of heat balance was adopted from [9]. The equation of the heat balance, during a time interval Δt, for the flame suppression is:

$$V_{fN} H_s = m_l \left(h_v - h_l\right) \cong m l_v \text{ [W]} \tag{1}$$

where: V_{fN} is the volume of gas burned in m^3_N, calculated as $V_{fN} = \Delta t \dot{V}_{fN}$, with the gas flowrate $\dot{V}_{fN}$ in $N\,m^3/s$, measured with the gas flowmeter and corrected in function of the gas temperature and pressure according with the manual instruction; H_s is the higher heating value of the fuel, in kJ/m^3_N; $m_l = \Delta t\, \dot{m}_l$ represents the liquid water mass, in kg, obtained from the mass flowmeter with $\dot{m}_l$ in kg/s; h_v – specific enthalpy of the generated vapour in the mixture, kJ/kg; h_l – specific enthalpy of liquids at the nozzle exit, in kJ/kg; l_v is the heat of vaporisation of water, in kJ/kg. At the exit of the nozzle the mass water flowrate $\dot{m}_w$ is measured with the Coriolis equipment

In the above equation was used the higher heating value of fuel, because in contact with the liquid phase the vapour issued from the combustion process, in contact with the liquid droplets is cooled until the liquid phase. This phenomenon is supposed to take place, considering that the final temperature of gases after the flame extinguish, arrives practically at the environmental temperature before the flame ignition. The water thermodynamic properties were calculated using [9].

For a time interval Δt, measured from the moment of the incidence between jet mist and flame at the flame extinguish, the heat balance may be written as:

$$\dot{V}_{fN} H_s \Delta t = \dot{m}_l \left(h_v - h_l\right) \Delta t \cong Q_{abs} \text{ [J]} \tag{2}$$

where Q_{abs} is the absorbed heat necessarily to be removed during the suppression flame process. The water quantity dispersed, during the time interval Δt, is $\dot{m}_l \Delta t = m_l$

This mass m_l from equation 1 differs of the mass flowrate measured by the Coriolis flowmeter $m_w = \Delta t\, \dot{m}_w$, since not all this flowrate is used in the flame extinguish, and because a part of this is unused. We may define the mass efficiency of liquid use for the flame extinguish as a ratio between the mentioned mass:

$$\varepsilon_{fe} = \frac{\dot{m}_l}{\dot{m}_w} \tag{3}$$

A similar equation of mass efficiency of liquid for time interval Δt is defined as the ratio between the liquid quantities, m_l and m_w. Concerning the heat capacity of the liquid water absorption, theoretically possible, to be captured by the mass m_w is:

$$Q_{t\,abs} = m_w \left(h_v - h_l\right) \cong m_w l_v \tag{4}$$

The real heat absorbed is obtained with the effective mass of water used:

$$Q_{r\,abs} = m_l \left(h_v - h_l\right) \cong m_l l_v \tag{5}$$

The ratio between the two heat represents the thermal efficiency of liquid use, expressed as:

$$\eta_{te} = \frac{\dot{V}_{fN} \Delta t H_s}{m_w \left(h_v - h_l\right)} = \varepsilon_{fe} \tag{6}$$

practically the same with the mass efficiency of liquid for the flame extinguish (eq. 2).

Because the liquid is heated from the inlet temperature in system until the exit temperature from the water heater, an added heat flux is necessary for this scope. The heat flux necessary for this aim is:

$$Q_{pr\,l} = m_w\left(h_{l\,ex} - h_{l\,in}\right) \tag{7}$$

where $h_{l\,ex}, h_{l\,in}$ are the liquid enthalpies at the exit and at the inlet in the water heater 7 (Figure 1).

Considering this additional heat, the effective thermal efficiency of liquid use may be well-defined as:

$$\eta_{t\,eff} = \frac{\dot{V}_{fN}\,H_s}{\dot{m}_w\left(h_v - h_l\right) + Q_{prl}} = \frac{\dot{V}_{fN}\,\Delta t\,H_s}{m_w\left[\left(h_v - h_l\right) + \left(h_{l\,ex} - h_{l\,in}\right)\right]} \tag{8}$$

We may define also the thermal power of flame as:

$$\dot{Q}_{t\,fl} = \frac{\dot{V}_{fN}\,H_s}{60}10^{-3}\ \text{[kW]} \tag{9}$$

with H_s in kJ/Nm^3, and $\dot{V}_{fN}$ in Ndm^3/min. The heat realised by the flame in the time Δt, is $Q_{t\,fl} = \Delta t\,\dot{Q}_{t\,fl}$.

Using our experimental data, and applying the above model, the main results are displayed on the table 1. The used fuel was butane with the higher heating value $H_s = 128\,000\,\text{kJ/m}^3_\text{N}$. For the heat absorption we consider the heat of vaporisation of water at triple point l_v= 2501 kJ/kg [10]. This point is a reference one for the moist air at 760 mm Hg. The superheated heat of vapour up to the environmental temperature is negligible.

Table 1. Results.

Measured parameters		Calculated quantities			
$\dot{m}_w$ [kg/min]	$\dot{V}_f$ [dm^3/min] *at 20°C, p=750 mm Hg	$\dot{V}_{fN}$ [Ndm^3/min]	$\dot{Q}_{t\,fl}$ [kW]	$\dot{m}_l$ [kg/min]	$\eta_{t\,ef} = \varepsilon_{fe}$
2.33	8.45	7.94	16.94	0.4064	0.174
2.33	8.4	7.89	16.83	0.4037	0.173
2.15	9.2	8.64	18.43	0.4421	0.205
2.15	11.1	9.2	19.63	0.4709	0.219

* the flowmeter used was calibrated for 23°C and 760 mmHg, the correction coefficient for butane is 0,93935 according to the apparatus guide.

We note that the thermal efficiency of liquid use is relatively reduced, in the range of 0,174 to 0,219. Consequently, an important liquid quantity is inefficient used, and the damage increases. These observations involve for the future to realise a better nozzle geometry in aim to have the fine droplets and the high rates of the liquid evaporation.

5. Conclusions

The paper put in evidence the importance of the temperature and of the nozzle geometry used in fire suppression system, supplied by a warm liquid having the domestic layout pressure in the range of (4-5) bar over the atmospheric pressure. So, a supplementary energy for water pump is not necessary.

The adequate geometry of the nozzle system is required in aim to improve the fineness of the liquid particles in the jet, and for the jet spreading. This characteristic allows to have a good efficiency in fire

IOP Conf. Series: Materials Science and Engineering **595** (2019) 012016 doi:10.1088/1757-899X/595/1/012016

extinguish. The Coandă effect is important in the jet spreading in function of the ceiling shape, and for the nozzle geometry.

The observation concerning the ceiling shape are very significant for the fire suppression system, because an additional artificial concave ceiling may create a high vapour concentration and a rapid extinguish. The geometrical dimensions of the system in accordance with the flame structure obtained are relevant (Figure 7).

From the heat balance, and from the extinguish effective thermal efficiency of liquid use, results that an unused liquid is important, with the negative consequences. This may represent an important parameter for the system operation.

References

[1] Andersson P, Arvidson M and Holmstedt G May 1996 *Small scale experiments and theoretical aspects of flame extinguishment with water mist*, Lund Institute of Technology, Lund University, Report 3080

[2] Chisacof A et al., *Clean Jet for the environment structure change*, grant CNCSIS ID_1708/ 2009-2011

[3] Liu Z and Kim A K 2000 *A review of water mist fire suppression systems – fundamental studies*, Journal of Fire Protection Engineering. **10(3)** pp 32-50

[4] Santangelo P E and Tartarini P 2010 *Fire Control and Suppression by Water-Mist Systems*, The Open Thermodynamics Journal. **4** pp 167-184

[5] Chisacof A, Panaitescu V, Pavel D and Poenaru M 2010 *The Two Phase Jet Use in Semi-open Space*, Proceedings of the ASME 2010 10th Biennial Conference On Engineering Systems Design and Analysis 2010, July 12-14, Istanbul, Turkey, paper ESDA2010-24961

[6] Panaitescu V, Pavel D, Chisacof A and Lazaroiu G 2012 *Free Jet of Mist Water use for Fire Heat Absorption,* Revista de Chimie. **63(3)** p 310-315

[7] Pavel D I and Chisacof A 2016 Experimental Aspects in Two-Phase Jet in Interaction with the Flame, Buletinul Institutului Politehnic din Iaşi, Universitatea Tehnică „Gheorghe Asachi" din Iaşi, Tom. Secţia Construcţii de Maşini.

[8] Coandă H M 1936, brevet nr 792 754 France.

[9] Chisacof A, Pavel and Dimitriu S 2018 *Geometrical shape nozzle and liquid temperature influence on the two-phase free jet characteristics.* Revista Termotehnica. **2** pp 9-15

[10] IAPWS-IF97 2008 *International Association Properties of Water and Steam (*editors Wagner W, Kretzschmar), Springer Verlag

IOP Conf. Series: Materials Science and Engineering **595** (2019) 012017 doi:10.1088/1757-899X/595/1/012017

Comparative analysis of the study of microclimate parameters in wooden houses in North-Eastern Bulgaria

P Zlateva and K Yordanov

Technical University of Varna, Bulgaria

E-mail: pzlateva@tu-varna.bg

Abstract. The subject of this study are some parameters of the microclimate in wooden houses in North-Eastern Bulgaria. This is necessary to identify the most influential parameters that affect human health and effectively improve energy efficiency in modern wooden residential buildings. Internal temperature, relative humidity and air speed and fine particulate matter were measured. External temperature and relative air humidity were also measured. The research results confirm that wood is a natural thermo-protective and moisture-regulating material. The analysis proposed can be applied to all energy efficiency projects in areas with similar climatic conditions.

1. Introduction

Prefabricated wooden buildings are an effective means of building private homes and are currently very popular in Europe and North America [1]. They have contributed to the transition of mankind from concrete structures to a life in a comfortable and natural environment [2]. Energy efficiency, healthy living environment, short time for construction and low prices are very important factors [3]. Wood is a natural heat-shrinkable [4] and moisture-regulating material [5]. Thus, an optimal balance of the natural air humidity is maintained in the house [6]. If the internal humidity level increases, the walls begin to absorb. If the air is too dry, the wood gives out the moisture absorbed. In this way the humidity in wooden houses is maintained steadily and permanently [7].

Figure 1. Floor layout of the Object 1.

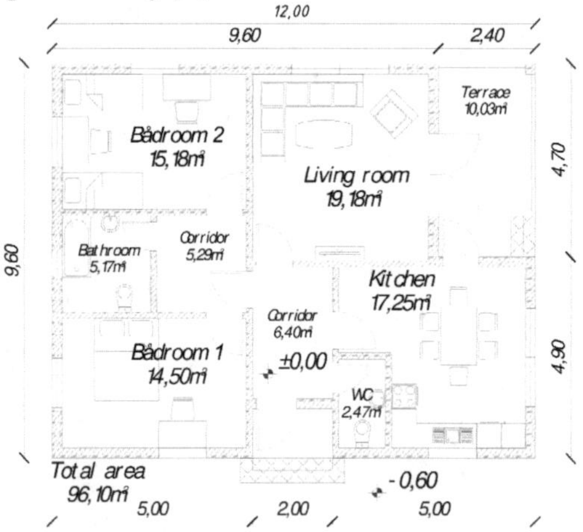

Figure 2. Floor layout of the Object 2.

Following various studies [8-9], it has been found that the complex air parameters in the rooms affect the lifestyle of the occupants and their comfort [10]. The research covers some of the microclimate parameters and some air quality indices in wooden buildings. The tests are carried out in two single-story houses with wooden elements incorporated in the façade walls located in the Black-Sea climatic region of North-Eastern Bulgaria as shown in Figure 1 and Figure 2 respectively. Object 1 is occupied by a family of two adults. The heating of Object 1 is a water jacket pellet stove developing a heat output of 13 kW and radiators connected to the collectors of the switchboards - a two-pipe system. Object 2 accommodates a three-member family. The heating in Object 2 is a water jacket stove producing a heat output of 13kW and a radiator arrangement identical to that in Object 1.

2. Research methodology

The measurements are made in two consecutive seasons - winter and spring, in randomly selected consecutive days for each season, thus meeting the minimum requirements for microclimate in buildings [11]. According to [12], the adopted fine particulate matter threshold limit values are as follows: PM10 average daily rate of 50 μg/m³; PM10 average annual rate of 40 μg/m³; PM2,5 average annual rate of 25μg/m³.

The following building microclimate parameters are measured: air temperature (°C); relative humidity (%); air velocity (m/s) as well as air quality indices - concentrations of PM10 (μg/m³) and PM2,5 (μg/m³). Temperature, humidity and airflow, PM10 and PM2,5 concentrations are measured in 12-hour series. Air velocity and PM10 and PM2,5 concentrations are measured per second. Air temperature and humidity are measured at 10 s. The external temperature and the relative air humidity are also measured simultaneously with the measurements in the residential buildings. All internal measurements are taken in a living room about 1 m above the floor. All external measurements are taken outdoors at a height of about 2 m from the elevation of the adjacent terrain, in a place that has maximal protection against direct sunlight [13].

3. Results and Discussion

The results obtained and illustrated in figures 3 to 10 are used for a comparative analysis of the two wooden houses.

3.1. Object №1 is located in a rural area

3.1.1. Winter Season Measurements

The air speed in a room in Object №1 for the period from 20:17:46 h on 05.03.2017 to 07:55:03 h on 06.03.2017 (winter season) has an average value of 0.092 m/s. The average PM2.5 concentration is 6.22μg/m³, and the average PM10 concentration is 11.23 μg/m³. The air velocity for this period is characterized by constant values – Figure 3. The registered PM2.5 concentrations are between 4 and 12 μg/m³, showing a decreasing trend and the registered PM10 levels are between 22 and 6 μg/m³. The humidity in the room is constant and is of about 40 %, while the outside air humidity fluctuates from 45 % to 72 %. The room temperature is constant between 20 – 21°C. The outside temperature varies between 7°C and 13°C – Figure 4.

The air speed in a room in Object №1 for the period from 08:57:46 h on 06.03.2017 to 19:57:07 h on 06.03.2017 (winter season) has an average value of 0.116 m/s. The average PM2,5 concentration is 8.50 μg/m³ and the average PM10 concentration is 15.18 μg/m³. Air velocity is characterized by constant values to 17:00 h, which is followed by variable air velocity values up to 19:57 h. The registered PM2.5 amounts are between 3 μg/m³ and 14 μg/m³ and the registered PM10 levels are between 5 μg/m³ and 26 μg/m³. Air humidity in the room is about 40 – 48 %, while the outside air humidity is 45 – 82 %. The temperature in the room is 20 – 21°C. The outside temperature is between 7°C and 19°C.

The air speed in a room in Object №1 for the period from 9:57:07 h on 06.03.2017 to 08:00:07 h on 07.03.2017 (winter season) has an average value of 0.137 m/s. The average PM2.5 concentration is

5.75 μg/m³ and the average PM10 concentration is 10.02 μg/m³. Air velocity to 05:00 h is highly variable, which is followed by a period of constant values. The registered PM2.5 concentration is between 3 μg/m³ and 8 μg/m³ and the registered PM10 levels are between 7 μg/m³ and 15 μg/m³. Air humidity in the room is 40 – 50 % and the outdoor air humidity is 60 – 92 %. The room temperature is between 21 – 23°C. The outside temperature is between 3°C and 15°C.

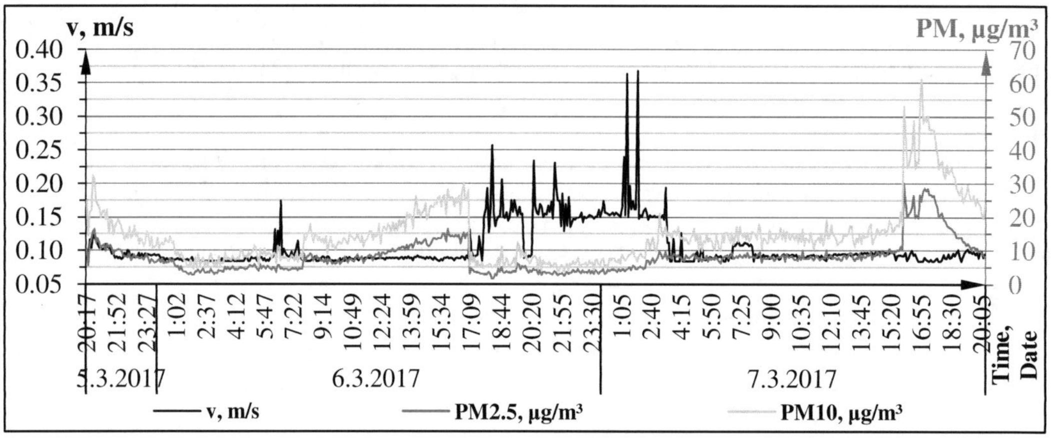

Figure 3. WINTER - Time distribution for 48h of air velocity and fine particulate matter – Object 1.

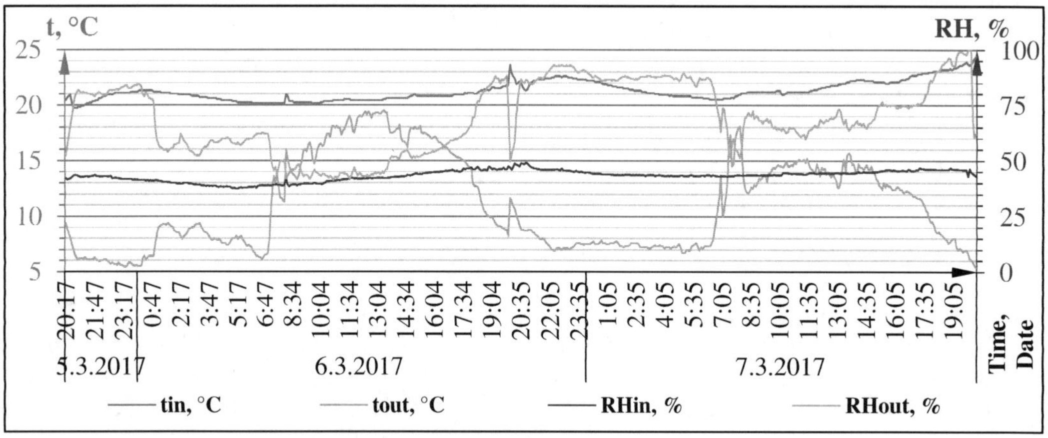

Figure 4. WINTER – Time distribution for 48h of temperature and relative humidity – Object 1.

The air speed in a room in Object №1 for the period from 08:00:07 h to 19:57:07 h on 07.03.2017 (winter season) has an average value of 0,098 m/s. The average PM2,5 concentration is 11,96 μg/m³ and the average PM10 concentration is 21,91 μg/m³. The air velocity is approximately constant. The registered PM2,5 concentration is between 3 μg/m³ and 25 μg/m³ and for the registered PM10 levels are between 14 μg/m³ and 45 μg/m³. The air humidity in the room is 40 – 42 %, and outdoor air humidity fluctuates from 62 % to 98 %. The room temperature is 21 – 23°C. The outside temperature varies between 7°C and 15°C.

3.1.2. Spring Season Measurements

The speed of the air in space of an object №1 for the period from 20:29:11 h on 08.05.2017 to 08:29:21 h on 09.05.2017 (spring season) has an average value of 0,126 m/s. The average quantity of

PM2.5 is 13.64 μg/m³ and average quantity of PM10 is 26.30 μg/m³. The air speed is slightly variable after which there is a strong increase due to the increased occupant's activity upon awakening from sleep – Figure 5. The recorded PM2,5 and PM10 levels increase by about 5 times after the tenth hour. The probable cause is the activity of the occupants. The room air humidity remains constant at about 60 %, while outside air humidity ranges from 88.4 % to 62 %. The internal temperature throughout the period is about 21 – 23°C. The outside temperature ranges from 11.2°C to 19.7°C – Figure 6.

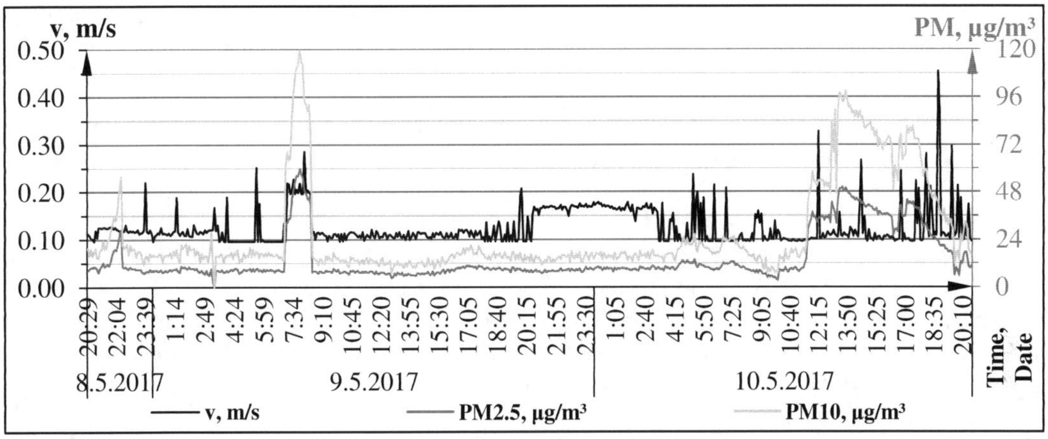

Figure 5. SPRING - Time distribution for 48h of air velocity and fine particulate matter – Object 1.

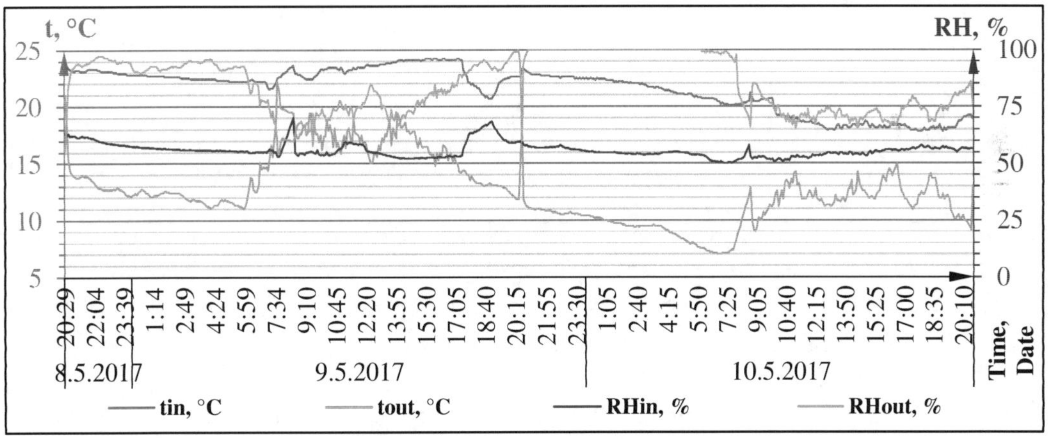

Figure 6. SPRING – Time distribution for 48h of temperature and relative humidity – Object 1.

The air speed in a room in Object №1 for the period from 08:35:09 h to 20:35:16 h on 09.05.2017 (spring season) averages 0.11 m/s. The average amount of PM2.5 is 7.95 μg/m³, the average amount of PM10 is 13.94 μg/m³. The air speed is constant. The registered PM2.5 concentration ranges between 6 μg/m³ and 10 μg/m³, and for PM10 it is between 10 μg/m³ and 18 μg/m³. Air humidity in the room is between 55 % and 65 %, and outside air humidity increases from 67.8 % to 92.2 %. The internal temperature increases from 20°C to 24°C. The outside temperature drops to 12.6°C.

The speed of the air in space of an object №1 for the period from 20:37:11 h on 09.05.2017 to 08:29:21 h on 10.05.2017 (spring season) has an average value of 0.121 m/s. The average quantity of PM2,5 is 13.28 μg/m³ and average quantity of PM10 is 25.58 μg/m³. The air speed is slightly variable after which there is a strong increase due to the increased occupant's activity upon awakening from sleep. The recorded PM2.5 and PM10 levels increase by about 5 times after the tenth hour. The probable cause is the activity of the occupants. The room air humidity remains constant at about 62 %, while outside air humidity ranges from 72.6 % to 61.5 %. The internal temperature throughout the period is about 21 – 23°C. The outside temperature ranges from 10.8°C to 18.9°C.

The air speed in a room in Object №1 for the period from 08:40:10 h on 10.05.2017 to 20:45:38 h on 10.05.2017 (spring season) averages 0.143 m/s. The average amount of PM2.5 is 26.90 μg/m³, the average amount of PM10 is 49.99 μg/m³. The air speed is variable. The registered PM2.5 amounts to between 6 μg/m³ and 44 μg/m³ and for PM10 the concentration is between 10 μg/m³ and 85 μg/m³. Air humidity in the room ranges from 51 % to 55 %, and the external air humidity drops from 80.8 % to 70.2 % and stays at these levels. The room temperature decreases from 17.5°C to 15°C during the 5th hour and stays at these levels. The outside temperature fluctuates between 10°C and 14°C.

3.2. Object 2 is located in a suburban area

3.2.1. Winter Season Measurements

The air speed in a room in Object №2 for the period from 20:13:07 h on 02.03.2017 until 08:00:07 h on 03.03.2017 (winter season) has an average value of 0.214 m/s – Figure 7. The average amount of PM2,5 is 28.84 μg/m³ and the average amount of PM10 is 51.77 μg/m³. The average values of the room air parameters are: humidity – 64.7 %, temperature – 18.3°C. The average values of outdoor air parameters are: humidity – 77.,9 %, temperature – 6.2°C – Figure 8. The internal air speed is variable due to effect of the heating system.

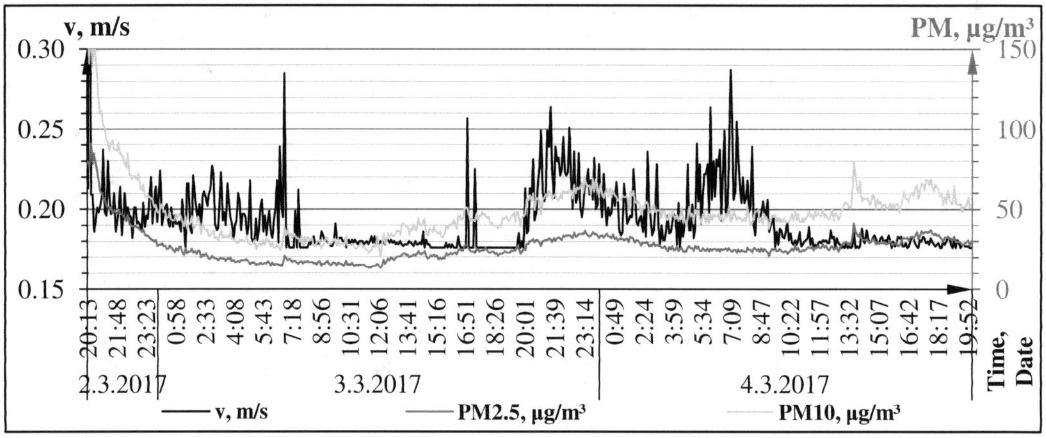

Figure 7. WINTER - Time distribution for 48h of air velocity and fine particulate matter – Object 2.

The air speed in a room in Object №2 for the period from 08:19:07 h until 20:00:07 h on 03.03.2017, (winter season) averages 0.179 m/s. The average concentration of PM2.5 is 28.69 μg/m³, that of PM10 is 52.45 μg/m³. The average values of the room air parameters are: humidity – 61.2 %, temperature – 20.3°C. The average values of outdoor air parameters are: humidity – 73.4 %, temperature – 5.7°C. The air velocity for this period is characterized by relatively constant values.

The air speed in a room in Object №2 for the period from 20:22:07 h on 03.03.2017 to 08:22:07 h on 04.03.2017 (winter season) has an average value of 0.206 m/s. The average amount of PM2.5 is 28.90 μg/m³ and the average amount of PM10 is 52.61 μg/m³. The average values of the room air parameters are: humidity – 51.5%, temperature – 18.4°C. The average values of outdoor air parameters are: humidity – 54.4 %, temperature – 13.3°C. The air velocity for this period is characterized by variable values within a short interval.

The air speed in a room in Object №2 for the period from 08:22:07 h to 20:17:07 h on 04.03.2017 (winter season) has an average value of 0.181 m/s. The average amount of PM2,5 is 28.8 µg/m³ and the average amount PM10 is 52.2 µg/m³. The average values of the room air parameters are: humidity – 51.5 %, temperature – 22.8°C. The average values of outdoor air parameters are: humidity – 54.6 %, temperature – 13.3°C. The air velocity for this period is characterized by constant values.

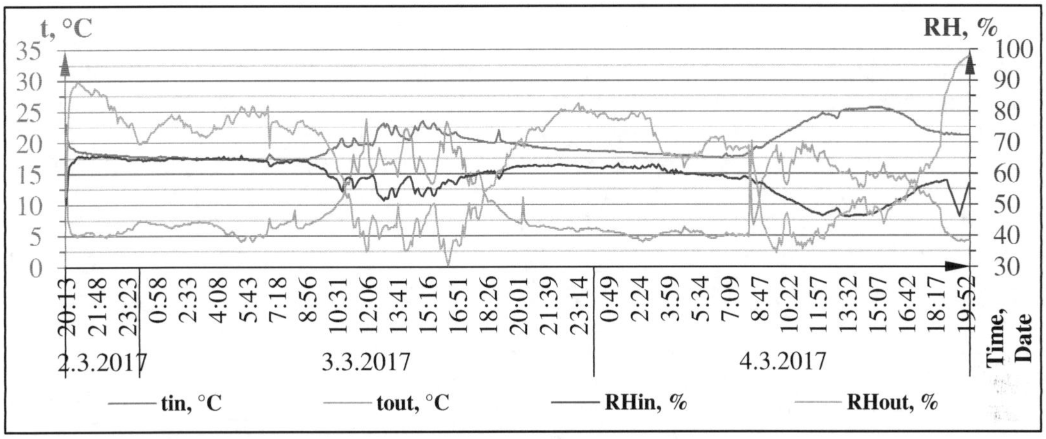

Figure 8. WINTER – Time distribution for 48h of temperature and relative humidity – Object 2.

3.2.2. Spring Season Measurements

The air speed in a room in Object №2 for the period from 08:08:07 h to 20:10:02 h on 28.04.2017, (spring season) averages 0.171 m/s – Figure 9. The average amount of PM2,5 is 25.15 µg/m³ and that of PM10 is 43.25 µg/m³. The average values of the room air parameters are: humidity – 52.5 %, temperature – 23.7°C. The average values of outdoor air parameters are: humidity – 47.3 %, temperature – 17.3°C – Figure 10. Air velocity is characterized by variable values.

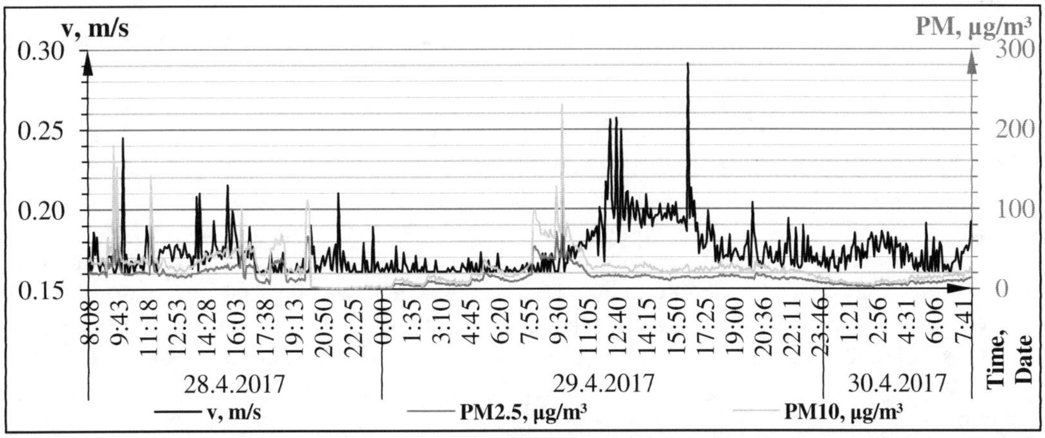

Figure 9. SPRING - Time distribution for 48h of air velocity and fine particulate matter – Object 2.

The air speed in a room in Object №2 for the period from 20:10:07 h on 28.04.2017 until 08:10:02 h on 29.04.2017, the spring season averaged 0.166 m/s. The average amount of PM2,5 is 6.12 µg/m³, PM10 is 10.8 µg/m³. The average values of the room air parameters are: humidity – 57.5 %, temperature – 21.6°C. The average values of outdoor air parameters are: humidity – 73.3 %, temperature – 8.7°C. Air velocity is characterized by relatively constant values.

The air speed in a room in Object №2 for the period from 08:10:02 h to 20:10:02 h on 29.04.2017 the spring season averages 0,186 m/s. The average amount of PM2.5 is 21.04 μg/m³ and the average amount of PM10 is 36.42 μg/m³. The average values of the room air parameters are: humidity – 57.2 %, temperature – 24.5°C. The average values of outdoor air parameters are: humidity – 54.12 %, temperature – 18.3°C. The air velocity for this period is variable.

The air speed in a room in Object №2 for the period from 20:10:02 h on 29.04.2017 to 08:11:20 h on 30.04.2017 (spring season) has an average value of 0.168 m/s. The average quantity of PM2,5 is 8.8 μg/m³ and the average amount of PM10 is 15.09 μg/m³. The average values of the room air parameters are: humidity – 58.2 %, temperature – 23.6°C. The average values of outdoor air parameters are: humidity – 67.4 %, temperature – 13.6°C. The air velocity for this period is variable.

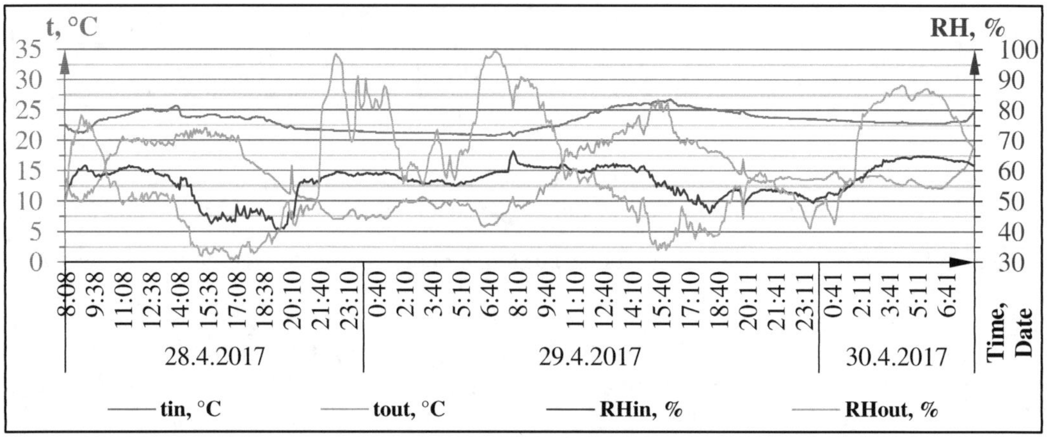

Figure 10. SPRING – Time distribution for 48h of temperature and relative humidity – Object 2.

4. Conclusions

The comparative analysis of the studies conducted shows that in both wooden houses the average quantity for PM2.5 is 21 μg/m³ and for PM10 is 38 μg/m³ is several times lower than the limit values according to the legislation of Bulgaria. Room air velocity is lower than the maximum acceptable for the minimum microclimate requirements of up to 0,187 m/s. The optimal air humidity is between 40 % and 60 %, and the maximum humidity is between 30 % and 75 %. There is a tendency for stable relative air humidity to be maintained in the investigated premises, irrespective of relative humidity fluctuations in the outside air, whereupon the wooden components are an "internal regulator" of relative humidity. The analysis of temperature dependences shows that the external temperature does not have a significant effect on the internal temperature, which proves the good insulation properties of wooden houses and their ability to maintain normal microclimate parameters. The analysis of results can be applied to all energy efficiency projects in regions with similar climatic conditions.

References

[1] Langley A 2014 Every building has a history *Heinemann Educational Books* p. 48

[2] Penkova N, Krumov K, Zashkova L and Kassabov I 2017 Heat Transfer and Climatic Loads at Insulating Glass Units in Window Systems *International Journal of Management and Applied Science* **5** pp. 22-28

[3] Valtchev G, Kalojanov N, Rasheva V, Minchev M and Tasheva S 2016 Analysis of results after implementation of energy saving measures in public buildings *Bulgarian Chemical Communications* **48** pp 283–289

[4] Nakaya T, Yamasaki M and Sasaki Y 2014 Influence of wall composition on thermal environment of wooden houses *Journal of Wood Science* **60** pp 117-126
[5] Beyer G 2012 Tackle climate change:build with wood *Working Group Sustainability* p 22
[6] Dominique G-M 2002 Sustainable Architecture and Urbanism: Concepts, Technologies, Examples *Springer Science & Business Media* p 64
[7] Glass S and TenWolde A 2007 Review of in-service moisture and temperature conditions in wood-frame buildings. General Technical Report FPL-GTR-174 (Forest Products Laboratory) p 53
[8] Patkova Z and Mastny P 2009 Microclimate of low-energy buildings in relation to applied heating systems International *Journal of Energy and Environment* vol 3 pp 85-93
[9] Polyakov M V and Polyakova AV 2012 Microclimate monitoring and conservation system for wooden architecture in Tomsk *XVIII Modern Technique and Technologies* p 30
[10] Joshi SM 2008 The sick building syndrome *Indian J Occup Environ Med* vol 12 pp 61-64
[11] Ordinance № РД-07-3 dated 18/07/2014 for the Minimum Microclimate Requirements
[12] Ordinance № 12 dated 15/07/2010 (promulgated in the State Gazette, issue 58/30th July 2010) for the Threshold Limit Values of Fine Particulate Matter
[13] Penkova N, Krumov K, Kassabov I and Zashkova L 2016 Solar energy gains and thermal loads at large scale transparent building envelopes in the presence of indoor solar ray tracing *Engineered Transparency Glass in Architecture and Structural Engineering* pp 373-381

Acknowledgments
The research received funding from the National Research Program "Low Carbon Energy for the Transport and Household (E+)" of the Bulgarian Ministry of Education and Science under grant agreement DO1-214/28.11.2018

The melting temperature of BCC perfect ternary alloy FeCrC under pressure

C D Tran[1], G Coman[2], H Q Nguyen[1], H T Nguyen[3] and V Q Dinh[1]

[1] Hanoi National University of Education, Hanoi, Vietnam
[2] "Dunarea de Jos" University of Galati, Romania
[3] University of Transport and Communications, Hanoi, Vietnam

Email: nthoaly@utc.edu.vn

Abstract. The analytic expressions of the mean nearest neighbour distance, the free energy, the absolute stabilizing temperature of alloy state and the melting temperature for substitution alloy AB with interstitial atom C and body-cantered cubic structure under pressure are derived by the statistical moment method. The theoretical results are applied to alloy FeCrC. Our calculated results are compared with other calculations and experimental data.

1. Introduction

Iron-based alloys are of fundamental importance in the development of science and technology due to their valuable mechanical and thermodynamic properties [1, 2]. Notably, the FeCrC alloy is a prominent physical system that attracts the attention of many researchers [3]. The FeCrC alloy is known as the basis for tool steels and stainless steels, so it is widely used in many industries. Besides, iron and its alloys are believed to be the main components of the Earth's core, so the information about their phase diagrams is of central importance for geophysics and planet physics [4].

Although the FeCrC alloy has been experimentally and theoretically investigated many times, our knowledge about its melting is still incomplete. On the theoretical side, Molecular Dynamic (MD) simulation and ab initio calculation are evaluated as two main methods to study the melting properties of the crystal. However, the increasing of alloying elements will lead to expensive computational cost and significant deviation from experiments in many cases. For example, in [5], the coexisting phases approach has been used to calculate melting phase diagrams of several Fe-C interaction potentials, such as Embedded Atom Method (EAM) potential [6, 7] and Analytic Bond Order (ABO) potential [8]. However, these obtained results are not consistent with Experimental Data (EXPT), and more work is required to solve this problem. To get a whole physical picture of the relationship between alloying elements and thermodynamic properties of the crystal, we often use the Calculation of Phase Diagram (CALPHAD) method [9]. However, using the CALPHAD method requires great support from experiments and the combination of many calculation software. Furthermore, the CALPHAD method mainly provides only melting information at zero pressure. As a result, building a general melting theory is a never-ending story but inspiring in modern physics.

For the reasons above, in this paper, we will determine the melting point of the FeCrC alloy with Body-Centered Cubic (BCC) structure by using the Statistical Moment Method (SMM). Nguyen Tang and Vu Van Hung introduced the SMM from the 1990s [10], and its melting theory is based on the condition of absolute stability for the crystalline state as

$$\left(\frac{\partial P}{\partial v}\right)_{T=T_S} = 0 \quad (1)$$

where P is the hydrostatic pressure, v is the atomic volume, and T_S is the critical temperature. The condition (1) corresponds to the infinite value of the isothermal compressibility, so the critical temperature T_S is just indeed a theoretical hypothesis. In the previous study of SMM theory [11, 12], the authors proved that the critical temperature T_S is close to the melting temperature T_m at the same physical conditions, so we can find T_m from T_s by using a simple correction. By following this path, we can easily obtain the melting curves of metals and binary alloys with only using the solid phase. It is the first time the SMM is expanded to investigate the melting of BCC perfect ternary alloy.

2. Method

The model of the ternary ABC alloy with BCC structure is shown in figure 1, where A is the main atom, B is the substitutional atom and C is the interstitial atom. For simplicity, we assume that the atom C only interacts with the main atoms A which are closest to it (A_1 and A_2). The analytic expressions for the cohesive energy u_{0X} and the alloying parameters k_X, ω_X, γ_{1X}, γ_{2X}, γ_X for the atom X (X = A, B, C, A_1, A_2) were determined in our previous works [13]. The atomic concentration c_X satisfy the following condition

$$c_A = 1 - c_B - 7c_C,\ c_{A_1} = 2c_C,\ c_{A_2} = 4c_C \quad (2)$$

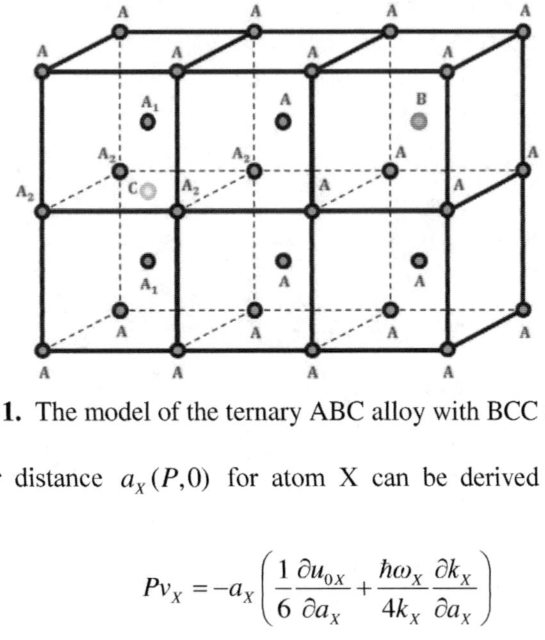

Figure 1. The model of the ternary ABC alloy with BCC structure.

The nearest neighbor distance $a_X(P,0)$ for atom X can be derived by solving the following equation

$$Pv_X = -a_X\left(\frac{1}{6}\frac{\partial u_{0X}}{\partial a_X} + \frac{\hbar\omega_X}{4k_X}\frac{\partial k_X}{\partial a_X}\right) \quad (3)$$

where v_X is the atomic volume. Then, the average atomic displacement $y_X(P,T)$ is given by

$$y_X(P,T) = \sqrt{\frac{2\gamma_X(P,0)\theta^2}{3k_X^3(P,0)}A_X(P,0)} \quad (4)$$

where $\theta = k_{Bo}T$, k_{Bo} is the Boltzmann constant and $A_X(P,0)$ was defined as in Ref. [14].

After that, the nearest neighbour distance $a_X(P,T)$ for atom X at pressure P and temperature T is determined by

$$a_A(P,T) = a_A(P,0) + y_A(P,T) \tag{5a}$$

$$a_B(P,T) = a_B(P,0) + y_B(P,T) \tag{5b}$$

$$a_C(P,T) = a_C(P,0) + y_{A_1}(P,T) \tag{5c}$$

$$a_{A_1}(P,T) \approx a_C(P,T) \tag{5d}$$

$$a_{A_2}(P,T) = a_{A_2}(P,0) + y_C(P,T) \tag{5e}$$

and we can calculate the mean nearest neighbour distance a_{ABC} between two main atoms A in the ternary ABC alloy as followsŞ

$$a_{ABC} = \frac{(1-c_B)B_{TAC}a_{AC} + c_B B_{TB} a_B}{(1-c_B)B_{TAC} + c_B B_{TB}} \tag{6a}$$

$$a_{AC}(P,T) = a_{AC}(P,0) + y_{AC}(P,T) \tag{6b}$$

$$a_{AC}(P,0) = (1-c_C)a_A(P,0) + c_C\sqrt{3}a_C(P,0) \tag{6c}$$

$$y_{AC}(P,T) = (1-7c_C)y_A(P,T) + c_C y_C(P,T) + 2c_C y_{A_1}(P,T) + 4c_C y_{A_2}(P,T) \tag{6d}$$

where B_{TAC} is the isothermal bulk modulus of the interstitial AC alloy and B_{TB} is the isothermal bulk modulus of the pure metal B [13].

The equation of state of the ternary ABC alloy with BCC structure has the form

$$P = -\frac{a_{ABC}}{6v_{ABC}}\sum_X c_X \frac{\partial u_{0X}}{\partial a_X} + \frac{3\gamma_G \theta}{v_{ABC}} \tag{7}$$

where γ_G is known as the Gruneisen parameter

$$\gamma_G = -\frac{a_{ABC}}{6}\sum_X \frac{c_X}{k_X}\frac{\partial k_X}{\partial a_X}\frac{\hbar\omega_X}{2\theta}\coth\frac{\hbar\omega_X}{2\theta} \tag{8}$$

By combining equations (1), (7), (8) we have

$$T_S = \frac{2Pv_{ABC} + \frac{a_{ABC}^2}{6}\sum_X c_X \frac{\partial^2 u_{0X}}{\partial a_X^2} - \frac{\hbar a_{ABC}^2}{4}\sum_X \frac{c_X\omega_X}{k_X}\left[\frac{1}{2k_X}\left(\frac{\partial k_X}{\partial a_X}\right)^2 - \frac{\partial^2 k_X}{\partial a_X^2}\right]}{\frac{k_{oB}a_{ABC}^2}{4}\sum_X \frac{c_X}{k_X^2}\left(\frac{\partial k_X}{\partial a_X}\right)^2} \tag{9}$$

Solving equation (9) will give us the value of the critical temperature T_S. Because T_S is close to T_m at the same physical condition, so we have $a_m = a_{ABC}(P,T_m) \approx a_S = a_{ABC}(P,T_S)$ and T_m can be expanded in term of $a_m - a_S$ as

$$T_m \approx T_S + \frac{a_m - a_S}{k_{Bo}\gamma_G(P,T_S)}\left\{\frac{Pv_{ABC}(P,T_S)}{a_S} + \sum_X \frac{c_X}{18}\left[\left(\frac{\partial u_{0X}}{\partial a_X}\right)_{T=T_S} + a_S\left(\frac{\partial^2 u_{0X}}{\partial a_X^2}\right)_{T=T_S}\right]\right\} \tag{10}$$

In 1939, Born [15] proposed that the melting initiates when the shear modulus vanishes and the crystal loses its ability to resist shear, so the melting curve of crystal under pressure can be given in the form [16]

$$T_m(P) = T_m(0)\frac{G(P)}{G(0)}\left(\frac{B_T(P)}{B_T(0)}\right)^{\frac{1}{b}} \tag{11}$$

where $G(P)$ and $G(0)$ are the shear moduli at pressure P and zero pressure, respectively, $B_T(0)$ is the isothermal bulk modulus at zero pressure, $b = \left(\frac{dB_T(P)}{dP}\right)_{P=0}$, $B_T(P)$ is the isothermal bulk modulus at pressure P.

3. Numerical results

For the Fe-Fe and Cr-Cr interactions, we use the Mie-Lennard-Jones pair potential as

$$\varphi = \frac{D}{n-m}\left[m\left(\frac{r_0}{r}\right)^n - n\left(\frac{r_0}{r}\right)^m\right] \tag{12}$$

where D is the dissociation energy, r_0 is the equilibrium value of r, m and n are taken empirically [17]. To describe the Fe-C interaction, we use the Finnis-Sinclair potential as follows

$$U = -A\sqrt{\sum_{i \neq j}\rho(r_{ij})} + \frac{1}{2}\sum_{i \neq j}\phi(r_{ij}) \tag{13}$$

where $\rho(r_{ij})$ is the embedded potential and $\phi(r_{ij})$ is the pair potential introduced in Ref. [6].

Firstly, we will check a limited case when $c_{Cr}, c_C \rightarrow 0$. Figure 2 shows the melting curve of the pure Fe metal obtained from the SMM correspond well with EXPT [18-21]. In the surveyed range, the differences are lower than 3 %. The melting temperature of the pure Fe metal is known as the increasing function of pressure. Specifically, when pressure increases from 0 to 5 GPa, the melting temperature of the pure Fe metal also increases from 1800 to 1990 K. The average melting slope $\frac{\Delta T_m}{\Delta P}$ is about 38 K.GPa^{-1} from the SMM and about 35 K.GPa^{-1} from EXPT [21].

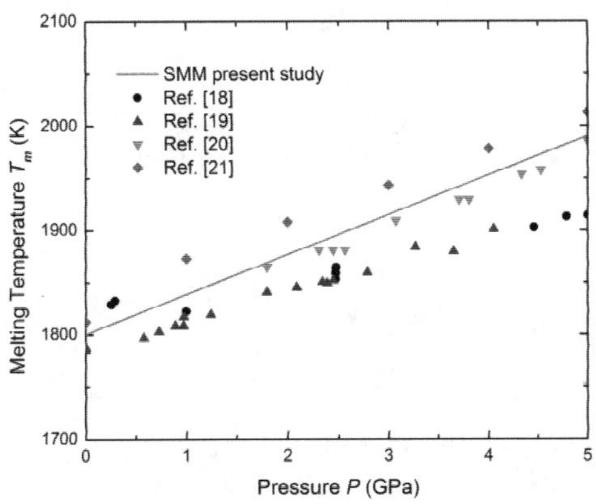

Figure 2. $T_m(P)$ for Fe obtained from the SMM and EXPT [18-21].

Figure 3 shows the effects of the interstitial atom C (figure 3a) and the substitutional atom Cr (figure 3b) on the melting temperature of Fe at zero pressure. We recognize that the melting temperature of the FeC alloy decreases rapidly with increasing atomic concentration of C. The average melting slope $\frac{\Delta T_m}{\Delta c_C}$ shown in figure 3a is about -125 K.(%)$^{-1}$ from the SMM and about -112.5 K.(%)$^{-1}$ from EXPT [22]. This value is rather high due to the substantial deformation of crystal structure when C is added into the lattice of the Fe metal [22]. Conversely, figure 3b shows that the melting temperature of the FeCr alloy does not change much when the atomic concentration of Cr ranges from 0 to 55%. It is clear to see that Fe and Cr approximately have the same atomic mass, the same interaction distance and the same dissociation energy [17], so the cohesive energy is almost unaffected by replacing Fe with Cr. However, the melting curve of the FeCr alloy at zero pressure has an extraordinary form. It has a minimum point at $c_{Cr} = 21\%$, and the value of the corresponding melting temperature is about 1769 K from the SMM, about 1782 K from the CALPHAD method [23] and about 1783 K from EXPT [24]. We notice that the deviation between EXPT [24] and the SMM's results for the melting temperature of the FeCr alloy becomes higher at Cr-rich side. The maximum error can be up to 10 % in the case of the pure Cr metal. The explanation for this result is that Cr melts at very high temperature, so the point defects including vacancies will have a significant influence on the physical properties of the crystal, while in this paper we only apply the SMM melting theory to the perfect model.

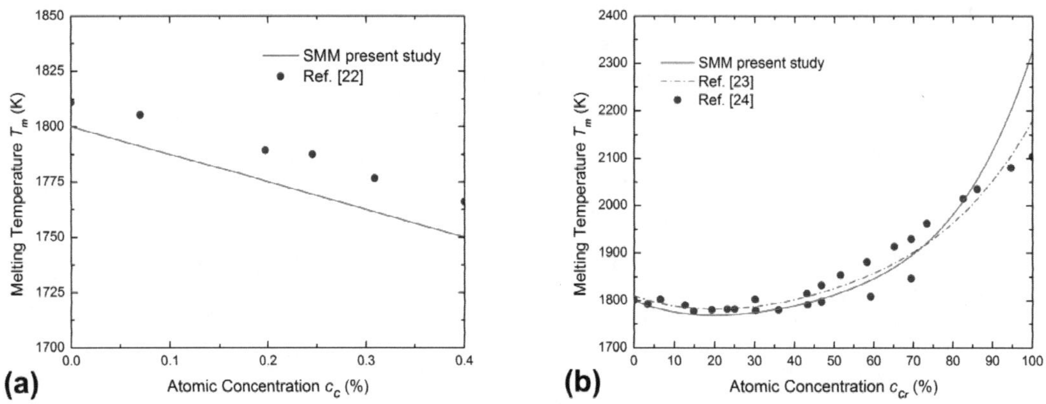

Figure 3. $T_m(c_C)$ for FeC (figure 3a) and $T_m(c_{Cr})$ for FeCr (figure 3b) obtained from the SMM are consistent with the CALPHAD's results [23] and EXPT [22, 24].

Next, we consider how the melting temperature of the FeCrC alloy depends on the weight percent of C at a few given weight percents of Cr (figure 4). At zero pressure, the maximum deviation between SMM's results and CALPHAD's results [9] for the melting temperature of the FeCrC alloy is lower than 1.5%. At $w_{Cr} = 13\%$ (figure 4a), the average melting slopes $\frac{\Delta T_m}{\Delta w_C}$ calculated by the SMM and the CALPHAD [9] are about -550 K.(%)$^{-1}$ and -515 K.(%)$^{-1}$, respectively. At $w_{Cr} = 17\%$ (figure 4b), the average melting slopes $\frac{\Delta T_m}{\Delta w_C}$ calculated by the SMM and the CALPHAD [9] method are about -620 K.(%)$^{-1}$ and -604 K.(%)$^{-1}$, respectively. Our numerical results are qualitatively and quantitatively consistent with CALPHAD's calculations [9]. However, the most exciting thing here is the expansion of maximum carbon solubility in the BCC Fe. In reality, when we substitute Cr for Fe, the crystal lattice is stretched a bit [1], so adding carbon to steel is easier to do.

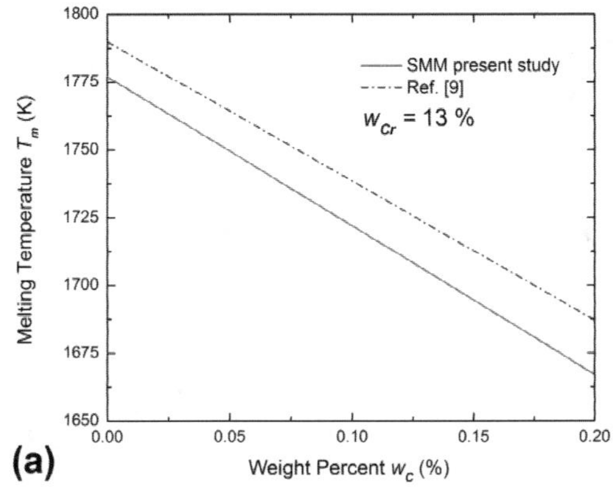

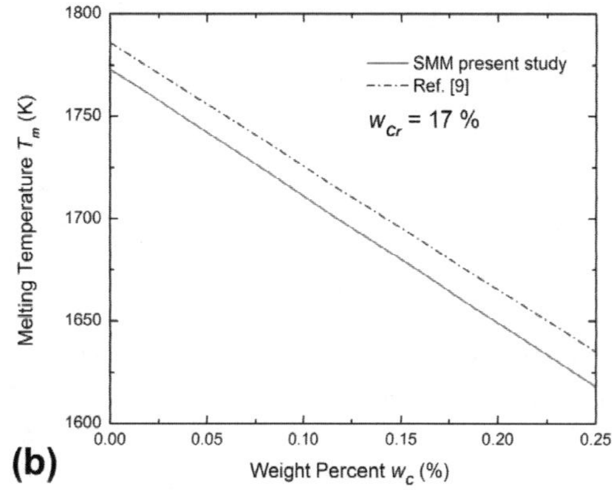

Figure 4. $T_m(w_C)$ for FeCrC at $w_{Cr} = 13\%$ (figure 4a) and $w_{Cr} = 17\%$ (figure 4b) obtained from the SMM correspond well with CALPHAD' results [9].

Similar to the main metal Fe, when pressure increases, the melting temperature of the FeCrC alloy also increases (figure 5). For example, in the range of pressure from 0 to 5 GPa, the melting temperature of $Fe_{82.95\%}Cr_{17\%}C_{0.05\%}$ increases from 1742 to 1926 K, the melting temperature of $Fe_{82.85\%}Cr_{17\%}C_{0.15\%}$ increases from 1680 to 1851 K, the melting temperature of $Fe_{82.75\%}Cr_{17\%}C_{0.25\%}$ increases from 1618 to 1776 K. Although the weight percent of C is very small, it has a significant contribution to the melting properties of the FeCrC system. It is an essential feature of the interstitial atoms that engineers need to pay close attention. Based on the presented results above, we believe that the melting curves of the FeCrC alloy shown in figure 5 have very high accuracy.

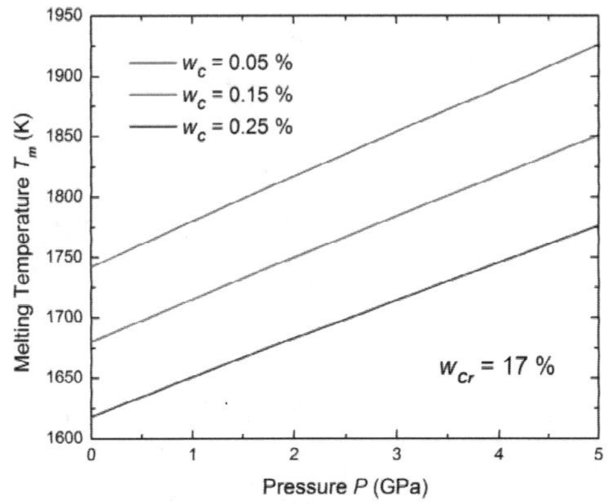

Figure 5. $T_m(P)$ of the FeCrC alloy obtained from the SMM..

4. Conclusion

By using the statistical moment method, we built the melting theory for the ternary ABC alloy with a body-centered cubic structure. The theoretical results have been performed for the FeCrC alloy in the interval of pressure from 0 to 5 GPa. Our numerical results correspond well with experiments and other calculations without large computational programs. These obtained results proved the effectiveness of the statistical moment method when studying on the melting properties of the crystal.

IOP Conf. Series: Materials Science and Engineering **595** (2019) 012018 doi:10.1088/1757-899X/595/1/012018

References

[1] Xiong W, Hedström P, Selleby M, Odqvist J, Thuvander M and Chen Q 2011 *Calphad* **35** 355
[2] Hadi I 2012 *J. Appl. Phys.* **112** 123106
[3] Andersson J O 1988 *Metall. Mater. Trans. A* **19** 627
[4] Lord O T, Walter M J, Dasgupta R, Walker D and Clark S M 2009 *Earth Planet. Sci. Lett.* **284** 157
[5] Melnykov M and Davidchack R L 2018 *Comput. Mater. Sci.* **144** 273
[6] Lau T T, Först C J, Lin X, Gale J D, Yip S and Van Vliet K J 2007 *Phys. Rev. Lett.* **98** 215501
[7] Henriksson K O and Nordlund K 2009 *Phys. Rev. B* **79** 144107.
[8] Hepburn D J and Ackland G J 2008 *Phys. Rev. B* **78** 165115.
[9] Khvan A V, Hallstedt B and Broeckmann C 2014 *Calphad* **46** 24.
[10] Tang N and Hung V V 1990 *Phys. Status Solidi (b)* **162** 379.
[11] Hung V V and Hai D T 2013 *Comput. Mater. Sci.* **79** 789.
[12] Hung V V and Thanh L D 2012 *Int. J. Mod. Phys. B* **26** 1250050.
[13] Tinh B D, Hoc N Q, Vinh D Q, Cuong T D and Hien N D 2018 *Adv. Mater. Sci. Eng.*
[14] Hieu H K 2014 *J. Appl. Phys.* **116** 163505
[15] Born M 1939 *J. Chem. Phys.* **7** 591
[16] Burakovsky L, Preston D L and Silbar R R 2000 *J. Appl. Phys.* **88** 6294
[17] Magomedov M N 1987 *J. Fiz. Khimic* **61** 1003
[18] Strong H M 1959 *J. Geophys. Res.* **64** 653
[19] Sterrett K F, Klement Jr W and Kennedy G C 1965 *J. Geophys. Res.* **70** 1979
[20] Strong H M, Tuft R E and Hanneman R E 1973 *Metal. Trans.* **4** 2657
[21] Guillermet A F and Gustafson P 1984 *High Temp.-High Press* **16** 591
[22] Okamoto H 1992 *J. Phase Equil.* **13** 543
[23] Jacob A, Povoden-Karadeniz E and Kozeschnik E 2018 *Calphad* **60** 16
[24] Adcock F 1931 *J. Iron Steel Inst* **124** 99

IOP Conf. Series: Materials Science and Engineering **595** (2019) 012019 doi:10.1088/1757-899X/595/1/012019

Thermodynamic properties of interstitial alloy FeC with BCC structure under pressure

H T Nguyen[1], C D Tran[2], G Coman[3], H Q Nguyen[2] and V Q Dinh[2]

[1] University of Transport and Communications, Hanoi, Vietnam
[2] Hanoi National University of Education, Hanoi, Vietnam
[3] "Dunarea de Jos" University of Galati, Galati, Romania

Email: nthoaly@utc.edu.vn

Abstract. The analytic expressions for the thermodynamic quantities such as the mean nearest neighbor distance, the free energy, the isothermal compressibility, the thermal expansion coefficient, the heat capacities at constant volume and at constant pressure of binary interstitial alloy with body-centered cubic (BCC) structure and the small concentration are derived by the statistical moment method. The theoretical results are applied to the interstitial FeC alloy in the interval of temperature from 50 to 1000K and in the interval of interstitial atom concentration from 0 to 5%. In special cases, we obtain the thermodynamic quantities of the main Fe metal. Our calculated results for the thermal expansion coefficient and the heat capacity under constant pressure of main metal Fe are in good agreement with experiments.

1. Introduction

Research on iron-based alloys is one of the topical issues for materials science and engineering. Especially, carbon is seen as the most importance alloying element in iron. The iron-carbon equilibrium phase diagram is experimentally and theoretically investigated more than any other binary systems [1]. The changes in the properties and microstructures at various temperatures shown in this diagram plays a crucial role in studying the behavior of a large variety of complex alloy steels. Many different potentials such as EAM [2, 3] and ABO [4] have been proposed to consider the physical characteristics of the iron-carbon system. Notably, in recent years, the high-pressure iron-carbon phase diagram is constructed to assess the role of carbon-rich phases during core crystallization in Earth [5].

In this paper, the high-pressure thermodynamic properties of the iron-carbon alloy (FeC) having body-centered cubic (BCC) structure is investigated thanks to the statistical moment method (SMM) [6]. By applying this method, we can easily consider the relationship between pressure, temperature, alloying element and thermodynamic properties of the FeC alloy.

2. Method

To analyze the physical properties of the FeC alloy, we use the interstitial alloy model illustrated in figure 1. In figure 1, A and B correspond to Fe and C, respectively. We assume that the atom B only interacts with the atoms A which closest to it (A_1 and A_2). If N_B atoms B are added into the crystal lattice of A, we will have $2N_B$ atoms A_1, $4N_B$ atoms A_2 and $(N-7N_B)$ atoms A, where N is the total atomic number of the system. The cohesive energy u_{0X}, the concentration of alloying element c_X and the alloying parameters k_X, ω_X, γ_{1X}, γ_{2X}, γ_X (X = A, B, A_1, A_2) were carefully defined in our previous works [7].

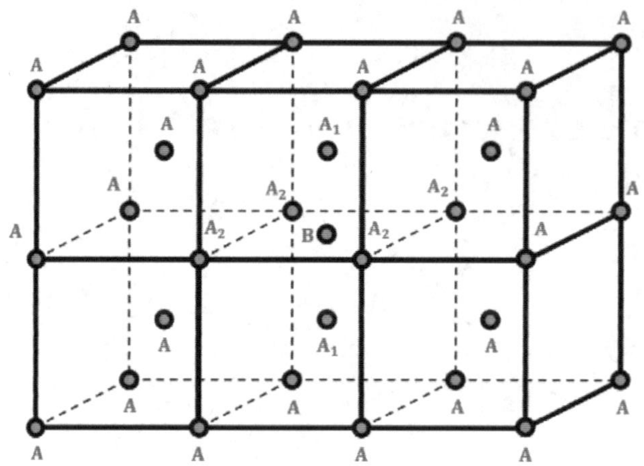

Figure 1. The model of the interstitial AB alloy with BCC structure.

The nearest neighbor distance $a_X(P,0)$ for atom X can be derived by solving the following equation

$$Pv_X = -a_X\left(\frac{1}{6}\frac{\partial u_{0X}}{\partial a_X} + \frac{\hbar\omega_X}{4k_X}\frac{\partial k_X}{\partial a_X}\right) \tag{1}$$

where v_X is the atomic volume. Then, the average atomic displacement $y_X(P,T)$ is given by

$$y_X(P,T) = \left(\frac{2\gamma_X(P,0)\theta^2}{3k_X^3(P,0)}A_X(P,0)\right)^{0.5} \tag{2}$$

where $\theta = k_{Bo}T$, k_{Bo} is the Boltzmann constant and $A_X(P,0)$ was defined as in [8]. After that, the nearest neighbour distance $a_X(P,T)$ for atom X at pressure P and temperature T is determined by

$$a_A(P,T) = a_A(P,0) + y_A(P,T) \tag{3a}$$

$$a_B(P,T) = a_B(P,0) + y_{A_1}(P,T) \tag{3b}$$

$$a_{A_1}(P,T) \approx a_B(P,T) \tag{3c}$$

$$a_{A_2}(P,T) = a_{A_2}(P,0) + y_B(P,T) \tag{3d}$$

Subsequently, the mean nearest neighbor distance in the interstitial AB alloy can be written as

$$a_{AB}(P,0) = a_{0AB} = (1-c_B)a_B(P,0) + c_B\sqrt{3}a_B(P,0) \tag{4a}$$

$$a_{AB}(P,T) = a_{AB}(P,0) + \sum_X c_X y_X(P,T) \tag{4b}$$

The isothermal compressibility of the interstitial AB alloy has the form

$$\chi_{TAB} = \frac{3\left(\frac{a_{AB}}{a_{0AB}}\right)^3}{2P + \frac{\sqrt{3}}{4a_{AB}}\frac{1}{N}\left(\frac{\partial^2\psi_{AB}}{\partial a_{AB}^2}\right)_T} \tag{5a}$$

$$\left(\frac{\partial^2\psi_{AB}}{\partial a_{AB}^2}\right)_T = \sum_X c_X\left(\frac{\partial^2\psi_X}{\partial a_X^2}\right)_T = \sum_X c_X\left\{\frac{N}{2}\frac{\partial^2 u_{0X}}{\partial a_X^2} + \frac{3N\hbar\omega_X}{4k_X}\left[\frac{\partial^2 k_X}{\partial a_X^2} - \frac{1}{2k_X}\left(\frac{\partial k_X}{\partial a_X}\right)^2\right]\right\} \tag{5b}$$

The thermal expansion coefficient of the interstitial AB alloy has the form

$$\alpha_{TAB} = -\frac{k_B \chi_{TAB}}{3}\left(\frac{a_{0AB}}{a_{AB}}\right)^2 \frac{a_{AB}}{3v_{AB}} \frac{\partial^2 \psi_{AB}}{\partial\theta \partial a_{AB}} \quad (6a)$$

$$\frac{\partial^2 \psi_{AB}}{\partial\theta \partial a_{AB}} = \sum_X c_X \frac{\partial^2 \psi_X}{\partial\theta \partial a_X} \quad (6b)$$

$$\frac{\partial^2 \psi_X}{\partial\theta \partial a_X} = \frac{3}{2k_X}\frac{\partial k_X}{\partial a_X}\frac{x_X^2}{\sinh^2 x_X} + \frac{6\theta^2}{k_X^2}\left[\frac{\gamma_{1X}}{3k_X}\frac{\partial k_X}{\partial a_X}\left(2 + \frac{x_X^3 \coth x_X}{\sinh^2 x_X}\right) - \right.$$

$$\left. -\frac{1}{6}\frac{\partial \gamma_{1X}}{\partial a_X}\left(4 + x_X \coth x_X + \frac{x_X^2}{\sinh^2 x_X}\right) - \left(\frac{2\gamma_{2X}}{k_X}\frac{\partial k_X}{\partial a_X} - \frac{\partial \gamma_{2X}}{\partial a_X}\right)\frac{x_X^3 \coth x_X}{\sinh^2 x_X}\right] \quad (6c)$$

The entropy of the interstitial AB alloy is determined by

$$S_{AB} = \sum_X c_X S_X \quad (7a)$$

$$S_X = S_{0X} + \frac{3Nk_B\theta}{k_X^2}\left[\frac{\gamma_{1X}}{3}\left(4 + x_X \coth x_X + \frac{x_X^2}{\sinh^2 x_X}\right) - 2\gamma_{2X}\frac{x_X^3 \coth x_X}{\sinh^2 x_X}\right] \quad (7b)$$

$$S_{0X} = 3Nk_B\left[x_X \coth x_X - \ln\left(2\sinh x_X\right)\right] \quad (7c)$$

The heat capacity at constant volumeof interstitial alloy AB is determined by

$$C_{VAB} = \sum_X c_X C_{VX} \quad (8a)$$

$$C_{VX} = 3Nk_B\left\{\frac{x_X^2}{\sinh^2 x_X} + \frac{2\theta}{k_X^2}\left[\left(2\gamma_2 + \frac{\gamma_{1X}}{3}\right)\frac{x_X^3 \coth x_X}{\sinh^2 x_X} + \frac{2\gamma_{1X}}{3} - \gamma_{2X}\left(\frac{x_X^4}{\sinh^4 x_X} + \frac{2x_X^4 \coth^2 x_X}{\sinh^2 x_X}\right)\right]\right\} \quad (8b)$$

The heat capacity at constant pressure of the interstitial AB alloy is determined by

$$C_{PAB} = C_{VAB} + \frac{9TV_{AB}\alpha_{TAB}^2}{\chi_{TAB}} \quad (9)$$

The adiabatic compressibility of the interstitial AB alloy has the form

$$\chi_{SAB} = \frac{C_{VAB}}{C_{PAB}}\chi_{TAB} \quad (10)$$

3. Numerical results

For the Fe-Fe interaction, we use the Mie-Lennard-Jones pair potential as

$$\varphi = \frac{D}{n-m}\left[m\left(\frac{r_0}{r}\right)^n - n\left(\frac{r_0}{r}\right)^m\right] \quad (11)$$

where D is the dissociation energy, r_0 is the equilibrium value of r, m and n are taken empirically [9].

To describe the Fe-C interaction, we use the Finnis-Sinclair potential as follows

$$U = -A\sqrt{\sum_{i\neq j}\rho(r_{ij})} + \frac{1}{2}\sum_{i\neq j}\Phi(r_{ij}) \quad (12a)$$

$$\rho(r) = t_1\left(r - R_1\right)^2 + t_2\left(r - R_1\right)^3 \quad \left(r < R_1\right) \quad (12b)$$

$$\Phi(r) = \left(r - R_2\right)^2\left(k_1 + k_2 r + k_3 r^2\right) \quad \left(r < R_2\right) \quad (12c)$$

where $\rho(r_{ij})$ is the embedded potential and $\Phi(r_{ij})$ is the pair potential. The Finis-Sinclair parameters were introduced in Ref. [2].

Our numerical results for the thermal expansion coefficient and the heat capacity at constant pressure of the FeC alloy are described by figures from figure 2 to figure 9.When the concentration we obtain thermodynamic quantities of the main Fe metal. Our calculated results illustrated in Figures 2, 3 and 7 are in relatively good agreement with experiments.

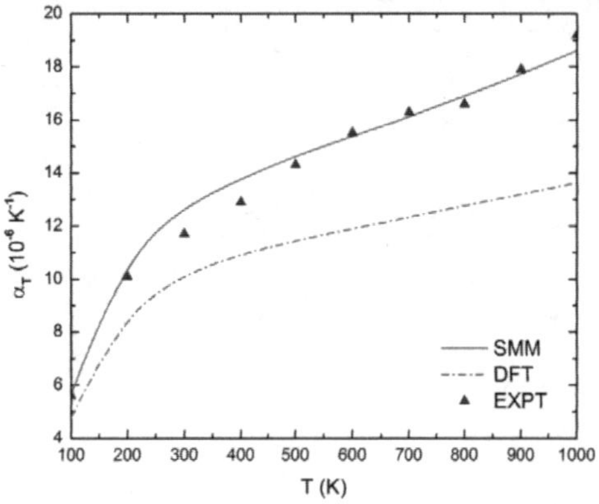

Figure 2. $\alpha_T(T)$ of Fe at zero pressure obtained from the SMM, DFT [10] and EXPT [11].

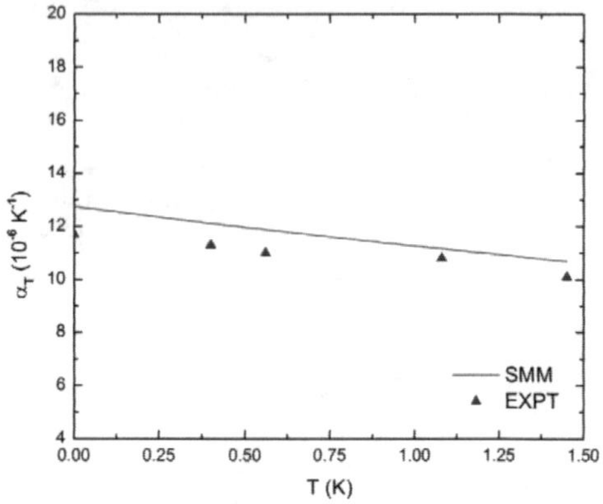

Figure 3. $\alpha_T(c_C)$ of FeC at 300 K and zero pressure obtained from the SMM and EXPT [12].

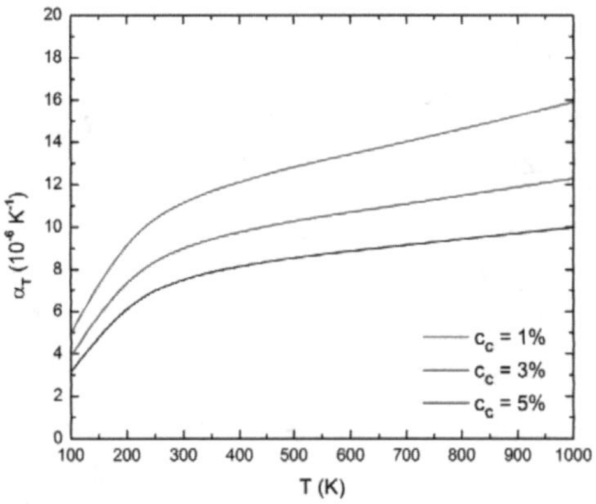

Figure 4. $\alpha_T(T)$ of FeC at c_C = 1, 3, 5 % and zero pressure obtained from the SMM.

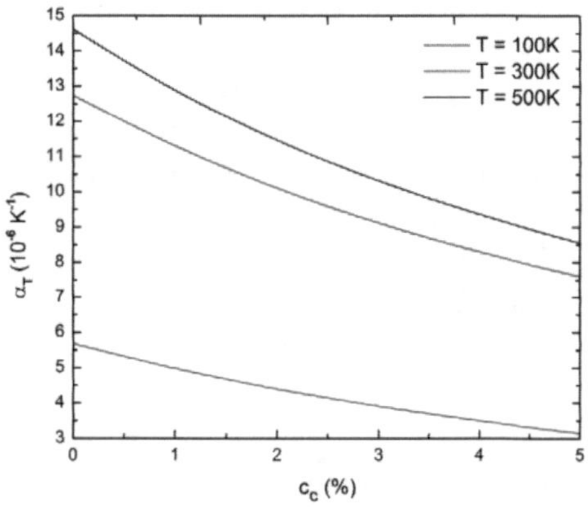

Figure 5. $\alpha_T(c_C)$ of FeC at T = 100, 300, 500 K and zero pressure obtained from the SMM.

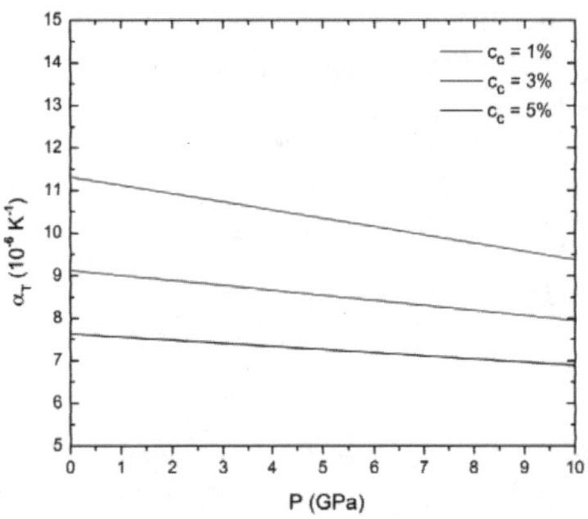

Figure 6. $\alpha_T(P)$ of FeC at T = 300 K and c_C = 1, 3, 5 % obtained from the SMM.

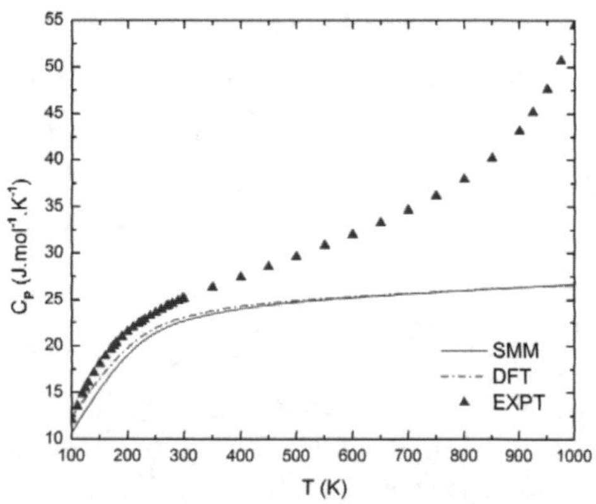

Figure 7. $C_P(T)$ of FeC at zero pressure obtained from the SMM, DFT [10] and EXPT

[13].

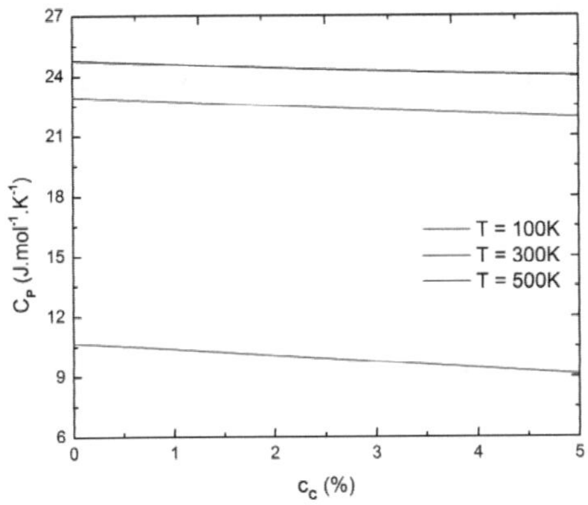

Figure 8. $C_P(c_C)$ of FeC at zero pressure and T = 100ş 300ş 500 K obtained from the SMM.

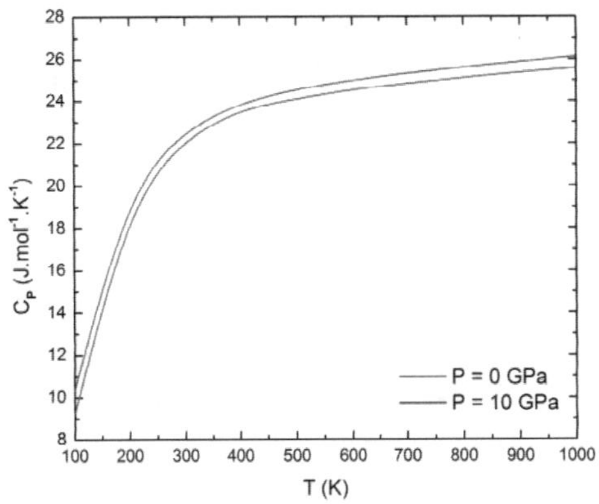

Figure 9. $C_P(T)$ of FeC at c_C = 1 % and P = 0, 10 GPa obtained from the SMM.

For the FeC alloy at the same temperature and pressure when the concentration of interstitial atoms increases, the thermal expansion coefficient α_T and the heat capacity at constant pressure C_P decrease. For example at T = 1000K, P = 0 when c_C increases from 0 to 5%, α_T decreases from 18.61.10^{-6} to 9.98.10^{-6} K^{-1}, C_P decreases from 26.68 to 25.07 J/ (mol.K).

For the FeC alloy at the same pressure and concentration of interstitial atoms when temperature increases, the thermal expansion coefficient α_T and the heat capacity at constant pressure C_P increase. For example at P = 0, c_C = 5% when T increases from 100 to 1000K, α_T increases from 3.14.10^{-6} to 9.98.10^{-6} K^{-1}, C_P increases from 9.13 to 25.07 J/ (mol.K).

For the FeC alloy at the same temperature and concentration of interstitial atoms when pressure increases, the thermal expansion coefficient α_T and the heat capacity at constant pressure C_P decreases. For example at T = 300K, c_C = 5% when P increases from 0 to 10 GPa, α_T decreases from 7.6.10^{-6} to 6.87.10^{-6} K^{-1}, C_P decreases from 21.92 J/ (mol.K) to 21.66 J/ (mol.K).

4. Conclusion

The statistical moment method is applied to investigate the high-pressure thermodynamic properties of the FeC alloy having a body-centered cubic structure. The analytic expressions for thermodynamic quantities of the FeC alloy are given in the explicit forms. Our numerical results in the limited cases are consistent with experimental data and DFT calculations. These obtained results prove the effectiveness of the statistical moment method in prediction high-pressure thermodynamic characteristics of the complex iron-based alloys.

References

[1] Okamoto H 1992 *J. Phase Equil.* **13** 543
[2] Lau T T, Först C J, Lin X, Gale J D, Yip S and Van Vliet K J 2007 *Phys. Rev. Lett.* **98** 215501
[3] Henriksson K O and Nordlund K 2009 *Phys. Rev. B* **79** 144107
[4] Hepburn D J and Ackland G J 2008 *Phys. Rev. B* **78** 165115
[5] Lord O T, Walter M J, Dasgupta R, Walker D and Clark S M 2009 *Earth Planet. Sci. Lett.* **284** 157
[6] Tang N and Hung V V 1990 *Phys. Status Solidi (b)* **162** 379
[7] Tinh B D, Hoc N Q, Vinh D Q, Cuong T D and Hien N D 2018 *Adv. Mater. Sci. Eng.* **2018**
[8] Hieu H K 2014 *J. Appl. Phys.* **116** 163505

[9] Magomedov M N 1987 *J. Fiz. Khimic* **61** 1003
[10] Dragoni D, Ceresoli D and Marzari N 2015 *Phys. Rev. B* **91** 104105
[11] Gray D E 1972 *American Institute of Physics handbook* (New York, Toronto, London: McGrawHill Companies)
[12] Bauccio M 1993 *ASM Metals Reference Book* (Ohio: Materials Park - ASM International)
[13] Desai P D 1986 *J. Phys. Chem. Ref. Data* **15** 967

IOP Conf. Series: Materials Science and Engineering **595** (2019) 012020 doi:10.1088/1757-899X/595/1/012020

Heat exchanger design based on minimum entropy generation

S Bucsa[1], D Dima[1], A Serban[1], M-F Stefanescu[1], V Popa[2] and A Dobrovicescu[1]

[1]Faculty of Mechanical Engineering and Mechatronics, University Politehnica of Bucharest, Romania
[2]Faculty of Engineering, "Dunarea de Jos" University of Galati, Romania

E-mail: adobrovicescu@yahoo.com

Abstract. The paper deals with the rating of the performance of heat exchangers. Each attempt in augmenting the overall heat transfer coefficient is accompanied by an increase in the pressure drop. Based on this observation the paper points out that both processes – heat transfer across a finite temperature difference and friction have a common destructive effect that can be measured by the entropy generation. The paper discusses the possibility of designing or simulating the behaviour of an existing heat exchanger based on the number of entropy generation unit NSU. The correlation between the thermal efficiency of the heat exchanger and its constructive characteristics is based on the NTU-ε method. The paper develops a correlation for calculating the entropy generation unit NSU function on the NTU number. The paper presents diagrams and methods for designing a heat exchanger for a specific number NSU or for simulating an existing design or physical heat exchanger.

1. Introduction

Any attempt in augmenting the heat transfer performance of a heat exchanger is accompanied by an increase in the pressure drop.

For a given thermal load and a specific heat transfer area

$$\dot{Q} = U \cdot A \cdot \Delta t_{ML} \tag{1}$$

The increase in the overall heat transfer coefficient leads to a decrease in the temperature difference between the hot and cold currents but in the same time the friction factor increases.

Apparently the temperature difference seems to be a thermal decisional parameter while the pressure drop has a mechanical effect. In fact both heat transfer across a finite temperature difference and pressure drop (friction) are irreversible processes and have a common effect – they destroy (consume) mechanical work.

The magnitude of the entropy generation that accompanies these irreversible processes can be used as a measure of the heat exchanger performance. The less is the entropy generation the more performing is the heat exchanger.

Scientists and engineers are preoccupied to design more performing heat exchangers but this overall performance is sometimes difficult to define.

Uttam Roy and Mrinmoy Majumder [1] used a Neural Network to evaluate and optimize the performance of a shell and tube heat exchanger looking for the maximization of the exergetic plant efficiency and to minimize the cost of the heat exchanger.

Bahmanyar and Talebi [2] used the entropy generation in the analysis of a shell and tube steam generator. They proposed a number for the generation of entropy to analyze the effect of the irreversibilities connected to heat transfer, friction and mixing. The conclusion of the work is that the decrease in the entropy generation of the heat exchanger leads to a higher performance for the overall system.

Deng Shengxiang et al. [3] enhanced the heat transfer by embedding fins in a phase change material. They constructed a numerical model based on the method of finite volume to estimate the efficiency of the heat exchanger for different fins arrangements.

Hara Shumpei et al. [4] performed an exergy analysis where water in which a surfactant is added is used as cooling medium.

An analysis of the techniques used for heat transfer enhancement in plate heat exchangers is presented by Zhang Ji et al. [5].

Navid Mahdavi and Shahram Khalilarye [6] introduced a new method for determining the pinch temperature of heat exchangers for refrigeration purposes.

In all these papers after a comprehensive analysis of the techniques of estimating the performance of heat exchangers the main interest is focused on defining the objective function for the extreme (optimum) search. The minimization of the overall entropy generation in the heat exchanger represents a valuable concept and strategy for finding the best heat transfer configuration.

2. Useful energy destruction in a heat exchanger

For a system in stationary regime the destruction of useful energy (exergy) is due to the irreversibility of the working processes:

$$\dot{I} = \dot{W}_{reversible} - \dot{W}_{real} \tag{2}$$

For power systems $\dot{I}$ represents the amount of power that cannot be produced due to irreversibility.

For refrigeration systems, $\dot{I}$ is the additional power supplementary to the minimum ideally required to accomplish the cooling task.

Quantitatively the useful energy destruction (exergy destruction) is equal to the product between the absolute temperature of the environment T_0 and the system overall entropy generation [7]

$$\dot{I} = T_0 \sum \dot{S}_{gen} \tag{3}$$

By estimating the entropy generation in a specific component of the system, i.e. a heat exchanger, one can determine the degree in which this component contributes to the destruction of the system useful energy (exergy).

The analyses focuses on the recuperative type gas to gas heat exchanger used in cryogenic refrigeration or liquefaction systems.

2.1. Thermal efficiency of a heat exchanger

The temperature – heat transfer surface diagram of the gas to gas heat exchanger is presented in figure1.

The thermal efficiency is defined as the ratio between the actual thermal load $\dot{Q}$ and the maximum possible one, $\dot{Q}_{max}$.

$$\varepsilon = \frac{\dot{Q}}{\dot{Q}_{max}} = \frac{C_h\left(t_i^h - t_o^h\right)}{C_{min}\left(t_i^h - t_i^c\right)} = \frac{C_c\left(t_o^c - t_i^c\right)}{C_{min}\left(t_i^h - t_i^c\right)} \tag{4}$$

where C_{min} is the lowest heat capacity between the hot stream and the cold one.

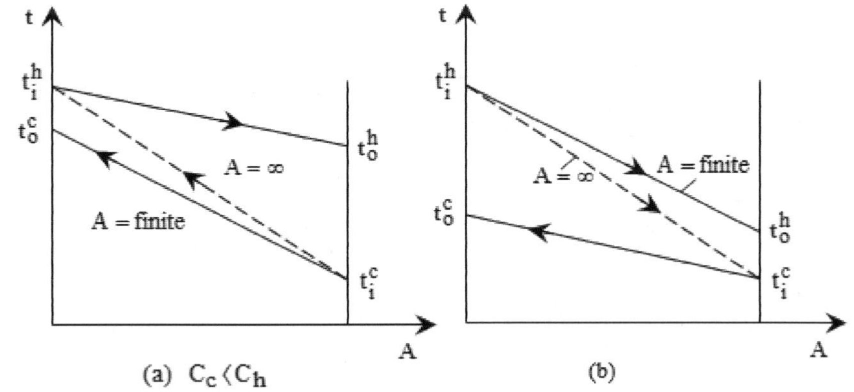

Figure 1. Variation of temperature across the heat transfer area. (a) $C_c \langle C_h$; (b) $C_h \langle C_c$

To respect the energy balance equation, the fluid experiencing the largest temperature difference should have the minimum thermal capacity. The maximum thermal load $\dot{Q}_{max}$ could be obtained through an infinite heat transfer area.

Depending on the value of C_{min}, one gets:

$$\varepsilon = \frac{t_i^h - t_o^h}{t_i^h - t_i^c}, \quad (C_{min} = C_h) \tag{5}$$

$$\varepsilon = \frac{t_o^c - t_i^c}{t_i^h - t_i^c}, \quad (C_{min} = C_c) \tag{6}$$

2.2. Number of thermal units (NTU)

By definition [8] the number of thermal units is calculated in the following way:

$$NTU = \frac{U \cdot A}{C_{min}} = \frac{1}{C_{min}} \int_0^A U(A)dA \tag{7}$$

Let us consider a counterflow heat exchanger (Figure 2).

Accounting for Eq. (3) the efficiency of the heat exchanger can be written such that:

$$\varepsilon = \frac{t_i^h - t_o^h}{\Delta t_0} = \frac{C_c}{C_h} \cdot \frac{t_o^c - t_i^c}{\Delta t_0} \tag{8}$$

where $\Delta t_0 = t_i^h - t_i^c$

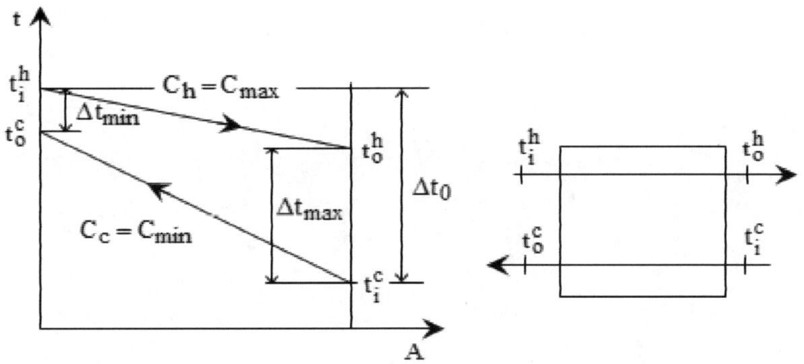

Figure 2. Counterflow heat exchanger. (a) Pipe diagram; (b) t-A diagram

If the inlet temperatures t_i^h and t_i^c are known and one considers that $C_{min} = C_c$ and consequently, $C_{max} = C_h$, equation (8) becomes:

$$\varepsilon = \frac{t_i^h - t_o^h}{\Delta t_0} = \frac{C_{min}}{C_{max}} \cdot \frac{t_o^c - t_i^c}{\Delta t_0} \tag{9}$$

From Figure 2 it follows that:

$$\Delta t_{max} = t_o^h - t_i^c = -\left(t_i^h - t_o^h\right) + \left(t_o^c - t_i^c\right) + \left(t_i^h - t_o^c\right) = -\varepsilon \frac{C_{min}}{C_{max}} \Delta t_0 + \varepsilon \Delta t_0 + \left(\Delta t_0 - \left(t_o^c - t_i^c\right)\right) =$$

$$= \left(1 - \varepsilon \frac{C_{min}}{C_{max}}\right) \Delta t_0 \tag{10}$$

$$\Delta t_{min} = t_i^h - t_o^c = \Delta t_0 - \left(t_o^c - t_i^c\right) = \Delta t_0 - \varepsilon \cdot \Delta t_0 = (1 - \varepsilon) \Delta t_0 \tag{11}$$

Accounting for Eqns. (10) and (11) one gets:

$$\Delta t_{mL} = \frac{\Delta t_{max} - \Delta t_{min}}{\ln \frac{\Delta t_{max}}{\Delta t_{min}}} = \frac{\Delta t_0 \cdot \varepsilon \left(1 - \frac{C_{min}}{C_{max}}\right)}{\ln \frac{1 - \varepsilon \frac{C_{min}}{C_{max}}}{1 - \varepsilon}} \tag{12}$$

Observing that for $C_{min} = C_c$

$$NTU = \frac{U \cdot A}{Cc} \tag{13}$$

$$\dot{Q} = C_c \left(t_o^c - t_i^c\right) = U \cdot A \cdot \Delta t_{mL} \tag{14}$$

the NTU relationship becomes:

$$NTU = \frac{t_o^c - t_i^c}{\Delta t_{mL}} \tag{15}$$

Equation (8) gives:

$$t_o^c - t_i^c = \varepsilon \cdot \Delta t_0 \tag{16}$$

By replacing Eqs. (16) and (12) into Eq. (15) gives:

$$NTU = \frac{\ln \frac{1 - \varepsilon \frac{C_{min}}{C_{max}}}{1 - \varepsilon}}{1 - \frac{C_{min}}{C_{max}}} \tag{17}$$

It follows that:

$$NTU\left(1 - \frac{C_{min}}{C_{max}}\right) = \ln \frac{1 - \varepsilon \frac{C_{min}}{C_{max}}}{1 - \varepsilon} \tag{18}$$

$$e^{NTU\left(1-\frac{C_{min}}{C_{max}}\right)} = \frac{1-\varepsilon\dfrac{C_{min}}{C_{max}}}{1-\varepsilon} \tag{19}$$

By solving Eq. (18) for ε one gets:

$$\varepsilon = \frac{\exp\left[NTU\left(1-\dfrac{C_{min}}{C_{max}}\right)\right]-1}{\exp\left[NTU\left(1-\dfrac{C_{min}}{C_{max}}\right)\right]-\dfrac{C_{min}}{C_{max}}} \tag{20}$$

2.3. Entropy generation in the heat exchanger

For the control volume of the heat exchanger (Figure 3) the entropy balance equation becomes:

$$\dot{S}_{gen} = \sum \dot{S}_o - \sum \dot{S}_i - \frac{\sum \dot{Q}}{T} \tag{21}$$

Considering that the heat exchanger is very well insulated $(\dot{Q}=0)$, Eq. (21) becomes:

$$\dot{S}_{gen} = \dot{m}_h (s_o - s_i)_h + \dot{m}_c (s_o - s_i)_c \tag{22}$$

The cryogenic recuperative heat exchanger is of a gas to gas type.

Accounting for the entropy variation relationship for an ideal gas, Eq. (22) becomes:

$$\dot{S}_{gen} = C_{max}\left(\ln\frac{T_o^h}{T_i^h} - \frac{R}{c_p^h}\ln\frac{p_o^h}{p_i^h}\right) + C_{min}\left(\ln\frac{T_o^c}{T_i^c} - \frac{R}{c_p^c}\ln\frac{p_o^c}{p_i^c}\right) \tag{23}$$

The mathematical model can be completed with the energy balance equation and the heat exchanger efficiency (Eq. 5)

$$C_{min}\left(T_o^c - T_i^c\right) + C_{max}\left(T_o^h - T_i^h\right) = 0 \tag{24}$$

$$\varepsilon = \frac{T_o^c - T_i^c}{T_i^h - T_i^c} \tag{5}$$

The outlet temperatures T_o^h and T_o^c can be calculated from Eqns. (5) and (24) function of the given inlet temperatures:

$$T_o^c = T_i^c + \varepsilon\left(T_i^h - T_i^c\right) \tag{25}$$

$$T_o^h = T_i^h - \frac{C_{min}}{C_{max}}\varepsilon\left(T_i^h - T_i^c\right) \tag{26}$$

Accounting for Eqs. (23), (25) and (26) and denoting by $NSU = \dfrac{\dot{S}_{gen}}{C_{max}}$ [9] the number of entropy generation units becomes:

$$NSU = \frac{\dot{S}_{gen}}{C_{max}} = \frac{C_{min}}{C_{max}}\left[\ln\left[1+\varepsilon\left(\frac{T_i^h}{T_i^c}-1\right)\right] - \left(\frac{R}{c_p}\right)_c \ln\left(1-\frac{\Delta p_c}{p_i^c}\right)\right] +$$
$$+\ln\left[1-\frac{C_{min}}{C_{max}}\varepsilon\left(1-\frac{T_i^c}{T_i^h}\right)\right] - \left(\frac{R}{c_p}\right)_h \ln\left(1-\frac{\Delta p_h}{p_i^h}\right) \tag{27}$$

IOP Conf. Series: Materials Science and Engineering **595** (2019) 012020 doi:10.1088/1757-899X/595/1/012020

Based on Eqs. (7), (20) and (27) one can design a heat exchanger for a specific number of entropy generation units.

3. Design a heat exchanger based on the NSU

Figures (3) and (4) show the variation of the Number of Entropy Generation Units (NSU) (Figure 3) and Number of Thermal Units (NTU) (Figure 4) with the heat exchanger efficiency.

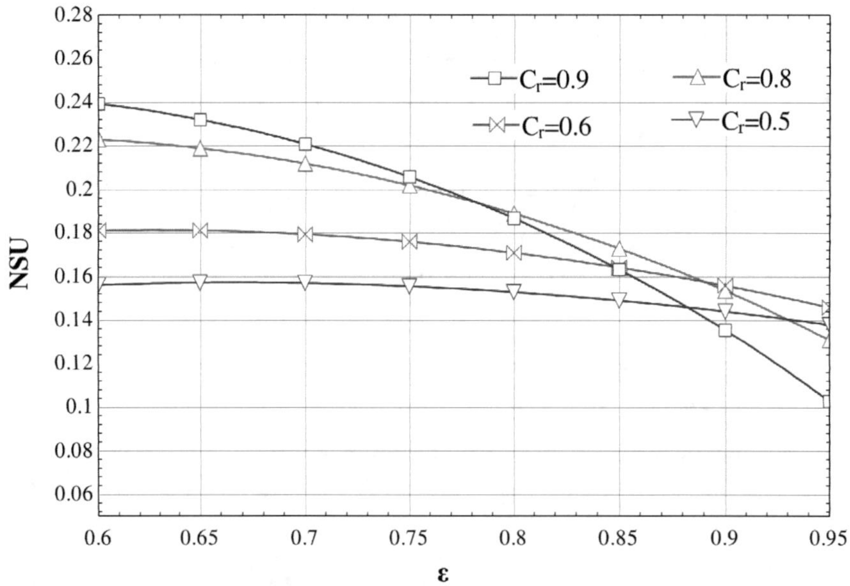

Figure 3. Variation of the NSU number with the heat exchanger efficiency ε

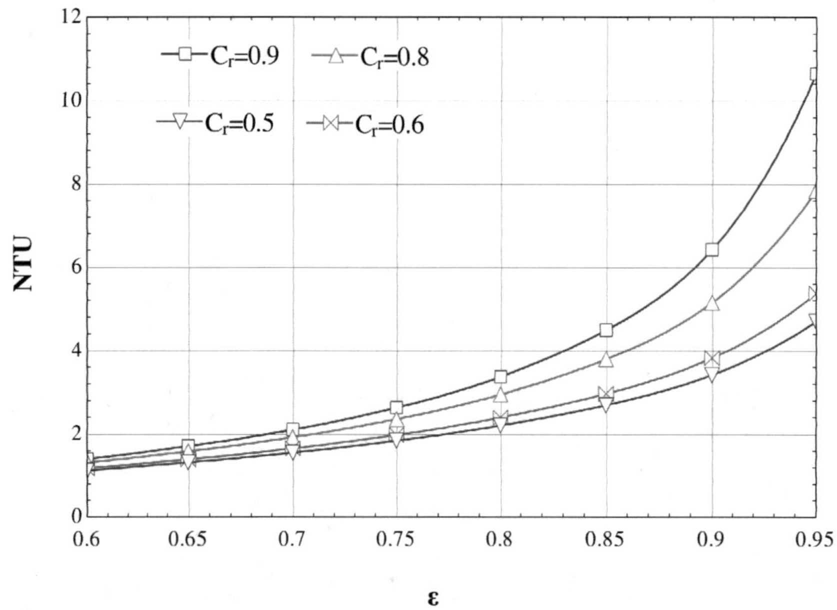

Figure 4. Variation of the NTU number with the heat exchanger efficiency ε

IOP Conf. Series: Materials Science and Engineering **595** (2019) 012020 doi:10.1088/1757-899X/595/1/012020

One can observe that:

$$NSU = \frac{\dot{S}_{gen}}{C_{\max}} = \frac{T_0 \cdot \dot{S}_{gen}}{T_0 \cdot C_{\max}} = \frac{\dot{I}}{T_0 \cdot C_{\max}} = NEU \tag{28}$$

where $\dot{I} = T_0 \cdot \dot{S}_{gen}$ is the exergy destruction caused by the irreversible processes in the heat exchanger.

Equation (28) defines the Number of Exergy Destruction Units (NEU) which is equal to the Number of Entropy Generation Units (NSU).

The operating and constructive optimum design of a thermal or cryogenic system may state a specific exergy destruction coefficient for a heat exchanger

$$\Psi_{HX} = \frac{\dot{I}_{HX}}{Fuel} 100 \tag{29}$$

where the Fuel represents the system exergy input.

Based on equations (28) and (29) one can calculate the NSU that corresponds to the recommended exergy destruction in the specified heat exchanger.

From figure (3), for a specific C_r, one can determine the efficiency ε for a desired NSU and use it in figure (4) to determine the NTU number and finally the necessary heat transfer area required to reach the specified magnitude of exergy destruction.

4. Conclusions

For rating both irreversible processes of heat transfer across a finite temperature difference and pressure drop due to friction, the Entropy Generation (Exergy Destruction) number is defined and used.

A method of designing a heat exchanger based on a stated NSU is presented.

The design method is based on the NTU-ε and NSU-ε correlations.

Nomenclature

A heat transfer area [m^2]
C caloric capacity [J/kgK]
Δt temperature difference [K]
p pressure [bar]
T temperature [K]
t temperature [^oC]
U overall heat transfer coefficient [W/m^2K]

Indices

HX heat exchanger
gen generated
i inlet
min minimum
max maximum
o outlet
0 equilibrium with the environment

Superscripts

c cold
h hot

Abbreviations

NTU Number of Thermal Units
NSU Number of Entropy Generation Units

NEU Number of Exergy Destruction Units

References

[1] Roy U and Majumder M 2019 Evaluating heat transfer analysis in heat exchanger using NN with IGWO algorithm *Vacuum* **161** pp 186-193
[2] Bahmanyar M E and Talebi S 2019 A performance analysis of vertical steam generator using an entropy generation method *Annal of Nuclear Energy* **125** pp 212 – 221
[3] Deng S X, Nie C, Wei G Y and Ye W B 2019 Improving the melting performance of a horizontal shell-tube latent-heat thermal energy storage unit using local enhanced finned tube *Energy and Buildings* **183** pp 161-173
[4] Hara S, Maxson A J and Kawaguchi Y 2019 Exergy transfer characteristics analysis of turbulent heat transfer enhancement in surfactant solution *International Journal of Heat and Mass Transfer* **130** pp 545 – 554
[5] Zhang J, Zhu X, Mondejar M and Haglind F 2019 A review of heat transfer enhancement techniques in plate heat exchangers, Renewable and Sustainable Energy Reviews *Elsevier* **101** (C) pp 305-328
[6] Mandavi N and Khalilarya 2019 S A novel method for assessment of pinch point characteristics of heat exchangers with different combinations of pure fluid and binary mixture as hot and cold streams *Energy Conversion and Management* **181** pp 342 – 358
[7] Moran M J, Shapiro H N, Boettner D D and Bailey M B 2018 Fundamentals of Engineering Thermodynamics (9th Edition, Wiley)
[8] Kays W M and London A L 1982 Compact Heat Exchangers (3rd Ed. Krieger Pub. Co. 1998)
[9] Bejan A 1982 Entropy Generation Through Heat and Fluid Flow (Wiley)

IOP Conf. Series: Materials Science and Engineering **595** (2019) 012021 doi:10.1088/1757-899X/595/1/012021

Analysis of a high-pressure screw compressor performances

C N Eparu[1], A Neacşa[2], A P Prundurel[1], R Rădulescu[1], C Slujitoru[3], N Toma[3] and M Niţulescu[3]

[1] Petroleum-Gas University of Ploiesti, Faculty of Oil and Gas Engineering, Drilling, Extraction and Transportation of Hydrocarbons Department, Romania
[2] Petroleum-Gas University of Ploiesti, Faculty of Mechanical Engineering, Mechanical Engineering Department, Romania
[3] National Turbines Development Research Institute (INCDT) - COMOTI Bucharest, Screw Compressors Assembly Department, Romania

E-mail: cristian.eparu@gmail.com

Abstract. Oil flooded screw compressors with discharge pressures of up to 26 bar have been available, commercially for some years. However, demands for higher pressure operation led to the design, construction and testing of a compressor operating with discharge pressures of up to 45 bar. To maximize the compressor efficiency, extensive performance simulations were carried out. Since 1980 oil flooded screw compressors have been used in many process gas compression applications and in the last year the higher pressure and larger capacity was realized. From 2010 to 2015 INCDT COMOTI manufactured screw compressors under GHH-Rand license (CU type, with max. delivery pressure 26 bar). Experience gain in this process gives us the power to develop the new type of screw compressors - based on license - with high delivery pressure - max. 45 bar. This paper will present details about the tests with a new high pressure screw compressor test in a closed loop test configuration. The machine was then tested in a closed loop test rig and, as is shown, the measured performance agrees well with that predicted by the numerical model. The parameters will be analysed starting from CFD modelling of the process and finally comparative analysis with actual parameters. The demonstrated advantages of this type of compressor, like high reliability, simple foundations, low operational cost, low initial cost, suitability for process fluctuation - gas pressure, gas composition - led to a significant demand for such compressors.

1. Introduction

In development strategy of INCDT-COMOTI, from 2016 to 2020, an important achievement was the development of a new range of screw compressors with high discharge pressures, up to 80 bar [1], for natural gas, in collaboration with the company GHH-Rand, for both internal market and export.

The good working relationships we developed with the GHH-RAND company ended by allowing us to manufacture under license - since April 2010 - the CU screw compressors family, also the replays from our potential customers require the need to design a new family of screw compressors, with oil injection, capable of developing a maximum discharge pressure of 45 bar [2, 3]. So, we develop the screw compressor type CU128 GM, with nominal suction pressure 4.5 bar, nominal discharge pressure 45 bar, and maximum compressing power 250 kW.

The paper is logically divided, following the developing stages for a new product, in our case a screw compressor with maximum discharge pressure of 45 bar (this compressor represents a development project of the licensed compressor type CU128G manufactured by GHH-Rand, for which the maximum discharge pressure is 26 bar). Thus, are drafted three main sections:

- Screw compressor flow simulation with CFD method [4];
- Configuration of test stand for screws compressors with maximum discharge pressure of 45 bar, suction pressure 4.5 bar, stand achieved in close-loop version [5];
- Testing of the compressor CU128 GM type on INCDT-COMOTI test stand. Data acquisition. Comparing theoretical parameters (computed in the first section of this paper) with the measured parameters. Conclusions.

2. Screw compressor flow simulation with CFD method

The CFD techniques in engineering domains became a usual approach in the last time. In our study we use a commercial CFD code, Ansys CFX, to predict the efficiency of a screw compressor [4, 6]. For this numerical investigation a CU type screw compressor class was studied [7, 8]. The fluid domain was divided in three subdomains, one rotating for the male and female rotors and two stators for suction side and pressure side. It is well known that the biggest challenge of a volumetric machine with positive displacement numerical analysis is represented by the rotors domains mesh generation. The rotors mesh was done with the TwinMesh software [4]. In the figure 1 is represented a 2D mesh section of our case. The URANS method with SST turbulence model with the BC (figure 2) was used to simulate the flow through the screw compressor [9, 10].

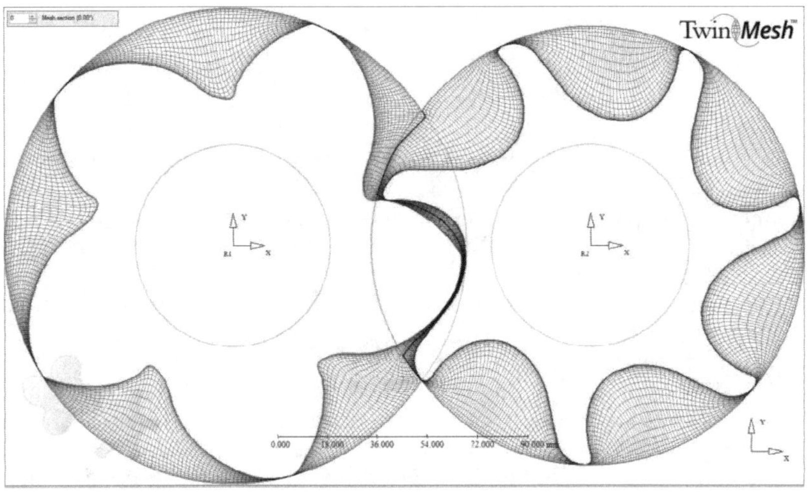

Figure 1. 2D cross section mesh.

The CFD techniques in engineering domains became common. In our study we use a commercial CFD code, Ansys CFX, to predict the efficiency of a screw compressor. For this numerical investigation a CU screw compressor class was studied, 5 lobes for the male (300° wrap angle) and 7 lobes for the female rotor. The rotors have a length of 266.7 mm with axis distance of 128 mm and rotation speed of 4500 rev/min of the male. The gap between the rotors is 42μm, between rotors and casing the gap is 65μm.

The fluid domain was divided in tree subdomains, one rotating for the male and female rotors and two stators for suction side and pressure side. It is well known that the biggest challenge of a volumetric machine with positive displacement numerical analysis is represented by the rotors domains mesh generation. The rotors mesh was done with the TwinMesh software. A good reference of the TwinMesh application on a screw compressor can be seen in A. Spille-Kohoff.et all paper [6].

In the figure 1 is represented a 2D mesh section of our case. In the rotors-casing clearance we have 20 elements and between rotors we have 40. The number of elements for the case, rotors domains and stator parts, is over 4.5 mil elements.

As results the velocity magnitude, torque, mass flow and absolute pressure variation by rotors position resulting from the numerical simulations are presented.

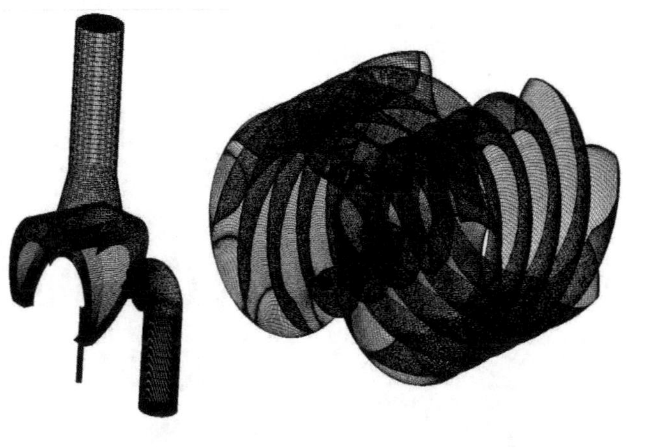

a) b)

Figure 2. The 3D mesh for:a) stationary fluid domains (1.9 elements); b) rotating fluid domains (2.6 elements).

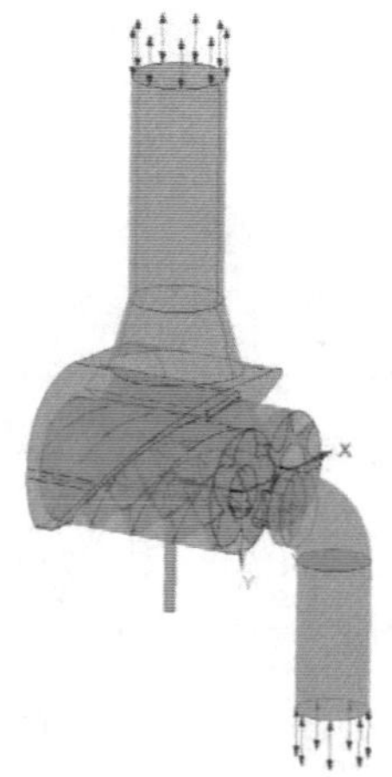

Figure 3. The Ansys CFX BC

The URANS method with SST turbulence model with the BC (figure 3) was use to simulate the flow through the screw compressor. For the suction and pressure side we choose the opening condition with 3bars pressure at the inlet and 30 bars at the outlet. In figure 3 is represented the boundary conditions for Ansys Cfx case. To simulate the oil injection, we use at sthe fluid models the multiphase with heat transfer and homogeneous model. At the oil boundary conditions, the pressure is 9.8 bars and the temperature 50ºC.

The aim of this study is to compare the numerical data with the date from the experimental tests. so, in Figure 4 is represented the absolute pressure contours on the screw compressor rotors. At the beginning in the screw compressor chamber the pressure is 3 bar because is connected with the suction side but after closing the compression starts through the chamber volume decreasing.

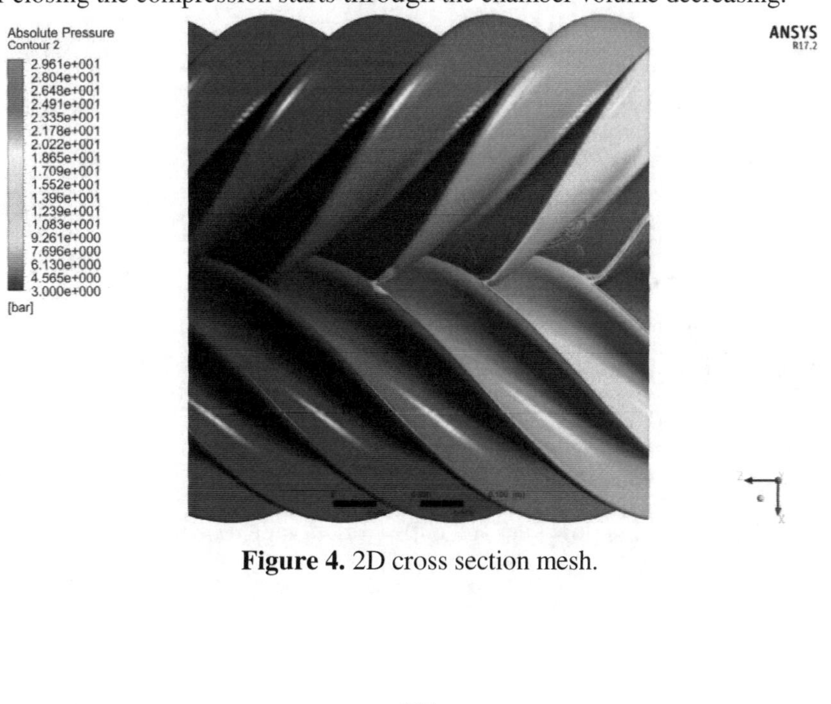

Figure 4. 2D cross section mesh.

Figure 5 shows absolute pressure and temperature in a cross plane for 72° rotation angle and in Figure 6 is represented the mass flow at inlet and outlet variation.

This study shows the CFD analyses results for an oil injected screw compressor, compressing air from 3 bars to 30 bars. The meshes for the rotating domains were done with a special meshing software and the unsteady simulations were performed with ANSYS CFX version 17.2 with multiphase, homogenous model and SST turbulence model.

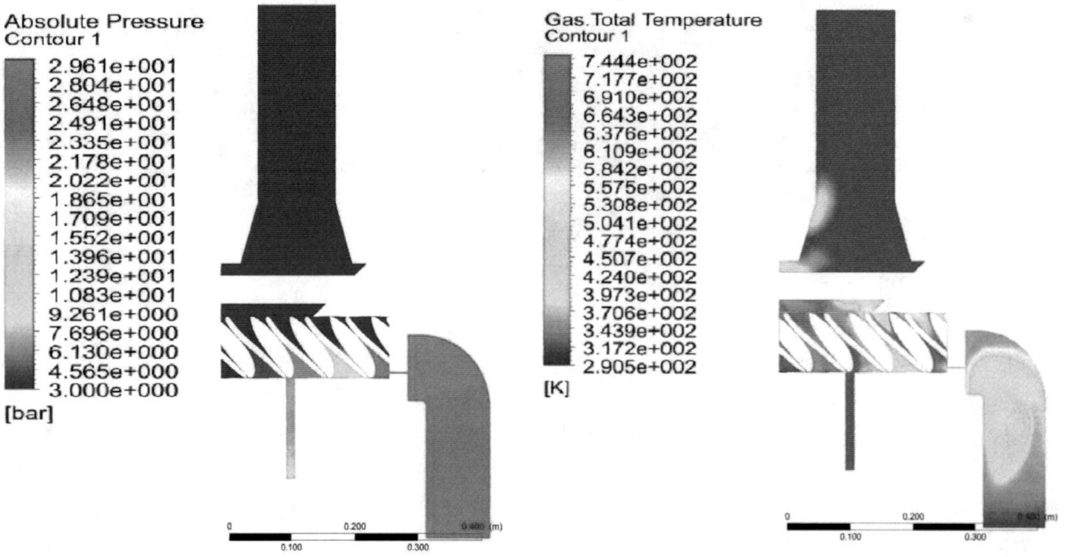

a) b)

Figure 5. Pressure (a) and temperature (b) on cross section at rotation angle 72°.

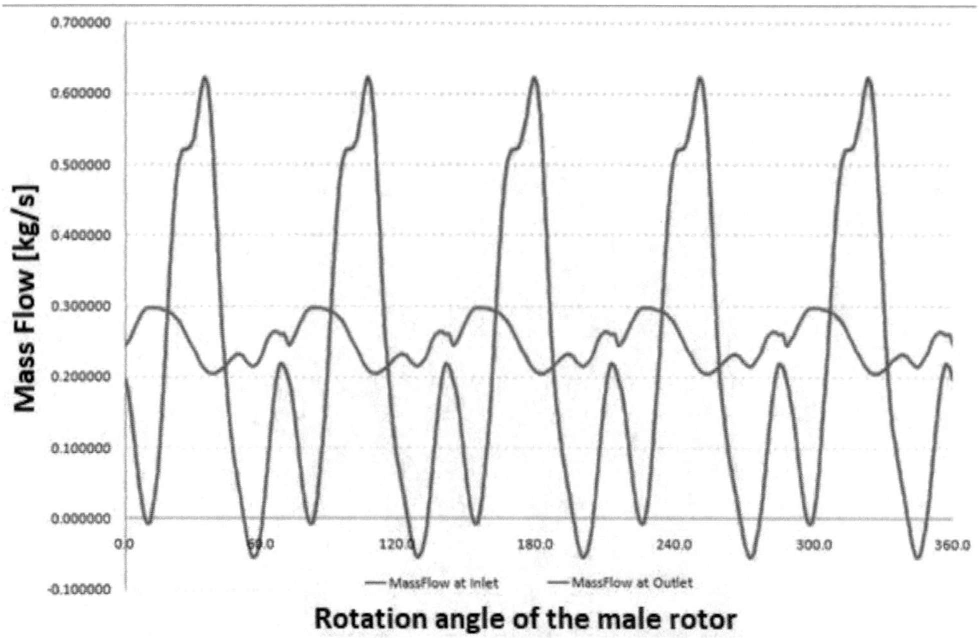

Figure 6. Mass flow inlet and outlet variation per one revolution.

3. Configuration of close-loop test stand for screws compressors

To achieve the configuration of test stand for screw compressor with maximum discharge pressure of 45 bar – for close loop version – were taken into account the existing equipment in the test stand, buying the new needed equipment and reconfiguring routes, in order to carry out the tests under the following conditions:

- rated suction pressure: maximum 4.5 bar
- rated discharge pressure: maximum 45 bar
- maximum compressing power: 250 kW

The stand configuration and data acquisition is achieved accordingly to the requirements of following standards [11, 12, 13, 14, 15]:

- API 619 Rotary type Positive displacement compressors for petroleum, petrochemical and natural gas industry;
- ISO 1217 Displacement compressors. Acceptance test (Annex C);
- PTC 9 Performance test code for displacement compressors, Vacuum pumps and blowers;
- ISO 5167-2 Measurement of fluid by means of pressure differential devices inserted in circular cross-section conduits running full.

The configuration/instrumentation layout of the test bench is shown in figure 7.

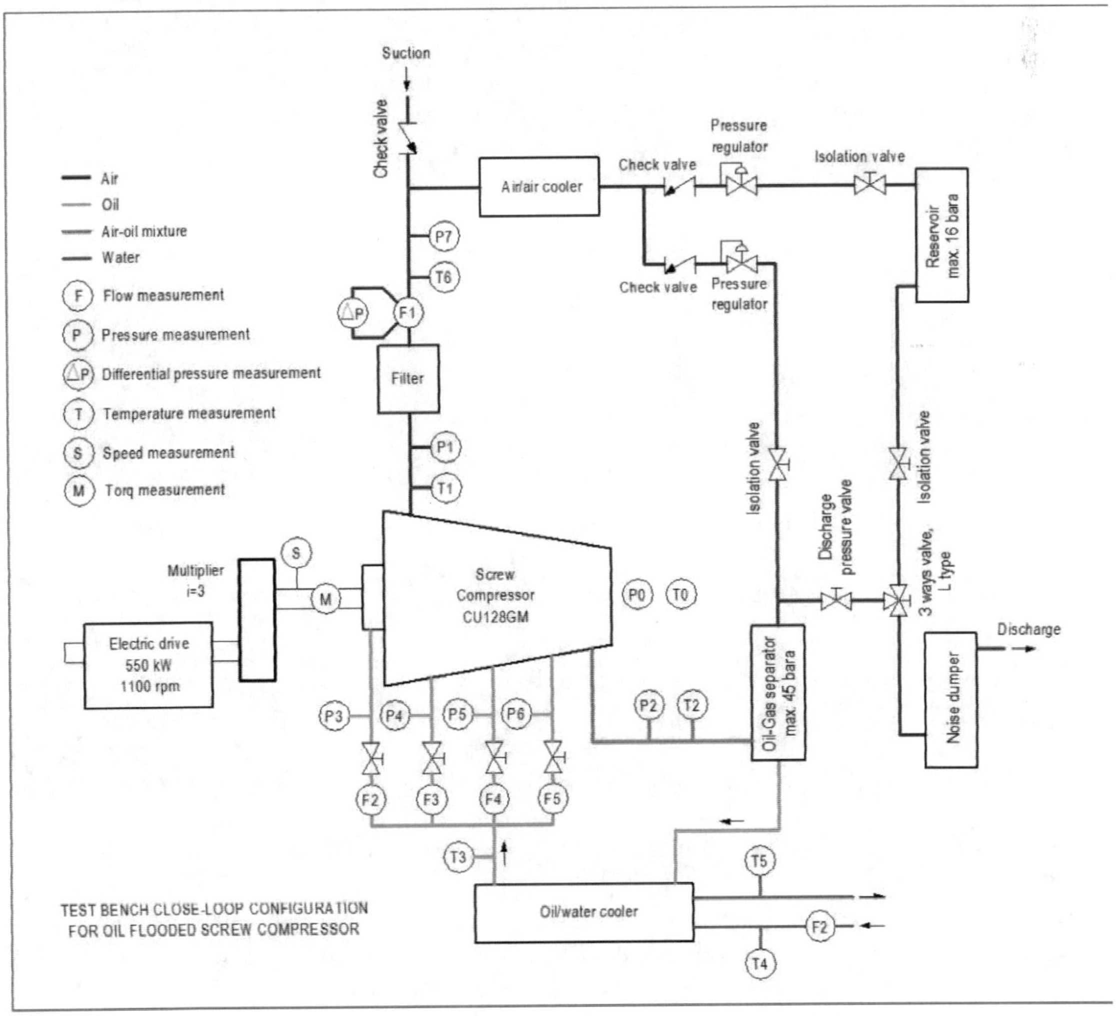

Figure 7. Configuration/Instrumentation layout of the test bench.

IOP Conf. Series: Materials Science and Engineering **595** (2019) 012021 doi:10.1088/1757-899X/595/1/012021

The parameters to be measured for the estimated performance of the compressor will be: compressor suction/discharge temperature and pressure; speed of the male rotor; power of the electric motor driver [16]; pressure/temperature of oil after the oil cooler; differential pressure, pressure and temperature across the differential devices (designed according to ISO 5167-2); in/out temperature for water, in the oil cooler; oil flow rates, for each oil point injection (seal, suction bearings, injection, discharge bearings); pressure of oil at each oil injection point; atmospheric pressure, atmospheric temperature and the relative humidity.

All measurements were obtained from electrically generated signals derived directly, respectively, from the speed, pressure, temperature and flow rates measurements. All the installation parameters (4-20 mA analogical, 0-24 Vcc digital) are supervised by a Programmable Logic Controller. The values of parameters are recorded at a certain rate of time in computer memory (25 μs).

The compressor process evolution was analysed through parameter values data acquisition [7]. Also, the parameters can be viewed in real time on the operating panel, which contains a series of screens that allow an operator to monitor the parameters.

In the figure 8 there are photos of the test bench, with the compressor installed on it.

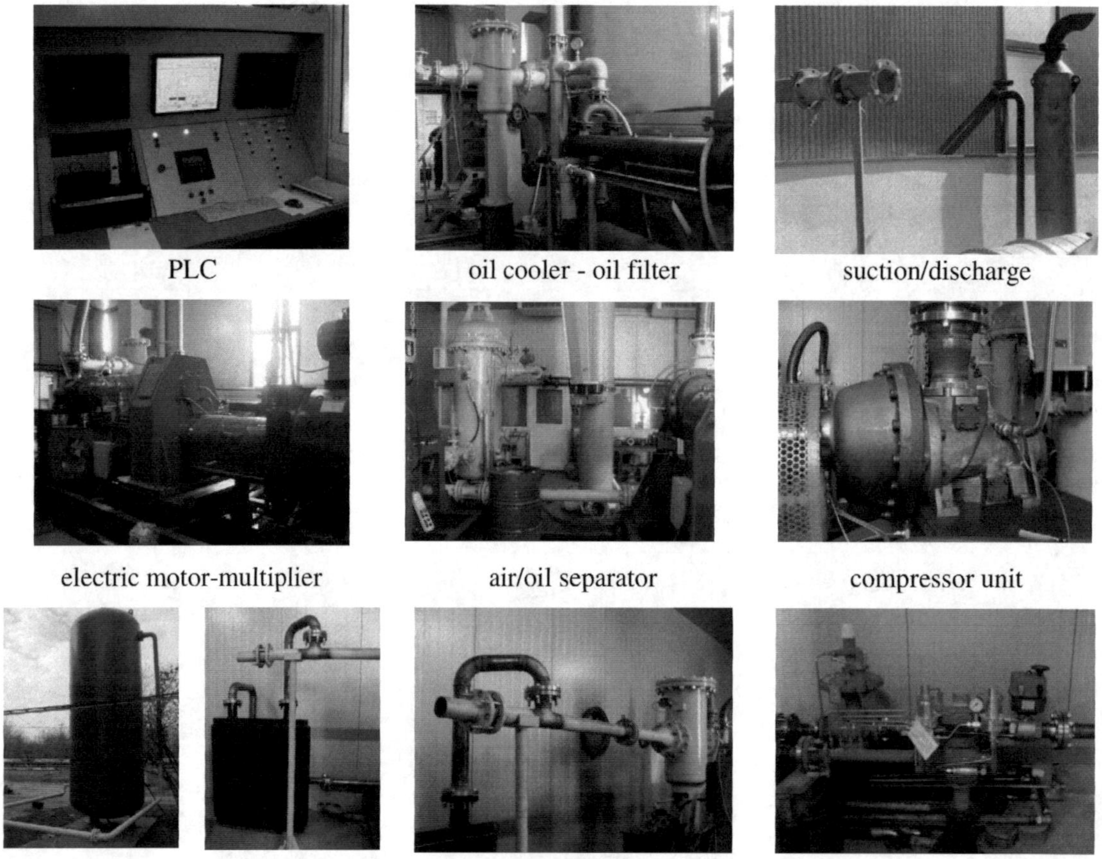

Figure 8. Test bench.

We can see: the electric motor driven; the multiplier; the compressor unit; the oil cooler/oil filter; the oil pipeline; the air pipeline suction and discharge; reservoir; pressure regulators; instrumentations for pressures, temperatures, flows, torque, rotation; air/oil separator etc.

During the course of the test, the temperatures of the compressor surfaces were measured automatically with a Fluke monitor - Figure 9.

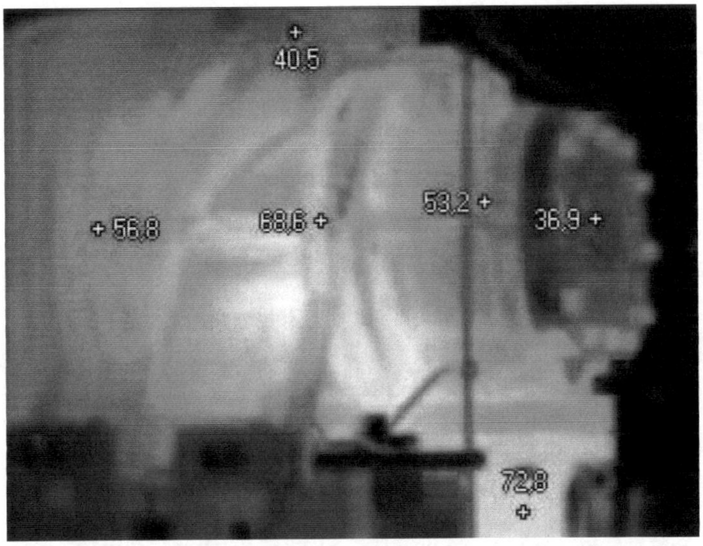

Figure 9. Temperatures of the compressor surfaces.

4. Testing of the compressor CU128 GM type on INCDT-COMOTI test stand.

The first tests were done by configuring the gas routes for suction pressure of 1 bar and discharge pressure of 13.5 bar (accordingly to the reception test regarding the GHH Rand schedule for testing of constructive version for CU128 G, with a compression ratio of 4.8). According to test reception, estimated parameters are as follows (for suction pressure - atmospheric pressure - and speed of 4500 rpm):

- discharge pressure: 13.5 bar
- volume flow related to suction condition: 847 m^3/h (1004 kg/h).

Following the compressor test, recorded data through data acquisition system enabled the drawing of diagrams with pressure suction / discharge pressure as parameters in figure 10 and flow rate/discharge pressure as parameters in figure 11.

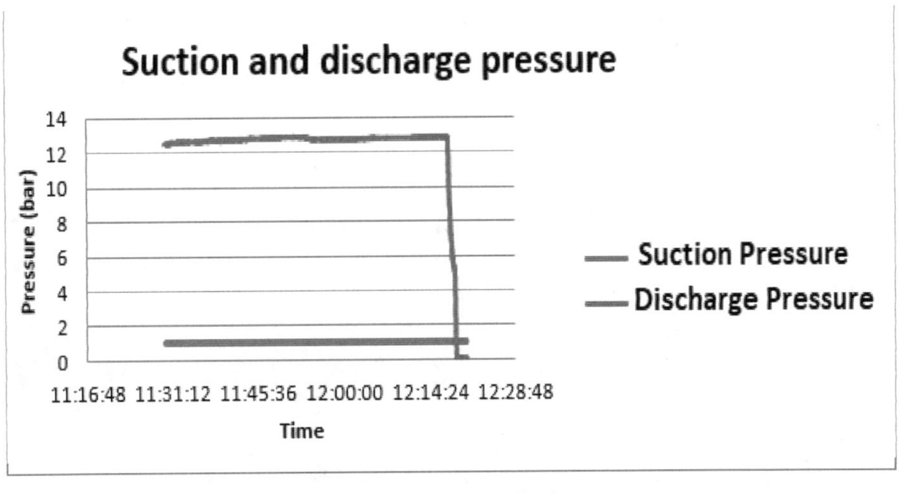

Figure 10. Suction and discharge pressure.

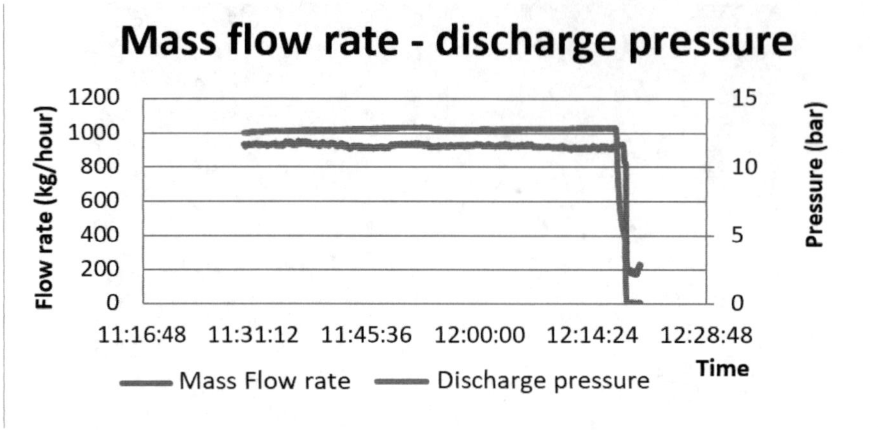

Figure 11. Mass flow rate - discharge pressure.

Table 1. The comparison between the GHH estimated values and the measured values from COMOTI's test bench is presented below (SS 1 bar and PS 13.5 bar).

	Suction pressure (bar)	Discharge pressure (bar)	Male speed Rpm	Suction temperature ^{0}C	Theoretical flow kg/h	Real flow kg/h	Estimated power kW	Real power kW
Section	1	13.5	4500	20	1108	974	121	-
Subsubsection	1	13.5	4500	21	-	1004	-	129

The experimental tests, with recirculation configuration, were carried out under the following conditions:

- the suction pressure was adjusted in (1 ... 4.5) bar range correlating it with the discharge pressure, respecting the optimal compressor compression ratio (air, K = 1.4 so πc = 8.99).

The following test regimes have been defined:

A) 4500 rpm for male rotor

- suction pressure 2 bar discharge pressure 20 bar;
- suction pressure 3 bar discharge pressure 30 bar;
- suction pressure 4 bar discharge pressure 40 bar.

B) 4000 rpm for male rotor

- suction pressure 4.5 bar discharge pressure 45 bar.

The regimes were thus chosen at suction pressures of 4 and 4.5 bar - as it does not exceed 300 kW maximum drive power. Finally, due to the fact that CFD simulation requires a long running time for a regime, was chosen as the test version with 3 bar for suction pressure and 30 bar for discharge pressure.

In figure 12 shows the data obtained through the experimental testing of the chosen regime.

Through the analyse of the temperature variation from the suction side and the pressure side of the screw compressor, a 53.8 ^{o}C temperature decreasing can be observed at the outlet route (air/oil separator, pressure regulators, air cooler, pipelines). Because of that, it wasn't needed to start the air cooler fan.

IOP Conf. Series: Materials Science and Engineering **595** (2019) 012021 doi:10.1088/1757-899X/595/1/012021

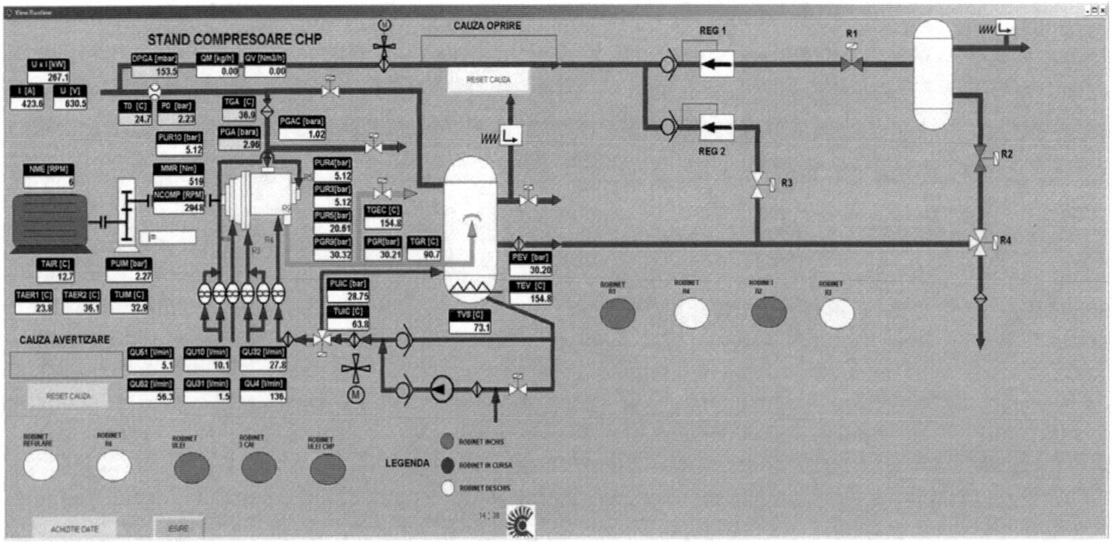

Figure 12. Test bench functional diagram.

Table 2. The comparison between the GHH estimated values and the measured values from COMOTI's test bench is presented below (SS 3 bar and PS 30 bar).

	Suction pressure (bar)	Discharge pressure (bar)	Male speed rpm	Suction temperature ^{0}C	Theoretical flow kg/h	Real flow kg/h	Estimated power kW	Real power kW
Estimate of GHH	3	30	4500	40	3113	2670	279	-
Test measurements	3	31.2	4500	36.9	-	2615	-	289

The diagram from the figure 10 was done by using the parameters from PLC.

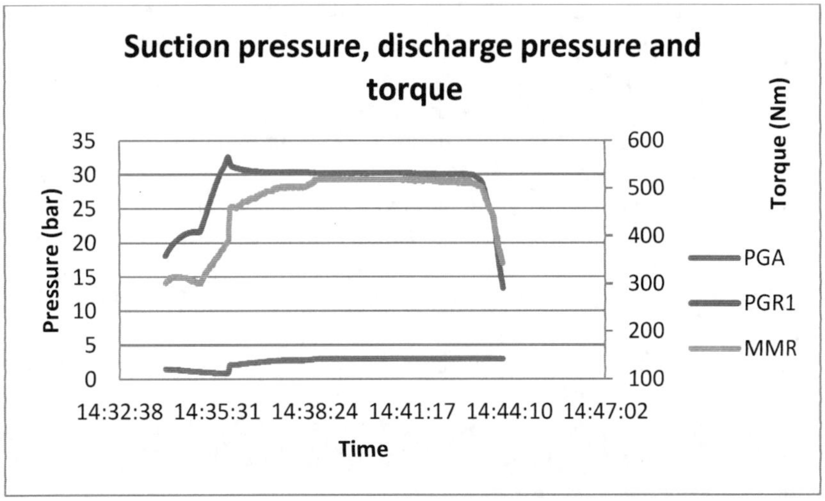

Figure 13. Suction pressure, discharge pressure and torque.

The comparison between the numerical data, obtained with CFD method, and experimental data will lead to an improvement of the design/testing procedure of the new screw compressor developed by NRDI Gas Turbine COMOTI.

IOP Conf. Series: Materials Science and Engineering **595** (2019) 012021 doi:10.1088/1757-899X/595/1/012021

5. Conclusion

The compressors are energy-consumer and a goal in the designing and the manufacturing of compressor is to reduce that energy consumption. Due to the specific process, the screw compressors can be designed so that the energy consumption is minimal for each process.

The oil injection used in technological processes for the minimization of frictions and to favour the two wounded wheel gear influences the thermodynamic process. The used oil flow may be varying between some limits which do not affect too much the compressor functioning.

The study presented in this paper shows results of a CFD analysis of an oil injected screw compressor which compresses air from 3 to 30 bar. The meshes for the rotating domains were generated by use of a special meshing software Twin-Mash and the unsteady simulations were performed with ANSYS CFX version 17.2 with multiphase homogenous flow model and embedded SST turbulence model. The aim was to compare the numerical data and the experimental test results.

Because the numerical investigation was not completed in time, finally we compared the test results with GHH Rand estimated schedule. We can see that the difference between estimated performance and measured performance-on the test bench- are small.

Finally, we believe that comparison between the numerical data, obtained with CFD method, and experimental data, will lead to an improvement of the design/testing procedure of the new screw compressor developed by NRDI Gas Turbine COMOTI.

References

[1] Eparu C, Suditu S, Prundurel A 2013 Ensuring security of local gas supply using pipeline gas storage *Ovidius Univ. An. Chem. Series* **24(2)** pp 98-108

[2] Neacşu S, Eparu C 2013 Informatics Platform for Managing the Natural Gas Transport Program Proc. of the 21st IBIMA Conf. pp 748-755

[3] Ştefănescu M, Silivestru V, Liviu A, Neacsu S, Florea I, Eparu C 2008 *Mentenanța turbomotoarelor* (Ploiesti: Editura Universității Petrol-Gaze din Ploieşti p. 138

[4] https://www.ansys.com/products/fluids/ansys-cfx.

[5] Amosov P E et al. 1977 Vintovie kompresornie mashinii - Spravochnik (Screw Compression Machines-Handbook) *Mashinostroeni*e Leningrad

[6] Spille-Kohoff A, Hesse J and Shorbagy A El 2015 CFD Simulation of a Screw Compressor Including Leakage Flows and Rotor Heating, *9th International Conference on Compressors and Their Systems, IOP Conf. Series: Materials Science and Engineering* **90**

[7] Eparu C, Neacşu S and Neacsa A 2019 The Optimization of Internal Processes from a Screw Compressor with Oil Injection to Increase Performances *IJHT* **37** 1 148-152

[8] Eparu C, Neacşu S, Neacsa A, Prundurel A 2019 The comparative thermodynamic analysis of compressor's energetic performance *MMEP* **6** 1 152-155

[9] Stošić N, Milutinovic L, Hanjalic K and Kovacevic A 1992 Investigation of the Influence of Oil Injection on the Screw Compressor Working Processes *Int. J. Refrig* **15** 206-220

[10] SKF - Explorer Catalogue

[11] Stošić N, Smith I K and Mujic E 2009 The Design of a High Efficiency Oil Flooded Gas Screw Compressor *City University London*

[12] API 619 - Rotary type Positive Displacement Compressors for Petroleum, Petrochemical and Natural Gas Industry (ISO 10440)

[13] ISO 1217 - Displacement Compressors. Acceptance Test (Annex C)

[14] ISO 5167-2 - Measurement of Fluid by Means of Pressure Differential Devices Inserted in Circular Cross-Section Conduits Running Full

[15] PTC 9 - Performance Test Code for Displacement Compressors; Vacuum Pumps and Blowers;

[16] Stošić N, Smith I K and Kovacevic A 2005 Improving Screw Compressor Performance *Fluid Machinery Group - International Conference on Compressors and their systems IMechE*

IOP Conf. Series: Materials Science and Engineering **595** (2019) 012022 doi:10.1088/1757-899X/595/1/012022

Thermodynamic analysis of irreversible closed Brayton engine cycles used in trigeneration systems

G Dumitrașcu[1], M Feidt[2], A R J Namat[1], Ş Grigorean[1] and B Horbaniuc[1]

[1] Mechanical and Automotive Engineering Department, Mechanical Engineering Faculty, "Gheorghe Asachi" Technical University of Iași, Romania
[2] University of Lorraine, LEMTA ENSEM - 2 Avenue de la Forèt de Haye, BP 90161, 54505 VANDOEUVRE CEDEX, Nancy, France

E-mails: gdum@tuiasi.ro; michel.feidt@univ-lorraine.fr; stockholm_61@hotmail.com; grigorean_stefan@yahoo.com; bogdan_horbaniuc@yahoo.com

Abstract. The paper presents a comparative thermodynamic analysis of irreversible closed Brayton engine cycles for two working fluids and for four imposed thermodynamic parameters. The chosen fluids are the air and the carbon dioxide. The four imposed thermodynamic constraints are the specific entropy variation during the heating process, the specific input heat, the specific output power and the energy efficiency. The irreversibility is quantified by the isentropic efficiencies of compressor and of gas turbine, and by the pressure drops coefficient during the heating and cooling processes of the working fluid. Each imposed thermodynamic parameter allowed to evaluate all corresponding cycle parameters, i.e. temperatures and pressures, and operational parameters, i.e. energy transfer values, function of pressure compression ratio and irreversibility magnitude. The thermodynamic analysis compares the selected results for the two working fluids.

1. Introduction

The most current researches of open or closed Brayton engine cycles follow to the new energy technologies trends, either new single or combined cycles with new working fluids or hybrid solar assisted engines or outer space applications. For instance, the specific research objectives refer to exhaustive exergy analysis regarding endogenous and exogenous lost exergy [1], exergy analysis following the maximization of power output and genetic algorithm optimization [2, 3], entropy generation minimization [4], performance assessments [5-10], ecological aspects [11] and for instance to specific reviews for solar energy technologies [12]. Recently proposed methods of thermal systems design [13, 14, and 15] apply the Finite Physical Dimensions Thermodynamics (FPDT) to design, optimize or perform sensitivity analysis of various thermal systems. This paper presents a comparative thermodynamic analysis of irreversible closed Brayton engine cycles for two working fluids and for four imposed thermodynamic parameters. The chosen fluids are the air and the carbon dioxide. The four imposed thermodynamic parameters are the specific entropy variation during the heating process, the specific input heat, the specific output power and the energy efficiency. The irreversibility is quantified through the isentropic efficiencies of compressor and of gas turbine, and through the pressure drops coefficient during the heating and cooling processes of the working fluid. Each imposed thermodynamic parameter allowed to evaluate all corresponding cycle parameters, i.e.

temperatures and pressures, and operational parameters, i.e. energy transfer values, function of pressure compression ratio and irreversibility magnitude. The thermodynamic analysis compares the selected results for the two working fluids.

2. Basic mathematical model

The analyzed scheme of the irreversible Brayton cycle is presented in T – s diagram, Figure 1.

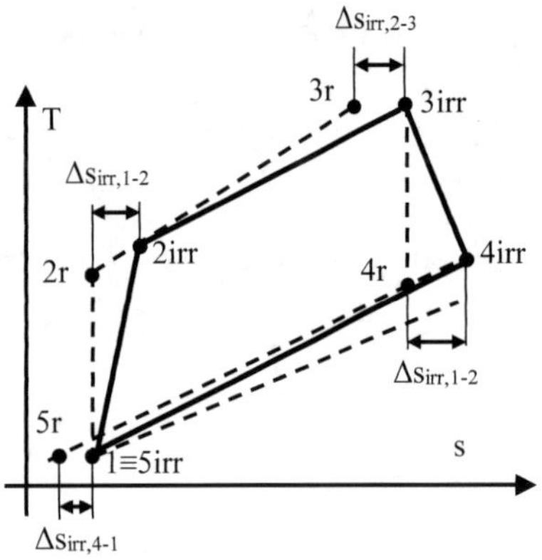

Figure 1. The scheme of irreversible Brayton cycle

2r, 3r, 4r, 5r: ending states of reversible processes
2irr, 3irr, 4irr, 5irr: ending states of irreversible processes

1 – 2r reversible adiabatic compression; 2irr – 3r reversible heating at constant temperature; 3irr – 4r reversible adiabatic expansion; 4irr – 5r reversible cooling at constant pressure

1 – 2irr irreversible adiabatic compression; 2irr – 3irr irreversible heating; 3irr – 4irr irreversible adiabatic expansion; 4irr – 1 irreversible cooling

Δs_{irr} – process additional irreversible entropy variations

2.1. General irreversible energy efficiency depending on the working fluid nature [16]

$$EE_{irr}=1-\frac{T_{mq,4irr-5r}}{T_{mq,2irr-3r}}\cdot\left(1+\frac{\Delta s_{irr,12}+\Delta s_{irr,2-3}+\Delta s_{irr,3-4}+\Delta s_{irr,4-1}}{s_{3r}-s_{2irr}}\right)=1-\frac{T_{mq,4irr-5r}}{T_{mq,2irr-3r}}\cdot N_{irr} \quad (1)$$

where:

- $T_{mq,4irr-5r}=\dfrac{T_{5r}-T_{4irr}}{\ln\left(\dfrac{T_{5r}}{T_{4irr}}\right)}\overset{T_{5r}=T_1}{=}\dfrac{T_1-T_{4irr}}{\ln\left(\dfrac{T_1}{T_{4irr}}\right)}$ is the mean thermodynamic temperature of the working fluid during the cooling at constant pressure;
- $T_{mq,2irr-3r}=\dfrac{T_{3r}-T_{2irr}}{\ln\left(\dfrac{T_{3r}}{T_{2irr}}\right)}\overset{T_{3r}=T_{3irr}}{=}\dfrac{T_{3irr}-T_{2irr}}{\ln\left(\dfrac{T_{3irr}}{T_{2irr}}\right)}$ is the mean thermodynamic temperature of the working fluid during the heating at constant pressure;
- N_{irr} is the number of internal overall irreversibility;
- s and Δs_{irr} are the specific entropy and process irreversible additional entropy variations depending on the working fluid nature;
- $T_{3irr}=T_{3r}$ is the maximum temperature on the irreversible cycle;
- $EE_{Carnot}=1-\dfrac{T_{mq,4irr-5r}}{T_{mq,2irr-3r}}$ is the energy efficiency of Carnot engine cycle operating between $T_{mq,4irr-5r}$ and $T_{mq,2irr-3r}$.

2.2. The hypotheses of comparative thermodynamic analysis

The comparative thermodynamic analysis is performed on the basis of below four restrictive requirements.

- Constant specific entropy variation during the reversible heating process:

$$\Delta s_q=s_{3r}-s_{2irr}=c_p\cdot\ln\left(\frac{T_{3r}}{T_{2irr}}\right)\overset{T_{3irr}=T_{3r}}{=}c_p\cdot\ln\left(\frac{T_{3irr}}{T_{2irr}}\right)=c_p\ln(2) \quad (2)$$

- Constant specific heat input, q_H:

$$q_H = c_p \cdot (T_{3r} - T_{2irr}) \overset{T_{3r}=T_{3irr}}{=} c_p \cdot (T_{3irr} - T_{2irr}) = 500 \frac{kJ}{kg} \quad (3)$$

- Constant specific power output, w:

$$w = q_H - |q_C| = c_p \cdot \left[(T_{3r} - T_{2irr}) - (T_{4irr} - T_{5r}) \right] \underset{T_{5r}=T_1}{\overset{T_{3r}=T_{3irr}}{=}} c_p \cdot \left[(T_{3irr} - T_{2irr}) - (T_{4irr} - T_1) \right] = 100 \frac{kJ}{kg} \quad (4)$$

- Constant energy efficiency, EE_{irr}:

$$EE_{irr} = \frac{w}{q_H} = 1 - \frac{|q_C|}{q_H} = 1 - \frac{T_{4irr} - T_{5r}}{T_{3r} - T_{2irr}} = 1 - \frac{T_{4irr} - T_1}{T_{3irr} - T_{2irr}} = 0.3 \quad (5)$$

The irreversibility is quantified by the isentropic efficiencies of compressor and of gas turbine, and by the pressure drops coefficient during the heating and cooling processes of the working fluid.

- The isentropic efficiency of irreversible adiabatic compression $1 - 2_{irr}$:

$$\eta_{sC} = 0.8 \text{ and } \eta_{sC} = 0.85 \quad (6)$$

- The isentropic efficiency of irreversible adiabatic expansion $3_{irr} - 4_{irr}$:

$$\eta_{sT} = 0.85 \text{ and } \eta_{sT} = 0.9 \quad (7)$$

- The pressure drop coefficient during the irreversible heating, $2_{irr} - 3_{irr}$ and cooling, $4_{irr} - 1$:

$$r_p = \frac{p_{3irr}}{p_{2irr}} = \frac{p_{4irr}}{p_1} = 0.975 \text{ and } r_p = \frac{p_{3irr}}{p_{2irr}} = \frac{p_{4irr}}{p_1} = 0.95 \quad (8)$$

The chosen working fluids are the air and the carbon dioxide regarded as perfect gases.
The imposed thermodynamic parameters are:

$$p_1 = 1 \text{ bar and } T_1 = 303 \text{ K} \quad (9)$$

The variable thermodynamic compression ratio:

$$\pi_C = \frac{p_{2r}}{p_1} = \frac{p_{2irr}}{p_1} \quad (10)$$

2.3. Equations of basic mathematical model

2.3.1. Ending states temperature of consisting processes of the cycle

$$T_{2r} = T_1 \cdot \pi_C^{\frac{k-1}{k}} \quad (11)$$

$$T_{2irr} = T_1 + \frac{T_{2r} - T_1}{\eta_{sC}} \quad (12)$$

$$T_{3r} = T_{3irr} = T_1 \cdot e^{\frac{\Delta s_q}{c_p}} \left[1 + \eta_{sC}^{-1} \cdot \left(\pi_C^{\frac{k-1}{k}} - 1 \right) \right] \text{ for imposed } \Delta s_q = c_p \cdot \ln(2) \quad (13)$$

$$T_{3r} = T_{3irr} = \frac{q_H}{c_p} + T_1 \cdot \left[1 + \eta_{sC}^{-1} \cdot \left(\pi_C^{\frac{k-1}{k}} - 1 \right) \right] \text{ for imposed } q_H = 500 \frac{kJ}{kg} \quad (14)$$

$$T_{3r} = T_{3irr} = \frac{c_p \cdot T_1 \cdot \left(\pi_C^{\frac{k-1}{k}} - 1 \right) + \eta_{sC} \cdot w}{c_p \cdot \eta_{sC} \cdot \eta_{sT} \cdot \left[1 - \left(r_p^2 \cdot \pi_C \right)^{\frac{1-k}{k}} \right]} \text{ for imposed } w = 100 \frac{kJ}{kg} \quad (15)$$

$$T_{3r} = T_{3irr} = \frac{T_1 \cdot \left[EE_{irr} \cdot \left(\eta_{sC} + \pi_C^{\frac{k-1}{k}} - 1 \right) + 1 - \pi_C^{\frac{k-1}{k}} \right]}{\eta_{sC} \cdot \left[\eta_{sT} \cdot \left(r_p^2 \cdot \pi_C \right)^{\frac{1-k}{k}} - \eta_{sT} + EE_{irr} \right]} \text{ for imposed } \eta = 0.3 \tag{16}$$

$$T_{4r} = T_{3irr} \cdot \left(r_p^2 \cdot \pi_C \right)^{\frac{1-k}{k}} \tag{17}$$

$$T_{4irr} = T_{3irr} - \frac{T_{3irr} - T_{4r}}{\eta_{sT}} \tag{18}$$

2.3.2. Energy transfers and energy efficiency

$$q_H = c_p \cdot \left(T_{3r} - T_{2irr} \right) = c_p \cdot \left(T_{3irr} - T_{2irr} \right) \tag{19}$$

$$q_C = c_p \cdot \left(T_{5r} - T_{4irr} \right) = c_p \cdot \left(T_1 - T_{4irr} \right) \tag{20}$$

$$w = q_H + q_C = q_H - |q_C| \tag{21}$$

$$EE_{irr} = \frac{w}{q_H} = 1 - \frac{|q_C|}{q_H} = 1 - \frac{T_{mq,4irr-5r}}{T_{mq,2irr-53}} \cdot N_{irr} \tag{22}$$

2.3.3. Irreversible entropy variations

$$\Delta s_{irr,1-2} = s_{2irr} - s_{2r} = c_p \cdot \ln\left(\frac{T_{2irr}}{T_{2r}} \right) \tag{23}$$

$$\Delta s_{irr,2-3} = s_{3irr} - s_{3r} = -R \cdot \ln\left(\frac{p_{3irr}}{p_{3r}} \right) \overset{p_{3r} = p_{2irr}}{=} -R \cdot \ln\left(\frac{p_{3irr}}{p_{2irr}} \right) \tag{24}$$

$$\Delta s_{irr,3-4} = s_{4irr} - s_{4r} = c_p \cdot \ln\left(\frac{T_{4irr}}{T_{4r}} \right) \tag{25}$$

$$\Delta s_{irr,4-1} = s_1 - s_{5r} = -R \cdot \ln\left(\frac{p_1}{p_{5r}} \right) \overset{p_{5r} = p_{4irr}}{=} -R \cdot \ln\left(\frac{p_1}{p_{4irr}} \right) \tag{26}$$

2.3.4. Number of internal overall irreversibility

$$N_{irr} = 1 + \frac{\Delta s_{irr,1-2} + \Delta s_{irr,2-3} + \Delta s_{irr,3-4} + \Delta s_{irr,4-1}}{s_{3r} - s_{2irr}} \tag{27}$$

3. Numerical results

The comparative selected numerical results are included in Figures 2 to 5. The selected results are:

- the maximum temperature on the cycle function of compression ratio, $T_{3irr} = T_{3r} = f_T(\pi_C)$, Figure 2;
- the specific output work function of compression ratio, $w = f_w(\pi_C)$, Figure 3;
- the irreversible energy efficiency function of compression ratio, $EE_{irr} = f_E(\pi_C)$, Figure 4;
- the number of internal overall irreversibility, $N_{irr} = f_N(\pi_C)$, Figure 5.

The maximum temperature on the cycle, the specific output work, and the irreversible energy efficiency are operational parameters used in the design of Brayton engines. The number of internal overall irreversibility puts a figure on overall internal irreversibility depending on the working fluid nature.

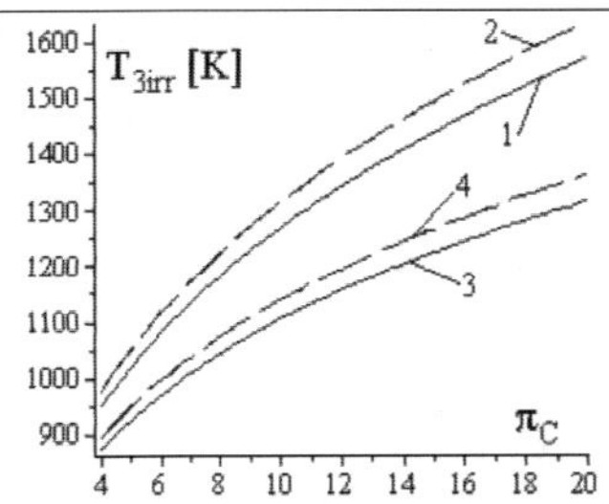

Fig. 2.a. case Δs_q = const. $=c_p \cdot ln(2)$

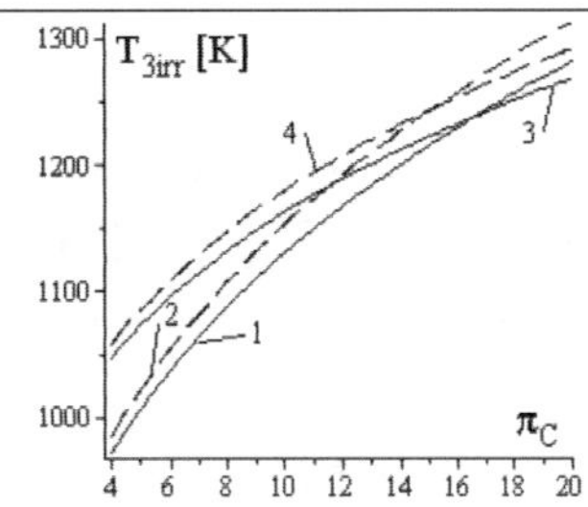

Fig. 2.b. case q_H = const. = 500,000 $J \cdot kg^{-1}$

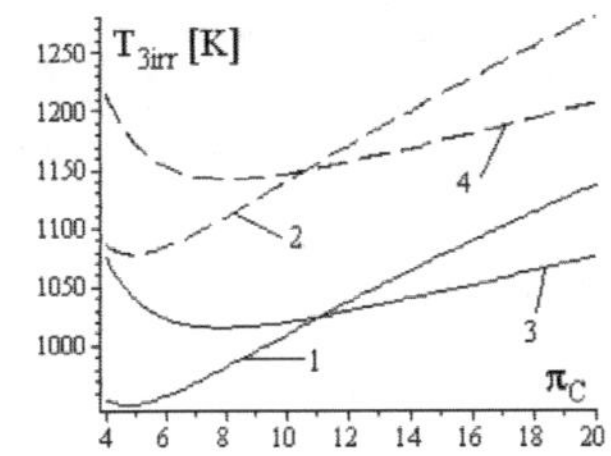

Fig. 2.c. case w = const. = 100,000 $J \cdot kg^{-1}$

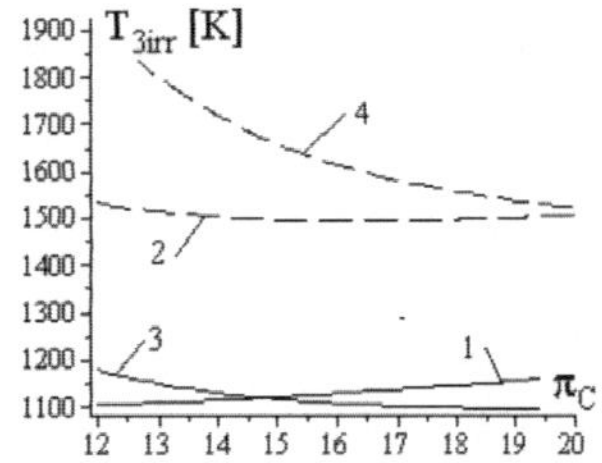

Fig. 2.d. case EE_{irr} = const. = 0.3

Figure 2. The dependences $T_{3irr} = f_T(\pi_C)$

1: air, $\eta_{sC} = 0.85$, $\eta_{sT} = 0.9$, $r_p = 0.975$; **2**: air, $\eta_{sC} = 0.8$ and $\eta_{sT} = 0.85$, $r_p = 0.95$

3: CO_2, $\eta_{sC} = 0.85$, $\eta_{sT} = 0.9$, $r_p = 0.975$; **4**: CO_2, $\eta_{sC} = 0.8$ and $\eta_{sT} = 0.85$, $r_p = 0.95$

The admissible maximum temperature on the cycle, currently from 1100 K to 1400 K, limits the maximum compression ratio, see Figures 2.a. and 2.d. These limitations depend on the fluid nature and on the internal irreversibility intensity. The imposed constraints emphasized different relationships $T_{3irr} = f_T(\pi_C)$. For all constraints a worsen internal irreversibility induces larger temperatures T_{3irr}. The constraint w = const. = 100 $J \cdot kg^{-1}$, Figure 2.c., revealed minimum temperatures T_{3irr}, the corresponding compression ratios are around five for air, and around eight for CO_2.

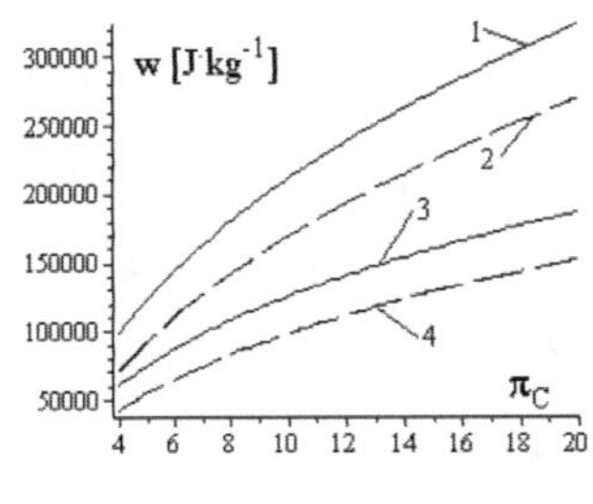

Fig. 3.a. case Δs_q = const. $=c_p \cdot ln(2)$

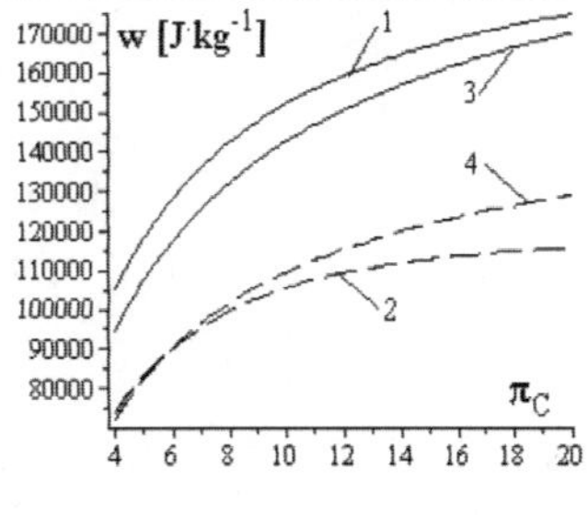

Fig. 3.b case q_H = const. = 500,000 $J \cdot kg^{-1}$

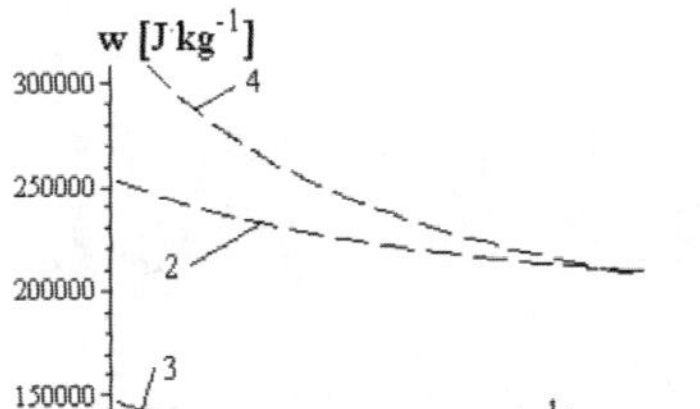

Fig. 3.c. case EE_{irr} = const. = 0.3

Figure 3. The dependences $w = f_w(\pi_C)$

1: air, $\eta_{sC} = 0.85$, $\eta_{sT} = 0.9$, $r_p = 0.975$;

2: air, $\eta_{sC} = 0.8$ and $\eta_{sT} = 0.85$, $r_p = 0.95$

3: CO_2, $\eta_{sC} = 0.85$, $\eta_{sT} = 0.9$, $r_p = 0.975$;

4: CO_2, $\eta_{sC} = 0.8$ and $\eta_{sT} = 0.85$, $r_p = 0.95$

Figure 3.a. shows that the specific output work is larger for air than for CO_2, and also the fact that the increased internal irreversibility diminishes it. Figures 3.b. exhibits smaller differences and the fact that the internal irreversibility has a stronger influence. Figure 3.c. exhibits also smaller differences but the influence of internal irreversibility has an opposite influence.

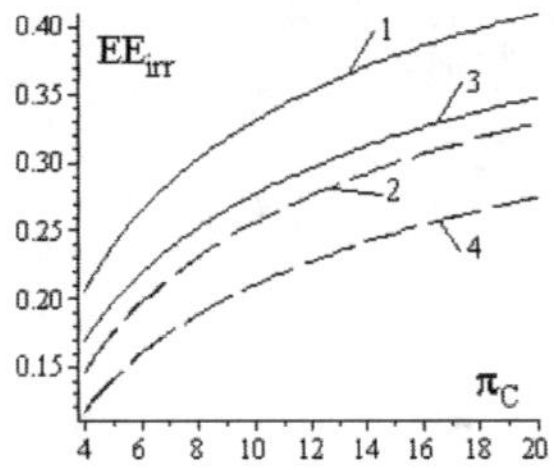

Fig. 4.a. case Δs_q = const. $=c_p \cdot ln(2)$

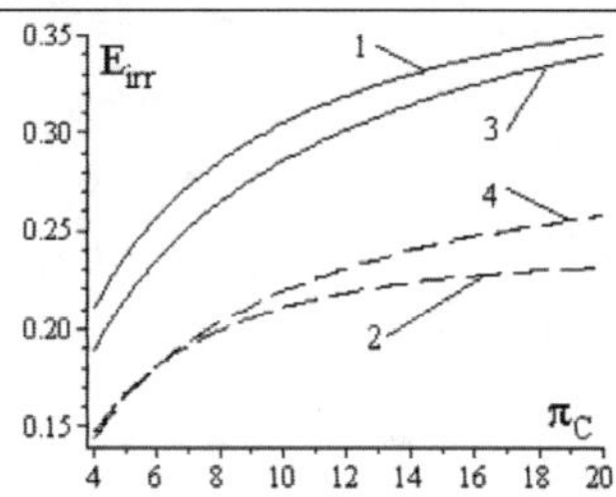

Fig. 4.b. case q_H = const. = 500,000 J·kg^{-1}

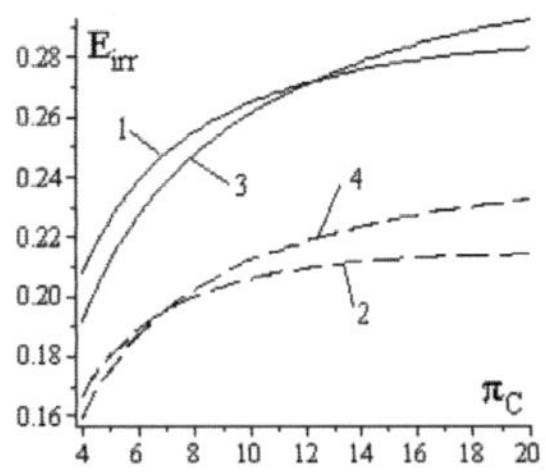

Fig. 4.c. case w = const. = 100,000 kJ·kg^{-1}

Figure 4. The dependences $EE_{irr} = f_E(\pi_C)$

1: air, $\eta_{sC} = 0.85$, $\eta_{sT} = 0.9$, $r_p = 0.975$;

2: air, $\eta_{sC} = 0.8$ and $\eta_{sT} = 0.85$, $r_p = 0.95$

3: CO_2, $\eta_{sC} = 0.85$, $\eta_{sT} = 0.9$, $r_p = 0.975$;

4: CO_2, $\eta_{sC} = 0.8$ and $\eta_{sT} = 0.85$, $r_p = 0.95$

Figure 4.a. shows that the energy efficiency is larger for air than for CO_2, and also the increased internal irreversibility drastically diminishes it. Figures 4.b. and 4.c. exhibit smaller differences, but the internal irreversibility has also a stronger influence.

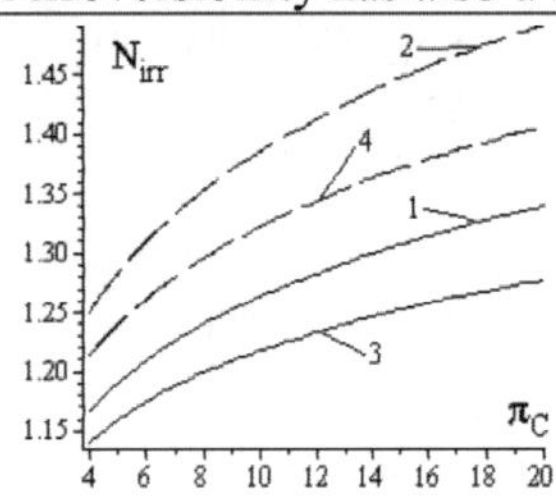

Fig. 5.a. case Δs_q = const. $=c_p \cdot ln(2)$

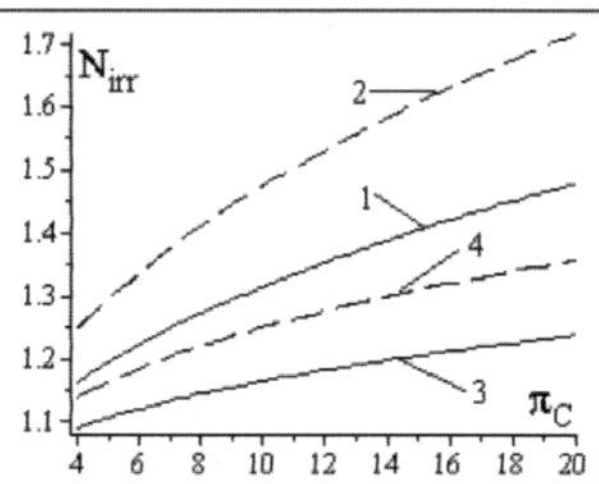

Fig. 5.b. case q_H = const. = 500,000 kJ·kg^{-1}

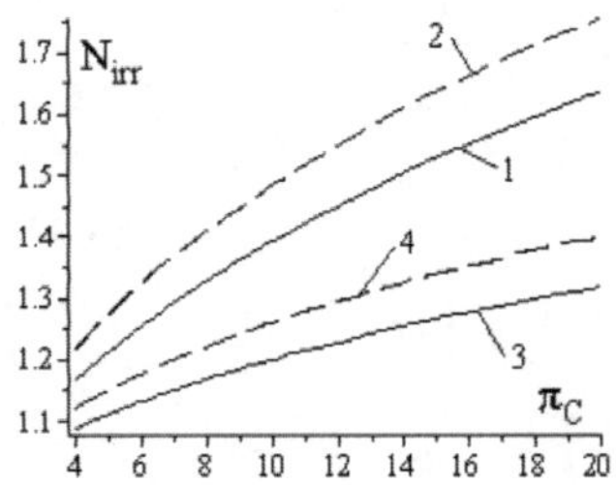

Fig. 5.c. case w = const. = 100,000 J·kg^{-1}

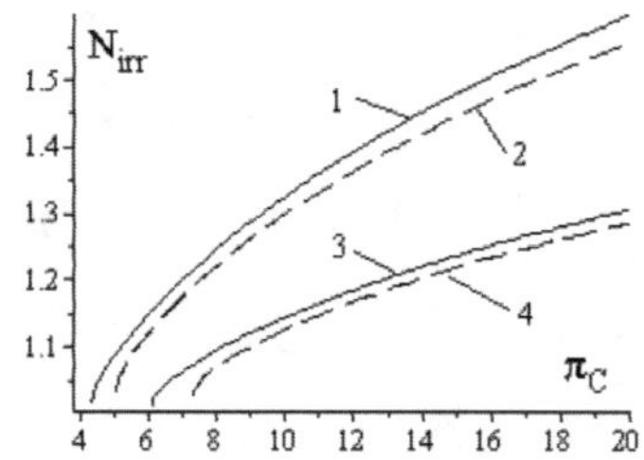

Fig. 5.d. case EE_{irr} = const. = 0.3

Figure 5. The dependences $N_{irr} = f_N(\pi_C)$

1: air, $\eta_{sC} = 0.85$, $\eta_{sT} = 0.9$, $r_p = 0.975$; **2**: air, $\eta_{sC} = 0.8$ and $\eta_{sT} = 0.85$, $r_p = 0.95$

3: CO_2, $\eta_{sC} = 0.85$, $\eta_{sT} = 0.9$, $r_p = 0.975$; **4**: CO_2, $\eta_{sC} = 0.8$ and $\eta_{sT} = 0.85$, $r_p = 0.95$

Figures 5.a. to 5.d. show numbers of internal overall irreversibility larger for air than for CO_2. Figures 5.a., 5.b. and 5.c. say that as smaller the irreversibility coefficients (isentropic efficiencies and pressure drops ratios) as larger N_{irr}. Figure 5.d. shows an opposite influence of irreversibility coefficients.

4. Conclusions

The design of Brayton engine cycles is performed on the basis of restrictive conditions imposed by the end users. In this paper were analyzed four possible scenarios to design the closed Brayton cycles in order to find out the operational and technological limits regarding:

- the maximum admissible temperature on the cycle, currently from 1100 K to 1400 K, and thus the maximum admissible compression ratio has different limitations depending on the fluid nature and on internal irreversibility;
- the specific output useful work, usually imposed in the designing process;
- the specific input heat, possible in the cases of limited thermal energy resources;
- the irreversible energy efficiency, main criterion in choosing of the convenient decision;
- the internal overall irreversibility evaluation depending on the working fluid nature.

The four scenarios to design Brayton engine closed cycles are:

- constant specific entropy variation of the working fluid during the reversible heating process;
- constant specific input heat on the cycle;
- constant specific output useful work;
- constant energy efficiency.

They were compared two working fluid, air and carbon dioxide, regarded as perfect gases. They were supposed two irreversibility cases. The first considered the best actual irreversibility coefficients: the isentropic efficiency of compressor of 0.85, and the isentropic efficiency of gas turbine of 0.9, and a constant pressure drop coefficient of 0.975, defined as the ratio of final pressure to initial one, for flows to heat exchangers.

The comparative thermodynamic analysis showed:

- for the case of specific constant entropy variation, the admissible maximum temperature T_{3r} = T_{3irr} = 1400 K imposes a maximum compression ratio of 14 for air, see Figure 2.a. – curves 1 and 2;
- for the case of constant energy efficiency, the admissible maximum temperature T_{3r} = T_{3irr} = 1400 K imposes the best irreversibility coefficients, see Figure 2.d. – curves 1 and 3;
- for the case of constant specific output work, the curve of maximum temperature $T_{3r} = T_{3irr}$ has a minimum for all cases, see Figure 2.c.;
- the internal overall irreversibility is smaller for carbon dioxide than for air, and it has similar features for specific constant entropy variation, specific input heat and specific output work, see Figures 5.a., 5.b., 5.c., but different for constant energy efficiency.

This comparative thermodynamic analysis might be completed by an exhaustive sensitivity analysis for all possible irreversibility cases and for all possible working fluids.

References

[1] Mossi I A K and Goni B K 2019 Advanced exergy analysis of a combined Brayton/Brayton power cycle, *Energy*, **166** pp 724-737
[2] Mohammad M N, Said F and Faramarz S 2016 Exergoeconomic multi objective optimization and sensitivity analysis of a regenerative Brayton cycle, *Energy Conversion and Management* **117** pp 95–105
[3] Mohammad M N, Said F and Faramarz S 2017 New exergy analysis of a regenerative closed Brayton cycle, *Energy Conversion and Management* **134** pp 116–124
[4] Haseli Y 2016 Efficiency of irreversible Brayton cycles at minimum entropy generation, *Applied Mathematical Modelling* **40** pp 8366–8376

IOP Conf. Series: Materials Science and Engineering **595** (2019) 012022 doi:10.1088/1757-899X/595/1/012022

[5] Xiaoli H and Guoqiang Z 2007 Maximum useful energy-rate analysis of an endoreversible Joule–Brayton cogeneration cycle, *Applied Energy* **84** pp 1092–1101
[6] Wissam S, Bou N, Charbel J M and Maroun G N 2018, Optimization of a Brayton external combustion gas-turbine system for extended range electric vehicles, *Energy* **150** pp 745-758
[7] Mohamed A S E, Khwaja M R and Solimana A M 2017, Performance analysis of different working gases for concentrated solar gas engines: Stirling & Brayton, *Energy Conversion and Management* **150** pp 651–668
[8] Seyed A S et al. 2015 Thermodynamic and thermo-economic analysis and optimization of an irreversible regenerative closed Brayton cycle, *Energy Conversion and Management* **94** pp 124–129
[9] Kunlin C et al. 2014 Performance assessment of a closed-recuperative-Brayton-cycle based integrated system for power generation and engine cooling of hypersonic vehicle, *Aerospace Science and Technology* **87** pp 278–288
[10] Merchán R P, Santos M J, Medina A and Calvo H A 2018 Thermodynamic model of a hybrid Brayton thermosolar plant *Renewable Energy* **128** pp 473-483
[11] Yasin U, Bahri S, Ali K and Ismail H A 2006 Ecological coefficient of performance analysis and optimization of an irreversible regenerative - Brayton heat engine *Applied Energy* **83** pp 558–572
[12] Le Roux W G, Bello-Ochende T and Meyer J P 2013 A review on the thermodynamic optimisation and modelling of the solar thermal Brayton cycle, *Renewable and Sustainable Energy Reviews* **28** pp 677–690
[13] Feidt M 2017 *Finite Physical Dimensions Optimal Thermodynamics 1*, Tuition House, 27-37 St. George'Road – London SW 4EU – UK, 272 p
[14] Feidt M 2017, *Finite Physical Dimensions Optimal Thermodynamics 2*, Tuition House, 27-37 St. George'Road – London SW 4EU – UK, 280 p
[15] Feidt M, Dumitrașcu Ghe, Horbaniuc B 2018 *Sensitivity analysis of and endoreversible Carnot engine using Finite Physical Dimensions Optimal Thermodynamics*, Colloque Francophone en Energie, Environnement, Economie et Thermodynamique, June 28 – 29, INSA Strasbourg, France
[16] Dumitrașcu Ghe 2008, The way to optimize the irreversible cycles, *Revista Termotehnica* **2**, 18-22, ISSN 1222-4057, online ISSN 2247-1871

Acknowledgments
This work was supported by UEFISCDI Romania – research grant PCCDI 32/2018.

IOP Conf. Series: Materials Science and Engineering **595** (2019) 012023 doi:10.1088/1757-899X/595/1/012023

Design analysis of low pressure distillation column for cryogenic air separation

M Pintilie[1], A Şerban[1], V Popa[2] and C L Popa[2]

[1]Faculty of Mechanical Engineering and Mechatronics, University Politehnica of Bucharest, Romania
[2]Faculty of Engineering, "Dunarea de Jos" University of Galati, Romania

E-mail: viorel.popa@ugal.ro

Abstract. Cryogenic air separation units are widely used in industry, which consume a large energy and need to produce a huge amount of need of nitrogen, oxygen and argon products with high purity. High purity distillation columns are critical unit operations in cryogenic air separation plants. In this paper the low pressure distillation column is investigated by using a steady state model. The first principles mathematic model of the column is firstly developed, and the pressure, temperature and static liquid nitrogen, oxygen and argon concentrations in the upper column are presented. The obtained results give important information on the variation of temperature, pressure, component concentrations and molar flows within the distillation column. The study highlighted the importance of considering air as ternary mixture. The separation process performance is significantly influenced by the location of side draw of argon and reflux ratio at the top of the column

1. Introduction

Products of air separation processes are very important in a variety of market sectors, such as petrochemical, metal, food processing and health care.

In the air separation industry, cryogenic distillation is the dominant technology for separating large quantities of air into individual high purity component products. Due to the complexity of the process, in addition to significant energy input, air separation units also have high degrees of material and thermal integration and low process agility.

Air separation process plants have the interesting characteristic that energy is considered the raw material. The real chemical raw material used is air, available in virtually limitless quantities, characterized by a uniform chemical composition and it is free.

Cryogenic distillation columns are operated at extremely low temperatures (−170 °C −190°C) to separate air components according to their different boiling temperatures. Thus air separation differs from typical distillation process, because it is not possible to cool the cryogenic columns with water or heat them with steam such as in common distillation processes. A refrigeration process is needed to achieve the low temperature level needed.

In addition, the plant contains two thermally integrated distillation columns that work at different pressures. If the argon columns are also counted, the complexity of the installation is greatly increased.

There are two important aspects that characterize cryogenic air separation systems:

- the products of cryogenic distillation are required in very high purity;
- cryogenic distillation consumes a large amount of energy (up to 85% of the whole air separation field).

A typical structure of a cryogenic air separation plant is presented in figure 1.

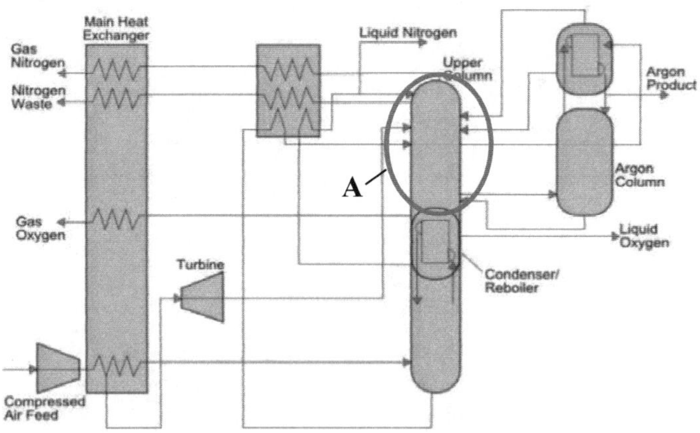

Figure 1. A cryogenic air separation plant that produces argon in addition to oxygen and nitrogen diagram.

The cryogenic distillation zone plays an essential role in the separation process.

In order to obtain the three main components of the air, it must contain the two thermally coupled columns by means of a condenser-evaporator (in which nitrogen and oxygen are obtained) and a crude argon column.

Modelling of a low pressure column (contour A, Figure 1) is required because this column strongly influences the operation of a cryogenic air separation process. This is because the effective separation of the three main air components (nitrogen, oxygen and argon) becomes possible as a result of the distillation process carried out in this column. Additionally, the working pressure in the column influences favorably the separation process, since the relative volatility of a component in a mixture increases with the pressure drop.

2. Low pressure air distillation column model

Steady state modelling is the key for process conceptualization, design and evaluation. However, the steady state is an idealistic definition used by engineers as a representation of "design" conditions which are not always accomplished due to change in raw materials, product specification, change in capacity due to marketing requirements, and the inherent dynamic behavior of processes.

Similar to a steady state model, dynamic models are based on first principles that cannot be violated. Conservation laws, phase equilibrium, heat and mass transfer, and kinetics are also applied in dynamic models. The most significant difference between steady state and dynamic simulation is that steady state assumes that variables are constant with respect to the time. This means that in steady state there is no accumulation in the system so the overall mass and energy input matches its output. Conversely, dynamic models take into account the mass and energy rate of accumulation within the system, which allows one to determine the how long it would take to reach a stable condition starting from a specified initial state.

Generally the steady state and dynamic modelling of air separation column are designed according to the pressure difference and mass flow rate.

2.1.Mathematical model and process simulation

Due to the continuous mass and heat transfer between the liquid and vapor phase, the model for each tray is generally comprised of equations for the total and components mass balances and energy balances, and a set of algebraic equations for vapor-liquid phase equilibrium relations and fraction summation equations (MESH equations) [1].

For each theoretical stage, the mass balance of individual components, enthalpy balance, and vapor liquid equilibrium equation can be written. The sum of these equations creates the mathematical model of a theoretical stage. The mathematical model of a column is composed of models of individual theoretical stages.

The following reasonable simplifying assumptions are considered:

- The liquid and vapor are ideally mixed on each stage, and the pressure and temperature on a tray are uniform;
- The vapor phase holdup on the stages is negligible;
- Vapor-liquid equilibrium is assumed to be ideal;
- Weeping and entrainment are ignored;
- Supposing air as a ternary mixture of nitrogen (78.1%), oxygen (20.95%) and argon (0.95%).

A schematic diagram of a theoretical stage is presented in Figure 2.

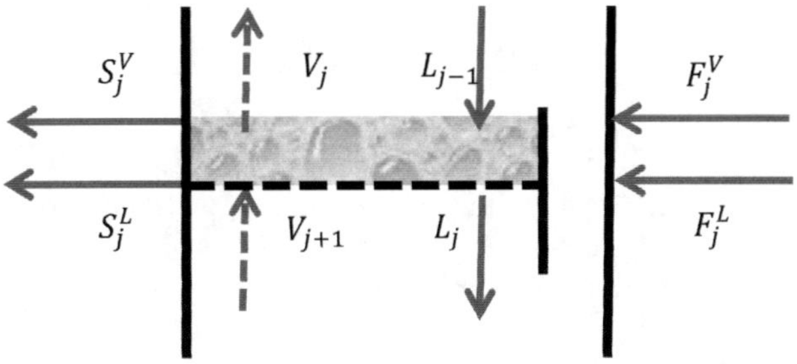

Figure 2. The schematic diagram of a theoretical stage.

The mass balance equation for each tray is of the form:

$$F_j^V + F_j^L + V_{j+1} + L_{j-1} - V_j - S_j^V - L_j - S_j^L = 0 \quad (1)$$

where: j is the number of theoretical stage, starting from the upper column; F_j^V, F_j^L is molar flow of vapor, respectively liquid, which enters in the column on stage j; S_j^V, S_j^L is molar flow of vapor and liquid side withdraw out of *j*-th stage respectively.

The mass balance per component is given by:

$$F_j^V \cdot z_{i,j}^V + F_j^L \cdot z_{i,j}^L + V_{j+1} \cdot y_{i,j+1} + L_{j-1} \cdot x_{i,j-1} - (V_j + S_j^V) \cdot y_{i,j} - (L_j + S_j^L) \cdot x_{i,j} = 0 \quad (2)$$

where

- i is the component index(1-nitrogen, 2-argon,3-oxygen)
- $x_{i,j}$, $y_{i,j}$ are vapor and liquid compositions
- $z_{i,j}^V$, $z_{i,j}^L$are composition of vapor and liquid which enter in the column on j stage

The energy balance for the ideal stage j is:

$$F_j^V \cdot H_j^{FV} + F_j^L \cdot H_j^{FL} + V_{j+1} \cdot H_{j+1}^V + L_{j-1} \cdot H_{j+1}^L - (V_j + S_j^V) \cdot H_j^V - (L_j + S_j^L) \cdot H_j^L - Q_j = 0 \quad (3)$$

All enthalpy calculations are employed Lee-Kesler method [2]:

$$H_j^V = \sum_{i \in CP} y_{i,j} \cdot \underline{H}_{i,j}^V(T_j, p_j) + \Delta H_{mix}^V(T_j, p_j) \quad (4)$$

$$H_j^L = \sum_{i \in CP} x_{i,j} \cdot \underline{H}_{i,j}^L(T_j, p_j) + \Delta H_{mix}^L(T_j, p_j) \quad (5)$$

$$H_j^{FV} = \sum_{i \in CP} y_{i,j} \cdot \underline{H}_{i,j}^{FV}(T_j^F, p_j^F) + \Delta H_{mix}^{FV}(T_j^F, p_j^F) \quad (6)$$

$$H_j^{LV} = \sum_{i \in CP} y_{i,j} \cdot \underline{H}_{i,j}^{LV}(T_j^F, p_j^F) + \Delta H_{mix}^{LV}(T_j^F, p_j^F) \quad (7)$$

$\underline{H}_{i,j}^V$, $\underline{H}_{i,j}^L$, $\underline{H}_{i,j}^{FV}$ and $\underline{H}_{i,j}^{LV}$ represent vapor and liquid enthalpies for any component on each tray, being calculated based on the temperature and the boiling pressure corresponding to each component on the tray where they are calculated.

ΔH_{mix} is the enthalpy of the mixtures and is calculated according to the compressibility factor of the interactive binary parameters.

The vapor–liquid equilibrium relation is written as:

$$y_{n,i} = k_i \cdot \gamma_{n,i} \cdot K_{n,i} \cdot x_{n,i} + (1 - k_i) \cdot y_{n,i+1} \quad 8)$$

$$n = N_2, O_2, Ar$$

where: k is the Murphee stage efficiency; K is the ideal temperature dependent vapor–liquid equilibrium constant; γ the activity coefficient.

The ideal vapor–liquid equilibrium constant is calculated as [3].

The activity coefficients were calculated using the Margules equation for multicomponent mixtures [1]:

$$\gamma_{N_2,i} = exp\left[\frac{A_{N_2,O_2} \cdot x_{O_2,i}^2 + A_{N_2,Ar} \cdot x_{Ar,i}^2 + (A_{N_2,O_2} + A_{N_2,Ar} - A_{O_2,Ar}) \cdot x_{O_2,i} \cdot x_{Ar,i}}{RT_i}\right] \quad (9)$$

$$\gamma_{O_2,i} = exp\left[\frac{A_{N_2,O_2} \cdot x_{N,i}^2 + A_{O,Ar} \cdot x_{Ar,i}^2 + (A_{N_2,O_2} + A_{O_2,Ar} - A_{N_2,Ar}) \cdot x_{N_2,i} \cdot x_{Ar,i}}{RT_i}\right] \quad (10)$$

$$\gamma_{Ar,i} = exp\left[\frac{A_{N_2,Ar} \cdot x_{N_2,i}^2 + A_{O_2,Ar} \cdot x_{O_2,i}^2 + (A_{N_2,Ar} + A_{O_2,Ar} - A_{N_2,O_2}) \cdot x_{N_2,i} \cdot x_{O_2,i}}{RT_i}\right] \quad (11)$$

where T is the absolute temperature, R is the ideal gas constant, $A_{j,k}$, accounts for liquid phase interactions between components j and k. Margules constants can be found in [3] and Antoin constants in [4].

For air and its components (oxygen, nitrogen and argon) the Peng-Robinson equation of state is applicable [5]:

$$p = \frac{RT}{v-b} - \frac{a}{V^2+3bV-2b^2} \quad (12)$$

The Chemsep programme was used to solve the system equations.

3. Results and discussions

In figure 3 the temperature and pressure variations in the low pressure column are presented.

Oxygen product at the bottom of the upper column boils at a determined pressure, and therefore the temperature is also determined.

The temperature in distillation column reduces from bottom to top. From stage 40 to 50 there is a high gradient of temperature difference. In dynamic mode it should be controlled with in a certain limit to stabilize the plant.

In the same time, the pressure decreases linearly from bottom to top. The pressure losses in the column are due to the upward movement of the vapor flux.

In figure 4 the products concentration variations along the column are presented.

The diagram highlights that although argon concentration is only 0.93% in ambient air, it has a strong impact on the low pressure column concentration profile.

Therefore, it is not indicated that the low pressure column separation process to be treated as a process for rectifying an oxygen-nitrogen binary mixture, in which the presence of argon produces small perturbations.

In the lower part of the low pressure column (oxygen zone), takes place a separation process of an oxygen-argon binary mixture. At the upper part of the oxygen zone, where the argon reaches a maximum concentration and where the side draw argon is located, the process of rectification of the oxygen-argon binary mixture turns into several stages in a process of rectification of the oxygen-argon-nitrogen ternary mixture.

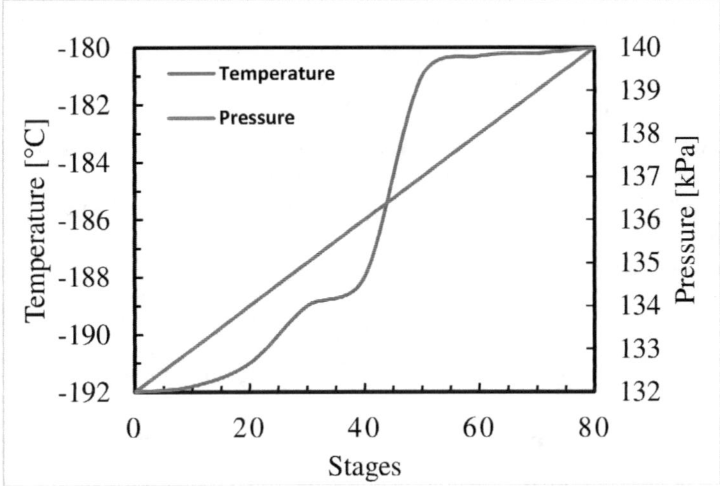

Figure 3. Temperature and pressure variation in low pressure distillation column.

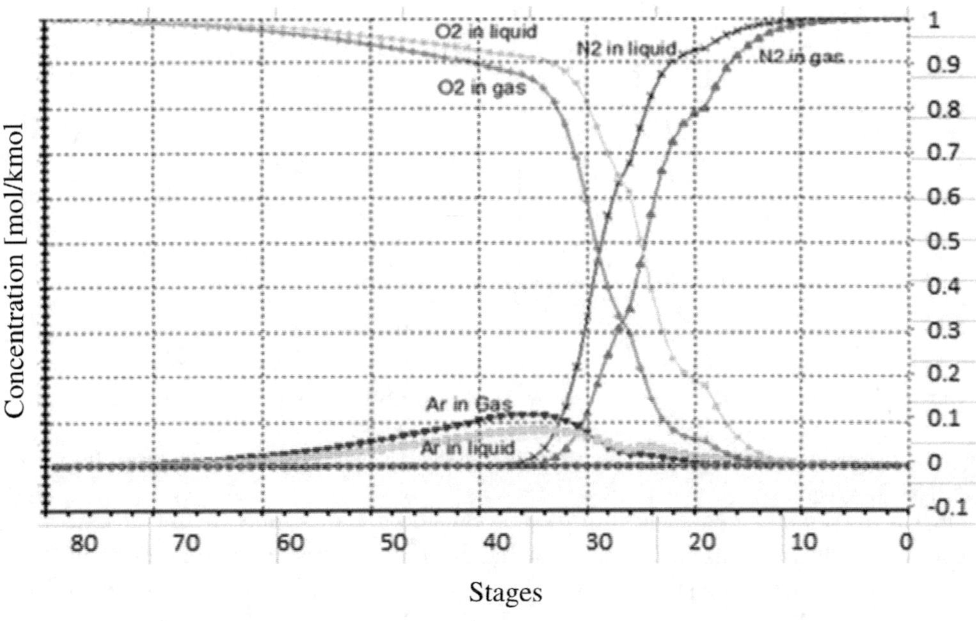

Figure 4. Products molar concentration variation in low pressure distillation column.

The LP column is the column which normally operates nearest to the minimum reflux ratio and hence the addition of stages beyond a certain point has little further effect.

Composition changes in the system are generally most extreme in the center of the column and sluggish at either end. At the bottom, or oxygen end, the third major component, argon, is very significant. Argon and oxygen have similar boiling points and volatilities, therefore most of the stages in the bottom section of the LP column are concerned with argon/oxygen separation.

The argon concentration changes relatively slowly from stage to stage, and generally reaches a local maximum at two different places. At the lower of these, the concentration is generally about 10% and may be as high as 15% argon, and this is the point at which the vapour stream to the crude argon side column is normally withdrawn.

The diagram shows a second argon accumulation point, with a much lower concentration, upper the feed point of the low pressure column with oxygen enriched liquid.

The location for side-draw of argon column must be low enough to prevent N_2 buildup in argon column, but y_{Ar} high enough.

In figure 5 the net molar flow variation in column is presented.

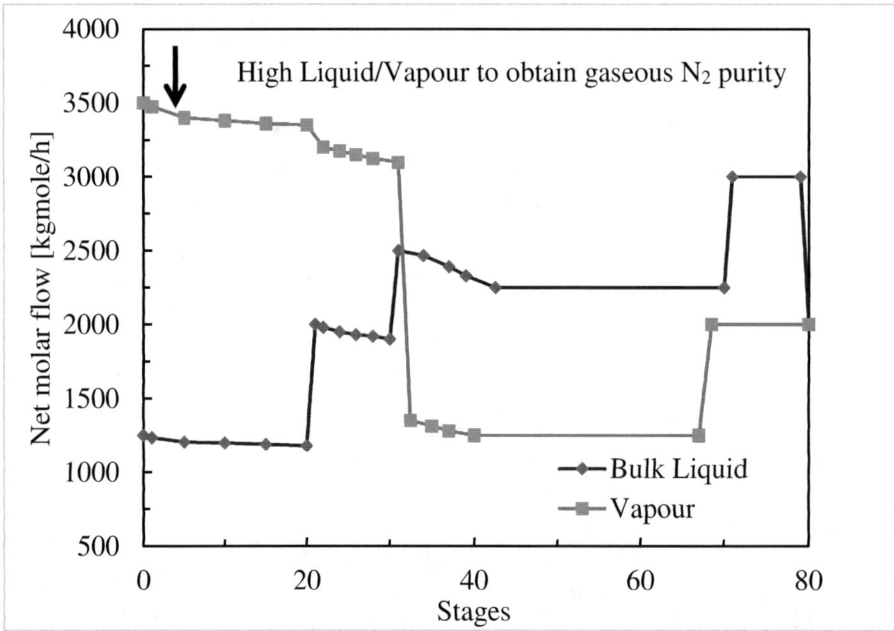

Figure 5. Products molar concentration variation in low pressure distillation column.

From this figure results that to obtain a high purity of gaseous nitrogen product a high reflux ratio L/V is necessary.

To produce a high quality gaseous nitrogen product, the reflux ratio in the top of the LP column must be improved significantly over that required for normal oxygen production. This is done by drawing off some of the vapour rising through the LP column as waste nitrogen at a point below the top of the column, leaving a smaller amount of the rising vapour to be stripped of oxygen by the full reflux flow.

4. Conclusions

By using a steady state model valuable information can be obtained on the design of the low pressure column. Column performances are affected by a number of factors like reflux ratio, feed conditions and products rate. The reflux ratio of the low pressure column may be increased by increasing the poor liquid flow. So doing this might improve the pure nitrogen purity, however, this will also tend to decrease the purity of the poor liquid, which in turn could cause a worse pure nitrogen purity. The location of the side draw stream to the argon column and the liquid-feed to the low pressure column, and the vapor–liquid feed ratio is critical to the purity of the final product. As the withdrawal of any product stream is increased then the purity of that stream will decrease. For example reducing the gaseous oxygen withdrawal increases the oxygen purity.

References

[1] Kumar L 2014 Analysis of steady state Cryogenic Air Separation unit of Rourkela Steel Plant and simulation of Fixed Bed Adsorption Separation of Air. Department of Mechanical Engineering National Institute of Technology Rourkela-769008, Odisha June 2014

[2] Prausnitz J M, Lichtenthaler R N and de Azevedo E G 1986. Molecular thermodynamics of fluid-phase equilibria. Englewood Cliffs, NJ: Prentice-Hall

[3] Lee B and Kesler M 1975 A generalized thermodynamic correlation based on three-parameter corresponding states, *AICHE J.* **21(3)** pp 510-527

[4] Harmens A 1970 Vapour-liquid equilibrium N_2-Ar-O_2 for lower argon concentrations. Cryogenics, **6** pp 201-213

[5] http://webbook.nist.gov/chemistry/.

[6] Lopez-Echeverry J S, Reif-Acherman S and Araujo-Lopez E 2017 Peng-Robinson equation of state: 40 years through cubic, *Fluid Phase Equilibria.* **447** pp 39-71

Design of air cooling unit with PCM

Ç Susantez[1], E Akyol[1], O Hacıhafızoğlu[1] and U Akyol[2]

[1] Mechanical Engineering Department, Engineering Faculty, Trakya University, Edirne, Turkey
[2] Mechanical Engineering Department, "Tekirdag Namik Kemal University", Çorlu, Tekirdag, Turkey

E-mail: cigdemsusantez@trakya.edu.tr

Abstract. In this paper a heat absorption unit has been designed to cool the entering hot air. For this aim, five slabs of PCM (Phase change material) have been used inside the cooling unit. The velocity and temperature values of entering hot air respectively are 0.1 m/s and 50°C. Three different cases for PCM design were investigated: in the first case RT 44HC, RT 35HC, RT 31, RT 28HC and RT 25HC as phase change materials were used in such a way that the PCM with higher number gets contact with the hot air first. In the second and third cases only one PCM type were used for the five slabs inside the cooling unit. They are RT 44HC and RT 25HC. Analysis performed for 10h. Results show that although maximum heat absorbed for the first 8h is in the case of only RT 25HC used in the cooling system, after 8h the cooling unit with mixed PCM shows a better performance and minimum exit air temperature has been observed for this case at the end of 10h.

1. Introduction

Phase change materials have been used in many applications such as latent thermal energy storage systems, building and automotive applications, thermal protection of electronic devices and foods, heat and cold therapy applications and logistics of pharmaceuticals [1-4]. It is also possible to see PCMs in many heat exchanger configurations. It is important to select the true PCM for the heat exchanger. In this respect the properties of latent heat, phase change temperature range and thermal conductivity of PCM play an important role [5].

There are many studies on different kinds of heat exchangers taking the advantage of PCM. Raj and Velraj [6] made an experimental and numerical study on their developed modular shell and tube type heat exchanger. They concluded that air spacers between the modules increase the heat transfer by increasing the retention time of air up to the frontal velocity of 2 m/s. Sun et al. [7] experimentally investigated the thermal behaviour of PCMs in rectangular slabs inside the wind tunnel. They showed that increasing the inclination angle of slab, air inlet temperature and velocity have positive effect on the energy charging speed. Saeed et al [8] performed an experimental work on plate type heat exchanger working with water and PCM. They obtained the result that the optimum plate-plate spacing is 1 in for reducing the PCM shelf-shielding and having a lower exit water temperature. Sharma et al. [9] made a numerical study on melting fraction of fatty acids as PCMs. They showed that capric acid is the best among investigated PCMs for latent heat storage system. Lopez et al. [10] numerically modelled heat exchanger with PCM slabs and exposing to air flow for building applications. They also validated their results with experiments. Hed and Bellander [11] presented a mathematical model for PCM air heat exchanger by taking into account the specific heat capacity of PCM as a function of temperature. They also verified their model with experiments. Dolado et al. [12] made an experimental and numerical study on PCM air heat exchanger. In their model they not only took into account temperature dependent enthalpy and thermal conductivity, but also convection inside the PCM by using effective conductivity. They concluded that an enhancement of the PCM thermal conductivity causes insignificant improvement on the thermal performance of thermal energy storage unit and 1 °C mismatching in the temperature of PCM enthalpy-temperature curve can cause relative errors up to 20 % in the power curve of thermal energy storage unit.

In this study a new design for the PCM air heat exchanger has been proposed. In the cooling unit five PCMs were used to cool the entering hot air. Three different cases were investigated. While in the first case five different types of PCMs were used, in other cases only RT 44HC and RT 25HC were used respectively.

2. Material and Method

The design of the investigated air cooling unit has been presented in figure 1 for three different investigated cases. The inlet air temperature is 50 °C. As it can be shown in figure 1, the PCM which has the highest melting temperature gets contact with the hot air first in the case of a, while in other cases only one type of PCM is used.

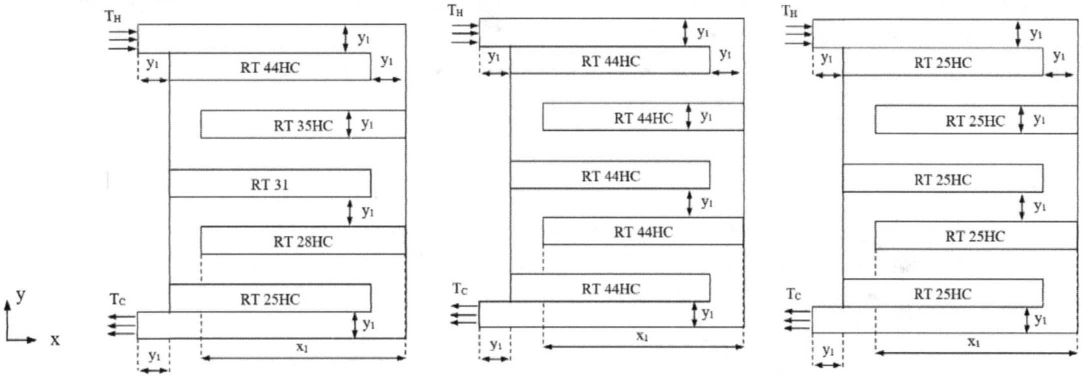

a) with mixed PCM **b)** with RT 44HC **c)** with RT 25HC

Figure 1. Cooling unit configurations.

Geometric dimensions of the cooling unit, thermophysical properties of air and PCMs are given in Table 1 and 2 respectively. Temperature dependent expressions of density, thermal conductivity and dynamic viscosity values of air have been obtained in the range of 20°C-50°C by cure fitting in Sigma Plot.

Table 1. Geometric dimensions and thermophysical properties.

x_1 (m)	0.11		
y_1 (m)	0.01		
c_p (J/kgK)	1007	[13]	
ρ (kg/m^3)	$1.2772\text{-}3.7\cdot10^{-3}$ T (°C)	[13]	$R^2 = 0.9991$
μ (kg/m s)	$1.7337\cdot10^{-5}+4.6\cdot10^{-8}$ T (°C)	[13]	$R^2 = 0.9999$
k (W/m K)	$2.37\cdot10^{-2}+7.3786\cdot10^{-5}$T (°C)	[13]	$R^2 = 1$

The governing equations are given below.

Table 2. Thermophysical properties of PCMs.

PCM type	ρ_{PCM} (kg/m³)	c_{PCM} (kJ/kgK)	k_{PCM} (W/mK)	L (kJ/kg)	T_s (°C)	T_l (°C)
RT 44HC [14]	750	2	0.2	220	41	44
RT 35HC [14]	825	2	0.2	210	34	36
RT 31 [14]	820	2	0.2	155	27	33
RT 28HC [14]	825	2	0.2	220	27	29
RT 25HC [14]	825	2	0.2	180	22	26

Energy equation for PCM:

$$\rho_{PCM} c_{PCM} \frac{\partial T}{\partial t} = k_{PCM} \left(\frac{\partial^2 T}{\partial x^2} + \frac{\partial^2 T}{\partial y^2} \right) + S_H \tag{1}$$

$$S_H = -\rho_{PCM} L \frac{\partial \beta}{\partial T} \frac{\partial T}{\partial t} \tag{2}$$

$$\beta = \begin{cases} 0 & if\ T \le T_s \\ \frac{T-T_s}{T_l-T_s} & if\ T_s < T < T_l \\ 1 & if\ T \ge T_l \end{cases} \tag{3}$$

where S_H is the source term, L is the latent heat, β is the melt fraction, T_s and T_l are solidified and liquefied temperatures respectively.

Continuity and momentum equations for air:

$$\frac{\partial \rho}{\partial t} + \frac{\partial \rho u}{\partial x} + \frac{\partial \rho v}{\partial y} = 0 \tag{4}$$

$$\rho\left(\frac{\partial u}{\partial t} + u\frac{\partial u}{\partial x} + v\frac{\partial u}{\partial y}\right) = -\frac{\partial P}{\partial x} + \frac{\partial}{\partial x}\left(\mu\frac{\partial u}{\partial x}\right) + \frac{\partial}{\partial y}\left(\mu\frac{\partial u}{\partial y}\right) - \frac{2}{3}\frac{\partial}{\partial x}\left(\mu\left(\frac{\partial u}{\partial x} + \frac{\partial v}{\partial y}\right)\right) \tag{5}$$

$$\rho\left(\frac{\partial v}{\partial t} + u\frac{\partial v}{\partial x} + v\frac{\partial v}{\partial y}\right) = -\frac{\partial P}{\partial y} + \frac{\partial}{\partial x}\left(\mu\frac{\partial v}{\partial x}\right) + \frac{\partial}{\partial y}\left(\mu\frac{\partial v}{\partial y}\right) - \frac{2}{3}\frac{\partial}{\partial y}\left(\mu\left(\frac{\partial u}{\partial x} + \frac{\partial v}{\partial y}\right)\right) \tag{6}$$

Energy equation for air:

$$\rho c_p\left(\frac{\partial T}{\partial t} + u\frac{\partial T}{\partial x} + v\frac{\partial T}{\partial y}\right) = \frac{\partial}{\partial x}\left(k\frac{\partial T}{\partial x}\right) + \frac{\partial}{\partial y}\left(k\frac{\partial T}{\partial y}\right) \tag{7}$$

Initial and boundary conditions are given bellow.

$$\left.\begin{matrix} T = T_H = 50^oC \\ V = V_{in} = 0.1m/s \end{matrix}\right\} \text{ for the inlet of the cooling unit} \tag{8}$$

$$\left.\begin{matrix} P = P_o = 0 \\ \frac{\partial T}{\partial x} = 0 \end{matrix}\right\} \text{ for the outlet of the cooling unit} \tag{9}$$

$$\left.\begin{matrix} u = 0 \\ v = 0 \end{matrix}\right\} \text{ on walls} \tag{10}$$

$$\frac{\partial T}{\partial \vec{n}} = 0 \text{ on the outer surfaces of the cooling unit} \tag{11}$$

$$\left.\begin{matrix} \vec{n}\cdot(\vec{q}_1 - \vec{q}_2) = 0 \\ T_1 = T_2 \end{matrix}\right\} \text{ on contact surfaces of PCM and air} \tag{12}$$

where $\vec{n}$ is unit normal vector, 1 and 2 represent air and PCM respectively.

$$\left.\begin{matrix} u = 0 \\ v = 0 \\ P = 0 \\ T = T_0 = 20^oC \end{matrix}\right\} \text{ if } t=0 \tag{13}$$

In this study, as the thickness of PCM slabs are thin, the effect of natural convection in the liquid phase in PCM is neglected. The average temperature of the cooled outlet air has been calculated and presented for all the investigated configurations of the cooling unit. Analysis has been performed in Comsol Multiphysics finite element and simulation software for 10h.

$$T_C = \frac{1}{A_{out}} \int_{A_{out}} T dA \tag{14}$$

Absorbed heat from the hot air during the process has been calculated according to equation (15).

$$\dot{Q}_{abs} = \dot{m}_{air} c_p (T_H - T_{out,ave}) = \bar{\rho} V_{in} A_c c_p (T_H - T_C) \tag{15}$$

where $\dot{m}_{air}$ is the mass flow rate of air, $\bar{\rho}$ is the density of air at 35 °C ($\bar{\rho}$ =1.145 kg/m^3) and A_c is the cross section area perpendicular to the air flow. Although we took temperature dependent air density expression in the analysis as it effects the exit air temperature, it was an obligation to use a constant density value for the air in the calculation of absorbed heat. The value of A_c is 0.001 m^2 as the depth perpendicular to the x-y plane was taken as 0.1 m.

3. Results and Discussion

Free triangular mesh has been used in the analysis as shown in figure 2. Six different mesh types with different number of elements were investigated in mesh dependency analysis as it can be seen on Table 3. Mesh type with number 6 has been used in the analysis. Validation of this code has also been shown in our previous study [15].

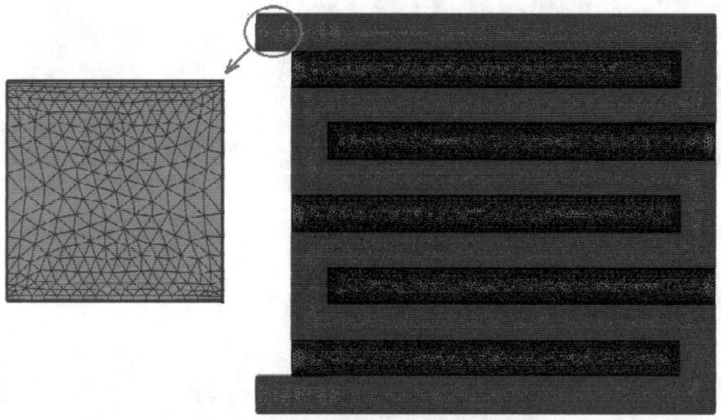

Figure 2. Geometry with mesh.

Table 3. Mesh dependency study.

Mesh type no	Number of Elements	$T_{out,ave}$ (°C) at t = 10 h
1	3253	28.4629
2	5005	28.8838
3	17121	29.2522
4	51355	29.5612
5	57371	29.6260
6	78682	29.9473

Simulational results for temperature distribution have been presented for cooling unit designs with mixed PCM, RT 44HC and RT 25HC in figure 3, 4 and 5 respectively. In the case of mixed PCM, the PCM plate, which has the highest melting temperature, gets in contact with the entering hot air first. It has been shown that RT 44HC, RT 35HC, RT 31, RT 28HC and RT 25HC start to phase change after 1 hour and they are continuing to phase change after 5 h. RT 35HC and RT 31 complete phase changing process while others are still continuing absorbing heat by phase changing at the end of the investigated time of 10h.

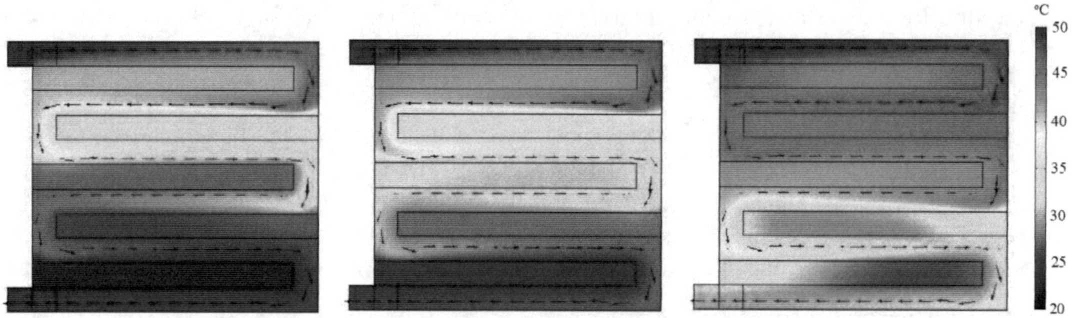

t = 1h t = 5h t = 10h

Figure 3. Temperature distribution in the cooling unit with mixed PCM.

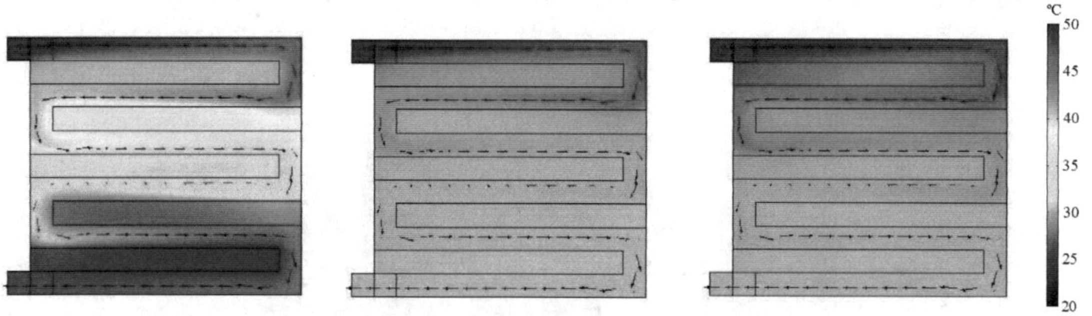

t = 1h t = 5h t = 10h

Figure 4. Temperature distribution in the cooling unit with RT 44HC.

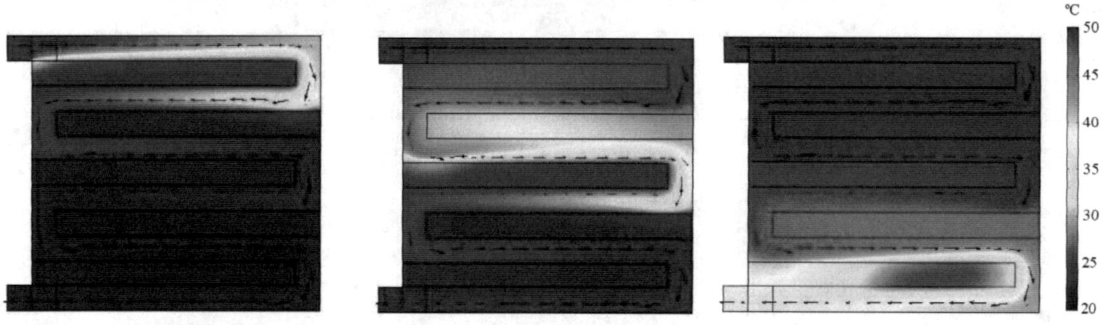

t = 1h t = 5h t = 10h

Figure 5. Temperature distribution in the cooling unit with RT 25HC.

In the case of only RT 44HC is used, only the first PCM which is close to the entrance of hot air starts to phase change after 1 hour. Only three PCM slabs close to the entrance of hot air are in the process of phase change, while others are at the temperature below solidified temperature after 5h. The first PCM which is close to the entrance of hot air almost completes phase change process and others are in the process of phase change at the end of 10h.

In the case of only RT 25HC is used, only three PCM slabs close to the entrance of hot air are in the process of phase change and others are at temperature below solidified temperature at the end of 1h. Only three PCM slabs, which are close to the exit, absorb heat by phase changing and other two PCM slabs are highly above the liquefied temperature at the end of 5h. The PCM slab close to the exit is about to complete phase change while others are highly above the liquefied temperature after 10h.

On the other hand, the average temperature of the air at the exit has presented for all the investigated configurations of the cooling unit as seen in figure 6. It has been observed that until 8h the values of exit air temperature for the cases of mixed PCM and RT 25HC are very close to each other and the minimum exit air temperature is seen for the case of mixed PCM at the end of 10h. The exit air temperature increases rapidly until solidified temperature and then stays constant around this temperature for the case of RT 44HC. In that case average value of the exit air temperature of the air is relatively higher than other cases. Because latent heat has more important effect on the cooling process than sensible heat, solidified temperature of RT 44HC is higher than that of other investigated PCMs and at the beginning of the process in the case of RT 44HC heat is absorbed from hot air in the form of sensible heat rather than latent heat. Increase on the average value of the exit air temperature has been observed after the process of phase change in the PCM slabs complete. Figure 7 shows the absorbed heat from the hot air as a function of time for all the investigated cooling unit designs. When solidified temperatures of PCMs have lower values, phase change process starts more quickly as it can be seen for the cases of mixed PCM and RT 25HC and during the phase change more heat is absorbed from the air. On the other hand, according to equation (15), absorbed heat is directly proportional to $(T_H - T_C)$. When phase change continues inside the PCM slabs in the case of RT 44HC, the exit air temperature is higher and as a result absorbed heat is lower.

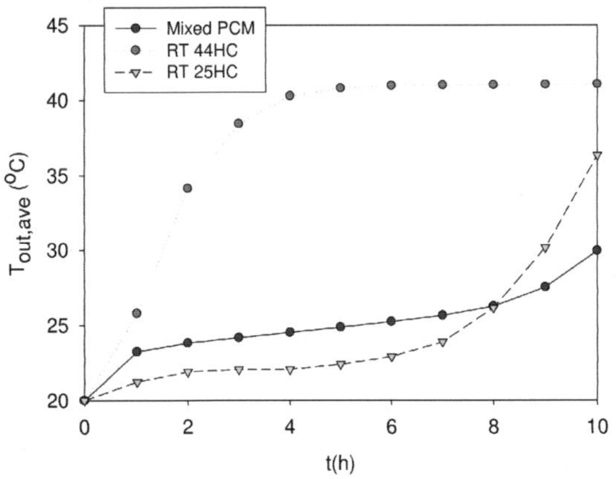

Figure 6. Average temperature of the exit air

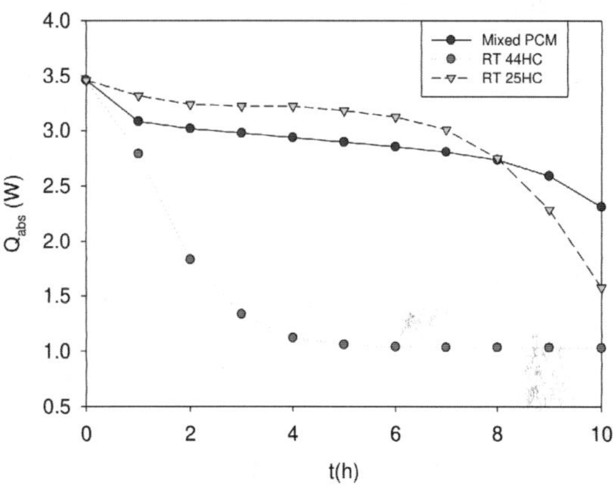

Figure 7. Absorbed heat

4. Conclusion

In this research, three different configurations of cooling unit design with PCMs have been investigated. Cooling units with mixed PCM, RT 44HC and RT 25HC were investigated. In the case of mixed PCM, increasing the absorbed heat was aimed by satisfying the PCM with the highest solidified temperature to get contact first with the hot air. It has been observed that during first 8h cooling unit performances with mixed PCM and RT 25HC are very close to each other and they are better than the case of only RT 44HC is used. The cooling unit design with mixed PCM shows a better performance by absorbing more heat after 8h.

Nomenclature

A: Area (m^2)
A_C: Cross section area perpendicular to the air flow (m^2)
A_{out}: Cross section area perpendicular to the air flow at the outlet (m^2)
c_p: Specific heat of air (J/kgK)
c_{PCM}: Specific heat of phase change material (J/kg·K)

k: Thermal conductivity of air (W/m K)
k_{PCM}: Thermal conductivity of phase change material (W/m K)
L: Latent heat (J/kg)
$\dot{m}_{air}$: Mass flow rate of air (kg/s)
$\vec{n}$: Unit normal vector
P: Pressure (Pa)
P_o: Pressure at the outlet (Pa)
$\dot{Q}_{abs}$: Absorbed heat (W)
$\vec{q}$: Heat flux vector (W/m^2)
S_H: Source term (W/m^3)
T: Temperature (^{o}C)
T_0: Initial temperature (^{o}C)
T_C: Cooled air temperature (^{o}C)
T_H: Hot air temperature (^{o}C)
T_s: Solidified temperatures (^{o}C)
T_l: Liquefied temperatures (^{o}C)
t: time (s, h)
u: Velocity in x direction (m/s)
V_{in}: Inlet velocity (m/s)
v: Velocity in y direction (m/s)
x, y: Coordinates
x_1: Length of the PCM plates (m)
y_1: Width of the PCM plates (m)

Subscripts
1, 2: Represent air and PCM respectively

Greek symbols
β: Melt fraction
μ: Dynamic viscosity of air (kg/m s)
ρ: Density of air (kg/m^3)
$\bar{\rho}$: Density of air at the mean temperature (kg/m^3)
ρ_{PCM}: Density of phase change material (kg/m^3)

References
[1] Ezan M A, Yüksel C, Alptekin E and Yılancı A 2018 Importance of natural convection on numerical modelling of the building integrated PVP/PCM systems *Solar Energy*. **156** 616
[2] Seddegh S, Joybari M M, Wang X and Haghighat F 2017 Experimental and numerical characterization of natural convection in a vertical Shell-and-tube latent thermal energy storage system, *Sustainable Cities and Society*. **35** 13
[3] https://www.rubitherm.eu/en/applications.html
[4] Zalba B, Marin J M, Cabeza L F and Mehling H 2003 Review on thermal energy storage with phase change: materials, heat transfer analysis and applications *Applied Thermal Engineering*. **23** 251
[5] Lazaro A, Dolado P, Marin J M and Zalba B 2009 PCM-air heat exchanger for free-cooling applications in buildings: Empirical model and application to design, *Energy Conversion and Management*. **50** 444

[6] Raj V A A and Velraj R 2011 Heat transfer and pressure drop studies on a PCM-heat exchanger modüle for free cooling applications, *International Journal of Thermal Sciences*. **50** 1573Sun X, Chu Y, Medina M A, Mo Y, Fan S and Liao S 2019 Experimental investigations on thermal behaviour of phase change material (PCM) in ventilated slabs, *Applied Thermal Engineering*. **148** 1359

[7] Saeed R M, Schlegel J P, Sawafta R and Kalra V 2019 Plate type heat exchanger for thermal energy storage and local shifting using phase change material, *Energy Conversion and Management*. **181** 120

[8] Sharma A, Won L D, Buddhi D and Park J U 2005 Numerical heat transfer studies of the fatty acids for different heat exchanger materials on the performance of a latent heat storage system, *Renewable Energy*. **30** 2179

[9] Lopez J P A, Kuznik F, Baillis D and Virgone J 2013 Numerical modeling and experimental validation of a PCM to air heat exchanger, *Energy and Buildings*. **64** 415

[10] Hed G and Bellander R, Mathematical modelling of PCM air heat exchanger 2006, *Energy and Buildings*. **38** 82

[11] Dolado P, Lazaro A, Marin J M and Zalba B 2011 Characterization of melting and solidification in a real scale PCM-air heat exchanger: Numerical model and experimental validation, *Energy Conversion and Management*. **52** 1890

[12] Çengel Y A and Cimbala J M 2008 *Akışkanlar Mekaniği Temelleri ve Uygulamaları* ed Engin T (İzmir: İzmir Güven Kitabevi)

[13] https://www.rubitherm.eu/en/index.php/productcategory/organische-pcm-rt

[14] Susantez Ç, Kahveci K and Akyol E, Effects of the number, thickness and non-uniform distribution of fins on the heat storage characteristics in a cylindrical thermal energy storage with phase change material 2018, *Fırat üniversitesi Mühendislik Bilimleri Dergisi*, **30** 225

IOP Conf. Series: Materials Science and Engineering **595** (2019) 012025 doi:10.1088/1757-899X/595/1/012025

Entropic model and optimization of a refrigeration machine

M Feidt[1], M Costea[2], C Stanciu[2] and D Stanciu[2]

[1]LEMTA, University of Lorraine, Vandoeuvre-lès-Nancy, France
[2]Department of Engineering Thermodynamics, University Politehnica of Bucharest, Bucharest, Romania

E-mail: liana5802@yahoo.fr

Abstract. The refrigeration machines are operating upon various reverse cycles. However, Blanchard, then Goth and Feidt have shown in the 1980s that unlike engines, there is no natural optimum in working fluid temperature, since these temperatures are not bounded in principle. This paper adds complements to the modeling of refrigeration machines on examples of machines derived from the Carnot one, endoreversible or not (model according to Chambadal; model according to Curzon – Ahlborn). The consequences on the machine operation optimization in the presence or absence of additional constraints are reported. The main constraints observed in the literature are (1) imposed refrigeration load, (2) imposed energy consumption, or (3) imposed coefficient of performance (COP). The present work reveals that while Chambadal model does not provide an optimal solution when constraints are not considered, the Curzon-Ahlborn model shows that working fluid temperatures depend on the heat transfer laws at the source and sink, but also on the extensity transfer entropy at the source, ΔS, taken as reference and accounting for the existence of the cycle. The results emphasize an optimal distribution of the physical properties of finite dimensions at the source and sink, as function of ΔS. This new result is the outcome of a sensitivity study. Extensions are under development.

1. Introduction

The equilibrium Thermodynamics has been for a long time the reference modelling for reverse cycle machines through the Carnot Coefficient of Performance (COP) defined as the useful effect divided by the work expenses. The well-known upper-bound of the refrigerating machine COP is:

$$COP = \frac{T_{CS}}{T_{HS}-T_{CS}} \tag{1}$$

where T_{CS} is the cold source temperature and T_{HS} is the hot source temperature.

These considerations have been overcome during the 1980s with the work of Blanchard [1], then Goth and Feidt [2], showing that there is no natural optimum in working fluid temperatures since they are outside of the source and sink temperatures.

The present paper proposes to complete and renew the modelling of reverse cycle machines by developing precise configurations that extend the Carnot one [3, 4]. In the literature, the endoreversible situation is well studied. Various objectives have been studied [5-12], mainly:

– the useful effect (refrigerating load), Q_C, aiming to its maximization;

– the work expenses, W, aiming to its minimization.

These optimizations are performed with or without constraints (for example, imposed refrigeration load, imposed energy consumption, imposed COP, see papers [13, 14]).

The present paper is a total reconsideration from a new point of view relating the heat source and sink heat transfer to the internal cycled fluid through ΔS_C, the extracted entropy at the source. This reference entropy appears as a necessary condition for the cycle existence and the coupling of the source and sink. Consequences of this modelling are carefully examined on some classical cases.

Section 2 of the paper explores the Chambadal and Curzon-Ahlborn models of the machines. The influence of irreversibilities appears gradually through internal or total entropy productions, but also through irreversibility degrees introduced by Novikov [15], and irreversibility ratios introduced by Ibrahim et. al. [16].

Section 3 reports on optimization of the refrigeration machine focusing on minimum work expenses. It is shown that the Chambadal model optimum coincides with the equilibrium thermodynamics one whatever the model refinement is used (basic one, linear heat transfer at the source). When moving to the Curzon-Ahlborn model (section 3.3), it appears that the external temperatures are related to the finite dimensions of the machine (G_H, G_C) that have to be optimally allocated.

Conclusions and perspectives are discussed in the last section.

2. Chambadal's model versus Curzon-Ahlborn's model of a refrigeration machine

2.1. Cycles representation in T –S diagram

The difference between the two machines is illustrated on figure 1.

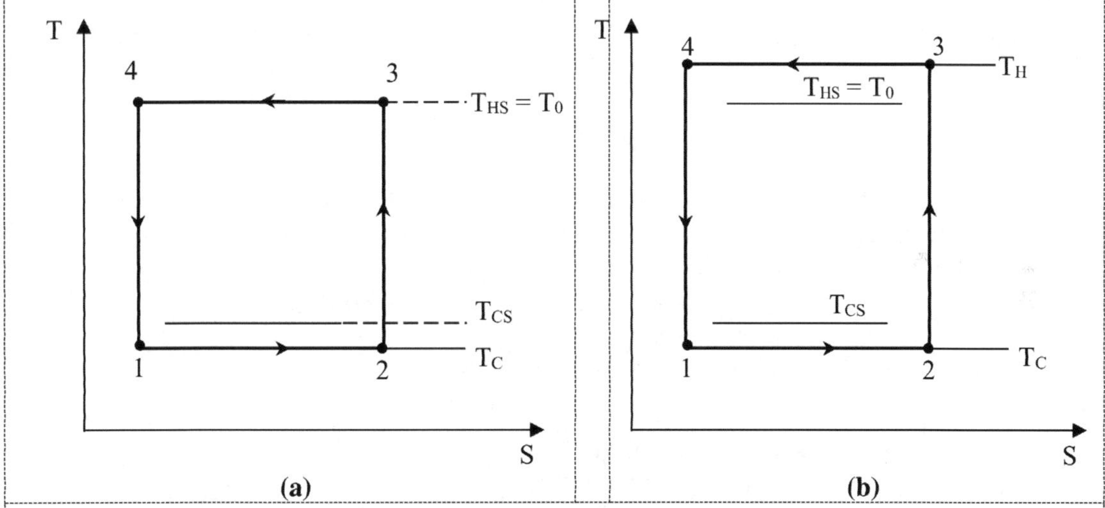

Figure 1. Comparison of the refrigeration machine according to Chambadal (a) and the one according to Curzon-Ahlborn's model (b).

For both machines, the cold side heat reservoir is a thermostat of temperature T_{CS}. The heat transfer at the cold source, Q_C, is done due to the temperature difference between the source and the working fluid, $\Delta T_C = T_{CS} - T_C$. Conversely, for the Chambadal's model, the heat transfer at the hot sink is considered perfect (reversible), at temperature $T_{HS} = T_0$. (T_0, ambient temperature, figure 1a).

Figure 1b corresponds to Curzon-Ahlborn's model that, in addition, considers the temperature difference $\Delta T_H = T_H - T_{HS}$, necessary for the heat removal to the environment.

Note that figures 1a and 1b suppose endoreversible cycles, and ignore the heat losses. Furthermore, one can see on fig. 1 that basically there are no natural bounds at the temperatures T_H and T_C. This fact is a characteristic of the reverse cycle machine [1, 2].

IOP Conf. Series: Materials Science and Engineering **595** (2019) 012025 doi:10.1088/1757-899X/595/1/012025

2.2. The basics of the thermodynamic model

The basics are valid for the two types of refrigeration machines presented in section 1 and also for the heat pumps of the two same types.

The energy balance of these machines when no heat losses are considered is expressed as:

$$|W| = |Q_H| - Q_C \tag{1}$$

where W is the mechanical energy consumed in the cycle, Q_H is the heat removal from the cycle hot side (and useful effect for a heat pump), and Q_C is the heat transferred to the cycle cold side that represents the useful effect for the refrigeration machine.

Generally, the heat transfers to and from the cycle can be differently expressed by using:

- the entropic form

$$Q_H = T_H \Delta S_H \tag{2}$$

$$Q_C = T_C \Delta S_C \tag{3}$$

- the heat transfer law involving a physical dimension G that represents a thermal transmittance, a general transfer law depending on temperature T_S (of hot or cold source), and the cycled fluid temperature T, leading to

$$Q_H = G_H \cdot f_H(T_{HS}, T_H) \tag{4}$$

$$Q_C = G_C \cdot f_C(T_{CS}, T_C) \tag{5}$$

Moreover, the entropy balance of the irreversible cycle of whatever machine is written as:

$$\Delta S_C + \Delta S_I = \Delta S_H \tag{6}$$

where ΔS_I represents the entropy production on the cycle, ΔS_C and ΔS_H are the transfer entropies at the cold and hot side respectively.

Note that a dissymmetry appears between refrigeration machine and heat pump regarding the entropy. Indeed, the input entropy is ΔS_C for the refrigeration machine, while ΔS_H is the one for the heat pump. Therefore, only the refrigeration machine approach will be presented hereafter, for which the reference entropy will be considered $\Delta S = \Delta S_C$ (simplified notation).

Note that the following inequality exits: $\Delta S_{CS} < \Delta S_C = \Delta S < \Delta S_H < \Delta S_{HS}$

By combining equations (2), (3) and (6) in equation (1) leads to:

$$|W| = (T_H - T_C)\Delta S + T_H \Delta S_I \tag{7}$$

The machine irreversibility rises the mechanical work consumption.

One considers also the coefficient of performance *COP*, as a measurement of the energetic efficiency of the refrigeration machine, as:

$$COP = \frac{Q_C}{|W|} \tag{8}$$

By combining equations (1) and (8) leads to:

$$\frac{1}{COP} = \frac{|Q_H|}{Q_C} - 1 \tag{9}$$

The above equation is expressed as a function of temperatures and entropy variation when equations (2), (3), and (6) are replaced in equation (9):

$$\frac{1}{COP} = \left(\frac{T_H}{T_C} - 1\right) + \frac{T_H}{T_C} \cdot \frac{\Delta S_I}{\Delta S} \tag{10}$$

Note that the first term in the right side of equation (10) corresponds to the reverse of the *COP* of the endoreversible refrigeration machine ($\Delta S_I = 0$). The second term accounts for the machine irreversibilities that decrease the real *COP*. Moreover, the first ratio of intensive terms, namely of temperature, is related to thermal transfers, while the second ratio of extensive type, ΔS_I / ΔS, corresponds to the irreversibility degree introduced by Novikov [15]. It corresponds as well to the entropic ratio considered by Ibrahim [16] for the thermal engines. For the refrigeration machine, this ratio is expressed as:

$$I_C = \frac{\Delta S_H}{\Delta S_C} = 1 + \frac{\Delta S_I}{\Delta S} \tag{11}$$

Note that other definitions are possible, and that there is a strong link between the ratio and the irreversibility degree.

Another description would consist of reasoning on the system (composed of the machine and the source and sink). The reader will easily find the energy consumption:

$$|W| = T_{HS}(\Delta S + \Delta S_S) - T_{CS} \cdot \Delta S \tag{12}$$

and the reverse *COP*:

$$\frac{1}{COP} = \left(\frac{T_{HS}}{T_{CS}} - 1\right) + \frac{T_{HS}}{T_{CS}} \cdot \frac{\Delta S_S}{\Delta S} \tag{13}$$

One can see in equation (13) the inverse of the reversible *COP* in the first term of the second member, then the term of total energy degradation, showing this time, the irreversibility degree of the system, $d_{IS} = \Delta S_S / \Delta S$.

3. Optimization options of the refrigeration machine by using the Chambadal's model

The study of the refrigeration machine reveals three main objectives leading the optimization, namely, maximum of useful effect, Q_C, minimum of energy consumption, W, and maximum of *COP*. Nevertheless, from an environmental point of view, minimum heat rejection, Q_H, is required.

To these typically engineering objectives corresponds an objective of the physicist, namely, the minimum of entropy production, that covers at least two aspects:

- entropy production of the machine, ΔS_I;
- entropy production of the system, ΔS_S.

These optimization aspects are summarized in table 1.

Table 1. Essential examples of objective functions.

Objective function	Refrigeration machine RM	Heat pump HP
Useful Effect U. E.	Max Q_C	Max$\|Q_H\|$
Energy Consumption E. C.	min $\|W\|$	min $\|W\|$
Energy Efficiency E. E.	$COP_{RM} = \frac{Q_C}{\|W\|}$	$COP_{HP} = \frac{\|Q_H\|}{\|W\|}$
Reject and Environment R	min $\|Q_H\|$	Max Q_C charge, no reject
Physical: entropy S	min ΔS_I min ΔS_S	min ΔS_I min ΔS_S
others (with or without constraint)		

Further, the paper will focus on the most common objective function, namely the search for the minimum of W, if it exists.

3.1. Refrigeration machine according to Chambadal, in the entropic model only

The combination of equations (1), (2), (3) and (6) allows to express the mechanical work as:

$$|W| = T_0(\Delta S + \Delta S_I) - T_C \cdot \Delta S \quad (14)$$

For ΔS as parameter of the model, other than zero so that the cycle exists, one can see that W is a decreasing function of T_C, whose minimum is zero when $T_C = T_0$ (thermodynamic equilibrium and machine absence).

3.2. Refrigeration machine according to Chambadal, in the entropic model with a linear heat transfer law at the source

The model of section 3.1 is now completed by the heat transfer equation at the source:

$$Q_C = G_C(T_{CS} - T_C) \quad (15)$$

By combining equations (15) and (3), the expression of the reference heat transfer entropy results as a function of T_C:

$$\Delta S = G_C\left(\frac{T_{CS}}{T_C} - 1\right) \quad (16)$$

Furthermore, equations (14) and (15) replaced in equation (16) lead to:

$$|W| = G_C\frac{(T_0 - T_C)(T_{CS} - T_C)}{T_C} + T_0 \cdot \Delta S_I \quad (17)$$

The derivation with respect to T_C confirms the decrease of W with T_C, whatever T_0, T_{CS}, G_C and ΔS are. This confirms the non-existence of a minimum of energy consumption for the irreversible refrigeration machine, contrarily to the engine, even if ΔS_I is a function of T_C (various cases are available near the authors). Moreover, Q_C and Q_H are also monotonous functions decreasing with T_C.

3.3. Refrigeration machine according to Curzon – Ahlborn, on an entropic basis

The model is still the one presented in section 3.2 further completed by the heat transfer equation at the machine hot side (see figure 1b):

$$Q_H = G_H(T_H - T_{HS}) \quad (18)$$

However, the original bias of the model is to maintain as parameter the reference entropy, ΔS, that ensures the cycle existence condition and consequently, of the machine.

The combination of equations (3) and (15), then of equations (2) and (18) leads to the following expressions of the temperatures T_C and T_H:

$$T_C = \frac{G_C T_{CS}}{G_C + \Delta S} \quad (19)$$

$$T_H = \frac{G_H T_{HS}}{G_H - (\Delta S + \Delta S_I)} \quad (20)$$

Then, the mechanical work expression can be derived as:

$$|W| = G_C\frac{(T_0 - T_C)(T_{CS} - T_C)}{T_C} + T_0 \cdot \Delta S_I \quad (21)$$

with ΔS, ΔS_I, T_{HS} and T_{CS} as parameters, and G_H and G_C as variables of finite dimensions related by:

$$G_H + G_C = G_T \quad (22)$$

The variational method [17] applied to the model provides the optimal distribution of G_H and G_C leading to the minimum energy expenses:

$$G_C^* = \frac{\Delta S\left[G_T\sqrt{T_{CS}} - (\Delta S + \Delta S_I)\left(\sqrt{T_{HS}} + \sqrt{T_{CS}}\right)\right]}{(\Delta S + \Delta S_I)\sqrt{T_{HS}} + \Delta S\sqrt{T_{CS}}} \quad (23)$$

$$G_H^* = \frac{(\Delta S+\Delta S_I)\left[G_T\sqrt{T_{HS}}+\Delta S\left(\sqrt{T_{HS}}+\sqrt{T_{CS}}\right)\right]}{(\Delta S+\Delta S_I)\sqrt{T_{HS}}+\Delta S\sqrt{T_{CS}}} \tag{24}$$

By replacing these expressions in equation (21), the minimum of the consumed mechanical work results as:

$$min|W| = \frac{1}{G_T-\Delta S_I}\left\{\sqrt{T_{HS}}(\Delta S+\Delta S_I)\left[G_T\sqrt{T_{HS}}+\Delta S\left(\sqrt{T_{HS}}+\sqrt{T_{CS}}\right)\right]-\right.$$
$$\left.-\sqrt{T_{CS}}\Delta S\left[G_T\sqrt{T_{CS}}-(\Delta S+\Delta S_I)\left(\sqrt{T_{HS}}+\sqrt{T_{CS}}\right)\right]\right\} \tag{25}$$

Note that equation (25) delivers the endoreversible case for $\Delta S_I = 0$. Also, it shows that min$W$ is an increasing function of ΔS, and a decreasing one of G_T.

4. Conclusions and perspectives

The present work deals with an approach of a refrigeration machine operating upon an irreversible Carnot cycle.

The proposed originality is methodological, since it considers a reference heat transfer entropy ΔS, corresponding to the heat input in the refrigeration machine cycle.

After a brief review of the useful concepts to the refrigeration machine according to Equilibrium Thermodynamics, we consider the equivalent of the Chambadal engine transposed into refrigeration machine, by focusing the search on a minimum of energy supply to the machine.

It appears that the proposed model, as the one from Equilibrium Thermodynamics, does not lead to optimum.

Conversely, the use of a Curzon - Ahlborn type model shows that considering the reference transfer entropy, that ensures the existence of the cycle and therefore, of the machine, is sufficient to lead to a minimum. Thus, there is no need to introduce an additional constraint, i.e. the imposed useful effect, or other types, as done in the literature [13, 14]. This model provides an optimum design by the optimal distribution of a finite physical dimension G_i.

One recalls that according to the model, the quantity G in steady state dynamic regime corresponds to:

- a heat transfer conductance, $K_i = k_i \cdot A_i$, at the source and sink;
- a product of the heat exchanger effectiveness, ε_i, multiplied by the heat capacity rate of the external fluid, $\dot{C}_i = \dot{m}_i c_{p,i}$, for an external transfer in sensible heat.

To the previous general case, one can add the typical specific cases where:
- the useful effect is imposed: $Q_C = Q_0$;
- the energy consumption is imposed: $W = W_0$;
- the machine COP is imposed: $COP = COP_0$.

These cases will be revisited in the future research, by looking for the particular values of the new variable ΔS, added to these constraints.

References

[1] Blanchard C H 1980 Coefficient of performance for finite speed heat pump *J. App. Phys.* **51** 2471–2472
[2] Goth Y and Feidt M 1986 Recherche des conditions optimales de fonctionnement des pompes à chaleur ou machines à froid asociées à un cycle de Carnot endoréversible *C. R. Acad. Sci.* **303** 113–122 (In French)
[3] Chambadal P 1957 *Les centrales nucleaires* (Paris: A Colin)
[4] Curzon F L and Ahlborn B 1975 Efficiency of a Carnot engine at maximum power output *Am. J. of Phys.* **1** 22-24
[5] Klein S A 1992 Design considerations for refrigeration cycles *Int. J. Refrig.* **15(3)** 181-185
[6] Chen L, Bi Y, Sun F and Wu C 1999 A generalized model of a combined refrigeration cycle

and its performance *Int. J. Therm. Science* **38(8)** 712-718

[7] Gordon J M and Ng K C 2000 *Cool Thermodynamics* (Cambridge: Cambridge International Science Publishing)

[8] Sieniutycz S 2003 Thermodynamic limits on production or consumption of mechanical energy in practical and industrial systems, *Progress in Energy and Combustion Science*. **29(3)** 193-246

[9] Calvo Hernández A, Medina A, Roco J M M and Velasco S 2004 Comparison of optimum thermal conductance distribution for an irreversible Carnot cycle under different optimization criteria *Advances in Finite Time Thermodynamics, Analysis and Optimization* (New York: Nova Science Publishers)

[10] Yu J, Zhou Y, Liu Y 2011 Performance optimization of an irreversible *Carnot* refrigerator with finite mass flow rate *Int. J. Refrig.* **34(2)** 567-572

[11] Grazzini G and Rocchetti A 2014 Thermodynamice optimization of irreversible refrigerators *Energy Convers. Mgmt* **84** 583-588

[12] Açıkkalp E 2014 Entransy analysis of irreversible Carnot-like heat engine and refrigeration cycles and the relationships among various thermodynamic parameters *Energy Convers. Mgmt* **80** 535-542

[13] Feidt M 2010 Thermodynamics applied to reverse cycle machines, a review *Int. J. of Refrig.* **33** 1327-1342

[14] Feidt M 2013 Evolution of thermodynamic modelling for 3 or 4 heat reservoirs reverse cycle machines: a review and new trends *Int. J. Refrig.* **36** 8-23

[15] Novikov I I 1958 The efficiency of atomic power plants (a review) *J. Nuclear Energy* **7** 125-128

[16] Ibrahim O M, Klein S A and Mitchell J W 1991 Optimum heat power cycles for specified boundary conditions *J. of Eng. for Gas Turb. Power* **113(4)** 514-521

[17] Feidt M 2017 *Finite Physical Dimensions Optimal Thermodynamics 1 – Fundamentals* (London: ISTE Press Elsevier) and *Thermodynamique optimale en dimensions physiques finies* 2013 (Paris: Hermes) (In French)

IOP Conf. Series: Materials Science and Engineering **595** (2019) 012026 doi:10.1088/1757-899X/595/1/012026

Analysis of a proportional pressure regulator

M Avram[1], M-F Ştefănescu[1], A Şerban[1], A Dobrovicescu[1] and B Grămescu[1]

[1]Faculty of Mechanical Engineering and Mechatronics, University Politehnica of Bucharest, Romania

E-mail: adobrovicescu@yahoo.com

Abstract. The paper deals with the theoretical and experimental analysis of a proportional pneumatic device. Such a piece of equipment can be integrated in the structure of an intelligent system that works with compressed gases. Different constructive variants of this type of equipment are analyzed. Authors focus their attention on a device that makes use of an electro-mechanic actuator operating with a proportional electromagnet. For the mathematical model some simplifying assumptions are identified. The mathematical model is numerically integrated. The behavior of the proportional pneumatic device at the variation of the inlet pressure and outlet mass flow rate consumption is pointed out.

1. Introduction

Modern pneumatic drive systems have in their structure besides "conventional" equipment of regulation and control also proportional equipment [1, 2]. What makes different the two families of equipment are first of all the command they use. For the first family the two parameters – pressure and flow rate can be modified only by manual adjustment, and the electric drive, when is used is of the kind" all or nothing". For the second family the parameter adjustment is continuously achieved, electronically following a pre-established program.

In the present paper authors intend to make a careful analysis of the existing solutions of pieces of equipment for the proportional control of pressure and to find out some new, original ones.

Any attempt in augmenting the heat transfer performance of a heat exchanger is accompanied by an increase in the pressure drop.

2. Constructive variants for proportional regulators

Nowadays there are several solutions to build up pneumatic proportional regulators.

A first possibility [3, 4] consists in using a piece of equipment with a cylindrical drawer with a translation movement (Figure 1).

This one can be used, with small changes, both for the flow and pressure regulation. When is used as a pressure regulator, from the left room Cs is missing the compression spring that exist when the structure is used as proportional distributor and through a circuit, the pressure is supplied from the consumer hole (2). In this case against the drawer acts to forces: the force developed by the proportional electromagnet F_{EMP} and the pressure force F_P.

IOP Conf. Series: Materials Science and Engineering **595** (2019) 012026 doi:10.1088/1757-899X/595/1/012026

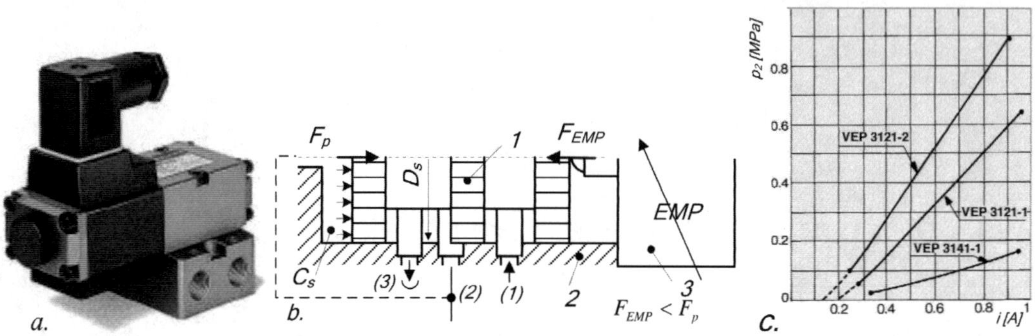

Figure 1. A three-hole equipment: a - an image of the equipment, b - the principal schematic, c – the static characteristic of the equipment.

This one can be used, with small changes, both for the flow and pressure regulation. When is used as a pressure regulator, from the left room C_s is missing the compression spring that exist when the structure is used as proportional distributor and through a circuit, the pressure is supplied from the consumer hole (2). In this case against the drawer acts to forces: the force developed by the proportional electromagnet F_{EMP} and the pressure force F_P.

If $F_{EMP} < F_P$ the consumer hole (2) is connected to the atmosphere, and pressure decreases until balance is achieved and the hole (2) is blocked. If $F_{EMP} > F_P$ the hole (2) is connected at the pressure source and pressure increases until the equilibrium is achieved. In this way pressure at the hole level (2) is kept constant at value proportional to the intensity of the driving current even when perturbations are present; pressure is calculated such that:

$$p_2 = k_{EP} \cdot i \tag{1}$$

where k_{EP} represents the constant of the proportional piece of equipment.

$$k_{EP} = \frac{4 \cdot k_{EMP}}{\pi \cdot D_s^2} \tag{2}$$

Based on the same idea five-hole equipment can be provided. These can control both the movement sense of the motor and the pressure value in one of its active rooms. The SMC Company produces pressure regulators that operate on the present principle.

In Figure 1 a three-hole equipment is presented (an image of it - Figure 1.a, the principle schematic – Figure 1.b and the pressure characteristic – intensity for three variants of proportional regulators, with different regulation fields – Figure 1.c).

Such a piece of equipment, with tree consumers is presented in Figure 1 (Figure 1.a - an image of the equipment, Figure 1.b - the principal schematic and Figure 1.c – the static characteristic of the equipment).

A second solution comes from the well-established constructive solution for a manual pneumatic pressure regulation [2, 3]. The only difference comes from the fact that in this case the discharge to the atmosphere valve is differently achieved. Such equipment (Figures 2 and 3) has in its structure two mains subassemblies:

- the main stage for amplification,
- the pilot subassembly.

Then main stage for amplification contains the following elements:

- The mobile assembles composed of a plain valve 1, the rod 2, membrane 3 and the two discs 4 that make up the rigid of the membrane;
- The plane valve 5 which can slide along the rod 2;
- The body of the equipment 8 that contains the holes (1), (2), (3).

The main stage for amplification contains the following elements:

- the mobile assembles composed of a plain valve 1, the rod 2, membrane 3 and the two discs 4 that make up the rigid center of the membrane;
- the plane valve 5 which can slide along the rod 2;
- the body of the equipment 8 that contains the holes (1), (2), (3).

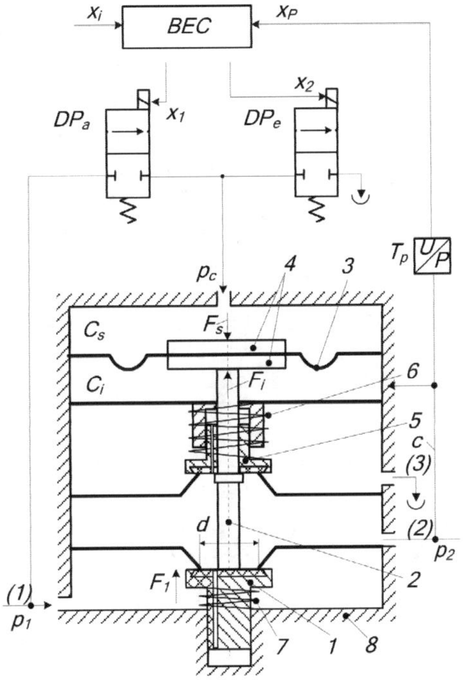

Figure 2. Pneumatic controlled pilot "all or nothing".

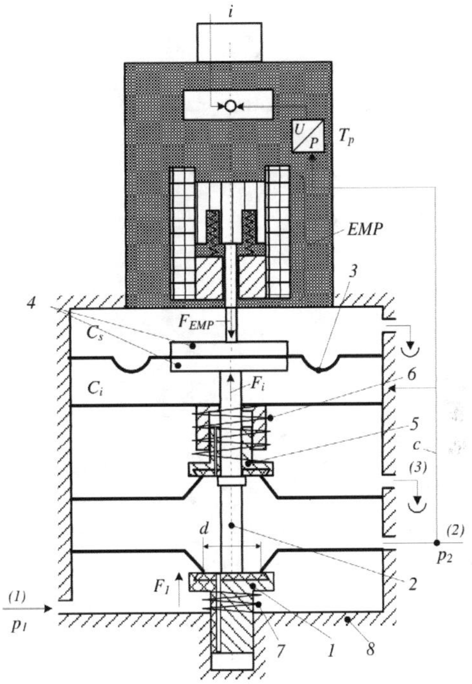

Figure 3. Proportional electromagnet pilot.

Regarding the pilot subassembly two variants may exist:

- with pneumatic controlled pilot ("all or nothing" Figure 2)
- with proportional electromagnet (Figure 3).

The equipment has a pressure transducer T_p that transforms pressure p_2 into a signal. At the level of the electronic block BEC permanently the seated pressure $p_{2,r}$ is compared to the effective one p_2 and depending on the comparison result, digital command signals are generated, or not, for the distributors DP_a and DP_e according to the logic presented in the Table 1.

Table 1. Logical signals command.

Condition	x_1	x_2
$p_2 < p_{2,r}$	24 V	0
$p_2 = p_{2,r}$	0	0
$p_2 > p_{2,r}$	0	24 V

When the command signal x_1 is null, the mobile assemble of the main stage of amplification is in the position presented in figure due to the forces developed by the tow helical springs 6, 7.

The increase in the command signal generates signal x_1 that switches distributor DP_a; in this way the inlet pressure p_1 reaches the upper room of membrane Cs and develops on its surface the pressure force F_s that will determine the moving down of the mobile assemble and the establishing of the connection (1) to (2) and in this way the increase in pressure p_2.

Through the reaction pipe c, pressure p_2 reaches the lower room of membrane C_i where it develops, on the lower surface, pressure F_i; this force is lower than force F_s due to the smaller surfaces against which pressure acts. In this moment, against the membrane acts the pressure force (F_s - F_i), under whose affect the principal valve opens. When pressure p_2 reaches the set value $p_{2,r}$, signal x_1 vanishes and the distributor DP_a moves to the preferential position.

If the inlet electrical signal decreases, signal x_2 is generated and distributor DP_e switches and makes contact to the atmosphere of the upper room of the membrane C_s. The mobile assemble is displaced up under the action of the elastic force developed by spring 7 and it will train in motion valve 5, what determines to establish connection (2) to (3), fact that leads to the decrease in pressure p_2. Company SMC produces this type of equipment [4]. Figure 4 shows such equipment with its characteristics.

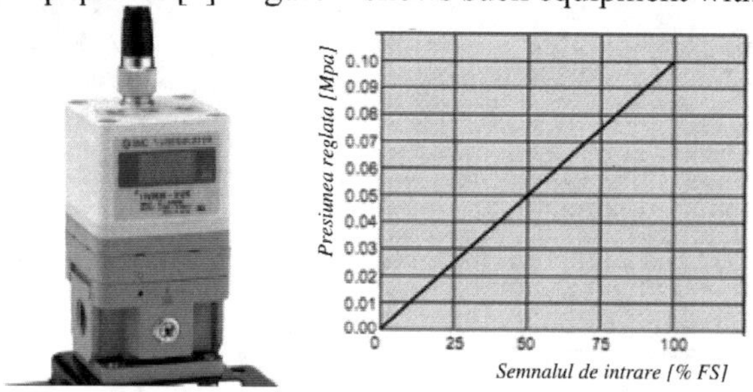

Figure 4. Company SMC equipment.

In the pilot variant with proportional electromagnet EMP (Figure 3), this one acts directly on the rigid center of the membrane. The moving equipment is under the action of two forces, the force developed by the proportional electromagnet F_{EMP} (directly proportional with the inlet measure i) and the force developed by pressure p_2, force F_i. In these conditions the regulated pressure at the equipment outlet is calculated with the relationship (1), where

$$k_{EP} = \frac{k_{EMP}}{S_m} \tag{3}$$

and S_m is the active surface of membrane 3.

3. Theoretical analysis of the proportional regulator

The theoretical analysis has been achieved on a model of a proportional regulator for which the electromagnetic actuator is a proportional electromagnet.

The principle schematic of this regulator is presented in Figure 5 and the mathematical model is composed of the following attached equations:

a. differential equation of the pressure in the room of volume V of the regulator

$$\frac{dP_r}{dt} = \frac{\chi \cdot R \cdot T}{V} \cdot (\dot{m}_t - \dot{m}_e) - \frac{\chi \cdot P}{V} \cdot \frac{dV}{dt} \tag{4}$$

where:

$$\dot{m}_t = \frac{K \cdot P_a \cdot S_c}{\sqrt{T_a}} \cdot N\left(\frac{P_r}{P_a}\right) \tag{5}$$

$$N\left(\frac{P_r}{P_a}\right) = \begin{cases} 1 & if\ 0 \le P_r/P_a \le 0.528 \\ a \cdot \left[\left(\frac{P_r}{P_a}\right)^{\frac{2}{\chi}} - \left(\frac{P_r}{P_a}\right)^{\frac{(\chi+1)}{\chi}}\right]^{\frac{1}{2}} & if\ \frac{P_r}{P_a} > 0,528 \end{cases} \tag{6}$$

$$K = \sqrt{\frac{\chi}{R} \cdot \left(\frac{2}{\chi+1}\right)^{\frac{\chi+1}{\chi-1}}} \tag{7}$$

$$a = \sqrt{\frac{2}{\chi-1} \cdot \left(\frac{\chi+1}{2}\right)^{\frac{\chi+1}{\chi-1}}} \tag{8}$$

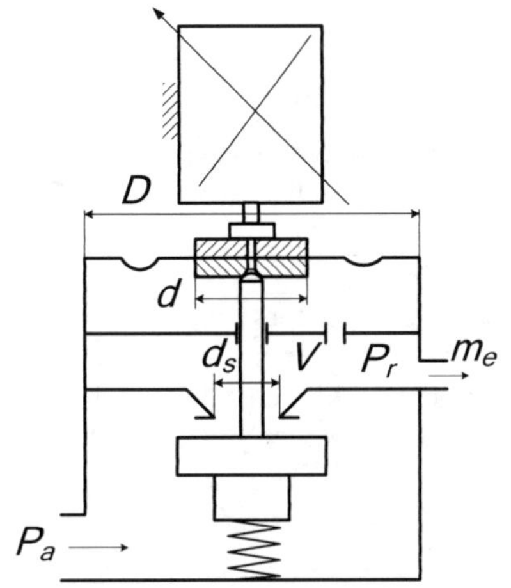

Figure 5. Principle schematic of this regulator.

Coefficients in Equations (4)-(8) have the following values [5]:

$$\chi = 1.4\ [-]$$

$$R = 287.04\left[\frac{m^2}{s^2 \cdot K}\right]$$

for which it follows:

$$K = 0.04042\left[\frac{\sqrt{K} \cdot s}{m}\right]$$

$$a = 3.8639\ [-]$$

The flow section has the following relationship:

$$S_c = \pi \cdot d \cdot x = k_1 \cdot x \tag{9}$$

b. the movement equation of the mobile subassembly:

$$m_r \cdot \frac{d^2x}{dt^2} + \mathrm{B} \cdot \frac{dx}{dt} = k_{\mathrm{EMP}} \cdot \mathrm{i} - A_{\mathrm{m}} \cdot (P_r - P_0) - P_a \cdot S_t - k_{arc} \cdot \mathrm{x} - F_{\mathrm{em}} \tag{10}$$

where [3]:

$$A_{\mathrm{m}} = \frac{\pi}{12} \cdot (D^2 + D \cdot d + d^2) - \frac{F_{em}}{P_r - P_0} \tag{11}$$

$$F_{em} = \frac{2 \cdot \pi \cdot E \cdot h \cdot x}{\sqrt{1 + \left(\frac{2 \cdot x}{D-d}\right)^2}} \cdot \frac{ln\sqrt{1 + \left(\frac{2 \cdot x}{D-d}\right)^2}}{ln\frac{D}{d}} \tag{12}$$

If the variation in volume V is neglected ($\frac{dV}{dt} = 0$), with the substitution

$$\frac{dx}{dt} = v$$

the mathematical model becomes:

$$\begin{cases} \frac{dP_r}{dt} = \frac{\chi \cdot R \cdot T}{V} \cdot \left[\frac{K \cdot P_a \cdot k_1 \cdot x}{\sqrt{T_a}} \cdot N\left(\frac{P_r}{P_a}\right) - \dot{m}_e\right] \\ \frac{dv}{dt} = \frac{1}{m_r} \cdot \left[k_{\mathrm{EMP}} \cdot \mathrm{i} - A_{\mathrm{m}} \cdot (P_r - P_0) - P_a \cdot S_t - k_{arc} \cdot \mathrm{x} - F_{\mathrm{em}} - B \cdot v\right] \\ \frac{dx}{dt} = v \end{cases} \tag{13}$$

The mathematical system of differential equations is solved numerically.

If one wants to solve it analytically, simplifying assumptions should be made [6, 7, 8].
The technique to simplify a mathematical model contains the following procedures:

- replacement of complex functions with more simple approximated relationships
- the use of mathematical development in series
- the use of empirical or statistical relationships instead of analytical functions
- linearization of non-linear functions around the average operation situations
- the modifying or omission of some factors
- consideration that certain variable factors are constant factors and use for them, average values (as friction coefficients, flow coefficients, flow velocities etc.).

In the following the mass m_s of the mobile ensemble and the elastic force of the membrane F_{em} are neglected [9,10]. For these conditions the mathematical model is simplified so fare:

$$\begin{cases} B \cdot \frac{dx}{dt} = k_{EMP} \cdot i - A_m \cdot (P_r - P_0) - P_a \cdot S_t - k_{arc} \cdot x \\ \dot{m}_t = \dot{m}_e + \frac{V}{\chi \cdot R \cdot T} \cdot \frac{dP_r}{dt} \end{cases} \tag{14}$$

This system can be solved based on the Laplace transform:

$$\begin{cases} B \cdot s \cdot x = k_{EMP} \cdot i - A_m \cdot (P_r - P_0) - P_a \cdot S_t - k_{arc} \cdot x \\ \dot{m}_t = \dot{m}_e + \frac{V}{\chi \cdot R \cdot T} \cdot s \cdot P_r \end{cases}$$

Based on that the following relationships could be determined:

$$x = \frac{k_{EMP} \cdot i - A_m \cdot (P_r - P_0)}{B \cdot s + k_{arc}} \tag{15}$$

$$\dot{m}_e = \dot{m}_t - \frac{V}{\chi \cdot R \cdot T} \cdot s \cdot P_r \tag{16}$$

On the other side the mass flow rate becomes:

$$\dot{m}_t = k_1 \cdot x \cdot \left\{ \frac{P_a}{R \cdot T_a} \cdot P_a \cdot \frac{2 \cdot \kappa}{(\kappa - 1) \cdot R} \cdot \left[\left(\frac{P_r}{P_a}\right)^{2/\kappa} - \left(\frac{P_r}{P_a}\right)^{(\kappa+1)/\kappa} \right]^{1/2} \right\} \tag{17}$$

The term in parenthesis on the right-hand term of Eq. (17) can be written such that:

$$T = \left(\frac{P_r}{P_a}\right)^{2/\kappa} - \left(\frac{P_r}{P_a}\right)^{\frac{(\kappa+1)}{\kappa}} = \left(1 - \frac{\Delta P}{P_a}\right)^{\frac{2}{\kappa}} - \left(1 - \frac{\Delta P}{P_a}\right)^{\frac{(\kappa+1)}{\kappa}} \tag{18}$$

and then, by developing each term in a binomial series and neglecting the terms that have greater power than three one gets:

$$T = \frac{\kappa - 1}{\kappa} \cdot \frac{\Delta P}{P_a} \cdot \left[1 - \frac{3}{2 \cdot \kappa} \cdot \frac{\Delta P}{P_a} - \frac{5 \cdot \kappa - 7}{6 \cdot \kappa^2} \cdot \left(\frac{\Delta P}{P_a}\right)^2\right] \tag{19}$$

Considering that the process is adiabatic, we can write:

$$\frac{P_a}{\rho_a{}^{\kappa}} = \frac{P_r}{\rho_r{}^{\kappa}}$$

and accounting for the state equation for a perfect gas

$$\rho_a = \frac{P_a}{R \cdot T_a}$$

It follows that:

$$\frac{P_a}{R \cdot T_a} = \frac{\rho_r}{\left(1 - \frac{\Delta P}{P_r}\right)^{-\frac{1}{\kappa}}} \tag{20}$$

By developing the numerator in a binomial series Eq. 16 becomes:

$$\frac{P_a}{R \cdot T_a} = \frac{\rho_r}{1 - \frac{1}{\kappa} \cdot \frac{\Delta P}{P_a} + \frac{1 - \kappa}{2 \cdot \kappa^2} \cdot \left(\frac{\Delta P}{P_a}\right)^2} \tag{21}$$

Replacing Equations (15), (19) and (21) into Equation (17) one gets:

$$\dot{m}_t = k^* \cdot k_1 \cdot \frac{k_{EMP} \cdot i - A_m \cdot (P_r - P_0)}{B \cdot s + k_{arc}} \cdot \sqrt{2 \cdot \rho_r \cdot (P_a - P_r)} \tag{22}$$

where:

$$k^* = \frac{1 - \frac{3}{2 \cdot \kappa} \cdot \frac{\Delta P}{P_a} - \frac{5 \cdot \kappa - 7}{6 \cdot \kappa^2} \cdot \left(\frac{\Delta P}{P_a}\right)^2}{1 - \frac{3}{2 \cdot \kappa} \cdot \frac{\Delta P}{P_a} - \frac{5 \cdot \kappa - 7}{6 \cdot \kappa^2} \cdot \left(\frac{\Delta P}{P_a}\right)^2} \tag{23}$$

The outlet mass flow rate becomes:

$$\dot{m}_e = k^* \cdot k_1 \cdot \frac{k_{EMP} \cdot i - A_m \cdot (P_r - P_0)}{B \cdot s + k_{arc}} \cdot \sqrt{2 \cdot \rho_r \cdot (P_a - P_r)} - \frac{V}{\chi \cdot R \cdot T} \cdot s \cdot P_r \tag{23.1}$$

or:

$$\dot{m}_e = k^* \cdot \frac{K_1 \cdot i - K_2 \cdot (P_r - P_0)}{\tau \cdot s + 1} \cdot \sqrt{(P_a - P_r)} - \frac{V}{\chi \cdot R \cdot T} \cdot s \cdot P_r \tag{23.2}$$

where:

$$K_1 = \frac{k_1 \cdot k_{EMP}}{k_{arc}} \tag{24}$$

$$K_2 = \frac{k_1 \cdot A_m}{k_{arc}} \cdot \sqrt{2 \cdot \rho_r} \tag{25}$$

$$\tau = \frac{B}{k_{arc}} \tag{26}$$

where τ is the time constant of the regulator.

The mathematical model (13) has been numerically integrated based on Matlab Simulink and AMESIM. In the same time experimental tests to verify the reaction in time of the equipment were performed. By closing the outlet hole and applying an inlet stage signal, one can observe the variation of the regulated pressure, like shown in Figure 6. This dynamic reaction is considered of the second order and was verified experimentally.

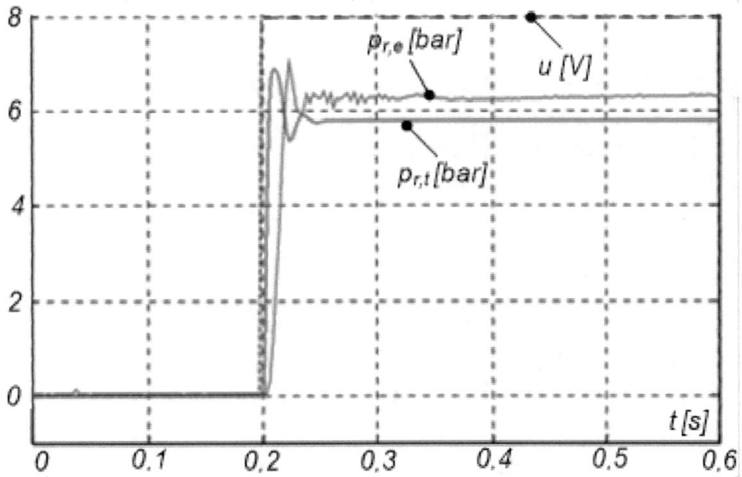

Figure 6. Set pressure variation at an applied signal strep.

The results of the numerical simulation with AMESIM are presented as follows:

- the reaction of the system for a stage variation of the input pressure, when the outlet flow passage of the regulator is 50% of the nominal flow one (Figure 7);
- the reaction of the system for a sinusoidal variation of the input pressure when the outlet flow passage of the regulator is 50% of the nominal flow one (Figure 8);

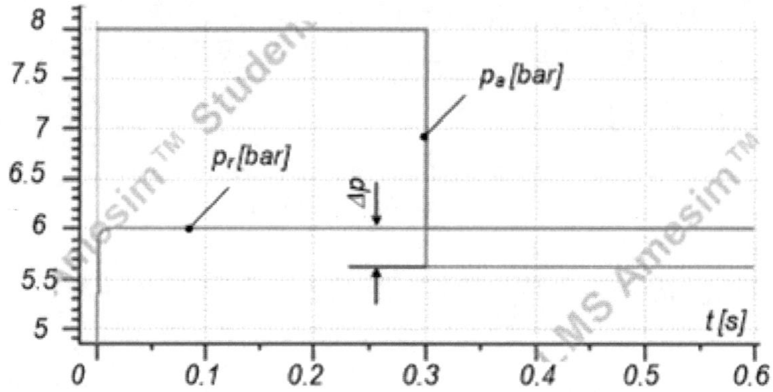

Figure 7. The reaction of the system for a stage variation of the input pressure

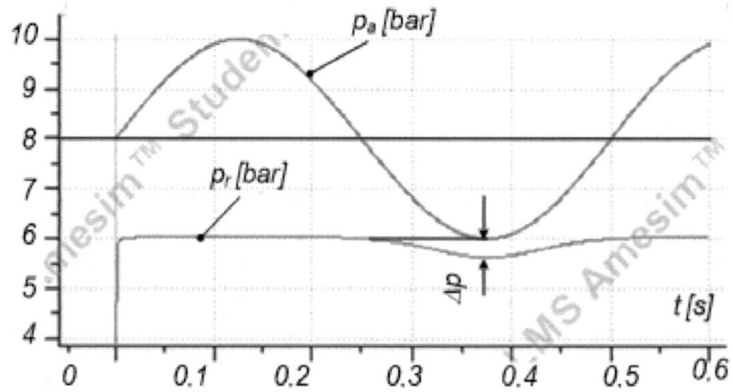

Figure 8. Reaction of the system for a sinusoidal variation of the input pressure

- the system response considering a stage variation of the outlet flow section of the regulator, when the supply pressure is constant (Figure 9).

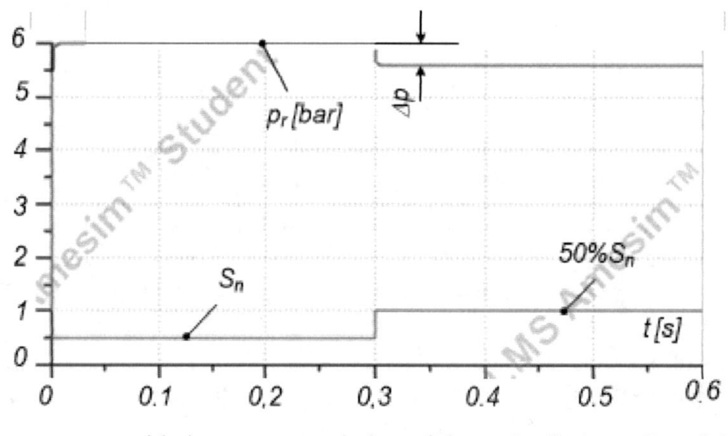

Figure 9. System response considering a stage variation of the outlet flow section of the regulator.

4. Conclusions

The current stage of knowledge and future prospects for development point out the importance of pneumo-automatics among top technologies.

Some of the negative aspects connected to the physical properties of the working medium as low viscosity and high compressibility should be solved.

The computerization of pneumatic systems represents an important qualitative gain for this field.

The pneutronics concept developed in a natural way as the result of the connection between three fields: pneumatics – electronics – informatics.

As a result of this connection systems with pneutronic drive have been developed. The base of a pneutronic system is represented by the proportional pneumatic equipment.

Besides them, in the system structure one can find: sensors, transducers, electronic circuits for signal processing, convertors A/D and D/A, controllers or microprocessors.

With such a configuration, a pneutronic system becomes more stable, much more accurate and rapid with a certain level of intelligence.

The paper presents a mathematical model for proportional equipment for the pressure regulation. With such equipment, in all the presented situations, pressure is regulated and stabilized and kept constant with a certain deviation around the regulated value.

References

[1] Foit K, Banaś W and Ćwikła G 2018 The pneumatic and electropneumatic systems in the context of 4 th industrial revolution *IOP Conference Series: Materials Science and Engineering* **400(2),** Article number 022024, ISSN: 17578981

[2] Avram M and Bucșan C 2014 Sisteme de acționare pneumatice inteligente (Editura Politehnica PRESS București ISBN 978-606-515-557-2)

[3] Banu G V 2011 Pneumatica - Noțiuni de bază *Departamentul tehnic al SMC Romania*

[4] *** https://www.smcworld.com/products/en/get.do?type=GUIDE&id=VEF-VEP-E

[5] Avram M 2005 Acţionări hidraulice şi pneumatice – Echipamente şi sisteme clasice şi Mecatronice (Editura Universitară, Bucureşti)

[6] Dolga V 2007 Proiectarea sistemelor mecatronice (Ed. Politehnica ISBN 978-973-625-573-1)

[7] Bishop H R 2002 The Mechatronics Handbook (CRC Press London-New York-Washington)

[8] Shetty D and Kolk A R 2000 Mechatronics System Design (Pws Publishing Company Boston)

[9] Cruz F B C, De Negri V J and Guenther R 2003 Mathematical Modelling of an Electropneumatic Pressure Regulator Servo Valve *ABCM Symposium Series in Mechatronics* **1** pp 725-734

[10] Vujić D and Radojković S 2001 Dynamic model of gas pressure regulator *FACTA UNIVERSITATIS Series: Mechanics, Automatic Control and Robotics* **3(11)** p. 269 – 276

IOP Conf. Series: Materials Science and Engineering **595** (2019) 012027 doi:10.1088/1757-899X/595/1/012027

Evaluation of the thermal comfort for its occupants inside a vehicle during summer

P Danca[1,2], I Nastase[1], F Bode[1,3], C Croitoru[1], A Dogeanu[1] and A Meslem[2]

[1]CAMBI Research Center, Technical University of Civil Engineering, Bucharest, Romania
[2]IUT Rennes, Laboratoire Génie Civil Génie Mécanique, France
[3]Faculty of Automotive, Mechatronics and Mechanical Engineering, Technical University of Cluj-Napoca, Romania

E-mail: ilinca.nastase@gmail.com

Abstract. Thermal comfort of occupants from the vehicle environment has gained more importance in the last decades because time spent by people in the vehicles is increasing. During each trip, driver's thermal comfort must to be ensured to reduce both thermal stress (thus reduces the risk of accidents) and for a healthy state of the occupants. Given the fact that the vehicle environment is non-uniform and fast transient, and because the actual standard (ISO 14505) proposes comfort evaluation methods developed for the steady state conditions in buildings, the purpose of the study is to evaluate a vehicle environment using the three standardised method presented in the above standard. All the three standardized evaluation indices were achieved through standardised methods. The equivalent temperature (t_{eq}) index was assessed with an advanced thermal manikin and its values were compared with the survey answers given by people in questionnaires (TSV Thermal Sensation Vote index) and with Predicted Mean Vote (PMV) index values calculated with Comfort Sense standardized evaluation equipment. A first conclusion is that the results of the three evaluation indices are different. Air temperature is different from a zone of the car to another and the thermal evaluation must be done in the place of each passenger.

1. Introduction

Thermal comfort has been studied for a long time, research in this domain leading to the emergence of several indices to assess the thermal comfort. These indices were developed initially for the buildings in a uniform controlled and steady environment in research laboratories. Although the comfort evaluation indices were developed for the steady-state environments, the same indices were adopted to evaluate the transient and non-uniform thermal environments as in the vehicle interior case. Contrary to the case of building, assessing thermal comfort in vehicles has several particularities [1]. The effect of solar radiation, poor interior insulation, the non-uniformity of the average radiant temperature, a very short time to ensure the comfort parameters are some of the characteristics of an automotive environment. Indoor Environmental Quality in buildings has gained importance in the last decade and now is developing a new direction of research, the environmental quality in vehicles. The ambience quality is an important criterion in marketing this type of products. It influences not only the thermal comfort inside the car, but it also reduces the risk of accidents [2-4] by reducing the driver's thermal stress and ensuring a good visibility (trough avoiding the fogging phenomena) , which leads to a safer trip.

The thermal comfort inside the vehicle is more difficult to be evaluated and controlled than in a building due to the increase influence of the heat exchange through convection, conduction, radiation [5-8]. Also, other influences are the environmental parameters and the non-uniformity of the vehicle environment resulted by the heating, ventilating and air conditioning system. Thus, ensuring and maintaining the thermal comfort inside the vehicle is one of the most important challenges for the researchers, designers and manufactures in the automotive industry.

The actual standard which, is proposing comfort evaluation methods for vehicles environment is ISO 14505 [9-11] standard "Evaluation of thermal environments in vehicles" being structured into three parts: 1. Principles and methods for assessment of thermal stress [9]; 2. Equivalent temperature determination [10]; 3. Evaluation of thermal comfort by the human subjects [11].

In the first part of the above-mentioned standard, is presented a comfort evaluation method based on the PMV and PPD model developed in the 1970 professor Fanger [12]. This comfort evaluation model was developed using data collected from human subjects evaluated in controlled laboratory conditions. Fanger's PMV and PPD indexes were adopted by international standardization organization such as ISO 7730 [13] and ASHRAE 55 [14]. Unfortunately, form the equations used for the calculation of these two indices we can observe that the results are generated particularly for homogeneous environments like the environments inside buildings.

The second part of the standard is presenting the t_{eq} index – the equivalent temperature index. This is a local and global index proposed by Nilsson following his PhD thesis [15-18]. The equivalent temperature uses the same method of calculation as the operative temperature for ambient air velocities under 0.1 m/s. This represents the average of the mean radiant temperature and the air temperature weighted respectively by the convection and radiation heat transfer coefficients for the occupant inside the vehicle. For values of the ambient air velocities greater than 0.1 m/s, the equivalent temperature is expressed as a function of the air temperature, the mean radiant temperature, the air speed and the thermal resistance of clothing [19-21].

In the third part of the standard ISO 14505 a direct method for assessing thermal comfort in automobiles is presented, the main index being the Thermal Sensation Vote (TSV) of human subjects that are surveyed. This subjective method quantifies and records the response of people about their thermal sensation in an environment, on the same scale of values as for the PMV. The TSV only considers the psychological and physiological factors. A study using this method involves standardized questionnaires for a controlled and representative sample of population. By centralizing the obtained thermal votes, results the thermal sensation from the particular investigated situation.

This article is a part of a larger study, intended to deepen the knowledge on thermal comfort inside vehicles, the purpose is to evaluate a vehicle environment using the three standardized method presented in the above standard. All the three standardized evaluation indices were achieved through standardized methods. The equivalent temperature (t_{eq}) index was assessed with an advanced thermal manikin and its values were compared with the survey answers given by people in questionnaires (TSV Thermal Sensation Vote index) and with Predicted Mean Vote (PMV) index values calculated with Comfort Sense standardized evaluation equipment. A first conclusion is that the results of the three evaluation indices are different. Air temperature is different from a zone of the car to another and the thermal evaluation must be done in the place of each passenger.

2. Experimental set-up

The measurement campaigns presented in this study were performed in a Renault Megane vehicle. The evaluation was conducted in the hot season, in Bucharest. Outside, the air temperature was 35°C according to the National Institute of Meteorology. During the measurement's sessions, the vehicle was placed in a hall (see **Error! Reference source not found.**) in order to ensure as more as possible constant thermal conditions in the vicinity of the car. Totally, were made 12 measurement sessions, in 10 of them human subjects were placed in the driver seat. All the tests were made for the second running position of the ventilation system Figure 1, this being the most used position in the summer conditions.

a)

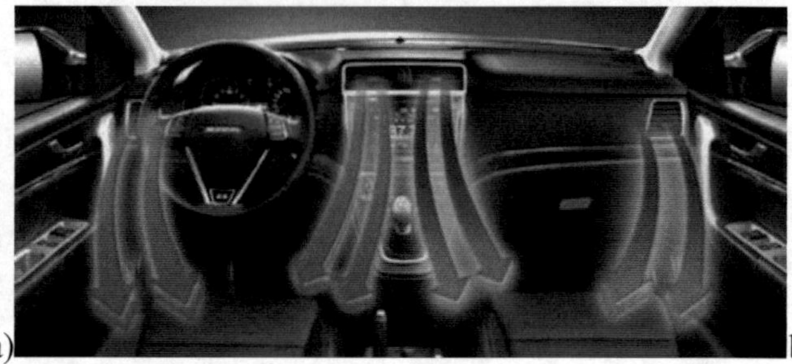
b)

Figure 1. Ventilation/air conditioning system set-up.

The air was introduced in the vehicle only trough dashboard diffusers as is showing in Figure 1. The air-conditioned system was turned on during all the tests. The mean velocity and the flowrates of the introduced air through the dashboard diffusers for this ventilation/air-conditioning running position are presented in Table 1.

Table 1. Values of the introduced air condition.

Diffuses/parameter	Mean velocity [m/s]	Temperature [ºC]	Q [kg/s]
Central diffuser	3.29	13	0.0316
Right side diffuser	1.58	9.4	0.0152
Left side diffuser	1.19	14	0.0114

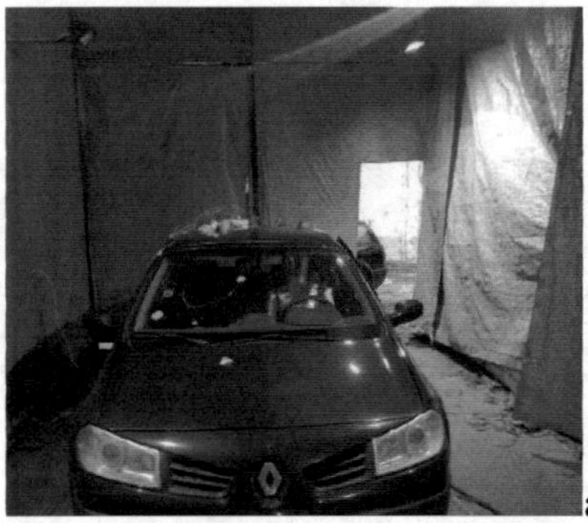
a)

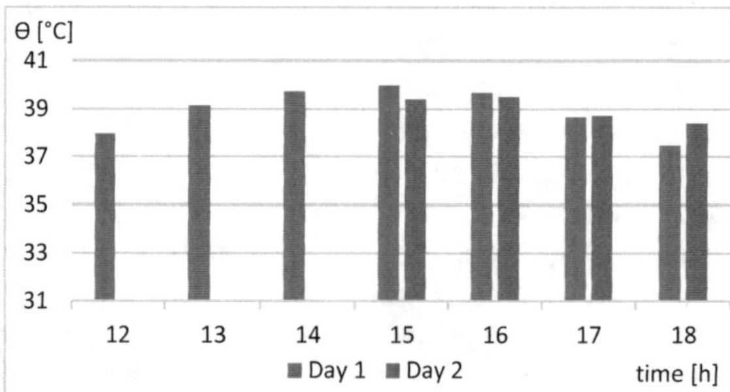

b)

Figure 2. a) experimental test car in the hall; b) temperature variation in the hall during the experimental tests.

2.1. *Thermal environment assessment*

To monitor temperature on the interior surfaces a network of 11 thermocouples was places on different surfaces as is shown in Figure 3. Three thermocouples were placed in the inlets: one in the centre (T19), one on the left side diffuser (T18) and another in the right-side diffuser (T20). Another nine thermocouples were placed on the interior surface: dashboard (T17), windshield (T16), sides windows (T21 and T22), ceiling (T24), floor (T23), top of the trunk (T25) and rear window (T26). These were connected to 4 data loggers manufactured by Ahlborn: two Almemo 710 each with 10 sensor connecting available ports.

IOP Conf. Series: Materials Science and Engineering **595** (2019) 012027 doi:10.1088/1757-899X/595/1/012027

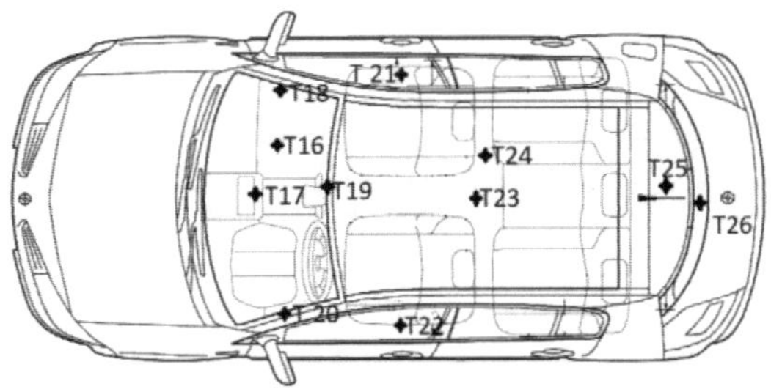

Figure 3. a) K-type thermocouples distribution on the surfaces.

Another 15 pieces of AX-DT 200 temperature and humidity wireless sensor (Figure 4 a) were placed in the places where are zones of the human body. For these, we built some special supports presented which can be seen in the following pictures. These zones are: foot, knee, abdomen, chest and head, five for each of the places of the other passengers. The distribution of all the AX-DT 200 sensors is presented in Figure 4. The temperature values were recorded after each minute.

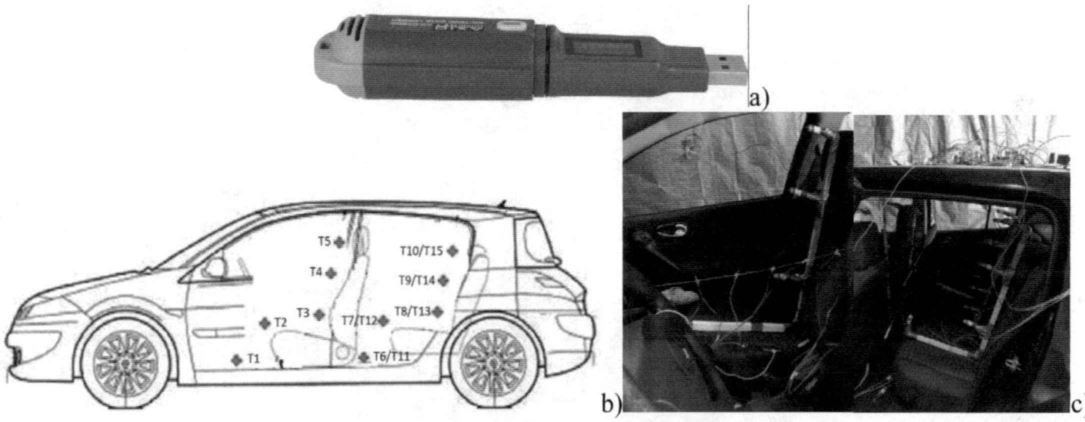

Figure 4. a) AX-DT 200 wireless sensor; b) sensors map distribution; c) images during measurement session.

2.2. Thermal comfort evaluation using PMV/PPD model – first part of ISO 14505

PMV and PPD indices are proposed by Fanger [12] over 30 years ago, and they are the result of a study in which subjects that had "standard" clothes performing a „standard" activity. During the assessment, the persons were exposed to various thermal conditions. According to the felt sensation, the subjects assessed the condition using a sensation scale with seven values (-3: cold, -2: cool, -1: slightly cool, 0 neutral, 1: slightly warm, 2: warm, 3: hot) [14]. Knowing the clothing insulation and metabolic rate resulting from the activity and introducing the external physical quantities that influence the phenomena of heat transfer (air temperature, radiant mean temperature, partial pressure of water vapor and relative air velocity) in the thermal equilibrium relationship of the human body, the PMV index is resulting from the equation:

$$PMV = (0.303e^{0.303}+0.028)\{(M\text{-}W)\text{-}3.05[5.73\text{-}0.007(M\text{-}W)\text{-}p_a]\text{-}0.42[(M\text{-}W)\text{-}58.15] \\ \text{-}0.0173M(5.87\text{-}p_a)\text{-}0{,}0014M(34\text{-}t_a)\text{-}3{,}96\cdot 10^{-8} f_{\mathrm{cl}}\,[(t_{cl}+273)^4-(t_{\mathrm{mr}}+273)^4]\text{-}f_{cl}\,h_c\,(t_{cl}\text{-}t_a)\,\} \quad (1)$$

where:

$$t_{\mathrm{cl}} = 35.7\text{-}0.0275(M\text{-}W)\text{-}\mathrm{I_{cl}}\{(M\text{-}W)\text{-}3.05[5.73\text{-}0.007(M\text{-}W)\text{-}\mathrm{p_a}] \\ \text{-}0.42[(M\text{-}W)\text{-}58.15]\text{-}0.0173M(5.87\text{-}\mathrm{p_a})\text{-}0.0014M(34\text{-}t_a)\,\} \quad (2)$$

M - the metabolic rate, in watts per square meter [W/m^2];
W - effective mechanical power, in watts per square meter [W/m^2];

t_{cl} - clothing surface temperature, in degrees Celsius [°C];
p_a - water vapour partial pressure, in Pascal [Pa];
t_a - air temperature, in degrees Celsius [°C];
v_a - relative air velocity, in metres per second [m/s];
I_{cl} - clothing insulation, in square meter Kelvin per Watt (m^2 K/W);
f_{cl} - clothing surface area factor;
t_{mr} - mean radiant temperature, in degrees Celsius [°C];
h_c - convective heat transfer coefficient, in watts per square meter Kelvin [W/m^2K]

NOTE: 1 metabolic unit = 1 met = 58.2 W/m^2; 1 clothing unit = 1 clo = 0.155 m^2 °C/W

The PPD index (Predicted Percentage of Dissatisfied) is associated with the *PMV* and is indicating the percentage of occupants under thermal discomfort. A value of 10% of the *PPD* index corresponds to the interval between -0.5 and +0.5 for *PMV* on Fanger's scale. Even for the *PMV*=0, about 5% of occupants are in discomfort.

$$PPD=100-95\exp[-(0.03353PMV^4+0.2179PMV^2)] \quad (3)$$

In our study the *PMV/PPD* indexes were calculated with a Comfort Sense measurement tool. It has various sensors which are measuring air temperature and velocity, relative humidity, operative temperature. Adding to this the clothing insulation and metabolic rate, the device is calculating automatically the values of *PMV* and *PPD*. For our study we imposed a value of 0.7 clo for the clothing insulation and 1.2 met for the metabolic rate. The Comfort Sense probes were placed on the driver place as is shown in Figure 5.

Figure 5. Comfort sense probes on the driver place.

2.3. Thermal comfort evaluation with t_{eq} index and thermal manikin – Second part of 14505
In our opinion the equivalent temperature is the most adapted for the evaluation in this environment. It offers the possibility to evaluate the thermal comfort at the local level taking into account the heat exchange between a specific body part with the surrounding environment. The equivalent temperature is using two relations function of the air velocity in the studied environment. In this case of velocities lower than 0.1 m/s, the equivalent temperature might be expressed as an average between the values of the air temperature and the radiant mean temperature weighted using the heat transfer coefficients (convection and radiation) between the occupant and the environment [22]. When the air speed values are larger than 0.1 m/s, the relation for the calculation of the equivalent temperature is:

$$t_{eq}=0.55 \cdot t_a+0.45 \cdot t_{mr} + \frac{0.24\text{-}0.75\sqrt{v_a}}{1+I_{cl}} (36.5\text{-}t_a) \quad (4)$$

In our study, the *teq* values are calculated with an advanced thermal manikin with 79 zone independently controlled and neuro-fuzzy control. The manikin was developed in our laboratory with the support of Mechatronics Department of the National Institute of Aerospace Research Elie Carafoli. The thermal manikin was designed for both seated and standing postures. The size of the manikin is defined by the standard skin surface of a human of 1.8 m^2 [23, 24]. The equivalent temperature that represents an indication of thermal comfort is obtained by evaluating the power consumption of a region of the manikin (see Equation 5). Due to the pwm (pulse with modulation) control signal which commutes on and off between maximum and minimum voltage, the power consumed by the thermostatic system was calculated by creating a calibration slope between pwm duty-cycle and the power calculated as a point by point mean of a single pulse period. The voltage drop on the patch was calculated differentially by measuring with the Hantek DSO5102P oscilloscope the voltage drops on the whole circuit from which it was subtracted the voltage drop on the transistors. The current consumed by the patch was measured with TH5A current transducer.

$$\theta_{ech} = \theta_{reg} - \frac{P}{S \cdot h_{calc}} \quad (5)$$

where:

θ_{ech} - equivalent temperature;

θ_{reg} - mean temperature of surface region calculated using a sliding average over a pre-set period;

S - surface area of manikin's region;

P - mean power consumption calculated using a sliding average over a pre-set period;

h_{cal} - convection coefficient calculated with equation (4) at constant environment temperature (θ_{ech}) of 24 °C and manikin's surface temperature controlled at 34 °C.

The 79 zones was grouped in 16 zones fallowing the prescription of the standard [10]. The manikin was placed on the place of the driver, with the left hand on the steering wheel and left hand on the gearshift as is showing in Figure 6. The imposed temperature of the manikin was 34 °C.

Figure 6. Thermal manikin placed in the driver place.

2.4. Thermal comfort evaluation using human subjects

The ISO 14505-3 gives guidelines and specifies a standard test method for the assessment, using human subjects, of thermal comfort in vehicles. The questionnaire used in our study made corresponding to from ISO 14505-3 and ISO 10551/2001 [25] which guide the construction and use of thermal sensation scales and proposes a set of specifications on direct assessment of thermal comfort/discomfort expressed by persons subjected to various degrees of thermal stress. In Figure 7 is presented an example of used questionnaires.

Evaluation of thermal environments in vehicles – subjective questionnaire

Session: ....
Date: ....
Time: ....
Age: ....
Sex: Male / Female
The place occupied in the car:

The driver seat	
Place the right driver	
Place behind the driver	
Place the right rear of the driver	

Please rate on these scales how you fell now:

1. Thermal sensation scale

		1	2	3	4	5	6	7	8	9	10	11
	Whole body	Head	Trunk		Arms	Upper legs		Lower legs		Feet	Ankles	Neck
			Front	Rear		Front	Rear	Front	Rear			
+3 Hot												
+2 Warm												
+1 Slightly warm												
0 Neutral												
-1 Slightly cool												
-2 Cool												
-3 Cold												

2. Thermal discomfort scale

		1	2	3	4	5	6	7	8	9	10	11
	Whole body	Head	Trunk		Arms	Upper legs		Lower legs		Feet	Ankles	Neck
			Front	Rear		Front	Rear	Front	Rear			
3 Very uncomfortable												
2 Uncomfortable												
1 Slightly uncomfortable												
0 Comfortable												

3. Stickiness scale

3 Very sticky	
2 Sticky	
1 Slightly sticky	
0 Not sticky	

4. Please rate on the scale how you would like to be now in vehicle inside:

Much warmer	Warmer	Slightly warmer	No change	Slightly cooler	Cooler	Much cooler
+3	+2	+1	0	-1	-2	-3

5. Please indicate how acceptable you find this thermal environment now:

☐ Acceptable ☐ Unacceptable

6. Please indicate how satisfied you are with this thermal environment now:

☐ Satisfied ☐ Dissatisfied

7. Please indicate other aspects that should be specified:

....

Figure 7. Example of used questionnaires.

Survey questionnaires made according to the standard ISO 10551 were given to the volunteers and 108 questionnaires were completed and the time necessary to complete a questionnaire is (30-60) sec. The general information require by questionnaire was: date and time, question about the person who had completed the questionnaire as age and sex. For thermal sensation assessment the subjects had to choose an option on the 7-points scale and they also had to select the thermal preference during the filling out the questionnaire. The volunteers had to answer about the acceptability of the thermal comfort and also to the local thermal discomfort question. Based on the response given in questionnaires we could calculate the Thermal Sensation Vote index (TSV). Before the tests the persons interviewed have been knowledgeable briefly about reasons for the questionnaire. The uncomfortable and thermal sensation scales were applied also for 11 parts of human body as: the head, the front torso, the back torso, the arms, the front and the back thighs, the front lower legs, the back lower legs, the foot, the ankles and the neck. As a particularity, on each questionnaire heart rate was noted.

3. Results and discussions

The temperature values recorded with de AX-DT 200 sensors at the level foot, knee, abdomen, chest and head are presented in Figure 8. The place where temperature is similar is the chest while the point where is found the biggest differences are foot and knees. We think that at the base of these differences is the fact that the introduced air is passing as a jet through the front part of the car, and in the rear part of the car, the velocity decreases favouring the air mixing. Also, a high difference can be noted if we look at the abdomen and foot temperature values for the front passenger. This difference may provide high uncomfortable states to the passenger.

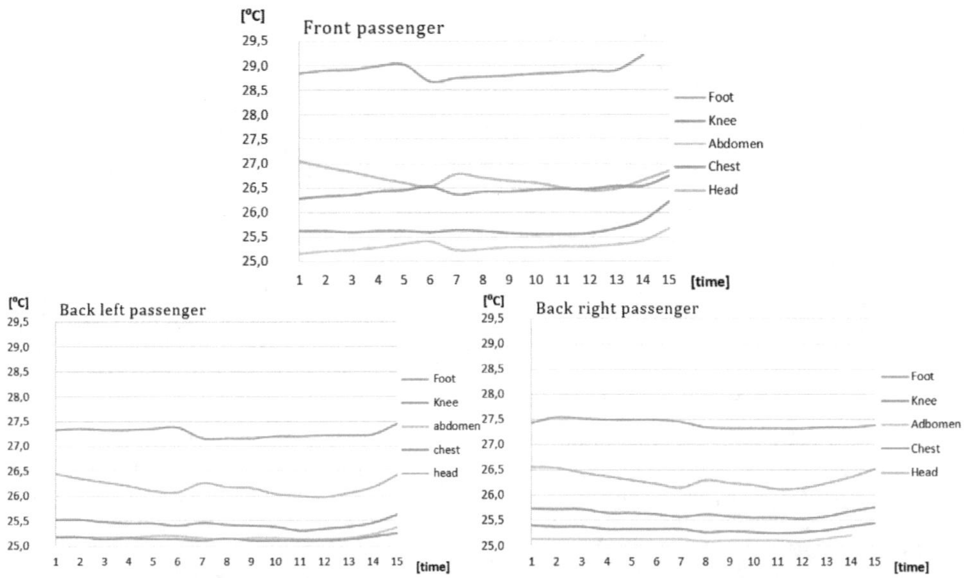

Figure 8. temperature variation in the place of the passengers.

Imposing the values of clothing insulation and of the metabolic rate and adding the measured values of the physical parameters, the PMV and PPD values were calculated. The mean value for each measurement session is presented in **Figure 9**. All the values are negative which means that there is a cold state tendency. An environment is considerate to be comfortable if the PMV values are within -0.5 and + 0.5 range. In our case on the driver seat, 6 of the values are in the neutral limits, while 4 are near to the border of this state.

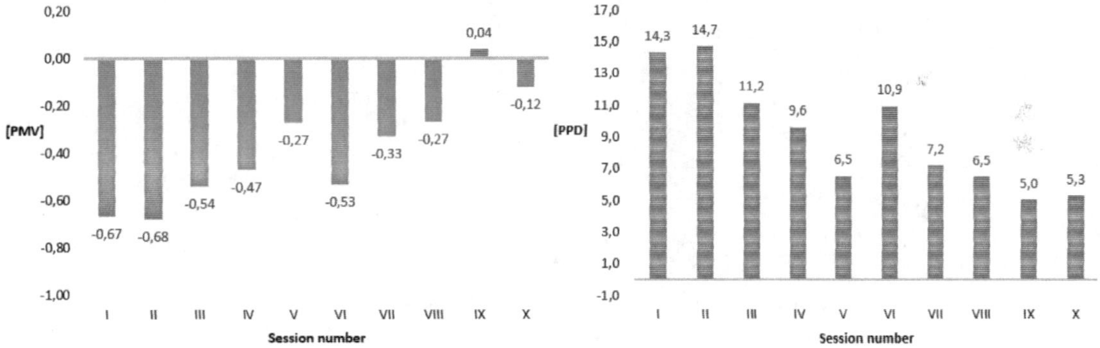

Figure 9. PMV and PPD values.

The results from the survey questionnaires study are presented in Figure 10. There are only three sensation voted by the volunteers. The highest percentages of the votes (80%) resulting from the questionnaires are in the neutral zone. The second biggest percentage from the votes (17%) are indicating that the passengers had a slightly warm sensation. While only 3% votes slightly cold. If we are going back to the PMV values we can see that excepting one vote, the values have negative values which means that there is a cold state tendency. From this we can see already a difference between the two indices, because their tendency is contradictory.

Comparing the TSV values at the local level we can note that for 6 of the persons surveyed voted a slightly cool sensation for the arms regions. This sensation may be produced by the cold air jets from the inlet diffusers which is passes near to these body parts. In fact, this is one of the observation of the volunteers.

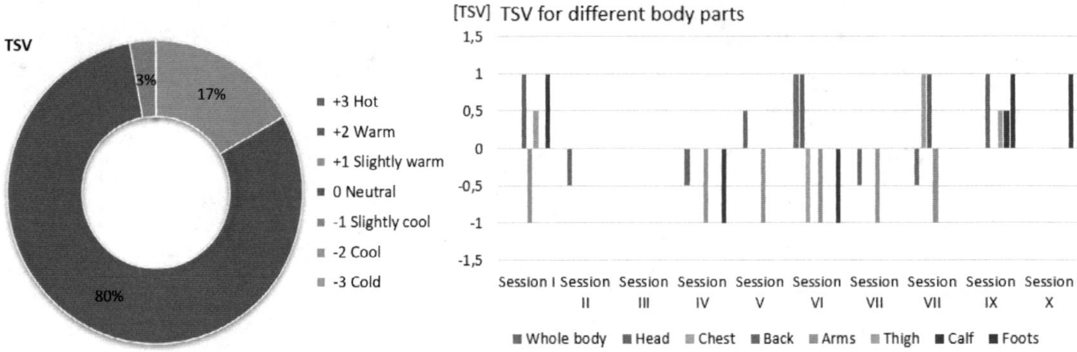

Figure 10. percentages of thermal sensations from the questionnaires.

Figure 11 shows better the differences of the two indices PMV and TSV. Although the majority of the votes of the two indices are showing a neutral thermal state, the majority of the TSV values are positive and the majority of the PMV values are negative.

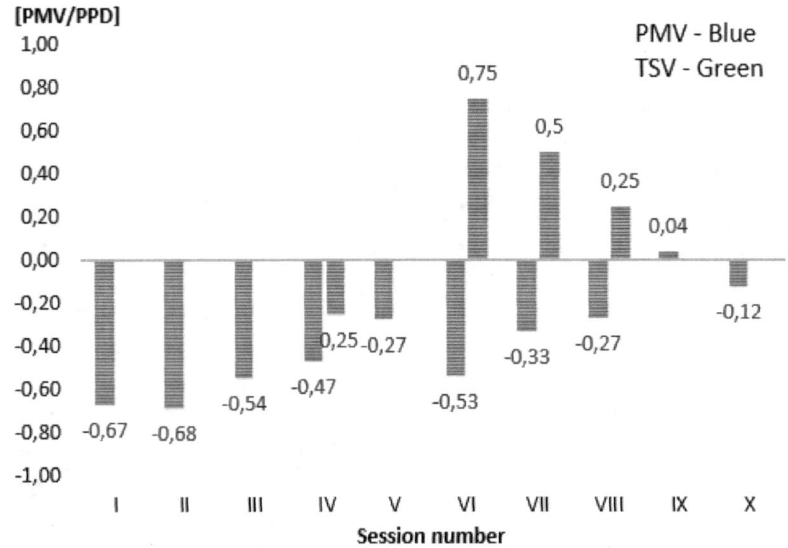

Figure 11. Comparation between PMV and TSV.

The results assessed with the thermal manikin are presented in Figure 12. In the shown diagrams with 1, 2, 3, 4 and 5 are represented comfort sensations corresponding to too cold, cold but comfortable, neutral, warm but comfortable, too hot thermal sensations. To each sensation is corresponding different ranges of t_{eq} values specific for each zone. This ranges are presented in ISO 14505-2 [10]. As it can be seen in Figure 12 the thermal manikin is assessing a high uncomfortable cold sensation for right upper arm placed in the direction of the inlet airflow which is coming from the central diffusers. Some of the regions as chest, left leg, left foot, right arm and left arm are near the region of cold sensation but still in the comfortable thermal sensation zone. The t_{eq} of the other body parts are in the zone 3 which is a neutral sensation.

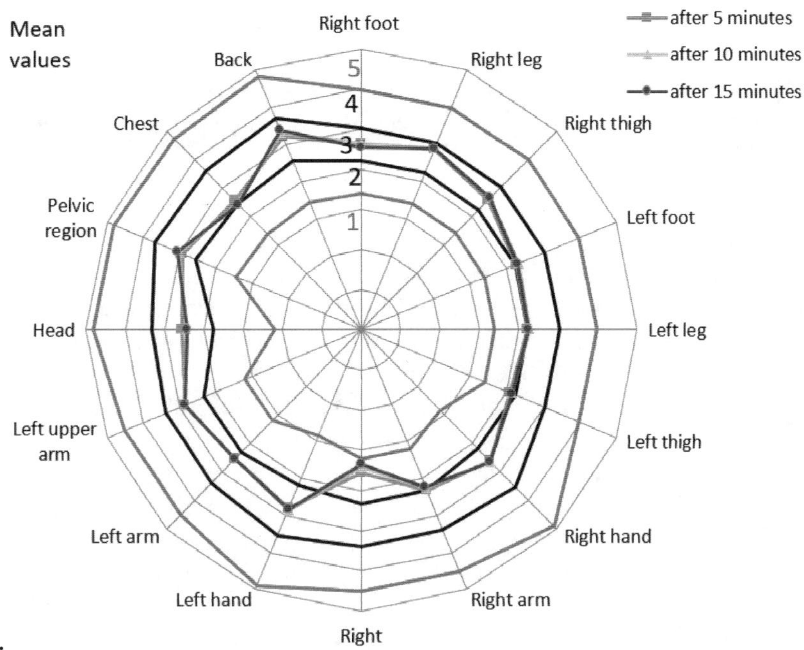

Figure 12. Distribution of the mean t_{eq} values.

In Table 2 are exposed mean values of the standardized indexes. The results of evaluation with all three standardized methods are predicting that most of the persons which are entering in the car should feel a comfortable thermal sate. Although, the majority of the votes of the three indices are showing a neutral thermal state, the majority of the TSV values are positive and the majority of the PMV values are negative. Thermal comfort assessment at the local level highlights that there are different thermal states felts for different body parts, for example arms which are in a cold thermal state due to the air jets from the inlet diffusers.

Table 2. Values of thermal comfort evaluation indexes.

Index	*teq* [°C]	TSV [-]	PMV [-]
V2	22.80 (cold but comfortable)	0.14 (neutral)	-0.41 (neutral)

4. Conclusions

This paper is focused on the comfort assessing in a transient non-uniform environment inside a vehicle. Determination of the vehicle occupants thermal comfort is very complicated due to the transient nature and non-uniformity of the vehicle interior. More than this, the actual standard is proposing three evaluation indexes and was developed for steady state and controlled conditions and some of the indexes are not adapted to this complex environment.

In this article are compared the values obtained for the three standardized indexes in term of thermal comfort, in a vehicle passenger in summer season.

The results are showing that the mean values PMV/PPD model calculated in a single point with Comfort Sense equipment are far from the TSV mean values resulting from questionnaires survey with human subjects, while the t_{eq} index calculated with an advanced thermal manikin reveals thermal states similar to the survey questionnaire study (TSV). This may be explained by the fact that the TSV and t_{eq} consider the sensation for each body part at the local level.

Another conclusion bay be, that for a correct evaluation of thermal comfort in non-uniform and transient environments like in the vehicles, measurement of the physically parameters in a single point from the environment is not relevant.

The main conclusion is that the PMV/PPD indexes are not very well adapted to the vehicle environment.

References

[1] Danca P, Vartires A and Dogeanu A 2016 An Overview of Current Methods for Thermal Comfort Assessment in Vehicle Cabin. *Energy Procedia.* **85** 162-9

[2] Chen C P, Hwang R L, Chang S Y and Lu Y 2011 Effects of temperature steps on human skin physiology and thermal sensation response. *Building and Environment.* **46(11)** 2387-97

[3] Zhou X, Lai D and Chen Q 2019 Experimental investigation of thermal comfort in a passenger car under driving conditions. *Building and Environment.* **149** 109-19

[4] Zhang W, Chen J and Lan F 2014 Experimental study on occupant's thermal responses under the non-uniform conditions in vehicle cabin during the heating period. *Chinese Journal of Mechanical Engineering*

[5] Danca P, Bode F, Nastase I and Meslem A 2018 CFD simulation of a cabin thermal environment with and without human body – thermal comfort evaluation. *E3S Web Conf.* **32**:01018

[6] Bode F, Nastase I, Danca P, Meslem A and Danca P 2017 *The influence of the Inlet angle of vehicle air diffuser on the thermal comfort of passengers*. 2017 International Conference on Energy and Environment (CIEM); 2017 19-20 Oct. 2017

[7] Danca P, Bode F, Nastase I and Meslem A 2017 On the Possibility of CFD Modeling of the Indoor Environment in a Vehicle. *Energy Procedia.* **112** 656-63

[8] Danca P and Vartires A 2015 *Thermal comfort assessment using human subjects.* Scientific Proceedings XXIII International Scientific-Technical Conference "trans & MOTAUTO'15. **XXIII** 3 pp. 22-25

[9] ISO 14505-1:2007 Ergonomics of the thermal environment - Evaluation of thermal environments in vehicles Part 1: Principles and methods for assessment of thermal stress.

[10] ISO 14505-2:2006 Ergonomics of the thermal environment - Evaluation of thermal environments in vehicles - Part 2: Determination of equivalent temperature

[11] ISO 14505-3:2006 Ergonomics of the thermal environment -Evaluation of thermal environments in vehicles Part 3: Evaluation of thermal comfort using human subjects

[12] Fanger PO 1970 *Thermal Comfort: Analysis and Applications in Environmental Engineering,* Danish Technical Press. Copenhagen, Denmark

[13] ISO 7730:2005 - Ergonomics of the thermal environment – Analytical determination and interpretation of thermal comfort using calculation of the PMV and PPD indices and local thermal comfort criteria

[14] ASHRAE Thermal environmental conditions for human occupancy, ASHRAE Standard 55–2004 American Society of Heating, Refrigerating and Air-Conditioning Engineers

[15] Nilsson H O 2004 *Comfort Climate Evaluation with Thermal Manikin Methods and Computer Simulation Models.* National Institut for Working Life

[16] Nilsson H O and Holmér I 2017 Definitions and Measurements of Equivalent Temperature, European commission cost contract no smt4-ct95-2017 Development of standard test methods for evaluation of thermal climate in vehicles

[17] Nilsson H, Holmér I, Bohm M and Norén O 1997 *Equivalent temperature and thermal sensation - Comparison with subjective responses.* Comfort in the automotive industry- Recent development and achievements; 1997; Bologna, Italy

[18] Nilsson H O 2007 Thermal comfort evaluation with virtual manikin methods. *Building and Environment.* **42(12)**:4000-5.

[19] Madsen T L 1984 Comparison between operative and equivalent temperature under typical indoor conditions. *ASHRAE Transactions.* **90(1)** 1077-90

[20] Madsen T L, Olesen B and Reid K 1986 New methods for evaluation of the thermal environment in automotive vehicles. ASHRAE Transactions. **92(part 1B)**:38-54

[21] Madsen T L 1976 Thermal comfort measurements. *ASHRAE Trans.* **82(1)**

[22] ASHRAE. *Thermal environmental conditions for human occupancy*, ANSI/ASHRAE Standard 55-2004, American Society of Heating, Refrigerating and Air-Conditioning Engineers, Atlanta, GA

[23] Dogeanu A, Iatan A, Croitoru C and Nastase I 2012 Conception of a real human shaped thermal manikin for comfort assesment. PhD & DLA Symposium Pesc, Hungary 2012

[24] Croitoru C 2011 *Studii teoretice şi experimentale referitoare la influenţa turbulenţei aerului din încăperile climatizate asupra confortului termic.* PhD thesis. Technical University of Civil Engineering of Bucharest

[25] EN ISO 10551:2001 Ergonomics of the thermal environment. Assessment of the influence of the thermal environment using subjective judgement scales

Acknowledgment
This work was supported by the grants: of the Romanian space agency ROSA, QUEST - Advanced air diffusion system of the crew quarters for the ISS and deep space habitation systems, STAR-CDI-C3-2016-577

IOP Conf. Series: Materials Science and Engineering **595** (2019) 012028 doi:10.1088/1757-899X/595/1/012028

Heat transfer characteristics of a gas cooler

D Dima[1], S Bucsa[1], C Ionita[1], E E Vasilescu[1], A Dobrovicescu[1] and E Niculae[2]

[1]Faculty of Mechanical Engineering and Mechatronics, University Politehnica of Bucharest, Romania
[2]Faculty of Entrepreneurship, Engineering and Business Management, University Politehnica of Bucharest, Romania

E-mail: adobrovicescu@yahoo.com

Abstract. The paper deals with the design and simulation of a gas-water heat exchanger used for heating purposes. The heat exchanger represents the gas-cooler/water heater of a transcritical CO_2 heat pump. The log-mean temperature difference and the NTU-ε methods are comparatively analyzed. The heat transfer area can be calculated through both methods but the true strength and what makes the difference when using the NTU-ε method is that based on it the simulation of the behaviour of an existing heat exchanger can be easily performed.

1. Introduction

A heat pump is an energy-efficient equipment because 75% of the total energy is derived from natural resources like water, air, soil, and the remaining 25% energy is the mechanical work converted into heat. The heat pump systems have been the subject of intensified research in recent years. Huseyin Benli [1] perform a thorough analysis of a heat pump with horizontal serpentines buried in the soil. Yingbai et al. [2] analyzes the destruction of exergy from the gas cooler in a carbon dioxide heat pump installation.

Although refrigerants with high ODP and GWP are currently being used, according to the EU Regulation [3], and in a relation with greenhouse gases policy it is obligatory that by 2050 these substances must be reduced by 80 to 95% compared to 1990's to limit climate changes.

Carbon dioxide has a major disadvantage compared to other working agents used in heat pump installations (R407A, R245fa, R600a, R717) and it has been found that the heat pump installations with carbon dioxide carrier operate in transcritical mode. It is in the researchers' attention because this fluid does not have a negative impact on the environment. Thermophysical properties and environmental impact factors for carbon dioxide are shown in Table 1 [4].

Table 1. Thermophysical properties and environmental impact factors for carbon dioxide.

Property	Carbon Dioxide
Normal boiling point [°C]	-78.4
Critical temperature t_k [°C]	30.97
Critical pressure p_k [bar]	73.33
ODP	0
GWP	1

This paper analyzes the behavior of the gas cooler built (or sealed) in plates, in terms of heat transfer process and the heat exchange area. Such heat exchangers are of a compact type (size) with 1500-2200 m^2/m^3 built from corrugated or smooth thin plates, forming fluid flow channels where heat exchange processes take place. On one side of the plate distributes the cooling fluid and on the other side the warming one [5]. This type of process is very well studied, as well as in various application aspects. Junyub and Kang [6] analyze the heat transfer characteristics to the condensation of R245fa refrigerant in a plate heat exchanger from a heat pump installation. Byung and Chung [7], Giovanni and Claudio [8] analyze the heat transfer and the pressure drop in heat exchangers with brazed plates during condensation of refrigerants R-1233zd and HFC1234yf.

Two methods are used to determine the heat exchange area. The first method is based on the logarithmic mean temperature difference, and it is the most known and used method for determining the heat exchange area. Major disadvantage of the method is that the inlet and outlet temperatures of the two fluids in the heat exchanger must be known. Also in this analyze, for the correct evaluation of the heat exchange area, must be determined according to each heat exchanger separately a correction coefficient, coefficient determined by a graphical method which most of the time introduces additional errors in the calculation [5, 9].

The second method used to determine the heat exchange area is the method based on the number of transfer units NTU. This method has a different approach to evaluate the heat exchange area by using two parameters, respectively, effectiveness and number of the heat transfer units. The equations applied for determining these parameters are different depending of the type of heat exchanger, so, the results obtained via this method are more reliable than one based on the logarithmic mean temperature difference approach [10, 11, 12].

2. Description of the heat pump installation

Figure 1.a shows the schematic diagram of an air-to-water heat pump installation composed of compressor Cp, gas cooler Gc, throttling valve Tv and evaporator Ev. Figure 1.b shows the transcritical evolution cycle of the heat pump installation in the pressure-enthalpy diagram. It can be seen (fig.1.b) that the 2-3 process is performed outside the saturation curve, so at the end of the gas cooling process (state 3) carbon dioxide is in the gaseous state.

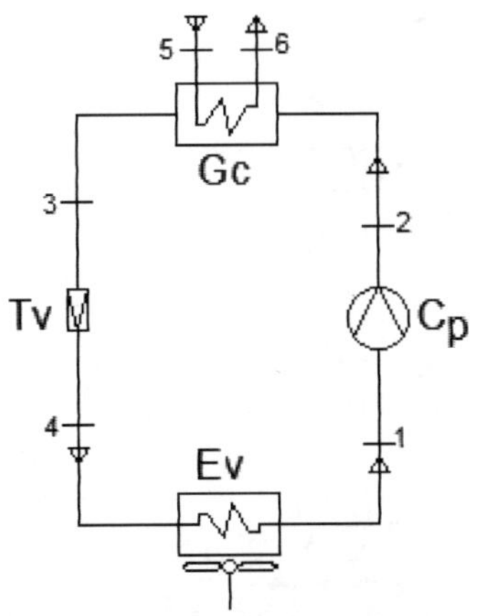

Figure 1.a. Heat pump diagram.

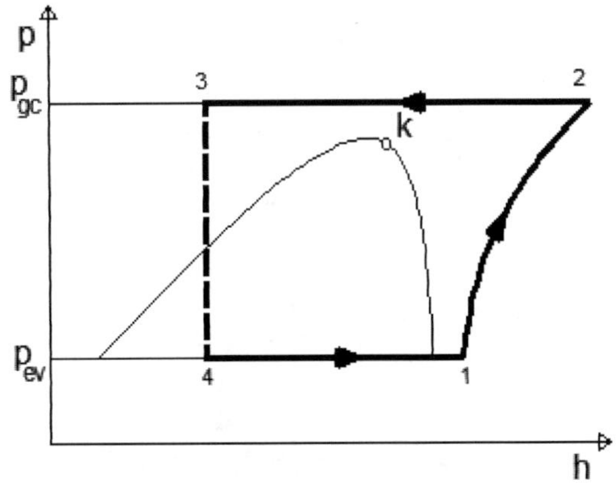

Figure 1.b. p-h diagram of the cycle.

The pressures and working temperatures at the characteristic points of the heat pump cycle are shown in Table 2. All data presented below was determined using the Engineering Equation Solver software [4].

IOP Conf. Series: Materials Science and Engineering **595** (2019) 012028 doi:10.1088/1757-899X/595/1/012028

Table 2. State parameters in the key points of the heat pump cycle.

No.	t[^{0}C]	p[bar]
1	7	36.73
2	108.3	120
3	65	120
4	2	36.73
5	55	1.5
6	60	1.5

3. Analysis of gas cooler operation

Figure 2 shows the temperature variations of the two fluids which are distributed in the gas cooler; so the Carbon Dioxide, is cooled from temperature t_2 to t_3 and water warms up from temperature t_5 to t_6.

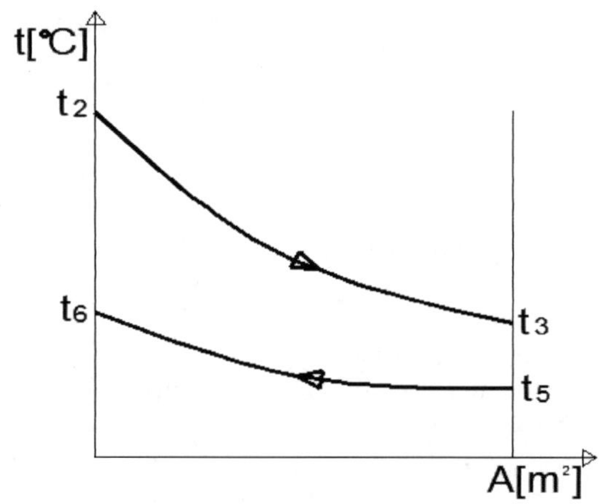

Figure 2. Counter flow gas cooler. Temperatures change across the heat exchanger.

The gas cooler presented in Figure 3 is equipped with plates [6].

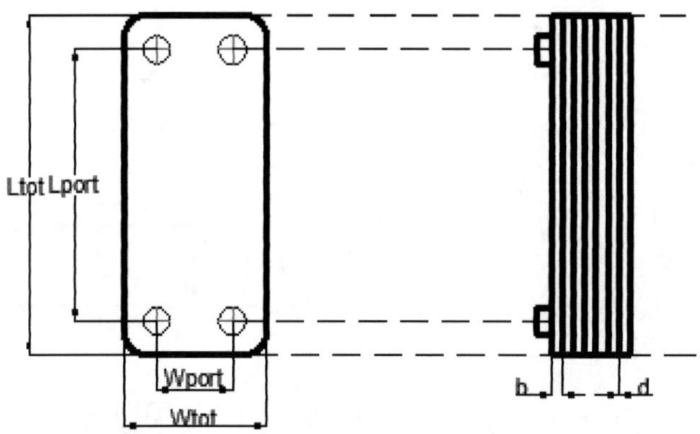

Figure 3. Configuration of the gas cooler.

In Table 3 are presented the geometric characteristics of the gas cooler from Figure 3.

Table 3. Geometry of the proposed gas cooler.

Total lenght	L_{tot}	[m]	0.3
Port to port lenght	L_{port}	[m]	0.24
Total width	W_{tot}	[m]	0.13
Port to port width	W_{port}	[m]	0.07
Port diameter	D_{port}	[m]	0.025
Distance between plates	b	[m]	0.007
Thickness of the plate	d	[m]	0.002
Hydraulic diameter	D_e	[m]	0.013
Plate area	A_e	[m^2]	0.037
Area of the flow channel section	A_f	[m^2]	0.0009

4. The mathematical model

The heat transfer characteristics of both fluids - carbon dioxide and water, were estimated at the mean working temperature of the fluid given by equations (1) and (2):

$$T_{m_{CO_2}} = \frac{t_2 - t_3}{\ln\left(\frac{T_2}{T_3}\right)} \tag{1}$$

$$T_{m_w} = \frac{t_5 - t_6}{\ln\left(\frac{T_5}{T_6}\right)} \tag{2}$$

The equivalent diameter was calculated with equation:

$$D_e = \frac{4 \cdot A_f}{Pe} \tag{3}$$

The convection heat transfer coefficient on the carbon dioxide side is calculated using the criterial equation [5]:

$$Nu_{CO_2} = 0.021 \cdot \mathrm{Re}_{fCO_2}^{0.8} \cdot \mathrm{Pr}_{fCO_2}^{0.43} \cdot \left(\frac{\mathrm{Pr}_{fCO_2}}{\mathrm{Pr}_{wall}}\right)^{0.25} \tag{4}$$

The convection heat transfer coefficient on the waterside is calculated using the criterial equation [6, 13]:

$$Nu_w = 0.0508 \cdot \mathrm{Re}_{fw}^{0.7304} \cdot \mathrm{Pr}_{fw}^{0.33} \cdot \left(\frac{\mu_{mw}}{\mu_{wall}}\right)^{0.14} \tag{5}$$

The overall heat transfer coefficient is given by the following equation:

$$U = \frac{1}{\frac{1}{h_{CO_2}} + \frac{d}{k_{wall}} + \frac{1}{h_w}} \tag{6}$$

Flow and heat transfer characteristics as well as convection coefficient values are presented in table 4.

Table 4. Flow and heat transfer characteristics.

Parameter	Carbon dioxide	Water
t_m [ºC]	86.2	57.5
t_{wall} [ºC]	71.8	71.8
k [W/m²K]	0.03697	0.6386
μ_f [m²/s	$8.862\cdot10^{-8}$	$4.924\cdot10^{-7}$
μ_{wall} [m²/s]	$7.945\cdot10^{-8}$	$4.037\cdot10^{-7}$
a_f [m²/s]	$6.435\cdot10^{-8}$	$1.551\cdot10^{-7}$
a_{wall} [m²/s]	$4.419\cdot10^{-8}$	$1.591\cdot10^{-7}$
$\dot{m}$ [kg/s]	0.2064	0.956
Re_f	123338	28803
Pr_f	1.377	3.175
Pr_{wall}	1.798	2.537
Nu	261.6	122.3
h [W/m²K]	742	5881
U [W/m²K]	655.2	

Two different methods for determining the heat transfer are used: the logarithmic mean temperature difference method and number of heat transfer units and efficiency method.

4.1. *Method based on the logarithmic mean temperature difference*

The relation of calculating the thermal power of the gas cooler is given:

$$\dot{Q} = U \cdot A \cdot LMTD \tag{7}$$

where, the mean logarithmic temperature difference is:

$$LMTD = \frac{\Delta t_{max} - \Delta t_{min}}{\ln\left(\frac{\Delta t_{max}}{\Delta t_{min}}\right)} \tag{8}$$

The necessary area for the heat exchange determined by the method based on the logarithmic mean temperature difference is:

$$A_{LMTD} = \frac{\dot{Q}}{U \cdot LMTD} \tag{9}$$

The mean-log temperature difference method is used mostly for designing or selecting a heat exchanger. In this case the flow rates, specific heat capacities, the inlet and outlet temperatures of both fluids and the overall heat transfer coefficient are known.

4.2. *Method based on the number of transfer units and heat exchanger thermal efficiency*

The NTU-ε method is very useful for simulations of heat exchanger's efficiency based on design and materials. For the case the heat transfer area and the overall heat transfer coefficient are known. The simulation for the heat exchanger are related to prediction of outlet temperatures at different inlet temperatures of the flows and thermal characteristics of the equipment.

For the heat pump heater (gas cooler) three equations for the rate of heat transfer could be written (Fig. 2):

$$\dot{Q} = C_{CO_2}\left(t_2 - t_3\right) \tag{10}$$

$$\dot{Q} = C_w\left(t_6 - t_5\right) \tag{11}$$

$$\dot{Q} = U \cdot A \cdot LMTD = U \cdot A \frac{(t_2 - t_6) - (t_3 - t_5)}{\ln \frac{t_2 - t_6}{t_3 - t_5}} \tag{12}$$

If C_{CO_2}, C_w, U, A (or $U{\cdot}A$) and the inlet temperatures t_2 and t_5 are known, Eqs. 10-12 contain as unknown parameters $\dot{Q}$ and the two outlet temperatures t_3 and t_6.

Combining Eq. 10 and 11, Eq. 13 results:

$$C_{CO_2}(t_2 - t_3) = C_w(t_6 - t_5) \tag{13}$$

$$C_{CO_2}(t_2 - t_3) = U \cdot A \frac{(t_2 - t_6) - (t_3 - t_5)}{\ln \frac{t_2 - t_6}{t_3 - t_5}} \tag{14}$$

Substituting t_6 from Eq. (13) into Eq. (14) gives:

$$\ln \frac{t_2 - \left[t_5 + \frac{C_{CO_2}}{C_w}(t_2 - t_3) \right]}{t_3 - t_5} = U \cdot A \left(\frac{1}{C_{CO_2}} - \frac{1}{C_w} \right) \tag{15}$$

Denoting P as:

$$P = U \cdot A \left(\frac{1}{C_{CO_2}} - \frac{1}{C_w} \right) \tag{16}$$

then Eq. (15) becomes:

$$\frac{t_2 - \left[t_5 + \frac{C_{CO_2}}{C_w}(t_2 - t_3) \right]}{t_3 - t_5} = e^P \tag{17}$$

In Eq. (17) the only unknown parameter is t_3 that can be written as:

$$t_3 = t_2 - (t_2 - t_5) \frac{e^P - 1}{e^P - \frac{C_{CO_2}}{C_w}} \tag{18}$$

$C_{CO_2} \langle C_w$ the heat exchanger efficiency becomes:

$$\varepsilon = \frac{t_2 - t_3}{t_2 - t_5} \tag{19}$$

Combining Eq. (18) and Eq. (19) the following results:

$$\varepsilon = \frac{e^P - 1}{e^P - \frac{C_{CO_2}}{C_w}} \tag{20}$$

where (Eq. 16) $P = \frac{U \cdot A}{C_{CO_2}} \left(1 - \frac{C_{CO_2}}{C_w} \right)$

Denoting by

$$NTU = \frac{U \cdot A}{C_{CO_2}} = \frac{U \cdot A}{C_{\min}} \tag{21}$$

it results that

$$P = NTU\left(1 - \frac{C_{min}}{C_{max}}\right) \tag{22}$$

Accounting for Eqs. (20) and (22) the efficiency ε becomes a function of NTU and $\frac{C_{min}}{C_{max}}$:

$$\varepsilon = f\left(NTU, \frac{C_{min}}{C_{max}}\right) \tag{23}$$

For the specific case C_{min}, C_{max}, U and the inlet and outlet temperatures are given. When the size of the heat transfer area is needed, the following steps have to be applied:

1. Calculate ε (Eq. 19)
2. Determine *P* (Eq. 20) and NTU (Eq. 22)
3. Determine *A* (Eq. 21)

Table 5. Heat transfer area calculated with LMTD and NTU-ε methods. ($\dot{Q}_{Gc}$= 19.98 kW; C_{CO_2}=0.4614 kW/K; C_w= 3.998 kW/K; C_r=0.1154; t_2=108.3 C; t_3=65 C ; t_5=55 C ; t_6= 60 C)

LMTD [K]	A_{LMTD} [m^2]	ε [-]	NTU [-]	A_{NTu} [m^2]
24.32	1.254	0.8124	1.78	1.254

Obviously (Table 5) with both methods the same value for the heat transfer area $A_{LMTD} = A_{NTU}$ is obtained.

For simulation the behaviour of a given (existing or in design phase) heat exchanger (*A* and *U* stated), the NTU-ε method is able to provide the outlet temperatures for different inlet temperatures and different sets of mass flow rates and fluid properties (C_{min}, C_{max}).

For the specific counterflow heat exchanger figures 4, 5 and 6 give the *NTU*-ε, $C_{CO_2} - t_3$ and t_6-t_3 diagrams.

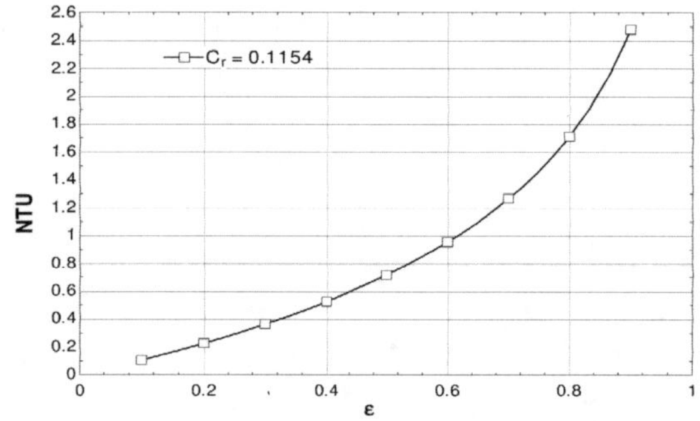

Figure 4. *NTU* – ε diagram.

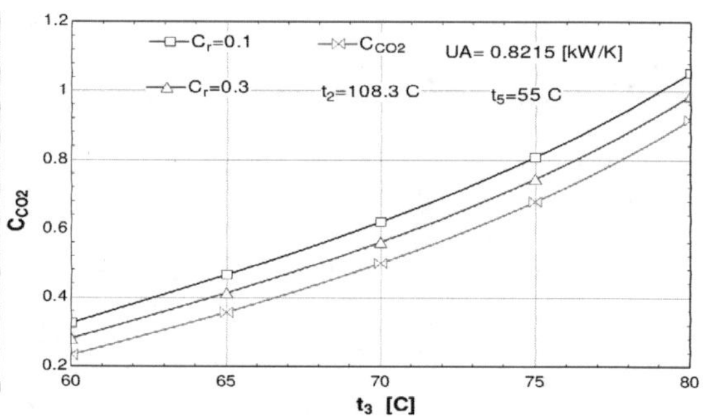

Figure 5. C_{CO_2} – t_3 diagram.

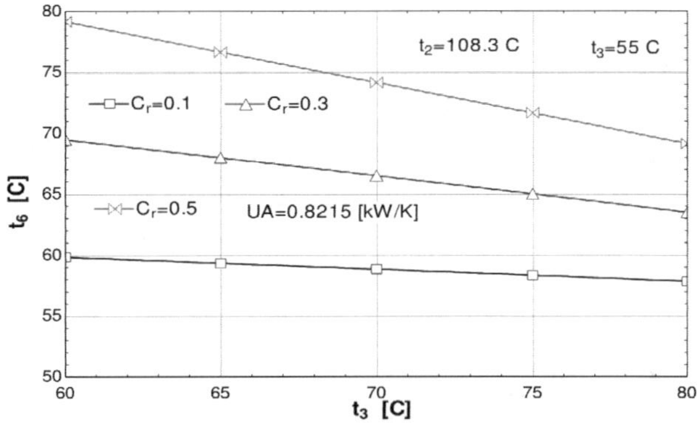

Figure 6. t_6 - t_3 diagram.

5. Conclusions

The study points out the differences in calculating the heat transfer area based on the log-mean temperature difference between two streams and the NTU-ε method.

The input data for the log-mean temperature difference method are the two streams thermal capacities, the inlet temperatures and one outlet temperature. The second outlet temperature is obtained through energy balance of the heat exchanger.

For the NTU-ε method are required the thermal capacities of the two heat streams and information about only the inlet temperatures. The efficiency of the heat exchanger gives one of the outlet temperatures and the energy balance gives the second one.

The heat transfer area can be calculated with the same accuracy by both methods but the true strength and what makes the difference when assessing an existing heat exchanger using the NTU-ε method less input parameters are needed in order to perform the calculations.

Nomenclature

A heat transfer area [m^2]
a thermal diffusivity factor [m^2/s]
C thermal capacity [J/K]
Δt temperature difference [°C]
h specific enthalpy [kJ/kgK]
k thermal conductivity [W/mK]
$\dot{m}$ mass flow rate [kg/s]
Nu Nusselt's number
p pressure [bar]
Pe perimeter [m]
Pr Prandtl's number [-]
Re Reynolds's number [-]
T temperature [K]
t temperature [°C]
U overall heat transfer coefficient [W/m^2K]

Greek symbols
ν kinematic viscosity [m^2/s]

Indices
gc gas cooling
CO_2 carbon dioxide

IOP Conf. Series: Materials Science and Engineering **595** (2019) 012028 doi:10.1088/1757-899X/595/1/012028

ev evaporator
f fluid
h hot
k critical point
m mean
max maximum
min minimum
w water

Abbreviations
GWP global warming potential
LMTD logarithmic mean temperature difference
NTU number of heat transfer units
ODP ozone depletion potential

References
[1] Benli H 2011 Energetic performance analysis of a ground-source heat pump system with latent heat storage for a greenhouse heating *Energy Conversion and Management.* **52**(11) 581
[2] Yingbai X, Zhichao W, Kuikui C and Xuedong Z 2011 The exergy analysis of gas cooler in CO2 heat pump system *Procedia Environmental Science* **11** 1555
[3] Regulation (EU) no 517/2014 of the European Parliament and of the Council of 16 April 2014 on fluorinated greenhouse gases and repealing Regulation (EC) no 842/2006
[4] https://www2.le.ac.uk/offices/itservices/ithelp/my-computer/programs/ess-10-2 Engineering Equation Solver (vers. 10.2 - 2017)
[5] Raducanu P 2006 Schimbatoare de caldura pentru instalatii termice si frigorifice (In Romanian) (ed. Printech, Bucuresti) pp 153
[6] Junyub L, Kang S, Dongwoo K, Dong Chan L and Yongchan K 2018 Condensation heat transfer characteristics of R245fa in a shell and plate heat exchanger for high-temperature heat pumps *International Journal of Heat and Mass Transfer.* **127** 730
[7] Byung H, Chung W, Oh J, Chang K and Yong T 2018 Characteristics on condensation heat transfer and pressure drop for a low GWP refrigerant and brazed plane heat exchanger *International Journal of Heat and Mass Transfer.* **22** 1272
[8] Giovanni A and Claudio Z 2013 Condensation of the low GDW refrigerant HFC234yf inside a brazed plane heat exchanger *International Journal of Refrigeration* **36**(2) 612
[9] Stamatescu C 1979 Tehnica Frigului Vol 2 Calculul si constructia masinilor si instalatiilor frigorifice industrial (In Romanian) (ed. Tehnica, Bucuresti) pp 119
[10] Randall F and Barron 1999 Cryogenic Heat Transfer (ed. CRC Press) 272
[11] Navarro H A and Cabezas-Gomez C 2007 Effectiveness-NTU computation with a mathematical model for cross-flow heat exchangers Brazilian Journal of Chemical Engineering. **24**(04) 509
[12] Kays W M and London A L 1998 Compact heat exchanger (ed. Malabar, Fla.: Krieger Pub. Co. Third edition, Repr. Ed. with corrections)
[13] Fernandez-Seara J, Uhia F J, Sieres J and Campo A 2007 A general review of the Wilson plot method and its modifications to determine convection coefficients in heat exchange devices *Appl. Therm. Eng.* **27** 2745

IOP Conf. Series: Materials Science and Engineering **595** (2019) 012029 doi:10.1088/1757-899X/595/1/012029

Kinematic Beta-Type Stirling Motor-Driven Compressor

V M Homutescu, D T Bălănescu, C E Panaite and A G Lupu

Faculty of Mechanical Engineering, Department of Mechanical and Automotive Engineering, "Gheorghe Asachi" Technical University, Iaşi, Romania

E-mail: balanescud@yahoo.com

Abstract. The theoretical functioning of a kinematic one-stage beta-type Stirling motor-driven compressor was analyzed. This machine comprises a kinematic Stirling engine and a reciprocating compressor in one single block. A double acting cylinder houses a piston that plays both the roles of the power piston of the Stirling engine and of the compressor's piston. The maximum performances of this machine are determined through numerical modelling. For a certain compression ratio of the compressor, a functioning regime for which the work produced by the Stirling engine is equal to the absolute value of the work consumed by the compressor can be found.

1. Introduction

A Stirling motor-driven compressor (SMDC) is a combination between a Stirling engine and a compressor. The Stirling engine directly drives the compressor. The SMDC analyzed here has a kinematic beta-type Stirling engine as prime mover. The power piston of the Stirling engine also acts as a compressor piston. It is placed inside a double-acting cylinder. The compression space of the Stirling engine is placed on one side of the shared piston and the working space of the compressor is placed on the other side. Suction and discharge valves and lines are conveniently placed. The working agent of the compressor and the working agent of the Stirling agent could be the same or different.

Several other devices that use Stirling engines for driving pumps and compressors for various uses are known.

Around 1860 W. H. Bailey fabricated at Manchester a Stirling-type hot air engine [1]. This Stirling engine was directly connected through the crankshaft with a water pump. The piston of the pump acts in a separate cylinder. Another Stirling engine, namely Robinson engine, uses a chain drive to transmit mechanical power to a water pump [1].

An "Apparatus for Compressing Gases" that uses a displacer and a constant total volume working space was patented in 1939 by Vannevar Bush [2]. The two variable volume chambers created by the displacer inside the cylinder were connected through a heater, a regenerator and a cooler, hence the resemblance of this device with a Stirling engine. The inventor called the device "direct thermal compressor".

A direct connection between a Stirling engine working as a prime mover performing as such clockwise thermodynamic cycle and a Stirling refrigerating machine performing a counterclockwise cycle is named a "duplex Stirling engine" [3]. The two machines are connected through a shared power piston, each of them having its own compression space and its line of heat exchangers (heater, regenerator and cooler). Another possibility is to directly connect a Stirling engine prime mover to a heat pump inside which the working agent performs a Rankine cycle [3]. The connection is made through a double-acting free-piston inertial compressor that shares one of its compression spaces with the buffer space of the Stirling engine.

Dineen [4] obtained in 1982 an US Patent for a hermetic oil-free Stirling engine compressor. An elastic diaphragm serves as power piston. The space outside the power piston form a "hydraulic chamber", and has another elastic diaphragm that serves as piston for the compressor. Inside this space a linear alternator is housed. The alternator uses some of the work yielded by the Stirling engine.

In 1985 Watanabe et al. obtained an US patent [5] for an "engine compressor having a Stirling cycle engine". An alpha-type Stirling engine with four cylinders is the prime mover. Each piston of the Stirling engine is directly connected through a guided piston rod to a compressor piston. The compressor cylinders could be single- or double-acting, and are fitted with the corresponding suction and discharge valves.

An US patent issued in 1992 to Momose [6] presents a gamma-type Stirling engine used as prime mover for driving the compressor of an air-conditioning equipment, or an air compressor. The pressure variation inside the Stirling engine is used for producing an axial displacement of a diaphragm. This oil-free compressor has symmetrical construction, and uses four membranes separating several chambers. Suction and discharge valves are placed inside the two working spaces of the compressor.

In 2004 Berkowitz obtained an US patent [7] for a free-piston Stirling engine that drives a heat pump with vapor compression. Both the Stirling prime mover and the compressor are placed inside the same hermetically sealed enclosure. The piston of the compressor is directly connected with the power piston of the Stirling engine, forming a single structural element. The device could use a single working agent, namely CO_2, or two working agents, helium for the Stirling engine and carbon dioxide for the heat pump. For the latter case, helium will leak inside the heat pump circuit, so a gas-liquid phase separator must be used for collecting a helium-rich gas mixture that will go back to the Stirling engine through a return line.

A scheme of a Stirling engine driving a compressor, attributed to William Beale, the inventor of the free-piston Stirling engine, was presented in [8]. A four-cylinder alpha-type Stirling engine was directly connected to four separate single- or double-acting compressor cylinders. The novelty over the Watanabe drawings [5] was that the compressed gas is used for driving a closed cycle gas turbine with intermittent feeding.

In a previous published work from 2005 [9] the authors presented and analyzed five possible schemes for SMDC, based on alpha, beta and gamma configurations. A beta-type Stirling engine could be transformed into a SMDC if the space below the power piston is organized as a working space for a reciprocating compressor, being fitted with suction and discharge valves. The mentioned paper [9] did not develop functioning models for the proposed SMDC.

In 2013 [10] authors developed a physico-mathematical model for a beta-type engine-driven Stirling compressor. This device is a Stirling engine fitted with suction and discharge valves and lines. When the pressure inside the engine rises, the discharge valve opens and some of the working agent leaves the compression chamber. Suction takes place when the pressure inside the machine drops below the pressure inside the suction line. So, it is compulsory that the gas needed to be compressed and the gas serving as working agent for the Stirling engine are the same.

In 2016 Tognarelli et al. obtained an Australian patent for a compressor train with a Stirling engine [11]. The shaft of the reciprocating compressor is connected to the shaft of the Stirling engine through a clutch. A reversible electric machine is connected to the same shaft line. The electric machine operates as an engine when the Stirling engine produces less work than the compressor needs, or as a generator otherwise. The Stirling engine should use the waste heat usually available in refinery plants.

The paper analyzes the functioning of a kinematic one-stage beta-type Stirling motor-driven compressor based on an isothermal physico-mathematical model of the Stirling engine and on a simplified physico-mathematical model for the reciprocating compressor.

2. Scheme and physical model of the beta-type Stirling motor-driven compressor

The beta-type SMDC schematic diagram is presented in figure 1. A SMDC unit comprises the cylinder 7 that houses the displacer piston 14 and the power piston 8. The space delimited by the cylinder heads 10 and 18 and by the cylinder itself is divided by the mentioned pistons in three variable-volume spaces: the expansion space 13 and the compression space 9 of the Stirling engine and the compressor working space 6. The expansion and compression spaces of the of the Stirling engine are connected through the cooler 16, the regenerator 15 and the heater 12. The working space of the compressor is equipped with suction valve 5 and with the discharge valve 19, housed inside the corresponding lines 4 and 20. The linkage between the pistons is made through a rhombic drive comprising four rods 2 and yokes 22 and 23.

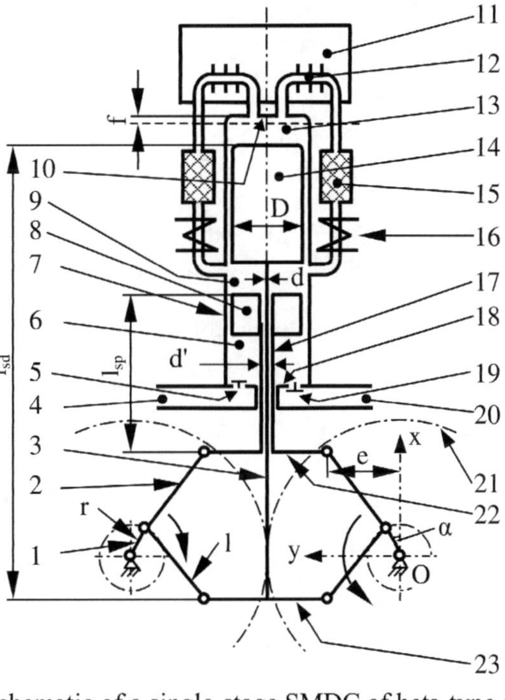

Figure 1. Schematic of a single-stage SMDC of beta-type arrangement:
1 – crankshaft; 2 – connecting rod; 3, 17 – stem; 4 – suction line; 5 – suction valve; 6 – compressor working space; 7 – cylinder; 8 – power piston; 9 – compression chamber; 10 – cylinder head of the Stirling engine; 11 – combustion chamber; 12 – heater; 13 – expansion chamber; 14 – displacer; 15 – regenerator; 16 – cooler; 18 – cylinder head of the compressor; 19 – discharge valve; 20 – discharge line; 21 – gear wheel; 22, 23 – yoke.

The Stirling engine acts as a prime mover, producing the work needed by the compressor. In a pure SMDC the compressor uses completely the work yielded by the Stirling engine. If the Stirling engine is producing more work that needed for the functioning of the compressor, the SMDC could be fitted with an electric generator providing electrical energy.

The Stirling engine functioning is described by the isothermal model with real volume variation. This model is well-known to those skilled in the art [1], [3], [8], and thus superfluous to describe here. The main hypotheses used are: agent is a perfect gas; processes inside each chamber are isothermal and quasistatic; instantaneous pressure inside the machine is the same in all chambers and the regenerator has a temperature that is a mean between heater and cooler temperatures.

The model used for describing the functioning of the reciprocating compressor uses polytropic processes for compression and expansion and isobaric processes for suction and discharge. Mean pressure losses at suction and discharge and mean polytropic indices for compression and expansion were adopted using the recommendations [12].

3. Mathematical model

A mathematical model describes the functioning of the entire SMDC and correlates the work yielded by the Stirling engine as prime mover with the work consumed by the reciprocating compressor. The model allows for calculating all thermodynamic parameters for any cycle point and for calculating the energies exchanged. The model also estimates the performances of the SMDC and calculates the dimensions of the needed flywheel.

The following dimensions needed for the model were mentioned in figure 1: r - crank radius; l - connecting rod; e - crankshaft - wrist pin offset; D - cylinder bore / power piston and displacer diameter; d - displacer stem diameter; d' - power piston stem diameter; l_{sd} - distance between displacer's yoke and displacer head; l_{sp} - distance between power piston's yoke and power piston head; f = distance between cylinder head and displacer head in its top dead center; α - crank position angle; xOy - coordinate system.

Based on the geometry the variations laws of the expansion chamber volume V_e and compression chamber volume V_c as functions of crank angle were determined. The volumes occupied by the working agent inside cooler, regenerator and heater are V_k, V_{reg} and V_h. Usually, these volumes are given as fractions of the volume swept by one of the pistons. The isothermal model of the Stirling engine uses constant temperatures for the agent inside the heat exchangers, T_k, T_{reg} and T_h, the temperature of the regenerator being calculated as a mean (e.g. a logarithmic one) of the temperatures of the cooler and heater.

Assuming that inside the Stirling engine evolves a constant total mass m_T of working agent having the specific gas constant R, the instantaneous pressure $p(\alpha)$ inside the prime mover is determined by the following relation:

$$p(\alpha) = \frac{m_T R}{\frac{V_c(\alpha)+V_k}{T_k} + \frac{V_{reg}}{T_{reg}} + \frac{V_e(\alpha)+V_h}{T_h}}. \quad (1)$$

The work yielded by a Stirling engine cycle is given by

$$W_{SE} = \int p(\alpha)\left[\frac{d}{d\alpha}(V(\alpha))\right]d\alpha, \quad (2)$$

where $V(\alpha)$ is the total volume of the working agent inside heater, regenerator, cooler, compression and expansion chamber.

The working space of the compressor has a variable volume $V_{ws}(\alpha)$. Its particular variation law is given by the chosen geometry. Due to the presence of the power piston stem, the maximum value of V_{ws} is always smaller than the volume swept by the power piston head (adjacent to the compression chamber of the Stirling engine) between its dead centers. The clearance volume V_{cl} value was adopted as a fraction of the volume swept by the power piston inside the compressor's working space.

The operating regime of the compressor is characterized by the pressure p_S of the suction tank and by the pressure p_D of the receiver tank, thus the external compression ratio is p_D / p_S. The suction mean pressure p_s and the discharge mean pressure p_d are calculated based on adopting suction and discharge relative pressure drop coefficients [12]:

$$\delta p_s = \frac{p_A - p_a}{p_A} \quad \text{and} \quad \delta p_d = \frac{p_d - p_D}{p_D}. \quad (3)$$

Mean values for polytropic indices n_c for compression and n_e for expansion were adopted. Temperatures at the end of the suction process and at the end of the discharge process were calculated by adopting the temperature differences ΔT_s during the suction gas heating process and ΔT_d during the discharge gas cooling process. Temperature at the beginning of the suction process is T_A.

The following functional crank angles were defined: α_{es} for the end of the suction process, α_{ec} for the end of the compression, α_{ed} for the end of the discharge process and α_{ee} for the end of the expansion process. With these notations, the pressure variation law inside the working space of the compressor is

$$p_{ws}(\alpha) = \begin{cases} p_s & if\ \alpha \in (\alpha_{ee}, 2\pi]\ and\ \alpha \in [0, \alpha_{es}] \\ p_s \left[\dfrac{V_{ws}(\alpha_{es})}{V_{ws}(\alpha)} \right]^{n_c} & if\ \alpha \in (\alpha_{es}, \alpha_{ec}] \\ p_d & if\ \alpha \in (\alpha_{ec}, \alpha_{ed}] \\ p_d \left[\dfrac{V_{ws}(\alpha_{ed})}{V_{ws}(\alpha)} \right]^{n_e} & if\ \alpha \in (\alpha_{ed}, \alpha_{ee}]. \end{cases} \tag{4}$$

The suction process is divided in two angle intervals because of the chosen coordinate system and of the origin for measuring the crank angle α.

In accordance with the above statements, the work W_{RC} consumed by the reciprocating compressor is given by

$$W_{RC} = \int p_{ws}(\alpha) \left[\frac{d}{d\alpha} (V_{ws}(\alpha)) \right] d\alpha . \tag{5}$$

By comparing the work W_{RC} consumed by the compressor with the work W_{SE} produced by the Stirling engine three possibilities appears. If $W_{RC} + W_{SE} < 0$, the Stirling engine output is not sufficient for powering the compressor. In this case the SMDC could not function by itself. For $W_{RC} + W_{SE} = 0$, the device will function as intended, as a Stirling motor-driven compressor. The net work exchange with the surroundings is null in this case. If $W_{RC} + W_{SE} > 0$, the Stirling engine produces more than the needs of the compressor. The excess work could be used by an electric generator, as in [11]. The possibility also arises to cope with the excess work by modifying the functioning regime of the Stirling engine, e.g. by reducing the mass of working agent m_T of the Stirling engine, or by reducing the temperature T_h of the heater.

When the power piston goes down (in figure 1), the total volume $V(\alpha)$ occupied by the working agent inside the Stirling engine increases, and the prime mover produces work. In the same time, inside the working space of the compressor the volume $V_{ws}(\alpha)$ decreases and the compression and discharge processes are taking place. So, it is compulsory to check if the Stirling engine delivers the right amount of work at the very moments when the compressor needs it. If a good compatibility between the work produced and the work consumed is achieved, there is a certain chance that the SMDC does not need a separate flywheel. If not the case, the SMDC needs a flywheel for accumulating appropriate energy during the excess work production periods and returning it during the periods when the device is producing less work than needed, i.e. during compression.

The torque variation of the SMDC against crank angle α could be calculated with a method using the pressure forces, the inertial forces of the reciprocating parts, masses and linear dimensions, apud [13]. The algorithm used calculates the pressure forces exerted by the working agents on the displacer and on the power piston, the forces of inertia for all parts performing reciprocating movements, the resultant forces that act on displacer and power piston and the component forces acting on the crankpin journal, perpendicular to the crank web. Torque M is finally expressed as product between crank radius and the mentioned component forces.

The masses of the main parts of the SMDC - displacer, power piston, connecting rods, stems, yokes, wrist pins - must be known, in order to calculate the inertial forces in the reciprocating parts. The mass of a connecting rod was divided in two, the first partial mass performing reciprocating movement together with the yoke, stem and piston, and the second partial mass performing a rotational movement with the crankshaft. Only the first partial mass is useful for torque calculations, and is usually expressed as a fraction of the connecting rod total mass [13]. The two-masses system has a dynamic behavior similar to the one of the connecting rod.

4. Results of the numerical simulation of the SMDC

A SMDC based on the configuration presented in figure 1 and described by the following particular values: r = 0.0175 m, l = 0.0525 m, e = 0.024 m, l_{sd} = 0.310 m, l_{sp} = 0.130 m, D = 0.040 m, d = 0.008 m, d' = 0.015 m, f = 0.0005 m was analyzed through numerical simulation.

Volumes of the heat exchangers of the Stirling engine are $V_h = V_k = 0.05\ V_{cmax}$; $V_{reg} = 1.2\ V_{cmax}$, where V_{cmax} is the maximum volume of the compression chamber. Temperatures of the heat exchangers are T_h = 823 K and T_k = 310 K. Regenerator mean temperature is T_{reg} = 525.4 K, the logarithmic mean of T_h and T_k.

The mass of working agent performing inside the Stirling engine is m_T = 0.000667 kg. This particular value was obtained by imposing a null net work output of the SMDC. Each crankshaft has a rotational speed of n = 1000 rpm.

The working agent of the Stirling engine and the working agent of the compressor is air, considered to be a perfect gas, with specific gas constant R_{air} = 287 J kg^{-1} K^{-1}.

The functional parameters of the compressor are p_S = 0.1 MPa, p_D = 0.9 MPa, δp_s = 0.05, δp_d = 0.10, n_c = 1.38, n_d = 1.34, ΔT_s = 10 K, ΔT_d = 15 K, T_S = 288 K, $V_{cl} = 0.0124\ V_{Sc}$, where V_{Sc} is the volume swept by the power piston inside the working space of the compressor.

The mass performing a reciprocating movement with the displacer (including the mass of the displacer, stem, yoke, wrist pins, and first portions of the mass of the two connecting rods) is 0.763 kg. The corresponding mass for the power piston is 0.791 kg. These masses were calculated using the particular design made for a small Stirling engine by the authors.

A dedicated program for the SMDC was developed in Mathcad. At the working regime characterized by the chosen values of p_S and p_D the following main results were obtained: W_{SE} = 12.83 J/cycle, W_{RC} = - 12.83 J/cycle. The power produced by the Stirling engine and consumed by the compressor is 213.9 W. The volumetric flow rate at the end of the suction process state is q = 0.041 m^3 min^{-1}, and the mass flow rate is $\dot{m}$ = 0.76 10^{-3} kg s^{-1}.

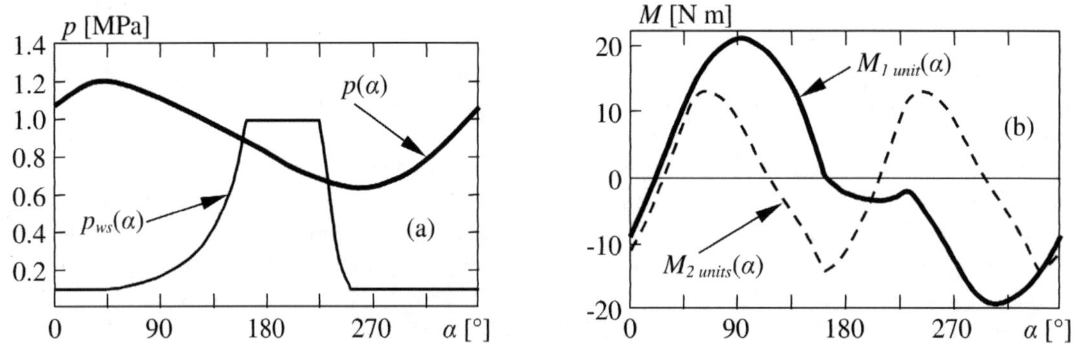

Figure 2. SMDC cyclical behavior:
(a) – variation of the pressure with crank angle inside the Stirling engine and inside the reciprocating compressor; (b) – torque variation with crank angle for an one-unit SDMC and for a SDMC with two-units with twin-cylinder with cranks at 180° apart configuration.

Figure 2 (a) shows the cyclical variations of the working agent pressure inside the Stirling engine and inside the working space of the compressor. The compression, discharge, expansion and suction processes inside the compressor are clearly visible. When compression and discharge processes take place the pressure inside the Stirling engine decreases. This confirms the assertion made in the previous chapter, that in the phase when the prime mover delivers work, the compressor uses some of the work directly.

In figure 3 (b) curve $M_{1\ unit}(\alpha)$ was calculated for a SMDC based on the scheme in figure 1. The curve shows that there is a part of the SMDC cycle when the torque is positive, and part when the torque is negative. I.e., there is a part of the cycle when SMDC produces work (the Stirling engine produces more that the compressor could consume), and a part of the cycle when SMDC needs work for the functioning. From here the necessity results of accumulating the excess work in a flywheel during the period when SMDC produces work. The flywheel restitutes the accumulated energy to the SMDC during the other period. The dimensions of the needed flywheel can be calculated by choosing a value for the coefficient of crankshaft speed non-uniformity δ. E.g., for $\delta = 1 / 200$ the flywheel must have a moment of inertia $I = 0.622$ kg m^2. A flywheel satisfying this condition and made of steel could have: internal radius $r_i = 0.250$ m, external radius $r_e = 0.2625$ m, width of 0.060 m and a total mass of 9.46 kg. The flywheel could be mounted on one of the crankshafts of the rhombic drive mechanism, or could be divided in two smaller flywheels, one for each crankshaft.

For obtaining a better value of the coefficient δ the SMDC could be realized with multiple units on the same crankshaft. In figure 3 (b) curve $M_{2\ units}(\alpha)$ was calculated for a SMDC having two functional units, in a twin-cylinder with cranks at 180° apart configuration. In comparison with the other curve in figure 3 (b), the reduction of the torque variation interval is clearly visible, so the dimensions of the flywheel will be also reduced. The needed moment of inertia in this case is $I = 0.223$ kg m^2.

5. Conclusions

A scheme for a motor-driven compressor comprising a Stirling engine directly coupled to a reciprocating compressor, in which the compressor is using in its construction the power piston of the Stirling engine, was proposed.

By applying an isothermal physico-mathematical model to the Stirling engine and a physico-mathematical model based on an equivalent p-V indicator diagram to the reciprocating compressor, it was found that, for a certain compression ratio of the compressor, there is a single functioning regime for which the net work exchanged with the surroundings by the SMDC is null.

The equality between the absolute value of the cyclical work required by the compressor and the cyclical work yielded by the Stirling engine could be achieved by various means, e.g. by regulating the total mass of working agent inside the Stirling engine.

Due to the nature of the physico-mathematical models used, the calculated performances of the SMDC have the character of maximum achievable performances.

A numerical example was used for obtaining the cyclical variations of the pressure inside the Stirling engine and inside the compressor, and the cyclical variation of the SMDC torque, as diagrams. The analysis of these diagrams shows that there is a period of the cycle in which the Stirling engine is producing work, and a period when it needs work. The same conclusion is valid also for the SMDC as a whole. This situation imposes the presence of an accumulator of mechanical energy into the componence of the SMDC - a flywheel. The flywheel stores the excess energy produced by the SMDC during the expansion of the working agent of the Stirling engine, and returns it to the SMDC when needed.

The numerical example allowed for establishing the maximum performances that a SMDC is capable of.

The SMDC could be an attractive solution for several applications, especially when conventional primary energy sources cannot be accessed. Possible uses of this motor-driven compressor are as:

- portable air compressor, for off-grid applications;
- for replacing a regular engine-driven compressor powered by an internal combustion engine;
- for recovering the waste heat of hot fluids from various industries;
- for harnessing the solar energy.

When properly designed, the SMDC could be also fitted with an electric generator, so the output could be compressed gas, electrical energy or a combination of both.

In our opinion, the SMDC is an unconventional heat engine worth to be further studied.

IOP Conf. Series: Materials Science and Engineering **595** (2019) 012029 doi:10.1088/1757-899X/595/1/012029

6. References

[1] Finkelstein T and Organ A 2001 *Air Engines: The History, Science, and Reality of the Perfect Engine* (New York: ASME Press)

[2] Bush V 1939 *Apparatus for compressing gases* US Patent US2,157,229

[3] Walker G 1980 *Stirling Engines* (Oxford: Oxford University Press)

[4] Dineen J 1982 *Stirling engine compressor with compressor and engine working fluid equalization* US Patent US4,361,008

[5] Watanabe T, Haramura S and Momose Y 1985 *Engine compressor having a Stirling cycle engine* US Patent US4,532,767

[6] Momose Y 1992 *Compressor integral with Stirling engine* US Patent US5,088,284

[7] Berchowitz D 2004 *Stirling engine driven heat pump with fluid interconnection* US Patent US6,701,721

[8] Urieli I Stirling cycle machine analysis https://www.ohio.edu/mechanical/stirling/ retrieved March 2019

[9] Homutescu V M Kinematic Stirling motor-driven compressors 2005 *Proceedings of CNEI 2005 Conference "Modelling and Optimization in the Machine Building Field", Bacău, Romania* MOCM-11 vol. 3 (Bacău: ALMA MATER Publishing House) pp 49-52

[10] Homutescu V M Maximum performances of the Stirling machine working as motor-driven compressor 2013 *Bul. I.P.I.* **LIX** 3 pp 27-38

[11] Tognarelli L, Bagagli R, Buffa F and Saltimi M 2016 *A compressor train with a Stirling engine* US Patent Application US20160348661A1, Australian Patent 2015212952

[12] Stamatescu C, Taşcă D and Grigoriu M 1965 *Compresoare volumice. Teoria, calculul şi construcţia* (Bucureşti: Editura Tehnică)

[13] Homutescu V M, Homutescu C A and Homutescu A 2001 *Maşini şi instalaţii termice şi hidraulice* (Iaşi: Editura CERMI)

Acknowledgments

This work was partly supported by the project POSCCE-A2-O2.2.1-2009-4-ENERED, ID nr. 911, co-financed by the European Social Fund within the Sectoral Operational Program "Increase of Economic Competitiveness".

IOP Conf. Series: Materials Science and Engineering **595** (2019) 012030 doi:10.1088/1757-899X/595/1/012030

Modelling of CO_2/acetone fluid mixture thermodynamic properties for compression/resorption refrigeration systems

R J B Moreira-da-Silva, D Salavera and A Coronas

Group of Applied Thermal Engineering – CREVER, Dep. Mechanical Engineering, Universitat Rovira i Virgili, Tarragona Spain

E-mail: alberto.coronas@urv.cat

Abstract. Compression/resorption refrigeration systems are similar to conventional vapor compression systems but using zeotropic mixtures with large boiling temperature difference, similar to the refrigerant/absorbent mixtures used in absorption technology. A resorber and a desorber replace the conventional condenser and evaporator, respectively. As the saturation temperature in both desorber and resorber change with the composition, there is a gliding temperature that can reduce the irreversibility of the heat transfer with external sensible heat and sink sources. Another advantage of these systems is that they operate at lower pressures and pressure ratios than the conventional ones. CO_2/acetone mixture was proposed for automotive air-conditioning with compression/resorption systems. In this paper, different methods are evaluated for modelling the thermodynamic properties, using experimental data. The selected thermodynamic model will be used to study the performance of the compression/resorption refrigeration cycle.

1. Introduction

Nowadays, vapour-compression refrigeration is the technology most commonly used for refrigeration purposes. The refrigerants typically used in vapour-compression cycles are hydrofluorocarbons (HFCs) that are potent greenhouse gases. Although vapour-compression refrigeration is a well-known technology and provides very good performance, the need to reduce environmental impacts has renewed interest in the use of natural refrigerants.

Compression/resorption refrigeration is a very interesting technological alternative that provides more flexible design options compared to conventional systems. The concept of compression/resorption is similar to the conventional vapour compression cycle but the refrigerant is a zeotropic binary mixture with a large boiling temperature difference, and the evaporating and condensing processes are not completed as in conventional vapour-compression cycles. The evaporation and condensation taking place at the evaporator and condenser at constant pressure and temperature (for pure refrigerants) in the conventional systems are replaced by desorption and absorption processes at non-isothermal conditions (Lorentz Cycle) in the desorber and the resorber, respectively. As the evaporation in the desorber of the refrigerant/absorbent mixture is partial, along with the vapor stream a residual solution stream is produced and once pumped returns to the resorber in the so-called "resorption circuit". The vapour stream leaving the desorber is compressed in the compressor and sent to the resorber after joining with the solution stream from the desorber. The two pressure levels in the conventional vapor-compression cycle are fixed by the refrigerant condensing and evaporating temperatures; however, in the resorption cycle, the operating pressures depend also on the composition of the refrigerant/absorbent mixture, providing great flexibility in terms of cycle design. Thus, these cycles can operate at lower high pressures and lower pressure ratio than the compression cycles working with the refrigerant fluid.

Ammonia/water is the most commonly used working pair in compression/resorption refrigeration systems, but due to the toxicity of ammonia that limits its use in some applications, other mixtures based on carbon dioxide were proposed by several researchers. A complete review on compression/resorption refrigeration technology was published by Groll [1] in 1997.

The first reference of compression/resorption cycles, also called vapor compression cycle with solution circuit, with CO_2 /acetone is from Groll and Kruse [2] based on the Ph.D. thesis of Groll [3] for refrigeration and air conditioning applications. In 2000, Spauschus and Hesse [4] published two patents about the use of CO_2 based mixtures and different absorbents for air conditioning in the automotive sector. The compression process performed in these documents is a two-phase compression, so the liquid and vapour streams leaving the desorber are compressed together by a wet compressor. As a result, the complexity and number of components in the cycle are reduced and due to the reduction of the discharge temperature at the compressor, the performance is improved. Mozurkewich et al. [5] studied by numerical simulation the performance of a wet compression/resorption cycle operating with CO_2/acetone and compared with other absorbents as N-methyl-2-pyrrolidone, neopentylglycol diacetate, and g-butyrolactone operating under the same pressure conditions as automotive air conditioning equipment with R134a. It was used property data of pure fluid and experimental data from Henderson [6]. They concluded that the performance with CO_2/acetone was better than with the other mixture. Yongming et al. [7] developed and tested a compression/resorption air conditioner working with mixtures based on CO_2. Wujek et al. [8] studied also the use of Ionic Liquids as absorbents of CO_2 for similar applications.

Most of these research works are based on thermodynamic properties of CO_2/acetone mixtures calculated with different methods, in general not based on experimental data because they were not available. For these reasons, in 2016 our research group launched a research project to complete the measurement and modelling of the thermodynamic properties of these mixtures for compression/resorption systems in refrigeration and air-conditioning applications.
The objective of this work is the evaluation of the methods for estimating the thermodynamic properties of CO_2/acetone in order to select the most convenient one to study the performance of the compression/resorption systems for refrigeration and air-conditioning applications.

2. Literature data and evaluation of thermodynamic property models

The first experimental data for CO_2/acetone found in the open literature are the vapour-liquid equilibrium data reported by Katayama et al. [9]. In 1991, Kato et al. [10] also measured the vapour-liquid equilibrium and compared with Kayatama's values obtaining similar results. Between 1996 and 1998, Day et al. [11], Chang et al. [12] and Ja et al. [13] measured vapour-liquid equilibria and densities. Day et al. [11] corrected some of the data published in a previous paper. Adrian and Maurer [14] determined the solubility of CO_2 in acetone. In addition, liquid phase density and the critical pressure were determined. The authors obtained deviations of 10 % between the density calculated with the equation of state and the measured values. Reaves et al. [15] measured the critical pressure and temperature for CO_2/acetone diluted mixtures.

Chen et al. [16] measured vapour-liquid equilibrium and densities. Bamberger and Maurer [17], Stievano and Elvassore [18], and Hsieh and Vrabec [19] measured vapour-liquid equilibrium. Wu et al. [20] and Chiu et al. [21] measured phase boundaries. Zahran et al. [22] determined by isothermal calorimetry, the excess enthalpies of CO_2/acetone. The results obtained show that the mixing is exothermic at all the conditions studied with large effects of pressure and temperature on the excess enthalpy. Höhler et al. [23] measured the solubility of CO_2 in acetone. In our group, Ramírez-Ramos [24] determined experimentally the vapour-liquid equilibrium (VLE) in a temperature range from 283.1 K to 353.1 K in a range of composition. VLE data were correlated using the RKS EoS, obtaining a good agreement between experimental and calculated data with deviations lower than 1.5%.

These experimental data can be used to select a suitable method, accurate and simple, to estimate the thermodynamic properties of CO_2/acetone. In the first step, the experimental vapour-liquid equilibria data collected from different sources are analysed in order to exclude those data sets presenting large deviations from the rest. The deviations between calculated and experimental data were obtained using the Peng Robinson and the Redlich-Kwong-Soave equations of state with the binary interaction parameter equal to zero. The results show that the deviations for most of the vapour pressure data sets analysed are below 5 % except for three references [18, 21, 28]. In Figure 1 can be seen that the selected experimental data sets cover almost completely the window for cooling applications using the compression/resorption refrigeration cycle, e.g., temperature range from -20 °C to 60 °C, and pressure up to 40 bar. In a second step, the binary interaction parameters were obtained by regression of the selected vapour-liquid equilibria data, in order to identify the method having lower deviations. The equations of state selected are Lee-Kesler-Plöcker (LKP), Peng-Robinson (PR) and Redlich-Kwong-Soave (RKS). Both cubic equations of state were evaluated with their original alpha-functions (PR and RKS) and with the Boston-Mathias one (PR-BM and RKS-BM), recommended for polar compounds (acetone in this case) and the supercritical region.

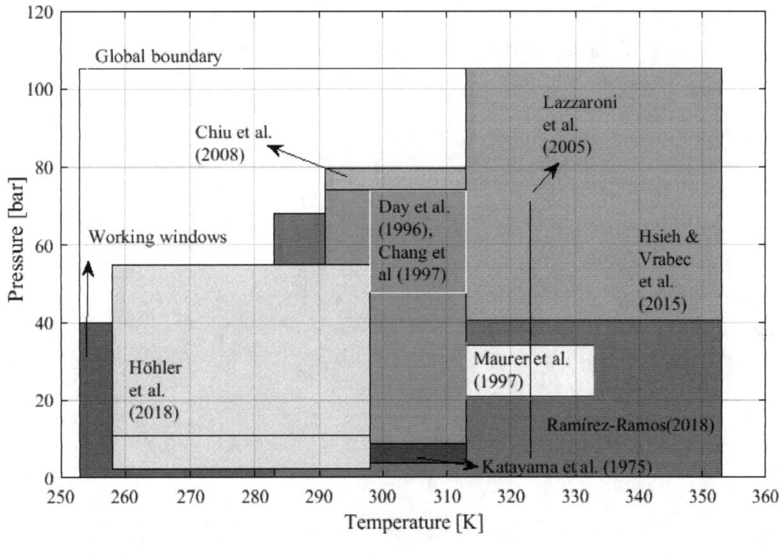

Figure 1. Experimental CO_2/acetone VLE data overlaying working window for cooling applications with compression/resorption refrigeration cycle.

In the case of Lee-Kesler-Plöcker equation of state [26] the mixing rule to calculate the pseudo-critical temperature for the mixture can be written as:

$$T_{cm} = \frac{x_1^2 \cdot V_{c_1}^{0.25} \cdot T_{c_1} + x_2^2 \cdot V_{c_2}^{0.25} \cdot T_{c_2} + 2 \cdot x_1.x_2.V_{c_{12}}^{0.25} \cdot T_{c_{12}}}{V_{cm}^{0.25}} \quad (1)$$

where sub-indexes 1 and 2 refer to CO_2 and acetone respectively, V_{cm} is the molar pseudo-critical volume for the mixture, which is based upon the pure components critical volume $V_{c1}$and V_{c2}, Equation (3), $V_{c_{12}}$is the critical volume term calculated using Equation (3), and T_{c12} is the critical temperature term calculated from the critical temperature of pure components, T_{c1} and T_{c2}, and a binary interaction parameter, k_{12}, Equation (4).

$$V_{c_m} = x_1 V_{c_1} + x_2 V_{c_2} \quad (2)$$

$$V_{c_{12}} = \frac{1}{8}\left[V_{c_1}^{\frac{1}{3}} + V_{c_2}^{\frac{1}{3}}\right]^3 \tag{3}$$

$$T_{c_{12}} = (1 + k_{12})\left(T_{c_1}T_{c_2}\right)^{0.5} \tag{4}$$

The binary interaction parameter k_{12} is obtained by regression of experimental VLE data. Regarding the cubic equations of state, they can be written in a polynomial form of the compressibility factor Z as:

$$0 = Z_m^3 - Z_m^2 + Z_m(A_m - B_m - B_m^2) - A_m B_m \tag{5}$$

$$0 = Z_m^3 - (1 - B_m)Z_m^2 + Z_m(A_m - 2B_m - 3B_m^3) - (A_mB_m - B_m^2 - B_m^3) \tag{6}$$

where Equation (5) corresponds to Redlich-Kwong-Soave EoS and Equation (6) for Peng-Robinson EoS. The coefficients A_m and B_m can be expressed with the following mixing rules:

$$A_m = x_1^2A_1 + x_2^2A_2 + 2x_1x_2(A_1A_2)^{0.5}(1 - k_{12}) \tag{7}$$

$$B_m = x_1B_1 + x_2B_2 \tag{8}$$

Coefficients A_1, A_2, B_1 and B_2 for the pure components are calculated as in a conventional way. The binary interaction parameter (k_{12}) can be expressed in a temperature dependent form as:

$$k_{12} = a + bT + \frac{c}{T} \tag{9}$$

where a, b and c are parameters obtained by regression of VLE data.

In this work, the Maximum Likelihood method is used to minimize the objective function OF and obtain the best values for k_{12}. The objective function OF is defined as

$$OF = \sqrt{\sum_{i=1}^{N}\left[\left(\frac{(P^{exp} - P^{cal})^2}{\sigma_P^2}\right)_i + \left(\frac{(T^{exp} - T^{cal})^2}{\sigma_T^2}\right)_i + \left(\frac{\left(x_1^{exp} - x_1^{cal}\right)^2}{\sigma_{x_1}^2}\right)_i + \left(\frac{\left(y_1^{exp} - y_1^{cal}\right)^2}{\sigma_{y_1}^2}\right)_i\right]} \tag{10}$$

where σ^2 is the statistical variance of associated measured variables (Pressure P, Temperature T, liquid mole fraction x, and vapor mole fraction y), and "*exp*" and "*cal*" indices refer to experimental and calculated values of P, T, x and y variables. Additionally, the Root-Mean-Square Deviation (RMSD) indicator, useful to evaluate the accuracy of the different methods in estimating the vapour-liquid equilibrium properties, is defined as follows:

$$RMSD = \frac{OF}{\sqrt{N - n}} \tag{11}$$

here N is the total number of experimental P, T, x, y data and n is the total number of parameters regressed.

In Table 1, the RMSD values for the selected equations of state and the values of the binary interaction parameter are shown. The deviations for the Lee-Kesler-Plöcker EoS with k_{12}=0 are much higher than the values obtained with other equations in the same conditions. However, when the binary interaction parameter is obtained by regression, in all the cases, the best results are obtained when k_{12} is independent of the temperature. It can be seen also that there is not any improvement in using the Boston-Mathias alpha-function.

Additionally, each of the proposed cubic equations of state has practically the same performance in estimating the VLE of CO_2/acetone. Reasonable results were achieved even without introducing a temperature dependence term in the binary interaction equation (7). As accuracy is not an issue anymore, either one of the cubic equations of state considered can be fairly used for modelling the thermodynamic properties necessary to study the performance of the cycle. Thus, we have selected the Redlich-Kwong-Soave equation to calculate the thermodynamic properties of the fluid mixture to study the performance and operating conditions of the cycle. Figure 2 shows the values of the maximum absolute deviation and the RMSD for the RKS equation.

Table 1. RMSD and binary interaction parameter values obtained for the different equations of state.

EOS	Parameter	RMSD
LK-PLOCK	k_{12} =0	373
	k_{12} =7.2 x 10^{-02}	81.7
PENG-ROB	k_{12} = 0	127
	a= -6.2 x 10^{-03}; b=c=0	81.4
	a= -1.0x10^{-01}; b=-3.5 x10^{-04}; c=0	81.3
	a=4.6; b=-7.5x 10^{-03}; c= -698.1	72.2
PR-BM	k_{12} =0	129
	a=-0.0052; b=c=0	76.5
	a= 0.15; b= -0.00053; c=0	77.4
	a=5.7; b= -0.0089; c=-896.3	71.0
RKS	k_{12} =0	125
	a=-0.008; b=c=0	79.2
	a=0.016; b= 0.00031; c=0	79.6
	a= 2.9; b=-0.0047; c= -460.8	70.4
RKS-BM	k_{12} = 0	125.46
	a=-0.0078; b=c=0	79.71
	a = 0.059; b=-0.00022; c=0	79.4
	a = 4.7; b = -0.0076; c= 721.7	79.5

In Figure 2(a), it can be seen that experimental data from [22] present the maximum absolute deviation for the temperature of 11 K with a RMSD of approximately 1.5 %. Regarding the liquid mole fraction, experimental data from [25] present the biggest deviation of 0.17 mol/mol that represents a value of 17% for RMSD (Figure 2(b)). Similarly, in Figure 2(c), it can be seen that the maximum deviation for pressure, about 7 bar, is found in [24,27,28,30] representing a RMSD value greater than 10 % for [23, 26]. Figure 2(d) shows that a maximum deviation value of about 0.03 mol/mol is obtained for vapour mole fraction representing a RMSD value lower than 1% in all cases.

3. Modelling of the thermodynamic properties

The thermodynamic properties of CO_2/acetone mixtures needed to study the performance and operating conditions of the compression/resorption system for refrigeration and air condition applications are vapour-liquid equilibrium (pressure, temperature, liquid, and vapour phase compositions) and vapour and liquid densities, enthalpies, and entropies. This section discusses the modelling of these properties. As explained in the cycle modelling section, the cycle design is based on the assumption that some states of the working fluid are at saturated conditions; thus, it is fundamental that the selected property calculation method computes adequately the vapour-liquid equilibrium. The VLE computation is a key factor to define the flow streams compositions throughout the cycle and may influence hugely on the cycle designing choices. Additionally, both liquid and vapour densities are essential in some component performance as the pump and compressor. Important to point out that the compressor energy requirement is a very important aspect for the cycle performance. Finally, thermodynamic properties such as enthalpy and entropy are crucial to foresee the energy performance in the cycle heat exchangers.

According to the results on the evaluation above presented, the Redlich-Kwong-Soave equation of state with the original alpha function, and k_{12} = -0.008 has been selected to perform the calculation of most of the thermodynamic properties needed.

The calculation procedure to calculate the VLE with equation of states is well-known [28]. Once the pressure and the composition of the vapor phase (P, y) at equilibrium conditions are fixed, the temperature and liquid phase composition (T, x) are calculated. In case that the input variables are the temperature and the liquid phase composition (T, x) at equilibrium conditions, the corresponding pressure and vapour-phase composition (P, y) are determined using the equation of the state.

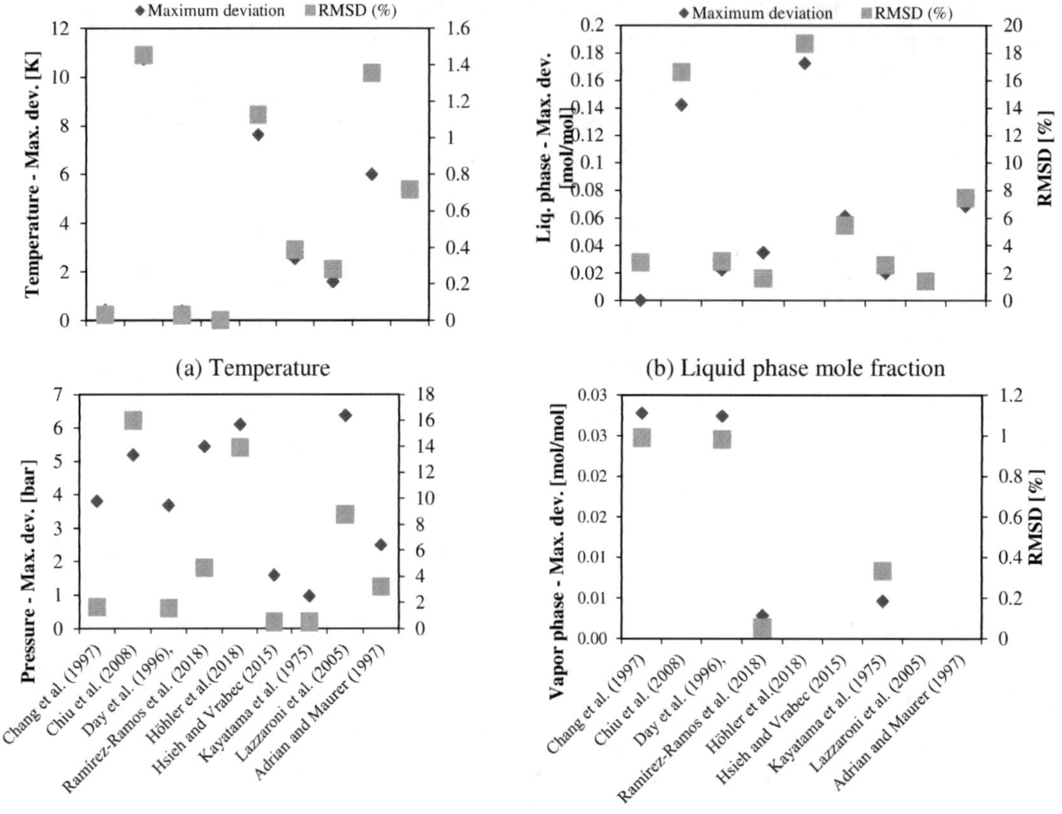

(a) Temperature (b) Liquid phase mole fraction

(c) Pressure (d) Vapour phase mole fraction

Figure 2. Maximum deviation and Residual Root Mean Deviation between calculated values using RKS-EOS and experimental data from different sources.

The density of the vapour phase can be computed by finding three roots of the cubic equation of state, Equation (5), for a given state: the largest one corresponds to the vapour phase, the smallest to the liquid phase [30]. Figure 3 shows a comparison between the vapour density calculated with the selected RKS-EoS and experimental data from [12] against the correspondent equilibrium liquid phase composition at different temperatures. It can be seen a good agreement between experimental and computed values through the EoS for the vapour phase density.

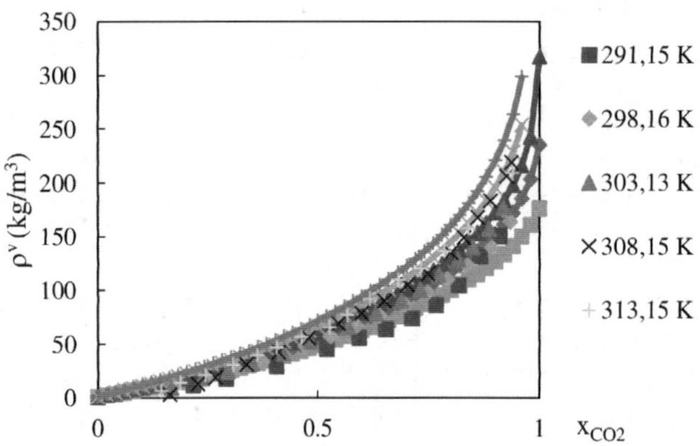

Figure 3 Comparison of experimental [14] and calculated saturated vapour density using Redlich-Kwong-Soave equation as a function of the correspondent equilibrium liquid-phase mole fraction at different temperatures.

Because the liquid density is

usually not accurately calculated by cubic equations of state [28], the modified Rackett model [30] is used in this work to estimate it. The corresponding mixing rule to calculate the critical temperature term ($T_{c,m}$) for the mixture has been calculated with the following equation:

$$T_{c_m} = \frac{x_1^2 V_{c_1}^2 T_{c_1} + x_2^2 V_{c_2}^2 T_{c_2} + 2x_1 x_2 V_{c_1} V_{c_2} \left(T_{c_1} T_{c_2}\right)^{0.5} (1 - k_{12})}{V_{c_m}^2} \tag{12}$$

where T_{ci} and T_{cj} is the critical temperature of the pure components, V_{c_m} is the critical volume term and k_{12} a binary interaction parameter. The critical volume term is calculated with Equation (2). The binary interaction parameter can be expressed in function of the temperature as:

$$k_{12} = a + b\,T + c\,T^2 \tag{13}$$

where *a*, *b* and *c* are obtained (Table 2) through experimental data regression from literature [11, 13, 16, 29].

Table 2. Coefficients of the binary interaction parameter for Equation (13).

Parameter	Value	Units
a	0.12	-
b	-0.00038	K^{-1}
c	3.272 x 10^{-06}	K^{-2}

The regression yields a maximum absolute deviation of 92.97 kg/m^3 and a RMSD of 2.3 % (Figure 4).

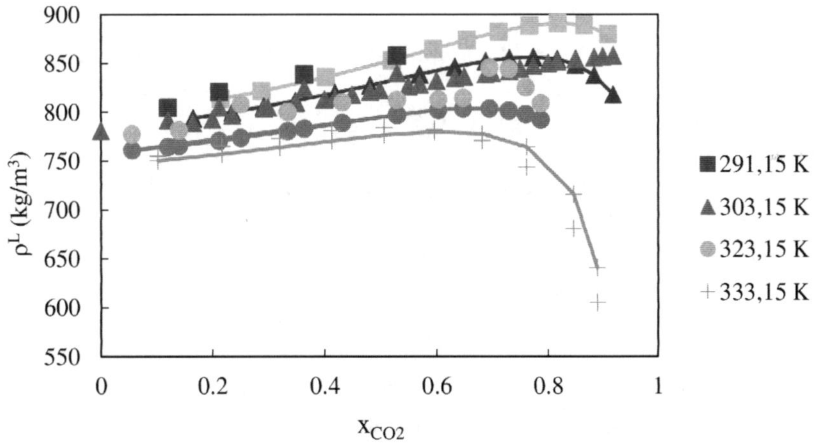

Figure 4. Comparison of calculated and experimental values of saturated liquid density using the modified Rackett model (lines) and literature data (symbols) [13, 14, 17, 30].

Regarding the energy-related properties, only Zahran et al. [22] measured excess enthalpy for CO_2/acetone. However, for a good understanding of the cycle behaviour, both the fluid mixture enthalpy and entropy are necessary: the former for the operation of heat exchangers such as desorber and resorber and the later for the compressor model, respectively. In this work, enthalpy and entropy are evaluated through departure function expressions corresponding to the RKS equation of state. Thus, the enthalpy of the binary fluid mixture *H* at a given state defined by the pressure P, temperature T, and total composition z is calculated as follows:

$$H(P,T,z) = H^0(T,z) - H^d(P,T,z) \tag{14}$$

where H^0 and H^d are the ideal gas enthalpy and the departure enthalpy, that can be obtained using the following equations:

$$H^0(T,z) = \int_{T_{ref}}^{T} \left[z\, C_{P,1}^0 + (1-z)C_{P,2}^0\right] dT \tag{15}$$

$$C_p^0 = A + B\left[\frac{\left(\frac{C}{T}\right)}{\sinh\left(\frac{C}{T}\right)}\right]^2 + D\left[\frac{\left(\frac{E}{T}\right)}{\cosh\left(\frac{E}{T}\right)}\right]^2 \tag{16}$$

$$H^d(P,T,z) = -\frac{A_m\alpha + \frac{dA_m\alpha}{dT}}{RT\, B_m}\ln\left(\frac{2V}{2V + B_m}\right) + 1 - Z_m \tag{17}$$

where $C_{P,1}^0$ and $C_{P,2}^0$ are the ideal gas heat capacity of the pure components, CO_2 and acetone respectively, calculated using equation (15) and the parameters A to E are pure fluid dependent taking the values indicated in Table 3. The enthalpy at the reference state is zero being the reference temperature T_{ref} equals to 298.15 K.

Table 3. Pure fluid coefficient values for the ideal gas heat capacity equation (15).

Fluid	A [J/mol K]	B [J/mol K]	C [K]	D [J/mol K]	E [K]
Carbon dioxide	29.37	34.54	1428	26.4	588
Acetone	57.04	163.2	1607	96.8	731.5

For the entropy calculation, the procedure is similar but using the following expressions:

$$S(P,T,z) = S^0(T,z) - S^d(P,T,z) \tag{18}$$

$$S^0(T,z) = \int_{T_{ref}}^{T} \frac{\left[z\, C_{P,1}^0 + (1-z)C_{P,2}^0\right]}{T} dT \tag{19}$$

$$S^d = \frac{\frac{dA_m\alpha}{dT}}{R\, B_m}\ln\left(\frac{2V}{2V + B_m}\right) - \ln\left(Z\left(1 - \frac{B_m}{V}\right)\right) \tag{20}$$

4. Modelling of the compression/resorption refrigeration cycle

Figure 5 shows the scheme of the compression/resorption cycle, being the main components: the resorber, the desorber, the compressor, the solution heat exchanger, the solution pump and the throttling valve. The CO_2 concentrated solution, also known as rich solution, with a mass CO_2 fraction x_{rich} and a mass flow rate $\dot{m}_{rich}$, is partially evaporated in the desorber at the low pressure P_{low} producing a cooling effect $\dot{Q}_D$, and leaving it with a vapor quality q_7 and a low temperature T_7. Then, the two-phase flow produced is separated into two streams. The solution stream, called weak solution, with a lower CO_2 mass fraction x_{weak} and a mass flow rate $\dot{m}_{weak}$ is pumped to the high pressure P_{high} and mixed with the compressed vapor stream, $\dot{m}_{vapor}$, before entering the resorber.

The absorption process of the vapor stream by the weak solution is completed in the resorber where the heat produced $\dot{Q}_R$ is delivered to the sink. The rich solution leaves the resorber at a high temperature T_4, and return to the desorber after being throttled in the expansion valve. The solution heat exchanger, with effectiveness ε_{SHX}, placed in the solution circuit, also known as resorption circuit, subcools the rich solution in order to increase the cooling effect in the desorber. The vapor stream from the separator is compressed by the compressor to the high pressure P_{high}, being $\dot{W}_C$ the mechanical power consumed in the process. The power consumed by the solution pump is $\dot{W}_P$. The efficiency of the refrigeration cycle is expressed in terms of the Coefficient of Performance, COP, defined as

$$COP = \frac{\dot{Q}_D}{\dot{W}_C + W_P} \tag{21}$$

The solution circulation ratio f is defined as the ratio of the rich solution and vapour mass flow rates.

$$f = \frac{\dot{m}_{rich}}{\dot{m}_{vapor}} \tag{22}$$

The compression pressure ratio is defined as the ratio of the high and low-pressure values.

$$CPR = \frac{P_{high}}{P_{low}} \tag{23}$$

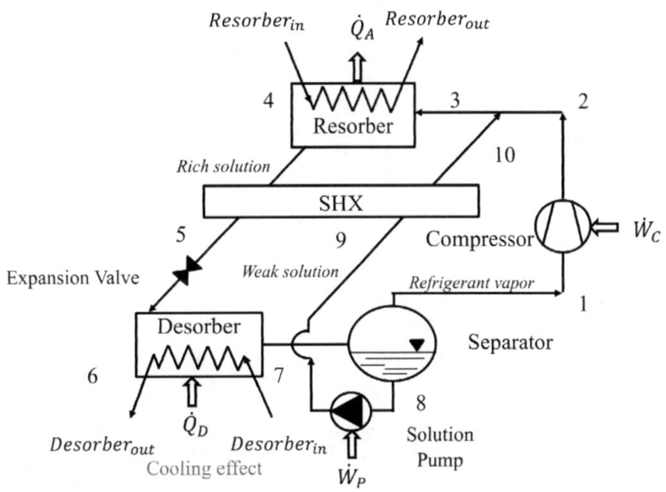

Figure 5. Compression/resorption refrigeration cycle scheme.

The cycle modelling is based on the mass and energy balances (Tables 4 and 5) and in the following assumptions:

- the two-phase flow leaves the desorber at the temperature T_7 and pressure P_{low} in equilibrium conditions;
- the rich solution with a CO_2 mass fraction x_{rich} leaves the resorber at the temperature T_4 and pressure P_{high} fluid mixture in saturated conditions;
- the processes in the solution pump and the compressor are isentropic;
- the throttling process in the expansion is considered isenthalpic;
- the process in the separator and the mixing process before the resorber are adiabatic;
- the effectiveness of the solution heat exchanger is 1;
- pressure drops and heat losses in pipes and other components are not considered.

The independent variables are:

- solution temperature at the resorber outlet: T_4;
- fluid mixture temperature at the outlet of the desorber: T_7;
- rich solution composition, x_{rich};
- difference between rich and weak solution compositions (in terms of CO_2 mass fraction) $\Delta x = x_{rich} - x_{weak}$;

- vapour mass flow rate $\dot{m}_{vapor}$ is taken as a reference equal to 1 kg/s.

Table 4. Energy balances in the main components of the compression/resorption cycle.

Component	Energy balances	
Compressor	$\dot{W}_C = \dot{m}_{vapor}(h_2 - h_1)$	(24)
Resorber	$\dot{Q}_R = \dot{m}_{vapor}[f\, h_4 - h_2 - (f-1)h_{10}]$	(25)
Solution heat exchanger SHX	$\dot{Q}_{SHX} = \dot{m}_{vapor} f(h_4 - h_5) = \dot{m}_{vapor}((f-1)(h_{10} - h_9)$	(26)
Expansion valve	$h_5 = h_6$	(27)
Desorber	$\dot{Q}_D = \dot{m}_{vapor}[h_1 + (f-1)h_{10} - f\, h_6]$	(28)
Solution pump	$\dot{W}_P = \dfrac{\dot{m}_{weak}(P_{high} - P_{low})}{\rho_8}$	(29)

Table 5. Mass balances in the separator.

Component	Mass balances	
Separator	$\dot{m}_{rich} = \dot{m}_{vapor} + \dot{m}_{weak}$	(30)
	$x_{rich}\dot{m}_{rich} = y_1\, \dot{m}_{vapor} + x_{weak}\dot{m}_{weak}$	(31)

The cycle can be represented graphically in the Dühring diagram for the CO_2/acetone mixture. This diagram, also known as PTXY diagram, is drawn using the RKS equation of state obtained in section 3. In Figure 6, has been represented the solution circuits corresponding to the typical temperature conditions of an air-conditioning application (e.g. T_4= 35°C, T_7=10°C based on [31]). These temperature levels are obtained considering a cooling water inlet temperature of 30°C in the resorber and chilled water returning temperature of 15°C in the desorber. In both cases, the composition of the rich solution is the same x_{rich}=50% in CO_2 mass fraction but the values of the rich and weak solution compositions are Δx =0.2 (case 1: blue circuit) and Δx =0.3 (case 2: red circuit). As can be seen in this figure, the high pressure is the same in both cases P_{high}= 40 bar, because it depends on x_{rich} and T_4, being the low pressure different P_{low}= 13.94 bar (case 1) and P_{low} = 9.4 bar (case 2), because it depends on the composition of the weak solution (x_{weak}= x_{rich} - Δx) and the temperature T_4. The corresponding pressure ratio values are 2.69 and 3.99 and the composition of the vapor stream in the compressor are y_1= 98.6% (case 1) and y_1=98% (case 2) in CO_2 mass fraction. The solution circulation ratio values are 3.43 (case 1) and 2.6 (case 2). Finally, the cooling capacity, the power consumed by the compressor and pump, and the COP are shown in Table 6:

Table 6. Performance indicators for typical air-conditioning application.

Indicators	Case 1	Case 2
$\dot{Q}_D(kW)$	230.7	240.4
$\dot{W}_P(kW)$	7.34	5.73
$\dot{W}_C$(kW)	57.1	82.9
COP	3.58	2.71

IOP Conf. Series: Materials Science and Engineering **595** (2019) 012030 doi:10.1088/1757-899X/595/1/012030

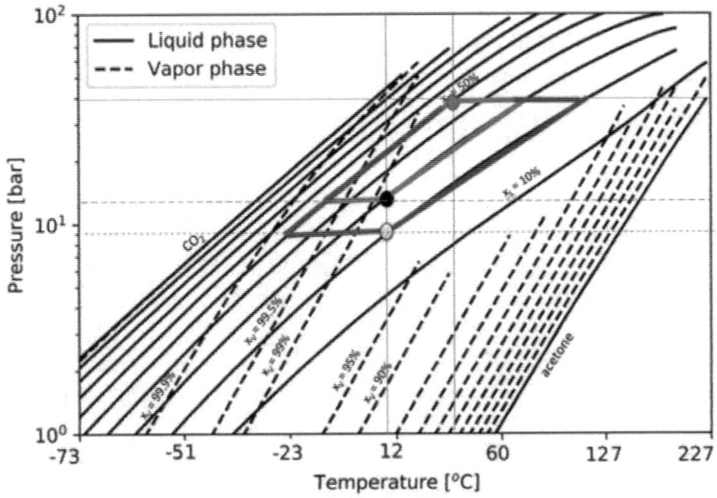

Figure 6. Compression/resorption cycle for air conditioning application in a pressure-temperature-composition (PTXY) diagram of CO_2/acetone mixture.

5. Conclusions

The open literature review on the experimental vapor-liquid equilibrium data of CO_2/acetone mixtures revealed the existence of enough data to cover the working window for the cooling applications of the compression/resorption system. These data sets were evaluated using the cubic equations of state of Peng Robinson and Redlich-Kwong-Soave with their original and the Boston-Mathias alpha-functions and the binary interaction parameter equal zero. The data sets with deviations lower than 5% in the pressure estimation, were used for the binary interaction parameter regression of the equations of state considered. The most suitable method, accurate and simple, found is the Redlich-Kwong-Soave with the original alpha-function and the binary interaction parameter k_{12}= -0.008.

Based on the estimation thermodynamic property method selected, it has been developed a thermodynamic model of the compression/resorption cycle. This model is useful to design the cycle and study the performance and operating conditions for cooling applications. The compression/resorption cycle have been represented graphically in the Dühring diagram for CO_2/acetone mixture in typical thermal conditions for air-conditioning.

References

[1] Groll E A 1997 Current Status of Absorption/Compression Cycle Technology. *ASHRAE Transactions*. **103(1)** p 1–13

[2] Groll E A and Kruse H 1992 *Kompressionkältemaschine mit Lösungskreislauf für umweltverträgliche Kältemittel*. Hannover: Institut für Kältetechnik und Angewandte Wärmetechnik

[3] Groll E A 1994 *Experimentelle und theoretische Untersuchungen von Kompressionkältermaschinen mit Lösungskreislauf*. Hannover: University of Hannover

[4] Spauschus H and Hesse U 2000 Reduced Pressure Carbon Dioxide-based Refrigeration System. US6112547. United States: United States Patent

[5] Mozurkewich G, Greenfield M L, Schneider W F, Zietlow D C and Meyer J J 2002 Simulated performance and cofluid dependence of a CO_2-cofluid refrigeration cycle with wet compression. *International Journal of Refrigeration*. **25(8)** p 1123–1136

[6] Henderson D R 1999 Properties of Working Fluids for Reduced Pressure Carbon Dioxide Systems. *SAE Transactions*. Vol. 108, Section 6: Journal of Passenger Cars, Part 2, p. 2208-

2215
[7] Yongming N, Jiangping C, Zhijiu C and Huanxin C 2007 Construction and testing of a wet-compression absorption carbon dioxide refrigeration system for vehicle air conditioner. *Applied Thermal Engineering*. **27(1)** p 31-36
[8] Wujek S S, Mccready M J, Mozurkewich G, Schneider W F and Elbel S 2014 Experimental and Modeling Improvements to a Co-Fluid Cycle Utilizing Ionic Liquids and Carbon Dioxide. *International Refrigeration and Air Conditioning Conference*. Paper 1510 http://docs.lib.purdue.edu/iracc/1510
[9] Katayama T, Ohgaki K, Maekawa G, Goto M and Nagano T 1975 Isothermal Vapor-Liquid Equilibria Acetone-Carbon Dioxide and Methanol-Carbon Dioxide Systems at High Pressures. *Journal of Chemical Engineering of Japan*. **8(2)** p 1–4
[10] Kato M, Aizawa K, Kanahira T and Ozawa T 1991 A new experimental method of vapor-liquid equilibria at high pressures *Journal of chemical engineering of Japan*. **24(6)** p 767-771
[11] Day C-Y, Chang C J and Chen C-Y 1996 Phase Equilibrium of Ethanol + CO_2 and Acetone + CO_2 at Elevated Pressures. *Journal of Chemical & Engineering Data*. **41(4)** p 839–843
[12] Chang C J, Day C-Y, Ko C-M and Chiu K-L 1997 Densities and P-x-y diagrams for carbon dioxide dissolution in methanol, ethanol, and acetone mixtures. *Fluid Phase Equilibria*. **131(1–2)** 243–258
[13] Ja C C, Kou-lunga C and Chang-yihb D 1998 A new apparatus for the determination of P – x – y diagrams and Henry's constants in high pressure alcohols with critical carbon dioxide. *Journal of Supercritical Fluids*. **12** p 223–237
[14] Adrian T and Maurer G 1997 Solubility of carbon dioxide in acetone and propionic acid at temperatures between 298 K and 333 K. *Journal of Chemical and Engineering Data*. **42(4)** p 668–672
[15] Reaves J T, Griffith A T and Roberts C B 2002 Critical Properties of Dilute Carbon Dioxide + Entrainer and Ethane + Entrainer Mixtures. *Journal of Chemical & Engineering Data*. **43(4)** p 683–686
[16] Chen J, Wu W, Han B, Gao L, Mu T, Liu Z, Jiang T and Du J 2003 Phase Behavior, Densities, and Isothermal Compressibility of CO_2 + Pentane and CO_2 + Acetone Systems in Various Phase Regions. *Journal of Chemical and Engineering Data*. **48(1)** p 1544–1548
[17] Bamberger A and Maurer G 2000 High-pressure (vapour + liquid) equilibria in (carbon dioxide + acetone or 2-propanol) at temperatures from 293 K to 333 K. *Journal of Chemical Thermodynamics*. **32(5)** p 685–700
[18] Stievano M and Elvassore N 2005 High-pressure density and vapor-liquid equilibrium for the binary systems carbon dioxide-ethanol, carbon dioxide-acetone and carbon dioxide-dichloromethane. *Journal of Supercritical Fluids*. **33(1)** p 7–14
[19] Hsieh C M and Vrabec J 2015 Vapor-liquid equilibrium measurements of the binary mixtures CO_2+ acetone and CO_2+ pentanones. *Journal of Supercritical Fluids* [online] **100**, p 160–166
[20] Wu W, Ke J and Poliakoff M 2006 Phase boundaries of CO_2 + toluene, CO_2 + acetone, and CO_2 + ethanol at high temperatures and high pressures. *Journal of Chemical and Engineering Data*. **51(4)** p 1398–1403
[21] Chiu H-Y, Lee M-J and Lin H 2008 Vapor− Liquid Phase Boundaries of Binary Mixtures of Carbon Dioxide with Ethanol and Acetone. *Journal of Chemical & Engineering Data*. **53(10)** p 2393–2402
[22] Zahran F, Renuncio J A R and Caban A 2010 Excess Molar Enthalpies of CO_2 + Acetone at Pressures from (9.00 to 18. 00)MPa and Temperatures from (313.15 to 333.15)K. Journal of Chemical & Engineering Data. **55** p 3649–3654
[23] Höhler F, Deschermeier R, Rehfeldt S and Klein H 2018 Gas solubility of carbon dioxide in methanol, acetone, mixtures of methanol and water, and mixtures of methanol and acetone. *Fluid Phase Equilibria*. **459** p.186–195
[24] Ramírez-Ramos G E 2018 *Experimental measurement of the Thermodynamic Properties of the*

CO_2 /acetone mixture for compression / resorption Heat Pumps. Universitat Rovira i Virigili

[25] Han F, Xue Y, Tian Y, Zhao X and Chen L 2005 Vapor-liquid equilibria of the carbon dioxide + acetone system at pressures from (2.36 to 11.77) MPa and temperatures from (333.15 to 393.15)K. *Journal of Chemical and Engineering Data.* **50(1)** p 36–39

[26] Plöcker U, Knapp H and Prausnitz J 1978 Calculation of High-Pressure Vapor-Liquid Equilibria from a Corresponding-States Correlation with Emphasis on Asymmetric Mixtures. *Industrial and Engineering Chemistry Process Design and Development.* **17(3)** p 324–332

[27] Lazzaroni M J, Bush D, Brown J S and Eckert C A 2005 High-pressure vapor-liquid equilbria of some carbon dioxide + organic binary systems. *Journal of Chemical and Engineering Data.* **50(1)** p 60–65

[28] Poling B, Prausnitz J and O'Connel J 1988 *The properties of gases & liquids.* Fifth edit. New York: McGraw-Hill. ISBN 0071499997

[29] Assael M J, Trusler M and Tsolakis T F 1996 *Thermophysical properties of fluids, An Introduction to their Prediction.* 1st. London: Imperial College Press

[30] Spencer C F and Danner R P 1972 Improved Equation for Prediction of Saturated Liquid Density. *Journal of Chemical and Engineering Data.* **17(2)** p 236–241

Acknowledgment
The authors are grateful to the Spanish Ministry of Economy and Competitiveness (DPI2015-71306-R) and FEDER/EU for financial support.

IOP Conf. Series: Materials Science and Engineering **595** (2019) 012031 doi:10.1088/1757-899X/595/1/012031

Theoretical and experimental features of the thermodynamic process in oil injection screw compressors

S Neacsu[1], C N Eparu[1], S Suditu[1], A Neacşa[2], N Toma[3] and C Slujitoru[3]

[1] Petroleum-Gas University of Ploiesti, Faculty of Oil and Gas Engineering, Drilling, Extraction and Transportation of Hydrocarbons Department, Romania
[2] Petroleum-Gas University of Ploiesti, Faculty of Mechanical Engineering, Mechanical Engineering Department, Romania
[3] National Turbines Development Research Institute (INCDT) - COMOTI Bucharest, Screw Compressors Assembly Department, Romania

e-mail: cristian.eparu@gmail.com

Abstract. For optimum performance of screw compressors, reduce wear and improve leakage, oil injection needs to be applied. The injected oil runs through the compressor together with the gas, taking a part of the heat of the gas resulting from compression during the process. The paper presents a theoretical analysis of the compression process affected by the cooling of the oil injection. The oil injection screw compressor is ideal for this process because the oil injected into the compressor along with the lubrication and sealing technology cools the compressed gas, which makes the compression process to be much closer to an isothermal process (minimum compression required). In order to calculate the thermodynamic of the gas lift process, the hydraulic calculation of the process, in which the extracted production, the amount of gas required, the injection pressures and the location of the valves, should be performed. These parameters were determined to correspond to the oil injection screw compressors manufactured in Romania by INCDT-COMOTI. The essential characteristics of those screw compressors that could be used considering the specificity of the location where the compressor will be mounted will be highlighted also.

1. Scientific and technical description of processes

Continuous artificial eruption or continuous gas lift involves the continuous injection of compressed gases directly into the column of fluid produced by the well in order to reduce their density and, implicitly, the dynamic bottom pressure, allowing the layer to properly flow according to this pressure.

Continuous artificial eruption installations are divided into two categories depending on the type of well completion, namely artificial eruption installations for simple completion and artificial dual-fill eruptions. The gas lift valves are used to discharge the well for its production and continuous gas injection in order to extract a certain fluid flow.

The flow of gases injected into a continuous artificial eruption well to produce a certain fluid flow is influenced by a number of factors such as: physical properties of the produced fluids, impurities, extraction pipes diameter, head pressure, the depth of the injection point and the productivity index of the well [1].

The layer-well working correlation for a continuous artificial eruption well involves determining the behaviour curve of the layer and the behaviour curve of the equipment, as well as establishing co-ordinates of the layer-well working correlation point, situated at the intersection of the two curves.

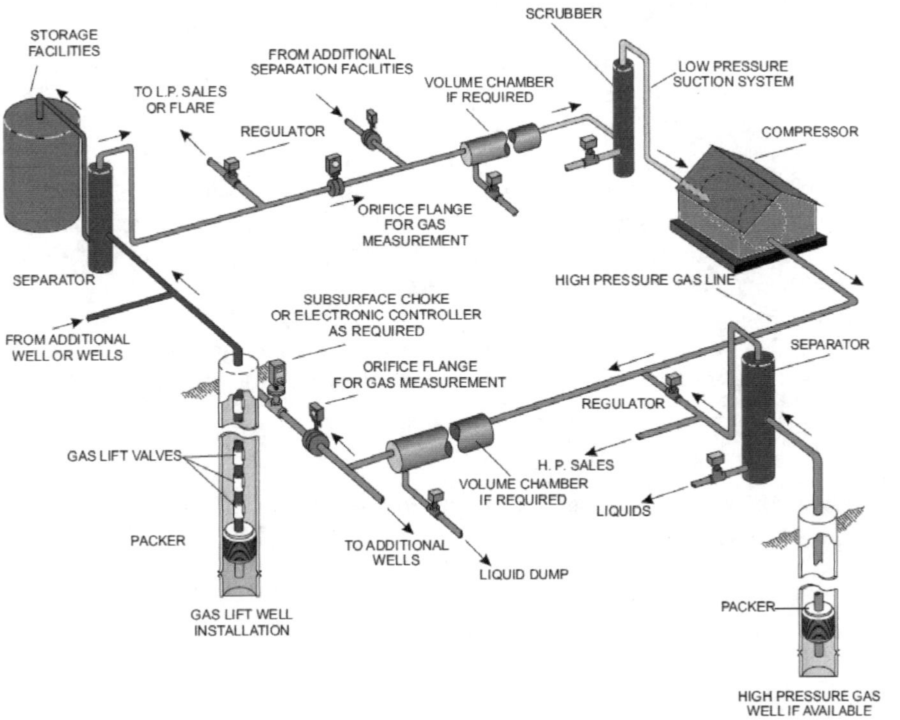

Figure 1. The gaslift installation with automated surface equipment (source Parveen Industries - https://parveen.in/includes/show_product.php?id=252).

In the case of continuous artificial eruption, unlike the natural eruption, the behaviour curve of the equipment is determined considering both the gas flow rate injected and the gas injection pressure [1].

The maximum flow rate produced by a continuous artificial eruption well depends on a number of factors such as:

- The pressure in the eruption head that can be considered constant or variable;
- The productivity index of the well, which over a short period of time can be considered constant, but which varies over time as a result of processes occurring in the reservoir;
- The capacity of the gas source (limited or unlimited);
- Injection line pressure (limited value);
- The diameter of the extraction pipes.

Of all these factors, the most important is the capacity of the gas source. If the gas source is very big, so as to be considered unlimited, the determination of the maximum flow produced by a continuous artificial eruption well requires the determination of the behaviour curve of a continuous artificial eruption well.

The screw compressor is part of the volumetric compressors category because the pressure increase is achieved by lowering the workspace volume. The major advantage of these types of compressors is that the compression process is continuous, the machine has no valves and is less sensitive to gas quality [2, 3, 4]. The screw compressors have been growing massively in popularity among users and this happens due to the high level of efficiency, the small gauge dimensions and because there are highly reliable [5]. Oil injection screw compressors are capable of reaching high performances by realizing major compression reports in one single stage, and, by upgrading the splitting technology/ using a better oil, the areas of applicability are continuously expanding [6].

The performance of the screw compressors is determined by the mass flow through it. From a thermodynamic point of view, the screw compressor is considered to be an open system because, during the operation, during the working process, a mass change with the external environment is made [1]. The oil injected into the compressor besides the mechanical effects also has a significant thermodynamic effect by cooling the compressed gas [6]. The compression forces required for all the loads being analysed are significantly lower for an oil-injected screw compressor than for the dry compressor [5].

Generally, from the technical point of view, in case of the screw compressors, the oil injection is done in aspiration, this operation goes to the wear minimization and the seal maximization of parts. The injected oil cools the compressed fluid partially, the physical phenomenon which leads to modifying the thermodynamic process, usually considered adiabatic.

The functional optimization of the screw compressor is driven in the direction of decreasing power necessary to compressing using the injected oil flow [6]. This process significantly modifies thermodynamic parameters of the compression. The oil flow rate for cooling the compressor is the variable parameter used for the optimization of the process.

2. Case study

The following data is known at a continuous artificial eruption well:

- Well depth: $H = 2800$ m
- Inner diameter of the tubing: $d_i = 63.5$ mm
- Inner diameter of the column: $D_i = 127$ mm
- Surface average temperature: $t_s = 10\ ^0$C
- Crude Oil density: $\rho_t = 830$ kg/m^3
- Reservoir water density: $\rho_a = 1100$ kg/m^3
- Gas relative density: $\rho_{rg} = 0.75$
- Surface tension of crude oil: $\sigma_t = 30\cdot 10^{-3}$ N/m
- Surface tension of water: $\sigma_a = 60\cdot 10^{-3}$ N/m
- Crude oil viscosity: $\mu_t = 2.2\cdot 10^{-3}$ Pa·s
- Gas viscosity: $\mu_g = 0.022\cdot 10^{-3}$ Pa·s
- Water viscosity: $\mu_a = 1\cdot 10^{-3}$ Pa·s
- The pressure in the eruption head: $p_2 = 4$ bar

Also, following the calibration of the well, the following data have result:

- Impurities: $i = 30\%$
- Bottom hole pressure: $p_d = 60$ bar
- Static pressure: $p_c = 90$ bar
- Liquid flow rate: $Q_l = 50$ m^3/day
- Well gas flow rate: $Q_g = 2500$ Nm^3/day
- Injected gas flow rate: $Q_{inj} = 30000$; 40000; 50000; 60000; 70000; 80000; 90000 Nm^3/day.

2.1. Thermodynamic analysis of the gas lift process

The gases used in the gas lift process are associated gases that are collected from the separator, which usually have low pressure. In order to be used, they must be compressed and inserted into the well column. Hence the gases are introduced through 7 valves, automatically into the tubing. They expand, mix with the oil and manage to raise it to the surface.

From the thermodynamic point of view, we identify the following processes:

A. Associated gas compression-process requiring an energy consumption with calculated parameters.

B. The introduction of gas from the column into the tubing is accomplished by means of valves. The corresponding thermodynamic process is an adiabatic expansion process. The compressed gas energy must be sufficient to be inserted into the tube.

2.2. Compression calculations

The compressor used as a model is an oil-injected volumetric screw compressor (CU128GM). The oil injection, besides the technological role, influences the thermodynamic compression process in that the oil cools the compressed gases. For this reason, it will be considered a polytrophic process whose exponent will be determined from the experimental data of the compressor [2, 7].

A functional dynamic model was designed so that the screw compressor process may be analysed. The main flows taken into consideration, according to those represented in Figure 2, are:

- Mass flows m_a, m_o, m_{a+o} for air and oil;
- Power flows for air and oil E_a, E_o, E_{a+o};
- Necessary power for compressor *Pc* and the lost power due to the irreversibility of thermodynamic processes;
- Lost heat flow by the machine to the environment Q_p and the correspondent entropy flow S_{Qp};
- The entropy flows that go through the machine for air and oil S_a, S_o and S_{a+o}.

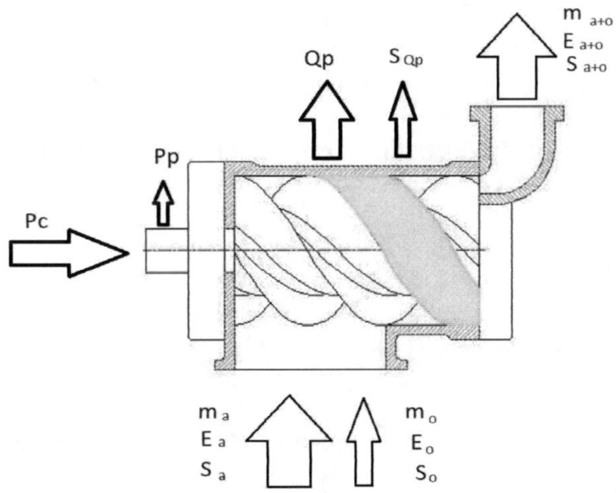

Figure 2. Functional dynamic model for a screw compressor.

For the model presented in figure 2 were used the following equations [1, 2, 8]:

- mass conservation law;
- energy conservation law;
- entropy balance for open systems;
- Gouy-Stodola´s theorem for lost power determination.

Based on the relations from above, a screw compressor with oil injection was created. For the accuracy of the obtained values, the numerical result obtained from the created model were calibrated, for specific regimes, with experimental data so that at the work regime, the outer and inner parameters must be identically.

Parameters to which the compression process is to be dimensioned result from the design of the gas lift process. The results of the calculation, made with a specialized software, are:

- *MAXIMUM LOAD* [100%]
- *The polytrophic exponent* = 1.04207
- *Gas flow rate* [m^3/day] = 50000 *Gas flow rate* [kg/s] = 0.4918
- *Input parameters:*

p_1 [bar] = 5.00; V_1 [m^3] = 0.12414775; T_1 [°C] = 20.0

- *Output parameters:*

p_2 [bar] = 38.00; V_2 [m^3] = 0.01772902; T_2 [[°C] = 45.0

- *Technical mechanics of compression* [J /kg] = 131192.682
- *Compression power needed* [kW] = 50326.038
- *Cooling heat* Q_r [kW] = 47681.930

From the analysis of the results one can conclude that due to the technology specific to the oil-injected screw compressor, the polytrophic compression process is approaching to the isothermal process in that the value of the polytrophic exponent is close to 1. This makes the compression power required minimum [9, 10].

The variable parameter that was taken into consideration for the process optimization was the oil flow of compressor cooling.

For the screw compressor with oil injection the temperature decreases according to the oil injected flow rate, fact that can be explained by the part of heat of compressed gas that the oil takes [11]. The thermodynamic process evolves from an adiabatic process to an isothermal one, fact that determines the significantly reduction of compression mechanical work [12].

Based on the obtained data can be observed that due to the oil cooling the value of the adiabatic exponent is reduced to values close to 1 so confirm that the real compression process is similar to an isothermal process fact that determines the maintaining of a low temperature regime of compressor during the functioning and also a reduced power necessary to machine entrainment.

Because the cooling air is stronger than oil heating so that per assembly it will result a decrease in entropy generation speed which means the increase of economy process.

Because the power depends on the mass flow, the oil having big density, concur at the increasing of masic specific air & oil flow results that the more oil is injected the more compressed necessary power increases.

2.3. *Valves process calculation*

The compressed gas is introduced into the well column and from there it automatically enters the tubing according to the gas lift process. When passing from the column to the tubing the gases suffer an adiabatic expansion. With each valve, depending on the depth and set pressure, the gas enters the column into the tubing.

This requires a consumption of the mechanical work made by the gas that expand from the column pressure to the tubing pressure. The graphical results are shown in figure 3. The results of the thermodynamic expansion calculation for each valve are shown in Table 1.

Table 1. The results of the thermodynamic expansion calculation for each valve.

Valve no.	p_col [bar]	p_tub [bar]	Specific expansion mechanical work [kJ/kg]	Generated power from gas expansion through the valve [kW]
1	34.5	8.5	30919.92	15206.42
2	33	11.2	23695.42	11653.41
3	31	14	17330.49	8523.13
4	29	16.3	12504.78	6149.85
5	27	18	8770.16	4313.16
6	25	19	5920.32	2911.61
7	22.7	19.8	2940.41	1446.09

If we add up the gas consumed powers when passing through the valves (50203.67 kW) the value is comparable to the one required to compress the gas (50326.038 kW), which shows that the system is balanced and that the power used for compression is consumed to introduce the gas into the pipe.

IOP Conf. Series: Materials Science and Engineering **595** (2019) 012031 doi:10.1088/1757-899X/595/1/012031

Considering the station parameters and the types of screw compressors manufactured in Romania by INCDT-COMOTI [13], before choosing the optimal solution, two types of compressors can be selected:

- Screw compressor with oil injection CHP220;
- Screw compressor with oil injection CU128GM.

Both the CHP 220 compressor and the CU128GM compressor show the following possible applications:

- Replacement of old GKN 10 piston compressors used in PETROM-OMV gas lift stations;
- High pressure compressors for national transport or distribution pipelines;
- Various applications required by users in the country or abroad, which require increasing the pressure and flow of the gas used.

Both compressor can be used in:

- Oil/gas extraction stations;
- Offshore stations due to their relatively small size, relatively low weight and low operating noise;
- Petrochemical industry or industrial applications;
- Pumping of gas into pipelines or storage tanks;
- Gas supply system for gas turbines (booster);
- Marine platforms exploitation.

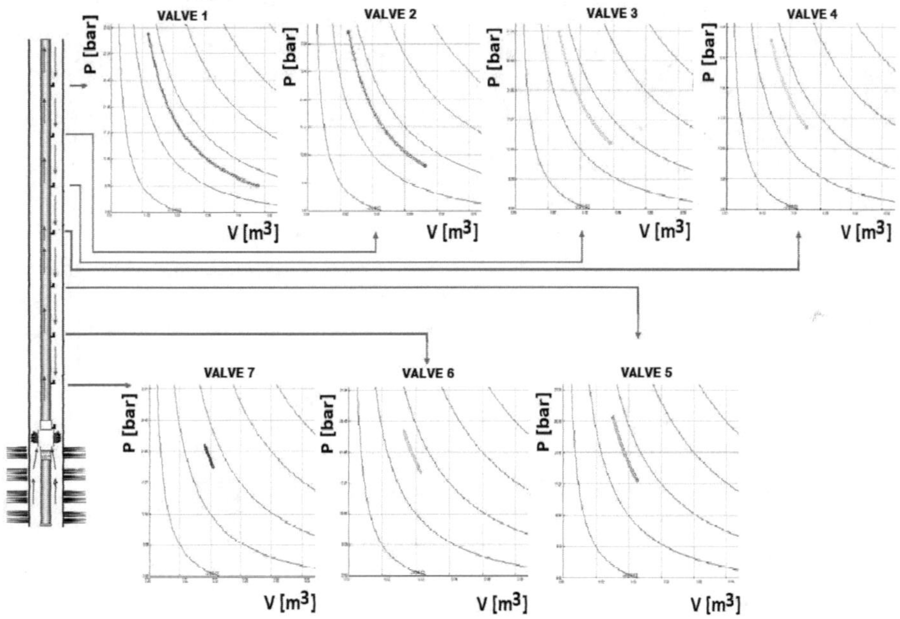

Figure 3. Gas injection mechanism through valves.

From the comparative analysis of the operating parameters of the compressor types specified above, it is advisable to choose as the compression unit for the gaslift process of the CU128 GM compressor. Operating parameters of the CU128GM high pressure discharge screw compressor (see figure 4) are as follows:

- Maximum Discharge pressure: 45 bar
- Suction pressure: 4.5 bar
- Volumetric ratio: 4.8
- By modulating the output pressure different volumetrically ratio can be obtained: 2.6; 3.5; 4.8 and the suction pressure can go to maximum 9 bar.
- Theoretical flow rate:

Minimum driving rotor speed 1148 - *Theoretical flow rate* 230 m^3/hour
Maximum driving rotor speed 4591 - *Theoretical flow rate* 921 m^3/hour

- Multiplier ratios:

Electrical engine drive 1500 RPM - *multiplier ratios* 1.2; 1.2407; 1.2830 ....2.5588
Electrical engine drive 3000 RPM - *multiplier ratios* 1.2; 1.2407; 1.2830 ....1.5208

Transmission ratios are obtained by correlating the number of toothed gear teeth on the motor shaft driven by the electric motor and the driving rotor shaft, and by varying the transmission ratio, the requested flow in the range is ultimately obtained.

Experimental samples for the CU128GM compressor were performed on the COMOTI stand (see figure 4), and after processing the generated data (see figure 5), the performance graph of the compressor (see figure 6) was obtained.

Figure 4. CU 128 GM V_i = 4.8 compressor ready for tests.

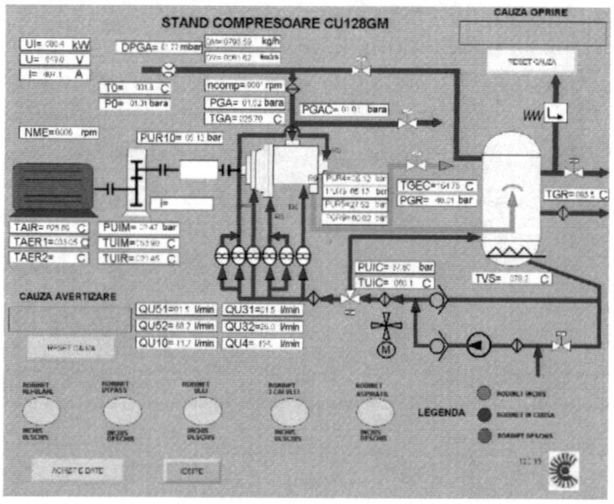

Figure 5. Print screen with the screw compressor test data.

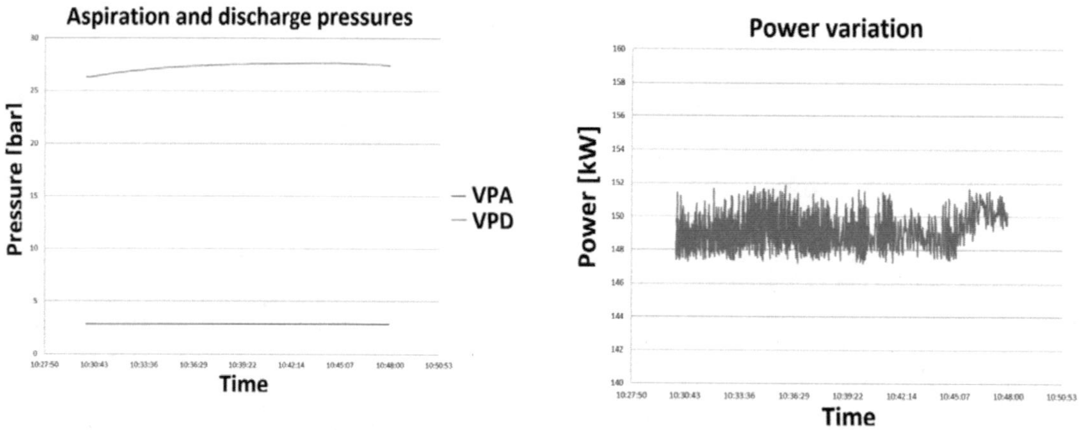

a) b)

Figure 6. CU 128 GM Vi= 4.8 compressor for the stand is tested: a) aspiration and discharge pressure; b) power variation.

3. Conclusion

To operate a well in gaslift, a quantity of gas is required (typically associated gases collected from the separator), which is compressed at the process pressure.

Gas compression is the only energy consumption (which translates into costs) for the gaslift process. For this reason, the dimensioning of the compression process must be rigorously made.

The paper presents the gas thermo-dynamic calculations made taking into account several gas compression variants for determining the optimum design solution and the maximum compression efficiency of the gas using an oil injection screw compressor. In this paper a compression process for an oil compressor has been dimensioned which allows a polytrophic process to be performed with a polytrophic exponent close to unit, which reduces the specific mechanical work of compression. The dimensioning of the compression process has been made correctly because the compressing power of the gas is equal to the power required to pass it through the well valves.

Due to the high flexibility and maintainability, oil-injected screw compressors can be used both in oil/gas extraction stations, offshore stations due to relatively small dimensions, relatively low weight and low operating noise, petrochemical industry or applications industrial piping, gas pumping in pipelines or storage tanks, the gas booster system, marine platforms. Another advantage of the oil-injected screw compressor is the low discharge temperature of compressed gas.

Several types of compressors have been studied for such a process and analysed for the operating parameters, the choice as compression unit of the CU128 GM compressor is advisable. It is underlined that this type of compressor will ensure the operation of the compressor station at optimal parameters.

Also, one of the reasons why this type of screw compressor can be considered is also the changes that may occur in the evolution of gas-lift conditions, for which the CU128GM compressor would operate at normal conditions and in these conditions unlike by the CHP220 compressor that would function with much diminished volumetric or adiabatic efficiency.

The minimal energy consumptions for compressors is achieved by optimizing the compression process fact that means a significant reduction in the operating costs.

References

[1] Neacsu S 2002 Comprimarea si lichefierea gazelor *Editura Romconvert* Ploiesti
[2] Neacsu S 2009 Termotehnica si mașini termice *Editura Printeh* Bucuresti
[3] Neacsu S 2001 Programul ″RG1″ pentru determinarea principalelor marimi termodinamice ale gazelor reale *Termotechnics National Conference* Galati **IV**
[4] Radcenco V 1994 Termodinamica generalizata *Editura Tehnica* Bucuresti
[5] [Eparu C, Neacşu S, Neacsa A and Prundurel A 2019 The comparative thermodynamic analysis of compressor's energetic performance *MMEP* **6** 1 152-155
[6] Eparu C, Neacşu S and Neacsa A 2019 The Optimization of Internal Processes from a Screw Compressor with Oil Injection to Increase Performances *IJHT* **37** 1 148-152
[7] API 619 - Rotary type Positive Displacement Compressors for Petroleum, Petrochemical and Natural Gas Industry (ISO 10440)
[8] Van Wylen G, Sonntag R and Borgnakke C 1994 Fundamentals of Classical Thermodynamics, *Jon Wiley & Sons New York,* USA
[9] Neacsu S 2003 Solution numeriques pour determiner les parametres d'etat des gaz reels, *Petroleum-Gas University Tehnical Bulletin* Ploiesti **LV 3**
[10] Stosic N, Smith I and Kovacevic A 2005 Screw Compressors Mathematical Modeling and Performace Calculation *Springer Verlag Berlin Heidelberg*
[11] ISO10440-1 2007 Petroleum and Natural Gas Industries/Rotary - Type Positive - Displacement Compressors Part 1/Process Gas Compressors
[12] Seshaiah N 2006 Experimental and Computational Studies on Oil Injected Twin Screw Compressors, *Mechanical Engineering Department National Institute of Technology Rourkela*
[13] Stefanescu M, Silivestru V, Liviu A, Neacşu S, Florea I and Eparu C 2008 Mentenanta turbomotoarelor *Editura Universitatii Petrol-Gaze* Ploiesti

Acknowledgments
The researches activities was financed by Executive Unit for Financing Higher Education, Research, Development and Innovation (UEFISCDI) within Bridge Grant (Knowledge Transfer to the Enterprise), Project code: PN-III-P2-2.1-BG-2016-0270.

Hydrogen a new fuel for internal combustion engines

D Gheorghe[1], D Tutunea[1], M Bică[1], A Gruia[1] and M Calbureanu[1]

[1] Road Vehicle and Transportation Department, University of Craiova, Craiova, Romania

E-mail: marinbica52@gmail.com

Abstract. The depletion of fossil fuels and the strict emissions regulations have motivated researchers to find and invest considerable resources in finding new renewable fuels. Hydrogen can be used in automotive industry to control the pollutant emissions and to improve engine performances. However, the producing of hydrogen is expensive, the supply infrastructure is not ready and the engines are affected by knocking phenomena and the increase of NOx emissions. The focus is now in the research of advance combustion technologies such dual fuel engines, homogeneous charge compression ignition (HCCI) and low-temperature combustion (LTC). The main limitations, challenges and perspectives are presented.

1. Introduction

The fuel depletion, environmental pollution and rapid increase in energy demand have pushed researchers to make studies to improve the process of combustion of internal combustion (IC) engines. The stringent regulations to reduce harmful emissions have promoted in the last years the use of low carbon or carbon-neutral alternative fuels combined with new solutions to enhance the charge flow, develop new emission controlling technologies and improve the mixture combustion process [1]. The idea of using hydrogen is not new and researchers around the world have been working to improve the technology in various aspects as production, storage, transport and utilization [2]. Also engine manufacturers have taken the initiative to develop hydrogen operated engines (Zetec 2.0 L from Ford, 6.0 L V12 bi-fuel from BMW). Ganesh et al. [3] investigated in a gasoline engine the effect of hydrogen on emissions and performance and found that the power output was 20% lower than gasoline and the NOx was four times higher than gasoline with negligible emissions of HC and CO. Park et al. [4] added hydrogen in a natural gas engine at various excess air ratios and found an improve engine combustion with reduce HC, CO and NOx emissions than the original engine. Saravanan and Nagarajan [5] tested different blends of hydrogen and found that engine brake thermal efficiency increased with 15% after hydrogen addition at 75% load, a dramatically decrease of Smoke, CO and CO2 with nearly the same NOx emissions as pure diesel engine. Wang et al. [6–9] studied the hydrogen addition at gasoline engines and found that hydrogen–gasoline blends could reduce HC and CO emissions at cold start and improve engine indicated thermal efficiency. Huang et al. [10-12] conducted experiments on the laminar burning velocities of the hydrogen-enriched fuels and found that the flame speed of hydrogen could reach 15 m/s, which were higher about five times than the speed of methanol. Bose et al. [13] investigated the effect on adding hydrogen in a diesel engine with a compression ratio of 17.5. The hydrogen was supplied constantly at 0.15 kg/h at all loads. The brake thermal efficiency increased with 3.9% comparative with petrodiesel fuel at 80% engine load. Dhyani and Subramanian [14] investigated on a multi-cylinder engine the effect of knock parameters such as injection timing, spark timing and equivalence ratio and found that knock and backfire were related to each other at higher equivalence ratio. Ceviz et al. [15] conducted experiments on hydrogen addition at a spark-ignition engine. The speed was kept constant at 2000 rpm using hydrogen fractions of 0%, 2.14%, 5.28% and 7.74% by volume as fuel. The hydrogen provided an improvement in the total brake-specific fuel consumption a reduction in HC and CO emissions, an increase in NOx emissions due to higher flame speed. The scope of this paper is to evaluate the possibility to use hydrogen as an alternative fuel for the internal combustion engine as primary fuel or in dual fuel operation.

Table 1. Main characteristics of hydrogen versus natural gas, gasoline and diesel liquid fuels [16].

		Hydrogen	Natural gas	Gasoline [(b)]	Diesel [(b)]
Density at NTP[(a)]	(kg/m^3)	0.09	0.7-0.9	737	820-950
Energy content	(MJ/kg)	120-142	53.6	46.4	48
Autoignition temperature	(K)	858	813	520-583	473
Flammability limits	(% gas-to-air volume ratio)	4-75	5-15	1.4-7.6	0.6-7.5
Minimum ignition energy	(mJ)	0.02	0.29	-	-
Quenching gap at NTP[(a)]	(mm)	0.64	2.1	-	-
Diffusion coefficient into air at NTP[(a)]	(cm^2/s)	0.61	0.24[(c)]	-	-

[a] Normal temperature and pressure conditions P = 1 bar, T = 293.15 K
[b] Liquid fuels.
[c] Diffusion coefficient of methane.

2. Hydrogen generation methods

The hydrogen quantity which exists on Earth is small and undeterminable and due to his very low density is pushed out from the gravitational attraction of the planet. It is necessary that hydrogen to be produced from other sources of raw material such as natural gas, oil, coal and water. To produce hydrogen an energy sources is needed as fossil fuels, nuclear or and renewable energy (solar, wind, geothermal and hydroelectric power) [16]. In recent years the majority of the hydrogen produced is from fossil fuels which is costly and produce high emissions. To produce hydrogen by renewable sources and economically is the key to a successful transition to a new hydrogen era. The hydrogen can be produced by several methods such as:

- Natural gas reforming. This method is the cheapest and efficient and is used frequently worldwide. Natural gas is reacted in the presence of synthesis gas and high temperature forming carbon monoxide (CO), a mixture of hydrogen and carbon dioxide (CO_2).

- Gasification. In this method coal, biomass or other materials are burned at high temperature between 1200 and 1500 °C to form gaseous components and then by a series of chemical reactions to hydrogen and carbon monoxide. By this process the hydrogen obtained from biomass is clean with zero greenhouse emission [17].

- Electrolysis. In this process an electric current decompose water in hydrogen and oxygen consuming the highest amounts of energy. It is clean and if the energy is from renewable sources is an emission-free process [18].

3. Hydrogen in CI engines

3.1 Hydrogen only operation

The idea of using hydrogen in compression ignition engines isn't new Homan et al. [19] try to use hydrogen and found that this fuel has high resistance to auto ignition even at compression ratio of 29:1. Later the same authors by using a glow plug ignition reveal that hydrogen ignition delay was short and indicated mean effective pressure is higher that petrodiesel fuel [20]. Naber and Siebers [21] conducted experiments in a constant-volume combustion vessel over a wide range of thermodynamic conditions and found that ignition delay of hydrogen has a high Arrhenius temperature dependence. Also ignition delay is affected by the temperature. Ikegami et al. [22, 23] used a single cylinder diesel engine to run with hydrogen as only fuel and found that operation range is limited.

3.2 Hydrogen/diesel dual-fuel engines

This solution was proven more realistic to use in compression engines, the hydrogen is combined in various ratio with petrodiesel fuel due to his lower auto ignition temperature. The hydrogen can be injected by carburetor, at the intake ports of the engines or at the intake manifold. Varde and Frame [24] found that the injection of small quantity of hydrogen in the intake of the engine at part load reduce smoke (between 10-15%) and increase oxides of nitrogen. Lilik et al. [25] experiments in a dual fuel engine the injection of hydrogen in the intake air and found an increase in NO_X emissions as is seen in Figure 1. Hydrogen can be mixed with diesel fuel at different rates depending on the speed and the load of the engine, in the literature the researchers use preponderant ratio between 10 and 40%.

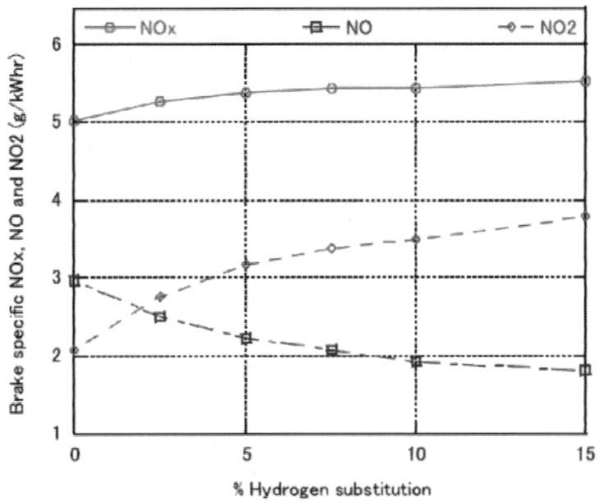

Figure 1. Brake specific NOx, NO and NO_2 emissions vs. energy percent from hydrogen fuel [25].

Taku and Yasumasa [26] used a single cylinder diesel engine with gas injectors at an intake port and the experiments shows that the engine could achieve higher thermal efficiency than diesel fuel at high engine load conditions. Also the emission of NO_X at higher loads are increased that one of the conventional diesel fuel. In Figure 2 is represented in-cylinder pressure, rate of heat release and diesel fuel injection signal. All rates of heat release have twin peaks, the firs peak is due to the combustion of a small amount of diesel fuel and hydrogen and the second is due to the main diesel injection and entrained hydrogen.

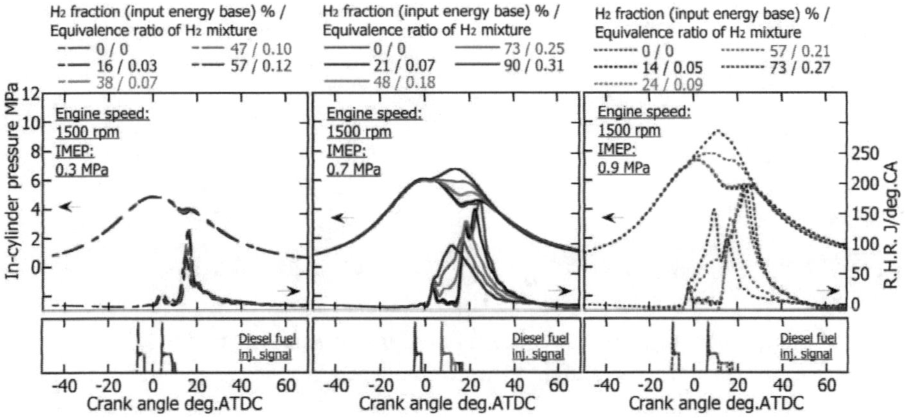

Figure 2. In-cylinder pressure, R.H.R. and diesel fuel injection signal under engine speed of 1500 rpm and IMEP of 0.3, 0.7 and 0.9 MPa condition (parameter: hydrogen fraction) [26].

Karagoz et al. [27] used for experiments a single cylinder, 4 stroke and variable compression ratio. Hydrogen/diesel blend was tested and compare with petrodiesel fuel at 40%, 60%, 75% and 100% engine loads. According to Figure 3, thermal efficiency decreases on each load condition because hydrogen's flame speed is nine times faster than diesel fuel and addition heat flow rises causing the increase of thermal loss. In the Figure 4 is shown the change in pressure due to the high flame speed resulting in fast combustion and increases in peak in cylinder pressure values.

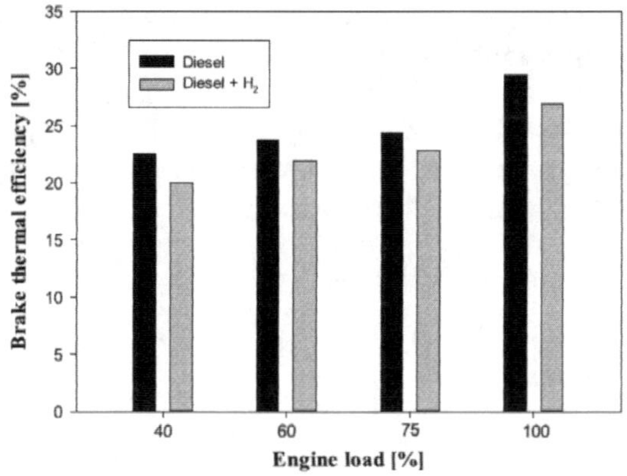

Figure 3. Effect of hydrogen enrichment on thermal efficiency at 40%, 60%, 75% and 100% engine load [27].

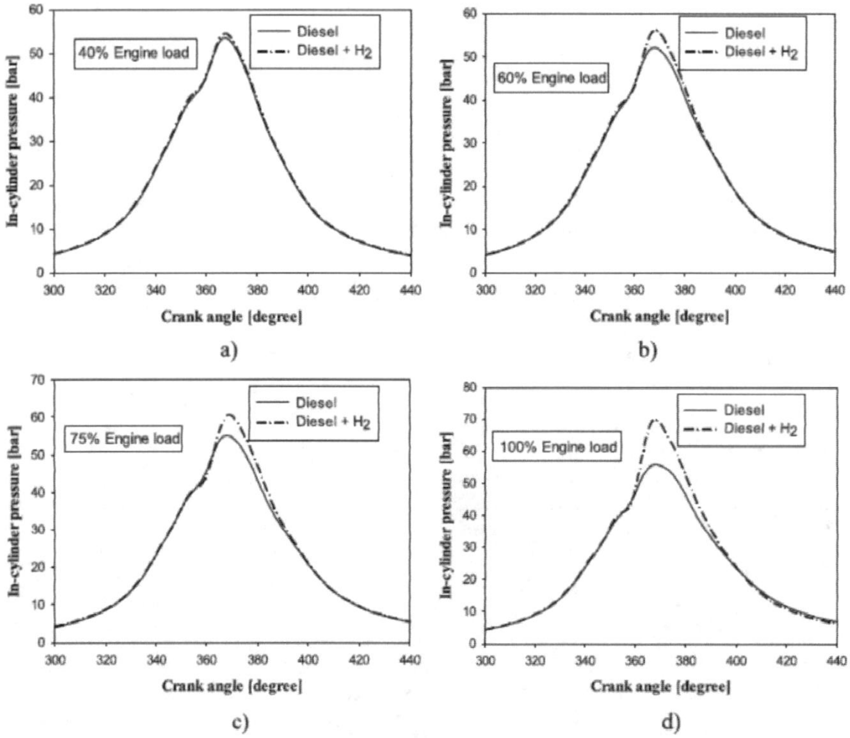

Figure 4. Effect of hydrogen enrichment on in-cylinder gas pressure related to CA at 40%, 60%, 75% and 100% engine loads

The EGR (Exhaust gas recirculation) system was used to reduce NO_X emissions and eliminate knocking of hydrogen-diesel dual fuel engines. Also this solution has many advantages the EGR have a negative effect on engine efficiency which rise with its percentage. The EGR have a dilution effect increasing smoke, CO and HC emissions due to the lower O_2 levels in the cylinder chambers [28].

3.3 Hydrogen and biofuels

Shirk et al. [29] studied the addition of hydrogen in blends of biodiesel (20%) and petroleum derived diesel fuel (80%) and found that small amounts of hydrogen up to 10% did not affect the steady-state operation of the engine. Kumar et al. [30] used the addition of hydrogen to vegetable oil and found that hydrogen fuel increased the brake thermal efficiency of the engine by 2% and decrease of smoke up to 20% at full load conditions. Also the hydrogen increases the NOx emissions due to the high combustion rates. Combustion duration was reduced and the ignition delay, maximum rate of pressure and peak pressure increases. Korakianitis et al. [31] conducted experiments on blends of diesel-hydrogen and rapeseed methyl ester-hydrogen and found that rapeseed methyl ester produces similar emission and thermal efficiency trends. Both blends exhibits increase in NOx emissions, HC, CO and smoke remaining relatively unchanged comparative with normal CI engine operation without hydrogen.

4. Hydrogen in gasoline engines

Ji et al. [32] studied on 1.6 L gasoline engine manufactured by Beijing Hyundai Motors hydrogen and gasoline at various load conditions. The indicated thermal efficiency increases due to the fact that wide flammability and short quenching distance of hydrogen improve combustion especially at idle and lean conditions. Unni et al. [33] investigated on a naturally aspirated single cylinder 396 cc gasoline engines with a carburetor performance and emissions of hydrogen and gasoline. The maximum power obtained with hydrogen (4.8 kW@ 3600 rpm) is lower than gasoline (6.16 kW@ 3600 rpm) and the brake thermal efficiency (BTE) of hydrogen is higher than that of gasoline due to the higher flame speed and very high rate of burning resulting in higher efficiency (Figure 5).

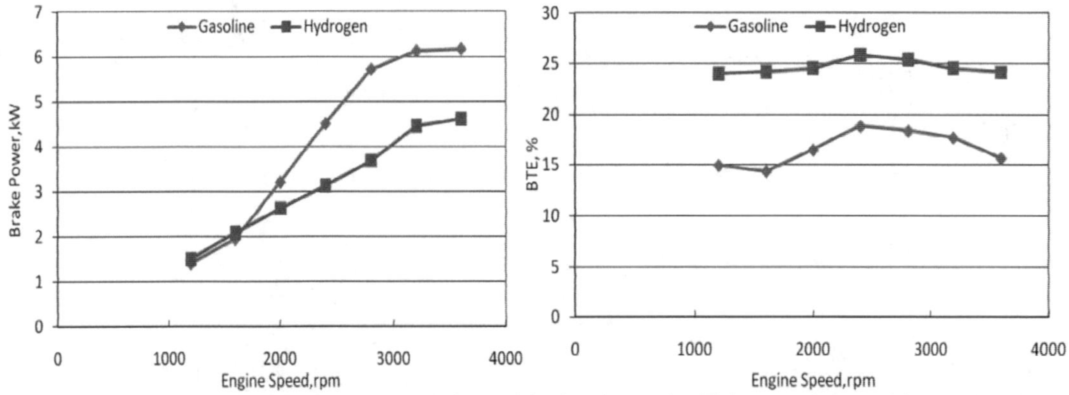

Figure 5. Variation of brake power (left) and brake thermal efficiency (right) with speed [33].

Sun et al. [34] investigated the effects of hydrogen direct injection on performance and engine stability. For the experiments used a four-cylinder SIDI engine with the injector situated between the intake valves and a centrally mounted spark plug. In Figure 6 is presented brake thermal efficiency with ignition timing under different excess air ratios for gasoline and a mixture of 10% hydrogen. Hydrogen addition increases BTE from 6% to 13% in four figures due to the complete combustion of hydrogen shorten combustion duration. Naruke et al. [35] investigated hydrogen addition on a single-cylinder 4-stroke SI engine. Hydrogen was supplied in the middle of the intake line and mixed with air using a static mixer. The value of λ was set to 1.80, 1.84, 1.89, 2.04, and 2.28, which are the lean limits for hydrogen concentrations of 0%, 2%, 4%, 10%, and 20%, respectively. In Figure 7 is presented the indicated thermal efficiency and knock at λ=1.8. The indicated thermal efficiency at the knock limit was increased by 0.3 point% at a fraction of hydrogen of 4%.

IOP Conf. Series: Materials Science and Engineering **595** (2019) 012032 doi:10.1088/1757-899X/595/1/012032

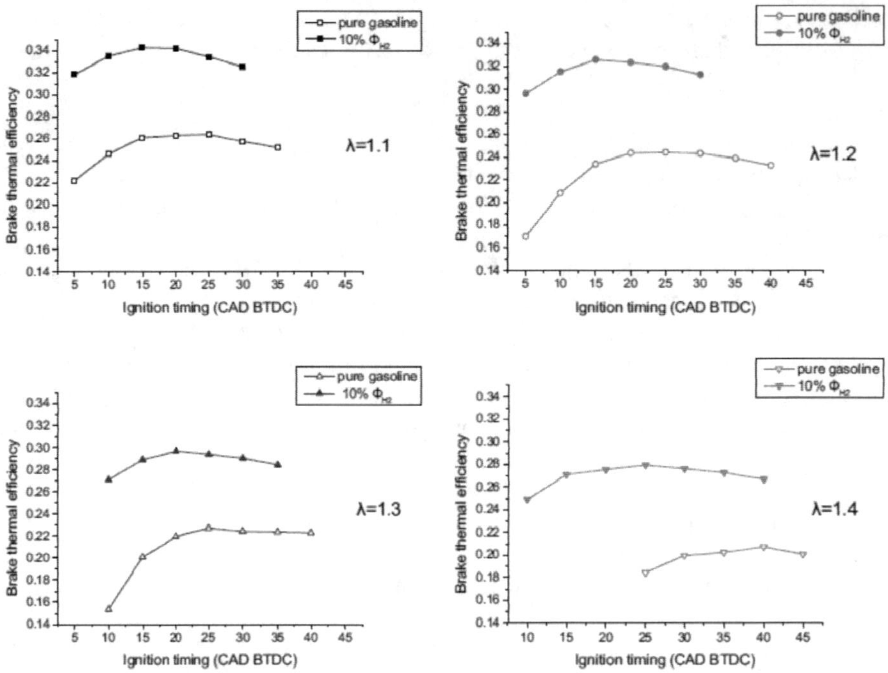

Figure 6. Brake thermal efficiency with ignition timing under different excess air ratios [34].

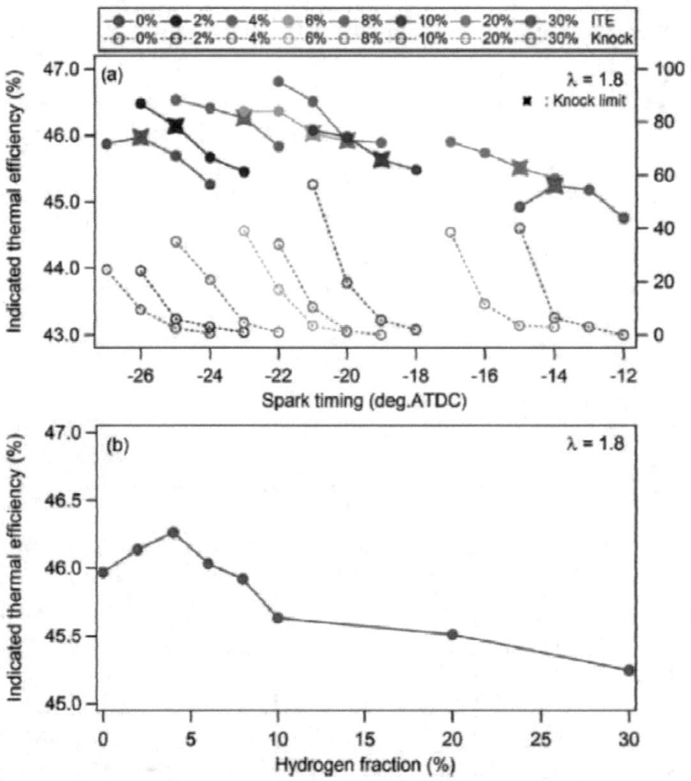

Figure 7. (a) Indicated thermal efficiency and knock percentage at λ=1.8 and (b) indicated thermal efficiency at knock limits against the hydrogen fraction in heating value at λ=1.8. Stars in graph (a) show the indicated thermal efficiency at the knock limit under each hydrogen fraction condition [35].

5. Conclusion

The future of diesel engine especial due the dieselgate scandal is considered ambiguous by many researchers and new actions are needed to advance the combustion performance and emissions production of the engine. The implementation of alternative fuels such as hydrogen with a carbonless structure with high energy content can be a solution. Hydrogen addition in internal combustion engines provide a decrease in HC, CO, CO_2 and smoke levels and an increase in heat release rate and brake thermal efficiency. However, the rise of the temperature in the combustion chamber brings an undesirable phenomenon the rise of NOx particularly at high load conditions. Applying EGR system can reduce the NOx formation but this system increases the smoke, CO and HC emissions do to the decrease of O_2 levels in the cylinder chambers. Depending on the type of engine used (burning of oil) and the air-fuel ratio (operating strategy used rich versus lean), a hydrogen fuelled engine can produce high NOx and significant carbon monoxide emissions to almost zero emissions. Hydrogen has high energy content by weight, but not by volume which is a challenge for storage and long term use.

References

[1] Changwei Ji, Shuofeng Wang and Bo Zhang 2012 Performance of a hybrid hydrogen–gasoline engine under various operating conditions, *Applied Energy* **97** pp 584–589

[2] Jayakrishnan K U, Premakumara G and Lalit M D 2017 Development of hydrogen fuelled transport engine and field tests on vehicles, *Int J Hydrogen Energy* **42** pp 643–651

[3] Ganesh R H et al. 2008 Hydrogen fueled spark ignition engine with electronically controlled manifold injection: an experimental study, *Renew Energy* **33** pp 1324–1333

[4] Park C, Kin C, Choi Y, Won S and Morivoshi Y 2011 The influences of hydrogen on the performance and emission characteristics of a heavy duty natural gas engine, *Int J Hydrogen Energy* **36** pp 3739–3745

[5] Saravanan N and Nagarajan G 2010 Performance and emission studies on port injection of hydrogen with varied flow rates with diesel as an ignition source, *Appl Energy* **87** pp 2218–2229

[6] Wang S, Ji C, Zhang M and Zhang B 2010 Reducing the idle speed of a spark-ignited gasoline engine with hydrogen addition, *International Journal Hydrogen Energy*. **35** pp 10580–10588

[7] Ji C and Wang S 2010 Experimental study on combustion and emissions performance of a hybrid hydrogen–gasoline engine at lean burn limits, *Int J Hydrogen Energy* **35** 1453–62

[8] Wang S, Ji C and Zhang B 2011 Starting a spark-ignited engine with the gasoline– hydrogen mixture, *Int J Hydrogen Energy* **36** pp 4461–4468

[9] Wang S, Ji C and Zhang B 2010 Effects of hydrogen addition and cylinder cutoff on combustion and emissions performance of a spark-ignited gasoline engine under a low operating condition, *Energy* **35** pp 4754–4760

[10] Zhang Z Y et al. 2008 Measurements of laminar burning velocities and Markstein lengths for methanole air nitrogen mixtures at elevated pressures and temperatures, *Combust Flame* **155(3)** pp 358 - 368

[11] Tang C L, Zhang Y J and Huang Z H 2014 Progress in combustion investigations of hydrogen enriched hydrocarbons, *Renew Sustain Energy Rev*. **30(2)** pp 195 - 216

[12] Tang C L, Huang Z H, Law C K 2011 Determination, correlation, and mechanistic interpretation of effects of hydrogen addition on laminar flame speeds of hydrocarbon air mixtures, *Proc Combust Inst*. **33(1)** pp 921- 928

[13] Bose P K, Maji D 2009 An experimental investigation on engine performance and emissions of a single cylinder diesel engine using hydrogen as inducted fuel and diesel as injected fuel with exhaust gas recirculation, *Int J Hydrogen Energy* **34** pp 4847 – 4854

[14] Dhyani V and Subramanian K A 2018 Experimental investigation on effects of knocking on backfire and its control in a hydrogen fueled spark ignition engine, *International Journal Hydrogen Energy* **43(14)** pp 7169 - 7178

[15] Ceviz M A, Sen A K, Kuleri A K and Oner I V 2012 Engine performance, exhaust emissions,

and cyclic variations in a lean-burn SI engine fuelled by gasoline hydrogen blends, *Appl Therm Eng*. **36** pp 314 - 324

[16] Pavlos D and Taku T 2017 A review of hydrogen as a compression ignition engine fuel, *Int J Hydrogen Energy* **42** pp 24470 – 24486

[17] Arribas L et al. 2017 Solar-driven pyrolysis and gasification of low-grade carbonaceous materials. *Int J Hydrogen Energy* **42(19)** pp 13598 - 13606

[18] Kato T 2009 Present status of hydrogen production by electrolysis *Nihon Enerugi Gakkaishi J Jpn Inst Energy*. **88** pp 371 - 377

[19] Homan H S, Reynolds R K, De Boer P C T and McLean W J 1979 Hydrogen-fuelled diesel engine without timed ignition, *Int J Hydrogen Energy* **4(4)** pp 315 - 325

[20] Homan H S 1989 An *experimental study of reciprocating internal combustion engines operated on hydrogen*, Cornell University

[21] Naber J and Siebers D L 1998 Hydrogen combustion under diesel engine conditions, *Int J Hydrogen Energy* **23(5)** pp 363 - 371

[22] Ikegami M, Miwa K, Shioji M and Esaki M 1980 A study on hydrogen fuelled diesel combustion, *Bull JSME*. **23(181)**

[23] Ikegami M, Miwa K and Shioji M 1982 A study of hydrogen fuelled compression ignition engines, *Int J Hydrogen Energy* **7(4)** pp 341 - 353

[24] Varde K S and Frame G A 1983 Hydrogen aspiration in a direct injection type diesel engine-its effects on smoke and other engine performance parameters, *Int J Hydrogen Energy* **8(7)** pp 549 - 555

[25] Lilik G K, Zhang H, Herreros J M, Haworth D C and Boehman A L 2010 Hydrogen assisted diesel combustion, *Int J Hydrogen Energy* **35(9)** pp 4382 - 4398

[26] Taku T and Yasumasa S 2017 The utilization of hydrogen in hydrogen/diesel dual fuel engine, *Int J Hydrogen Energy* **42** pp 14019 - 14029

[27] Karagoz Y T, Sandalci L, Yuksek A and Dalkilic S 2015 Engine performance and emission effects of diesel burns enriched by hydrogen on different engine loads, *Int J Hydrogen Energy* **40** pp 6702 – 6713

[28] Singhyadav V, Soni S L and Sharma D 2012 Performance and emission studies of direct injection C.I. engine in duel fuel mode (hydrogen-diesel) with EGR, *Int J Hydrogen Energy* **37(4)** pp 3807 - 3817

[29] Shirk M G, McGuire T P, Neal G L and Haworth D C 2008 Investigation of a hydrogen-assisted combustion system for a light-duty diesel vehicle, *Int J Hydrogen Energy* **33(23)** pp 7237 - 7244

[30] Senthil K M 2003 Use of hydrogen to enhance the performance of a vegetable oil fuelled compression ignition engine, *Int J Hydrogen Energy* **28(10)** pp 1143 - 1154

[31] Korakianitis T, Namasivayam A M and, Crookes R J 2011 Diesel and rapeseed methyl ester (RME) pilot fuels for hydrogen and natural gas dual-fuel combustion in compression-ignition engines, *Fuel* **90(7)** pp 2384 - 2395

[32] Changwei Ji, Shuofeng W and Bo Z 2012 Performance of a hybrid hydrogen–gasoline engine under various operating conditions, *Applied Energy* **97** pp 584–589

[33] Jayakrishnan K U, Premakumara G and Lalit M D 2017 Development of hydrogen fuelled transport engine and field tests on vehicles, *Int J Hydrogen Energy* **42** pp 643 – 651

[34] Yao S, Xiumin Yu, Wei D and Yang T 2018 Effects of hydrogen direct injection on engine stability and optimization of control parameters for a combined injection engine, *Int J Hydrogen Energy* **43** pp 6723 – 6733

[35] Masaki N, Kohei M, Satoshi S, Kotaro T and Mitsuru K 2019 Effects of hydrogen addition on engine performance in a spark ignition engine with a high compression ratio under lean burn conditions, *Int J Hydrogen Energy* **44** pp 15565 – 15574

IOP Conf. Series: Materials Science and Engineering **595** (2019) 012033 doi:10.1088/1757-899X/595/1/012033

A review of hydrogen/diesel fuel blends in internal combustion engines

D Gheorghe, D Tutunea, M Bică, A Gruia and M Calbureanu

Road Vehicle and Transportation Department, University of Craiova, Craiova, Romania

E-mail: marinbica52@gmail.com

Abstract. A hydrogen fuelled internal combustion engine has great advantage on exhaust emission in comparison with a conventional engine from fossil fuel. Despite diesel engines are indispensable for the transport sector, the decreasing emission regulations limit values in recent years' force researches to find other alternative fuels to use in addition or straight without major engines modifications. A comprehensive review of the recent research activities on the use of hydrogen/diesel fuel blends is documented. Also the challenges and perspectives and main limitations are analysed in this paper

1. Introduction

Since the introduction by the Rudolf Diesel in 1982 diesel engines are considered one of the most efficient types of engine used worldwide. The recent diesel scandal regarding the pollutant emissions have increase attention regarding the future of diesel engines. Many countries have limited the access of diesel cars in the next years forcing industry and researches to investigate and to research new alternatives. The use of hydrogen is prepared by the automotive manufacturers as a method to enhance performance and improve harmful emission. However, the production of hydrogen is still a very costly process and is expected that this obstacle to be overtaken in future years. The reduction of diesel emission could bring back diesel engine's reliability [1]. The energy content of hydrogen is one of the highest among the commonly used fuels providing at 1 kg almost three times more energy than petrodiesel fuels (diesel and gasoline). However, the low density specific to hydrogen restricts his use do to the storage space required for powering a car for an adequate driving range. Antunes et al. [2] investigated on a single cylinder diesel engine the hydrogen addition directly into a combustion chamber and reported an increase in peak power by 14% comparative with petrodiesel fuel. Miyamoto et al. [3] studied the 16% addition by volume of hydrogen in a direct injection, single cylinder diesel engine and found a reduction in HC and NOx emissions. Saravanan and Nagarajan [4] used three methods of hydrogen addition in a Kirloskar AV1 diesel engine and found that at 1500 rpm and 25%, 50%, 75% and 100% load conditions are no difference between injecting from port and injecting from manifold. The NOx emissions increase with 8% and 17% depending on the method used. Karagoz et al. [5] studied the effect of engine load to the performance of hydrogen addition 0-30% and found a decrease of BTE and an increase of BSFC with 8.4%. Wu and Wu [6] investigated the performance of a single cylinder diesel engine fueled with 20% hydrogen and found a reduction of NOx and smoke emissions compared with petrodiesel fuel. Szwaja et al. [7] studied the addition of 5% hydrogen by energy share to fossil diesel and found that combustion duration did not change significantly but the peak combustion pressure increased. In Table 1 are compared a few of the combustion properties of hydrogen and diesel fuel. According to the data the high diffusivity and auto-ignition temperature of hydrogen make this fuel for compression ratio (CI) engine compared to gasoline engines [8].

Table 1. Important properties of diesel and hydrogen [8].

Properties	Diesel	Hydrogen
Lower Heating Values (MJ/kg)	43	120
Stoichiometric air fuel ratio	14.5	34.2
Energy Density at 15 °C and 100 kPa, MJ/m^3	35.8	10.3
Autoignition temperature, K	530	858
Laminar burning velocity, m/s	0.3	2.65 – 3.25
Flammability limits (% volume in air)	0.7 to 5	4 to 75
Density at 15 °C and 100 kPa, kg/m^3	848	0.0083
Diffusivity in air, cm^2/s	0.038	0.63

2. Effect of hydrogen addition on performance

2.1. Power output

The use of different ratios of hydrogen (e.g. 5 - 50% by volume) to the diesel engine and its effect on emission and performance was evaluated by various researchers. Shin et al. [9] investigated the hydrogen addition and found that the blend hydrogen/diesel improves the process of combustion and leads to the increase of engine power for all loads. Ghazal [10] investigated in his research the relation between air–fuel ratio and engine speed and hydrogen addition. He found for a ratio of 15 air–fuel (A/F) and a percentage of hydrogen between 5% and 10% for different speeds the maximum power output improves. At higher air–fuel ratio the hydrogen addition affects the engine maximum power if the proportion of hydrogen is over 30–40%. In Figure 1 it is observed an increase with 14% in power for all engine speeds comparative with petrodiesel fuel.

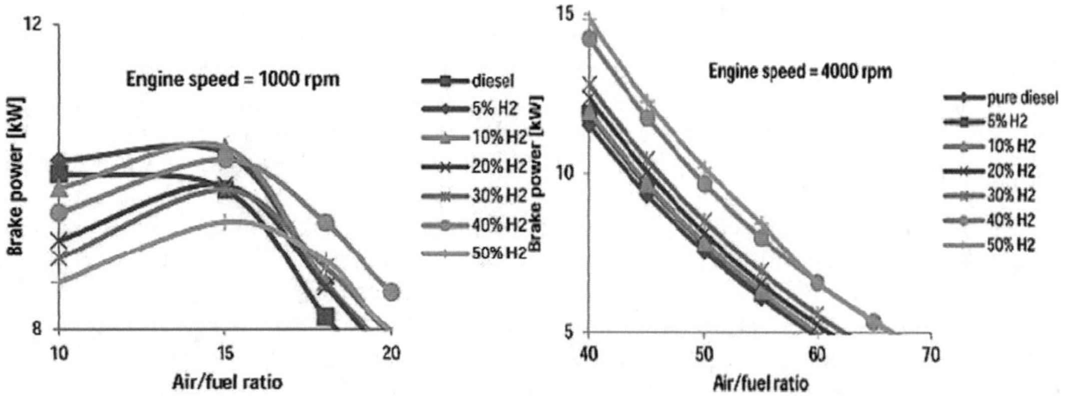

Figure 1. The brake power versus A/F ratio for different H_2 concentrations at 1000 and 4000 rpm [10].

2.2. Duration of combustion

Comparative with petrodiesel fuel hydrogen has a higher flame speed which affects the combustion process and engine performance. Liew et al. [11] showed that the addition of 2% hydrogen reduce the combustion duration revealing the effect of the fuel in the process of diffusion combustion. At a percentage of 6% a 23% was observed a reduction of combustion duration comparative with neat biodiesel (Figure 2). Grab-Logarinski and Szwaja [12] studied the hydrogen addition in a diesel engine in reduce proportion (maximum 5%) and found that hydrogen reduce ignition delay and improve performance of the engine. If the hydrogen addition exceeds 15% in energy ratios the diesel engine exhibits severe knock.

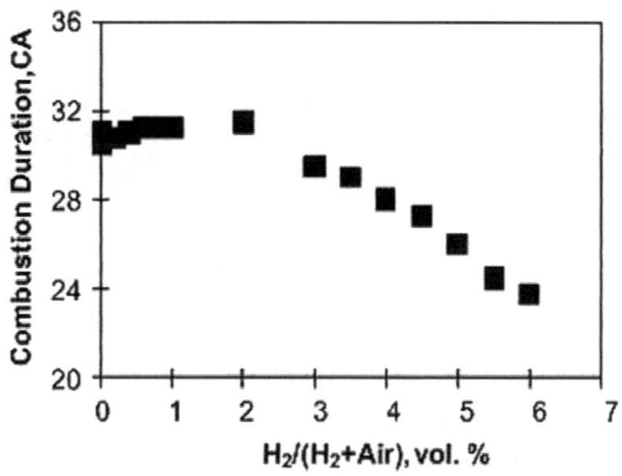

Figure 2. Effect of H_2 addition on combustion duration, 70% load [11].

2.3. In cylinder pressure

At higher loads the hydrogen addition increases the pressure as is shown in Figure 3 due to the higher energy content. Saravanan and Nagarajan [4] found out that a percentage of 30% hydrogen at higher loads produce the best results. Masood et al. [13] show in his researches that the combustion process tends to increase as the percentage of hydrogen is higher due to higher flame velocities.

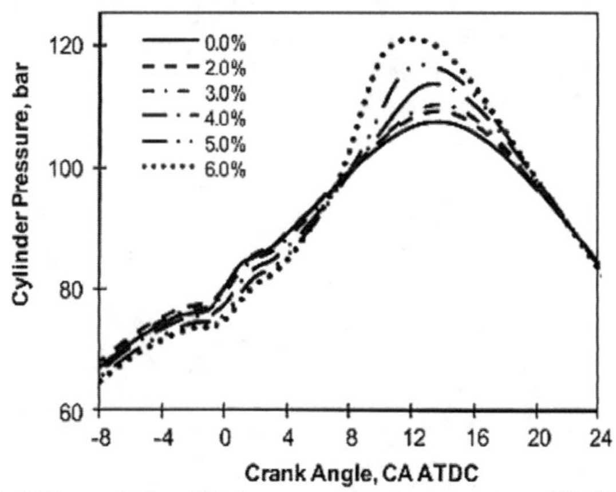

Figure 3. Effect of H_2 addition on cylinder pressure, 70% load [11].

2.4. Heat release

The heat release rate is influenced by the addition of hydrogen. Figure 4 shows that the addition of hydrogen increases the heat release. It is observed a rapid rise of heat different that of highlighted diffusion combustion of diesel engines due to the high quantity of hydrogen used.

3. Conclusions

The high flammability range of hydrogen in comparison with all other fuels can allow to be combusted in engines over a wide range of fuel-air mixtures. The high flame speed is an order of magnitude higher (faster) than petrodiesel fuels meaning that hydrogen fuelled engines can approach thermodynamically ideal engine cycle. Also hydrogen is a desirable fuel for IC engines it can produce combustion anomalies such as rapid pressure rise, backfiring, pre ignition (surface ignition) and auto ignition. The power output of hydrogen internal combustion engine can increase with 15-20% more than gasoline or diesel engine depending of the method used for fuel injection. Due to the higher pressure and temperature associated with hydrogen usage the engine parts are solicited at higher values than petrodiesel fuels. Regarding the effects of hydrogen on engine parts its effects are the decrease in ductility and true stress at fracture.

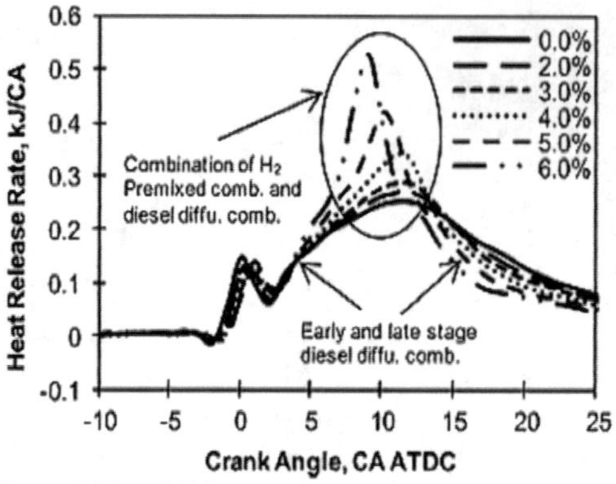

Figure 4. Effect of H_2 addition on heat release process, 70% load [11].

References

[1] Pavlos D and Taku T 2017 A review of hydrogen as a compression ignition engine fuel, *International Journal of Hydrogen Energy* **42** pp 24470-24486

[2] Gomes A J M, Mikalsen R, Roskilly A P 2009 An experimental study of a direct injection compression ignition hydrogen engine, *Int J Hydrogen Energy* **34(15)** 6516-6522

[3] Miyamoto T, Mikami M, Kojima N, Kabashima H and Urata Y 2009 *Effect of hydrogen fraction in intake mixture on combustion and exhaust emission characteristics of a diesel engine*, SAE Paper No. 2009-24-0086

[4] Saravanan N and Nagarajan G 2009 *Experimental investigation on performance and emission characteristics of dual fuel DI diesel engine with hydrogen fuel* SAE Paper No. 2009- 26-032

[5] Karagoz Y, Sandalci T, Yuksek L and Dalkılıc A S 2015 Engine performance and emission effects of diesel burns enriched by hydrogen on different engine loads, *Int J Hydrogen Energy* **40** pp 6702 -6713

[6] Wu H W and Wu Z Y 2012 Investigation on combustion characteristics and emissions of diesel/hydrogen mixtures by using energyshare method in a diesel engine. *Appl Therm Eng* **42**(Suppl. C) pp 154-162

[7] Szwaja S and Grab-Rogalinski K 2009 Hydrogen combustion in a compression ignition diesel engine. *Int J Hydrogen Energy* **34** 4413-21

[8] Priybrat S and Atul D 2018 Effect of hydrogen supplementation on engine performance and emissions, *Int J Hydrogen Energy* **43** pp 7570 – 7580

[9] Shin B, Cho Y, Han D, Song S and Chun K M 2011 Hydrogen effects on NOx emissions and brake thermal efficiency in a diesel engine under low-temperature and heavy-EGR conditions. *Int J Hydrogen Energy* **36** pp 6281–6291

[10] Ghazal O H 2013 Performance and combustion characteristic of CI engine fueled with hydrogen enriched diesel, *Int J Hydrogen Energy* **38** pp 15469–15476

[11] Liew C et al. 2010 An experimental investigation of the combustion process of a heavy-duty diesel engine enriched with H_2. *Int J Hydrogen Energy* **35** pp 11357–11365

[12] Adnan R, Masjuki H and Mahlia T 2012 Performance and emission analysis of hydrogen fueled compression ignition engine with variable water injection timing. *Energy* **43** pp 416–426

A review of new technologies in valve systems of internal combustion engines and their effects

D Akal[1], U Akyol[2] and K Uzuneanu[3]

[1] Mechanical Engineering Department, "Trakya University", Edirne, Turkey
[2] Mechanical Engineering Department, "Tekirdag Namik Kemal University", Çorlu, Tekirdag, Turkey
[3] Thermal Systems and Environmental Engineering Department, "Dunarea de Jos" University of Galati, Galati, Romania

E-mail: uakyol@nku.edu.tr

Abstract. Efficient use of energy is becoming increasingly crucial due to global warming and depleting fossil fuel sources. Automotive manufacturers aim for fuel economy, reduction of exhaust emissions, durability, volume and weight reduction, and power and torque increase in internal combustion engines. One of the most important developments in this field is the new technologies in classic valve systems. There are still ongoing studies in this field as the gas exchange processes affects all engine parameters. In this review; the principles of new technologies used on classic valve systems and their effects on engines have been investigated. The research has shown; new mechanism technologies used on valve systems have positive effects on engine performance, power, torque and fuel consumption.

1. Introduction

Efficiency in engines has become a primary objective in recent years due to the rapid development of automotive industry, depleting oil derivative fuels, increasing exhaust gas emissions and as a result of cost analysis and supply and demand ratios. The valves in internal combustion engines are opened and closed by the camshaft. The biggest problem in standard use of the camshaft is that the valve opening duration and lift are fixed. In these type of engines, the camshaft delivers an appropriate volumetric efficiency in low speeds, but it's negatively affected in high speeds. As speed increases in internal combustion engines, the gas intake volume and speed continues to change in relation to the piston speeds. Thus, the valve timing should also be changeable.

2. Innovations in valve technology

Many studies have been made in the automotive industry for developing valve systems that will increase the volumetric efficiency, and thus improve the internal combustion engines. There is still ongoing research in this field. In conventional engines, the intake valves within the cylinders let the gases into the cylinder, and the exhaust valves dispose the burnt gases out of the cylinder. Larger valves increase the volumetric efficiency. As the valves get larger, the intake and exhaust of the gases get easier. However, bigger valves also mean that the spark plug gets farther away from the combustion chamber, which decreases efficiency. For better volumetric efficiency, it makes more sense to increase the number of valves, rather than the size. Initially, the engine manufacturers have increased the intake and exhaust numbers per cylinder to create multi-valve systems. Multi-valve technology has increased the gas efficiency within the cylinder [1]. In internal combustion engines, variable valve technologies are now being used to ensure the power and torque increase at wider speed ranges. In recent years, the gas intake can be increased even more with variable valve timing coupled with variable valve lift controls. Thanks to all these improvements, internal combustion engines have better fuel economy, higher power and torque [1].

In the automotive industry, the studies on valve technologies show that, engines with high efficiency, more flexible valve timing and valve lift are now being produced. With today's technology, it's possible to increase fuel economy and reduce exhaust emissions in multi-cylinder engines, by deactivating some cylinders when the additional power is not needed. In this context, there is still on-going research on the valve mechanisms in electromechanical, electrohydraulic, electro-pneumatic and electric motor driven systems [1].

In electromechanical valve systems, there is an electromagnet for each valve and an electronic control unit that controls the valve system. In these systems, technologies such as variable valve timing, variable valve lift and the disabling of some cylinders can be used more easily [2]. In engines designed with this technology, the problems of noise and valves hitting against the covers have also been somewhat eliminated [3-5]. Additionally, increased fuel economy and reduced exhaust gas emissions have been observed.

When the electrohydraulic valve systems are examined, the stem of the valves works in connection with a hydraulic cylinder. The system works with filling and discharging a fluid into the hydraulic cylinder in the valve stem. Opening and closing of valves is managed by directional control valves of the fluid. The required pressure difference for the system can be provided either by the lubrication system of the engine or by an external electric motor [6].

Electro-pneumatic valve systems show fundamental similarities with the electrohydraulic systems. In this system, air is used as a fluid. It is preferred especially in valve systems where greater power is needed. This system operates efficiently at normal speeds but suffers when operating at low speeds [7]. Additionally, variable valve lift control and contact problems sometimes occur in the engines where electro-pneumatic valve systems are used.

Recently, some studies have been conducted on valve systems driven by electric motors. This system is also based on variable valve timing and variable valve lift. The engines with electric motor driven valve systems offer lower power consumption [8].

3. Variable valve mechanisms

Two of the biggest parameters affecting the engine efficiency the most are valve timing and lift. In modern engines, the intake valves are opened before the piston reaches the top dead centre (TDC) and closed after it passes the bottom dead centre (BDC) to increase volumetric efficiency. Similarly, exhaust valves are opened before the piston reaches BDC and closed after it passes TDC. In this way, maximum performance and efficiency can be achieved. In the past, the camshafts were designed with a set working range. In these type of engines, the biggest setback of using a standard camshaft is not having the ability to change the valve timings and lift. Even though these engines provide a viable performance in low speeds, the power drops in high speeds. Full efficiency can only be achieved within a small revolutions range in the engines in which a standard camshaft is used. Conversely, the valve lift in high speeds causes a drop in power in low speeds. High angle camshaft offer a high performance in high speeds, but causes rattling and noise when operating in low speeds. In order to avoid all these problems, the automotive industry has developed engines with variable valve mechanisms with high efficiency, which will ensure perfect performance in varied working conditions.

3.1. Audi valvelift system

In this system, the valves are controlled by the low speed cam or the high speed cam on the camshaft, depending on the engine speed. When the engine is running at low speeds, the low profile cam mechanism is activated for a narrower lift. In increased engine speeds, the high cam is activated to increase the lift. The activation of the low and high cam profiles on the camshaft is made by two electromagnetic mechanisms. In the Figure 1, the activation of low and high cams is indicated by red and green arrows. This increases the efficiency in relation to the amount of air entering the engine cylinder [9]. Friction is also reduced because there are fewer components in variable valve systems. About an additional 7% fuel economy can be expected in the engines where this system is used [10].

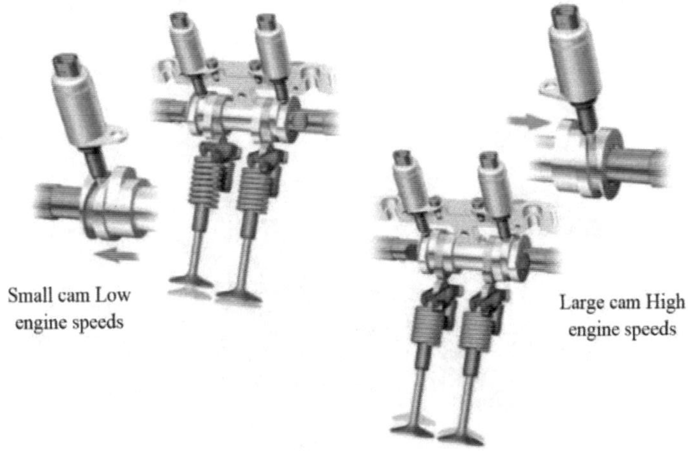

Figure 1. Audi Valvelift System [11].

3.2. BMW Valvetronic System

With this technology, there is no need for a throttle valve because it has a different type of software and control system compared to other variable valve mechanisms. It boasts a fully equipped intake valve that adjusts the amount of air required for the engine, and eliminates pumping loss. Developed by BMW, this system consists of an additional camshaft, which allows the opening and closing of the valves, a monitoring sensor, a spiral gear system and an electric motor. The timing of the cam shaft and the opening duration of the valves are regulated by the spiral gear system, which is driven by the stepper motor. One of the most important features that distinguish this system from others is that the cooling water temperature at the cylinder top surface is much lower [12]. Due to the lower temperature of the cooling water, a smaller volume circulation is used and less power is required from the engine. Motors using this system operate easily and quietly in cold weather conditions. In addition, fuel consumption and harmful exhaust emissions are lower compared to other engines [13].

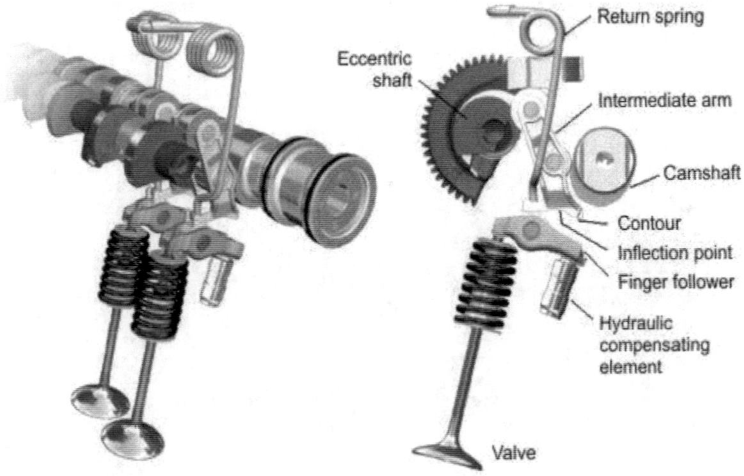

Figure 2. Valvetronic system as developed by BMW [14].

3.3. Porsche VarioCam System

Similar to other variable valve timing mechanisms, this system changes the angle of the camshaft according to the position of the crankshaft. The difference is, this system not only controls the intake valves, but also the exhaust valves. When the system is activated, the intake valves are closed earlier because the lapping duration of intake and exhaust valves are longer. As a result, more torque is obtained from the engine [15]. As in other systems, the camshaft is composed of low and high cam, which controls the valves according to the speed of the engine. The drive system, which changes the angular position of the camshaft, operates with hydraulic oil.

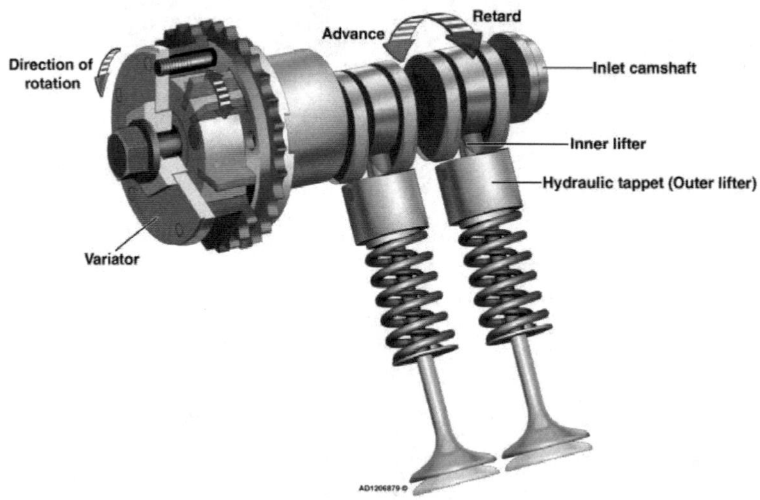

Figure 3. Porsche's VarioCam System [16].

3.4. Toyota VVTL-I system

Developed by Toyota, this technology aims for creating more momentum in low speeds and more power in high speeds by controlling the timing and lift of the valves [17]. Like other variable valve mechanisms, there are high and low cam profiles on the camshaft. The rocker has a two-piece design for making the switch between these cam profiles easier. The transition between the cam profiles on the camshaft is made by hydraulic oil pressure working the locking pin. The low cam is active in low speeds. When the engine speed reaches 6000 rpm, the oil pressure is increased and the high cam profile gets activated, resulting in higher power [17].

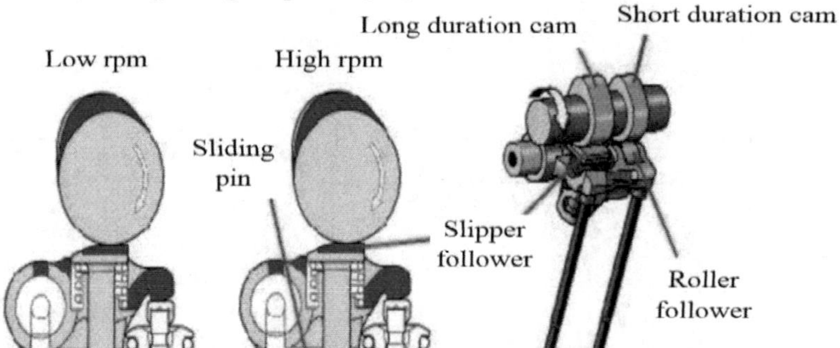

Figure 4. Toyota VVTL-I System [18].

3.5. Fiat MultiAir System

The MultiAir system offers better combustion control and a faster response to torque requests. This means concretes advantages in terms of:

- Lower fuel consumption
- CO_2 emission reduction
- Power and torque increase
- Higher driving responsiveness
- Easier start-up.

The MultiAir system is managed by injection/ignition control unit which controls the operation of the intake valves. In this way the fluid inside the engine can be changed without resorting to the traditional expansion on the throttle valve [19].

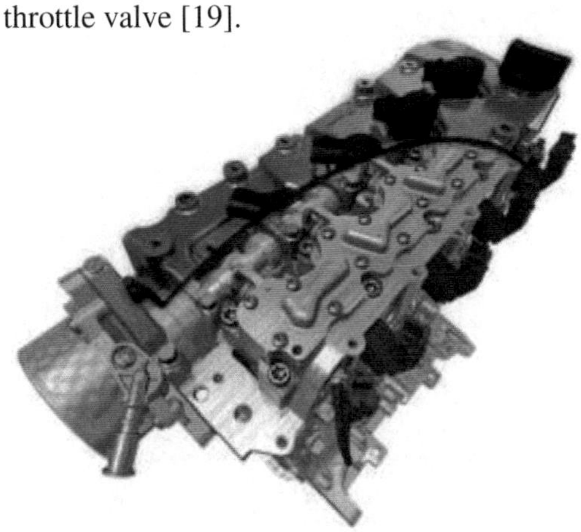

Figure 5. MultiAir System [19].

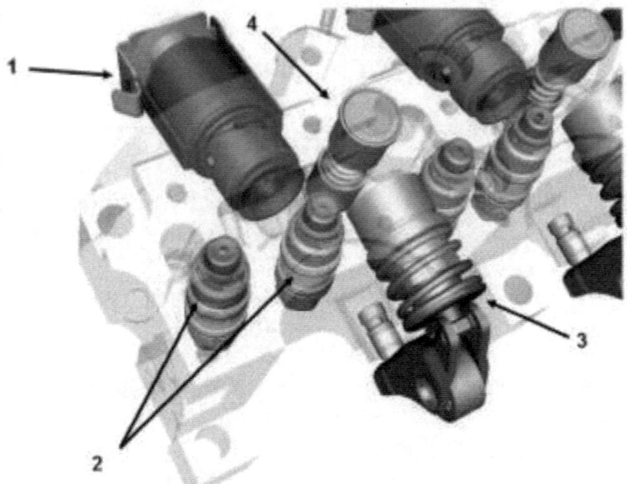

Figure 6. Uniair module components (1- Solenoid valve, 2- Hydraulic brake, 3- Upper pumping element, 4- Oil accumulator) [19].

MultiAir technology manages the torque and power delivered by the engine by varying the lift profile of the intake valves, without direct use of the throttle body. This application consists of following components:

- Single camshaft (exhaust side) with hydraulic tappets
- Uniair module (electro-hydraulic) with integrated intake tappets
- Use of the brake servo vacuum unit [19].

The MultiAir engine is characterised by a camshaft controlling the exhaust valves and an electro-hydraulic "Unair" module controlling the intake valves. The steel camshaft is fitted on the camshaft housing by means of five caps and is driven by a toothed belt. Cams are suitably oriented and profiled on the shaft. Their number corresponds to the exhaust valves plus a further four cams to control the unair module pumping elements. The front side of the shaft is pre-set for fitting a toothed pulley, connected to the crankshaft by means of a toothed belt. The timing sensor faces the toothed wheel fitted on the rear side of the shaft, while the shank drives the brake servo vacuum unit [19].

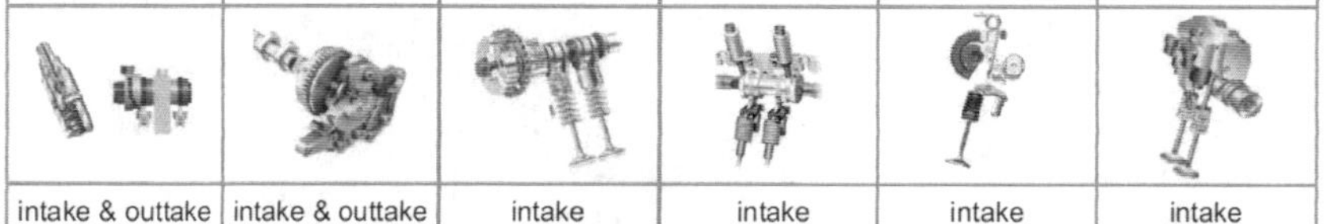

	• cylinder deactivation (switchable lost motion element or shiftung cam)	• cam phaser	• cam phaser • lost motion element (switchable tappet, cam follower, ...)	• cam phaser • shifting cam	• cam phaser • actuated intermediate shaft	• electro-hydr. system
	intake & outtake	intake & outtake	intake	intake	intake	intake
fuel reduction	approx. 7-8 %	approx. 4 %	approx. 7 %	approx. 8 %	approx. 8 %	approx. 8-15 %
Variability						
- phaising		continuous (intake and outtake)				
- valve lift	discrete: 2 stages standard and zero lift		2-3 stages	2-3 stages	continuous	continuous
- shape			compatible cam contours	independent cam contours		free shape
series production	since 1981 Active Fuel Managment (GM), ACC (Daimler), VCM (Honda), MDS (Chrysler), CoD (Audi,VW), ...	since 1983 commonly used by car manufacturers	since 1989 VTEC (Honda), MIVEC (Mitsu.), VVTL-i (Toyota), VarioCam Plus (Porsche), ...	since 2006 Audi valvelift system	since 2001 Valvetronic (BMW)	since 2009 MultiAir (Fiat)

Figure 7. Different variable valve train systems in series production [20].

4. Conclusions and evaluations

The studies carried out to increase the volumetric efficiency and performance of the engines show that the multi-valve technology and variable valve timing systems are of great importance. This trend will continue in the future due to higher taxation and increasing pressure to achieve greater fuel economy and performance, while keeping lower emission rates. However, the use of high quality materials, precision surface treatment and the high costs of these systems can lead to a search for alternative replacements of valve technologies. Conducted studies show that the engines designed with a variable valve timing system have improvements in power, torque, fuel consumption and exhaust emissions. Example studies and their effects are listed in the table below.

Despite many innovations in valve systems, the automotive industry and especially the Swedish and Chinese companies continue to work on engines with variable timing and no camshaft, where each valve is controlled separately. Manufacturing engines without a camshaft is expected to bring the following advantages: lighter engine; lower friction; silent operation; lower fuel consumption; lower emissions; higher power and torque.

References

[1] Birgül N and Sekmen P 2013 Investigation of The Dynailc Performance of an Electro Mechanical Valve (EMV) System at Different Lifting Operations and Determination of THE Support Limits of an Internal Combustion Engine *Journal of the Faculty of Engineering and Architecture of Gazi University* p 607-16

[2] Atzler F 2001 On the Future of the Piston Engine with Internal Combustion an Overview *Marie Curie Fellowship Conference*

[3] Shiao Y and Pan W 2011 Design of A 3-Stage Voltage Controller for EMV Actuation in SI Engines *11th International Conference On Control, Automation and Systems*
[4] Peterson K S and Stefanopouluo A G 2004 Extremum Seeking Control for Soft Landing of An Electromechanical Valve Actuator *Automatica* **40** p 1063-69
[5] Lua C A Toledo B C and Benedetto M D and Gennaro S D 2007 Output Feedback Regulation of Electromagnetic Valves for Camless Engines *Proceedings of the European Control Conference* p 1-26
[6] Pournazri M Khajepour A and Fazeli A 2011 *An Efficient Lift Control Technique in Electrohydraulic Camless Valvetrain Using Variable Speed Hydraulic Pump* (USA: Society of Automotive Engineers (SAE)) p 0940
[7] Trajkovic S Tunestål P and Johansson B 2007 Introductory Study of Variable Valve Actuation for Pneumatic Hybridization *Society of Automotive Engineers (SAE)* p 0288
[8] Yihui Q 2010 A Custom-designed Limitedangle Actuator for an Electromechanical Engine Valve Drive. Part I: Conceptual Design *Proceedings of the 5th IET International Conference on Power Electronics, Machines and Drives (PEMD)*
[9] Gerzeli I 2009 Design of Skip Cycle Mechanism for A Four Stroke Internal Combustion Gasoline Engine *Master's Theses İstanbul Technical University*
[10] http://www.autozine.org/technical_school/engine/vvt_31.htm> 2008
[11] McCharty J and Stretch D 2012 Compact, electro-hydraulic, variable valve actuation system providing variable lift, timing and duration to enable high efficiency engine combustion control *US Department of Energy, High-Efficiency Engine Technologies*
[12] http://autospeed.com/cms/A_1083/article.html> 2007
[13] http://www.usautoparts.net/bmw/technology/valvetronic. htm> 2008
[14] Günkan M F 2014 To Devolop Cycle Mechanism and Drive System for an Internal Combustion Engine *Master's Theses İstanbul Technical University*
[15] http://www.porsche.com/uk/aboutporsche/porscheandenvironment/technology/variocamplus/ 2013
[16] http://www.workshopmanuals.co.uk 1999 (Germany : Technical Manual 911 Carrera)
[17] http://www.celicas.co.uk/Technology-Vvtli.html>2009
[18] Karakayis A 2014 Investigation of Control System Strategies for Hydraulic Valve Actuation in an IC Engine *MEM47 Automotive Project, University of Brighton School of Computing, Engineering and Mathematics, Division of Engineering & Product Design*
[19] Tofaş academy education book Fiat 1.4 Turbo Multiair timing System
[20] Huber R 2012 Dynamics of Variable Valve Trains and Extrapolation Methods for Time-Stepping Schemes *Doctoral Theses Munich Technical University*

IOP Conf. Series: Materials Science and Engineering **595** (2019) 012035 doi:10.1088/1757-899X/595/1/012035

Experimental study of NO_x and CO_2 emissions at load regimes for a passenger car powered by direct injection diesel engine

S M Burciu and G Coman

[1]Department of Thermal Systems and Automotive Engineering,
Dunarea de Jos University of Galati, Galati, Romania

E-mail: Gelu.Coman@ugal.ro

Abstract. The paper presents a study of NO_x and CO_2 emissions for a standard diesel direct injection engine, which equips the Renault-Dacia Solenza 1.9D-SDI cars. The different pollutions levels are analysed between idling regimes and medium and high loads at different speeds and transmission gearboxes. To simulate the load regimes, chassis dynamometer V-Tech Dyno VT-2/B1 was used. The exhaust gases were taken to various operating modes on the chassis dynamometer, specific regimes in urban and extra-urban mode. In the paper, some recommendations are made on operating regimes that would ensure the functioning of the car at a lower polluting level.

1. Introduction

The main legislative instrument to achieve the 2030 objectives of the Clean Air Programme is Directive 2016/2284/EU on the reduction of national emissions of certain atmospheric pollutants which entered into force on 31 December 2016 [1]. Air pollution is an important public health problem in most cities [2]. Without effective measures to mitigate the adverse impacts of motor vehicle use, the living environment in the cities will continue to deteriorate and become increasingly unbearable, [2, 3, 4].

According with ACEA (European Automobile Manufacturers Association), the global vehicle fleet continues to grow. The EU passenger car fleet grew by 5.7% over the last five years. The EU has a total fleet of 259.7 million passenger cars. According to the same Association, Romania had in 2017 the oldest fleet, with vehicles older than 16 years. Out of the total cars running in Romania, over 4 million are 10 years old, [5]. The new Romanian legislation on pollution standards for second-hand cars has as a consequence the import of more than 500,000 used cars in 2018 [6]. The increase in the number of passenger automotives with diesel engine and the intensive traffic in urban areas have a negative effect on environment and human health, [7]. Chana describes in his study the emissions from actual operations and their dependence to the speed or age of the vehicle, [8].

The most important pollutant emissions of diesel engine are nitrogen oxides (NO_x) and smoke, the carbon dioxide CO_2 being considered a greenhouse effect gas, [9, 10]. Regarding the impact of pollutant emissions on environment [2, 3], special measures are taken in order to reduce as much as possible the pollution produced by diesel engines, especially in the urban areas, [10, 11].

In order to limit the pollutant emissions level at diesel engines, especially NO_x and soot, the pollution legislations have become more severe, this fact leading to the applying of new active and passive methods for fuels combustion control [12, 13], and using alternative fuel too, [14, 15].

The present paper aims precisely to show the difference between the NO_x and CO_2 emissions determined at idling speed and the one determined at the load regimes specific to the normal operating conditions in urban and extra-urban mode, in case of a passenger car powered by direct injection diesel engine.

In the case of real operating regimes (operating regimes in load), the NO_x and smoke level may exceed the maximum smoke and NO_x level imposed by European anti-pollution legislation, but this thing depends largely on the engine's technical condition, fuel quality and operating regimes. According to this legislation, the level of smoke is measured by free acceleration at idling speed between the minimum and the maximum rotation of engine and the determination of the pollution level for NO and NO_x is carried out under idling regimes with specified conditions (oil temperature, speed engine), so there are no provisions for the measurement of pollutants in the gasses with the engine in load.

2. Equipment used for experimental studies

The equipment used for experimental determinations includes a car, V-Tech Dyno VT-2 / B1 Chassis dynamometer, an analyzer of exhaust gases and a smoke meter.

The car used in the study, was Renault-Dacia-Solenza 1.9D-SDI with the characteristics indicated in the table 1, from [16].

Table 1. Vehicle characteristics.

Maximum power	Maximum rotation of engine	Engine type	Engine volume	Cylinders number	Type of fuel and fuel system	Environmental standard	Maximum authorized weight
HP	rpm]	-	cm^3	-	-	-	kg
63	4500	F8Q	1870	4	Diesel/Standard diesel injection-SDI	EURO III	1460

The total emission of nitrogen oxides NO_x(%vol) and the CO_2 (vol%) in the exhausts gasses were measured with the Multilyzer STx gas analyser. The analyzer used for the opacity coefficient measure is Johnson Controls – Ultimax 600, 800-65 model.

The roller stand enables the dynamical determination of the engine's power and torque, as well as of the power and torque of the vehicle's engine wheels.

The V-TechDyno VT-2/B1 Chassis dynamometer (figure 1) has been used [17], which has the following features:

- maximum load on the axis 3000 kg;
- number of axes - 1 axis;
- maximum speed 300 [km/h];
- maximum engine power measured 600 [HP];
- minimum external wheel diameter 400 [mm];
- the TELMA eddy-current electric brake with a maximum torque of 800 Nm/1000 rpm, with a precision of 0.1% for inertial mode and 1% for brake mode respectively.

The data acquisition system that equips the V-TechDyno VT-2/B1 Chassis dynamometer is shown in figure 2.

To perform the load tests, the road test method was used (figure 3). This mode simulates road condition, driving and changing gears at the same way as on the road. Basic data should be written to the software as weight, aerodynamic coefficient C_x value and front area. During acceleration, the brakes will apply the proper load on dyno rolls – this will enable to simulate air and traction resistances. This mode also gives the possibility to simulate hills.

Figure 1. The V-TechDyno VT-2 / B1Chassis dynamometer

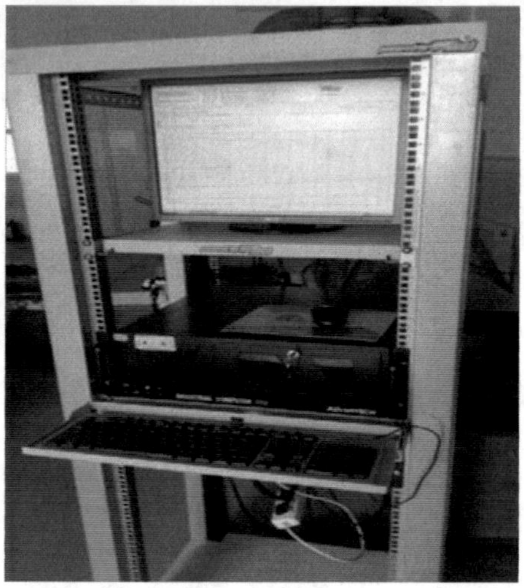

Figure 2. The V-TechDyno VT-2 / B1 data acquisition system

The determinations were made for the horizontal road, in the 2nd, 3rd, 4th, and 5th gears of the gearbox, at the usual travel speeds of (25 – 150) km/h and in the range of (1500 – 4000) rpm, which is the specific operating range for the diesel engine and the car powered with diesel engine.

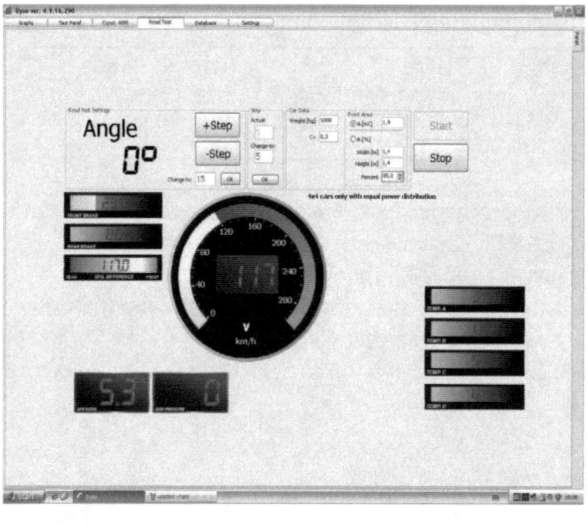

Figure 3. Tuning panel of road test mode.

In accordance to the European provisions, for diesel-fuelled cars, at the periodic technical inspection [18], the smoke index is checked, which must fit as follows:

- For natural aspirated diesel engines, the smoke index $k \leq 2.5\ m^{-1}$.
- For turbocharged diesel engines with pollution norms up to and including EURO 3, the smoke index $k \leq 3\ m^{-1}$.
- For turbocharged diesel engines that have the pollution standard EURO 4, EURO 5, EURO6, the smoke index $k \leq 1.5\ m^{-1}$.
- Carbon dioxide is not yet legislated.

IOP Conf. Series: Materials Science and Engineering **595** (2019) 012035 doi:10.1088/1757-899X/595/1/012035

In terms of the limit for NOx, according to the Euro standard, that is 0,5 [g/km] for EURO 3, 0.25 [g/km] for EURO 4, for EURO 5 is 0.18 [g/km], respectively 0,08 [g/km] for EURO 6, but these are only determined when homologate a new automotive model for establish and verify the EURO norm.

The paper aims to show the difference between the smoke index, NOx and CO2 emissions determined in the idling regimes and the ones determined at the load regimes of engine, in urban and extra-urban mode, for a passenger car powered by direct injection diesel engine.

The roller stand enables the dynamical determination of the engine's power and torque, as well as the power and torque of the vehicle's engine wheels.

3. Experimental determinations

Using the equipment presented, experiments were carried out for measuring the CO_2 and NO_x emissions and also the opacity coefficient at idling regimes and load regimes.

For engine idling at various speeds, a regime which is not representative of the operation of engine because it represents a maximum of 5% of the total running time, the following were found, according to the graphs in figure 4, 5, 6:

- At speeds above 2500 rpm, CO_2 emissions increase by up to 37% compared to those measured at speeds below the above-mentioned value, (figure 4);
- At speeds above 2500 rpm, NO_x emissions increase by up to 66% as compared to those measured at speeds lower than the value mentioned, figure 5;
- The maximum value of the measured smoke index of 0.18 m^{-1} (figure 6), is 14 times smaller than the maximum admissible limit accepted by European legislation, under the specified conditions and 7.4 times lower than the one measured directly by acceleration between minimum and maximum at idling regime. However, this value was measured in a stabilized idle mode, not through acceleration from a minimum to a maximum speed;

The maximum value of the smoke index obtained, measured according to legislative provisions and presented in figure 7, was 1.8 times lower than the maximum admitted limit for this euro category and type of engine.

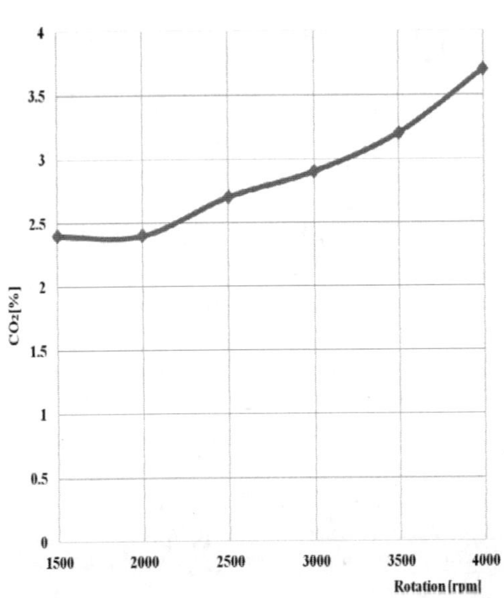

Figure 4. Emissions of CO_2 for idling regimes.

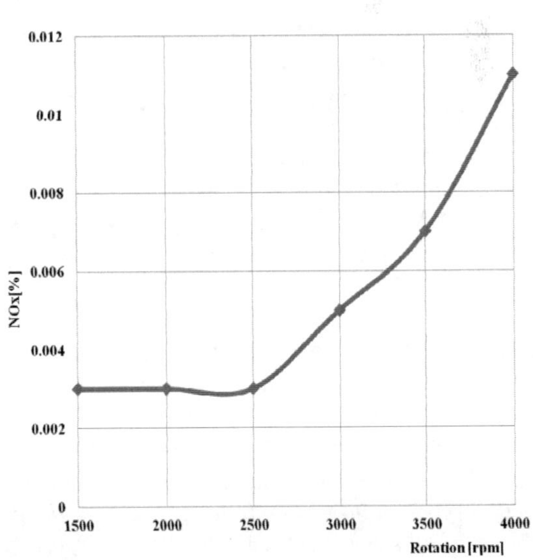

Figure 5. Emissions of NO_x for idling regimes.

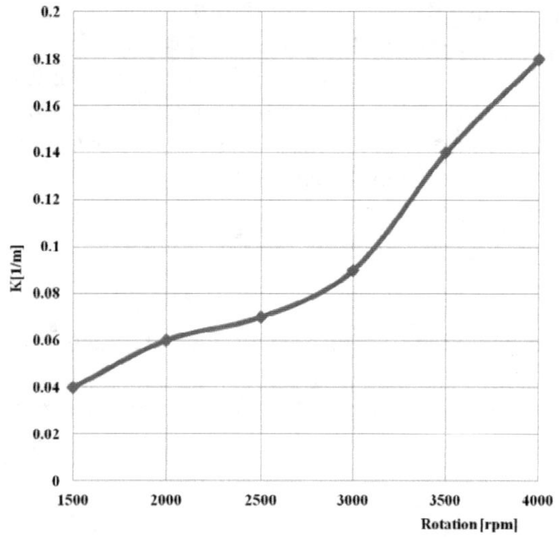

Figure 6. Smoke emission for idling regimes

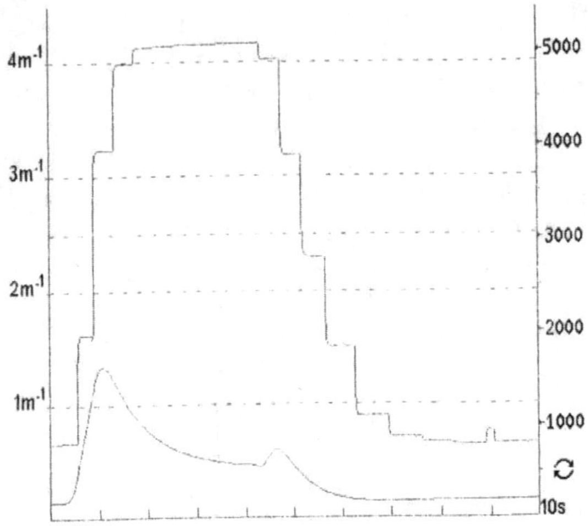

Figure 7. Maximum opacity index for fast acceleration between minimum and maximum speed.

For the idle mode, which is not specific to the operation of cars because it represents a maximum of 5% of the service life, the conclusion is as follows: in situation of good technical condition of the diesel engine (with proper adjustment of the fuel system and with a normal compressing in the engine mechanism), the engine having the appropriate thermal regime (coolant temperature of at least 80^0C), the level of CO_2, NO_x and smoke pollution is low, below the limits imposed by the legislation and in order to reduce them, it is recommended the idling speed in the range of (800 – 1500) rpm, which also holds true for real life situations.

For the functioning of the indicated diesel engine in the load regime, ie the urban and off-road operation of the car, based on the analysis of the variations shown in figures 8, 9, 10, the following findings can be mentioned:

- *as far as CO2 emissions are concerned:*

• The smallest emissions were recorded for 2nd gear operation in the range of (25-40) km/h, corresponding to (1500-2500) rpm for the engine, and around the speed of 70 km/h for 3rd gear operation, ie 3000 rpm for the engine.

• The CO_2 emission reduction tendency was also found for operation within the (2500-3000) rpm range, which corresponds to the (50-70) km/h speed range for the 3rd gear, (70-90) km/h for the 4th gear and (90-110) km/h for 5th gear operation. This aspect of reducing CO_2 emissions is explained by the fact that within the speed range (2500-3000) rpm, the economic zone is located, the engine having a lower consumption than for other speed ranges.

• When the engine runs above 3200 rpm, CO_2 emissions increase too as it exits the economic zone. The highest CO_2 emissions are encountered in the 5th gear drive, at speeds over 120 km/h, respectively above (3200-3500) rpm. When running in the 5th gear at speeds above 120 km/h, the CO_2 emissions are of (12.5-14) %, being (35-55)% higher than emissions from the a 4th or a 3rd gear run respectively.

- *with regard to the smoke index (opacity of the exhaust gas), according to figure 10*:

• For the full speed range of (25-100) km/h, for the run in the 2nd, 3rd and 4th gear at stabilized speeds, the smoke index does not exceed 0.22 m^{-1}, which represents a reduced value, 6 times lower than the index of smoke measured in acceleration (figure 7), according to the legislative provisions.

• When running in the 5th gear at stabilized speeds above 100 [km/h], the smoke index increases up to 7 times as compared to the less polluting regimes indicated previously, however it never exceeds the smoke index measured in acceleration (figure 7), according to the legislative provisions.

• In situation of good technical condition of the diesel engine, the engine having the appropriate thermal regime (coolant temperature of at least 80^{0}C), the smoke level is low if run at approximately constant speeds (stabilized regimes), the pollution level being just below the one recorded at idle and acceleration from minimum to maximum regimes, according to the legislative provisions [18].

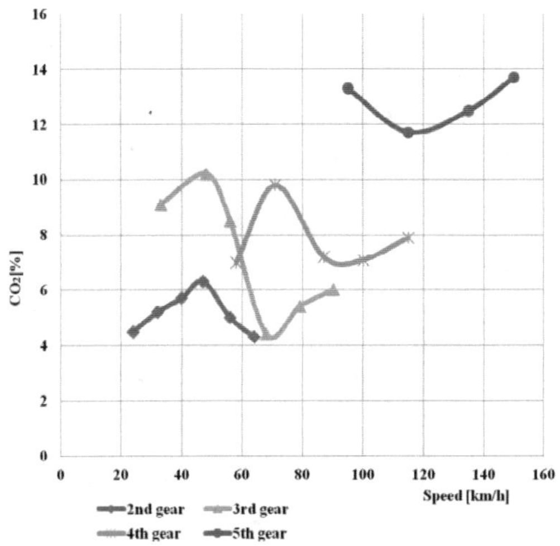

Figure 8. CO_2 emissions depend of speed and gear.

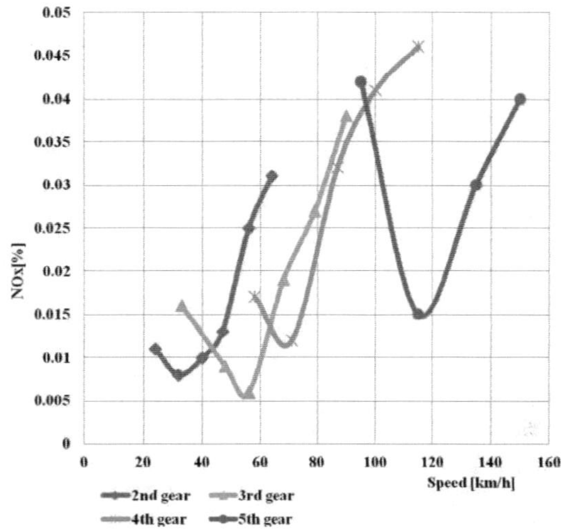

Figure 9. NO_x emissions depend of speed and gear.

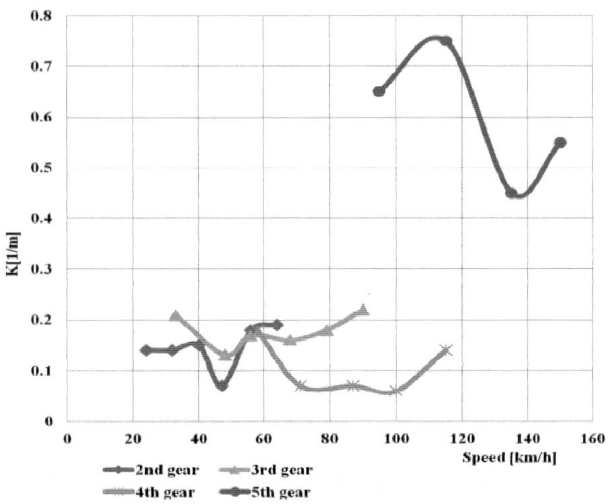

Figure 10. Smoke emissions depend of speed and gear.

- *with regard to NO_x emissions:*

• The smallest emissions of (0.005 - 0.015) % are achieved for 2nd, 3rd, and 4th gear operation, for engine operation within the (2000-3000) rpm range. Speed ranges that produce minimal pollution and which are recommended for operation are: (25-45) km/h in the 2nd gear, (45-65)km/h for the 3rd gear operation, and (60-75) km/h for the 4th gear, respectively.

• Running at approximately constant speeds, within this speed range of (25-75) km/h,which is specific to urban driving, with engine at the appropriate heat regime, lead to quite low NO_x emissions.

• This aspect of NO_x reduction is explained by the fact that, in the (2000-3000) rpm range, the economical operating area is situated, with the economic pole around the speed of 2500 rpm, the engine having a lower consumption than for other speed intervals. For the indicated speed range, even if the amount of O_2 is quite high, ranging from 7 to 15 % as shown in figure 12, however, due to lower gas temperature, average loads being considered (the exhaust gas temperature did not exceed 200^{0}C, as shown in Figure 11), the NO_x emissions are at minimum values.

• Running at steady speeds above 75 km/h in the 4th gear and above 120 km/h respectively in the 5th gear, NO_x emissions may increase three times as compared to the regimes with low pollution indicated above. This is explained by the increased fuel consumption and flue gas temperature (exhaust gas temperature exceeded 220^{0}C, as shown in figure 11). The influence of temperature has a more important role in NO_x formation than oxygen does, because these emissions increase even if the amount of O_2 is reduced by about 50% for the 4th gear and (3-5) times for running in the 5th gear, as shown in figure 12.

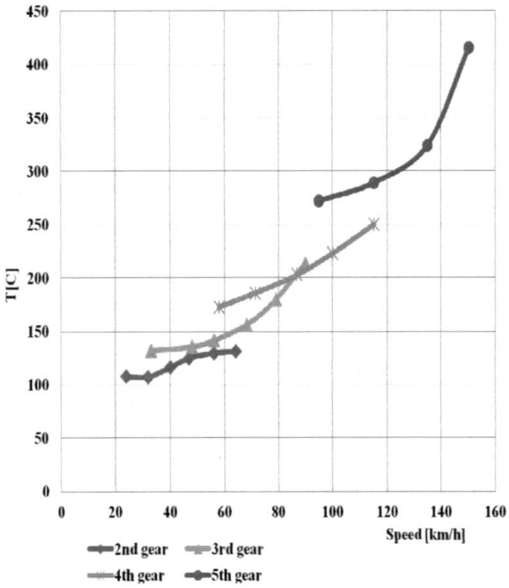

Figure 11. Exhaust gas temperature depends of speed and gear.

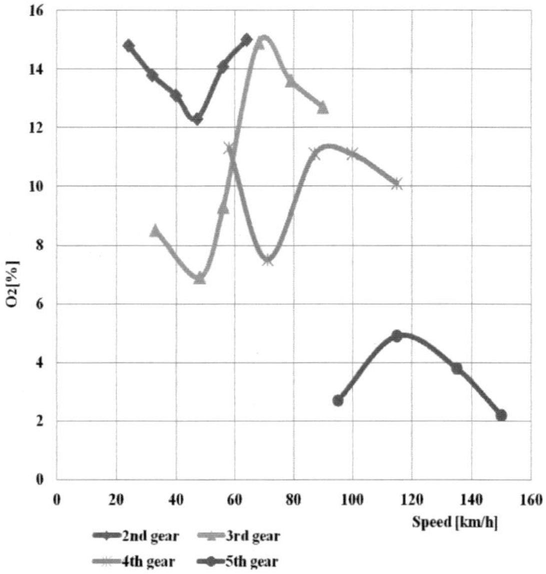

Figure 12.The O_2 exhaust depends of speed and gear

4. Conclusions

According to the experimental results, conclusions are as following:

• In order to ensure the lowest pollutant values, it is recommended to operate the engine in the range of (2000-3000) rpm, which is the economic pole in the speed range of 2500 rpm. Engine operation in the recommended range allows you to run at speeds of (25-75) km/h on the 2nd, 3rd and 4th gear, which is specific to urban operation.

• Speed ranges that produce minimal pollution and which are recommended for operation are: (25-45) km/h on 2nd gear of the gearbox, (45-65) km/h for running on the 3rd gear of the gearbox and (60-75) km/h for the 4th gear.

• When running at constant speeds above 75 km/h on the 4th gear and above 120 km/h respectively on the 5th gear, the NO_x emissions will increase up to 3 times as compared to the low pollution regimes indicated, and the smoke index increases up to 7 times.

• The highest emissions of CO_2, NO_x and smoke have been recorded for the 5th gear at speeds above (100-120) km/h, running regimes specific to extra-urban operation.

• In situation of good technical condition of the diesel engine, the engine having the coolant temperature of at least 80^0C, if run at approximately constant speeds (stabilized regimes) in urban mode, the pollution is reduced compared to the one achieved in dynamic driving.
• It is advisable to avoid sudden accelerating, running at approximately constant speeds in the (25-75) km/h speed range, which will ensure a minimum level of CO_2, NO_x and smoke pollution, below the values obtained for acceleration at idle regime when running between the minimum and maximum possible speeds.

References

[1] Directive (EU) 2016/2284 of the European Parliament and of the Council of 14 December 2016 on the reduction of national emissions of certain atmospheric pollutants, amending Directive 2003/35/EC and repealing Directive 2001/81/EC
[2] Oka S 2006 Energy Efficiency in Serbia: Research and Development Activity, *Thermal Science.* **2** pp 5-32
[3] Oka S 2006 Energy Efficiency in Serbia National Energy Efficiency Program: Strategy and Priorities for the Future, *Thermal Science*, **4** pp 7-16
[4] Faiz A 1996 *Air Pollution from Motor Vehicles. Standards and Technologies for Controlling Emissions*, The World Bank Washington, D.C
[5] European Automobile Manufacturers' Association, http://www.acea.be/.
[6] Roumanian Automobile Manufacturers' Association, http://www.acarom.ro/.
[7] Petrović V S 2008 Particulate Matters from Diesel Engine Exhaust Emission, *Thermal Science.* **2** pp 183-198
[8] Chana T L 2004 On-road remote sensing of petrol vehicle emissions measurement and emission factors estimation in Hong Kong. *Atmospheric Environment.* **38(14)** pp 2055–2066
[9] Xiao H 2017 An Experimental Study of the Combustion and Emission Performances of 2,5-dimethylfuran Diesel Blends on a Diesel Engine, *Thermal Science.* **21(1B)** pp 543-553
[10] Vellaiyan S and Amirthagadeswaran K S 2017 Multi-response Optimization of Diesel Engine Operating Parameters Running with Water-in-Diesel Emulsion Fuel, *Thermal Science.* **21(1B)** pp 427-439.
[11] Premkartikkumar S R 2014 Effectiveness of Oxygen Enriched Hydrogen-HHO Gas Addition on DI Diesel Engine Performance, Emission and Combustion Characteristics, *Thermal Science.* **18** pp 259-268
[12] Madhujit D 2015 An experimental study on combustion, performance and emission analysis of a single cylinder, 4-stroke DI-diesel engine using hydrogen in dual fuel mode of operation, *International Journal of Hydrogen Energy*. **27** pp 8586–8598
[13] Marko J 2017 Impacts of low temperature combustion and diesel post injection on the in-cylinder production of hydrogen in a lean-burn compression ignition engine, *International Journal of Hydrogen Energy*. **2** pp 1276–1286
[14] Cernat A 2017 Experimental aspects of hydrogen use at diesel engine by diesel gas method, *Thermal Science.* **22(00)**, pp 138-138
[15] Kotek M 2017 Comparison of PM production in gasoline and diesel engine exhaust gases, *Agronomy Research.* **15(S1)** pp 1041–1049
[16] http:// www.auto-data.net/ro.
[17] Burciu S M 2016 Laboratory works: First part–Car dynamics. Second part – Car diagnosis, Galati University Press, Galati, Romania.
[18] RNTR 7, Romanian Auto Register, http://www.rarom.ro

IOP Conf. Series: Materials Science and Engineering **595** (2019) 012036 doi:10.1088/1757-899X/595/1/012036

Prediction of spark ignition engine fuel consumption for different functional conditions using a theoretical model

V Vornicu, T Ulian, E Rakosi, G Manolache and S Talif

Department of Mechanical and Automotive Engineering, "Gheorghe Asachi" Technical University of Iasi, Iasi, Romania

E-mail: tsorinel@yahoo.com

Abstract. The theoretical model estimates the characteristic parameters of a car spark ignition engine by using complex functions. Analyzing the variation of the efficiencies and the fuel consumption at various operating regimes it is possible to optimize them according to the operating regime of the car. It is also studied a method for estimation of the specific fuel consumption of the spark ignition engine by considering engine operation in variable mode. The theoretical study presented by authors contributes actively to the optimization of the interaction between vehicle and propulsion system by identifying low fuel consumption modes.

1. Introduction

The study of the combustion process in internal combustion engines shows that the actual quantity of air, L_G, in kg/kg of fuel, differs from the minimum quantity of air, L_{minG}, in many operating regimes. The minimum quantity of air is also known as the theoretical (stoichiometric) quantity of air, required for the complete combustion. The ratio of these two parameters defines the air excess ratio [4, 8]

$$\lambda = \frac{L_G}{L_{\min G}} = \frac{L_M}{L_{\min M}}. \quad (1)$$

In formula (1), L_M and L_{minM} are the molar quantities of air to fuel, in kmol/kg of fuel.

2. Theoretical model

The theoretical model developed by authors is a complex one, proposing an analysis of the energy parameters of the engine. Preliminary variants of this model have been presented by authors in other articles [9-12]. Starting from the variations of the volumetric efficiency, η_v, and the indicated mean pressure, p_i, obtained in the model, the indicated specific fuel consumption and the brake specific fuel consumption functions were developed. This was possible by assuming a variation of the mechanical efficiency, η_m. Complexity of the problem also comes from the fact that, as well known, engine operation involves variation of λ in a very narrow range around the stoichiometric value. Hence, the quantity of fresh air drawn into the cylinders of a spark ignition engine, in kmol/kg of fuel, is given by

$$M_a = \lambda \cdot L_{minM} + \frac{1}{M_C}, \quad (2)$$

where the term M_c, in kg/kmol, represents the combustion molecular mass. In the case of spark ignition engines operating with gasoline, the recommended values of M_C are in the range 110...120

kg/kmol. Since gasoline is a complex mixture of hydrocarbons, it is recommended to consider that it consists entirely of a single and representative pure hydrocarbon, which is n-octane (C_8H_{18}). The molecular mass of this chemical compound is M_c = 114 kg/kmol [1, 2].

By considering that M_r, in kmol, is the quantity of residual exhaust gas retained inside the cylinders from the last combustion, it results that cylinders charge, comprising of the fresh air and the residual gas, is

$$M_i = M_a + M_r. \quad (3)$$

The parameter M_r is expressed function by the coefficient of residual gases, which is defined as

$$\gamma_r = \frac{M_r}{M_a}. \quad (4)$$

From equations (4) and (2) it results

$$M_r = \gamma_r \cdot M_a = \gamma_r \cdot \left(\lambda \cdot L_{minM} + \frac{1}{M_C} \right). \quad (5)$$

In these conditions, equation (3) becomes

$$M_i = M_a \cdot (1+\gamma_r) = \left(\lambda \cdot L_{minM} + \frac{1}{M_C} \right) \cdot (1+\gamma_r). \quad (6)$$

Variation of γ_r with the engine load, χ, is given by

$$\gamma_r(\chi) = a_{\gamma r} \cdot \chi + b_{\gamma r}, \quad (7)$$

where the coefficients $a_{\gamma r} = -0.3$ and $b_{\gamma r} = -0.35$ were determined by authors on the basis of data from the literature [2, 4].

By considering $L_{minM} \cong \frac{1}{28.96} \cdot L_{minG}$ for liquid petroleum fuels, from equations (6) and (7) it results

$$M_i(\chi) = \left(\lambda \cdot \frac{L_{minG}}{28.96} + \frac{1}{M_C} \right) \cdot \left[1 + \left(a_{\gamma r} \cdot \chi + b_{\gamma r} \right) \right]. \quad (8)$$

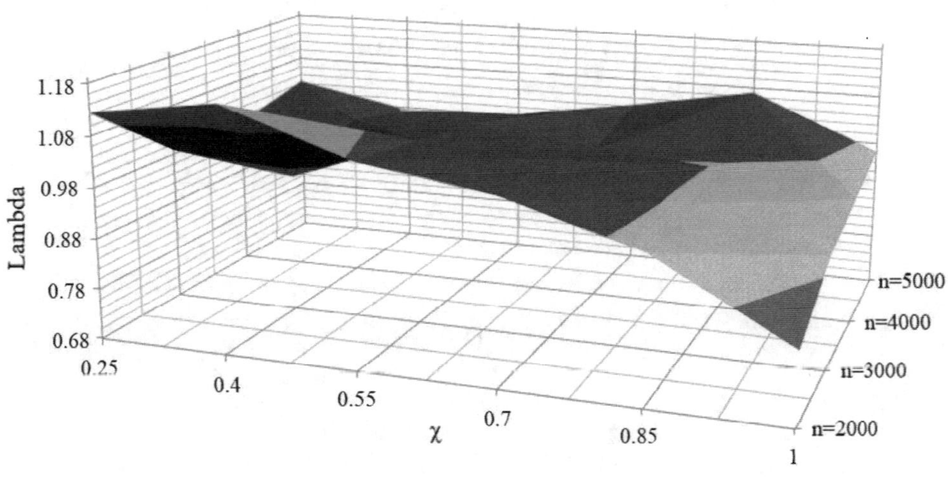

Figure 1. Variation of the air excess coefficient.

If there are also taken into account the mass fractions of the fuel components (C = 0.854; H_2 = 0.142; O_2 = 0.004) as well as $L_{minG} \cong 14.82$ kg of air/kg of fuel and $L_{minM} \cong 0.5118$ kmol, then the cylinder charge in the initial state is expressed as a function of engine load and speed, as follows:

$$M_i(\chi) \cong (0.5118 \cdot \lambda_{n,\chi} + 0.00877) \cdot \left[1 + (a_{\gamma r} \cdot \chi + b_{\gamma r})\right]. \tag{9}$$

In equation (9), the values of the air excess coefficient, $\lambda_{n,\chi}$, changes with the operating regime according to the matrix $\mathcal{L} = \left\| \lambda_{n,\chi} \right\|$, which was determined by processing experimental data corresponding to medium-capacity spark ignition engines and collected from literature [6, 7]. The values in this matrix correspond to the diagram $\lambda(n, \chi)$, shown in Figure 1.

On the other hand, the indicated efficiency, η_i, as known, is given by the ratio of indicated work, L_i (mechanical work generated by gas expansion in cylinders), to fuel's input energy, Q_c:

$$\eta_i = \frac{L_i}{Q_c}. \tag{10}$$

It should be mentioned that cycle losses can be estimated on the basis of the indicated efficiency, so the capacity of the actual cycle to convert heat into mechanical work can be predicted. This can characterize the economy of the cycle.

According to the literature [3-5], the indicated efficiency can be expressed as

$$\eta_i = 8.315 \cdot \frac{M_i}{H_i} \cdot \frac{p_i}{\eta_v} \cdot \frac{T_0}{p_0}, \tag{11}$$

where H_i is the lower calorific value of the fuel, while p_0 and T_0 are the initial pressure and the initial temperature of the environment. In the developed model, the initial environmental conditions are taken into consideration by introducing the parameter

$$\phi_0(p_0, T_0) = \frac{p_0}{T_0}. \tag{12}$$

Variation of the mean indicated pressure, described by matrix $\Pi_i = \left\| p_{i_{n,\chi}} \right\|$,

$$\Pi_i = \begin{pmatrix} 2.52 & 3.08 & 3.10 & 2.76 \\ 3.84 & 4.64 & 5.02 & 6.79 \\ 4.89 & 5.89 & 6.01 & 6.88 \\ 5.96 & 6.85 & 7.24 & 7.88 \\ 6.96 & 7.72 & 8.18 & 8.44 \\ 7.30 & 8.41 & 8.90 & 9.00 \\ 7.84 & 8.89 & 9.41 & 9.81 \\ 8.37 & 9.52 & 9.92 & 10.33 \\ 9.85 & 10.53 & 10.87 & 10.90 \\ 10.54 & 11.02 & 11.11 & 11.13 \end{pmatrix}, \tag{13}$$

was expressed as

$$p_i(\chi)=\left[\varphi_t\cdot\phi_0\cdot\left(1-\varepsilon^{-1}\right)\cdot\frac{1}{\varepsilon-1}\cdot\left(1-\varphi_{pu}+\gamma_r\right)\cdot\begin{cases}-278\cdot\chi^2+473\cdot\chi+225\\-167\cdot\chi^2+284\cdot\chi+263\\-28\cdot\chi^2+48\cdot\chi+311\\-56\cdot\chi^2+95\cdot\chi+299\end{cases}\cdot\right.$$

$$\cdot\begin{cases}0.69\cdot\chi+0.13\\0.656\cdot\chi+0.212\\0.594\cdot\chi+0.277\\0.475\cdot\chi+0.4\end{cases}\cdot\begin{cases}\varepsilon^{1.3775}\\\varepsilon^{1.3883}\\\varepsilon^{1.3937}\\\varepsilon^{1.397}\end{cases}\quad for:\begin{cases}n=5\cdot10^3\ rpm\\n=4\cdot10^3\ rpm\\n=3\cdot10^3\ rpm\\n=2\cdot10^3\ rpm\end{cases}.$$

$$\cdot\left\{\frac{\mu_t\cdot\tau_z}{e_e-1}\cdot\left(1-\frac{1}{\varepsilon^{e_e}-1}\right)\cdot\left[\begin{cases}-278\cdot\chi^2+473\cdot\chi+225\\-167\cdot\chi^2+284\cdot\chi+263\\-28\cdot\chi^2+48\cdot\chi+311\\-56\cdot\chi^2+95\cdot\chi+299\end{cases}\cdot\begin{cases}\varepsilon^{0.3775}\\\varepsilon^{0.3883}\\\varepsilon^{0.3937}\\\varepsilon^{0.397}\end{cases}\quad for:\begin{cases}n=5\cdot10^3\ rpm\\n=4\cdot10^3\ rpm\\n=3\cdot10^3\ rpm\\n=2\cdot10^3\ rpm\end{cases}\right]^{-1}-\right.$$

$$\left.-\frac{1}{e_c-1}\cdot\frac{1}{\varepsilon^{e_c}-1}\right\}\quad, \tag{14}$$

while the volumetric efficiency was calculated as

$$\eta_v=a_n\cdot\chi+b_n=\begin{cases}0.69\cdot\chi+0.13\\0.656\cdot\chi+0.212\\0.594\cdot\chi+0.277\\0.475\cdot\chi+0.4\end{cases}\text{for engine speed: }\begin{cases}n=5\cdot10^3\ rpm\\n=4\cdot10^3\ rpm\\n=3\cdot10^3\ rpm\\n=2\cdot10^3\ rpm\end{cases}. \tag{15}$$

Notations used in equation (14) are:

- ε - compression ratio;
- e_c, e_e - polytrophic compression exponent, polytrophic expansion exponent;
- T_z - combustion temperature;
- φ_{pu} - overfilling degree;
- φ_t - rounding factor of the main indicated diagram.

Taking into account the mean indicated pressure, the volumetric efficiency and the parameter describing initial environmental conditions as defined by formulas (14), (15) and (12), equation (11) can be written as

$$\eta_i(\chi)=\frac{8.315}{H_i}\cdot\frac{1}{\phi_0}\cdot\frac{\left\|p_{i_{n,\chi}}\right\|}{a_n\cdot\chi+b_n}\cdot\left(0.5118\cdot\left\|\lambda_{n,\chi}\right\|+0.00877\right)\cdot\left[1+\left(a_{\gamma r}\cdot\chi+b_{\gamma r}\right)\right]. \tag{16}$$

Hence, the dependence of the indicated efficiency on the engine operating mode can be analyzed.

By considering formula (16), the indicated specific fuel consumption can be computed with formula

$$c_i(\chi)=\frac{3600}{8.315\cdot\frac{1}{\phi_0}\cdot\frac{\left\|p_{i_{n,\chi}}\right\|}{a_n\cdot\chi+b_n}\cdot\left(0.5118\cdot\left\|\lambda_{n,\chi}\right\|+0.00877\right)\cdot\left[1+\left(a_{\gamma r}\cdot\chi+b_{\gamma r}\right)\right]}. \tag{17}$$

Similar to the case of the air excess coefficient, the mechanical efficiency has associated the matrix

IOP Conf. Series: Materials Science and Engineering **595** (2019) 012036 doi:10.1088/1757-899X/595/1/012036

$$\eta_m = \begin{pmatrix} 0.500 & 0.732 & 0.796 & 0.841 \\ 0.735 & 0.822 & 0.874 & 0.936 \\ 0.792 & 0.860 & 0.895 & 0.936 \\ 0.829 & 0.880 & 0.913 & 0.944 \\ 0.854 & 0.893 & 0.923 & 0.948 \\ 0.861 & 0.902 & 0.929 & 0.951 \\ 0.870 & 0.907 & 0.933 & 0.955 \\ 0.878 & 0.913 & 0.936 & 0.958 \\ 0.897 & 0.922 & 0.942 & 0.960 \\ 0.903 & 0.925 & 0.943 & 0.961 \end{pmatrix}. \quad (18)$$

The brake specific fuel consumption can be estimated using the formula

$$c_e = \frac{c_i}{\eta_m}. \quad (19)$$

From formulas (17) and (19) it results

$$c_e(\chi) = \frac{3600}{8.315 \cdot \frac{1}{\phi_0} \cdot \frac{\left\| p_{i_{n,\chi}} \right\|}{a_n \cdot \chi + b_n} \cdot \left(0.5118 \cdot \left\| \lambda_{n,\chi} \right\| + 0.00877\right) \cdot \left[1 + \left(a_{\gamma r} \cdot \chi + b_{\gamma r}\right)\right]} \cdot \frac{1}{\left\| \eta_{m_{n,\chi}} \right\|} \quad (20)$$

3. Results

The model developed by authors contributes to the study of the indicated power of a spark ignition engine. The results presented below are obtained by simulation with this model and correspond to an existing engine with the following characteristic parameters:

- $\varepsilon = 9.5$ (compression ratio);
- $i = 4$ (number of cylinders);
- $S = 80.5$ mm (stroke);
- $D = 79.5$ mm (bore);
- $V_s = 0.3995$ dm^3 (displacement);
- $n = 5500$ rpm (nominal engine speed);
- $\tau = 4$ (engine's strokes).

The variation of the indicated specific fuel consumption with the engine speed at different loads is graphically represented in Figure 2.

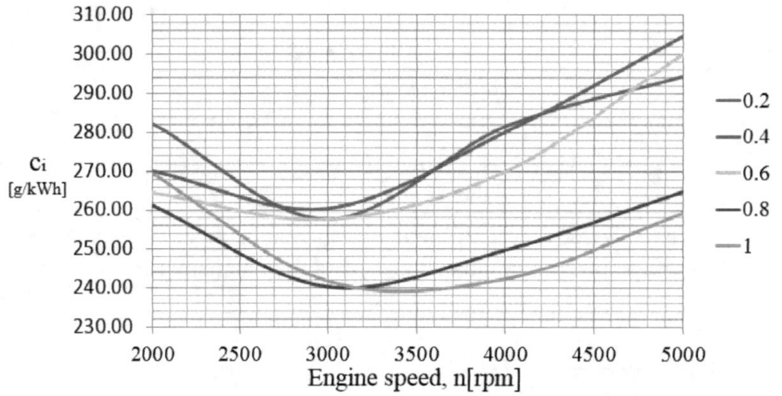

Figure 2. Variation of the indicated specific fuel consumption with the engine speed.

It is also possible to use the developed model in order to study the variation of c_i with the engine load for various preset speeds, as shown in figure 3.

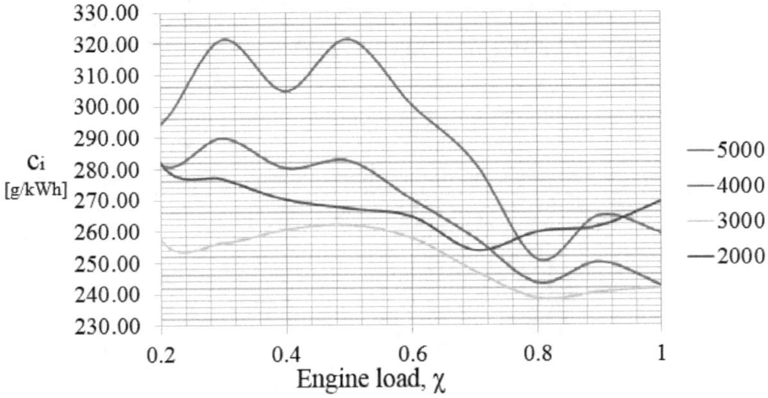

Figure 3. Variation of the indicated specific fuel consumption with the engine load.

Besides, by using the values of the mechanical efficiency, η_m, it is possible to predict the evolution of the actual brake specific fuel consumption with the engine speed and with engine load, as shown in figure 4 and figure 5, respectively.

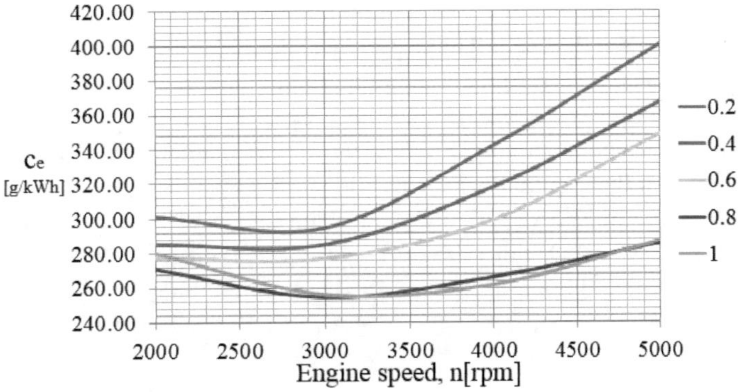

Figure 4. Variation of the brake specific fuel consumption with the engine speed.

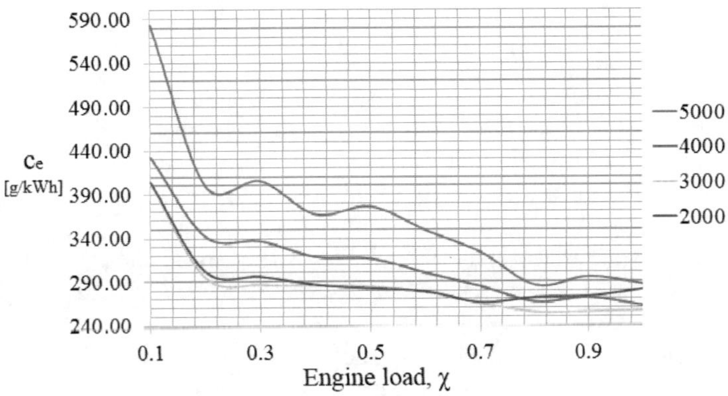

Figure 5. Variation of the brake specific fuel consumption with the engine load.

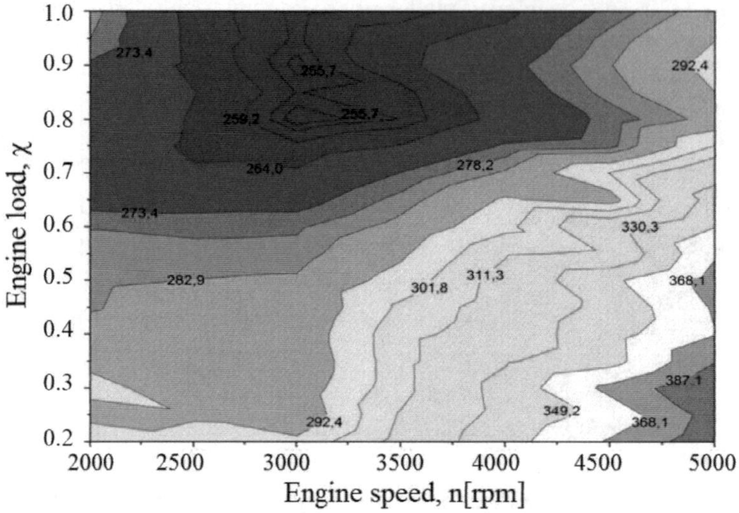

Figure 6. Map of the brake specific fuel consumption, in g/kWh.

4. Conclusions

By using the proposed theoretical model, which represents a further development of a model previously developed by authors, the analytical expressions of some characteristic parameters of the engine, i.e. p_i, c_i, η_i and c_e, are obtained. These parameters, which affect the operation of the engine, are defined within the model as functions of the engine load, χ.

The model also offers the possibility to create the brake specific fuel consumption map, as shown in figure 6, which highlights the economical operating modes of the engine. Besides, it is possible to study the variation of c_i with the engine load for various preset speeds.

It can be considered that the theoretical study performed by authors on the basis of the developed model makes an important contribution to the optimization of the interaction between car and engine, by identifying low fuel consumption regimes. The theoretical model is about to be validated with laboratory experimental data.

References

[1] Heisler H 1995 *Advanced Engine Technology* (Oxford: Butterworth-Heinemann)
[2] Heywood B J 1988 *Internal Combustion Engine Fundamentals* (New York: McGraw-Hill)
[3] Heisler H 2002 *Advanced Vehicle Technology – 2nd edition* (Oxford: Butterworth-Heinemann)
[4] Grunwald B 1980 *Theory, calculation and construction of motor vehicles for road vehicles* (Bucharest: Ed. Didactică şi Pedagogică)
[5] Grotstollen H 2011 Optimal design of motor and gear for drives with high acceleration by consideration of torque-speed and torque-acceleration product *IEEE Appl.* **47 (1)** 144-52
[6] Rakosi E, Golgotiu E, Roşca R and Manolache G 2004 Analysis of Procedures Used in Improving the Performance of The Spark Ignition Engines Iasi *Bull. Inst. Pol. Iași* **X (LIV) 1-2** 135-44
[7] Rakosi E, Roşca R and D. Mojilian 1999 Ways of Improving the Performances of the Spark Ignition Engines by Direct Injection *Proc. of the 2nd Int. Coll. Fuels,* Esslingen, Germany 413-22
[8] Roşca R, Rakosi E and Manolache G 2005 *Automotive technology elements* (Iași: Ed. Politehnium)
[9] Vornicu V, Ulian T, Rakoși E and Talif S 2018 Contributions on modeling the intake and compression processes of the spark ignition engine *Procedia Manufacturing* **22** 683-90
[10] Vornicu V, Ulian T, Rakoși E, Manolache G and Gaiginschi L 2017 Theoretical model for

determination of the spark ignition engine thermo-gasodynamic parameters on various functional conditions *IOP Conf. Ser.: Mater. Sci. Eng.* **252 (7)** 012073

[11] Vornicu V, Ulian T, Rakoşi E and Talif S 2017 Theoretical calculation model of the propulsion engine admission parameters for different functional conditions *MATEC Web of Conf.* **112** 10007

[12] Vornicu V, Ulian T, Rakoşi E, Manolache G, Talif S and Gaiginschi L 2018 Analysis of the spark ignition engine performances by using a theoretical model *IOP Conf. Ser.: Mater. Sci. Eng.* **444 (7)** 072014

Reducing CO_2 emissions from tractor engines by using the diesel fuel and rapeseed oil blend

S M Burciu[1], M Lupchian[1] and M Neacşu[2]

[1]Department of Thermal Systems and Environmental Engineering, “Dunarea de Jos” University of Galati, Galati, Romania
[2]Department of Materials Science and Engineering, “Dunarea de Jos” University of Galati, Galati, Romania

E-mail: mariana.lupchian@ugal.ro

Abstract. The paper presents the results of some experimental research that led to the conclusion that diesel engines that fit agricultural tractors can reduce CO_2 emissions by using a fuel oil mixture made up of 50% diesel and 50% refined rapeseed oil. Following the experimental determinations made and presented in the paper, both in load and idle operation, using a fuel mixture consisting of diesel and rapeseed oil, a reduction of up to 25% of CO_2 emissions was obtained. Such a reduction in CO_2 emissions is important because this gas belongs to the greenhouse effect category, and at the moment, a large number of U650 tractors are still being used in agriculture.

1. Introduction

Since the 1970s, the interest in the use of vegetable oils as alternative fuels for diesel engines has been renewed [1]. The rapid increase in the number of diesel vehicles has led to an increase in the "greenhouse effect" due to the pollution of the surrounding air with CO_2, CO, NO_x, HC, PM. Plant oil based fuels have been used as diesel fuel extension agents, which could mitigate environmental pollution, contributing to reducing CO_2 emissions in a closed cycle [2].

CO_2 emissions from transport will continue to increase as the world's number of vehicles is expected to double by 2030 [3]. Under the Kyoto Protocol, developed countries have pledged and succeeded in reducing greenhouse gas emissions over the period 2008-2012 by about 5% compared to 1990 levels [4]. To achieve this objective, it was necessary to further reduce CO_2 emissions from cars [5], therefore very demanding targets were set around the world. Emissions from diesel engines are related to diesel quality, engine operating regimes and also to its design.

Legislation on emissions, the need to reduce CO_2 emissions and fuel consumption will lead to more and more complex engines compared to those in the past. New passive and active technologies have led to the reduction of pollutant emissions (CO (carbon monoxide), HC (hydrocarbons), NO_x (nitrogen oxides) and particles emissions) [6]. CO_2 emissions are not classified as harmful pollutant emissions but they favour the "greenhouse effect". The CO_2 concentration of the exhaust gases depends on the operating mode of the engine and is directly proportional to the fuel consumption caused by the engine load. When running the engine with pure rapeseed oil, the CO concentration is higher, which characterizes the incomplete combustion process of the rapeseed oil-air fuel mixture [7].

The main purpose of the widespread use of vegetable oils fuels should be seen as a measure taken to resolve environmental problems, create new jobs, and suspend the rise in prices for mineral fuels [8].

Significant progress has been made over the last 20 years within the European Union in the area of emissions and air quality. Directive 2001/81 / EC has made a key contribution to this progress by setting maximum emission limits from 2010 [9].

Vegetable oil produced from oilseeds, crude or refined but not chemically modified, may be used as biofuel in cases where its use is compatible with the type of engine involved and with the corresponding emissions requirements [10].

At the University of Hohenheim, rapeseed fuel was used in a direct injection engine between 1992 and 1995. The performance of the test engine was unsatisfactory for a low engine load that can be encountered frequently to agricultural tractors. At a higher load, satisfactory results were obtained [11]. Cultivation of rape has expanded quite a lot in Romania too; the surface on which the oilseed plant has come to be grown is at this moment the highest recorded in our country since 2010, reaching 460 thousand hectares. In Romania, more and more farmers include rapeseed in their mix of crops. The rapeseed is growing steadily on the national market, and the upward trend is still maintained, the harvested area exceeding 590,000 hectares [13].

2. Equipment used for experimental studies

All the experimental measurements presented in this study were performed on a 4 cylinder, naturally aspirated, watercooled, D103, four-stroke, diesel engine. U 650 type tractors are equipped with this diesel engine and it has a power output of 65 HP at 1800 rpm. Although they are from an older generation, U 650 tractors are still widely used in agriculture in Romania.

The engine laboratory equipment (figure 1 and figure 2) was equipped with a Froude hydraulic brake. The coupling between hydraulic brake and diesel engine is directly, without transmission gear.

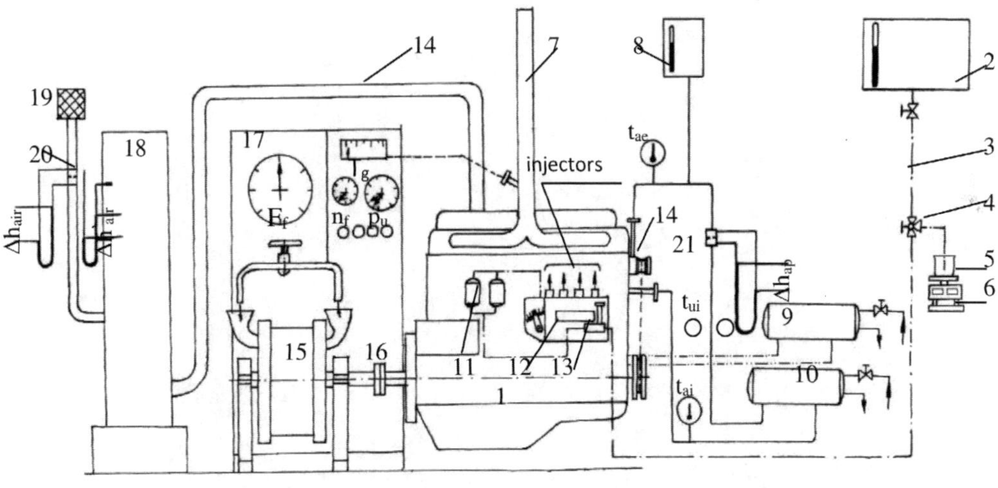

Figure 1. The scheme of experimental laboratory equipment.

The elements specified in Figure 1 are: 1 – Diesel engine D 103; 2 - fuel tank; 3 - fuel supply pipe; 4 - three-way valve; 5 - Calibrated fuel consumption measurement; 6 - electronic weighing; 7 - exhaust gases outlet pipe; 8 - expansion tank for coolant; 9 - oil cooler; 10 - coolant cooler; 11 - fuel filter battery; 12 - injection pump with speed regulator; 13 - supply pump; 14 - cooling liquid pump; 15 - hydraulic brake; 16 - coupling; 17 - signaling and warning panel; 18 - suction air tank; 19 - air filter; 20 - diaphragm used for measuring air flow; 21 - Diaphragm used for measuring the flow rate of the coolant.

The fuel mix that was used consists of 50% diesel with cetane number (CC) 51 and 50% refined rapeseed oil (straight oil) commercially available. In Table 1 are presented the main physical-chemical properties of rapeseed oil and diesel fuel [2, 9].

Table 1. Physical and chemical properties of refined rapeseed oil and diesel fuel.

Parameter	Rapeseed oil	Diesel fuel
Kinematic viscosity, 20^0 C [mm^2/s]	60-72	4-6
Lower Calorific value -LCV [MJ/kg]	37.4-38	41.5-42
Density, 20 °C [kg/dm^3]	0.91-0.92	0.82-0.84
Cetane number	40	51
Flash point, [°C]	270-321	65-80
Freezing point [°C]	-18 … 0	-12… 0

Fuel viscosity directly influences the spraying of the fuel into the combustion chamber, the efficiency of blending and vaporising the fuel together with its calorific value, determining the maximum power the engine can develop at a given injection pump rate [14].

In the present case, no modifications were made to the diesel fuel injection system and no equipment was installed to correct the viscosity of the fuel mixture. The viscosity of the fuel mixture (diesel and refined rapeseed oil at 50%) measured with the Hoppler viscometer was 18.4 mm^2/s at 20^0C, about 3.8 times higher than diesel fuel and 3.5 or less than refined rapeseed oil. The values measured with the Hoppler viscometer at 20^0C were for diesel fuel of 4.8 mm^2/s respectively for the refined rapeseed oil of 64.5 mm^2/s.

Figure 2. Laboratory equipment with diesel engine D103.

Using the Multilyzer STx gas analyzer, the CO_2 emissions (%vol.), CO (%vol.) and the residual oxygen content O_2 (vol%) of the exhaust gases were measured. The measurement of the temperature of the combustion gases was carried out on the exhaust pipe at a distance of about 1.5 m from the engine exhaust manifold, a distance similar to the displacement of the exhaust manifold of the engine the U650 tractor is equipped with.

3. Experimental determinations and results

A significant reduction in the effective engine power was noticed at loads greater than 0.75. According to the results presented in figure 3, the reduction is between (6-20)%, the higher values being at the full load and close to the nominal speed, and the lower values at lower loads and revolutions ranging between (1200-1700) rpm. The reduction in power is mainly due to a drop in the calorific value of the fuel mix by about (8-10)%, but this can be accepted under conditions where a reduction in CO_2 emissions is found using the indicated fuel mix, taking into account that for most of the operating time it is not exceeded 0.8 of the full load.

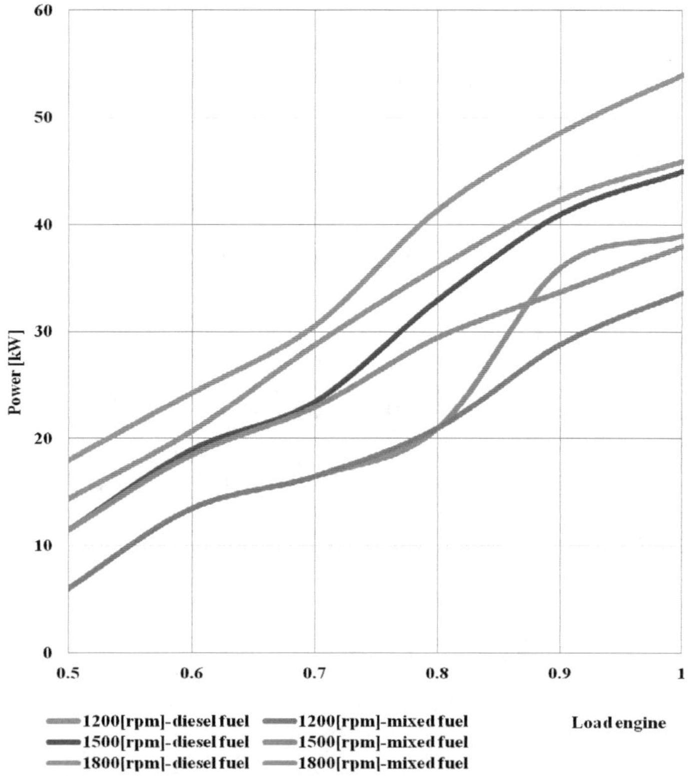

Figure 3. Effective Power with diesel fuel compared with mixed fuel.

For the idle regimes, there is a clear reduction of CO_2 emissions, which is a positive aspect, especially as it will also be confirmed in the case of medium and high load operation. CO_2 emissions are reduced by 12-24% for idling, with lower cuts at lower speeds, according to the variations shown in Figure 4.

At the same time, CO emissions increased 4-23% at revolutions above 1200 rpm, according to the variations in Figure 5, the larger increase being around the nominal speed of 1800 rpm. This is explained by the reduced combustion rate of the fuel mixture relative to the engine speed and a shift of the combustion on the expansion, which leads to the formation of a larger amount of CO emissions.

If, in the case of the use of diesel fuel, CO emissions account for (1-2)% of CO_2 emissions, in the case of 50% diesel fuel and 50% rape oil, CO emissions represent (1.5-2.5)% of the CO_2 emissions.

In the case of engine idling, there is a reduction of up to 24% of CO_2 emissions but also an increase of up to 23% of CO emissions. However, taking into account that CO emissions account for only a maximum of 2.5% of CO_2 emissions, we can assume that the use of the specified fuel mixture will have a positive effect.

For load regimes, at least for loads ranging from 0.6 to 0.8, the CO_2 reduction is noteworthy and it ranges from 5 to 25%, with higher values at low speeds and lower values of (4-8) % at loads above 0.8, as shown in Figure 6.

IOP Conf. Series: Materials Science and Engineering **595** (2019) 012037 doi:10.1088/1757-899X/595/1/012037

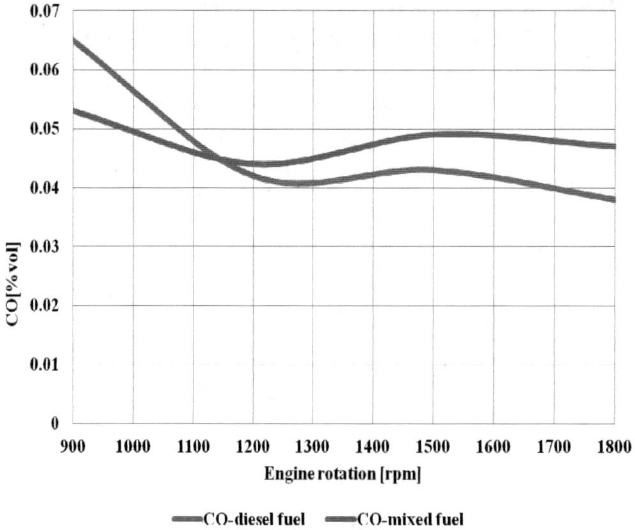

Figure 4. Emissions of CO_2 for idling regimes.

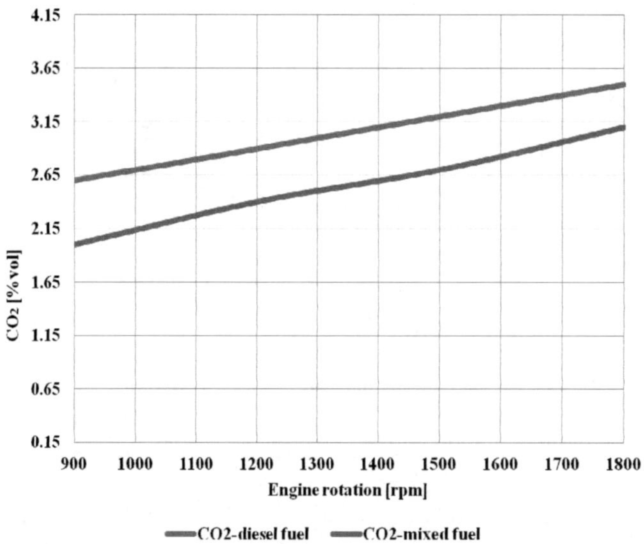

Figure 5. Emissions of CO for idling regimes.

At the same time, the emission of CO in the exhaust gases, according to Figure 7, increases on average by (30-40)% at partial loads and has an increase of less than 8-10% in the full load when using the diesel fuel with rapeseed oil blend compared to the classic diesel fuel.

Reduction of CO_2 emissions occurs even when the exhaust gas temperature is slightly increased by up to 8-10% (especially at lower loads) as shown in Figure 8, because the combustion moves to the releasing cycle (mixture used have a diminished cetane number as against diesel fuel) and that increase the CO emissions. Certainly, a reduction in the temperature of the gas could be achieved in the case of increased injection rate when using the diesel fuel and rapeseed oil blend and possibly a decrease of CO emissions, but these will be verified in experimental determinations which will follow.

In case of load engine running, there is a reduction of up to 25% of CO_2 emissions but also an increase of up to 40% of CO emissions; however, taking into account that CO emissions represent only (1.5- 2.5)% of CO_2 emissions, we can assume that when using the specified fuel mixture, the effect is still a positive one, especially considering that most of the time the engine will work at loads up to 0.8, in which case we have set that CO_2 emissions drop by as much as 25%.

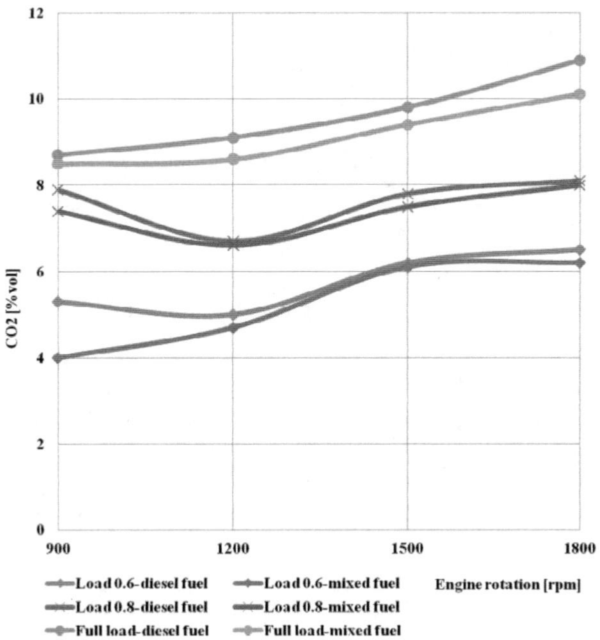

Figure 6. Emissions of CO_2 depend of load and rotation.

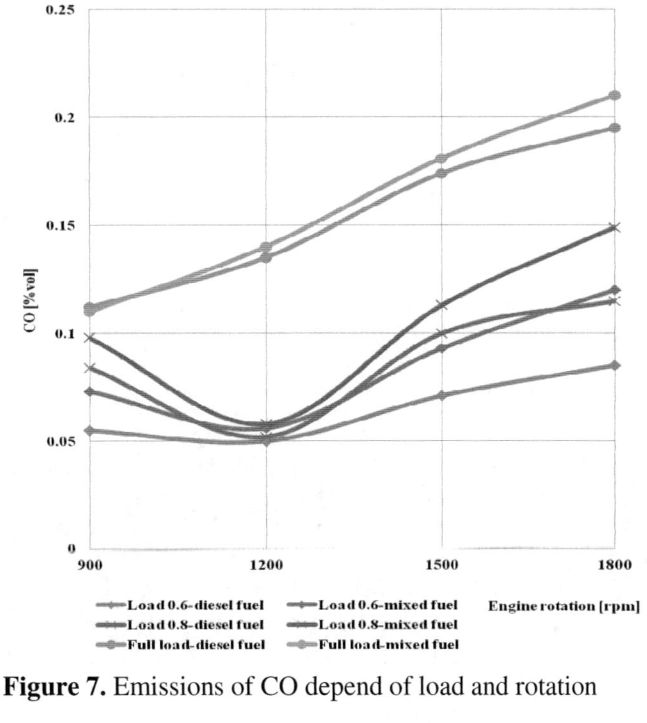

Figure 7. Emissions of CO depend of load and rotation

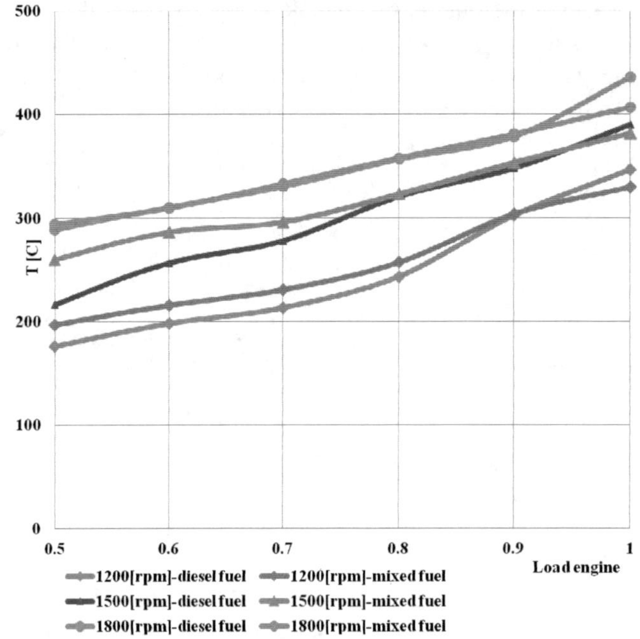

Figure 8. Exhaust gas temperature depends of load and rotation.

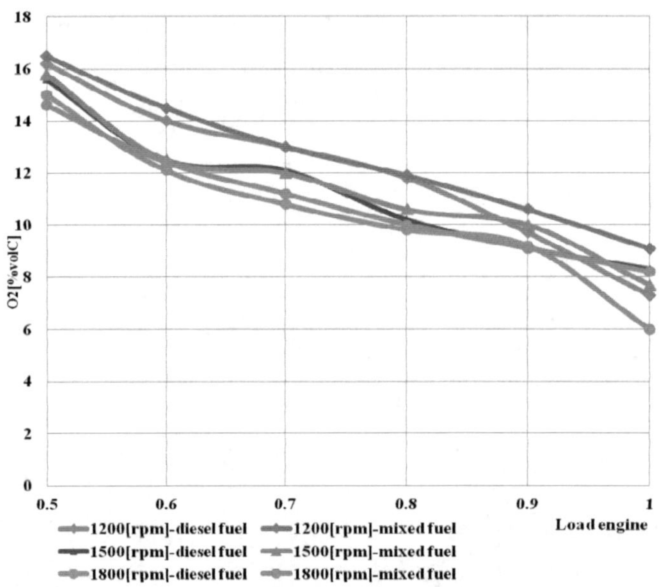

Figure 9. Oxygen in the exhaust gases depends of load and rotation.

In principle, the use of the diesel fuel with rapeseed oil blend increases the amount of O_2 in the exhaust gases due to the presence of the oxygenated compounds in the refined rapeseed oil used in the fuel mixture in a ratio of 50%, the amount of O_2 in the combustion gases being anyway higher at the partial loads of 0.5-0.8 irrespective of the type of fuel, as shown in figure 9.

The experimental results, carried out on the laboratory stand (within the Department of Thermal Systems and Automotive Engineering at the Faculty of Engineering, Galati), with the D103 diesel engine used to equip U650 tractors, led to the following results regarding the CO_2 and CO emissions from the combustion gases in the situation of using the fuel mixture made up of diesel and refined rapeseed oil in a ratio of 50%, compared to the normal operation with diesel fuel with a cetane number of 51:

• For idling, there was a 12-24% reduction in CO_2 emissions, with lower cuts at lower engine speeds.
• CO emissions increase by 4-23% at revolutions above 1200 rpm, with higher rises around the nominal speed of 1800 rpm.
• For load operation, at 0.6-0.8 loads most frequently encountered in operation, the CO_2 emissions reduction is in the range of 5-25% with higher values at low speeds.
• For workloads above 0.8, the reduction of CO_2 emissions is lower, being only 4-8%, with the lowest values at full load.
• CO emissions from combustion gases increase on average by 30-40% at partial loads of 0.6-0.8 and have a smaller increase of only 8-10% on the full load when using the fuel mixture compared to the classic fuel (diesel). As we can see, CO emissions represent only 1.5-2.5% of CO_2 (which is a greenhouse gas) emissions.

4. Conclusion

Following the experimental determinations carried out and presented in the paper, both in load and idle modes, using diesel fuel with rapeseed oil blend (diesel and refined rapeseed oil in a ratio of 50%), a decrease of up to 25% of CO_2 emissions was obtained, higher values being obtained at 0.6-0.7 loads of the engine, and lower values of up to 8% at loads higher than 0.8. Such a reduction in CO_2 emissions is important precisely because at the moment there is still a large number of U650 tractors used in Romanian agriculture.

In the determinations made, it was found that the CO emissions in the combustion gases increase on average by 30-40% at the partial loads of 0.6-0.8 of the engine and they have an increase of only 8-10% at the load full when using the fuel mixture compared to classic fuel (diesel). Taking into account that CO emissions represent only 1.5-2.5% of CO_2 emissions, we can assume that if the diesel fuel with rapeseed oil blend is used, the effect is still a positive one, especially considering that in most operating times, the engine will work at loads up to 0.8, in which case we have established that CO_2 emissions decrease by as much as 25%.

Given that CO_2 is considered a greenhouse gas, a reduction of up to 25% CO_2 emissions to partial load operation when using the 50% fuel oil and refined rapeseed oil blends can be considered an important one.

Taking into account the price of refined rapeseed oil, which is about 0.9 Euro / litre compared to diesel fuel of 1.24 Euro / liter, the authors recommend the use of this fuel mixture made up of 50% diesel with the cetane number 51 (CC 51) and 50% refined rapeseed oil, as it has been found to have a reduced decrease of up to 10% of power in the payload of 0.6-0.8 most commonly encountered in operation, but at the same time a reduction of up to 25% of CO_2 emissions.

References

[1] McDonnell K, Ward S, Leahy J J and McNulty P. 1999 Properties of rapeseed oil for use *as a diesel fuel extender*. JAOCS. **7** pp 539–43

[2] Labeckas G and Slavinskas S 2006 *Performance and exhaust emissions of direct-injection diesel engine operating on rapeseed oil and its blends with diesel fuel*. VGTU. **20(5)** 186-94

[3] Labeckas G, Slavinskas S and Pauliukas A 2007 *The effect of rapeseed oil blending with ethanol on engine performance and exhaust emissions* Journal of Kones Powertrain and Transport. **14(3)** pp 331-38

[4] Directive 2005/55 / EC of the European Parliament and of the Council of 28 September

[5] Azwan B S 2013 *Optical Analysis of Characteristics of Rape-seed Oil Diesel Spray Applied to Direct Injection Diesel Engine* (University of Tokushima, Graduate School of Advanced Technology and Science)

[6] Engeljehringer K 2013 *The future of automotive testing of emission measurement* (Rev:

Ingineria Automobilului SIAR. **7(2)** 11-30

[7] Lacusta I, Novorojdin D and Besleaga I 2016 *Practical use of alternative fuels in motor vehicles and environmental impact*. Rev: Ingineria Automobilului SIAR **41** 15-7

[8] Labeckas G and Slavinskas S 2005 *Performance and exhaust emissions of direct-injection Diesel engine operating on rapeseed oil and its blends with Diesel fuel*. VGTU **20(5)** 186-94

[9] Cordoş N and Mariaşiu F 2011 Hydrocarbon emissions from biofuel based on rapeseed oil compared to diesel fuel in diesel engine D-2402 *Research Journal of Agricultural Science* **43**

[10] *** Directive 2003/30 / EC of the European Parliament and of the Council of 8 May 2003 on the promotion of the use of biofuels and other renewable fuels for transport

[11] Ammerer J and Rathbauer M W 2003 *Rapeseed oil as fuel for farm tractors* (IEA Bioenergy)

[12] Burciu M 2016 Laboratory guide: Diagnosing and car dynamics. Galati University Press

[13] http://www.economica.net

[14] Beşleagă I G, Lagusta I and Popuşoi V 2009 *Study of the physico-chemical properties of fuels for compression-ignition engines* (Moldova vol 45) no 1 (ISSN 1857-0003) pp 311-15.

Reducing NO_x emissions from tractor engines by using the diesel fuel and rapeseed oil blend

G Coman, S M Burciu and M Lupchian

Department of Thermal Systems and Environmental Engineering, "Dunarea de Jos" University of Galati, Romania

E-mail: mugurel.burciu@ugal.ro

Abstract. The paper presents the influence of the use of rapeseed oil and diesel mixture on diesel engine performance and pollutant emissions. The main focus of the study is to identify the possibility of using rapeseed oil mixed with diesel at tractor engines. The fuel mix used consists of 50% diesel with cetane number (CC) 51 and 50% refined rapeseed oil. The experimental results have led to the conclusion that, in the case of tractor engines, NO and NO_x emissions can be substantially reduced compared to the situation when diesel fuel is used. Following the experimental determinations carried out and presented in the paper, both in load and idle operation, by using a fuel mixture consisting of diesel and rapeseed oil, a NO_x and NO_x emission reduction of up to 25% was achieved. Such a reduction in NO and NO_x emissions is important to emphasize precisely because of the fact that at the moment there is still a large number of U650 tractors in agriculture.

1. Introduction

The purity and quality of air is vitally important to the health of the population. Climate change is a global challenge which demands collective action and international cooperation. The main legislative instrument to achieve the 2030 objectives of the Clean Air Programme is Directive 2016/2284/EU on the reduction of national emissions of certain atmospheric pollutants which entered into force on 31 December 2016 [1]. The usage of biofuels is increasing yearly due to their lower environmental impacts compared to fossil fuels. Besides the minimal impact on the environment, the interest in using biofuels has also increased due to the high price of diesel [2]. Today, there are some 6.5 million trucks in circulation throughout the EU and the average age of Europe's trucks is 12 years [3]. An important sector in which diesel engines are used is agriculture. In recent years, the development of agriculture in Romania has led to the renewal of the agricultural machinery park. However, a very large number of old-generation agricultural machinery, equipped with diesel engines are still used, which has a negative impact on the environment. A possibility to reduce the emissions generated by these engines is the use of rapeseed oil and diesel blends [4]. The physical and chemical properties of biofuels highly depend on the raw materials used for biofuel production and have a significant influence on combustion and emission formation processes [5]. In the EU, biofuel production has largely been focused on biodiesel, and its constituent is chiefly rapeseed oil [6]. Worldwide, there is a tendency to increase the area planted with rape. In the European Union, rapeseed oil represents approximately 77% of the total biodiesel [7]. There are many studies on the use of rapeseed oil and rapeseed oil blends in diesel engines. Thus, the influence of biodiesel properties on engine operating parameters and the amount of pollutant emissions is studied in [8,9]. Qi et al. [10], studied how the use of rapeseed oil

IOP Conf. Series: Materials Science and Engineering **595** (2019) 012038 doi:10.1088/1757-899X/595/1/012038

biodiesel fuel blended with conventional diesel fuel influences the combustion and emission characteristics in two-cylinder agricultural diesel engines. Due to the viscosity and high density of crude rapeseed oil, depending on the amount of rapeseed oil in the fuel mixture, modifications to the diesel engine feed system are required [11]. Labeckas et al. [12] investigated the effect of rapeseed oil blending with ethanol on engine performance and exhaust emissions. Thus are presented the testing results of a four stroke, four cylinder, direct injection, unmodified, naturally aspirated diesel engine operating on neat rapeseed oil (RO) and its 2.5 vol% (ERO 2.5) and 7.5vol% (ERO 7.5) blends with ethanol [12].

2. Equipment used for experimental studies

All the experimental measurements presented in this study were performed on a 4 cylinder, naturally aspirated, water-cooled, D103, four-stroke diesel engine. This diesel engine equips U 650 tractors and has a power output of 65 HP at 1800 rpm. Although they are older generation, U 650 tractors are still widely used in agriculture in Romania. The engine laboratory equipment (Figures 1 and 2) was equipped with a Froude hydraulic brake. The coupling between hydraulic brake and diesel engine is directly, without transmission gear.

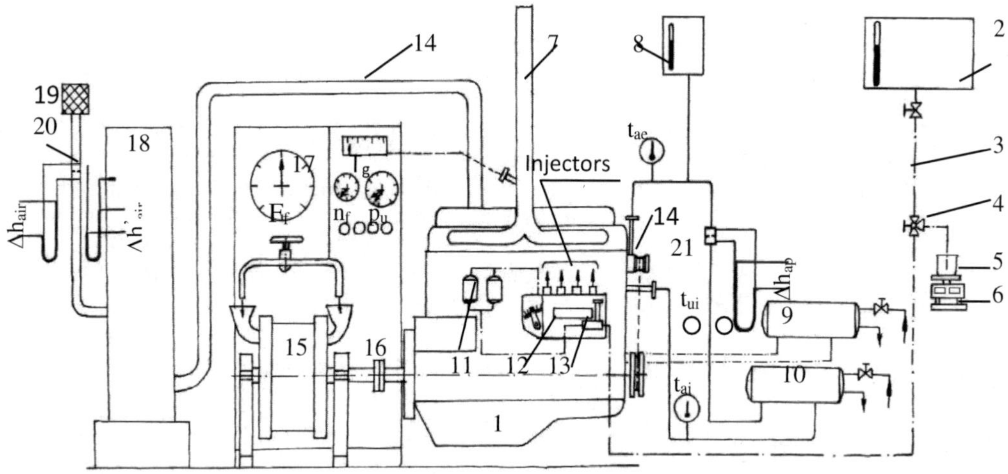

Figure 1. The scheme of experimental laboratory equipment.

The elements specified in figure 1 are: 1 – diesel engine D103; 2 - fuel tank; 3 - fuel supply pipe; 4 - three-way valve; 5 - calibrated fuel consumption measurement; 6 - electronic weighing; 7 - flue gas outlet pipe; 8 - expansion tank for coolant; 9 - oil cooler; 10 - coolant cooler; 11 - fuel filter battery; 12 - injection pump with speed regulator; 13 - fuel pump; 14 - cooling liquid pump; 15 - hydraulic brake; 16 - coupling; 17 - signaling and warning panel; 18 - suction air tank; 19 - air filter; 20 - diaphragm used for measuring air flow; 21 - diaphragm used for measuring the flow rate of the coolant.

The fuels used in experimental research was mixture of 50 % refined rapeseed oil - 50% diesel fuel and 100% diesel fuel as reference fuel.

In the Table 1 are presented the main physical chemical properties of refined rapeseed oil and diesel fuel.

In the present case, no modifications have been made to the diesel fuel injection system and no equipment has been installed to correct the viscosity of the fuel mixture. The viscosity of the fuel mixture (diesel and refined rapeseed oil at 50%) measured with the Hoppler viscometer was 18.4 mm^2/s at 20 ^0C, about 3.8 times higher than diesel fuel and 3.5 or less than refined rapeseed oil. The values measured with the Hoppler viscometer at 20 ^0C were for diesel fuel of 4.8 mm^2/s respectively for the refined rapeseed oil of 64.5 mm^2/s.

Table 1. Physical and chemical properties of refined rapeseed oil and diesel fuel.

Parameter	Rapeseed oil	Diesel fuel
Kinematic viscosity, 20^0 C [mm^2/s]	60-72	4-6
Lower Calorific value -LCV [MJ/kg]	37.4-38	41.5-42
Density, 20 °C [kg/dm^3]	0.91-0.92	0.82-0.84
Cetane number	40	51
Flash point, [°C]	270-321	65-80
Freezing point [°C]	-18 … 0	-12… 0

Figure 2. Laboratory equipment with diesel engine D103.

The amounts nitric oxide NO (%vol), the total emission of nitrogen oxides NO_x(%vol) and the residual content of oxygen O_2(vol%) in the exhausts were measured with the Multilyzer STx gas analyser. The measurement of the flue gas temperature was carried out on the exhaust pipe at a distance of about 1.5 [m] from the engine exhaust manifold, that is to say, at a distance similar to the displacement of the U650 on the exhaust of the tractor engine.

3. Experimental determinations and results

There was a significant reduction in the effective engine power at loads greater than 0.75. According to the results presented in figure 3 the reduction is between (6-20)%, the higher values being at the full load and close to the nominal speed, and the lower values at lower loads and revolutions ranging between (1200-1700) rpm.

The reduction in power is mainly due to the lower calorific value of the fuel mixture by about (8-12)%, but this can be accepted under conditions where a reduction in pollutant emissions is found when using the indicated fuel mixture, considering that most of the time of use, we do not exceed 0.8 of the full load.

For the idle regimes, there is a clear NO and NO_x emission reduction, which is a positive aspect, especially that it will also be confirmed in the case of medium and high load operation. NO and NO_x emissions are reduced by (10-20)% for idling, with higher cuts at lower engine speeds, according to the variations shown in Figure 4 and 5.

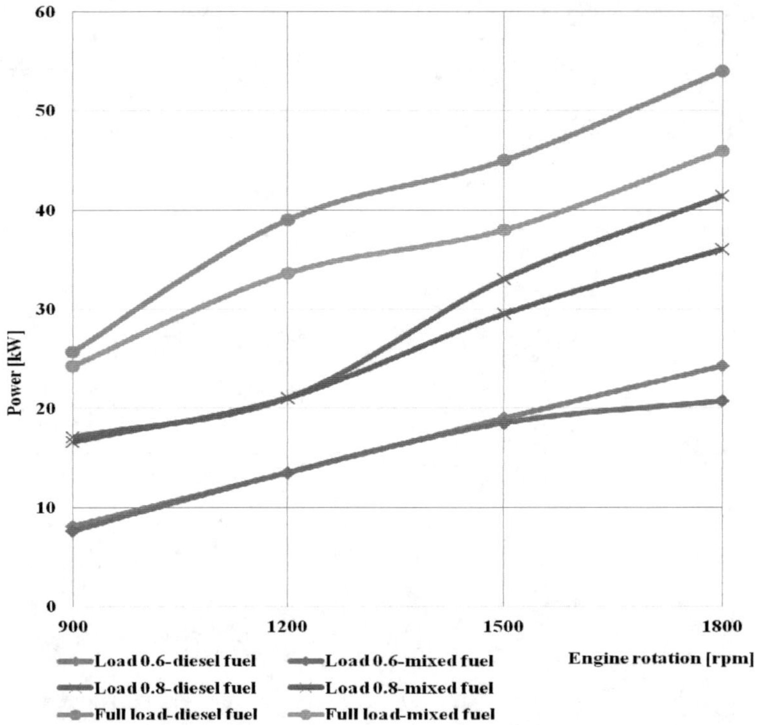

Figure 3. Effective Power with diesel fuel compared with mixed fuel.

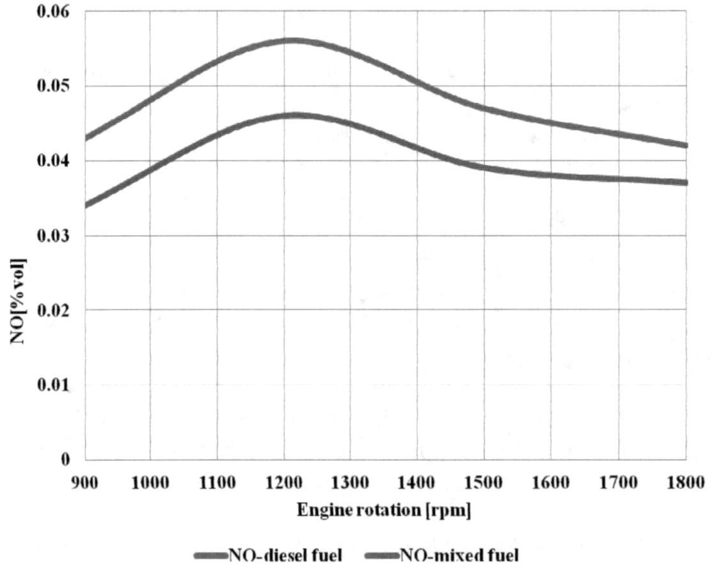

Figure 4. Emissions of NO for idling regimes.

For load performance, at least for loads ranging from 0.6 to 0.9, the NO and NO_x emission reduction is noteworthy and it ranges from 8 to 25%, with higher values for loads of 0.6-0,7 and lower values at loads above 0.8, as shown in figure 6 and figure 7.

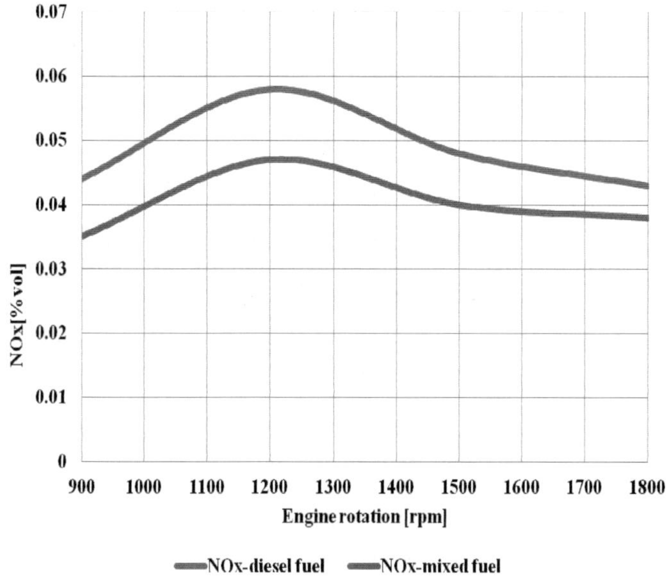

Figure 5. Emissions of NO_x for idling regimes

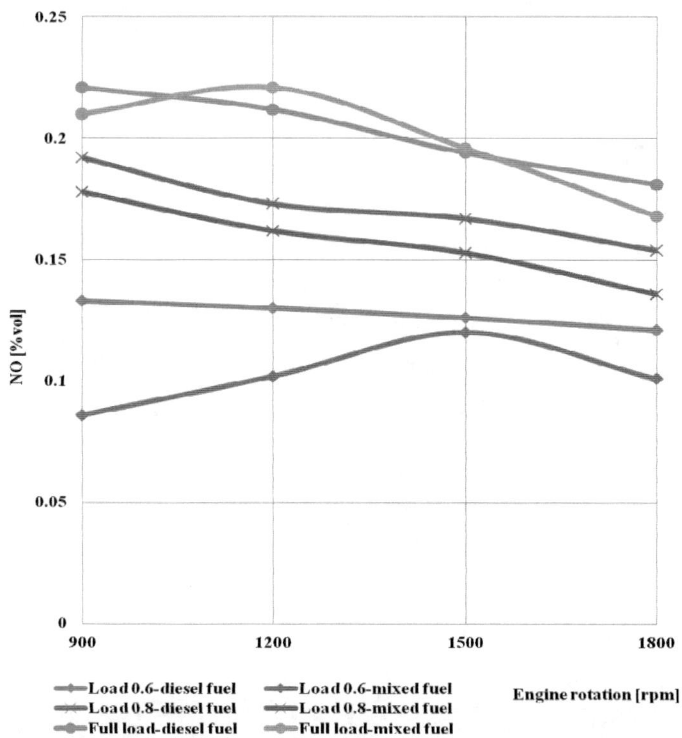

Figure 6. Emissions of NO depend of load and rotation.

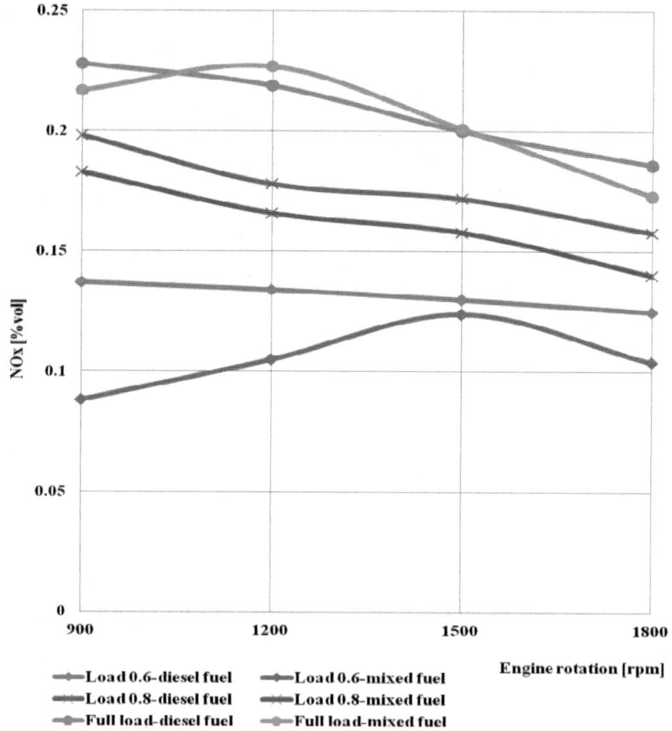

Figure 7. Emissions of NO_x depend of load and rotation.

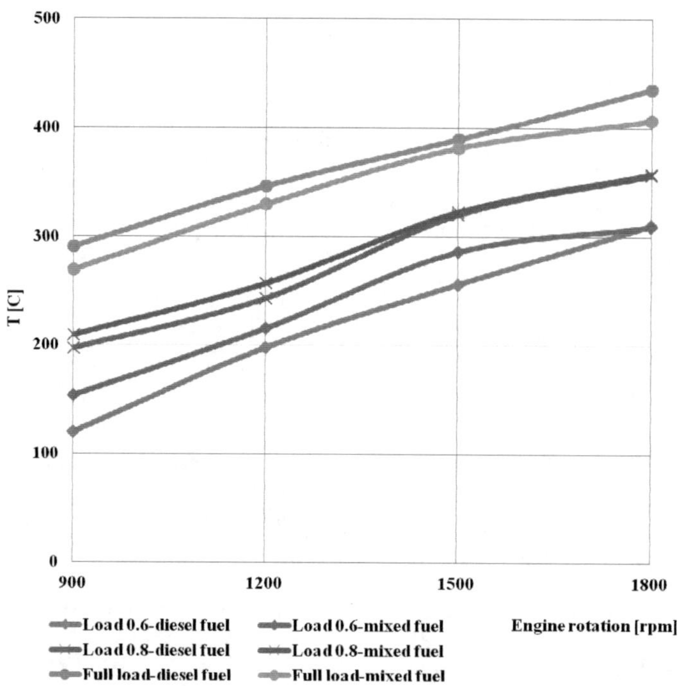

Figure 8. The temperature of the exhaust gases depends of load and rotation.

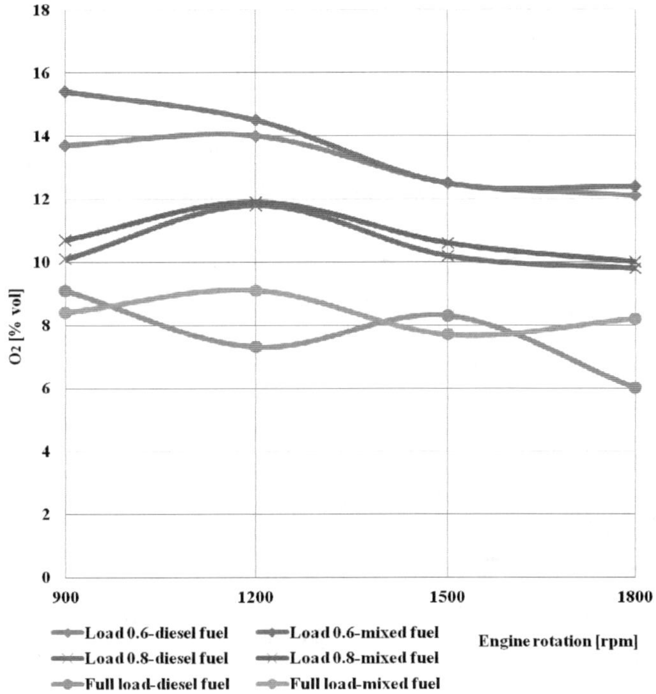

Figure 9. Oxygen in the exhaust gases depends of load and rotation.

This reduction of NO and NO_x emissions occurs even when the exhaust gas temperature is slightly increased by up to 10% (especially at lower loads), as shown in figure 8, as the combustion moves to detention, the mixture has the cetane number diminished as compared to classic diesel. There would certainly be a reduction in the temperature of the gas in the situation of increasing the injection advance when using the fuel mixture and possibly a further reduction of the pollutant emissions with this, which will be verified in the framework of extensive experimental determinations.

In principle, the use of the fuel mixture increases the amount of O_2 in the flue gas due to the presence of the oxygenated compounds in the rapeseed oil, the amount of O_2 in the combustion gases being anyway higher at partial loads regardless of the type of fuel according to Figure 9.

In constant load and variable speed regimes, the major influence on NO and NO_x emissions reduction with the increase in engine speed is the decrease in the amount of O_2 in the combustion gases according to Figure 9.

The experimental results, carried out on the laboratory stand (within the Department of Thermal Systems and Automotive Engineering at the Galati Engineering Faculty) with diesel engine D103, which equips the U650 tractors, led to the following positive results regarding the reduction of the NO and NO_x emissions in the case of using a fuel mixture made up of diesel oil and refined rapeseed oil in a 50% ratio, compared to normal diesel fuel with a cetane number of 51:

- for idling regimes, NO and NO_x emissions are reduced by 10-20%, with higher cuts at lower engine speeds.
- for operation regimes at average loads of 0.6-0.7, the NO and NO_x emission reduction reaches 20-25% and drops to 8-10% at loads above 0.8.
- taking into account the price of refined rapeseed oil of about 0.9 Euro/ litre compared to that of diesel fuel which is of 1.24 Euro / litre, the authors recommend the use of this fuel mixture consisting of 50% diesel with CC 51 and 50% refined rape seed oil, as there was no noticeable

decrease in power in the average 0,6-0,8 load range, which was most frequently encountered in operation.

- in constant speed and variable load modes, the increase in the gas temperature has a major influence on the increase in NO and NO_x emissions with increasing engine load.
- for constant load and variable speed regimes, a major influence on the NO and NO_x emissions reduction with increasing engine speed has the decrease of O_2 in the flue gas.
- regardless of the type of fuel used, 100% diesel or mixed fuel with 50% refined rapeseed oil, according to experimental results it is recommended that at loads above 0.6 the engine should operate at higher rpm, ie in the range (1500-1800) rpm because the NO and NO_x emissions will be reduced by up to (10-25)%.

4. Conclusion

Following the experimental determinations carried out and presented in the paper, both in load and idle modes, using a fuel mixture consisting of diesel and refined rapeseed oil in a ratio of 50%, a decrease of up to 25% of NO and NO_x emissions was noticed, with higher values at loads of 0.6-0.7 and lower values at loads of the engine over 0.8. Such a reduction in NO and NO_x emissions is important to emphasize precisely because of the fact that at the moment there is still a large number of U650 tractors in Romania's agriculture. At the same time, regardless of the type of fuel used, 100% diesel or 50% rape oil mixture, according to the experimental results, it is recommended that at loads above 0.6 the engine should operate at higher rpm, ie in the range (1500-1800) rpm as NO and NO_x emissions will be reduced by up to 25%.

References

[1] ***European Union, Directive 2016/2284/EC of the European Parliament and of the Council of 14 December 2016 on the reduction of national emissions of certain atmospheric pollutants, amending Directive 2003/35/EC and repealing Directive 2001/81/EC, Brussels.

[2] Demirbas A 2009 Progress and recent trends in biodiesel fuels. *Energy Convers Manage.* **50**

[3] European Automobile Manufacturers' Association, https://www.acea.be.

[4] Labeckas G and Slavinskas S 2005 Performance and exhaust emissions of direct-injection Diesel engine operating on rapeseed oil and its blends with Diesel fuel, Transport: *Journal of Vilnius Gediminas Technical University and Lithuanian Academy of Sciences*, **20(5)** p 186-194

[5] Torres Jiménez E et al 2010 Physical and chemical properties of ethanol–biodiesel blends for diesel engines, *Energy Fuels.* **24(3)** 2002–9

[6] Deborah B et al 2014 European biodiesel market and rapeseed oil: what impact on agricultural food prices? *International Journal of Global Energy Issues.* **37(5/6)**

[7] Zentková I and Cvengrošova E 2013 The utilization of rapeseed for biofuels production in the EU, *Visegrad Journal on Bioeconomy and Sustainable Development.* **2(1)** 11-14

[8] Kegl B 2008 Effects of biodiesel on emissions of a bus diesel engine, *Bioresour Technol* **99** 863-73

[9] Hribernik A and Kegl B 2007 Influence of biodiesel fuel on the combustion and emission formation in a direct injection (DI) Diesel engine, *Energy Fuels.* **21** 1760–7

[10] Qi D H, Lee C F, Jia C C, Wang P P and Wu S T 2014 Experimental investigation of combustion and emission characteristics of rapeseed oil–diesel blends in a two-cylinder agricultural diesel engine. *Energy Convers Manage.* **77** 227–32

[11] Cordoş N and Mariaşiu F 2011 Hydrocarbon emissions from biofuel based on rapeseed oil compared to diesel fuel in diesel engine D-2402, *Research Journal of Agricultural Science.* **43(1)**

[12] Labeckas G, Slavinskas S and Pauliukas A 2007 The effect of rapeseed oil blending with ethanol on engine performance and exhaust emissions, *Journal of KONES.* **14(3)** 331-338

[13] Burciu M 2016 Laboratory Guide - Diagnosis and car dynamics, *Galati University Press*, ISBN 978-606-696-054-0

IOP Conf. Series: Materials Science and Engineering **595** (2019) 012039 doi:10.1088/1757-899X/595/1/012039

Study of the thermal stability of biodiesel and diesel fuel

D Gheorghe, D Tutunea, M Bică, A Gruia and M Călbureanu

[1] Road Vehicle and Transportation Department, University of Craiova, Craiova, Romania

E-mail: marinbica52@gmail.com

Abstract. In this study the physico-chemistry characterization and the thermal behaviour of biodiesel and diesel fuel were carried out. Biodiesel is currently the most suitable fuel as alternative for diesel engines due to the environmental and technical advantages. In Romania the use of blends of biodiesel/diesel as fuels is growing every year mainly to the implementation of Kyoto protocol and to the similar properties to diesel fuel. The tests were realized under atmospheric conditions of pure air flow at various heating rates. The thermogravimetric profile indicates that diesel has three mass losses events and biodiesel of sunflower oil has five mass loss events.

1. Introduction

The demand of energy and the rising costs of the petroleum-derived fuels have triggered the research to develop new type of fuels that can reduce environmental pollution and global warming [1]. The global needs in energy are assured by fossil fuels reducing the reserves of petroleum and increasing the emission of CO_2 [2]. Biodiesel is commonly obtained by the transesterification of triglycerides (TG) in the presence of short-chain alcohol of methanol or ethanol as catalysts to obtain fatty acid methyl esters (FAMEs) [3]. 75% of the total investment is constituted by the vegetable oils or animal fats that are primary resources for biodiesel production [4]. Non-edible oils stand as new promising sources of raw materials (waste cooking oils, crude jatropha oil, algae) for biodiesel production in various countries to satisfy the energy demands [5]. The biodiesel obtained is susceptible to the degradation process when is displayed to heat, light or air due to the chemical composition specific to each feedstock [6]. The thermos-analytical applications have been use with success to the characterization of biodiesel and petroleum derivatives determining the pyrolysis and combustion patterns for the evaluation of thermal properties of heavy petroleum fractions [7, 8]. In generally the kinetics of these processes are monitored by thermogravimetric methods using multiple heating rates to evaluate the complex chemistry of diesel fuel stability [9]. The most important parameters of TG analysis are initial and final decomposition temperature and the maximum temperature of conversion [10]. Dantas et al. [11] investigated the thermal stability of biodiesel of corn and found that the fuel is stable until 225°C. Chand et al. [12] studied the thermal behaviour of the biodiesel of soybean oil and found that the fuel is thermal stable until 150 °C. Li et al. [13] studied the kinetic parameters and thermal degradation of peanut, palm and waste cooking oil biodiesel in a nitrogen atmosphere using various heating rates between 20 – 40 °C min-1 in a temperature range of 25 – 600 °C and found a single mass loss step for all samples. The range of temperatures varies between 155.8–329.8 °C (peanut biodiesel), 164.2–333.5 °C (palm biodiesel) and 142.2–325.5 °C (waste cooking oil biodiesel) with final carbon residues of 4.07%, 3.77% and 3.49%. Santos et al. [8] investigated the thermal stability of palm, cotton and sunflower oil biodiesel and B10 in a helium atmosphere with a heating rate of 10 °C min-1 in a temperature range of 30 – 600 °C and found that almost all the samples showed a single step mass loss with the exception of palm oil which had two events. The temperature ranges for cotton, sunflower and palm oil were 304–490 °C, 300–495 °C and 325–489 °C and the carbon residue was 97.5%, 98.7% and 88.6%. Silva et al. [14] studied the thermal stability of pequi oil biodiesel obtained by methanol and ethanol in a nitrogen and air atmosphere, with a heating rate of 10, 20 and 40 °C min-1 in a temperature range of 27 – 600 °C and found for all samples a single step mass loss representing the volatilization processes in both used gases. The mass loss starts at around 110 °C. Rodrıguez et al. [6] compared the thermal stability of higuereta and soybean oil with diesel fuel using various heating rates between 5, 10, 20 and 30 °C min-1 in a temperature range of 25 – 450 °C and found that in diesel fuel the mass loss is between 50 – 250 °C and in the case of biodiesel varies from 185 to 300 °C. The aim of this study was to investigate biodiesel of sunflower oil determining physical–chemical properties and its thermal stability and compare with diesel fuel.

2. Materials

Sunflower oil was purchased from local shop. Methanol with a purity of 99.5% and NaOH were purchased from S.C. Laborex Romania. Diesel fuel was purchased from a local gas station from Craiova, Romania on November 5th 2017. Euro L Diesel is a diesel fuel from LuKoil Company obtained at Petrotel Ploiesti Refinery.

3. Equipments

In this study, the thermal degradation characteristics of biodiesel blends were studied using a Diamond TG/DTG Analyzer from PerkinElmer Instruments. The physico-chemical tests on biodiesel were performed according to standardized methods ASTM in the laboratory of Fuels at Faculty of Science Craiova. The oil was transformed in biodiesel in a small reactor of 30 L capacity at the Faculty of Mechanics.

4. Methyl ester production

The biodiesel was produced using a ratio of 6:1 of methyl alcohol and sunflower vegetable oil, using NaOH as catalyst (4%), under constant agitation at 60°C. In the first phase, methanol and sodium hydroxide were placed to react under stirring until achieving a complete homogenization of the base catalyst, resulting in sodium methoxide. After 2 hours the transesterification reaction was completed and mixture was withdrawn from the reactor and poured into a separator to remove the glycerine layer. The biodiesel was washed out 5 times to remove impurities and unreacted agents. Finally, biodiesel was dried completely by using silica jell.

5. Results and discussion

Physical–chemical properties. The results for biodiesel samples were tabulated in Table 1. It was seen that the flash point of biodiesel is 168 which helps the transportation of biodiesel but it should be mixed with diesel fuel for better combustion in engine. Biodiesel presents a lower heating value due to the presence of chemically bound oxygen in vegetable oils which lowers their heating values. Kinematic viscosity of biodiesel at 40°C is higher than diesel fuel and suggests a reduced atomization and incomplete combustion in the engine. The biodiesel analyzed has a copper strip corrosion of which indicate that this fuel is not corrosive. The cloud point and pour point of biodiesel is higher than diesel fuel because biodiesel produced from vegetable oils has free fatty acids which cause a higher value of these properties. Methanol used for production was produced from corn and corn has a small quantity of sulphur; hence, biodiesel has a negligible amount of sulphur.

Thermal analysis. The samples were heated at five heating rates of 5, 10, 15, 20 and 25 °C min^{-1} in a temperature range of 30 – 700 °C. All thermal conversion data were recorded in a computer coupled to the thermo analytical equipment and evaluated by Pyris software.

Table 1. Physical–chemical properties of biodiesel.

Test	Method	Unit	Biodiesel	ASTM
Flash point	D93	°C	168	130 min.
Kinematic Viscosity at 40 °C	D445	cSt.	3.6	1.9 – 6.0
Total sulfur	UOP 357	ppm	3.0	15
Copper Strip	D130	-	1a	No. 3 max
Cetane Index	D976	-	49.2	47
Cloud point	D2500	°C	-5	N/A
Pour point	D97	°C	-7	N/A
Carbon residue	D189	Wt%	0.002	0.050 max
Acid number	D974	Mg KO H/g	0.270	0.800
Heating value	D240	MJ/kg	39.59	N/A
Specific gravity	D1298	-	0.884	N/A

N/A Not Available

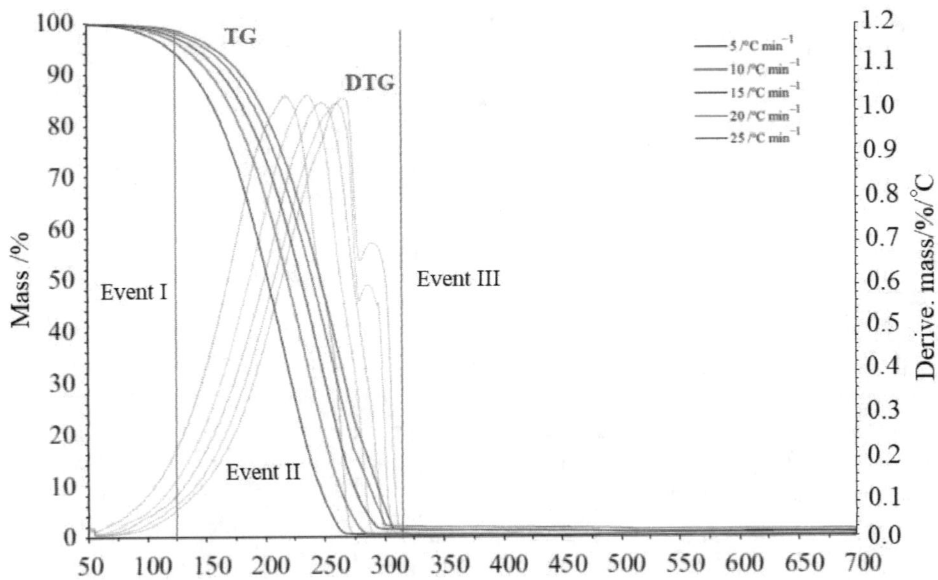

Figure 1. TG and DTG profiles of diesel fuel.

Figure 1 shows the TGA (thermogravimetric analysis) and DTG (derivation thermogravimetric) of diesel fuel. The TG curves can be divided in three events; the first represents the evaporation of hydrocarbons with shorter carbon chains, the second represents the devolatilization where the samples loss the 97–98% of the total mass and the third with the carbonization of the sample. The second event takes place between 82.5–273.8 (5 °C min^{-1}, mass loss 98.18%), 95.3–294.5 (10 °C min^{-1}, mass loss 98.11%), 101.4–313.82 (15 °C min^{-1}, mass loss 97.21%), 112.03–317.44 (20 °C min^{-1}, mass loss 96.73%) and 119.03–326.68 (25 °C min^{-1}, mass loss 96.11%).

Figure 2 shows the TGA and DTG for biodiesel samples. The biodiesel has five events, the first is the evaporation, the second associated with the highest mass loss, the third associated with thermal degradation of the residuals of the different triglyceride components, the forth with the loss of 3–6% of the sample mass and the fifth with the carbonization of the sample. The main mass loss occurs in the second event and takes place between 54.7–300.8 (5°C min^{-1}, mass loss 87.658%), 168.2–319.6 (10 °C min^{-1}, mass loss 88.69%), 176.9–325.31 (15 °C min^{-1}, mass loss 88.22%) and 185.21–334.63 (20 °C min^{-1}, mass loss 88.11%). Both for diesel and biodiesel samples the higher heating rate create another rate of degradation peak shifting the degradation process to a higher temperature.

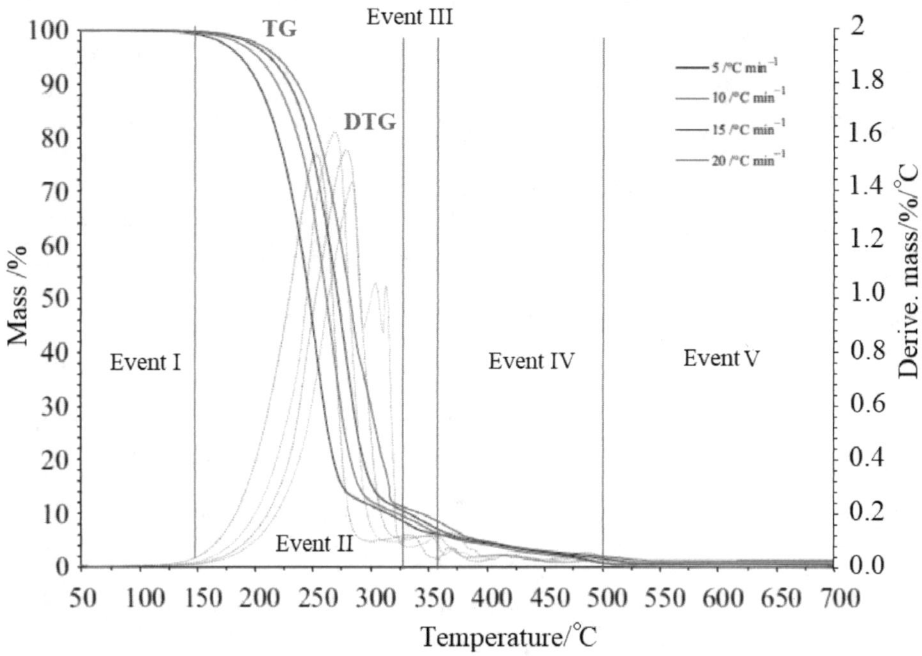

Figure 2. TG and DTG profiles of sunflower oil biodiesel.

6. Conclusion

The thermal behaviours of diesel fuel and biodiesel blends were evaluated using TG and DTG, using thermogravimetric analysis under pure air atmospheric conditions. From the TG curves it was observed that sunflower oil biodiesel has a higher thermal stability than diesel fuel around 130-150 °C. Such temperatures are significant in the storage process. Biodiesel has five events of decomposition whereas the diesel fuel was divided into three events.

References

[1] Jain S and Sharma M P 2010 Prospects of biodiesel from Jatropha in India: A review. *Renew Sustain Energy Rev.* **14 (2)** pp 763–771

[2] Mathimani T, Kumar TS, Chandrasekar M, Uma L and Prabaharan D, 2017 Assessment of fuel properties, engine performance and emission characteristics of outdoor grown marine Chlorella vulgaris BDUG 91771 biodiesel. *Renewable Energy.* **105** pp 637–646

[3] Rashid U and Anwar F 2008 *Production of biodiesel through optimized alkaline-catalysed transesterification of rapeseed oil.* Fuel **87** pp 265–273

[4] Avinash A, Sasikumar P and Murugesan A 2018 Understanding the interaction among the barriers of biodiesel production from waste cooking oil in India-an interpretive structural modeling approach, *Renew. Energy* **127** pp 678-684

[5] Balat M, 2011 Potential alternatives to edible oils for biodiesel production – a review of current work, *Energy Convers Manage.* **52** pp 1479–1492.

[6] Rodrigues RP, Verhelst S, 2009 Thermal and kinetic evaluation of biodiesel derived from soybean and higuereta oil. *J Therm Anal Calorim.* **96** pp 897–901

[7] Maciel STA, Wisniewski A Jr and Souza MJB 2015 Use of micropyrolysis and TG to study the thermal catalytic conversion of onshore crude oil using the zeolite catalysts type Y and

ferrierite. *J Therm Anal Calorim.* **122** pp 369–77

[8] Santos AGS, Caldeira VPS, Souza LD, Oliveira DS, Araujo AS, Luz GE Jr 2016 Study of termal stability by thermogravimetry for oil, biodiesel and blend (BX) of different oil seeds. *J Therm Anal Calorim.* **123** pp 2021–2028

[9] Leiva C R M, Crnkovic P M and Santos A M 2006 The use of thermogravimetry to determine the activation energy of the combustion process of combustible oils, *Quim Nova* **29** 940–46

[10] Freire L M S et al. 2009 Thermal investigation of oil and biodiesel from Jatropha curcas L. *J Therm Anal Calorim.* **96** pp 1029–33.

[11] Dantas M B et al. 2007 Thermal and kinetic study of corn biodiesel obtained by the methanol and ethanol routes. *J Therm Anal Calorim.* **87(3)** pp 835–839

[12] Chand P, Reddy CV, Verkade JG, Wang T and Grewell D 2015 Thermogravimetric quantification of biodiesel produced via alkali catalyzed transesterification of soybean oil. *Energy Fuels.* **23** pp 989–992

[13] Li H, Niu SL, Lu CM and Cheng SQ 2015 Comparative evaluation of thermal degradation for biodiesels derived from various feedstocks through transesterification. *Energy Convers Manag.* **98** pp 81–88

[14] Silva T A et al. 2014 Methylic and ethylic biodiesels from pequi oil (Caryocar brasiliense Camb.): production and thermogravimetric studies. *Fuel.* **136** pp 10–18

IOP Conf. Series: Materials Science and Engineering **595** (2019) 012040 doi:10.1088/1757-899X/595/1/012040

Study on performance of an innovative power system for hybrid propulsion consisting in a gas turbine with heat exchanger and steam turbine

D T Balanescu, V M Homutescu, C E Panaite and A Popescu

Department of Mechanical and Automotive Engineering, "Gheorghe Asachi" Technical University of Iasi, Iasi, Romania

E-mail: mariohomutescu@gmail.com

Abstract. Hybrid electric vehicles are currently considered a viable solution for fossil fuel saving and pollutant emissions reduction in transportation sector. The typical configuration of a hybrid electric propulsion system contains a reciprocating internal combustion engine – gasoline or diesel. Aiming to increase performance and to reduce pollution in road transport, a hybrid electric propulsion system based on a micro gas and steam turbine system operating with compressed natural gas is proposed in the paper. The micro gas turbine has a recuperative heat exchanger and the steam cycle has one pressure level. The study analysed the performance of the proposed system as a function of gas turbine engine pressure ratio and temperature difference of the gas turbine engine heat exchanger. Results indicate that, at equal power, the analysed system is more performant than internal combustion engines currently used in hybrid electric propulsion, so they could represent a viable solution in this field.

1. Introduction

Fossil fuels shortage and concern for clean power generation solutions in the propulsion systems sector have led to a growing interest for the electric traction. Excluding the vehicles with continuous connection to the power network (such as trolleys and trams), which require specific infrastructure facilities, all electric vehicles incorporate some sort of energy storage systems. In the case of pure electric vehicles, the most stringent problem is to ensure the drive range need. As indicated in [1], identifying of the range needs requires studies using individual real travel data, which also indicates the adequate locations of charging stations. Besides the drive range problem, pure electric vehicles are not cost competitive [2], so they need substantial subsidies support to be financially attractive [3]. The drive range is significantly extended in the case of hybrid electric vehicles (HEV), which typically combine an electric motor and an internal combustion engine (ICE).

For both pure electric and hybrid electric technology the mobility range improvement and vehicle cost reduction are inherently related to the development of the electric energy storage; presently, the supercapacitors are the most promising energy storage devices due to a short charging time and long lifespan [4]. Obviously, the energy management is also very important. Accordingly, management strategies for maximum braking energy recovery were studied [5, 6], while optimum energy distribution between the electric source and ICE was analysed in [7].

In addition to extending the drive range, the performance improvement represents a major concern in the field of hybrid electric vehicles. Aiming to higher performance, several solutions have been studied lately as alternative to ICE for integration in hybrid electric vehicles. Fuel cell appears to be one promising solution. A hybrid fuel cell electric vehicle has the advantage that uses only the electric motor. The experimental study on a hybrid propulsion system consisting in a 90 kW proton exchange membrane fuel cell and a supercapacitor, performed in [8], indicates impressive performance such as a drive range of 435 km and a maximum speed of 158 km/h. It was shown that performance can be improved by using a lithium-ion battery together with a supercapacitor [9]; in this case, the drive range extended to 545 km and the maximum speed was 161 km/h.

Gas turbine and combined cycle units based on gas turbine also represent an alternative to ICE, not only in conventional approach [10], but also for hybrid propulsion systems. The use of the gas turbine for waste heat recovery from ICE exhaust gases of a hybrid propulsion system improves fuel economy with more than 5% [11]. It was also investigated the use of the gas turbine as range extender of a hybrid vehicle – in simple cycle [12] or in combined cycle [13, 14]. Results of these studies indicate the turbine as an attractive solution to improve the performance of hybrid propulsion systems. A study on performance of a micro gas-steam turbine combined cycle (MCC) unit operating with natural gas (compressed natural gas – CNG – application) was developed by authors and results are presented in the paper. In the transportation field, natural gas represents a very attractive and promising alternative fuel to the gasoline, methanol or diesel due to cleaner combustion qualities and low cost [15].

The opportunity of the study on MCC implementation in the area of hybrid propulsion systems, which implies a low power level, is supported by the impressive performance of high and medium power combined cycles power stations, which currently are the most efficient and least polluting power units operating with fossil fuels. Compared to studies conducted in [13] (gas turbine without heat exchanger and steam cycle with one pressure level) and [14] (gas turbine with heat exchanger and steam cycle with two pressure levels), the analysed configuration is based on a gas turbine with heat exchanger and a steam cycle with one pressure level.

2. Description of the hybrid propulsion system

Schematic of the analysed MCC hybrid system is shown in figure 1 and has a series-parallel topology, which was chosen due to a good flexibility and high efficiency. This configuration not only ensures flexibility [16], but allows the turbines to operate mainly in optimum conditions, at rated rotation speeds. Accordingly, efficiency is maintained high in each operating mode. Also, when compared to other powertrain architectures, the series-parallel topology makes possible regenerative braking [17].

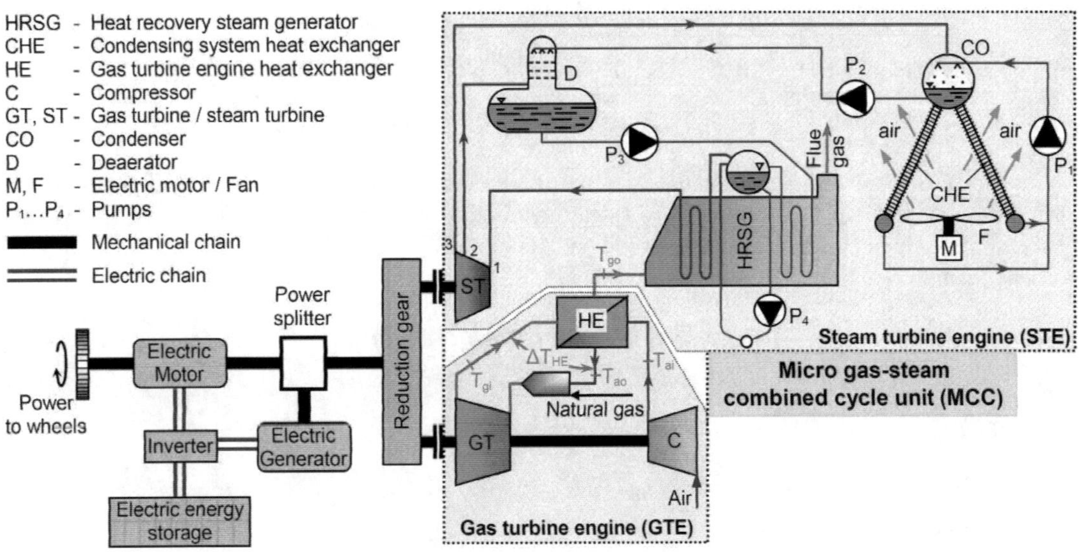

Figure 1. Layout of hybrid system based on micro gas-steam turbine combined cycle unit.

IOP Conf. Series: Materials Science and Engineering **595** (2019) 012040 doi:10.1088/1757-899X/595/1/012040

The flue gas exiting the gas turbine passes through heat exchanger HE preheating the air entering the combustion chamber of the gas turbine engine (GTE). After HE, the flue gas passes through heat recovery steam generator (HRSG) – the link between GTE and the steam turbine engine (STE) – and then are released into the atmosphere. Thus, the waste heat from GTE is partially recovered in STE, which produces additional power; hence, MCC converts fuel into mechanical energy more efficiently than GTE in single operation. But the use of STE implies the presence of the condensing system. The typical condensing system, with cooling tower, induces water losses, which can-not be accepted in the case of a propulsion system of an on-road vehicle. That is why the condensing system type included into the analyzed scheme operates in closed cycle: condensate is subcooled by air in a heat exchanger CHE, then a fraction of the subcooled condensate flow is sprayed into the steam turbine exhaust flow and thus steam condensation occurs.

3. Method

The performance of MCC hybrid system presented in figure 1 was analysed by computing the output power, efficiency and specific fuel consumption according to [18], as described below.

Output power of MCC represents the sum of the output powers produced by GTE and STE, in kW,

$$P_{MCC} = P_{GTE} + P_{STE} = \eta_m \cdot \left\{ w_{GTE} \cdot \dot{m}_a \cdot \left[\left(AER \cdot AFR \right)^{-1} + 1 \right] \cdot \left(1 - \theta_{ca} \right) + \left(\dot{m}_S \cdot \Delta h_{ST} - \dot{m}_{ES} \cdot \Delta h_{ES} \right) \right\} \quad (1)$$

where: η_m is the mechanical efficiency of turbines; w_{PT} represents the specific work output of GTE, in kJ/kg; $\dot{m}_a$ is the inlet mass flow rate of the compressor, in kg/s; *AER* is air excess ratio; *AFR* is air-fuel ratio, in kg air/kg fuel; θ_{ca} represents the fraction of air mass flow rate used for cooling of turbine nozzles; $\dot{m}_S$ and $\dot{m}_E$ are the steam turbine inlet mass flow rate (point 1 in figure 1) and extraction steam mass flow rate (point 2), respectively, in kg/s (they result from the thermal balances of HRSG and deaerator); Δh_{ST} represents the total change of specific enthalpy in the steam turbine (from point 1 to point 3), in kJ/kg; Δh_{ES} is the change of specific enthalpy in the steam turbine from the extraction section (point 2) to the turbine exhaust (point 3), in kJ/kg.

Efficiencies of the two turbine engines of the MCC unit, i.e. GTE and STE, were calculated with

$$\eta_{GTE} = 3.6 \cdot P_{GTE} \cdot \left(FC_{MCC} \cdot LHV \right)^{-1} \cdot 10^5 \quad (2)$$

$$\eta_{STE} = P_{STE} \cdot \left(\dot{m}_g \cdot \Delta h_{HRSG} \right)^{-1} \cdot 10^2 \quad (3)$$

while overall efficiency of the MCC was expressed as

$$\eta_{MCC} = 3600 \cdot P_{MCC} \cdot \left(FC_{MCC} \cdot LHV \right)^{-1} \cdot 10^5 \quad (4)$$

The terms FC_{MCC}, *LHV*, $\dot{m}_a$ and Δh_{HRSG} in formulas (2), (3) and (4) denote fuel (natural gas) consumption of MCC unit, in Nm^3/h, lower heating value of fuel, in kJ/Nm^3, flue gas mass flow rate, in kg/s, and the change of the flue gas specific enthalpy in HRSG, in kJ/kg.

Fuel consumption, in Nm^3/h, was given by

$$FC_{MCC} = 3600 \cdot \dot{m}_a \cdot \left(\rho_f \cdot AER \cdot AFR \right)^{-1} \cdot \left(1 - \theta_{ca} \right) \quad (5)$$

where ρ_f is fuel density, in kg/Nm^3.

Specific fuel consumption, in Nm^3/kWh, was calculated with the classic formula

$$SFC_{MCC} = FC_{MCC} \cdot P_{MCC}^{-1} \quad (6)$$

In order to assess the environmental impact of MCC hybrid system unit, CO_2 emission rate, expressed in g_{CO2}/kWh, was calculated with formula

$$E_{CO2} = 1000 \cdot SFC_{MCC} \cdot g_C \cdot \rho_f \cdot M_{CO2} \cdot M_C^{-1} \tag{7}$$

where g_C is the carbon fraction in the fuel, in kg carbon/kg fuel, while M_{CO2} and M_C are the molar masses of carbon dioxide and carbon, respectively.

The study was conducted by assuming that inlet mass flow rate of the compressor ($\dot{m}_a$), gas turbine inlet temperature and temperature of flue gas exhausted from HRSG are 0.45 kg/s, 1400 K and 393 K, respectively. A minimum temperature difference of 20 K was accepted between the temperature of flue gas exhausted from HRSG and temperature of superheated steam. Pressure loss in the combustion chamber of GTE and combustion efficiency were assigned 3% and 98%, respectively. Mechanical efficiencies of gas turbine and steam turbine were assumed 99% while isentropic efficiencies of these two turbines as well as isentropic efficiency of compressor were assigned 86%. The fraction of air mass flow rate used for cooling of turbine nozzles (θ_{ca}) was assumed 3%. It was considered a condensing pressure of 0.4 bar, which induces a saturation (condensing) temperature of 75.9°C.

The following composition (volume fractions) of natural gas was assumed: CH_4 = 95.1%, C_2H_6 = 2.8%, CO_2 = 0.9% and N_2 = 1.2%. Accepting this composition (representative for Romania), the following values were determined: AFR = 16.57 kg air/kg fuel, LHV = 34062.9 kJ/Nm^3, ρ_f = 0.714 kg/Nm^3 and g_C = 0.726 kg C/kg fuel.

The study was undertaken to evaluate MCC unit performance and effectiveness of GTE heat exchanger, defined as (see figure 1):

$$\sigma_{HE} = \frac{T_{ao} - T_{ai}}{T_{gi} - T_{ai}}, \tag{8}$$

as a function of GTE compressor pressure ratio, π_c, and characteristic temperature difference of GTE heat exchanger, ΔT_{HE}, which is given by

$$\Delta T_{HE} = T_{gi} - T_{ao}. \tag{9}$$

The terms T_{ai}, T_{ao} and T_{gi} in equations (8) and (9) denote air inlet temperature, air outlet temperature and flue gas temperature at GTE heat exchanger inlet, expressed in K.

4. Results and interpretation

The results of the analysis are graphically represented in figure 2 and figure 3. It can be seen that maximum value of π_c is 14 when ΔT_{HE} = 60 K and decreases gradually (down to 9) when ΔT_{HE} increases (up to 240 K). This is because T_{ai} increases with π_c while T_{gi} decreases as long as gas turbine inlet temperature is constant, so less and less heat is transferred from flue gas to air in GTE heat exchanger until it becomes zero, which means that the use of GTE heat exchanger becomes meaningless at a certain value of π_c (the maximum possible one).

Figure 2a clearly shows that both P_{GTE} and P_{STE} increase with π_c, but decrease with ΔT_{HE}. As consequence, P_{MCC} (the sum of P_{GTE} and P_{STE}) exhibits the same trends.

By analysing figure 2b and figure 3b, it can be observed that the rate of heat exchanged in HE, emphasized by σ_{HE}, influences decisively the value of η_{GTE}. Thus, lower values of π_c and/or ΔT_{HE} lead to the increase of σ_{HE}, which induces the increase of η_{GTE}. But lower values of π_c and/or ΔT_{HE} also induce lower temperature of the flue gas exhausted from GTE (t_{go}), so less heat is introduced in the steam cycle where this heat is converted less efficiently into power (see variations of P_{STE} and η_{STE} in figure 2a and figure 2b). Consequently, STE performance has the lowest influence on the performance of MCC unit when π_c and ΔT_{HE} are the lowest. In an opposite approach, the increase of π_c and/or ΔT_{HE} leads to STE performance improvement. Thus, both P_{STE} and η_{STE} increase and compensate more effectively the GTE performance loss when π_c and ΔT_{HE} increase. Hence, η_{MCC} varies in a quite narrow range compared to η_{GTE} and η_{STE}.

It must be noted that, in spite of the effective compensation, the performance of MCC unit is higher when π_c and ΔT_{HE} exhibit low values. This emphasizes that the rate of heat conversion into power should be maximized in GTE and not in STE. Besides, it is highlighted the role of STE as waste heat recovery unit.

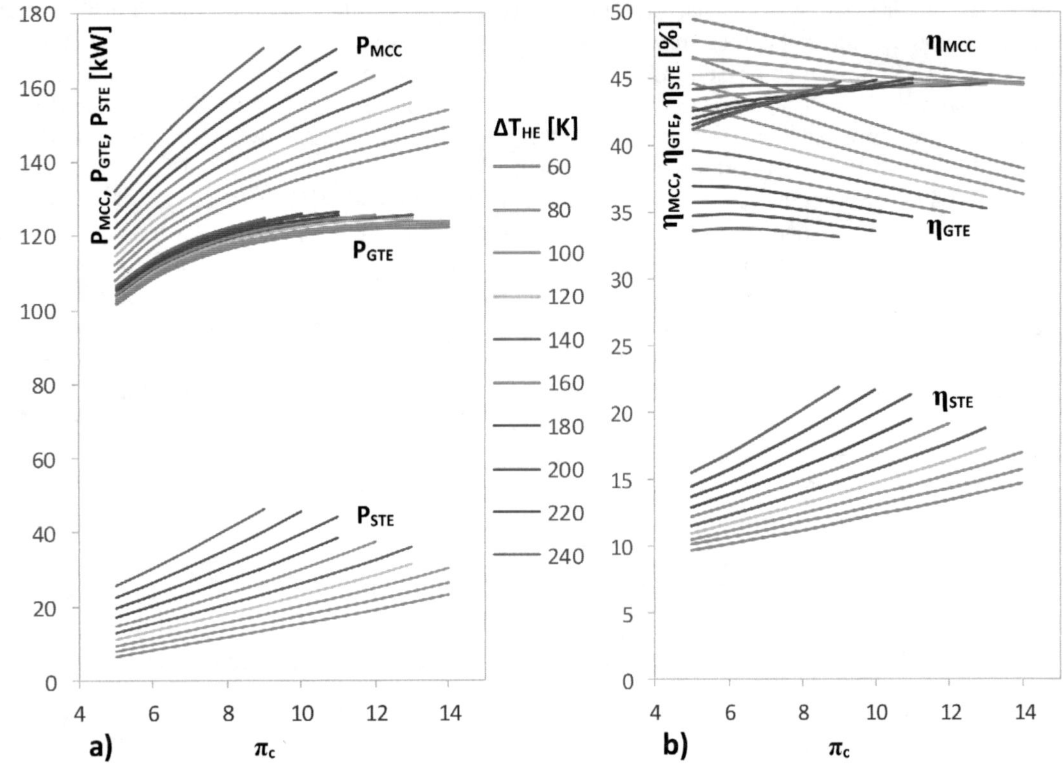

Figure 2. Output powers (a) and efficiencies (b) versus GTE compressor pressure ratio for several temperature differences of GTE heat exchanger.

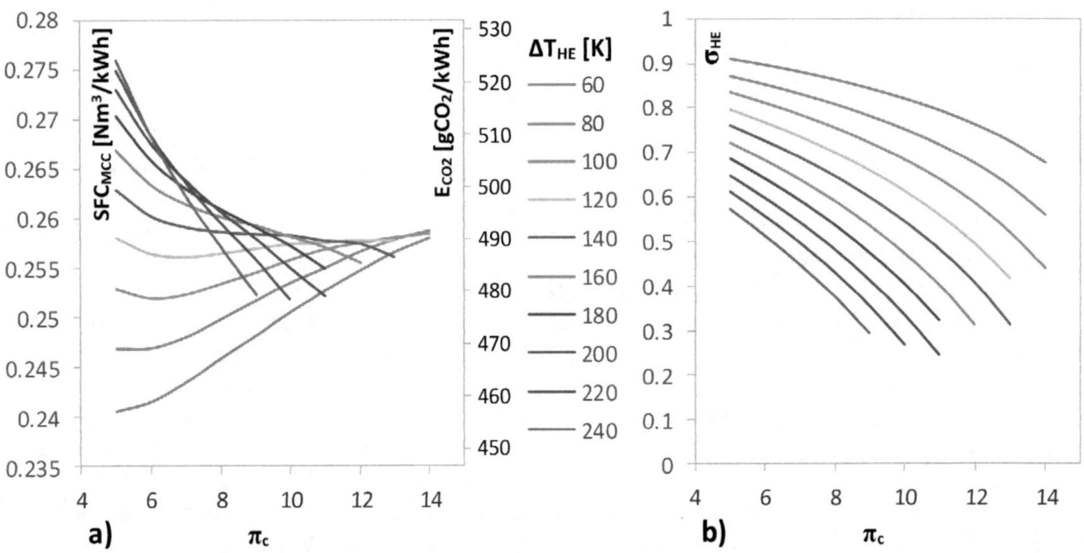

Figure 3. Specific fuel consumption / CO_2 emission rate (a) and effectiveness of GTE heat exchanger (b) versus GTE compressor pressure ratio for several temperature differences of GTE heat exchanger.

Taking into account the discussion above, yields that highest performance level of MCC unit is achieved when π_c and ΔT_{HE} have minimum values, namely 5 and 60 K, respectively. In this case, σ_{HE}, η_{GTE} and η_{MCC} attain their maximum values, of 0.91, 46.7% and 49.6%, respectively, η_{STE}, *SFC* and E_{CO2} the minimum values, i.e. 9.8%, 0.241 Nm^3/kWh and 457.1 g_{CO2}/kWh, while P_{MCC} = 408.5 kW. These values indicate the significantly higher performance potential of the analysed MCC unit compared to the most advanced diesel ICE (currently the most efficient typical solution for hybrid propulsion), which do not exceed 42% efficiency [19] and have CO_2 emission rates higher than 617 gCO_2/kWh [20]. Compared to MCC based on GTE with heat exchanger and steam cycle with two pressure levels, studied in [13], with η_{MCC} = 49.6% and SFC = 0.242 Nm^3/kWh, MCC unit analysed in this paper offers similar performance but in a simpler version, with a simpler construction of both HRSG and steam turbine, which is possible due to the single pressure level in the steam cycle. Accordingly, the analysed hybrid propulsion system based on MCC unit represents an interesting alternative solution for conventional hybrid propulsion systems with ICE.

5. Conclusions

In order to maximize the performance of MCC unit, the performance of GTE should be maximized since it converts heat into power more efficiently than STE, which plays the role of a waste heat recovery system.

The performance of the analysed MCC unit operating with CNG – characterized by maximum efficiency, minimum SFC and minimum CO_2 emission rate values of 49.5%, 0.241 Nm^3/kWh and 457.1 g_{CO2}/kWh, respectively – is considerably higher compared to the most advanced diesel ICE that are currently used as typical instantaneous power generation source for hybrid vehicles. Besides, it offers the same performance with MCC based on STE with two pressure levels, which has a more complex construction. Due to these aspects, it can be assumed that hybrid propulsion system consisting of GTE with heat exchanger and STE with one pressure level represents a noticeable alternative to the typical hybrid system, based on ICE.

The main concern in further studies on MCC units for hybrid on-road propulsion should be related to the development of micro steam turbines with output powers less than 50 kW and to micro gas turbines operating with high temperature and pressure of flue gas.

References

[1] Shi X, Pan J, Wang H and Cai H 2019 Battery electric vehicles: what is the minimum range required? *Energy* **166** 352-8

[2] Weiss M, Zerfass A and Helmers E 2019 Fully electric and plug-in hybrid cars – an analysis of learning rates, user costs and costs for mitigating CO_2 and air pollutant emissions *Journal of Cleaner Production* **212** 1478-89

[3] Breetz H L and Salon D 2018 Do electric vehicles need subsidies? Ownership costs for conventional, hybrid and electric vehicles in 14 U.S. cities *Energy Policy* **120** 238-49

[4] Horn M, MacLeod J, Liu M, Webb J and Motta N 2019 Supercapacitors: A new source of power for electric cars? *Economic Analysis and Policy* 61 93-103

[5] Li Y, Huang X, Liu D, Wang M and Xu J 2019 Hybrid energy storage systems and energy distribution strategy for four-wheel independent-drive electric vehicles *Journal of Cleaner Production* **220** 756-70

[6] Luin B, Petelin S and Al-Mansour F 2019 Microsimulation of electric vehicle energy consumption *Energy* **174** 24-32

[7] Maamria D, Gillet K, Colin G, Chamaillard Y and Nouillant C 2018 Computation of eco-driving cycles for Hybrid Electric Vehicles: Comparative analysis *Control Engineering Practice* **71** 44-52

[8] Fathbadi H 2018 Fuel cell hybrid electric vehicle (FCHEV): Novel fuel cell/SC hybrid power generation system *Energy Conversion and Management* **156** 192-201

[9] Fathbadi H 2018 Novel fuel cell/battery/supercapacitor hybrid power source for fuel cell hybrid

electric vehicles *Energy* **143** 467-77

[10] Uzuneanu K 2018 Theoretical study on performance of a combined gas and steam turbine propulsion system for road transport Proceedings of the 8th International Conference on Advanced Concepts in Mechanical Engineering, IOP Conf. Series: Materials Science and Engineering **444** 082026

[11] Nader W B, Mansour C, Dumand C and Nemer M 2018 Brayton cycles as waste heat recovery systems on series electric vehicles *Energy Conversion and Management* **168** 200-14

[12] Karvountzis-Kontakiotis A, Andwari A M, Pesyridis A, Russo S, Tuccillo R and Esfahanian V 2018 Application of Micro Gas Turbine in Range-Extended Electric Vehicles *Energy* **147** 351-61

[13] Balanescu D T and Homutescu V M 2018 A New Approach in Combined Cycles – Low Power Unit for Hybrid Terrestrial Propulsion. Performance Assessment *Proceedings of the 11th International Conference Interdisciplinarity in Engineering, Inter-Eng 2017, Procedia Manufacturing* **22** 730-37

[14] Balanescu D T, Homutescu V M and Popescu A 2018 Micro gas and steam turbines power generation system for hybrid electric vehicles *Proceedings of the 8th International Conference on Advanced Concepts in Mechanical Engineering, IOP Conf. Series: Materials Science and Engineering* **444** 082022

[15] Chen Y, Ma J, Han B, Zhang P, Hua H, Chen H and Xin Su 2018 Emissions of automobiles fueled with alternative fuels based on engine technology: A review *Journal of traffic and transportation engineering* **5 (4)** 318-34

[16] Li L, Yan B, Song J, Zhang Y, Jiang G and Li L 2016 Two-step optimal energy management strategy for single-shaft series-parallel powertrain *Mechatronics* **36** 147-58

[17] Wang E, Guo D and Yang F 2015 System design and energetic characterization of a four-wheel-driven series–parallel hybrid electric powertrain for heavy-duty applications *Energy Conversion and Management* **106** 1264-75

[18] Kehlhofer R, Hannemann F, Stirnimann F and Rukes B 2009 *Combined-Cycle Gas & Steam Turbine Power Plants - 3rd edition* (New York: PennWell)

[19] Chaim M B, Shmerling E and Kuperman A 2015 A Fuel Efficiency Evaluation of Gas-Turbine-Engine-Based Hybrid Vehicles International Journal of Green Energy **12 (4)** 328-32

[20] Quiros D C, Smith J, Thiruvengadam A, Huai T and Hu S 2017 Greenhouse gas emissions from heavy-duty natural gas, hybrid, and conventional diesel on-road trucks during freight transport *Atmospheric Environment* **168** 36-45

Acknowledgments

This work was partly supported by the project POSCCE-A2-O2.2.1-2009-4-ENERED, ID nr. 911, co-financed by the European Social Fund within the Sectoral Operational Program "Increase of Economic Competitiveness".

IOP Conf. Series: Materials Science and Engineering **595** (2019) 012041 doi:10.1088/1757-899X/595/1/012041

About Structural and Thermal Analysis of Diesel Engine Piston Using Ansys Software

M Frățița, F Popescu, K Uzuneanu, I V Ion and C M Anghelușă

Thermal Systems and Environmental Engineering Department, "Dunărea de Jos" Universitaty of Galați, Romania

E-mail: michael.fratita@ugal.ro

Abstract. The main role of the piston is to vary the dimensions of the combustion chamber in internal combustion engine. The piston is in contact with the hot gases resulting from combustion. In addition to thermal stresses, depending on the operating mode of the engine, the piston is required to alternate the inertial force. Therefore, the steel piston undergoes due to high inertial forces, and lightweight aluminium pistons cannot operate at high temperatures. In order to combine the advantages of the two types of materials it has been chosen the constructive design of the bimetal pistons. This paper aims to compare the performances of steel pistons and bimetal pistons using the Ansys software.

1. Introduction

The piston is an important component of the internal combustion engines. It has the role to shut off the combustion chamber and to convert the energy of the burned gases into mechanical work. Also, the piston creates a variable space in the combustion chamber contributing to gas exchange. Using the connecting rod - crank shaft mechanism, the translation movement of the piston is converted to rotation movement. The piston is subjected to cyclical actions of the pressure resulted from the burned gases and the inertial forces. The intensity of the cycles depends on engine parameters (rotation/min and load). The design of the piston crown depends on by the engine type (two stroke or four stroke engine) and the fuel (gas, compressed natural gas, and diesel).

The electronic computer unit of the engine can process a large number of sensors. The manufacturers of turbo-charged diesel motor started to use more sensors. These are temperature, pressure and actuator position sensors in order to obtain an accurate control of the engine. To avoid reaching a high temperature on the piston crown that causes melt or damage, engine manufacturers have installed exhaust gas temperature (EGT) sensor. When the temperature increases to the threshold of 800°C, the electronic computer unit reduces the value of the fuel injected by limiting the maxim power of the vehicle. The sensor can spot malfunction in engine operation when the injectors do not work properly and doses more fuel, preventing accidental melting of the piston.

Diesel engines working at high pressures and high exhaust gas temperatures up to 1200°C use bimetal piston. The piston crown is made of high strength forged steel, and skirt is made of aluminium alloys [1]. Thus, it can be made a piston with high temperature resistance and a less total weight than a piston made by steel or cast. Therefore, lower inertial forces allow a better operation of the engine at higher speed.

IOP Conf. Series: Materials Science and Engineering **595** (2019) 012041 doi:10.1088/1757-899X/595/1/012041

2. Materials for pistons

The most commonly used materials for manufacturing the pistons are: Aluminium alloys and Steel Cast.

The engine type and the type of fuel used allows the material to be classified according to the maximum temperature of the piston crown, such as [2]:

- Aluminium alloys for spark ignition engine with the maximum temperature of the piston crown of 280°C;
- Aluminium alloys for compression ignition engine with the maximum temperature of the piston crown of 350°C;
- Cast for compression ignition engine with the maximum temperature of the piston crown of 450°C.

3. The main size of the piston

Following the size of the piston has been released the data presented in Table 1 [2, 3, 4].

Table 1. Piston dimensions.

Parameter	Value
L_p – piston length	68.8 mm
L_m – skirt length	45.5 mm
H_c – Ring belt height	45.8 mm
h – crown piston to the top compression ring	10 mm
h_c – distance between the ring grove	2.5 mm
δ – head rib	12.5 mm

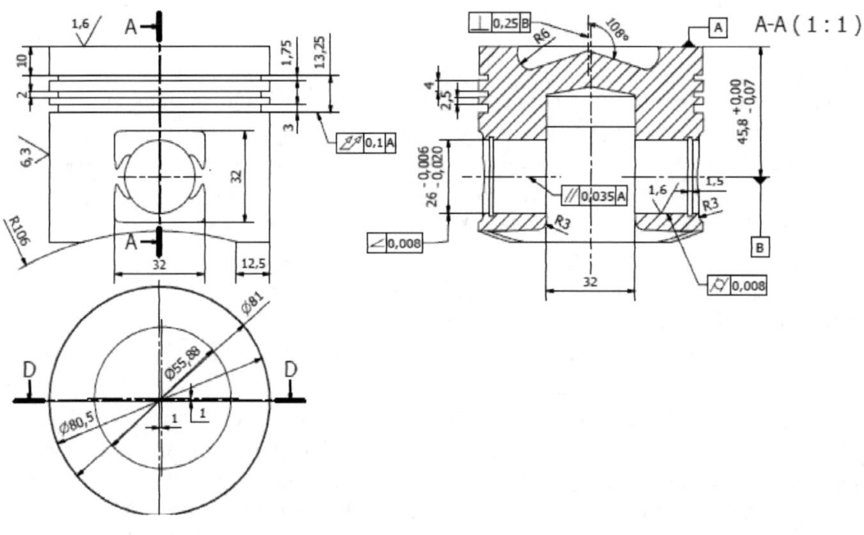

Figure 1. Drawing of piston.

4. 3D modelling of the piston

The 3D modelling of the piston has been made using Autodesk Inventor after creating the theoretical size. In the following figure it can be seen a 3D steel piston.

Figure 2. Steel piston.

The second 3D model presented in figure 3 is a bimetal piston. It is composed of two parts. The piston crown is made by steel and the skirt is made by aluminium alloy. This model has the advantage of using the steel high resistance and light weight of aluminium alloy.

Figure 3. Bimetal piston.

The general properties of 3D models can be seen in figure 4 a) and b). It can be seen that the steel piston assembly and the wrist pin has a total weight of 1418 grams, while the bimetal piston assembly has a total weight of 1119 grams. Also, the inertial moments of bimetal piston are lower than the steel piston.

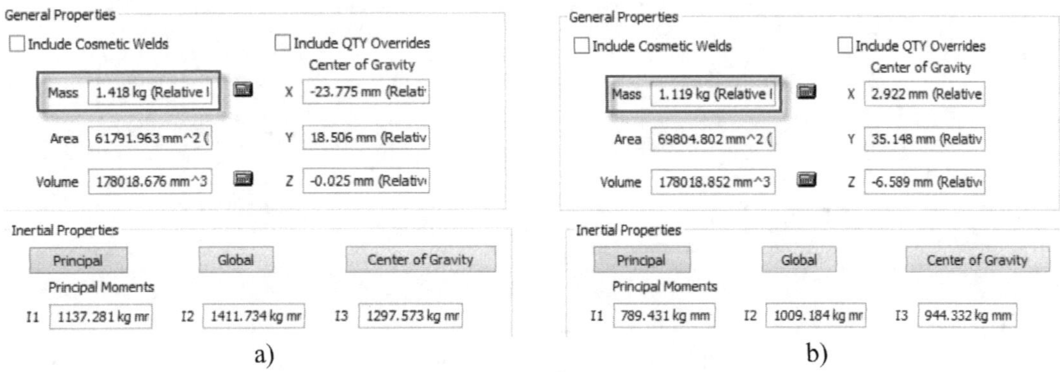

Figure 4. General properties of piston.

IOP Conf. Series: Materials Science and Engineering **595** (2019) 012041 doi:10.1088/1757-899X/595/1/012041

5. Structural and Thermal Analysis

5.1.Meshing

The static analysis is performed using Ansys software. For generating the mesh, it was used uniform network [5, 6] with the minimal surface of 0.02 mm. In the case of the steel piston and wrist pin, shown in figure 5, the model has been meshed in 149 180 elements by 241 492 nodes. In the second case, the assembly of the bimetal piston has been meshed in 155 048 elements by 253 912 nodes.

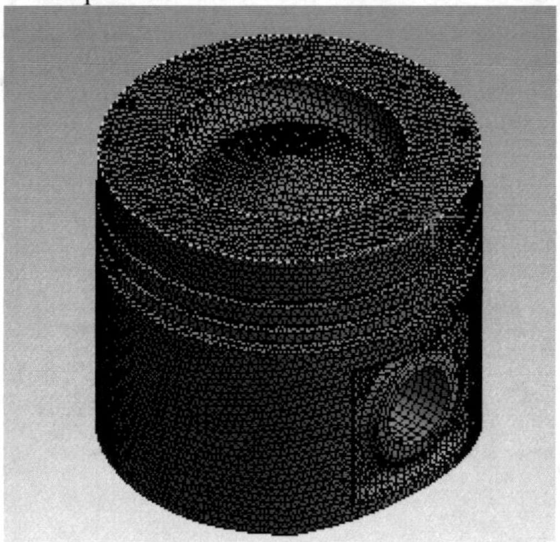

Figure 5. Meshing of piston.

5.2 Constrain

In this study, the pressure of gases has been applied on the piston crown surface having a maximum value of 19 MPa. This value results after thermal calculation of the engine. Also, the wrist pin has been considered fixed element. The second simulation uses the same constrain.

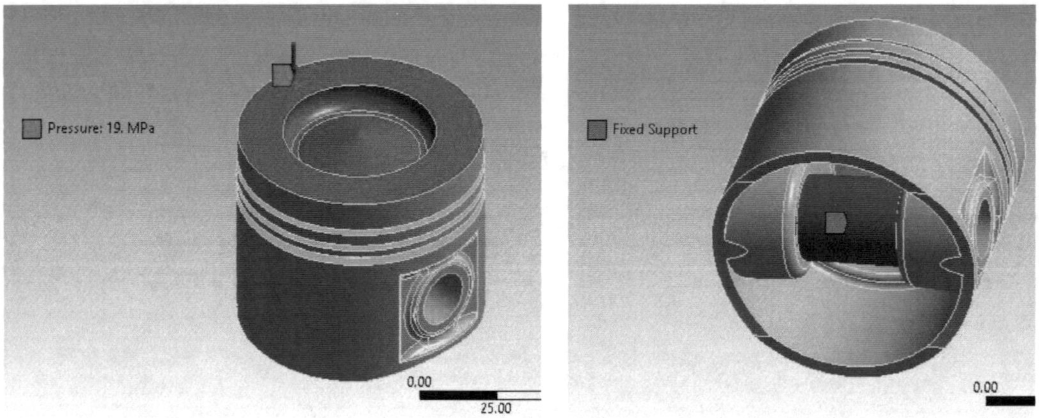

a) b)

Figure 6. Constrains of piston.

5.3 Static Structural

The static structural analysis has been performed in Static Structural Module of Ansys Software. The maximum stress recorded for steel piston is 302 MPa. It can be seen the maximum stress is recorded between the head rib and the piston pin boss reinforcement area.

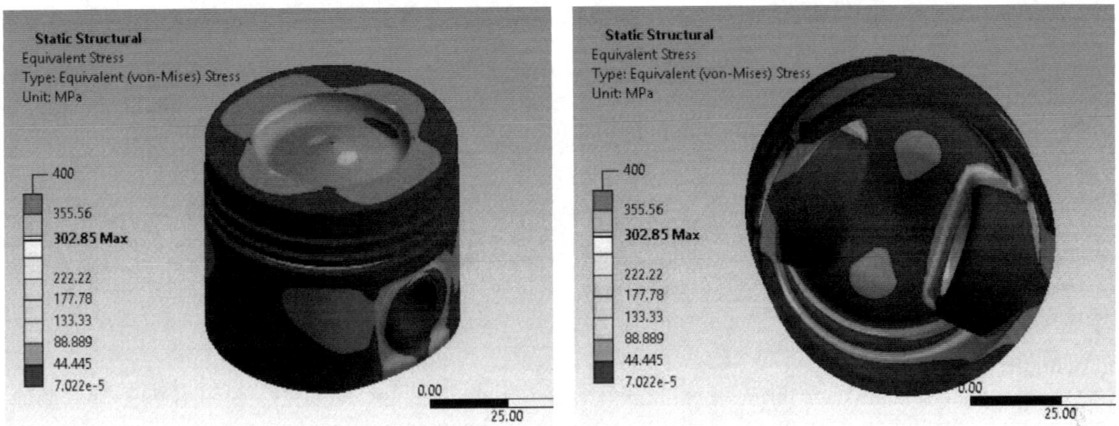

Figure 7. Static Structural Steel Piston.

For bimetal piston it has been recorded a higher von-Misses stress of 358 MPa. This value is also recorded between the head rib and the piston pin boss reinforcement area.

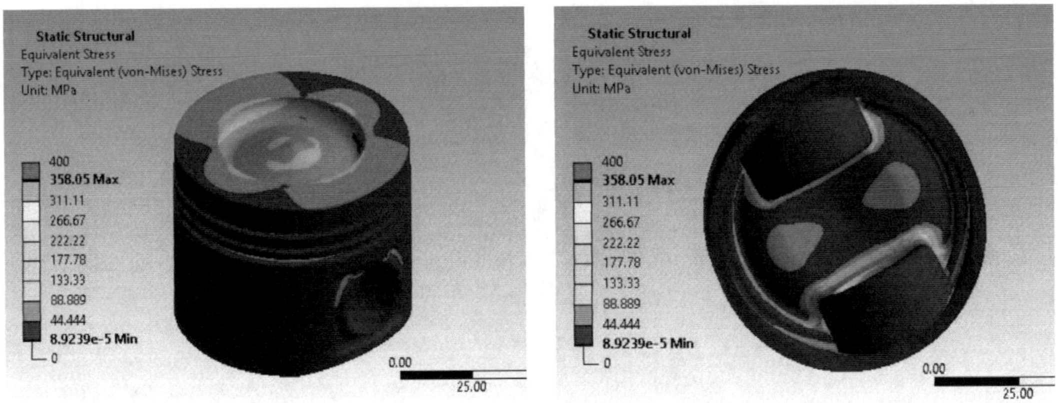

Figure 8. Static Structural Bimetal piston.

5.4 Thermal Study

In terms of thermal study [7], it has been considered the temperature of the combustion chamber in the piston crown area of 450°C, in both situations. It can be observed the temperatures of piston skirts are different due to different materials. Since the thermal conductivities are different, the temperature of aluminium alloy skirt is 30°C higher than steel skirt temperature.

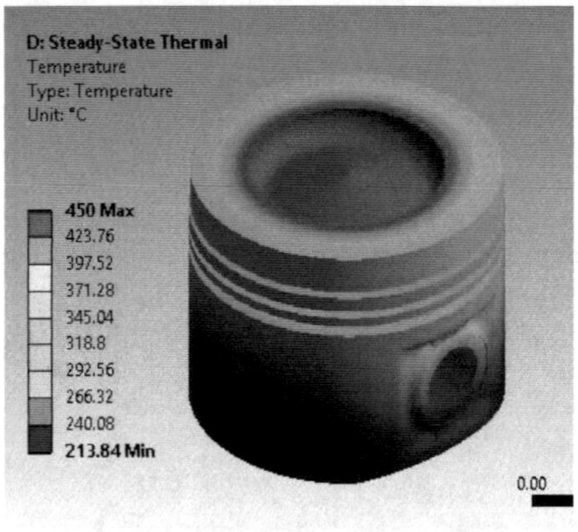

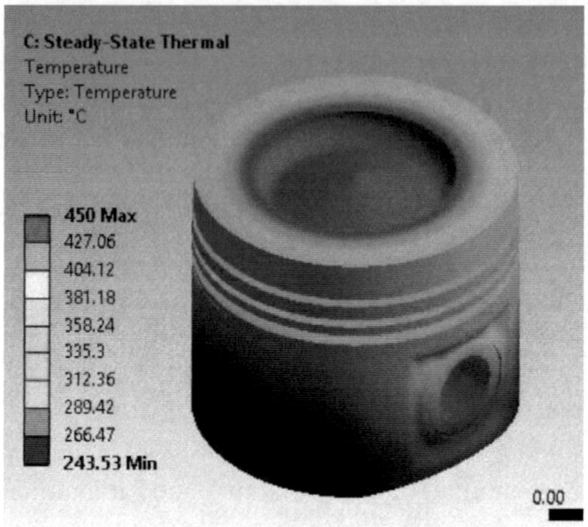

Figure 9. Thermal Study of pistons.

6. Conclusions

Creating a piston by two different mechanical assembled materials increases the complexity of its design as well the complexity of manufacturing. According to the static structural analysis, it is necessary to consider the design of the piston pin boss reinforcement to reduce the stress concentrators. Also, the bimetal piston weight, for the analysed piston, is lower by 299 grams, resulting lower inertial forces. Due to lower thermal conductivity coefficient of the steel, it can be used lower tolerance at the piston crown, resulting a much higher efficiency at cool start of the engine. Since the bimetal pistons have only the piston crown made of steel, it can be saved high-quality steel consumption.

References

[1] Reddy K A 2016 *Design and Analysis of Toroidal Piston Using Ansys* Journal of Mechanical Engineering and Applied Mechanics. **1(2)** p 9-21

[2] Uzuneanu K and Gheorghiu C 2006 *Internal Combustion Engines* (in Romanian) (Ed Fundaţiei Universitare "Dunărea de Jos", Galaţi 2006)

[3] Uzuneanu K and Panait T 2011 *Analysis of heat transfer in the combustion chamber of an internal combustion engine using thermal networks* (Recent Advances in Fluid Mechanics and Heat & Mass Transfer – 9th IASME/WSEAS International Conference on Heat transfer, Thermal Engineering and Environment (HTE '11), Florence, Italy, Aug. 23-25, 2011, ISBN 978-1-61804-026-8, p 257 – 262)

[4] Nimbarte V R and Khamankar S D 2015 *Stress Analysis of Piston using Pressure Load and Thermal Load* IPASJ International Journal of Mechanical Engineering. **3(8)** p 1-8

[5] Vasudev K R, Rajesh K R and Nageswara T R 2018 *Design and Analysis of Bimetallic Piston* International Journal of Latest Technology in Engineering, Management & Applied Science (IJLTEMAS) VII(VI) p 134-138

[6] Gopala G and Kumarb L S 2017 *Analysis of Piston, Connecting rod and Crank shaft assembly* Materials Today: Proceedings. **4(8)** p 7810–19

[7] Yadav M V, Sridhar A V and Krishna R V V 2016 *Thermal Analysis of Bimetallic Piston Using Different Material Combinations* International Journal & Magazine of Engineering Technology, Management and Research. **3(3)** p 831-37

IOP Conf. Series: Materials Science and Engineering **595** (2019) 012042 doi:10.1088/1757-899X/595/1/012042

Aerodynamic and structural analysis of a small-scale horizontal axis wind turbine using QBlade

D Zahariea[1], D E Husaru[2] and C M Husaru[2]

[1]Fluid Mechanics, Fluid Machinery and Fluid Power Systems Department, "Gheorghe Asachi" Technical University of Iasi, Iasi, Romania
[2]"Gheorghe Asachi" Technical University of Iasi, Iasi, Romania

E-mail: dzahariea@tuiasi.ro

Abstract. In this paper the aerodynamic and structural analysis of a small-scale horizontal axis wind turbine using QBlade software will be presented. QBlade software will be used to validate the wind turbine design, according with the Blade Element Momentum theory. The influence of the tip and root losses on the power curve will be analysed. The pitch angle control curve will be obtained. The annual energy production will be analysed on the basis of Weibull parameters. The structural analysis in terms of modal and static structural will be presented for three different internal structures: solid blade, hollow blade without spar, and hollow blade with spar. It has been found that the natural frequencies obtained for lateral translations (flapwise and edgewise), chord rotation around the blade longitudinal axis and longitudinal translation are in a safety range from the harmonics of wind turbine rotational speed at the operating condition, thus avoiding the resonance.

1. Introduction

One of the most important open source software for design and simulation of wind turbines with horizontal and vertical axis is QBlade, [1]. QBlade software can be used for airfoils design and analysis [2], due to integrated XFOIL code [3]. QBlade software is based on the Blade Element Momentum (BEM) theory, thus is widely used for simulation [4], optimization [5], and validation of horizontal axis wind turbine design [6]. CFD analysis is used with good results for validation of QBlade computations for wind turbine design in [7]. Some other results that are used for validation of QBlade analysis against other similar software, and against experimental data are presented in [8]. QBlade is also used with good results for vertical axis wind turbine simulation [9].

In this paper, QBlade software will be used for aerodynamic and structural analysis of a 10-kW horizontal axis wind turbine. The wind turbine has been designed using BEM method with NREL's S823 airfoil; QBlade software being used in this paper for design validation considering as design parameters the power coefficient and the power of the wind turbine. The influence of root and tip losses on the power and power coefficient will be also analysed. Using QBlade software, the distribution of the pitch angle of the rotor blade for which the wind turbine power will be limited to the rated power for high winds will be computed. The influence of the site wind potential on the wind turbine annual energy production will be also analysed.

From the point of view of blade structure, three constructive types will be considered: the solid blade, the hollow blade without spar, and the hollow blade with longitudinal spar. QBlade will be used for computing the blade mass, the modal frequencies for flapwise, edgewise, torsional and longitudinal directions, as well as for evaluating the von Misses stress.

2. Wind turbine characteristics

The wind turbine that will be analysed in this paper has been designed and presented as Bachelor's thesis at "Gheorghe Asachi" Technical University of Iasi in 2017, [10]. It is about a wind turbine with horizontal axis and z=3 blades, with rated power P_r=10 kW, rated wind velocity U_r=10 m/s, and rated rotational speed n_r=175 rpm. The wind turbine has been designed for a site where the wind potential can be described by the following Weibull parameters: the shape parameter k=1.85 and the scale parameter A=7.35 m/s. The rated tip speed ratio of the wind turbine is TSR=6.5 while the rated power coefficient is CP_r=0.4. The wind turbine has been designed using BEM theory.

3. Airfoil characteristics

The starting point of the designing process has been the selection of an airfoil and the analysis of the airfoil characteristic parameters (pressure coefficient, boundary layer shape, polars). For this wind turbine, the NREL's S823 airfoil has been selected, for which the maximum thickness is 21.2% at 24.3% of the chord, and the maximum camber is 2.4% at 70.5% of the chord [11]. This airfoil is not included in the QBlade default airfoil database, therefore, has to be imported from an external file. The external file with airfoil geometric characteristics is obtained from online airfoil database, [11]. Once the airfoil is imported in the QBlade project, the distribution of the pressure coefficient and the boundary layer shape on the lower and the upper surfaces of the airfoil can be analysed. For example, in Figure 1 the pressure coefficient C_p is presented for the optimum angle of attack α_{opt}=7°.

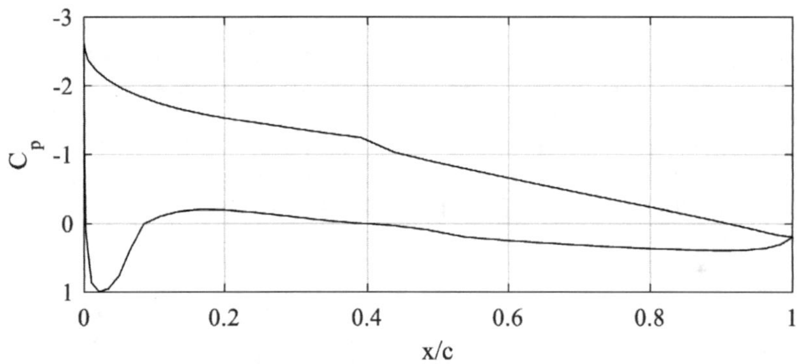

Figure 1. The pressure coefficient.

The angle of attack has a strong influence on the pressure coefficient, but also on the boundary layer shape. The boundary layer shape is comparatively presented in Figure 2, a) for the optimum value of angle of attack α_{opt}=7°, while figure 2, b) corresponds for the angle of attack of 11°. This kind of analysis shows that for the optimum angle of attack the boundary layer separation point on the upper side of the airfoil is very close to the trailing edge of the airfoil, Figure 2, a). Increasing the angle of attack value, the separation point is moving away from the trailing edge towards the leading edge, while the thickness of the boundary layer is increasing with respect to the optimum angle of attack case.

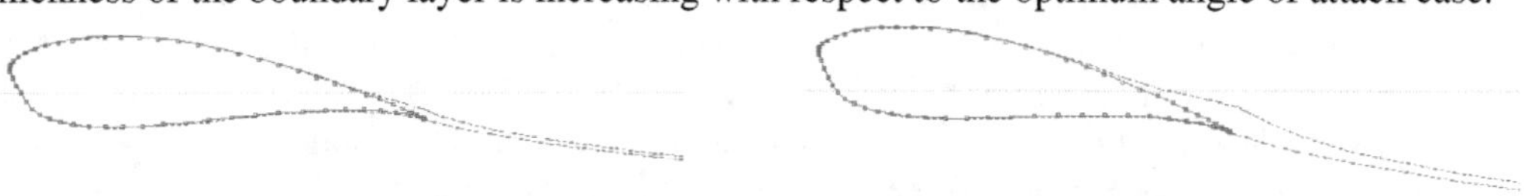

a) angle of attack = 7° b) angle of attack = 11°

Figure 2. Boundary layer and the separation point.

Finally, the airfoil polars can be analysed, no matter the source of these polars: imported from external files with experimental data, or generated inside QBlade. These polars are obtained for a usual range of angle of attack, for example -5°÷15°. Most likely is that during different working stages of the wind turbine, the airfoil will work under angles of attack outside the initial range, therefore the polars must be extrapolated on the full range -180°÷180°. QBlade can perform this extrapolation using Montgomerie and Viterna methods [12, 13]. For this analysis, the Montgomerie method has been used and the lift, the drag and the lift to drag coefficients can be obtained. For example, the lift coefficient for full angles of attack range is presented in Figure 3.

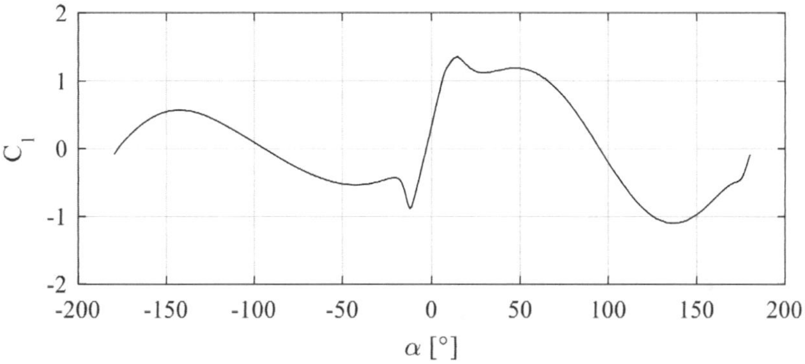

Figure 3. Lift coefficient for -180°÷180° with Montgomerie extrapolation method.

Analysing the airfoil polars, has been observed that the angle of attack for which the lift to drag ratio has the maximum value is α_{opt}=7°, and this will be the angle of attack for all blade sections starting from hub until the tip. For this specific angle of attack, the lift coefficient is 1.082, the drag coefficient is 0.01256, while the lift/drag ratio has its maximum value of 86.1265, [11].

The characteristic curves of the airfoil are strongly dependent of the Reynolds number, therefore the Reynolds number must be chosen before starting the aerodynamic design. This Reynolds number, known as polar Reynolds number, has been chosen to be Re_p=5x10^5. During the aerodynamic design, based on the velocity resulted in the computation process, another Reynolds number will be computed, the blade Reynolds number. For this case, the blade Reynolds number resulted in the range Re_b=(5.07÷5.32)x10^5 with smaller value close to the hub. Thus, there is a relative error between these two Reynolds numbers in range of ε_{Re}=1.4÷6.4%, with smaller value close to the hub, figure 4. The graph presented in Figure 4 has been obtained using a fourth-degree polynomial fit. This error, noticed at the Reynolds number for the airfoil characteristic curves, indicates that the real wind turbine will have different global characteristics. However, in this particular case, the maximum relative error (6.4%) is small enough in order to consider that the computed global characteristics will be close to the real wind turbine characteristics.

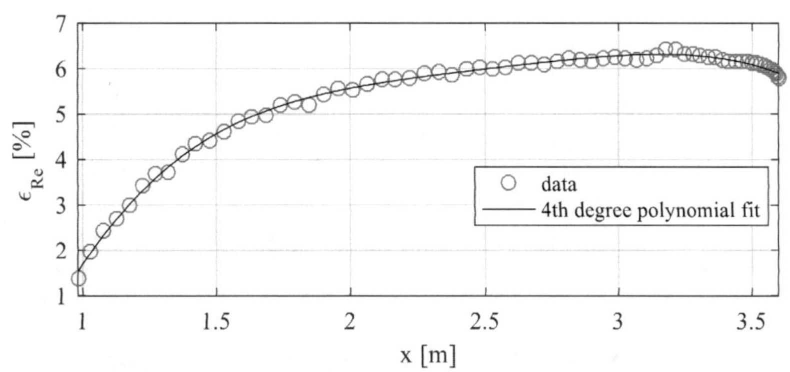

Figure 4. Relative error between the polar and the blade Reynolds numbers.

4. Blade characteristics

The main geometric parameters of the wind turbine blade are: the tip radius R_t=3.6 m, the hub radius R_h=0.15 m, the intermediate radius which will define the active part of the blade until the tip, as well as the inactive part of the blade until the hub R_a=0.972 m. For this study, the intermediate radius is defined by the radius ratio $k = R_a/R_t$=0.27. Thus, the blade length will be L=2.628 m. The chord length c_{opt}[m] (computed with the Schmitz relationship, [12]) and the blade twist angle β_{opt}[°] along the blade are presented in figure 5 and figure 6.

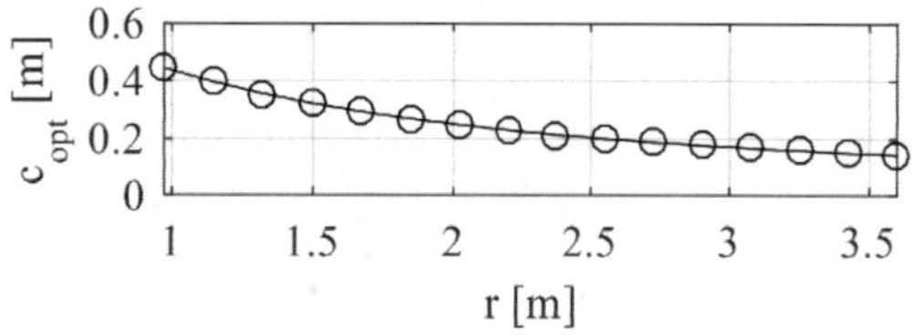

Figure 5. The chord length.

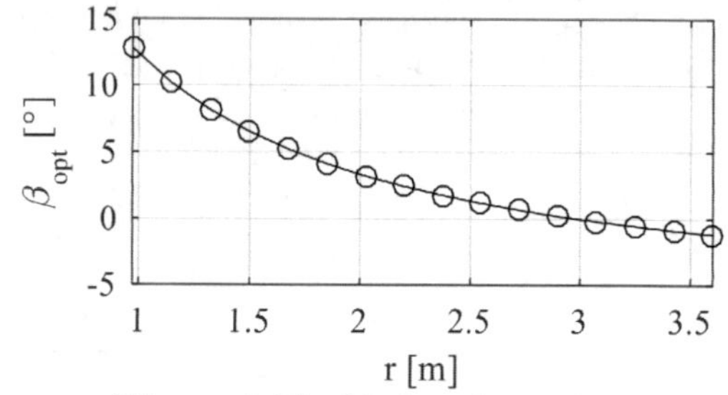

Figure 6. The blade twist angle.

The wind turbine blade is presented in Figure 7. There are indicated the position and the airfoil name for all sections along the blade. The active part of the blade is between the tip radius and the intermediate radius; for this part, a single airfoil is used in all sections. The inactive part of the blade is between the intermediate radius and the hub radius and is composed by two parts: the cylindrical part for connection with the hub, and the interpolated part between airfoil section and the circular section.

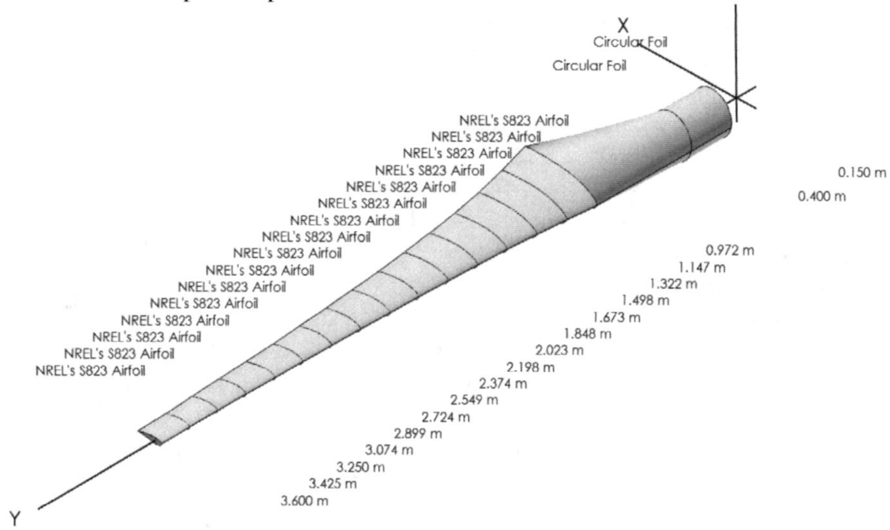

Figure 7. The blade.

5. Aerodynamic analysis

After the blade and rotor definition several aerodynamic analyses can be performed. First of all, the power coefficient and the power of the wind turbine can be comparatively analysed for three cases: without tip and root losses, only with tip losses, and only with root losses. The tip and root losses are defined by the Prandtl's loss factor, [14]. The Prandtl's loss factors for the tip and for the root are presented in Figure 8. Figures 9, a) and 9, b) shows that there is a small influence of the root losses on both the power coefficient, and the power of wind turbine. On the other hand, the influence of the tip losses is very important and should never be neglected on the designing process of wind turbines. For example, from figure 9, b), the wind turbine power without any losses is 11.3 kW; if only the tip losses will be considered, then the power will be 10.3 kW with a decrease in power of around 8.85%, and finally, if only the root losses will be considered, the power, basically, does not change.

The point marked with a black circle represents the wind turbine rated point in terms of power coefficient (figure 9, a) and wind turbine power (figure 9, b). There is a good agreement between the wind turbine rated data (TSR=6.5⇒CP_r=0.4 and P_r=10 kW) and the values computed in QBlade (TSR=6.5⇒CP=0.4123 and P=10.3 kW).

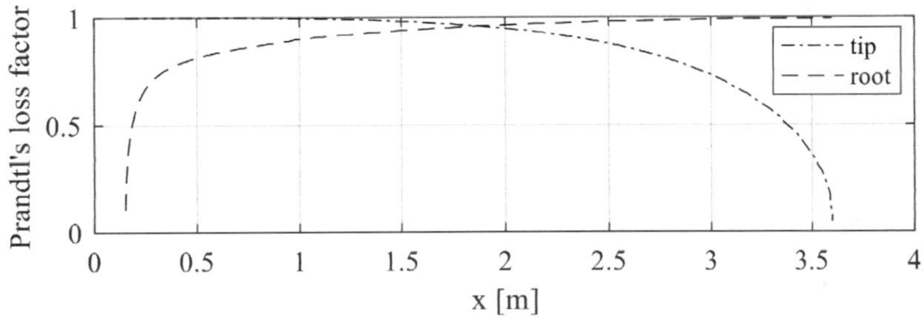

Figure 8. Prandtl's loss factor for the root and for the tip.

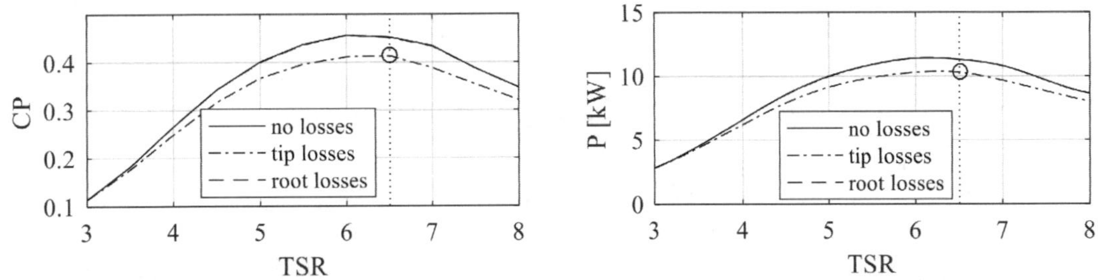

a) power coefficient, $CP(TSR)$ b) power, $P(TSR)$

Figure 9. The power coefficient and the power of wind turbine.

Another aspect regarding the aerodynamic analysis is related with the power control of the wind turbine by pitch regulation, which is one of the most important method for limiting the rotor power and speed, [15]. Pitch regulation represents the process of mechanically adjusting of the blade pitch angle in order to maintain the rated power of the wind turbine for high wind speeds, above the wind speed rated value, which in this case is U_r=10 m/s. QBlade is able to obtain the required distribution of the pitch angle for which the wind turbine power will be limited to the rated value (P_r=10 kW) in case of high winds. This pitch angle curve will be further use for controlling the actuator which will rotate the blade around its longitudinal axis. The pitch angle curve control and the wind turbine power curve are presented in figure 10. Analysing these curves, can be observed the cut-in wind speed (the minimum wind speed at which the wind turbine will generate power) which is 4 m/s, and the cut-off wind speed (the maximum wind speed at which the wind turbine will generate power) which is 25 m/s. Between the cut-in wind speed and the rated wind speed there is the domain of partial-loads, where the wind turbine power depends with the wind speed based on a non-linear relationship. For this domain, the pitch angle is set to zero. Between the rated wind speed and the cut-off wind speed there is the domain of full loads, where the wind turbine power should have the constant value of the rated power. In order to maintain this constant value, despite the increase in wind speed, the aerodynamic angle of attack should change according with the pitch angle curve, starting from -6.5° at 11 m/s until -28.2° at 25 m/s, figure 10.

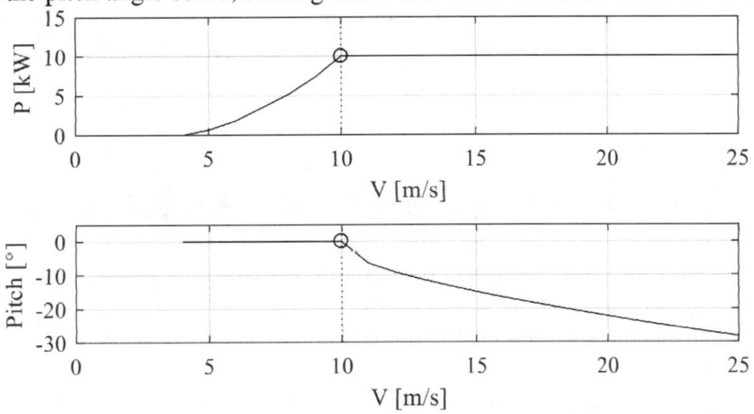

Figure 10. The pitch angle control curve.

The annual energy production (AEP, [MWh/year]) of the wind turbine is the total amount of energy produced during a year and can be obtained by multiplying the power from the power curve with the wind speed frequency distribution. This wind turbine has been designed for a wind speed frequency defined by the shape parameter k=1.85 and the scale parameter A=7.35 m/s. However, the wind turbine can operate in different sites, for different wind potentials. For example, the influence of the Weibull parameters on the annual energy production is presented in Figure 11, a) for the scale parameter, and figure 11, b) for the shape parameter.

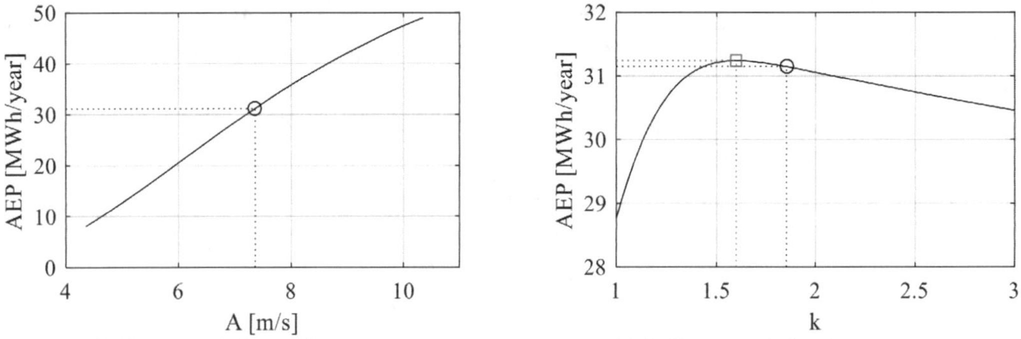

a) influence of the scale parameter b) influence of the shape parameter

Figure 11. Annual energy production.

The scale parameter is related with the mean wind speed, thus, increasing the scale parameter, the annual energy production will increase as well, Figure 11, a). The point parked with a black circle represents the annual energy production for the wind turbine, if will operate on the designed site. For this case, the annual energy production is 31.149 MWh/year. On the other hand, the dimensionless shape parameter which is related with the breadth of the wind speed distribution, has a more complex influence on the annual energy production. The maximum annual energy production, which is 31.242 MWh/year can be obtained on a site characterized by the shape parameter 1.597, Figure 11, b). The red square on the Figure 11, b) represents the maximum annual energy production in terms of shape parameter.

6. Modal analysis

Three types of blades have been analysed: solid blade, hollow blade without spar, and hollow blade with a longitudinal spar. The blade is made by a generic double bias material with density of 1750 kg/m^3 and Young's modulus 1.2e+10 Pa, while the spar is made by a generic triaxial material with density of 1850 kg/m^3 and Young's modulus 2e+10 Pa. For the hollow blades, the shell thickness is variable along the blade with 2% of the chord thickness. The spar thickness is also variable along the blade with 8% of the chord thickness. The spar position is defined by 25% of the chord, and the spar angle is 0°. Considering all these characteristics, the blades mass computed with QBlade are: 92.4527 kg for the solid blade, 18.5303 kg for the hollow blade without spar, and 27.9298 kg for the hollow blade with spar.

Modal analysis consists in determining the modal frequencies (natural vibration frequencies) and the mode shapes for different rotor speed. In order to compare the modal frequencies with the excitation frequencies generated by the rotation of the rotor blades, the data computed with QBlade are presented in Campbell plot format, figure 12. Two types of excitation frequencies have been considered: the blade passing frequency associated with once-per-revolution (1P), and the rotor passing frequency associated with three-per-revolution (3P), as the wind turbine analysed in this paper has three blades. QBlade can compute modal frequencies for different bending direction: flapwise, edgewise, torsional and longitudinal. The results presented in figure 12 shows that the most dangerous bending direction is the flapwise for which the modal frequencies are closer to the excitation frequencies 1P and 3P than the modal frequencies in edgewise bending direction. However, even the flapwise modal frequencies are in a safety range from the excitation frequencies 1P and 3P for all the wind turbine operating conditions. For example, at rated speed of 175 rpm the excitation frequencies are 2.9167 Hz (1P) and 8.75 Hz (3P), while the modal frequencies in flapwise direction are 21.1694 Hz (solid blade), 19.1978 Hz (hollow blade without spar), and 20.0929 Hz (hollow blade with spar).

Because of the order of magnitude, only flapwise and edgewise modal frequencies has been presented in Figure 12. This decision is justified, because, for example, for the solid blade, the 1st mode of the torsional modal frequencies are around 273 Hz, while for the longitudinal direction the 1st mode of modal frequencies are around 326 Hz; these values are far from the excitation frequencies 1P (2.9167 Hz) and 3P (8.75 Hz), thus, will not be presented. Moreover, only the 1st mode of modal frequencies will be presented, as for the 2nd, the 3rd, and so on, the modal frequencies are higher than for the 1st mode. For example, for the solid blade and the flapwise direction, the modal frequencies for the 2nd, the 3rd, and the 4th modes are: 62.5463 Hz, 102.095 Hz and 144.288 Hz (at rated speed of 175 rpm).

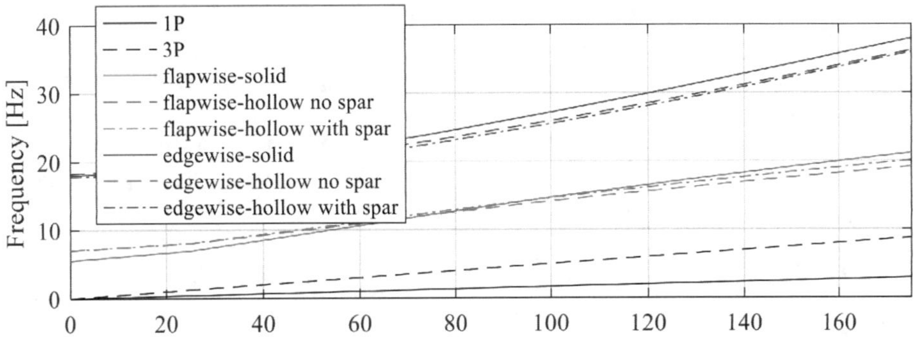

Figure 12. Campbell diagram.

7. Structural analysis

The von Misses stress has been computed for all three blades (solid blade, hollow blade without spar and hollow blade with spar) for aerodynamic loading defined by the rated wind speed of 10 m/s, figure 13. From the point of view of structural behaviour, the best option is the solid blade; however due to the high mass which is 92.4527 kg, other options should be considered. The smallest value of the mass is for the hollow blade without spar; however, for this case the highest von Misses stress values occur. On the other hand, the solid blade with spar is in-between the other two constructive solutions in terms of mass and von Misses stress. Another interesting thing that can be observed from figure 13 is about the position of maximum von Misses stress, which is at the hub radius R_h=0.15 m for the hollow blades, and around the intermediate radius R_a=0.972 m for the solid blade.

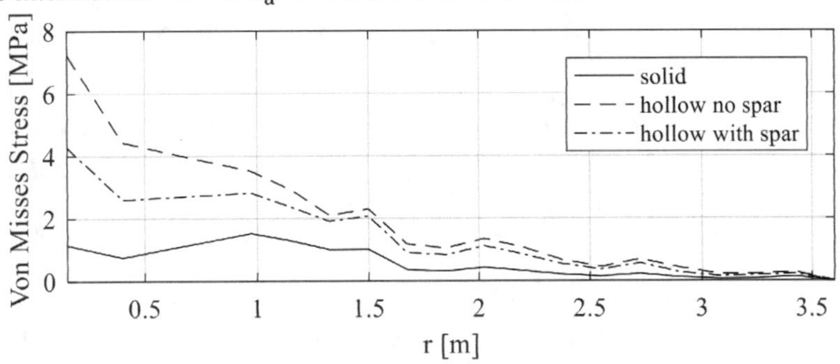

Figure 13. Von Misses stress.

8. Conclusions

QBlade is a very useful tool for performing different design and validation tasks in the field of wind turbines: airfoil and polar design, blade and wind turbine design, static structural and modal analysis. Several aerodynamic analyses can be made at detailed level, for example the influence of tip and root losses on the wind turbine power parameters. Good agreement has been observed between the rated and computed data: 3% for the power and 3.075% for the power coefficient.

The pitch angle control curve is another extremely important result which establishes the required range for the pitch angle 0÷-28.2° for maintaining the rated power of 10 kW for wind speeds is the domain 10÷25 m/s.

The annual energy production obtained for the designed site (k=1.85, A=7.35 m/s) is 31.149 MWh/year. However, better annual energy production can be obtained for sites with smaller shape parameter, in range k=1.43÷1.85, with a maximum value of 31.242 MWh/year for k=1.597.

The modal analysis shows that the most dangerous situation is defined by the hollow blade without spar and the flapwise bending direction; however, even for this case there is a safety margin of around 120%. A better behaviour has been noticed for the hollow blade with spar; the safety margin in this case is around 130%. The hollow blade with spar is the better option from the point of view of the relationship between the mass blade and the von Misses stress. However, because of the QBlade material limitation, the results obtained for the modal and static structural analyses can be considered only as a first approximation.

Due to the specific approximations of the BEM theory, which represents also default limitations of the QBlade software, these results, even if they are encouraging and obtained with very low computational effort, should be used only as preliminary design. As consequence, further investigations using more accurate methods, like CFD analysis, should be performed for obtaining results more closely to the wind turbine real behaviour, which of course, can be exactly determined only by experiment.

References

[1] Hermann Fottinger Institute of TU Berlin, QBlade. Wind turbine design and simulation, http://www.q-blade.org/, accessed at 5.02.2019

[2] Dhurpate P R, Sutar K B and Kale S 2016 Numerical analysis of different airfoils using QBlade software, *Imp. J. of Interdiscip. Res.* **2** 6

[3] Drela M, XFoil Subsonic Airfoil Development System, http://web.mit.edu/drela/Public/web/xfoil/, accessed at 15.02.2019

[4] Raut S, Shrivas S, Sanas R, Sinnarkar N and Chaudhary M K 2017 Simulation of micro wind turbine blade in QBlade, *Int. J. for Res. in Appl. Sci. & Eng. Technol.* **5** IV 256-62

[5] Tenghiri L, Khalil Y, Abdi F and Bentamy A 2008 Optimum design of a small wind turbine blade for maximum power production, *IOP C. Ser. Earth Env.* **161** 012008

[6] Faisal M 2017 Rotor blade performance analysis with blade element momentum theory, *Energy Procedia* **105** 1123-29

[7] Koc E, Gunel O and Yavuz T 2016 Comparison of QBlade and CFD results for small-scaled horizontal axis wind turbine analysis, *IEEE Int. Conf. on Renewable Energy Research and Applications (Birmingham)* 204-9

[8] Marten D, Wendler J, Pechlivanoglu G, Nayeri C N and Paschereit C O 2013 QBLade: an open source tool for design and simulation of horizontal and vertical axis wind turbines, *Int. J. of Emerging Technol. and Adv. Eng.* **3** 3 264-9

[9] Sharma R and Patel B 2015 Design and simulation of Darrieus (Eggbeater) type vertical axis wind turbine using open source software QBlade, *Int. J. for Innov. Res. & Technol.* **I** 12 162-9

[10] Husaru C M 2017 *Designing of a wind turbine on-grid connected power supply system for two consumers*, Bachelor's Degree Thesis, scientific supervisor Zahariea D, Technical University "Gheorghe Asachi" of Iași, Romania

[11] Airfoil Tools, http://airfoiltools.com/, accessed at 25.02.2019

[12] Marten D and Wendler J 2013 QBlade guidelines v0.6, http://q-blade.org/project_images/files/guidelines_v06.pdf, accessed at 26.02.2019

[13] Faisal M, Syerly K, Husni S and Surya H 2017 Airfoil lift and drag extrapolation with Viterna and Montgomerie methods, *Energy Procedia* **105** 811-6

[14] Hansen M O L 2008 *Aerodynamic of Wind Turbines*, 2nd ed. (London: Earthscan)

[15] Hau E 2006 *Wind Turbines, Fundamentals, Technologies, Application, Economics*, 2nd ed. (Berlin: Springer)

Acknowledgments

The authors would like to acknowledge the technical resources offered by the Laboratory of Computer-Aided Fluid Engineering (equipped with technical resources with the financial support of the grant ENERED, POSCCE-A2-O2.2.1-2009-4, ID 911), from the Department of Fluid Mechanics, Fluid Machinery and Fluid Power Systems, "Gheorghe Asachi" Technical University of Iasi, Romania.

IOP Conf. Series: Materials Science and Engineering **595** (2019) 012043 doi:10.1088/1757-899X/595/1/012043

Analysis of the terrain specifics and roughness factor on the wind shear over complex terrains

A K Terziev

Deptartment of "Power Engineering and Power Machines", Technical University of Sofia, Bulgaria

E-mail: aterziev@tu-sofia.bg

Abstract. Wind power production depends mainly on the distribution of wind speed over the terrain. Wind behaviour is a function of terrain features, the roughness of the terrain - the complexity of the relief (the presence of mountain ranges, hills or valleys) and the presence of natural or artificial obstructions along the wind's path (shrubs, trees, small and large buildings). In order to assess the factors influencing the distribution of the wind shear, long-term measurements with high meteorological masts installed on complex terrain were carried out. The masts are equipped with calibrated equipment to measure wind parameters. A numerical solution has been carried out with specialized software, which also supplied information about the velocity field on the site. A comparison was made between the results of the numerical solution and the experimental ones. An adjustment was proposed to change the initial conditions in order to refine the results of numerical solutions.

1. Introduction

Determining the wind shear for a complex terrain is a difficult task. It depends not only on the roughness factor of the relief, but also on its orography. Details such as "speed up" of winds over valleys, mountain peaks, natural strains between mountain ranges, etc., should be included here. Figure 1 schematically presents the types of boundary layers formed by wind flow over complex terrains.

Figure 1. Boundary layers and sublayers over complex terrains.

The presence of a complex surface leads to large horizontal differences in the structure of the boundary layer. The impact of different mountain structures on the behavior of the boundary layer is well illustrated in [1, 2].

The strength of winds along mountain ranges, apart from orography, is influenced by local features (such as temperature inversions, for example). As stated in [3], several hours are required to establish a stable wind profile in mountains and valleys. The windy part of the mountains is 3 hours after sunset and in the valleys 3 hours after sunrise [4]. In addition, wind direction in the valleys is influenced by wind warming and cooling, which is more efficient in the high and narrow sections compared to the lower parts [5].

Unlike flat terrain, wind behavior over complex terrain can vary, so a common analytical methodology cannot be proposed. However, analytical approximations are known in the literature, which can even predict such behavior. The non-linear nature, such as detaching the eddy from the main wind flow and vortexing, cannot be predicted with regular mathematical models, and nonlinear numerical relations have to be used [6]. A simplified model for a numerical solution is proposed in [7]. In conclusion, one of the most famous commercial products used to analyze wind flows over flat and low complex terrain uses linear analytical approximation [8].

2. Modeling of terrain orography

The exact interpretation of topography and orography is an essential prerequisite for correct determination of the wind shear.

The most primary and labor-intensive way of building a relief surface is by using topographic maps. For the territory of Bulgaria, such maps are available both in the web and in the specialized institutes. Depending on the desired accuracy of the relief surface, maps are available at a scale of 1: 5 000, 1:10 000, 1:25 000, 1:50 000. In order to determine the wind potential parameters for a given location, it is advisable to use the most detailed maps (1: 5 000), while for areas over 5 km from the site in question, less detailed maps can be used.

For the purpose of the present study, the chosen site is located in the municipality of Ruen, Central Eastern Bulgaria (Figure 2). The 3D map (Figure 2) is prepared using Surfer software. Due to the complex topography of the relief, measurements of the wind parameters are made in three points, using three meteorological masts - 30, 70 and 80 m.

The area is characterized by a complex terrain displacement of 100 to 600 m, which is a prerequisite for a significant change in the wind behavior above the ground.

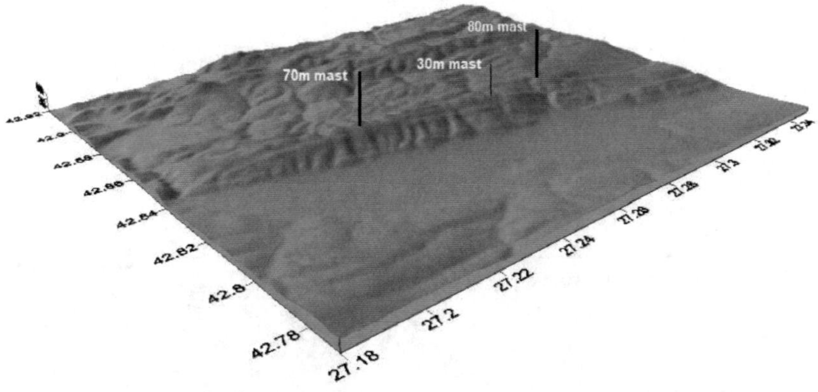

Figure 2. Topology of terrain and position of measuring masts.

In order to analyze the potential, a measurement of 30 m mast was initially carried out in the period December 2007 - September 2009. Measurements with a 70 m mast start in March, 2010 and end January, 2011. Measurements with a 80 m mast start in November, 2010 and end January, 2012.

Five different sensors have been used during the on-site measurements – anemometers, wind vanes, temperature, humidity and pressure sensors. Summarized technical information concerning the measuring equipment is presented in table 1. The equipment is calibrated in certified laboratory and it is installed on the mast according to the current standards for performing wind measurements.

Table 1. Technical specification of the measuring equipment.

Sensor	Model	Range	Accuracy
Anemometer	Thies Clima 4.3350	0÷75 m/s	<1% of means value (0.3÷50 m/s) or ±0.2 m/s
Wind vane	Thies Clima 4.3150	0-360°	± 1°
Humidity sensor	Vaisala HMP 50	0-90 %	± 3%
Temperature sensor	Vaisala HMP 50	-40 ÷ +60 °C	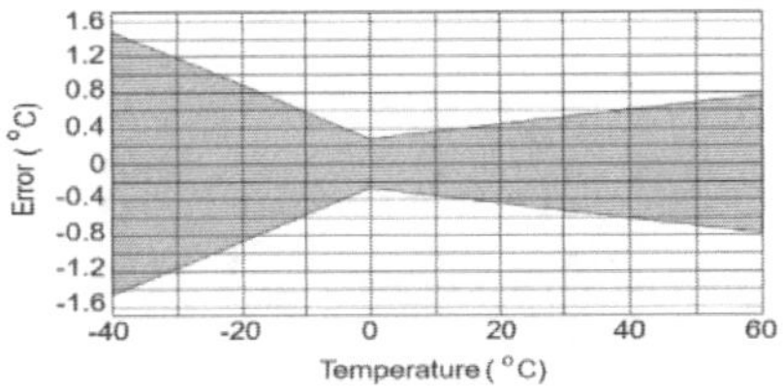
Pressure sensor	Setra 276	600 ÷ 1 100 hPa	± 0.25%

Obviously, for both tall masts, the measurement period overlaps, so an analysis of the wind shear can be made. In the 30 m mast the measurements started earlier, so it is necessary to make a correlation with the wind rose data. Due to the above, in Figure 3 is information about the wind rose for the three masts. It can be seen that the prevailing directions are two: north-northeast and south-southwest. The correlation between data is very good, which is an indicator of reliability in data analysis.

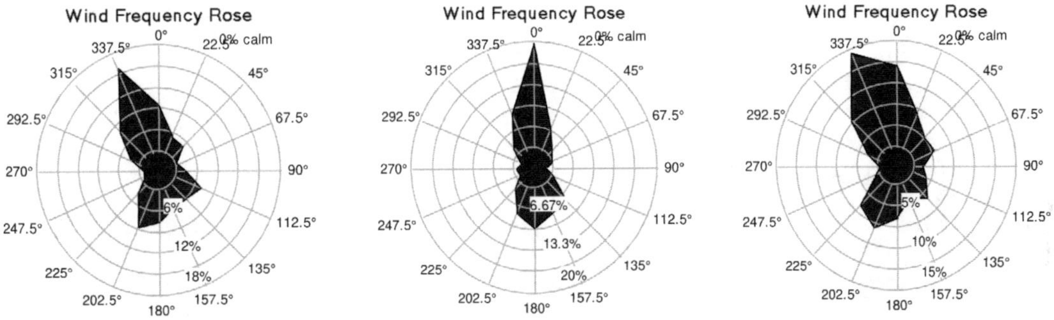

a) b) c)

Figure 3. Wind rose for: a) 30 m mast, b) 70 m mast, c) 80 m mast.

Figure 4 presents information on the wind speed at height obtained using power law and logarithmic law wind shear profiles.

According to Figure 4 roughness factor of the terrain in front of the 70 m mast is 0.862 while for the 30 m mast is 0.485. The position of the 80 m mast also implies a lower roughness factor, namely 0.102. Given that 70 and 30 m masts are installed on the same plateau, the behaviour of wind speed at altitude is very close. The orography of the 80 m mast differs from that of the two masts, which is why the wind shear is also different (sharper), which implies the lower roughness factor. The power law and logarithmic profiles (Figure 4) are obtained on the basis of the following equations 1 and 2:

$$U(z)=\begin{cases}\dfrac{u_*}{k}\ln\left(\dfrac{z}{z_0}\right), & if\, z>z_0\\ 0, & if\, z<z_0\end{cases} \qquad (1)$$

where: $U(z)$ - wind speed at some height above ground z, m/s; z_0 - surface roughness, m; k - von Karman's constant (0.4); u_* - friction velocity, m/s.

$$U(z) = u(z_1)\left(\frac{z}{z_1}\right)^{\alpha} \tag{2}$$

where: $U(z)$ - wind speed at some height above ground z, m/s; z_1 - height of the known wind speed, m; α - power law exponent.

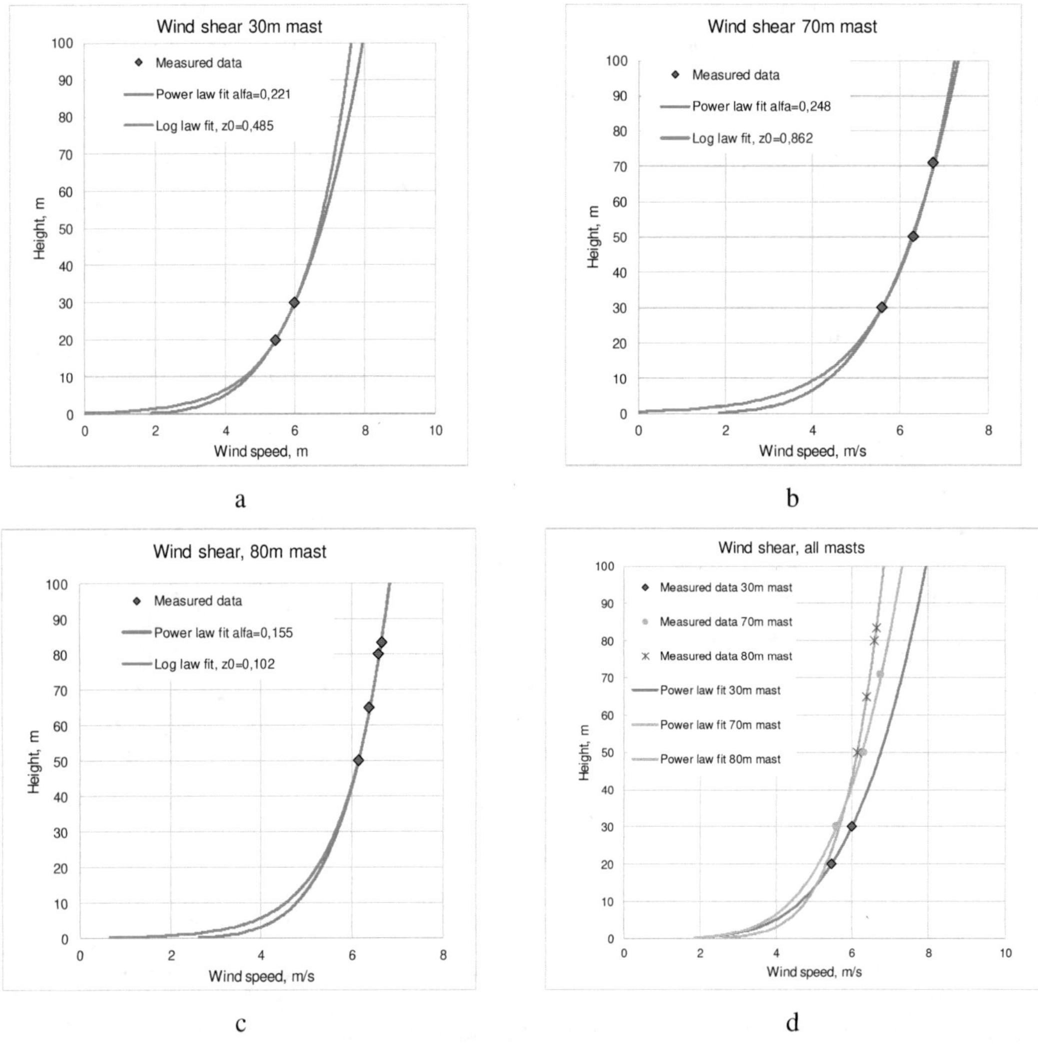

Figure. 4 Wind profiles for all masts.

The visualization of the wind profile in view of the orography of the terrain is presented in Figure 5. It shows the height profiles for the 70 and 80 m masts for the North direction, which coincides with both the prevailing wind directions and the highest sector frequency. The difference between the two wind profiles is clearly visible - at the 70 m mast, the terrain ahead of the mast in the direction of the wind is significantly more complex (with greater displacement in altitudes), which is also the reason for the greater velocity gradient. However, for the 80 m mast, due to the smaller difference in displacement, the wind profile is significantly "sharper". The main conclusion that can be drawn from the comparison between the two profiles is that at the same altitude on the hill, at a significantly more complex terrain before the hill, the wind speed can be increased by up to 10%, only due to the orographic features of the terrain.

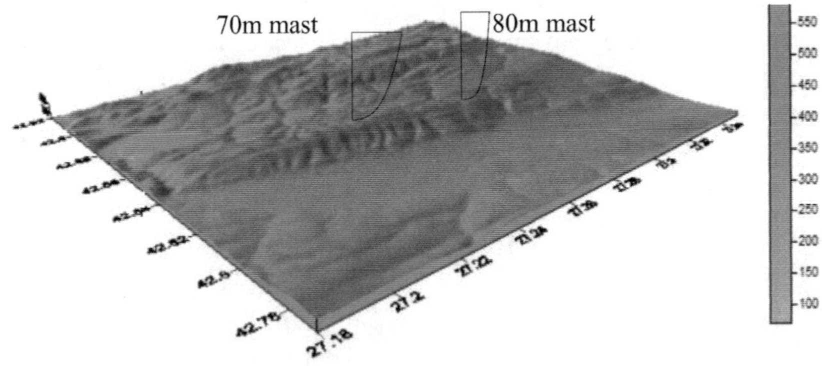

Figure 5. Wind profiles at a height of 70 and 80 m in prevailing north wind direction.

A cross-sectional view of the topography of the terrain in the vicinity of 70 and 80 m masts is shown in Figure 6.

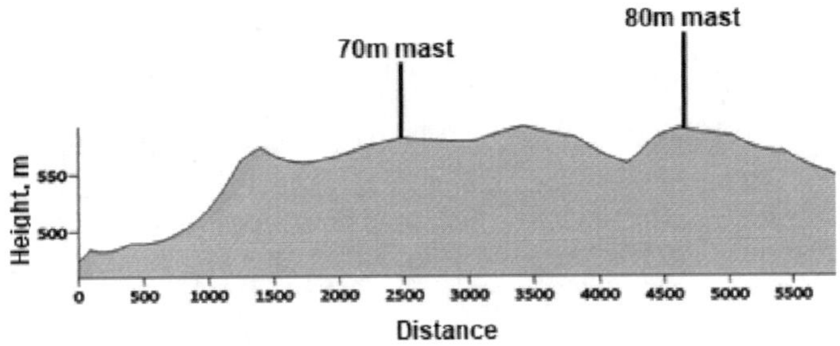

Figure 6. Cross-sectional topography of the terrain in the vicinity of 70 and 80 m masts.

3. Numerical study

Various commercial software products are used for engineering calculations. Some of them are specialized, focusing on wind flow study (WAsP WindPro, WindSim), while others can be "tailored" to wind power (Fluent / Ansys). WAsP commercial software [9] is a linear model that combines two models: physical model (atmospheric stability, factor roughness, natural change in geography of relief, etc.) and statistical model (Weibull wind parameters distribution). The physical model is used to determine the wind profile.

With the development of computing techniques, in recent years there has been an increasing interest in modelling of flows in semi-confined spaces [10 - 15]. In [16] is presented modelling of flows over complex terrains using different turbulent models.

In order to study the influence of orography and the roughness factor of the relief on the wind velocity profile, the following approaches are used:

- **Case 1:** A 30 m mast is used as the basis for the numerical simulations, after which the results of the numerical solutions are compared with those obtained from the 70 m mast.
- **Case 2:** A 70 m mast is used as the basis for numerical simulations, then the results are compared with those obtained from a 30 m mast;

This way an analysis will be made regarding the interpretation of data obtained from measurements with the high masts and their interpretation near the site through numerical simulations.

Case 1. Using a low mast to predict the wind flow.

Figure 7 shows information about the input meteorological data used in the numerical simulations for a 30m mast and in Figure 8 for a 70 m mast. The numerical simulations are performed using the specialized software product WAsP.

Comparison of data shows that there is a good correlation (60%) between data in terms of directional distribution. In addition, the correlation between the frequency distribution in directions for 30 and 80 m masts is over 87%.

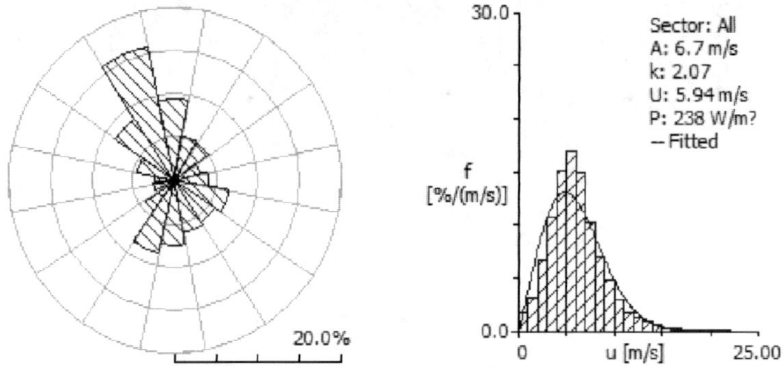

Figure 7. Input meteorological data of a 30m mast.

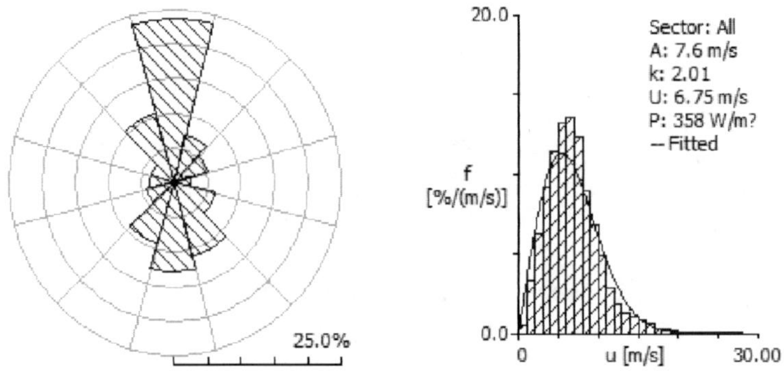

Figure 8. Input meteorological data of a 70m mast.

The results of the numerical solution using two masts as main ones are presented in Figure 9 and 10, there is a very good match between the numerical results of using the two masts.

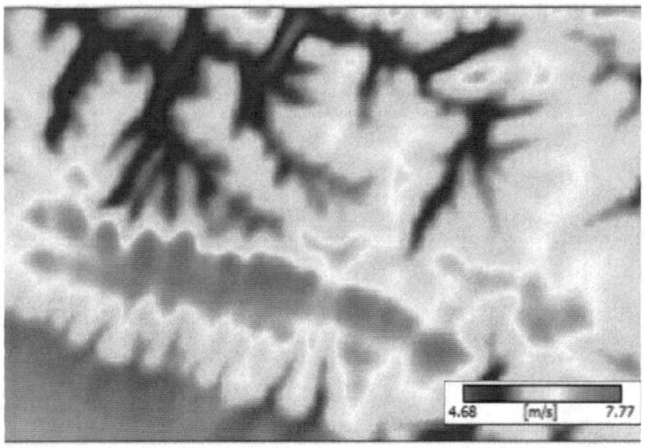

Figure 9. Wind speed near terrain using 30 m masts.

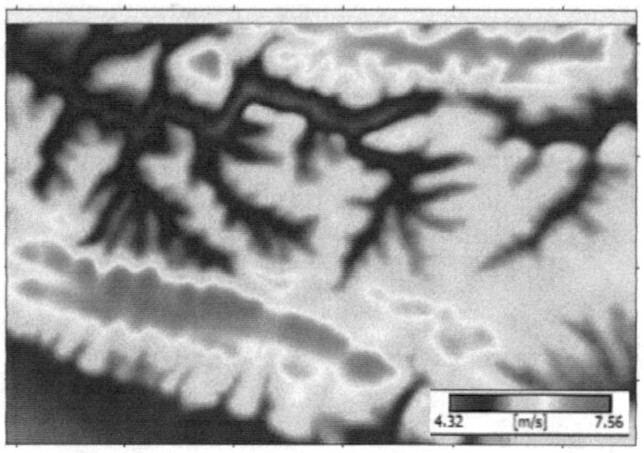

Figure 10. Wind speed near terrain using 70 m masts.

An analysis of the results of the numerical simulations for case 1 is presented below. Table 2 is about the frequency distribution of wind parameters as well as average wind speed by sectors for a 30 m mast (on-site measurements) as well as the same wind characteristics obtained by numerical simulations near a 70m mast. The last column also shows the RIX index (Roughness index). This index shows the probability of turbulence of the wind flow by sector, resp. accelerating or delaying the flow caused by the roughness factor of the relief.

The information presented in Table 3 is of great interest. A comparison was made between the results of a numerical solution obtained with the use of a 30 m mast and real ones, using a high mast. Correlation of wind speed data by direction is 54.5%, which is not very satisfactory.

Table 2. Main mast and numerical simulation results.

		30m mast (main data)		70m mast (numerical simulation)		
Sector	Angle, °	Frequency, %	Average speed, m/s	Frequency, %	Average speed, m/s	RIX,%
1	0	13.9	7.17	13.9	7.98	11.2
2	30	6.8	5.39	6.1	5.80	13.2
3	60	5.7	4.72	5.5	5.49	3.7
4	90	4.5	4.76	4.8	5.56	3.0
5	120	7.7	5.39	7.6	6.27	1.3
6	150	6.2	4.87	7.8	6.33	2.4
7	180	9.9	6.33	9.2	6.98	7.0
8	210	12.0	6.58	9.9	7.18	7.9
9	240	4.8	5.63	5.9	6.79	5.2
10	270	3.5	6.35	3.9	7.07	0.6
11	300	7.5	5.93	10.2	7.42	8.6
12	330	17.5	6.74	15.2	7.84	7.5
ALL			6.10		6.98	6.0

Table 3. Comparison between results from numerical and experimental solutions.

		70m mast (numerical simulation)		70m mast (on site measurements)	
Sector	Angle, °	Frequency, %	Average speed, m/s	Frequency, %	Average speed, m/s
1	0	13.9	7.98	23.6	7.87
2	30	6.1	5.80	7.2	5.91
3	60	5.5	5.49	6.3	5.66
4	90	4.8	5.56	5.9	5.25
5	120	7.6	6.27	7.4	6.57
6	150	7.8	6.33	7.6	6.76
7	180	9.2	6.98	7.9	7.06
8	210	9.9	7.18	8.1	7.18
9	240	5.9	6.79	6.2	5.53
10	270	3.9	7.07	6.1	5.40
11	300	10.2	7.42	6.6	5.89
12	330	15.2	7.84	7.3	6.43
ALL			**6.98**		**6.76**

Meanwhile, however, the difference in average speeds is 3.2%, which is not significant in value. More importantly, however, is the significant difference in frequency distribution of velocities, which is important in the precise determination of energy production from a wind park.

The difference between numerical and experimental data is clearly visible in figure 11. At 10 m above the ground, the difference in wind speeds obtained from the numerical and experimental data is 28.5%. At 20 m height, this difference drops to 16.8%, and at 30 m - to 11.1%, which is within the allowable range. The main conclusion that can be made here is that the software product does not take sufficient account of the orography of the terrain, the size of the boundary layer, which is why there is also an error in the wind shear. At a height of 100 m, however, the difference between numerical and experimental studies is 0.5%. Along with the above the software package has the potential for user intervention by setting the corresponding roughness factor of the relief. However, this factor may be established if there are at least two speed measurements within and outside the boundary layer in the vicinity of the site. Experience shows that the error in numerical research decreases in determining the roughness.

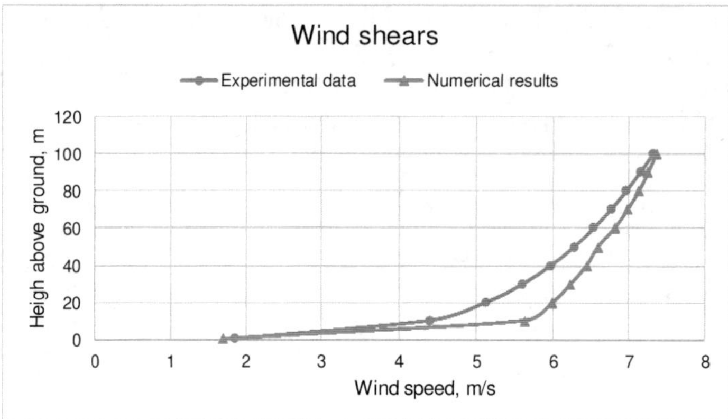

Figure 11. Wind shear (numerical results and experimental studies).

Case 2. Using a tall mast to predict the wind flow.

The main purpose of the study is based on data from the 70 m mast measurement to predict the wind parameters measured with a 30 m mast. Wind input parameters are similar to those shown in Figure 8.

Table 4 presents information on the wind parameters of 30 m mast - experimental and numerical surveys.

Table 4. Comparison between results from numerical and experimental solutions.

		70 m mast (numerical simulation)		70 m mast (on site measurements)	
Sector	Angle, °	Frequency, %	Average speed, m/s	Frequency, %	Average speed, m/s
1	0	21.0	6.95	13.9	7.17
2	30	7.8	5.59	6.8	5.39
3	60	5.0	5.18	5.7	4.72
4	90	3.2	4.84	4.5	4.76
5	120	6.9	5.64	7.7	5.39
6	150	9.8	5.96	6.2	4.87
7	180	12.7	6.34	9.9	6.33
8	210	8.9	6.43	12.0	6.58
9	240	5.1	5.52	4.8	5.63
10	270	3.8	4.77	3.5	6.35
11	300	4.9	5.24	7.5	5.93
12	330	10.7	6.00	17.5	6.74
ALL			**6.03**		**6.10**

Correlation of wind speed data in directions is 61%, which is no better than when using a 30 m mast. Meanwhile, the difference in average speeds is 1.2%, in favour of the higher mast.

Figure 12 shows the wind shear in the vicinity of 30 m masts (numerical and experimental surveys). At 10 m above the ground, the difference in wind speeds obtained by the numerical and experimental research is 15.5% (the speed is higher in the numerical studies). At 20 m height, this difference decreases to 6.1%, and to 30 m - it is 0.3%, i.e. as the high of the taken measurements increases, the interpolation of data to the ground is more accurate. Velocity values at height greater than 30 m are extrapolated using the logarithmic velocity profile and numerical simulations. It is seen that at a height of 70 m the difference in speeds is 9.3%, the real one closer to the numerical research.

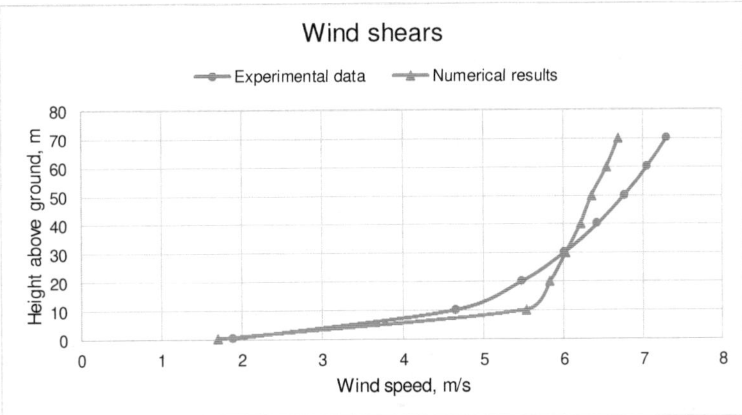

Figure 12. Wind shear (numerical results and experimental studies).

4. Conclusions

On site measurements carried out with meteorological high masts installed in the vicinity of the site provide reliable information on the velocity distribution. The specialized software used in the experiment does not take into account the peculiarity of the relief, so the results obtained may differ from the real ones.

The results from the performed study shows that at least two meteorological masts are needed in measurements on complex terrain and should be equipped with the respective for the site equipment.

It is important to determine the thickness of the boundary layer, because it has the most significant influence on the wind shear. For this type of complex terrains, the thickness of the boundary layer exceeds 20 m, indicating that measurements of the wind parameters at the specified height are required. Together with measurements at a height of 20 meters, measurements of at least two points are required.

The use of a second measuring mast serves as a reference source of information to analyze the results obtained from the numerical solution. As can be seen from the analysis made, due to the use of linear models in the numerical solution, the error between numerical and experimental results may exceed 10%. The specified error can be reduced by making adjustments in roughness factor of the terrain adopted in numerical model.

The terrain orography characteristics obtained can be used to configure the turbines (micro-sitting) in the park as well as to fine-tune energy production.

References

[1] Smith R B 1978 The influence of mountains on the atmosphere. In: Landsberg HE, Saltzman B (Eds) *Adv. Geophys*. **21** pp 87–230

[2] Atkinson B W 1981 Meso-scale Atmospheric Circulations. Academic Press, London 495

[3] Vergeiner I and Dreiseitl E 1987 Valley winds and slope winds—observations and elementary thoughts. Meteorol. *Atmos. Phys*. **36** pp 264–286

[4] Heimann D, De Franceschi M, Emeis S, Lercher P and Seibert P (Eds) 2007 Air pollution, traffic noise and related health effects in the Alpine space—a guide for authorities and consulters, ALPNAP comprehensive report. Università degli Studi di Trento, Trento

[5] Steinacker R 1984 Area-height distribution of a valley and its relation to the valley wind. Contr. *Atmos. Phys*. *57* pp 64–71

[6] Zenman O and Jensen N O 1987 Modification of Turbulence Characteristics in Flow over Hills. Quart. J. Roy. *Meteorol. Soc*. **113** 55–80

[7] Taylor P A 1977 Numerical studies of neutrally stratified planetary boundary layer flow over gentle topography, I: Two-dimensional cases. Bound.-Lay. Meteorol. **12** pp 37–60

[8] Petersen E L, Mortensen N G, Landberg L, Højstrup J and Frank H P 1998 Wind Power Meteorology. Part II: Siting and Models. *Wind Energy*. **1** pp 55–72

[9] Troen I and Petersen E L 1989 *European Wind Atlas*, Risø National Laboratory, Roskilde
[10] Mirkov N, Rasuo B and Kenjeres S 2015 On the improved finite volume procedure for simulation of turbulent flows over real complex terrains, *Journal of Computational Physics.* **287** pp 18-45
[11] Balogh M, Parente A and Benocci C 2012 RANS simulation of ABL flow over complex terrains applying an Enhanced k-ε model and wall function formulation: Implementation and comparison for fluent and OpenFOAM *Journal of Wind Engineering and Industrial Aerodynamics*. **104-106** 360-368
[12] Makridis A and Chick J 2013 Validation of a CFD model wind turbine wakes with terrain effects, *Journal of wind engineering and industrial aerodynamics*. **123** pp 12-29
[13] Castellani F, Astolfi D, Burlando M and Terzi L 2015 Numerical modelling for wind farm operational assessment in complex terrain, *Journal of Wind Engineering and Industrial Aerodynamics*. **147** pp 320-329
[14] Antonov I S, Petkov N P and Elenkov L A 1992 Numerical model of a two-dimensional two-phase turbulent jets, *Proceeding of University of Mining and Geology.* **38**, pp 94-104
[15] Antonov I S, Lien H D, Angelov M S 1997 Numerical modeling of two-phase non-isothermal turbulent jet, *10th international conference "Numerical methods for laminar and turbulent flow*, Swansea, U.K. 28.07-01.08 1997
[16] Abdi D and Bitsuamlak G. 2014 Wind flow simulation on idealized and real complex terrain using various turbulence models, *Advance in Engineering Software*. **75** pp 30-41

Acknowledgments
The author would like to thank the Research and Development Sector at the Technical University of Sofia for the financial support.

Basic geometric and physical parameters of pumpkin seeds

D Atanasov, M Georgieva, I Kiryakov and V Rulev

University of Food Technologies – Plovdiv, Bulgaria

E-mail: mutk0a@abv.bg

Abstract. The granulometric composition of Pumpkin seeds of the "Samson" variety used as feedstock for food and vegetable oils producing industry are studied. Their geometric dimensions and physical indices are obtained experimentally. Results and dependencies necessary for designing and dimensioning of drying ovens from installation for vegetable oils production are obtained.

1. Introduction

The technological processes from the food and flavour industries and the materials used in them have a strongly expressed specificity in comparison with the processes and materials from the remaining branches. Most often these materials are not isometric which imposes in measuring drying processes and installations to know such characteristics of the disperse materials as shape and dimensions of the particles, their surface, granulometric composition, density, bulk density, etc. parameters.

There are data of pumpkin seeds parameters in the literature [4, 10], but they are incomplete and refer to narrow interval of the material moisture content.

2. Purpose of this study

The purpose of this study is to obtain experimentally the geometric dimensions and the main physical parameters of pumpkin seeds, necessary for drying installations projection.

3. Materials and Methods

The object of this study was the pumpkin seeds (*Cucurbita maxima*) of the "Sampson F1" variety were cultivated in a field situated close to Plovdiv (Bulgaria) – harvest 2017. Pumpkin seeds have been rich in protein, glyceride oil, phytosterols and vitamins. The seeds have been a source of carotenoids, tocopherols and trace elements (K, Mg, Mn, Zn, Se, Co, Cr and Mo). Seeds have been a major product for the oil-producing industry in the production of vegetable oil after cold pressing. The resulting oil has been a high content of unsaturated fatty acids. All these qualities have determined the application of pumpkin seeds to the food industry and medicine [8, 9].

For the purpose of projection and measurement of drying equipment from installations used for essential oils production, the following characteristics were determined for the material investigated [3,7,11]:

1. Density ρ_M, kg/m^3- it is the material weight in a volume unit of its own:

$$\rho_M = \frac{m_p}{V_p}, \tag{1}$$

where m_p is a pumpkin seeds quantity, kg;

V_P - a pumpkin seeds volume, m^3.

The material density has been determined picnometrically [1, 5] in the following sequence:

- the picnometer has been filled with distilled water, then tempered at temperature t = 20 °C, and then it has been weighed on an analytical balance having an accuracy of 0.0001 g. The measured mass – m_1 is recorded, g;
- the mass – m, g of randomly selected 100 - 200 particles of one and the same type pumpkin seeds has been measured;
- the material with the mass m has been carefully placed in the picnometer, then the picnometer has been dried up with a filter paper, tempered at temperature t = 20 °C and weighed. The measured mass - m_2, g has been recorded.

The material density has been calculated by the equation:

$$\rho_M = \frac{m}{m+m_1+m_2}\rho_{H_2O,t} \tag{2}$$

where $\rho_{H_2O,t}$ is water density at t = 20 °C, kg/m^3.

2. Bulk density ρ_H, kg/m^3 of the particles - density of loose bulk layer of disperse material:

The bulk density of the material has been determined in the following sequence [1,5]:

- a definite quantity of the material has been weighed -m_p, g;
- it has been poured down loosely in the measuring glass and the level has been equalized;
- the material volume - V, dm^3 has been recorded with an accuracy of up to 0.0002 dm^3;
- the bulk density has been calculated by the equation:

$$\rho_H = \frac{m_p}{V} \tag{3}$$

The measurements have been processed statistically by the method for indirectly measured values [12].

3. Mean linear dimensions. The linear dimensions of the pumpkin seeds have been determined by means of direct measurements of the three characteristic sizes of the particles - length l, mm; width b, mm; thickness h, mm.

The measurements have been done by means of a calliper-gauge with an accuracy of 0.1 mm, measuring the linear sizes of 100 randomly selected particles, so that a representative mean characteristics have been obtained.

The data obtained for each linear dimension have been subjected to statistical processing according to the method for directly measured values [12].

4. Average equivalent diameter $\bar{d}_e$, mm defined as the diameter of the equivalent volumetric sphere. The equivalent diameter of polydisperse material of big particles that can be counted down, could be calculated from the expression:

$$\bar{d}_e = \sqrt[3]{\frac{6V_{PT}}{n\pi}} \tag{4}$$

where: n - the number of particles in the sample that should be not less than 200 [11]. When measuring the average equivalent diameter, 300 particles have been counted down at each measurement; V_{PT} - total volume of the number of particles, mm^3.

5. Average surface F_p, mm^2 of the particles:

$$F_P = \pi\bar{d}_e^2 \tag{5}$$

6. Average volume V_p, mm^3 of the particles:

$$V_P = \frac{V_{PT}}{n} \tag{6}$$

7. Factor of the shape φ_S, defined as a ratio on the surface of the equal volumetric sphere and of the particle:

$$\varphi_S = 4.84\frac{V_P^{2/3}}{F_P} \quad (7)$$

8. Coefficient of non-sphericity f - reciprocal value of the factor of shape. It has been calculated by means of the equation [2].

$$f = 0.207\frac{F_P}{V_P^{2/3}} \quad (8)$$

9. Dispersity a_f - the surface corresponding to a volume unit from the particle:

$$a_f = \frac{6}{d_e} \quad (9)$$

10. Initial porosity ε_0 - the ratio of the pores volume in the immovable layer to the volume of the layer:

$$\varepsilon_0 = \frac{V - V_0}{V} = 1 - \frac{\rho_H}{\rho_M} \quad (10)$$

where V_0 is a pumpkin seeds volume in the layer, m^3.

11. Archimedes criterion:

$$Ar = \frac{g d_e^3}{\nu^2}\frac{(\rho_M - \rho_{air})}{\rho_{air}} \quad (11)$$

where: ρ_{air} - air density at $t = 20°C$, kg/m^3;
ν - air cinematic viscosity at temperature of 20°C, m^2/s;
g - acceleration of gravity, m/s^2.

The data processing from the directly measured values has been done in the following sequence [12]:
- evaluating the measurements and gross blunder removal;
- calculating the average value $\bar{x}$ and the mean squire deviation S_x of the measured values:

$$\bar{x} = \frac{\sum x_i}{n} \quad (12)$$

$$S_x = \sqrt{\frac{1}{n-1}\sum(x - x_i)^2} \quad (13)$$

where x_i and $\bar{x}$ are respectively the values of the separate measurement and the mean arithmetical value:
n - the number of measurements;
-calculating the confidential interval Δx for the series of measurements:

$$\Delta x = t(\upsilon, \tfrac{\alpha}{2})\frac{S_x}{\sqrt{n}} \quad (14)$$

where $t(\upsilon, \frac{\alpha}{2})$ is Student's coefficient, determined according to the degrees of freedom $\upsilon = n-1$ and the level of significance α;
- calculating the relative error in the measurement:

$$\varepsilon = \frac{\Delta\bar{x}}{\bar{x}}.100, \% \quad (15)$$

The data processing for the indirectly measured values was done in the following sequence [12]:
-establishment of the differential dependence of the indirectly measured value y, from the directly measured x_i:

$$S_y = \frac{\partial f(\Delta x_i)}{\partial x_i} \quad (16)$$

- calculating the absolute error in the measurement:

IOP Conf. Series: Materials Science and Engineering **595** (2019) 012044 doi:10.1088/1757-899X/595/1/012044

$$y = \sqrt{\sum S_y^2} \tag{17}$$

- calculating the relative error:

$$\varepsilon = \frac{\Delta \bar{y}}{\bar{y}}.100,\% \tag{18}$$

4. Result and Discussion

Table 1 shows the values of pumpkin seeds parameters calculated after the above formulae and methods. The pumpkin seeds density investigation depending on its moisture content W (total bases) within the limits of moisture content $W = 0 \div 38.6\%$ (at temperature of $t = 20$ °C and atmospheric pressure) showed that the dependence is a linear one (Figure 1) and can be described by means of the equation

$$\rho_M = 458.26 + 354.23\text{W} \tag{19}$$

by a correlation ratio of r^2=0.8536.

Similar dependence is also obtained for bulk density (Figure 2). The obtained equation has a correlation coefficient r^2=0.8692 and has the following appearance

$$\rho_H = 293.8 + 261.18W \tag{20}$$

Table 1 Physical characteristics of pumpkin seeds.

Parameters	Moisture Content (Total Bases) of pumpkin seeds, %					
	38.6	35.0	26.8	10.6	4.9	0
1. Density - ρ_M, kg/m^3	595.0	582.2	553.2	495.8	475.6	458.3
2. Bulk Density - ρ_H, kg/m^3	394.6	385.2	363.8	321.5	306.6	293.8
3. Average Equivalent Diameter of the Particle - $\bar{d}_e$, **mm**	10.54	10.45	10.25	9.82	9.66	9.52
4. Average Sizes of the Particle:						
- length $\boldsymbol{l}$, **mm**;	17.26-17.57	16.92-17.31	16.66-17.06	16.29-16.63	15.93-16.29	15.91-16.24
- width $\boldsymbol{b}$, **mm**;	11.02-11.31	10.8-11.06	10.76-11.03	10.29-10.54	10.10-10.35	10.04-10.28
- thickness $\boldsymbol{h}$, **mm**.	5.02-5.22	4.96-5.18	4.74-4.95	4.57-4.75	4.43-4.65	4.52-4.75
5. Average Surface of the Particle $\boldsymbol{F_P}$, $\mathbf{mm^2}$	349.0	343.1	330.1	302.9	293.2	284.7
6. Average Volume of the Particle $\boldsymbol{V_P}$, $\mathbf{mm^3}$	612.9	597.9	563.6	495.8	471.9	451.42
7. Factor of the Shape - $\boldsymbol{\varphi_S}$	1.00064	1.00117	1.00042	1.00096	1.00057	1.00041
8. Coefficient of Non-Sphericity - $\boldsymbol{f}$	0.99936	0.99883	0.99958	0.99904	0.99943	0.99959
9. Dispersity - $\boldsymbol{a_f}$	0.5693	0.5742	0.5854	0.6110	0.6211	0.6303
10. Initial Porosity - $\boldsymbol{\varepsilon_0}$	0.337	0.338	0.342	0.352	0.355	0.359
11. Archimedes Criterion – **Ar** ($\boldsymbol{t} = \mathbf{20}$°C and $\boldsymbol{p} = \mathbf{1.01x10^5}$ **Pa**)	1.2074x10^{-3}	1.1513x10^{-3}	1.0323x10^{-3}	8.1331x10^{-4}	7.4257x10^{-4}	6.8483x10^{-4}

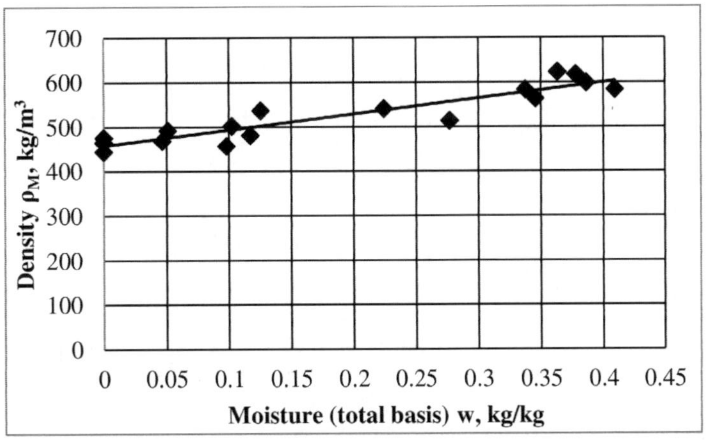

Figure 1. Change of pumpkin seeds density depending on its moisture content.

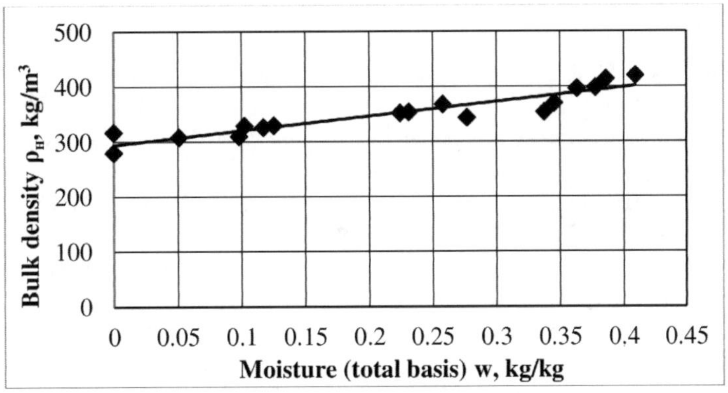

Figure 2. Change of the bulk density of the pumpkin seeds depending on its moisture content.

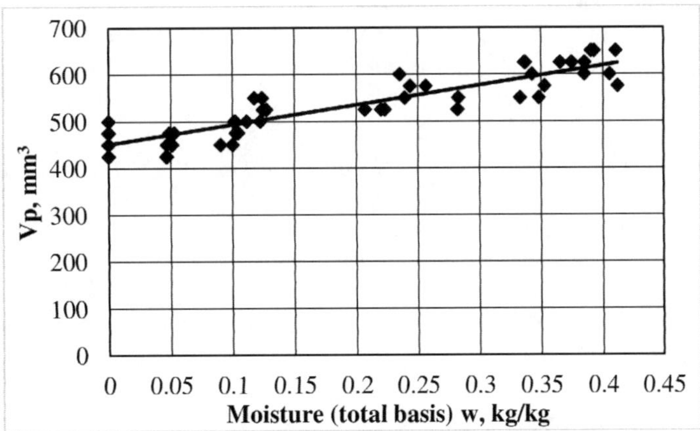

Figure 3. Change of pumpkin seeds volume depending on its moisture content (dry bases).

The pumpkin seeds volume investigation confirms that the variation of according to humidity its linear and can be described by means of the equation

$$V_p = 451.42 + 418.52W \tag{21}$$

by a correlation ratio of r^2=0.8207.

IOP Conf. Series: Materials Science and Engineering **595** (2019) 012044 doi:10.1088/1757-899X/595/1/012044

5. Conclusion

1. The basic geometric and physic parameters (density; bulk density; average: linear sizes, equivalent diameter, surface and volume; factor of shape; coefficient of non-sphericity; dispersity; initial porosity and Archimedes criterion) of pumpkin seeds (*Cucurbita maxima*) of the "Sampson F1" variety, Plovdiv 2017 at atmospheric conditions and wide range of material moisture content (from 0 to 40 %) have been obtained experimentally (Table 1).

2. Analytical equations describing the dependencies of the density and bulk density of the pumpkin seeds from their moisture content (total bases) for the same conditions have been obtained (Eqs. 19 and 20).

3. Analytical equation describing the dependence of the pumpkin seeds volume shrinking from their moisture content for the same conditions has been obtained. The equation confirms that the dependence between the volume shrinking and pumpkin seeds moisture content as well as other materials is a linear one (Eq. 21).

4. The results obtained could be useful at designing of drying equipment and installations for pumpkin seeds oil production.

References

[1] Bonev B and others 1989 *Practical Guidance of Physics* (HIFFI Plovdiv)

[2] Elenkov V R 1988 *Drying and Drying Equipment* (Zemizdat, Sofia)

[3] Ginzburg A 1973 *Fundamentals of the theory and technology of drying foods* (in Russ.) (Food Industry, Moscow)

[4] Hadjiiski Tz, Boiadjiev B and Rangelov P 1980 Industrial Drying of Grape Seeds. Oil - Soap Industry, *Issue of Institute of Vegetable Oils, Proteins and Washing Agents Book* **1** 10

[5] Nevenkin S 1993 *Drying and drying equipment* (in Bulg.) (Engineering Sofia)

[6] Hristozov D V and others 1990 *Laboratory Guidance of Physics* (Sofia Science and Art)

[7] Likov A V 1968 *Theory of Drying* (Moscow, Energy)

[8] Mushtaev V I and Ulianov V M 1988 *Drying of Dispersive Materials* (Moscow, Chemistry)

[9] Petkova Zh Y, Antova G A and Angelova-Romova M Y 2018 Development of lipid damage of pumpkin seed oil stabilized with different antioxidants during long-term storage, *Bulgarian Chemical Communications* **50** Special Issue C 112

[10] Petkova Zh Y and Antova G A 2014 Phospholipid composition of Cucurbitaceae seed oils *Bulgarian Chemical Communications* **46** Special Issue A 100

[11] Razuvaev N I 1975 *Complex Processing of By-Products from Vine Production Industry* (Moscow, Food Industry)

[12] Sendov S H 1987 *Application of Fluidized Systems in Industry* (Sofia, Technique)

[13] Vuchkov I H and Stoianov S 1986 *Mathematical Simulation and Optimization of Technological Objects* (Sofia, Technique)

IOP Conf. Series: Materials Science and Engineering **595** (2019) 012045 doi:10.1088/1757-899X/595/1/012045

Correlations between kinetic and thermodynamic parameters of the extraction of tobacco leaves

S Tasheva[1], V Gandova[2], V Popova[3,5], T Ivanova[3], S Damyanova[4] and A Stoyanova[3]

[1] Department of Heat Engineering, University of Food Technologies, Plovdiv, Bulgaria
[2] Department of Analytical and Physical Chemistry, University of Food Technologies, Plovdiv, Bulgaria
[3] Department of Tobacco, Sugar, Vegetable and Essential Oils, University of Food Technologies, Plovdiv, Bulgaria
[4] Department of Biotechnologies and Food Technologies, University of Russe, Branch Razgrad, Bulgaria

E-mail: vpopova2000@abv.bg

Abstract: The correlations between thermodynamic and kinetic parameters in the extraction of tobacco leaves (solid – liquid system) were determined. The values of the parameters internal molecular diffusion coefficient, activation energy and entropy were used in the calculation procedure presented in the study, with regard to the extraction process of leaves of three types of tobacco (Virginia, Burley and Oriental), carried out with 50% ethanol as a solvent at different temperatures (20, 30, 40, 50, 60 and 70°C) and duration (10, 20, 30, 40, 50 and 60 min). On the basis of the Arrhenius law the effective diffusion coefficient in solid – liquid system and the reaction rate constant of the three types of tobacco were computed.

1. Introduction

Extraction is a process of transferring one or more components of a complex solid with a porous structure into a liquid medium (solvent). According the Benda [1] this process is performed on the basis of selective solubility of each substance(s) of the solid material.

The extraction process is often called diffusion, as is the driving force of transfer. The diffusion of extracted components of the starting mixture in the extractant is based on the difference in their concentration [1-6].

The diffusion coefficient is a major characteristic of the diffusion setting and is an important parameter in the theory of all diffusion processes [1-12]. The ideal thermodynamic system would represent an ensemble of non-interacting particles, which diffuse as a result of the chaotic movement. The theory of chaotic motion model is based on the strict laws of static mechanics and is sufficiently accurate. In this reason, the diffusion coefficient in a given case can be calculated accurate enough, so as to create the conditions for the development of theories of the complex diffusion process.

For example, the molecular diffusion coefficient represents only a physical constant characterizing the ability of a substance to penetrate through diffusion into an immobile environment, not dependent on the hydrodynamic conditions in which the process runs. The molecular diffusion coefficient is a function of the properties of the extracted substance and the setting, in which it diffuses, as well as of temperature and pressure. Usually the molecular diffusion coefficient increases with increasing the temperature and decreasing the pressure (for gases). In some specific cases, the molecular diffusion coefficient was determined by experimental data or in theoretical or semi-empirical equations, taking into account the influence of the temperature and the pressure at which the diffusion proceeds [3-7, 13-15]. According to several authors [6, 7, 16-18] the molecular diffusion coefficient was also called a conductivity coefficient for the solid-liquid system.

Often an effective diffusion coefficient was used instead of an internal molecular diffusion coefficient, which gives the optimal value of the diffusion coefficient for the solid-liquid extraction and can be used for its application in electronic devices for controlling the extraction process.

According to literature data [16, 18-22] there are results for determining an effective diffusion coefficient depending on the value of the internal molecular diffusion coefficient. In some of the publications Arrhenius law was used for the determination of the effective diffusion coefficient based on a constant value of the diffusion coefficient. Some authors [22-24] prepared mathematical models for calculating the effective diffusion coefficient in dependence with the structure of the solid, the temperature and the type of solvent used. Castillo-Santos et al. [25] and Simeonov et al. [9] calculated an effective diffusion coefficient based on the different porosity of the raw material.

Some authors [11, 12, 20, 21, 26, 27] represent the influence of temperature on the process by the Arrhenius law to determine the reaction rate constant (k) in the extraction process. On this basis were presented the graphical dependencies between ln*k* and the extraction temperature $1/T$ and the activation energy of the system was determined as well as the entropy and enthalpy.

The effective diffusion coefficient calculated according to the Arrhenius law establishes the relationship between the kinetic (coefficient of internal molecular diffusion and temperature) and thermodynamic (activation energy and temperature) parameters of the extraction process.

The reaction rate constant calculated for a given thermodynamic system provides information to determine the order of the reaction, its speed, and the mechanism of the reaction. It can be used to optimize a process.

Ivanova et al. [13-15] presented the results of the investigations of the extraction process of leaves from three types of tobacco (Oriental, Virginia and Burley) with 50% ethanol solvent, namely: the calculated values of the internal molecular diffusion coefficient and the activation energy of the thermodynamic liquid extraction system.

The purpose of the present work is to calculate the effective diffusion coefficient and the reaction rate constant for liquid extracts from three types of tobacco leaves - Virginia, Oriental and Burley, with 50% ethanol solvent, as well as to establish a relationship between the kinetic and thermodynamic parameters of the extraction process of the three types of tobacco.

2. Materials and methods

In the extraction of capillary-pore material the process is carried out by molecular diffusion. The density of the diffuse stream in the material relative to a unit of its surface was described by Fick's law using the effective diffusion coefficient D_e (coefficient of mass conductivity). The dependence of the effective diffusion coefficient on the parameters of the porous structure was characterized by the equation (1):

$$D_e = D \cdot \varepsilon \cdot \Pi = D \cdot m \quad (1)$$

where: D is the coefficient of diffusion in a homogeneous setting (in a pure liquid); Π - coefficient of diffusion permeability; m - parameter depending on the porous structure [4, 9].

According to Castillo-Santos et al. [25] and Simeonov et al. [9] the porosity of the raw material can be used.

Effective diffusion coefficient can be determined, which connects the kinetic and thermodynamic parameters of the extraction process according to equation (2) [28]:

$$D_e = D_0 \cdot e^{-\frac{E_a}{R.T}} \quad (2)$$

where: D_e is the effective diffusion coefficient, m^2/s; D_0 - pre-exponential factor or internal molecular diffusion coefficient, m^2/s; E_a - activation energy in the system, J/mol; R - universal gas constant, J/mol·K; T- extraction temperature, K.

The effective diffusion coefficient in dependence with the temperature of the Arrhenius law was calculated based on the values of the internal molecular diffusion coefficient for liquid extracts of Oriental, Virginia and Burley tobacco types, provided by Ivanova et al. [13-15], according to equation (2).

Temperature is a parameter that affects the diffusion properties of substances being extracted, as well as the running of the process. Its influence was expressed by a reaction rate constant, according to equation (3) [11, 12, 20, 21, 26, 29, 30]:

$$k = A \cdot e^{-\frac{E_a}{R.T}} \quad (3)$$

where: k is the reaction rate constant; E_a - activation energy, J/mol; R - universal gas constant, J/mol·K; T - absolute temperature, K.

To determine the reaction rate constant and Arrhenius constant value the activation energy and entropy of the thermodynamic system were used. Calculations prepared for liquid extracts obtained from the extraction of three types of tobacco, extracted with ethanol solvent at concentration 50%, provided by Popova et al. [19, 31, 32].

Arrhenius constant (A) or frequency factor can be calculated according to equation (4) [11, 26]:

$$A = \frac{R \cdot T}{N \cdot h} e^{\frac{\Delta S}{R}} \quad (4)$$

where: N is the Avogadro number; h - Plank constant; ΔS - activation entropy in the system; R - universal gas constant, J/mol·K.

3. Results and discussion

In Table 1 were presented results obtained after calculation of effective diffusion coefficient on the basis of the Arrhenius law, which takes into account the temperature influence on the extraction process. The values obtained for effective diffusion coefficient were calculated with parameters of an internal molecular diffusion of the three types of tobacco (Virginia, Burley and Oriental) with ethanol solvent at a concentration 50% according to Ivanova at al. [13-15].

As it can be seen from the presented results, the determined effective diffusion coefficient of the three types of tobacco exhibited values between 3.68 and $888.78.10^{-9}$ m^2/s (Table 1). It was observed good correlation between results in this work and literature data.

The values of the effective diffusion coefficient D_e depended on the type of the raw material, the solvent used, the particle size, as well as on the raw material: solvent ratio. After reaching a maximum the effective diffusion coefficient decreases (increases the resistance in the solid phase), because of the changes occurring in the structure of the solid phase. From the data presented in Table 1 it can be seen that the values of effective diffusion coefficient D_e for a certain temperature also reach the maximum, and then decrease.

At 30 and 60°C temperature, the maximum value of effective diffusion coefficient D_e coincided with the end of the process for all three types of tobacco. Consequently, it can be concluded that these values were optimal in terms of the temperature and duration of the solvent extraction process with 50% ethanol for tobacco. For the solid phase used under these conditions there was no change in structure as a result of the extraction process.

Scientific literature describes values of the effective diffusion coefficient of different plant materials, which vary considerably depending on the characteristics of the material, the extractant, the extraction conditions and the calculation procedure (Table 2).

IOP Conf. Series: Materials Science and Engineering **595** (2019) 012045 doi:10.1088/1757-899X/595/1/012045

Table 1. Effective diffusion coefficient (D_e) for extraction of tobacco leaves with solvent 50% ethanol

t, °C	τ, min	$D_e \cdot 10^9$, m^2/s		
		Oriental	Virginia	Burley
20	10	3.68	3.72	87.01
20	20	5.10	5.99	310.76
20	30	5.90	5.63	279.69
20	40	6.76	6.15	296.78
20	50	5.25	5.91	377.58
20	60	5.52	5.93	367.62
30	10	4.26	4.05	128.97
30	20	5.75	5.26	295.22
30	30	5.10	6.28	298.33
30	40	4.95	5.87	315.42
30	50	5.40	6.40	296.78
30	60	6.19	8.52	378.51
40	10	4.97	5.71	152.28
40	20	7.72	6.52	259.48
40	30	10,12	6.56	289.00
40	40	9.31	6.60	337.18
40	50	9.92	7.25	400.88
40	60	10.70	7.22	547.91
50	10	5.80	13.08	220.63
50	20	9.72	19.43	441.28
50	30	10.25	22.67	428.85
50	40	8.87	25.99	427.30
50	50	9.16	47.37	626.18
50	60	10.54	46.69	612.74
60	10	7.65	7.81	413.31
60	20	11.03	10.57	551.60
60	30	10.55	12.39	658.82
60	40	11.68	12.87	649.49
60	50	12.44	12.35	596.66
60	60	14.50	13.80	749.96
70	10	38.96	20.49	170.92
70	20	62.51	37.13	888.78
70	30	50.44	23.12	739.61
70	40	5049	17.05	680.57
70	50	67.52	13.48	560.93
70	60	69.65	14.62	621.52

Table 2. Effective diffusion coefficient values for different plant raw materials.

Plant material	Conditions	D_e (m^2/s)	Reference
Vanilla beans	60% (w/w) ethanol-water solvent; at 30, 40 and 50°C; fixed particle size	$1.22 - 2.43.10^{-11}$	[25]
Birch bark	Ethanol as a solvent; at boiling temperature; with mixing; for different solid particle sizes	$7.12 - 47.2.10^{-11}$	[5]

Mangrove barks	Water as a solvent; isothermal extraction; temperature range 20-75°C; a simple pore diffusion model with constant diffusivity	5.0 - 7.5.10^{-11}	[10]
Oak barks	Water as a solvent; isothermal extraction; temperature range 20-75°C; a simple pore diffusion model with constant diffusivity	4.35 - 5.10^{-11}	[10]
Geranium	Ethanol as a solvent, with 70% concentration:		
	– depending on the calculated function of the effective coefficient	1.04 - 3.70.10^{-11}	[33]
	– depending on the method of the regular calculated regime	1.58 - 5.25.10^{-11}	[24]
	– taking into account the influence of other process parameters	1.2097 - 5.0567.10^{-10}	[24]
Tobacco	Water as a solvent	0.28064 - 0.47784.10^{-10}	[24]
Peanut kernels	Hexane as a solvent; at 25°C; for different thickness of the solid particles (seed slices)	0.49 - 1.26.10^{-13}	[28]
Tung seeds	Different petroleum distillation products as solvents; at 30, 50, and 70°C; for different thickness of the solid particles (seed slices)	15.0 - 44.5.10^{-13}	[28]
Chickpea	Grounded chickpea, a mixture of hexane and isopropanol (75:25, v/v) as a solvent; at 60°C	0.13 - 2.04.10^{-13}	[28]

Depending on the calculated values of the reaction rate constant (k) by equation (3), in Figures 1-3 were presented its variation on the temperature and the duration of the process for the three types of tobacco.

It is determined that reaction rate constant values increased with the increase of process duration and temperature. These results can be explained by the exponential type of the Arrhenius law (equation (4)). In future investigations the values of the reaction rate constant will be used to determine the order of the reaction, velocity and mechanism of the reaction.

In Figure 4 was presented the graphical dependence ln$K$ versus $1/T$ for the three types of tobacco (Oriental, Virginia, Burley). These graphical presentations were used to determine the angle of linear dependence, from where the activation energy and enthalpy of the process can be determined, in the absence of system data. The linear dependences were confirmed from literature data [11, 12, 20, 21, 26, 30].

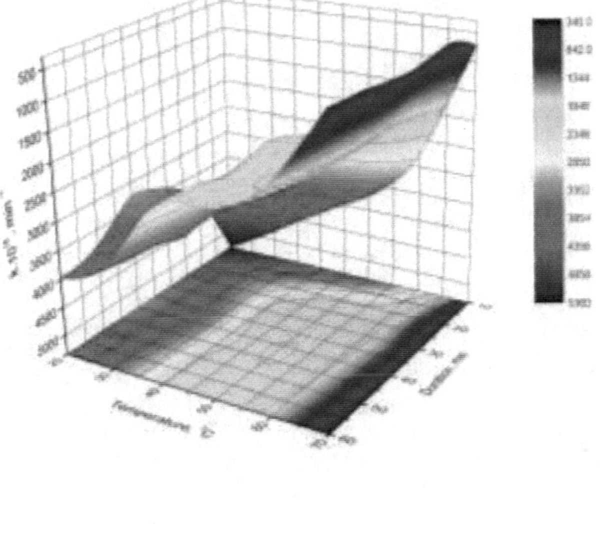

Figure 1. Reaction rate constant (k) in dependence with temperature and duration of the extraction process for Oriental tobacco.

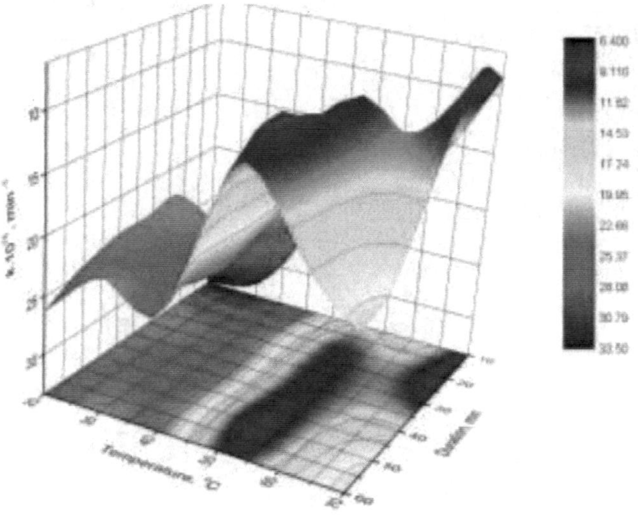

Figure 2. Reaction rate constant (k) in dependence with temperature and duration of the extraction process for Virginia tobacco.

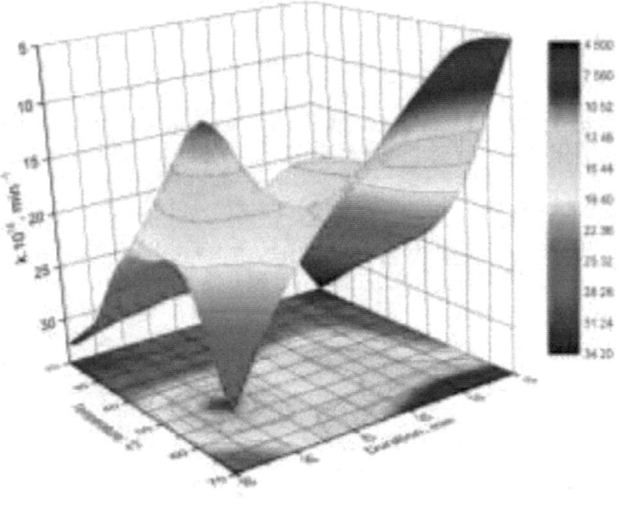

Figure 3. Reaction rate constant (k) in dependence with temperature and duration of the extraction process for Burley tobacco.

As it can be seen from the presented results the temperature had a very significant influence on the process of the extracted substances. Beside this factor, the structure of the solid, its size and the raw material: solvent ratio influences the process. All of these parameters will be taken into account in future studies of the solid-liquid extraction process.

4. Conclusions

There were determined relations between kinetic and thermodynamic parameters involved in extraction process, related to the influence of temperature on the diffusion of substances and the degree of their extraction from the solid. The values of the effective diffusion coefficient for leaved from three tobacco types were calculated, which were within the following ranges: for Oriental tobacco - between 3.68 and $69.65 \cdot 10^{-9}$ m^2/s, for Virginia flue-cured tobacco – between 3.72 and $47.37 \cdot 10^{-9}$ m^2/s and for Burley tobacco - between 87.01 and $888.78 \cdot 10^{-9}$ m^2/s. The maximum values of the effective diffusion coefficient D_e can be used in practice for the characterization of the extraction process. Moreover, they can be applied for optimization of the process and its intensification against parameters influencing it.

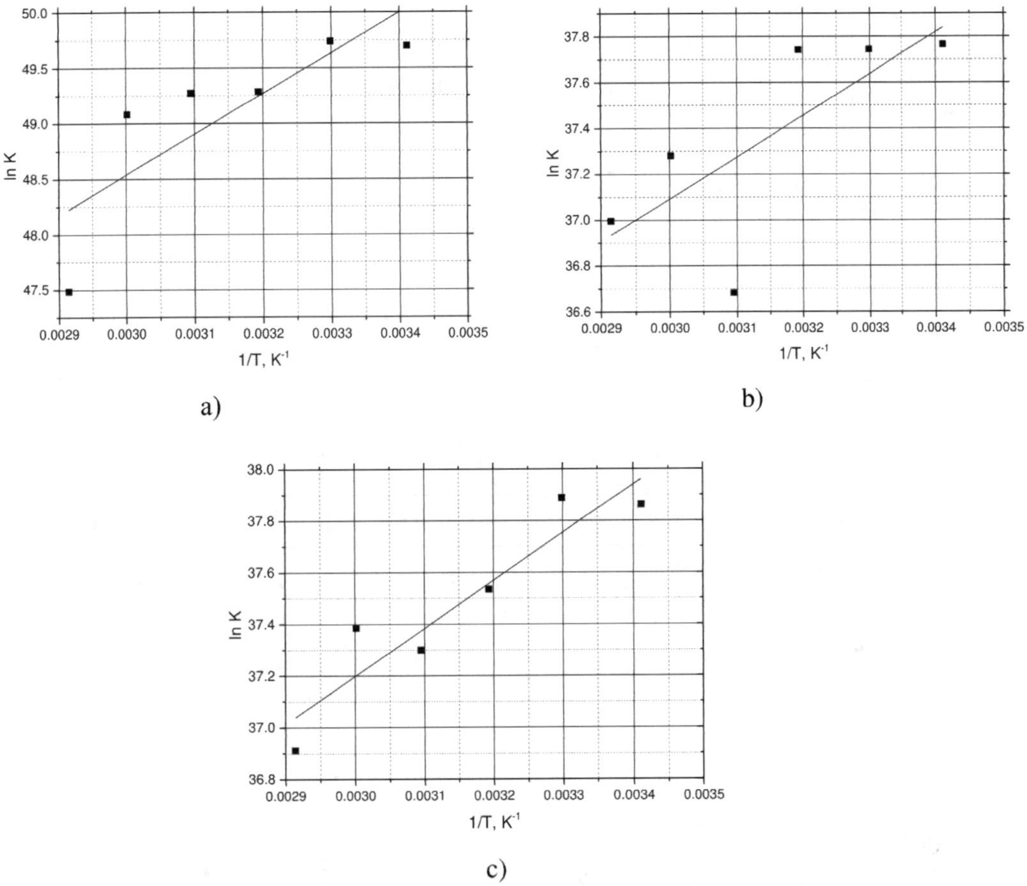

Figure 4. Graphical dependence between ln$K$ versus $1/T$ for three tobacco types: a) Oriental, b) Virginia, c) Burley.

References

[1] Benda A 2014 *Heat and mass transfer in materials and processes* (Moscow: MGUP) *(in Russian)*

[2] Gricenko V 2014 *Processes and devices of food production* (Rubtsovsk: RIIAltGTU) *(in Russian)*

[3] Ivanec V and Borodulin D 2006 *Processes and devices of chemical technology* (Kemerovo: KTIPP) *(in Russian)*

[4] Szytula A and Leciejewicz J 1989 *Handbook on the physics and chemistry of rare earths, vol 12,* Editors K A G Schneidner Jr and L Erwin (Amsterdam: Elsevier) pp 133

[5] Koptelova E, Kutakova N and Tretyakov S 2013 *Lesnoy Zhurnal (Forestry Journal)* **4** 119

[6] Makarevich N, Bogdanovich N, Tretiakov S, Koptelova E and Kutakova N 2014 *Chemistry of plant materials* **4** 251

[7] Ostroushko V and Patchenko V 2012 *East - European Journal of Advanced Technology* **4/6(58)** 12

[8] Parahonia AM and Pushanko MM 2012 *Ukrainian Food Journal* **1(2)** 78 *(in Ukrainian)*

[9] Simeonov E, Tzibarska I and Minchev A 1999 *Chemical Engineering Journal* **73** 255

[10] Tsibranska I 2000 *Communications of Bulgarian Academy of Sciences* **53(6)** 71

[11] Tusek A, Benkovic M, Cvitanovic A, Valinger D, Jurina T and Kljusuric J 2016 *Industrial Crops and Products* **91** 205

[12] Uzoh C, Onukwuli O and Nwabanne J 2014 *Materials for Renewable and Sustainable Energy* **3(38)** 1

[13] Ivanova T, Popova V, Damyanova S, Tasheva S, Atanasova T and Damyanov D 2009 *Scientific Works of the UFT* **56** 225
[14] Ivanova T, Popova V, Damyanova S, Tasheva S, Atanasova T and Damyanov D 2010 *Scientific Works of the Union of Scientists in Bulgaria – Plovdiv* **B12** 67
[15] Ivanova T, Popova V, Damyanova S, Tasheva S, Atanasova T and Damyanov D 2010 *Scientific Works of the University of Russe* **49(9.2)** 86
[16] Promtov M, Stepanov A and Aleshin A 2017 *TSTU Bulletin* **23(2)** 265
[17] Rudobashta S and Kosheleva M 2015 *Textile Technology* **6** 175
[18] Hou K, Zheng Q, Li Y, Shen J and Hu S 2000 *Computers and Chemical Engineering* **24** 1343
[19] Popova V, Tasheva S, Damyanova S and Stoyanova A 2014 *Scientific Works of the University of Russe* **53(10.2)** 102
[20] Nwabanne J 2012 *International Journal of Multidisciplinary and Engineering* **3(6)** 11
[21] Paunovic D, Mitic S, Kostic D, Mitic M, Stojanovic B and Pavlovic J 2014 *Advanced Technologies* **3(2)** 58
[22] Simeonov E and Chilev C 2015 *Journal of Chemical Technology and Metals* **50(5)** 597
[23] Simeonov E and Koleva V 2012 *Chemical and Biochemical Engineering* **26(3)** 249
[24] Simeonov E, Yaneva Z and Chilev C 2017 *Bulgarian Chemical Communications* **49(2)** 399
[25] Castillo-Santos K, Ruiz-Lopes I, Rodrigues-Jimenes G, Carrillo-Ahumada J and Garcia-Alvarado M 2017 *Journal of Food Engineering* **192** 36
[26] Amin S, Hawash S, Diwani G and Rafei S 2010 *Journal of American Science* **6(11)** 293
[27] Philibert J 2006 *Defect and Diffusion Forum* online **249** 61
[28] Ruiz R, Martinez C and Vizcarra M 2011 *Revista Mexicana de Ingenieria Quimica* **10(3)** 387
[29] Chan C, Yusoff R and Ngoh G 2013 *Chemical Engineering Research and Design* **92(6)** 1169
[30] Charpe T and Rathod V 2016 *Brazilian Journal of Chemical Engineering* **33(4)** 1003
[31] Popova V, Tasheva S, Damyanova S and Stoyanova A 2014 *Vestnik MGUP* **1(16)** 67
[32] Popova V, Tasheva S, Damyanova S and Stoyanova A 2014 *Journal of Eurasian Technological University* **2(16)** 9
[33] Saha D and Mukherjee A 2017 *Journal of Chemical Sciences* **129(7)** 825

IOP Conf. Series: Materials Science and Engineering **595** (2019) 012046 doi:10.1088/1757-899X/595/1/012046

Determining the expenses for heating of a residential building using different energy sources

G Komitov[1], V Rasheva[2] and I Binev[3]

[1]Dept. of Agricultural Mechanization, Agricultural University of Plovdiv, Bulgaria
[2]Dept. of Industrial Thermal Engineering, University of Food Technologies, Plovdiv, Bulgaria
[3]Dep. of Energetics, Trakia University St. Zagora, Faculty of Technics and Technology, Jambol, Bulgaria

E-mail: v_rasheva@abv.bg

Abstract: The heating of residential and industrial buildings is important for creating a good microclimate in them and for their normal functioning. This article defines the thermal losses of an energy model of a residential building according to the energy efficiency legislation in force in the Republic of Bulgaria. The advantages and disadvantages of using different energy sources for buildings heating are considered. Various types of energy sources- conventional and alternative (electricity, pellets, HHO gas, LPG, natural gas – CNG, gas oil) are used. Comparison and analysis of the expenses, energy and environmental impact of their implementation has been done.

1. Introduction

The heating of residential and industrial buildings is important for creating a good micro-climate in them and for their normal functioning. The analysis shows that the largest share of household energy consumption is represented by heating (50%) and hot water (23%) (Figure 1) [1, 2].

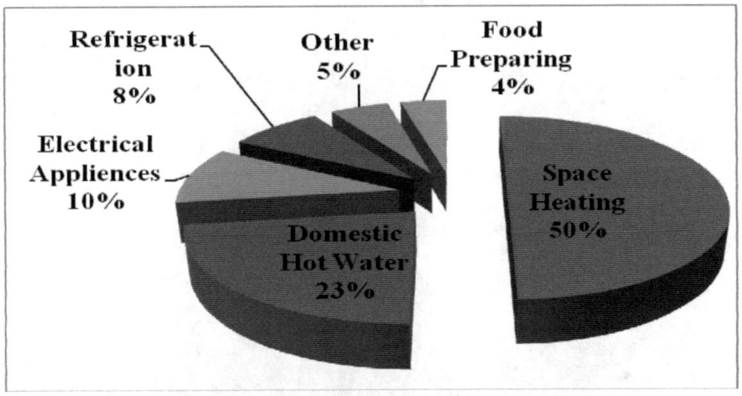

Figure 1. Structure of energy consumption in a household.

Energy consumption for heating and domestic hot water accounts more than 70% of the total energy consumption of the household, so the type of energy source and the efficiency of energy

IOP Conf. Series: Materials Science and Engineering **595** (2019) 012046 doi:10.1088/1757-899X/595/1/012046

conversion systems is of utmost importance.

The final consumption of energy sources for heating in Bulgaria is as follows (Figure 2): wood and coal - 52.6%, electricity - 38.5%, natural gas - 2.4%, natural gas for district heating - 6.5% [2]. The large share of wood and coal use (around 52.6%) is clearly visible. In domestic heating with non-efficient stoves and solid fuel boilers and the use of coal containing ash and sulphur and wet wood, pollutants such as sulphur dioxide, nitrogen dioxide, carbon monoxide and fine particulate matter are the main problem in cities [3-7]. Analyses indicate that this type of heating causes at least 85% of fine particulate emissions, so measures to improve air quality are mostly targeted in this area [2].

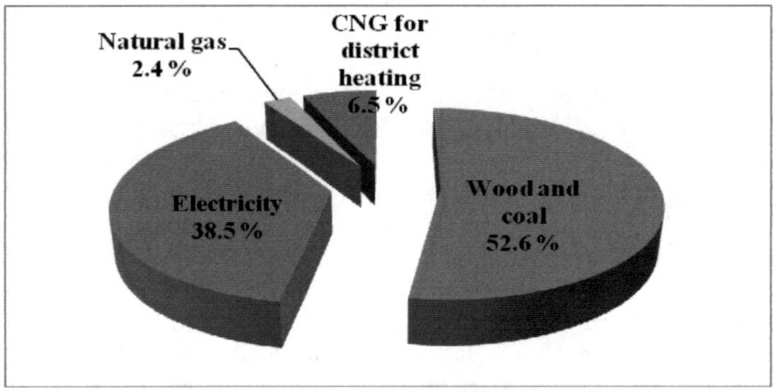

Figure 2. Energy consumption for heating in Bulgaria.

Fig. 2 shows also that the use of electricity for home heating stands in second place with a share of 38.5%. The problem in this case is that nearly 1/3 of electricity is lost, until it reaches the end user. Except of that the electricity distribution system is significantly burdened and, at the same time, there is energy dependence of households on the electricity distribution companies [1-2]. The use of renewable energy sources is negligible, but more and more people living at houses choose pellets as heating fuel for their homes.

The European Union is addressing the issue of introducing of renewable energy as a means of reducing greenhouse gas emissions and compliance with the Paris agreement on climate change. Increased use of renewable energy also plays an important role in sustainable energy production at affordable prices, technological development and innovation [8].

In many countries, more rational use of energy and development of renewable energy sources (RES) is planned to be replaced part of the fossil fuels used to satisfy their energy needs (Figure 3) [1-2, 8-11].

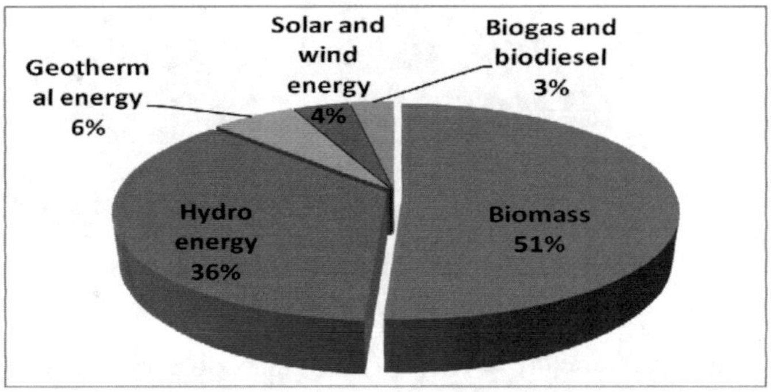

Figure 3. Distribution of energy from renewable energy sources (RES).

The perspectives for energy saving and the development of new energy sources are related to economic models geared to sustainable development at regional and local level [1].

The most efficient and secure way to meet the energy needs of households is decentralized generation of renewable energy, including in our homes. It concentrates complex benefits because: avoids losses in energy transportation, avoids costs for building the transmission network, and uses clean and inexhaustible resource [12].

An alternative to houses heating is the use of oxy-hydrogen gas (HHO gas). Yull Brown has patented an electrolyze technology for HHO gas generation [13]. A number of researchers have investigated the existing generators of HHO gas and their applicability in practice [14-16], the possibility of using HHO gas as a fuel in internal combustion engines [17-20], in food industry [21], for electricity generation [22] or for heating purposes [23-24].

The purpose of this paper is to determine the costs for heating of a residential building using 7 different sources of energy – conventional and alternative and to compare them.

2. Materials and Methods

In our case, a residential building with a heated area of 514 m^2 and a heated volume of 1285 m^3 was chosen for a model. The building is located in the city of Plovdiv and depending on its energy consumption falls in energy class "B" [25]. To determine the energy required for heating the model building, a methodology based on the energy efficiency legislation in force in Bulgaria is used [25]. For the purpose of the comparative study seven different energy sources for building heating are used: HHO gas heating (Brown gas), electric hot water boiler [26], pellet boiler - class A1 [27], gas oil boiler [28-29] and natural gas condensing boiler [30]. The calculated energy demand for heating of the model building is Q_Y = 36,788 kWh per year. The number of heating days per a year in the region of Plovdiv is d = 165 days and it is determined according to [25]. The energy distribution according to the number of heating days per a year $Q_{H,d}$ is obtained after dividing the required amount of energy Q_Y by the number of heating days according the equation (1).

$$Q_{H,d} = \frac{Q_Y}{d} = 223 kWh/day \quad (1)$$

The energy required to heat the building for 1 hour is obtained after dividing the required energy for one day $Q_{H,d}$ by the number of hours per day (2):

$$Q_h = \frac{Q_{H,d}}{h} = 9.29 kWh \quad (2)$$

Table 1 presents data for the calorific value and the prices of the investigated energy sources [31].

Table 1. Data on the calorific value and the prices of the investigated energy sources.

№	Energy sources	Calorific value - Q_{PS}			Price of energy source - p_{PS}		
1	HHO gas	Q_{HHO}	6.26	Wh/l	p_E	0.11	€/kWh
2	Electricity	p_E	1.00	kWh	p_E	0.11	€/kWh
3	Wood pellets	Q_W	5.28	kWh/kg	p_W	0.26	€/kg
4	Agro pellets	Q_A	3.33	kWh/kg	p_A	0.13	€/kg
5	Natural gas - CNG	Q_{CNG}	9.89	kWh/m^3	p_{CNG}	0.44	€/m^3
6	LPG	Q_{LPG}	13.9	kWh/kg	p_{LPG}	1.12	€/kg
7	Gas oil	Q_G	11.63	kWh/kg	P_G	1.09	€/l

Bulgaria has a fixed exchange rate of BGN to the EURO and 1 EURO = 1.95583 BGN [32].

Considering the wide variety of energy sources and the different price of each of them, the comparison between the energy sources from table 1 consists in determining the value that the end user pays for model building heating for one hour. The HHO gas, also known as Brown's gas, is an alternative source of energy while at the same time it is nearly ecological. The Brown's gas is produced by electrolysis of distilled water and a certain amount of potassium hydroxide solution (Fig. 4). In course of an electrochemical reaction, HHO gas is released. The energy potential of the HHO gas according to different sources is Q_{HHO} = 6.26 Wh/l [33].

Figure 4. Principal scheme for generating of HHO gas.

The quantity of HHO gas required to ensure the heating of the building for 1 hour is calculated according to the equation (3):

$$D_{HHO} = \frac{Q_h}{Q_{HHO}} = 1484 l_{HHO} \tag{3}$$

On the other hand, according to our experiments for generation of 1 liter HHO gas is required E_{HHO} = 1.006 Wh electric energy. Therefore, for heating of the model building for 1 hour, the required amount of energy is (4):

$$q_{HHO} = D_{HHO}.E_{HHO} = 1.49 kWh \tag{4}$$

According to the electricity prices of electricity company EVN-Bulgaria, the average price of electricity is 0.21 BGN/kWh (to 01.01.2019) [31] or 0.11 €/kWh. The cost of the model building heating can be expressed by equation (5):

$$C_{PS} = q_{PS} \cdot p_{PS} \tag{5}$$

where q_{PS} is the amount of energy required for building heating with the particular energy source.

In the option of heating with hot water electric boiler, the boiler "Ecotermal", model „10 MRL“ is selected [26]. It is an electric boiler with a power of 10 kW and a power supply of 400 V. The value of electricity used for heating in this case is determined by dependence (6):

$$C_E = Q_h \cdot p_E \tag{6}$$

The amount of energy q_{PS} required for building heating with energy sources - pellets, gas oil, CNG or LPG is determined by equation (7):

$$q_{PS} = \frac{Q_h}{Q_{PS}} \tag{7}$$

where Q_{PS} is the calorific value of the particular energy source.

The cost of heating by using different energy sources C_{PS} for one hour is determined from the equation (8):

$$C_{PS} = q_{PS} \cdot p_{PS} \tag{8}$$

where P_{PS} is the price of each energy source.

According to the measures taken by the Paris Greenhouse Gas Reduction Agreement, the use of coal is limited to all EU countries and therefore the authors do not recommend their use.

Heating with automated pellet boilers can be considered as renewable energy source heating, especially if the pellets are made from waste from agricultural produce, such as branches, plant stems, scales, etc. The boiler burning pellets used in this publication is "Burnit" boiler, model "UB 27" with a rated output of 27 kW [27]. The boiler is automated and is suitable for burning wood and agro pellets as well as solid fuel.

3. Results and Discussion

The results for determining the cost of building heating using different energy sources are summarized in Table 2 and graphically presented in Figure 5.

Table 2. Value of heating per an hour at different energy sources.

№	Energy sources	Energy requirement - q_{PS}			Cost of heating C_{PS},
1	HHO gas	q_{HHO}	1.49	kWh	0.16 €/h
2	Electricity	q_E	9.29	kWh	1.00 €/h
3	Wood pellets	q_W	1.76	kg	0.45 €/h
4	Agro pellets	q_A	2.79	kg	0.36 €/h
5	Natural gas - CNG	q_{CNG}	0.94	m^3	0.41 €/h
6	LPG	q_{LPG}	0.67	kg	0.73 €/h
7	Gas oil	q_D	0.70	l	0.77 €/h

The results of Table 2 only make sense to determine the required for building heating amount of a specific energy source in natural terms in order to limit its use or its preservation in nature. Considering that electricity is one of the cleanest energies, the use of an energy source different from electricity can lead to a significant increase in the fine dust particles. At the same time, in the production of a unit of electricity, 3 times more primary energy is consumed and, therefore, more is the CO_2 released into the atmosphere. Therefore, the result of Table 2 could be used for environmental and economic research on the use of the type of energy carrier.

Figure 5 shows the cost of heating of model building at seven different energies.

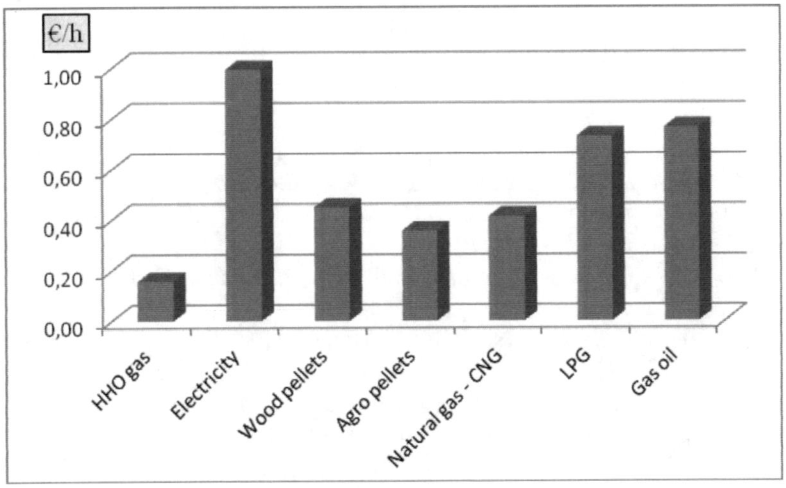

Figure 5. Cost of heating at different energy sources.

According to data from Table 2 and Figure 5 heating the building using HHO gas is cheapest. In this case one-hour heating of the model building costs 0.16 €/h. There is almost no harmful emission and the costs are for water and electricity necessary for the decomposition of water. Therefore, heating with HHO gas is suitable for all heated buildings with a separate boiler room.

On the other hand, the most expensive heating of the model building is heating with an electric hot water boiler - 1.00 €/h. This type of heating is suitable for use of buildings with limited space.

The use of green energy from biomass for heating is quite acceptable option at a relatively low price of heating - 0.36 ÷ 0.45 €/h depending on the type of pellets. The cost of pellets can be reduced by properly planning the purchase of the pellets. At the end of the heating season, the price of pellets drops by almost half. The emissions from the boiler for automatic pellet burning are minimal and are controlled by the combustion process. This type of heating is suitable for all heated buildings.

Economical option for building heating is heating with a boiler burning CNG (0.41 €/h). The heating with CNG relies on a gas supplier and this will lead to energy dependence. The heating with the CNG is simply to the type of the installation and is a practical option in the absence of free spaces for the boiler.

The value of building heating using LPG and gas oil as energy source varies within the range of 0.73 ÷ 0.75 €/h. The prices of these types of energy source directly depend on the oil price. Again there is energy dependence on oil companies. It is practical and justified to use these energy sources in hard-to-reach areas and those that do not require year-round heating. Such buildings can be mountain chalets and villas.

4. Conclusions

This article defines the thermal losses of an energy model of a residential building according to the energy efficiency legislation in force in the Republic of Bulgaria. The advantages and disadvantages of using different energy sources for buildings heating are considered. Various types of energy - conventional and alternative (electricity, pellets, HHO gas, natural gas, LPG and gas oil) are used. Comparison and analysis of the energy and environmental impact of their implementation has been done.

The most economical heating of the model building with the required heat output of 9,29 kW is that of HHO gas. The value of this heating is 0.16 €/h if electricity is used to generate the HHO gas. This type of heating is suitable for all heated buildings that have a boiler room.

Biomass energy for heating the model building shows relatively good results, depending on the type of biomass (pellet). The value of heating the model building with biomass varies from 0.36 ÷ 0.45 €/h, but it can be reduced by properly planning its purchase.

The use of natural resources leads to energy dependence on the supplier of the energy sources.

References

[1] Legal framework for the use of biomass in the countries of the European Union and the Republic of Bulgaria *http://www.abea-bg.org/files/Biomass_pravna%20ramka.pdf*

[2] Guidance for Household Energy Auditors *http://reach-energy.eu/wordpress/wp-content/uploads/2015/01/D3.4-EAP.pdf*

[3] Valtchev G. 2012 Combustion equipment and technology, *2nd Edition, Academic Publishing House of UFT, Plovdiv*

[4] Valtchev G 2003 Heat Transfer Systems with Mineral Oils. *Scientific works of UFT - Plovdiv,* Volume XLIX, 182-187

[5] Angelov B, Kunev D and Totev T 2018 Contamination of the environment with mercury in the burning of Bulgarian lignite, *Energy Forum,* IHS Varna, 10 – 18

[6] Totev T, Bonev B and Stanoev V 2004 Modeling of combustion technologies, *Journal Mechanics of machines.* **54**, year XII, Vol. 5 pp 3-8

[7] Directive (EC) 2018/2001 *of the European Parliament and of the Council of 11 December 2018 on the promotion of the use of energy from renewable sources*

[8] Ion V I, Ciocea G and Popescu F 2016 Waste Heat Recovery Technologies from Heating Boiler Flue Gas. *TERMOTEHNICA* Supliment 1

[9] Ion V I, Popescu F and Dimofte E 2017 Integration of Biomass Resources into Existent District Heating System. *Proceedings of the 6th Intern. Conf. on Thermal Equipment, Renewable Energy and Rural Development TE-RE-RD 2017, Romania,* 57 - 62

[10] Zlateva P 2012 Effective Use of Biomass for Energy Production in Bulgaria, *Scientific Announcements of the Union of Scientists - Varna,* **1** pp 26-30

[11] Mitkov I, Ivanov I and Dallev M 2016 Comparative analysis of biomass fuels. *Agricultural Science,* Vol. VIII Issue 20, Doi: 10.22620/agrisci.2016.20.021

[12] Energy strategy of the republic of Bulgaria until 2020. *http://www.mi.government.bg/files/useruploads/files/epsp/22_energy_strategy2020_.pdf*

[13] Yull Brown *US patent number 4,014,777 issued on March 29, 1977, and US patent number*

4,081,656 issued on March 28, 1978.

[14] Deltchev D, Terziev A and Iliev I 2017 Review on existing Brown's gas (HHO) production's systems and analysis of capabilities for its use in practice. *Proceedings of the 6th International Conference on Thermal Equipment, Renewable Energy and Rural Development TE-RE-RD 2017,* Moieciu de Sus – Romania, pp 7 - 12

[15] King M B 2011 Water Electrolyzers and Zero-Point Energy. *Physics Procedia.* **20** pp *435-445*

[16] Venkata R M et all. 2018 Production of Brown's Gas using Hydroxy Generator. *International Journal of Engineering & Technology.* **7(4.5)** pp 428-457

[17] Bambang S, Sudjud D and Djoko SK. 2016 Application of Dry Cell HHO Gas Generator with Pulse Width Modulation on Sinjai Spark Ignition Engine Performance. *Intern. J. of Research in Engineering and Technology.* **05(02)**

[18] Sowba P T, Ravichandran N and Senthil R 2013 Utilization of Brown Gas As A Suplemental Fuel IN THE Diesel Engine AS Pre-Combustion Exhaust Emission reduction Method. *International Journal of Engineering Research & Technology.* **2(6)** pp 2983-2986

[19] Dhananjay B, Jay O, Bhavin H and Gaurang A 2015 An Experimental Analysis of S.I Engine Performance with HHO Gas as a Fuel. *International Journal of Research in Engineering and Technology.* **04(04)** pp 608-615

[20] Gohar G A and Raza H 2017 Comparative Analysis of Performance Characteristics of CI Engine with and without HHO Gas (Brown Gas). *Adv Automob Eng* **6:172**. doi: 10.4172/2167-7670.1000172

[21] Heinz J S 2017 Hydrogen Economy for the Food Sector. *4/2017 eFOOD-Lab international 20-23*

[22] Chia-Nan W, Min-Tsong C, Hsien-Pin H, Jing-Wein W and Sridhar S 2017 The Efficiency Improvement by Combining HHO Gas, Coal and Oil in Boiler for Electricity Generation. *Energies.* **10, 251**; doi:10.3390/en10020251

[23] Tabazah T, Hamdan M A, Abo D O and Abdelhafez E 2014 Utilization of Water Produced Hydrogen for Domestic Heating Purposes. *International Journal of Thermal & Environmental Engineering.* **7(2)** pp 95-99

[24] Paul E D et all. 2015 Hydrogen and fuel cell technologies for heating: A review. *International Journal of Hydrogen Energy.* **40** pp 2065-2083

[25] Regulation № 7 of 2004 on Energy Efficiency of Buildings. *Available at: https://www.mrrb.bg/bg/normativni-aktove/naredbi/page/5/*

[26] Electric boiler "ecothermal" *http://www.ecotermal-bg.com/index.php*

[27] Burnit UB *https://burnit.bg/bg/project/BD-BB-burnit-advance-ub/*

[28] Sime Rondo 3 *http://www.sime.it/int/en/products/navbasamen/rondo/*

[29] Burner "Mack 3" *http://sime.co.uk/wp-content/uploads/2017/02/Mack-GB-Brochure.pdf*

[30] Immergas Victrix 12 *https://www.immergas.com/product?title=VICTRIXErP*

[31] EVN *https://www.evn.bg/ Ceni-el-energia.aspx* EVN *https://www.evn.bg/ Ceni-el-energia.aspx*

[32] http://www.bnb.bg/

[33] Simov G 2014 Brawn's gas. Brawn's gas will change world. (in Bulgarian). *http://grigorsimov.blog.bg/politika/2014/11/13/braunov-gaz-hho-braunoviiat-gaz-shte-promeni-sveta-avtor-qhi.1313295*

IOP Conf. Series: Materials Science and Engineering **595** (2019) 012047 doi:10.1088/1757-899X/595/1/012047

Effect of yaw angle on the global performances of Horizontal Axis Wind Turbine - QBlade simulation

D E Husaru[1], P D Bârsănescu[2] and D Zahariea[3]

[1]Mechanical Engineering," Gheorghe Asachi" Technical University of Iaşi, Iaşi, Romania
[2]Mechanical Engineering, Mechatronics and Robotics Department, "Gheorghe Asachi" Technical University of Iaşi, Iaşi, Romania
[3]Fluid Mechanics, Fluid Machinery and Fluid Power Systems Department, "Gheorghe Asachi" Technical University of Iasi, Iasi, Romania

E-mail: husarudorin@yahoo.com

Abstract. The yaw angle has a great importance in the wind turbine working. Even though most of the large turbines have yaw mechanisms, they do not have an instant response. The aerodynamic forces and torque on the blades fluctuate, depending on the yaw angle. The design and the numerical simulation of the wind turbine were performed with the Blade Element Momentum method in open source QBlade software. The power coefficient, torque, thrust and power output generated by wind turbine in non-yawed flow were analysed. The simulations have been performed for non-yawed flow, in rotational speed range between 2100 and 3300 rpm and wind velocity of 15 m/s. Simulations for yaw angle range have been performed between ± 60°, with 5° step at rated rotational speed of 2700 rpm. The results are presented through charts for global parameters in both non-yawed flow, and yawed flow. The effect of yaw angle on global performances of the wind turbines is more important after the value of 25° when the power output decrease with about 15% from power output in non-yawed flow. The average value of the exponent from conventional relation of power coefficient in yawed flow is 1.77 in good concordance with experimental tests.

1. Introduction

Nowadays, the wind energy knows a continuous increase most of all because it is a renewable and non-polluting energy. The kinetic energy of the wind is converted into mechanical energy by wind turbines. The two main types of wind turbines are classified according rotational axis in HAWT (Horizontal Axis Wind Turbine) and VAWT (Vertical Axis Wind Turbine) [1]. The HAWT are used in the most of windy areas around the world due to higher efficiency, but in recent years in the most favourable areas already are installed wind turbines, so new research directions have been identified in order to increase the percentage of energy provided by wind turbines: installation of small wind turbines, expansion of offshore areas and increase of rotor dimensions [2].

Several methods to design and predict aerodynamic performances of wind turbines such as vortex methods, 3D viscous-inviscid interaction technique, computational fluid dynamics algorithms and Blade Element Momentum (BEM) method are used. The most used method by the designers of wind turbines is the BEM method. This method provides a good aerodynamic performance analysis with a relatively simple procedure in different flow regime such as different wind speeds, yaw angles, setting angles of blades [3,4]. The classical BEM method was first developed by Glauert [5] by combining momentum

IOP Conf. Series: Materials Science and Engineering **595** (2019) 012047 doi:10.1088/1757-899X/595/1/012047

theory and blade element theory. The BEM method involves discretization in N annular elements with height dr of the stream tube introduced in the 1-D momentum theory and no flow across the elements. The annular elements are considered independent one to another (no radial dependency) and the force on the blades are considered constant in each annular element due to hypothesis of a rotor with infinite number of blades [6].

The classical BEM method is a two-dimensional method extrapolated into the third dimension through applied semi-empirical corrections, tip and root losses induced by finite number of the blades and 3D corrections, derived from experimental tests or CFD computations. The BEM method have a main advantage in comparison to CFD computation due to less computational time, very low cost and possibility to develop and test rapidly different rotor design, reaching at a rotor model that will be studied with CFD technique in details [7].

Most of the working time, HAWT operate in yawed flow because the direction of wind varies continuously over time. Wind turbines working in yawed flow leads to difficult problems of aerodynamics, control and aeroelasticity. Some of these for the moment are unresolved, [8]. Yaw angle, figure 1, represent the angle between wind turbine rotor plane and wind direction. Several researchers have conducted numerical and to experimental test in order to predict the operation and to determinate the global performances of the HAWT in yawed flow [9, 10].

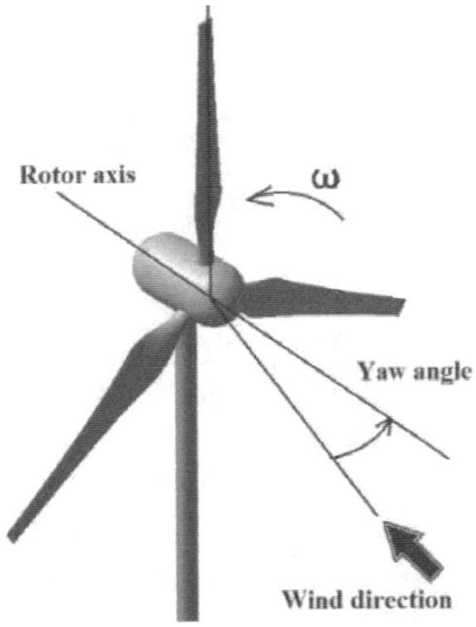

Figure 1. Yaw angle of HAWT.

Some researchers, based on the assumption that induction is constant with yaw angle, with x=3 (aerodynamics of the helicopter rotor in a forward flight) do not consider that aerodynamics of wind turbines has additional difficulties and differences from the aerodynamics of the helicopter rotor.

It is proposed that the power coefficient, and implicitly, power output in yawed flow have a cosinusoidal behaviour expressed in the conventional form in equation (1), [11].

$$C_P(\psi_0) = C_P(0^o) \cdot cos^x(\psi_0) \tag{1}$$

Experimental tests in wind tunnel and on field [12], shows that x varies from 1.88 and 5.15 while in measurement on the rotor with diameter of 4.5 m in German Dutch Wind Tunnel [13], x is about 1.8.

QBlade is an open source software dedicated to design and predict the aerodynamic performances of Horizontal Axis Wind Turbines and Vertical Axis Wind Turbines blades using BEM method and the integration in XFOIL enables to design custom airfoils and compute their polars and also allow to import experimental polars.

In this paper, power coefficient and power curve generated for a HAWT with three blades, with NACA 4415 airfoils and 0.58 m rotor diameter obtained with QBlade software in the range ± 60° for yaw angle have been analysed.

The relatively small geometrical dimensions of the wind turbine rotor are due to the fact that this research is part of a more complex analysis in which, in a further step, some experimental tests will be carried out. For experimental tests the MF-TA4 wind tunnel from the Department of Fluid Mechanics, Fluid Machinery and Fluid Power Systems will be used, for which the cross-section area of the test section is 2.853 m^2. In order to obtain a blockage ratio under 10% it was chosen that the swept area of the rotor to be equal to 0.2642 m^2, which correspond to the rotor diameter of 0.58 m.

2. Wind turbine rotor model

2.1. Airfoil characteristics

One of the important points in the design of wind turbine rotors is the selection of the blade airfoil. A wind turbine blade can contain one or more airfoils. There are a multitude of bidimensional airfoils that can be used in the design of HAWT blades. At low Reynolds numbers (<10^5), which is the case of this research, there is difficult to find experimental polars. In [14], Eastman N. J. and Sherman A., have conducted experimental tests for different airfoils at low Reynolds numbers, including NACA 4-digit series of airfoils. These series of airfoils have been used since the first half of the 20th century in different applications, including wind turbines [15]. In this paper a NACA 4415 airfoil was used for the entire blade length, being one of the airfoils that have been experimentally characterized for different Reynolds numbers. In figure 2 are presented the dimensionless coordinates of the NACA 4415 airfoil, having a maximum camber of 4% located at 40% from the leading edge and having a maximum thickness of 15% with respect to the airfoil chord length.

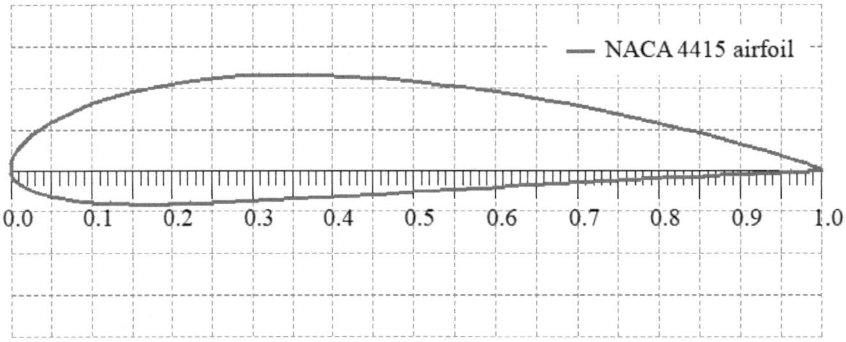

Figure 2. NACA 4415 airfoil dimensionless coordinates.

2.2. Polars of NACA 4415 airfoil

In this paper, the variation of the lift coefficient (C_L) and drag coefficient (C_D) in the range of the incidence angle (α) between -7° and 21.2° corresponding to a Reynolds number equal with 83000 for NACA 4415 airfoil were taken from charts resulting from experimental tests. Due to the small number of points on the charts of incidence angle range, an interpolation of 100 point with 5th degree polynomial was made. The polars of the airfoil was created in XFOIL and then was imported in QBlade software. In figure 3, (a) lift coefficient and 3, (b) drag coefficient for incident angle in the range -7 and 21.2 degrees are presented. In figure 3, (c) are presented the polar curves for NACA 4415 airfoil in QBlade. The optimum incidence angle is 6.387 degrees conditioned by maximum ratio between lift coefficient and drag coefficient equal with 40.055. At this value of the incidence angle, optimum lift coefficient and drag coefficient are equals with 1.0386 and respectively 0.0259.

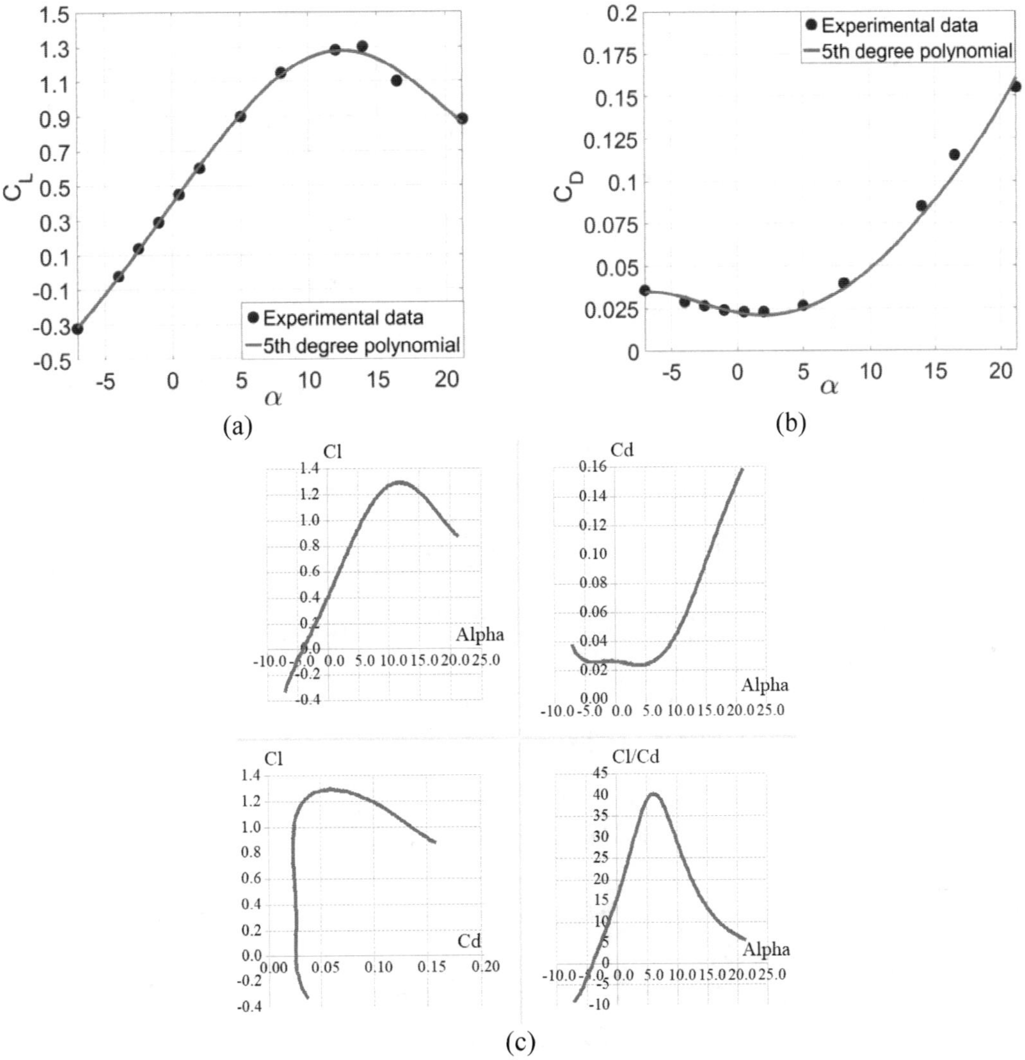

Figure 3. NACA 4415 Polars.

2.3. Blade design and Blade Element Momentum method

The design of the blade was accomplished by using BEM method considering the rated parameters of the wind turbine rotor presented in table 1.

Table 1. Wind turbine rated parameters.

Wind velocity V [m·s^{-1}]	Rotational speed n [rpm]	Tip speed ratio TSR [-]	Rotor diameter D [m]	Reynolds Number Re [-]
15	2700	5.46	0.58	83000

The total length of the blade is 0.27 m. The blade was divided in 23 sections with 2 circular airfoils with 0.016 m diameter, at radial positions 0.02 m and 0.03 m. Between section 2 (circular foil) and section 3 (NACA 4415 airfoil) automatic linear interpolation was performed by QBlade.

The NACA 4415 airfoil chord varies from 0.044874 m in section 3 and 0.016946 m in the final section. The chord along the blade is presented in figure 4, (a). The total twist of the blade is 13.43 degree between 13.957 degrees at first section of the active zone of the blade and 0.5234 degrees at tip of the blade. The variation of the twist angle along the blade is presented in figure 4, (b). Global solidity

IOP Conf. Series: Materials Science and Engineering **595** (2019) 012047 doi:10.1088/1757-899X/595/1/012047

of the blade is 6.1189 %. Geometry of the blade and airfoils used at each radial position are presented in figure 4, (c).

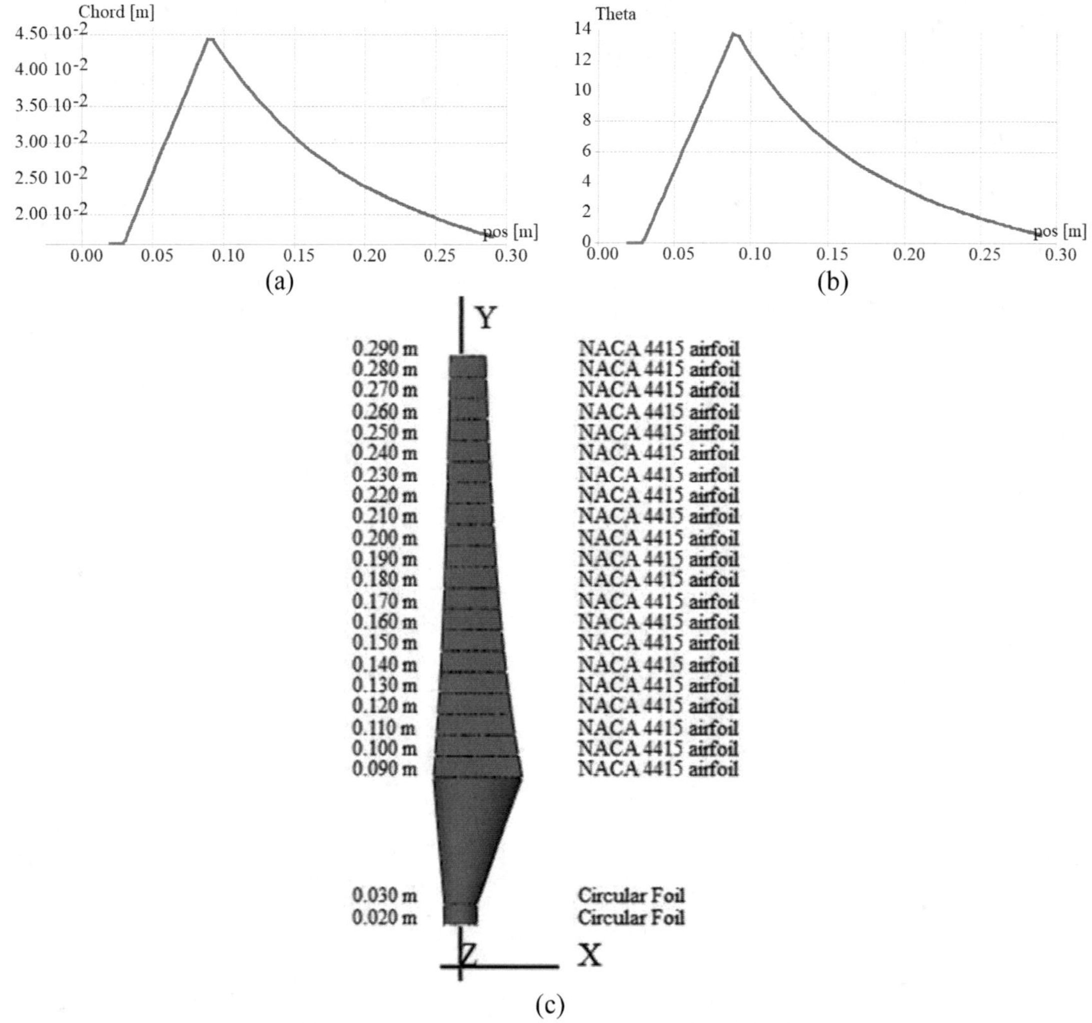

Figure 4. Geometrical dimensions of the blade.

3. QBlade simulations and results

The functional performance analysis of wind turbine rotor was performed in *BEM Simulation* menu for tip speed ratio range between 2 and 8 with 0.1 step. The air density and air viscosity are considered 1.225 kg/m^3 and respectively 1.647e-05 Pa·s. Prandl correction, 3D correction and Reynolds number correction have been used. Figure 5, (a) shows the variation of Reynolds number along the blade and figure 5, (b) shows the absolute error between the Reynold number of imported polar and Reynolds number calculated by QBlade.

The largest absolute error is recorded at the hub area and at the inactive part of the blade (16504 at radius 0.03 m), while at active part of the blade the maximum absolute error for the Reynolds number is -3275 (at radius 0.0902 m), which corresponds to a relative error of less than 4%.

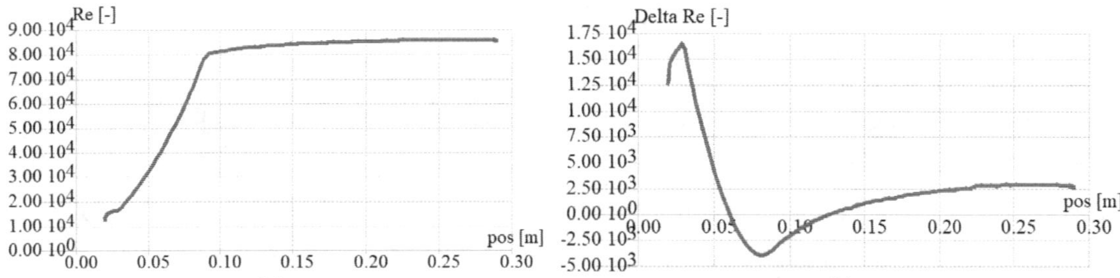

(a) (b)

Figure 5. Reynolds number and absolute error.

The multi-parametric analysis of functional performances was performed in *Multi-Parameter BEM Simulation* menu. The simulations were performed for rotational speed range between 2100 and 3300 rpm with 50 rpm step, and for wind velocity range between 5 and 25 m/s with 1 m/s step.

Global performances of the wind turbine rotor are showed in figure 6, (a) for power coefficient, (b) for power output, (c) for torque and (d) for thrust according to rotational speed. For rated rotational speed of 2700 rpm and rated wind velocity of 15 m/s, the values of the calculated parameters are presented in table 2 (red point from figure 6).

Table 2. Wind turbine global performances parameters.

Power coefficient CP [-]	Power P [W]	Torque T [Nm]	Thrust S [N]
0.40408	220.69	0.7805	30.08

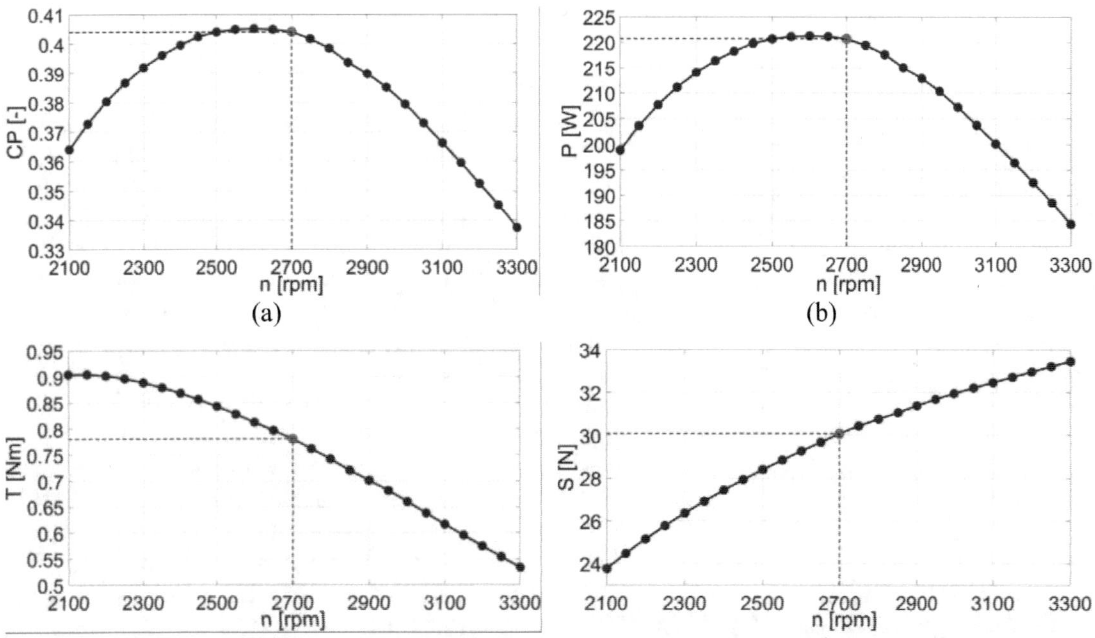

(c) (d)

Figure 6. Global performances of wind turbine rotor.

The influence of yaw angle on global performances of wind turbine rotor was studied using *LLT (Lifting Line Theory) HAWT Simulation* menu, [16] for the yaw angle range between ±60° with 5° step size under an windfield defined by the mean wind velocity of 15 m/s, by the hub height of 0.88 m, and by the turbulence intensity of 5 %. The LLT simulations were accomplish with 360 time steps, for 10 complete rotor revolution. In figure 7,(a) and 7, (b) are presented the positions of the wind turbine rotor and the near wake for positive and respectively negative values of yaw angle.

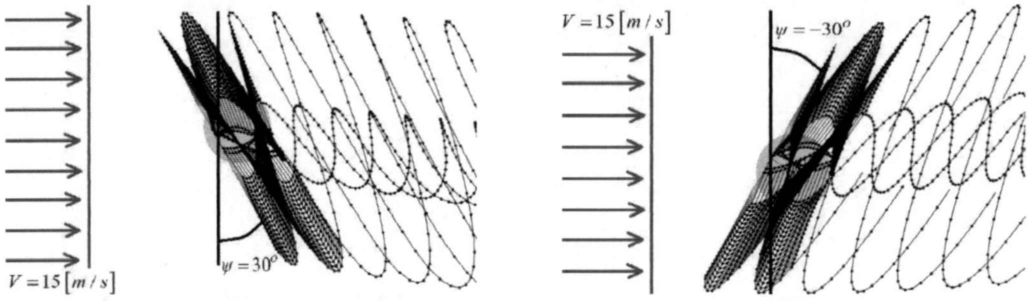

Figure 7. Position of the wind turbine rotor.

The results of the LLT simulations are in particular influenced by two free parameters used to adjust the initial core size and its growth rate in time, vortex time offset (S_C) and turbulent vortex viscosity (δ_v). These parameters affect significant the performances of the wind turbine rotor analyzed. Too large values for them reduces the induction in the near wake and leads to over predictions of rotor power output, leading to significant discrepancies between the BEM results and LLT results, [16]. In this paper, two cases with different values of these two parameters, presented in table 3, were analyzed.

Table 3. The significant parameters used in LLT simulations

	Vortex time offset S_C [-]	Turbulent vortex viscosity δ_v [-]
Case 1	2	1
Case 2	0.2	0.2

Power coefficient and power output according to yaw angle range are presented in figure 8, (a) and respectively in figure 8, (b). The yaw angle increase causes a significant decrease of global parameters. Maximum values are obtaining at 0°, 0.504 for power coefficient and 275 W for power output in case 1 and 0.453 for power coefficient and 248 W for power output in case 2.

The effect of yaw angle on global performances of the wind turbines is more important after the value of 25° when the power output decrease with about 15% with respect to the power output in non-yawed flow in both cases. At yaw angles greater than 45° the power output generated by wind turbine decrease to half. Power coefficient and power output at equal positive and negative value of the yaw angle have small differences, with greater values obtained at positive yaw angles more important after 50 degrees. For example, power output is 115.75 W at -50° and 117.29 W at 50° in case 1 and respectively 98.7 W at -50° and 101.5 W at 50° in case 2.

At zero yaw angle, significant discrepancy between BEM results and LLT results has been noticed. This discrepancy is mainly due to the chosen values of the two free parameters (vortex time offset and turbulent vortex viscosity), the simulations time steps number and some limitation of the LLT method. The relative error between BEM results and the two cases of LLT simulations for power coefficient is 24.7 % in case 1 and 12.28 % in case 2.

IOP Conf. Series: Materials Science and Engineering **595** (2019) 012047 doi:10.1088/1757-899X/595/1/012047

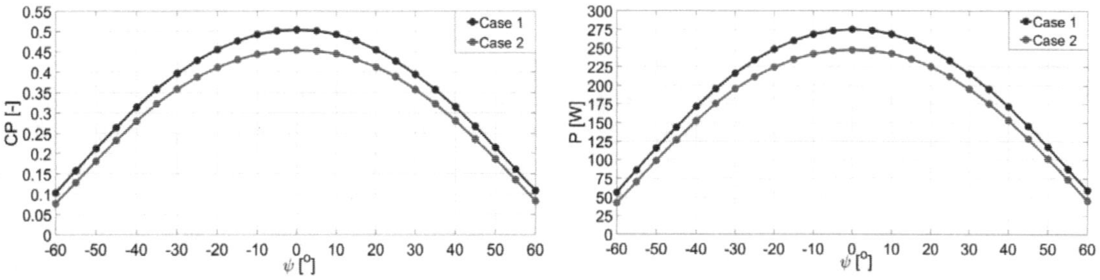

(a) (b)

Figure 8. Influence of yaw angle on power coefficient and power output.

Further, using the results obtained in case 2, the x exponent of conventional expression for power coefficient from equation (1) in yawed flow was determined. In figure 9, the x exponent values (green dots) according to yaw angle range from 5° to 5° are presented. The LLT computed data was approximated with a 4th degree polynomial (brown line), described by equation (2):

$$x = 1.5581 \cdot 10^{-8} \cdot \psi^4 + 2.4558 \cdot 10^{-8} \cdot \psi^3 + 0.00024239 \cdot \psi^2 - 0.00074046 \cdot \psi + 1.3874 \quad (2)$$

The maximum and minimum values of the x exponent, in case 2, equal to 2.56, and 1.18 have been obtained at -60°, and respectively at 5°.

The mean of the exponent (red line), equal with 1.77 has been computed with data presented in figure 9, being in a good concordance with the value obtained from experimental tests conducted in German Dutch Wind Tunnel, which is 1.8, [13], and also with the lower value, equal to 1.88, of the x exponent range presented in [12].

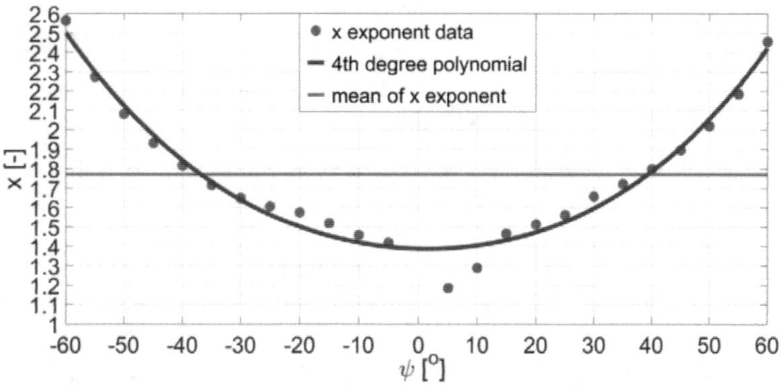

Figure 9. Exponent of conventional equation for power coefficient.

Relative deviation (ε) of the x exponent with respect to its mean value is showed in figure 10, for case 2. Maximum deviation of 45% has been observed for yaw angle of -60°. The relative deviation decrease in the range 0°÷ ±40°, reaches a minimum point at ±40°, and then increase till ±60°.

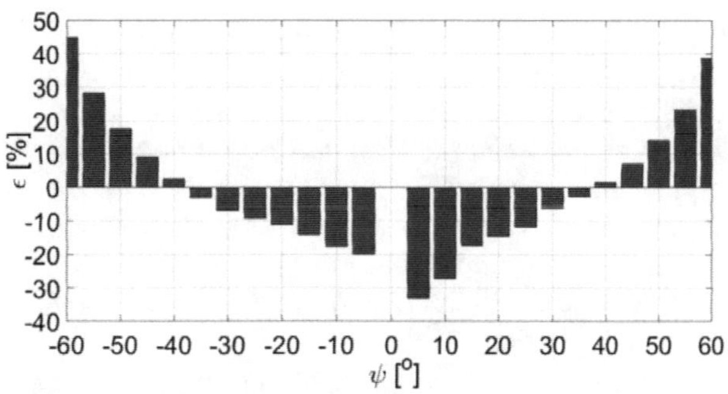

Figure 10. Relative deviation for power coefficient.

IOP Conf. Series: Materials Science and Engineering **595** (2019) 012047 doi:10.1088/1757-899X/595/1/012047

4. Conclusion

In this study, the global performances of HAWT in non-yawed and yawed flow were analyzed using QBlade open source software. The design of the blades was accomplished in QBlade using BEM method. The most important design steps are the selection of optimum incidence and the calculation of airfoil polars for a good approximation of the Reynolds number.

The yaw angle affects the global performances of the wind turbines no matter how small it is. In this study has been observed that the effect of yaw angle is more important after the value of 25°, when the power output decrease with about 15% from the power output in non-yawed flow. Moreover, at yaw angles greater than 45° the power output generated by wind turbine decrease to half.

Researchers opinions on the value of the x exponent from conventional equation for power coefficient at different yaw angles are divergent. In this paper, mean value of the x exponent of conventional equation for power coefficient in the range ±60° is 1.77, being in good concordance with the mean value obtained in several experimental tests.

The LLT results for those two cases analyzed in this paper have significant differences compared to BEM results at zero yaw angle. In case 1, values of the global parameters for wind turbine model are over predicted with 24.7 % compared to BEM results. In case 2, the relative error between LLT results and BEM results decrease at 12.28 %. This discrepancy between the two methods results is influenced by chosen values of the two free parameters used in LLT simulations and the simulations time steps number.

In order to obtain good prediction of the global parameters of wind turbines in *LLT HAWT Simulation* menu it is necessary to adapt the values of the vortex time offset (S_C) and turbulent vortex viscosity (δ_v) to the dimension of computed wind turbine rotor. Large values for these parameters reduces induction in the near wake and leads to over predictions of global parameters of wind turbine.

In the future, some experimental tests (using MF-TA4 wind tunnel) and new numerical simulations (using ANSYS Fluent) are necessary to improve the knowledge about the effect of the yaw angle on the aerodynamic performances of HAWT and the value of the exponent from conventional expression for yaw angle.

References

[1] Marinic-Kragic I Vucina D and Milas Z 2018 Numerical workflow for 3D shape optimization and synthesis of vertical-axis wind turbines for specified operating regimes *Renewable Energy* **115** 113-127

[2] Husaru D E Barsanescu P 2018 *Numerical analysis for prototype blade of Horizontal Wind Axis Turbine in ANSYS Static Structural* Conf Ser Science and Engineering **444** 062010

[3] V Dehouck M Lateb J Sacheau H Fellouah 2018 Application of the Blade Element Momentum Theory to Design Horizontal Axis Wind Turbine Blades *J. Sol. Energy Eng* **140** / 014501-1

[4] Faisal M 2017 Rotor Blade Performance Analysis with Blade Element Momentum Theory *Energy Procedia* **105** 1123 – 1129

[5] Glauert H 1935 *Airplane Propellers* (New York: Julius Springer) 169–360

[6] Hansen M 2008 *Aerodynamics of wind turbines* Second Edition (London: Earthscan)

[7] D Marten J Wendler 2013 *QBlade Guidelines_v06* TU Berlin

[8] Micallef D Sant T 2016 *Wind Turbines: Design, Control and Applications: A Review of Wind Turbine Yaw Aerodynamics*, chapter 2 (edited by Aissaoui A G Tahour A)

[9] Bartl J Mühle F Sætran L 2018 Wind tunnel study on power output and yaw moments for two yaw-controlled model wind turbines *Wind Energ. Science* **3**, 489–502

[10] Schepers J G Snel H 2007 *Model Experiments in Controlled Conditions (MEXICO)* Energy Research Centre of the Netherlands ECN-E-07-042

[11] Schepers J G Schreck S 2013 *The importance of aerodynamics and the role of aerodynamic measurements* ECN-M--13-002

[12] Dahlberg J A and Montgomerie B 2005 *Research program of the Utgrunden Demonstration Offshore Wind Farm* Final report Part 2 Wake effects and other loads Swedish Defense Research Agency FOI 2005-02-17

[13] J G Schepers 2001 *EU project in German Dutch Wind Tunnel* ECN-RX--01-006

[14] Eastman N J Sherman A 1937 *Airfoil section characteristics as affedcted by variations of the Reynolds number* Report no. **586** National Advisory Committee for Aeronautics

[15] J F Herbert-Acero J Martínez-Lauranchet O Probst S Méndez-Díaz 2014 *Aerodynamic Optimization of Small Wind Turbine Rotors Based on NACA 4-Digit Airfoils through Computational Intelligence* EWEA European Wind Energy Conf & Exhibition 2014 10-13 March 2014, Fira de Barcelona Gran Via, Spain.

[16] Marten D 2016 *QBlade v0.95-Guidelines for Lifting Line Free Vortex Wake Simulations* TU Berlin

Acknowledgments

The authors would like to acknowledge the technical resources offered by the Laboratory of Computer-Aided Fluid Engineering (equipped with technical resources with the financial support of the grant ENERED, POSCCE-A2-O2.2.1-2009-4, ID 911), from the Department of Fluid Mechanics, Fluid Machinery and Fluid Power Systems, "Gheorghe Asachi" Technical University of Iasi, Romania.

IOP Conf. Series: Materials Science and Engineering **595** (2019) 012048 doi:10.1088/1757-899X/595/1/012048

Modelling and simulation of a hydraulic system used for wind turbines

R-I Pelin[1] and I Tița[2]

[1] Technical University "Gheorghe Asachi" Iași, Department of Mechanical Engineering, Mechatronics and Robotics, Iasi, Romania
[2] Technical University "Gheorghe Asachi" Iași, Department of Fluid Mechanics, Hydraulic and Pneumatic Machines and Systems, Iasi, Romania

E-mail: iddtita@yahoo.com

Abstract. For wind turbines, pitch control is essential. In some cases, especially in the case of large scale wind turbines, hydraulic control is used. In this paper a hydraulic system used in wind turbine pitch control is analysed. Modelling and simulation of hydraulic systems is necessary in order to acknowledge the drawbacks and advantages before starting the design of the system. By acknowledging, the differences between the model of a physical system and a system in a virtual environment, one can handle the improvement, in terms of price or time-consumption. Considering the proposed diagram of the pitch control hydraulic system, it is presented the block diagram created in MATLAB Simscape Fluids. Some results of the simulation are presented. There are included conclusions and plans for future work.

1. Introduction

The power extracted by the rotor of the wind turbine is essential determined by the blade pitch control [1]. Pitch control is also important in order to protect the turbine in the case of high values of the wind speed, reducing extreme loads to wind turbine rotor and structure.

In the past few years' active pitch control, collective or individual, gains ground in wind turbine construction. Collective pitch control is the traditional method and most of the wind turbines are provided with. In this case, the commanded value for the pitch is sent simultaneous to all the blades. In this purpose hydraulic or electrical systems are used.

Researches presented in [2] demonstrate that a very significant reduction in operational loading can be achieved by means of individual pitch action. As it regards very large wind turbines with rating up to 15MW a research on this case [3] confirm individual pitch control as a future reliable solution.

2. Modelling for a hydraulic pitch control system

Modelling a hydraulic system is important in order to identify the parameters which influence the behaviour of the system in certain functioning conditions.

Pitch systems might be either electrical or fluid power systems, the distribution among the two types being equal. The rotation of the blade along their longitudinal axis is done by a double acting cylinder, controlled with a servo valve. Accumulator in pitch systems are equipped with pressure sensor on fluid side, which may be used as a feedback for controlling the accumulator fluid pressure. The amount of gas may be estimated by a combination of pressure signal and current volume of fluid contained in accumulator [4].

Figure 1 shows the diagram for the hydraulic pitch control system. In the case of three blades rotor, there are three similar circuits with common source. The common source is located in the nacelle of the wind turbine. Considering that the actuation cylinders are connected to the blades it is necessary to use flexible hoses. Proportional valve controls the position angle of the blade. Maintaining the blade in the position is important. It may be done either mechanically with a locking circuit or hydraulically using a pilot-operated check valve. Accumulator is the safety equipment and acts in the event of an important fault of the system. The system also assures protection of the rotor when the rods of the cylinders are fully extended and is realized stalling position which stops the rotation.

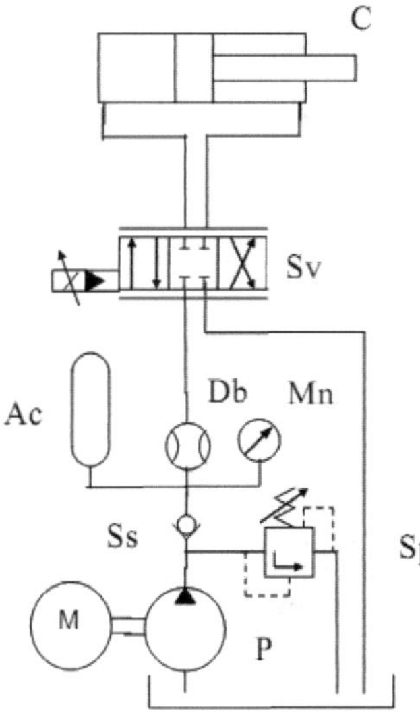

Figure 1. Hydraulic diagram for the pitch control system, for a single blade control circuit.

P - fixed displacement pump
M - electric motor
Sp – relief valve
Ss – check valve
Ac – piston type accumulator
Db – Flow meter
Mn – Pressure gauge
Sv – Proportional solenoid valve
C – Double acting differential cylinder

The model for servo cylinder is well known, see e.g. [5] and [6]. Equations (1) - (5) describe the functioning of the system with the notations given in table 1:

$$Q_1 = A\frac{dx}{dt} + \frac{V_1}{\beta_o}\frac{dp_1}{dt} \tag{1}$$

$$-Q_2 = -\alpha A\frac{dx}{dt} + \frac{V_2}{\beta_o}\frac{dp_2}{dt} \tag{2}$$

$$m\frac{d^2x}{dt^2} + c\frac{dx}{dt} + kx + F = p_1 A - p_2\alpha A \tag{3}$$

$$Q_1 = \mu A_1\sqrt{\frac{2}{\rho}(p_0 - p_1)} \tag{4}$$

$$Q_2 = \mu A_2\sqrt{\frac{2}{\rho}(p_2 - p_{atm})} \tag{5}$$

Laplace transforming and linearizing equations (1) - (5) results equations (6) - (10):

$$Q_1 = AsX + \frac{V_1}{\beta_o}sP_1 \tag{6}$$

$$-Q_2 = -\alpha AsX + \frac{V_2}{\beta_o}sP_2 \tag{7}$$

$$ms^2X + csX + kX + F = P_1A - P_2\alpha A \tag{8}$$

$$Q_1 = K_1P_1 \tag{9}$$

$$Q_2 = K_2P_2 \tag{10}$$

Table 1. Nomenclature.

Notation	Meaning
V_1	Volume of forward chamber
V_2	Volume of return chamber
A	Area of the piston
p_1	Pressure in forward chamber
P_2	Pressure in return chamber
m	Total mass
k	Load spring gradient
F	Load force at the blade
x	Displacement of the piston
β_o	Bulk modulus of the oil
Q_1	Flow in forward chamber
Q_2	Flow from return chamber
α	Rod influence on area
ρ	Density of the oil
c	Viscous damping coefficient
$A_{1,2}$	Area of the orifice
μ	Flow coefficient

3. Block diagram MATLAB Simscape Fluids

The simulation in Simscape Fluids supposes to configure the bloc diagram corresponding to hydraulic diagram [7]. In Figure 2 one can see the corresponding block diagram for the hydraulic diagram in Figure 1. In the diagram one may change rotation speed pf the pump, pump displacement, pressure limited by the relief valve, piston area, volume of the accumulator, precharge pressure of the accumulator and the command low for the proportional valve.

In each case of simulation, it is possible to see the behaviour for the piston position, piston speed, flow in accumulator line, command for directional control valve, pump flow, flow to the cylinder, flow through pressure valve and pump discharge pressure.

IOP Conf. Series: Materials Science and Engineering **595** (2019) 012048 doi:10.1088/1757-899X/595/1/012048

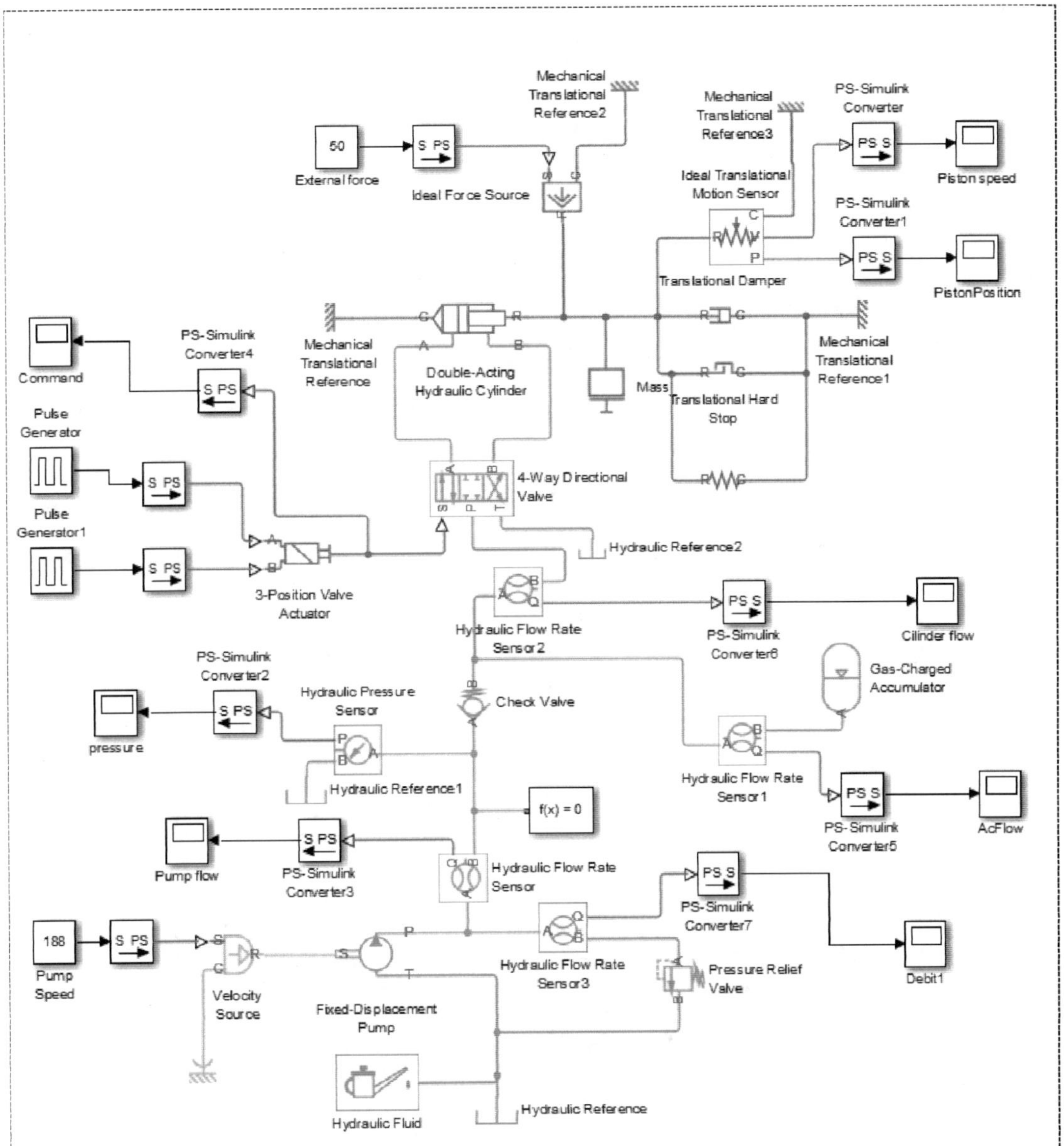

Figure 2. Block diagram for the hydraulic pitch control system.

4. Simulation results

There are two sets of results presented in Figures 3 and 4. First is done for: piston area $A = 12\cdot 10^{-3}$ m^2, $\alpha\cdot A$=9$\cdot$10^{-3} m^2, V=20$\cdot$10^{-6}m^3/rad, V_{Ac}=2$\cdot$10^{-3}m^3, p_0=5$\cdot$10^{5}N/m^2, p_{Sp}=160$\cdot$10^{5}N/m^2. The second is done for: piston area A=12$\cdot$10^{-3}m^2, $\alpha\cdot A$=9$\cdot$10^{-3}m^2, V=20$\cdot$10^{-6}m^3/rad, V_{Ac}=2$\cdot$10^{-3}m^3, p_0=3$\cdot$10^{5}N/m^2, p_{Sp}=160$\cdot$10^5 N/m^2.

The command law for the proportional valve is presented in Figure 5.

IOP Conf. Series: Materials Science and Engineering **595** (2019) 012048 doi:10.1088/1757-899X/595/1/012048

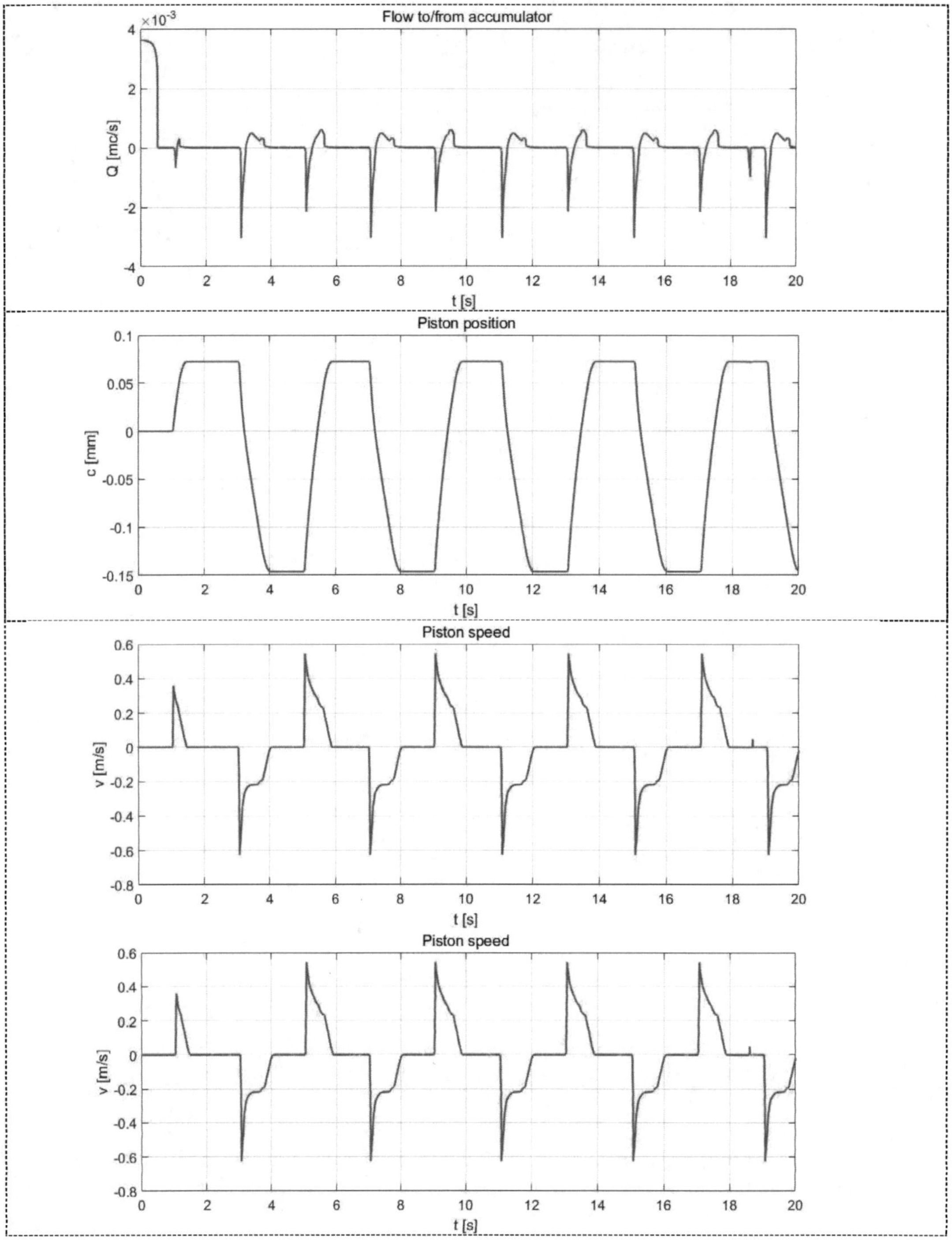

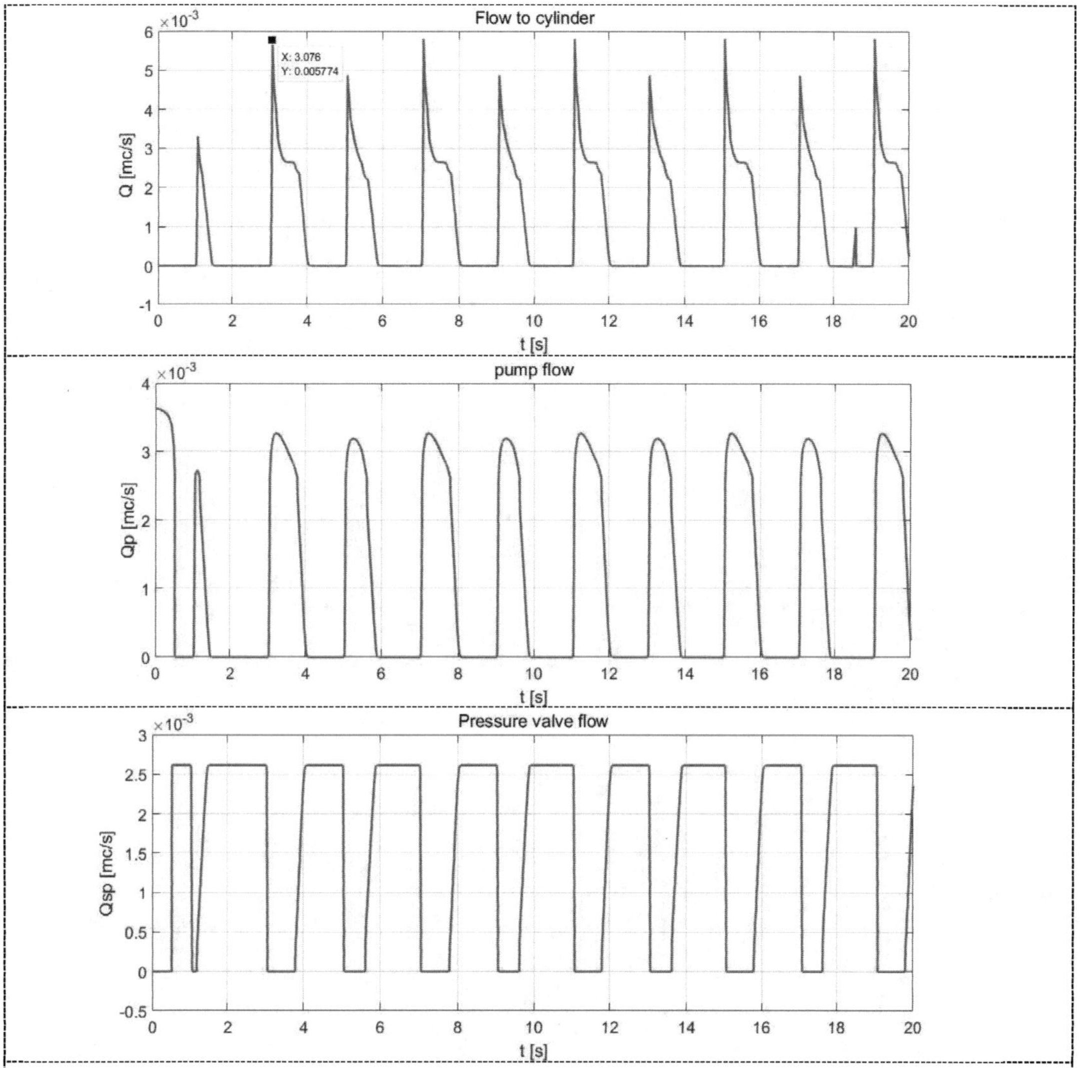

Figure 3. Simulation results for first set of values.

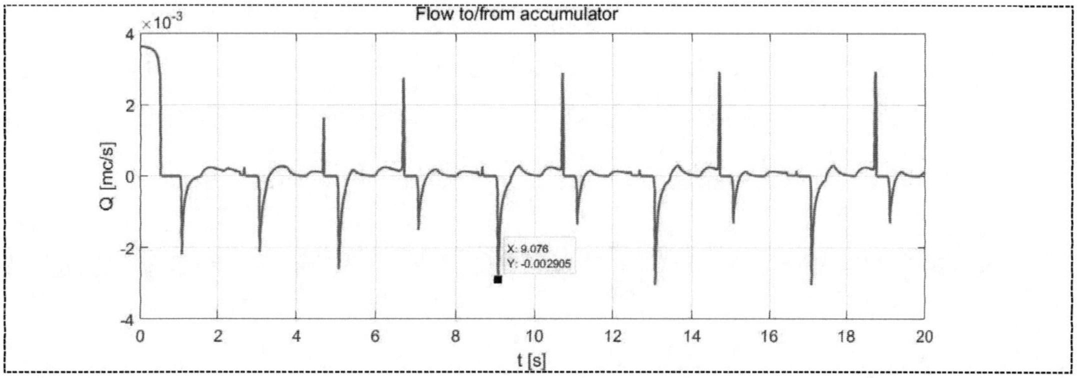

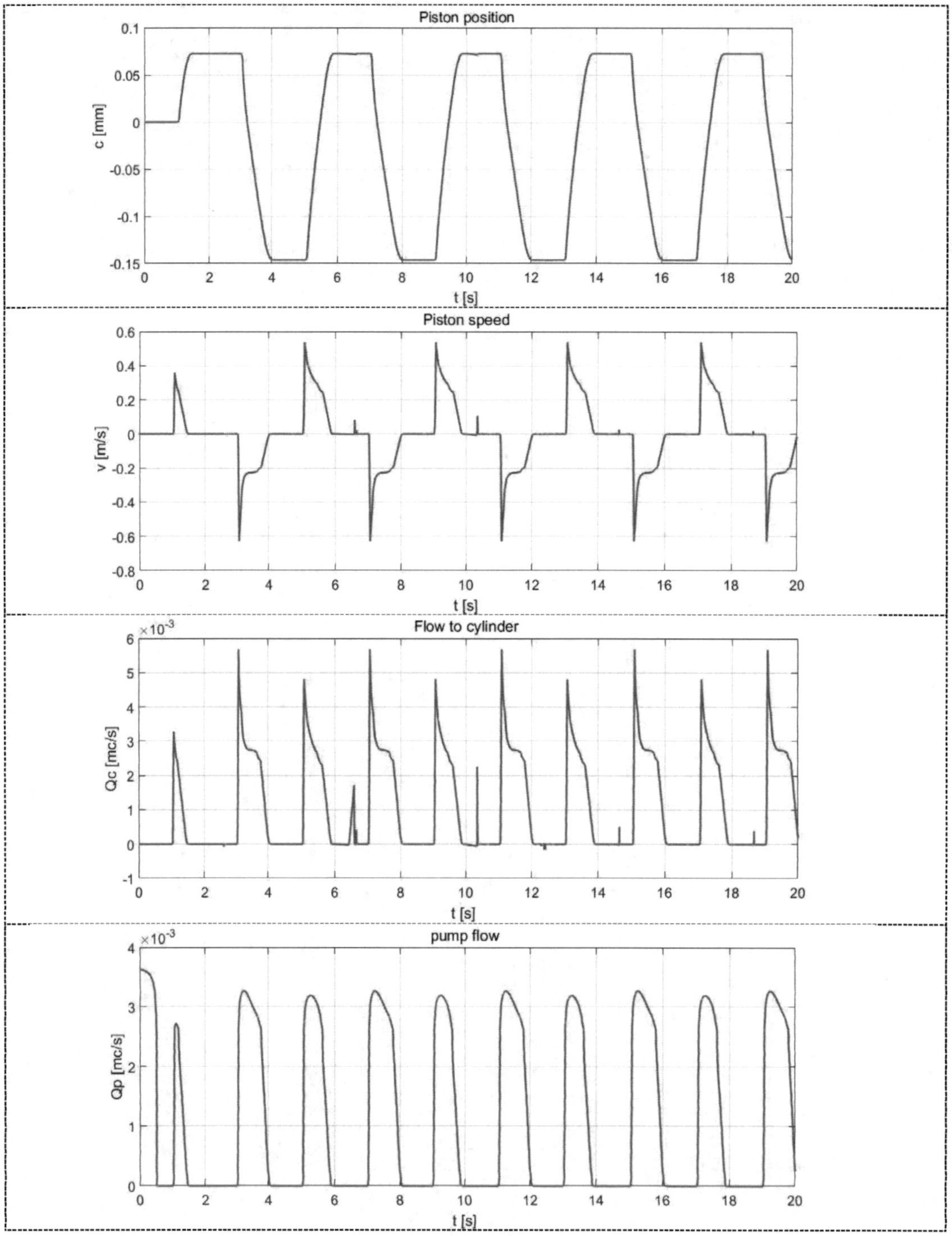
Piston position
c [mm]
t [s]
Piston speed
v [m/s]
t [s]
Flow to cylinder
×10^-3
Qc [mc/s]
t [s]
pump flow
×10^-3
Qp [mc/s]
t [s]

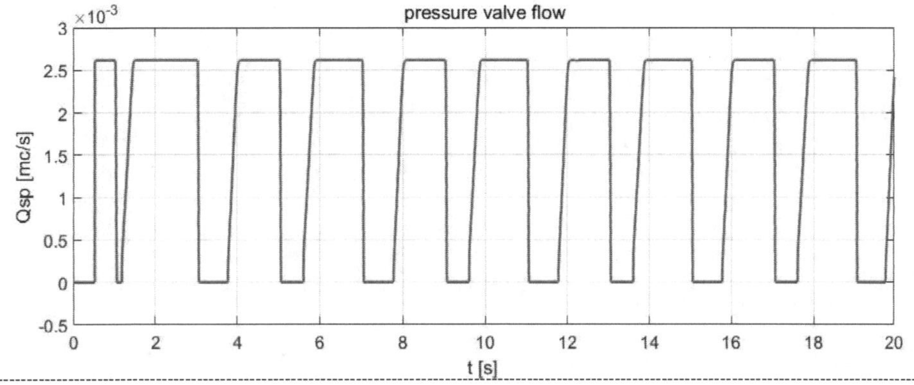

Figure 4. Simulation results for the second set of values.

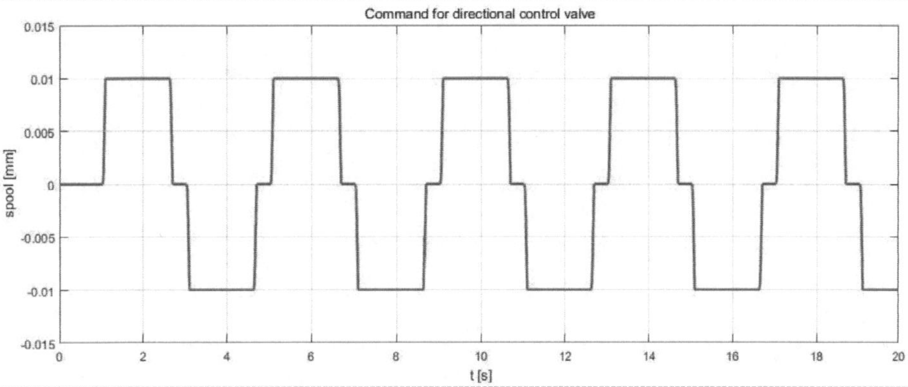

Figure 5. Command for directional control valve.

5. Conclusions

For research of fluid power systems MATLAB Simscape Fluids is an efficient instrument and recent development makes this instrument even more useful.

For the general structure of a pitch control system it was configured the block diagram. It offers as result of simulation the piston position, piston speed, flow in accumulator line, command for directional control valve, pump flow, flow to the cylinder, flow through pressure valve and pump discharge pressure. Simulation was done in two cases study. Between these two the differences regarded the accumulator namely the volume and the precharge pressure. Results of simulation reveal the influence of parameters of the accumulator.

In the future work it will be considered pressure feedback and system regulator suitable in this case. It will be also considered and other parameters to be modified in the system in order to reveal their influence. It is also for interest if the hydraulic pitch control system is suited for medium and low power wind turbines for isolated consumers.

References

[1] Pelin R-I, Bârsănescu P D and Tita I 2018 Hydraulic systems used for pitch control of wind turbines: a literature overview *IOP Conf. Ser.: Mater. Sci. Eng.* **444(4)**:042013

[2] Bossanyi E A 2003 Wind Turbine Control for Load Reduction *Wind Energy* **6** pp 229-244

[3] Chen Z J and Stol K A 2014 An assessment of the effectiveness of individual pitch control on upscaled wind turbines *Journal of Physics: Conference Series* **524**, conference 1

[4] Liniger J, Sepehri N, Soltani M and Pedersen H 2017 Signal-based gas leakage detection forfluid power accumulators in wind turbines *Energies* **10(3)**:331
[5] Watton J 2009 *Fundamentals of Fluid Power Control* (New York Cambridge University Press)
[6] Merritt H E 1967 *Hydraulic Control Systems*, (New York John Wiley and Sons)
[7] Matlab & Simulink Simscape Fluids User's Guide https://uk.mathworks.com/help/pdf_doc/physmod/hydro/hydro_ug.pdf (accesed at 12.09.2018)

Acknowledgments
The authors would like to acknowledge the technical resources offered by the Laboratory of Wind Turbine Control Systems, from the Department of Fluid Mechanics, Fluid Machinery and Fluid Power Systems, "Gheorghe Asachi" Technical University of Iasi, Romania. The Laboratory of Wind Turbine Control Systems has been equipped with technical resources with the financial support of the grant ENERED, POSCCE-A2-O2.2.1-2009-4, ID 911.

IOP Conf. Series: Materials Science and Engineering **595** (2019) 012049 doi:10.1088/1757-899X/595/1/012049

Numerical analysis of the influence of the lower hull angle of a round skirtless air cushion vehicle

M S Pavăl[1], A Popescu[1] and D Zahariea[2]

[1]Mechanical Engineering and Road Automotive Engineering Department, "Gheorghe Asachi" Technical University of Iaşi, Iaşi, Romania
[2]Fluid Mechanics, Fluid Machinery and Fluid Power Systems Department, "Gheorghe Asachi" Technical University of Iaşi, Iaşi, Romania

E-mail: mihai-silviu.paval@tuiasi.ro

Abstract. Ever since 1716, when the first concept of building a vehicle to operate with an air cushion was presented, the world remained fascinated by this concept. This article presents a study on important parameters of the flow into the inner cavity of a simple plenum hull, the first configuration of the air cushion vehicle. In order to achieve this, six constructive models were chosen, the only major difference being the angle of inclination at the bottom of the hull. The paper presents the specific mathematical model adapted for this type of configuration, the imposed constructive dimensions, the steps of numerical analysis and results. An important step of this analysis is the estimation of the total weight of each prototype according to the lower hull angle. The estimate was performed using Catia software and the actual analysis was completed using Matlab.

1. Introduction

Air cushion vehicle represent a unique type of transportation due to the fact that these vehicles can operate across a variety of surfaces in a relatively short time. The main factor that makes this movement possible is the air cushion.

For current constructive models, this air cushion is generated between the running surfaces and the flexible skirt of the vehicle. This was not always the case, because the flexible skirt appeared after an initial concept of this vehicle.

Two main types of air cushion vehicle have been featured over the years, open plenum chamber and peripheral jet. The major difference between the two concepts is the buoyancy tank, [1].

The open plenum chamber was the first configuration of this kind of vehicle. An example of such a vehicle is the first experimental model made by Chinese, named Craft '33', [2]. The open plenum chamber configuration has two alternatives: simple plenum (figure 1a) and skirted plenum (figure 1b). Both concepts are represented in figure 1, where: 1 – air source, 2 – hull, 3 – running surface and 4 – flexible skirt, [3].

The purpose of this paper is to study the important parameters describing the flow of fluid only for the simple plenum constructive configuration.

A literature study on these types of vehicles, reveals that most article mention a Swedish philosopher Emanuel Swedenborg, being considered the first person to present a vehicle concept based on an air cushion. His idea was published in Daedalus Hyperboreus, [4], and presented in figure 2, [5].

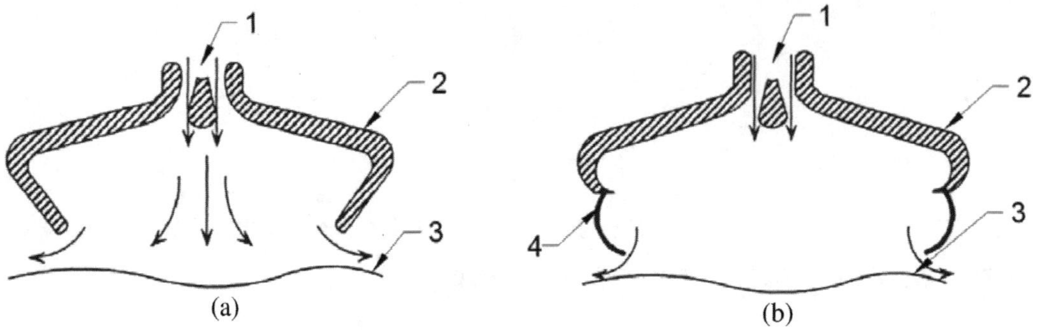

Figure 1. Open plenum chamber configuration (adapted from [3])

This concept was described as an elliptical shape made of a wooden frame covered with sailcloth. In the middle there was a cabin in which the crew used two paddles to introduce air under the hull. The main problem of this vehicle was the inability to create the air cushion due to the limitation of human effort, [5].

2. Description of the mathematical model of the simple plenum configuration

The emergence of this concept in academic space, triggered scientific research in the area. Scientists developed mathematical relationships to calculate the governing parameters for this vehicle concept, relationships that are still valid today.

In terms of mathematical modeling describing the flow of a fluid into the inner cavity for such an air cushion vehicle configuration, many researchers were concerned with theoretical issues of this flow, the most important work being carried out by Cocksedge [6], Dukkipati [7] or Wong [8].

The proposed mathematical model is adapted to the examples presented by the authors mentioned above, where the important fluid flow parameters for this configuration are described.

The first step establishes the simplifying assumptions:

- body forces generated at speed will not be considered
- to define a mathematical equation describing the fluid exit velocity, the air is considered to be at rest in the air cushion
- to define a mathematical equation describing the discharge coefficient for the exit flow, it is considered the case of a long wall and ideal fluid flow.

The second step is the description of the mathematical equations. Figure 3 shows a sectional view of this configuration that specifies the parameters to be used in mathematical modeling.

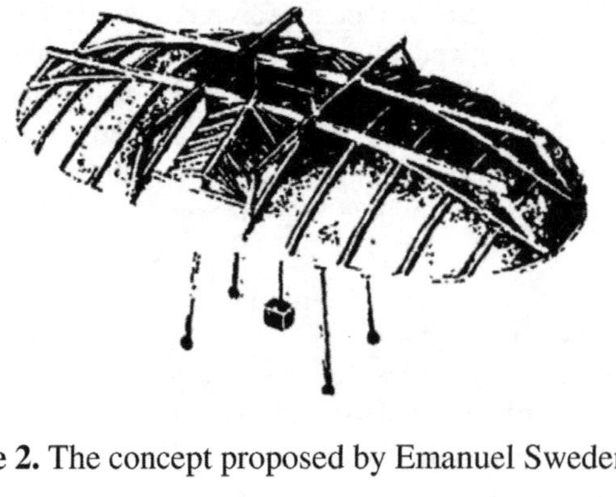

Figure 2. The concept proposed by Emanuel Swedenborg

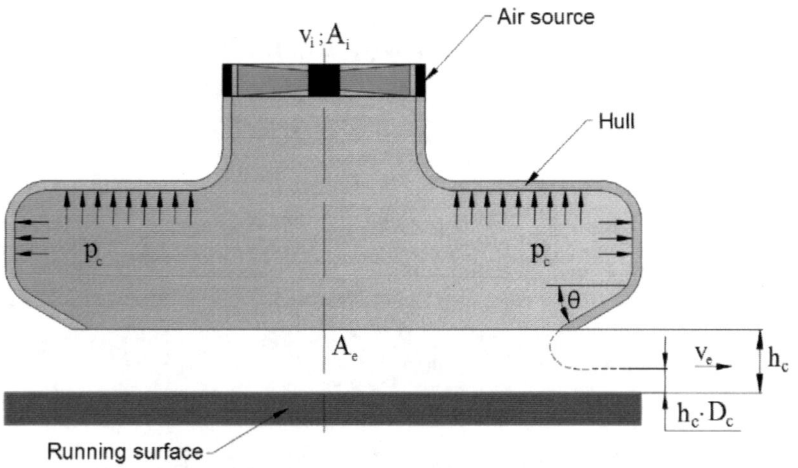

Figure 3. Main fluid flow parameters that describe the mathematical equations (adapted from [8])

The most important parameters describing this configuration are: the lift force (L_f) that can be expressed using equation (1), where: T_f – fan thrust [N], $p_c \cdot A_c$ – cushion lift [N], L_b – body forces generated at speed [N] and clearance height (h_c) that can be expressed by the equation (2), where: Q_f – the flow through the fan [m^3/s], D_c – discharge coefficient for the exit flow [-], S – the peripheral length of the cushion [m] and v_e – exit velocity [m/s].

$$L_f = T_f + p_c \cdot A_c + L_b \tag{1}$$

$$h_c = Q_f \left(D_c \cdot S \cdot v_e\right)^{-1} \tag{2}$$

As far as the calculation of the discharge coefficient is concerned, it can be determined by using equations (3) and (4). Regarding the equation (3), this was obtained using five values for θ (i.e. 0°, 45°, 90°, 135° and 180°), that were used for a polynomial approximation for experimental graphs. Regarding the equation (4), this is a theoretical expression that yields lower numerical values when compared to equation (3), due to the fact that the air velocity is neglected.

$$D_{c1} = 0.5 + 0.4 \cdot \theta \cdot 10^{-3} + 0.109 \cdot \theta^2 \cdot 10^{-4} - 0.494 \cdot \theta^3 \cdot 10^{-7} + 0.345 \cdot \theta^4 \cdot 10^{-9} \tag{3}$$

$$D_{c2} = \frac{1}{2}\left[1 + \frac{\sin\theta}{\left(\frac{\pi+2}{\pi-2}\right)(1+\cos\theta) - \sin\theta \cdot \cos\theta}\right] \tag{4}$$

To determine the pressure of the air cushion (p_c), equation (5) is used, where: G – the hull weight of the air cushion vehicle [N] and A_c – air cushion area [m^2].

$$p_c = \frac{G}{A_c} \tag{5}$$

Considering the second assumption and using Bernoulli's theorem, equation (6) shows the exit velocity (v_e) under the peripheral gap, where: p_c – cushion pressure [N/m^2] and ρ_{air} – mass density of air [kg/m^3].

$$v_e = \left(\frac{2 \cdot p_c}{\rho_{air}}\right)^{1/2} \tag{6}$$

IOP Conf. Series: Materials Science and Engineering **595** (2019) 012049 doi:10.1088/1757-899X/595/1/012049

Applying the expression of continuity of mass, equation (7) is obtained by which the fan thrust (T_f) can be determined, where: A_e – the outlet area of the fluid [m^2] and A_i – the inlet area of the fluid [m^2].

$$T_f = \frac{2 \cdot p_c \cdot A_e^2}{A_i} \tag{7}$$

The last two important parameters of this mathematical model are nozzle power (P_N), computed using equation (8) and the augmentation factor (K_a), equation (9):

$$P_N = p_c \cdot \left(\frac{2 \cdot p_c}{\rho_{air}} \right)^{1/2} \cdot h_c \cdot S \cdot D_c \tag{8}$$

$$K_a = \frac{A_c}{2 \cdot h_c \cdot S \cdot D_c} \tag{9}$$

Using equations (1), (7) and taking into account the first assumption, yields equation (10) which represents lift force:

$$L_f = \frac{2 \cdot p_c \cdot A_e^2}{A_i} + A_c \cdot p_c \tag{10}$$

3. The dimensional characteristics of the constructive model proposed for the study

In order to be able to apply the mathematical relationship described above, it is imperative to impose dimensional characteristics. Therefore, in figure 4 are presented the main dimensions of the proposed model of study.

As mentioned earlier, this study refers to six constructive forms in which the only difference is the angle of inclination of the hull which can have θ_1=30°, θ_2=36°, θ_3=42°, θ_4=48°, θ_5=54° and θ_6=60° values. In terms of clearance height, usually the range in which it can vary is between 200 to 600 millimeters [9]. For the configuration presented in this article, these values are very high and for this study a value of 50 millimeters has been adopted.

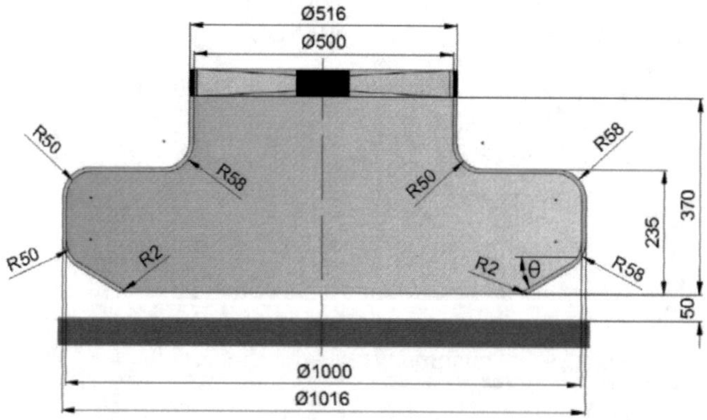

Figure 4. Main dimensions of the proposed model

4. Diagram of steps taken to perform numerical analysis

For the actual numerical analysis, the diagram presented in figure 5 was used. Analysing this diagram, the constructive dimensions and the imposed initial parameters, one can notice that the only unknown is the weight of the prototype. In the following it will be presented the weight analysis of the prototype for all six proposed cases, in which different types of material are used. The analysis was performed using Catia software.

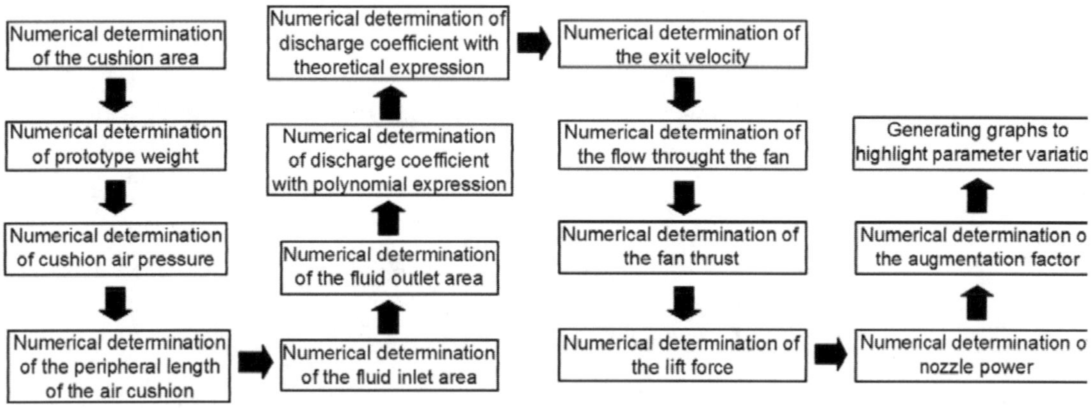

Figure 5. Succession of stages necessary to carry out the actual analysis

5. Weight analysis of the proposed design model for different types of material

Studying the literature on these types of vehicles, one can notice that several authors have used certain materials in various studies on the hull of an air cushion vehicle. These materials are: plywood [10][11][12][13], sandwiching polystyrene sheets between plywood sheets [14], styrofoam with balsa wood [15], Aluminum [16], Al 6061-T6 [17], fiber reinforced epoxy composite (60 % S – glass fibre and 40 % epoxy resin) [18][19] or Dyneema epoxy composite material [20]. It should be noted that these materials were considered hypothetical only where numerical calculation and / or numerical simulation were made or even used in the actual construction of an air cushion vehicle.

Two important factors that are considered when selecting the hull material of an air cushion vehicle are as follows: the structure should have a relatively light weight and must provide a higher strength. For air cushion vehicles that are also used on water, the corrosion resistance of the material must also be considered, due to the long period in which the hull is in salty water.

Taking into consideration the materials presented above, table 1 presents the main materials used for the hull weight analysis of the chosen prototype where new types of materials can be observed.

Figure 6. Isometric view of hull made of aluminum for θ_1=30°

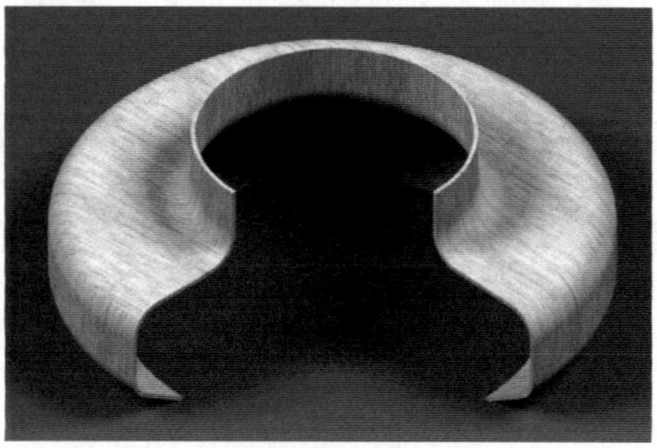

Figure 7. Isometric view partially sectioned for wooden hull with θ_1=30°

To perform this weight prototype analysis for the various materials presented in table 1, the Catia software was used due to the complexity of this structural shape. The 3D models of all six constructive designs were created and figure 6 shows an isometric view of the hull having the aluminum material. In order to have a picture of the interior of this hull, figure 7 shows a partially sectioned view of the hull made of wood.

IOP Conf. Series: Materials Science and Engineering **595** (2019) 012049 doi:10.1088/1757-899X/595/1/012049

Table 1. The final materials chosen for the study of weight

No. of case	Type of material	Density [kg/m^3]
1	Aircraft plywood [21]	580 - 650
2	Marine plywood [22]	450 - 720
3	Bendy plywood [23]	400
4	Aluminum / Al 6061 – T6 [24]	2700
5	Balsa wood [20]	160
6	Fiber reinforced epoxy composite (60 % S-glass fibre and 40 % epoxy resin) [18][19]	2490 – S-glass fibre 1200 – epoxy resin
7	Dyneema epoxy composite material [20]	970

Due to the realization of these 3D models, it was easy to determine the volume, thus resulting in the mass of each prototype in part for each of the previously mentioned materials. Table 2 lists these values where the centre of gravity coordinates is also included.

To analyse the materials used, figures 8, 9 and 10, respectively, show the variation of the mass in relation to the surface for the three types of plywood. Referring to figure 8 and 9, this analysis is performed for two densities, practically the extremes of the density range.

Table 2. The constructive features of the models proposed for the study

No.	θ [°]	Volume [m^3]	Center of Gravity x [mm]	y [mm]	z [mm]
1	30	0.01265	-4.242e-014	-2.616e-013	169.553
2	36	0.01231	3.001e-013	5.808e-014	173.172
3	42	0.01208	-1.481e-014	7.848e-013	175.900
4	48	0.01190	-1.603e-013	1.252e-013	177.987
5	54	0.01177	7.445e-013	-2.709e-013	179.593
6	60	0.01167	2.656e-013	8.861e-013	180.824

Also, figure 11 shows the variation in mass depending on the surface of the prototype for aluminum and balsa wood and in figure 12 shows the variation in mass in relation to the surface of the prototype for the two composite materials, namely: fiber reinforced epoxy composite and dyneema epoxy composite.

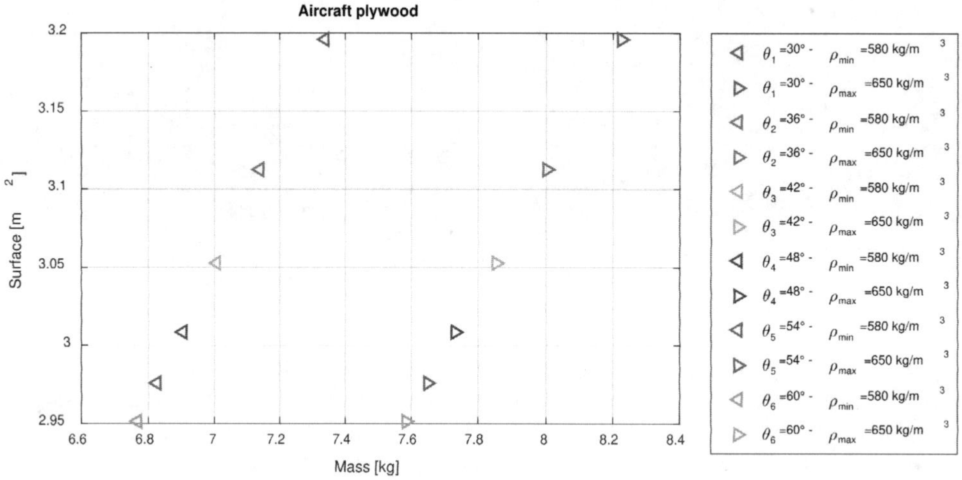

Figure 8. Variation of mass according to the surface of the prototype for aircraft plywood

IOP Conf. Series: Materials Science and Engineering **595** (2019) 012049 doi:10.1088/1757-899X/595/1/012049

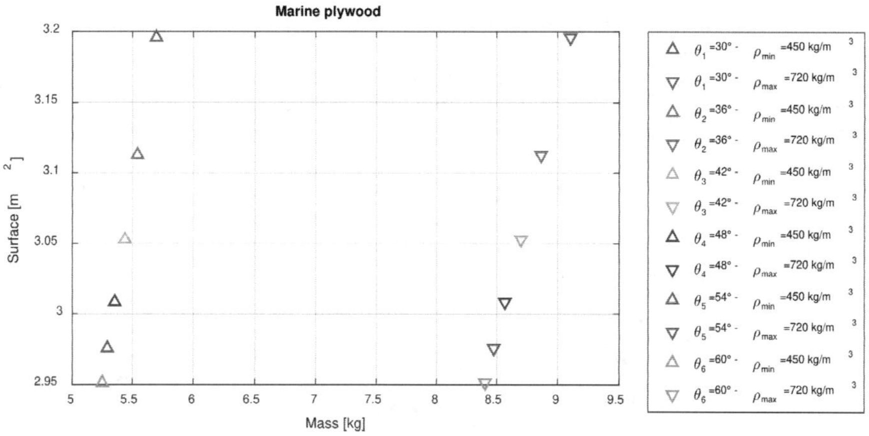

Figure 9. Variation of mass according to the surface of the prototype for marine plywood

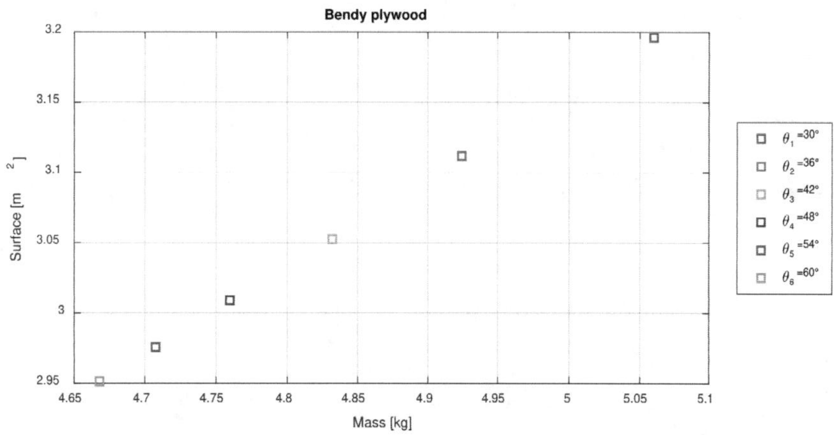

Figure 10. Variation of mass according to the surface of the prototype for bendy plywood

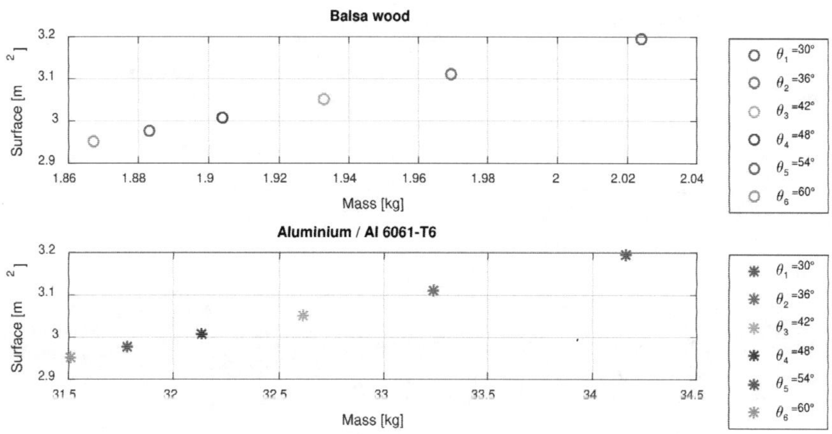

Figure 11. Variation of mass according to the surface of the prototype for balsa wood and aluminium

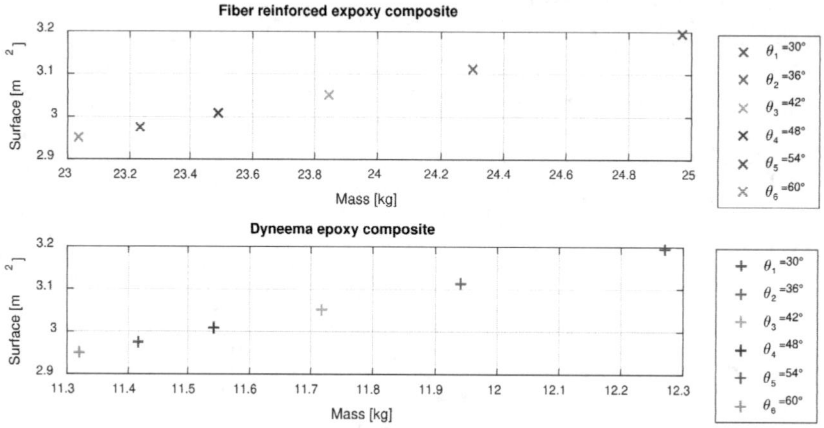

Figure 12. Variation of mass according to the surface of the prototype for the two types of composite materials

6. Numerical calculation

Using the mathematical model described above, a numerical analysis was performed for all required angles. It should be noted that for aircraft plywood and marine plywood materials, the calculations were made with the lowest density. Regarding the parameters: discharge coefficient and augmentation factor, their numerical values are presented in table 3. For the θ_1=30° and θ_6=60°, figure 13 shows the variation of the fan flow in relation to the lift force, figure 14 shows the variation of the nozzle power in relation to the lift force and figure 15 shows the variation of cushion pressure in relation to the lift force.

Table 3. Numerical values of discharge coefficient and augmentation factor

θ [°]	D_{c1} [-]	D_{c2} [-]	K_{a1} [-]	K_{a2} [-]
30	0.5208	0.4077	4.8775	6.2304
36	0.5268	0.3695	4.8216	6.8737
42	0.5334	0.3038	4.7615	8.3605
48	0.5407	0.1597	4.6978	15.900
54	0.5485	-0.4150	4.6305	-6.1209
60	0.5570	2.5059	4.5598	1.0136

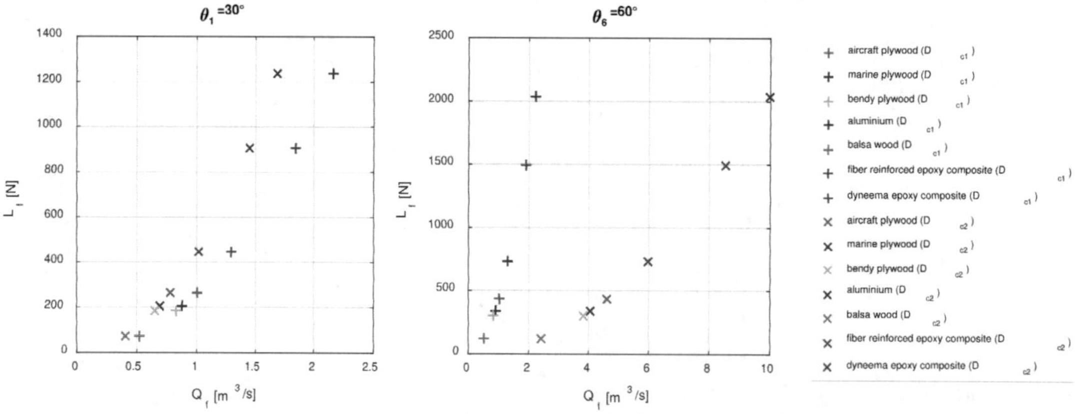

Figure 13. Variation of the fan flow in relation to the lift force for θ_1=30° and θ_6=60°

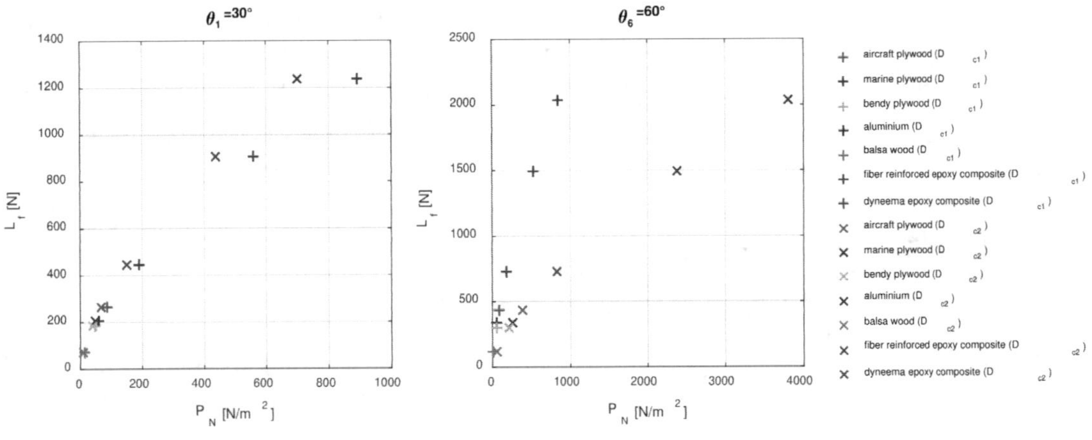

Figure 14. Variation of the nozzle power in relation to the lift force for θ_1=30° and θ_6=60°

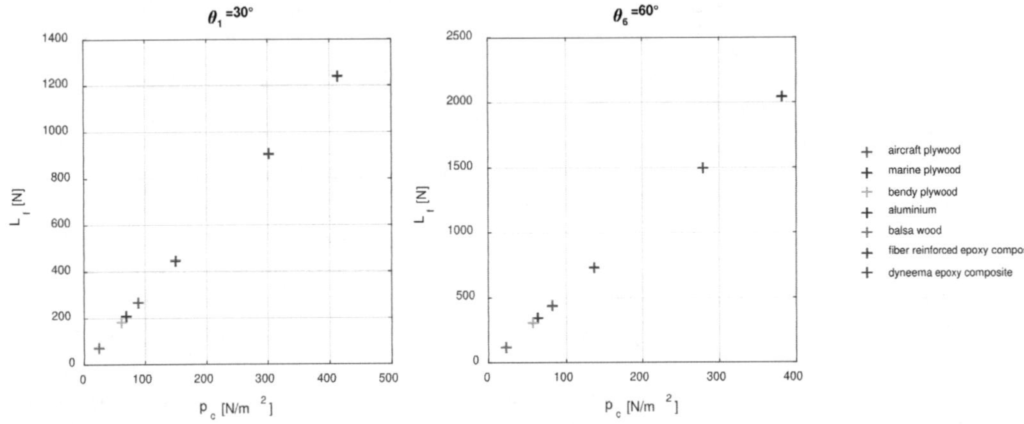

Figure 15. Variation of cushion pressure in relation to the lift force for θ_1=30° and θ_6=60°

7. Conclusions

The purpose of this article was to perform a numerical analysis using a theoretical model that describes simple plenum configuration and applying this model for a constructive form to obtain numerical results that can be interpreted and may provide an image of the important parameters that characterize this configuration.

Regarding the types of materials used, three types of plywood were used for comparison, i.e. aircraft plywood, marine plywood and bendy plywood. These materials have a relatively low density and can be used to construct such a hull.

Analyzing the values presented in table 2 and considering that the main dimension that varies is the value of the lower hull angle for all six prototypes, it can be inferred that the θ angle has a major influence on the constructive features. Therefore, it can be noticed that as the value of θ angle increases, the weight of each prototype gradually decreases. Calculations show that for each material, the mass of the prototype with θ_6=60° is less than 7.74% of the prototype mass with θ_1=30°.

Performing an analysis of material types, it can be noticed that for aircraft plywood the difference is 0.5684 kg, for the marine plywood the difference is 0.441 kg, for the bendy plywood the difference is 0.392 kg, for the aluminum the difference is 2.646 kg, for the balsa wood the difference is 0.1568 kg, for fiber reinforced epoxy composite the difference is 1.9345 kg and for the dyneema epoxy composite the difference is 0.9506 kg. It should be noted that the numerical study did not take into account the weight of the air source.

A thorough literature review performed by the authors of this paper revealed that similar studies have not been performed or published. This analysis is particularly important in the design stage of such a vehicle because it estimates the weight of the prototype. Using the graphs presented in this article, it is possible to estimate with some approximation depending on the surface, the weight of the model that is to be built / studied.

Another important observation is that the discharge coefficient calculated with polynomial equation is larger than the one calculated using the theoretical expression.

Given that aluminum has the highest density of all materials used for this numerical analysis, it follows that the highest numerical value obtained are for this material.

References

[1] Pavăl M S and Popescu A 2018 Hovercrafts – An Overview. Part II: Basic Construction Principles, Classification, Advantages, Disadvantages *Buletinul Institutului Politehnic din Iaşi* **64** pp 51 – 61.

[2] Yun L and Bliault A 2000 *Theory and Design of Air Cushion Craft* (Great Britain: Arnold), pp 7 – 8.

[3] Tae-Cheol J 2003 *Design of Air Cushion Vehicles Using Artificial Intelligence: Expert System and Genetic Algorithm* Ryerson University, p 2.

[4] Pavăl M S and Popescu A 2018 Hovercrafts – An Overview. Part I: Basic Concepts of Advanced Marine Vehicles – Fundamental Elements, Short History, Patents *Buletinul Institutului Politehnic din Iaşi* **64** pp 39 – 50.

[5] Benya Y Y and D'yachenko V K 1972 *Basic Theories of Air Cushion Vehicles* National Technical Information Service pp 7– 8.

[6] Cocksedge G G 1969 *Thrust-Cushion Vehicles. A Preliminary Analysis* McMaster University pp 21 – 26.

[7] Dukkipati R V 2000 *Vehicle Dynamics* (New Delhi: Narosa Publising House) pp 425 – 428.

[8] Wong J Y 2001 *Theory of Ground Vehicle. Third Edition* (Canada: John Wiley & Sons) pp 485 – 488.

[9] Pavăl M S, Popescu A, Popescu T, Zahariea D and Husaru D E 2018 Numerical study on the movement of air inside the inner cavity of a hovercraft model *IOP Conf. Ser.: Mater. Sci. Eng.* **444** 082005.

[10] Okafor B E 2013 Development of a Hovercraft Prototype *International Journal of Engineering and Technology* **3** pp 276 – 281.

[11] Anandhakumar A, Ganesan S, Goutham S and Pasupathi K 2015 Design and Fabrication of Hovercraft *International Journal of Innovative Research in Science, Engineering and Technology* **4** pp 597 – 601.

[12] Patel M, Raval D, Parmar K, Darji H and Parikh K 2016 Air Suspended Vehicle for Internal Transportation Purpose *International Journal for Scientific Research & Development* **4** pp 372 – 375.

[13] Shrirao R R, Kale S P, Rakhde D K and Khot A S 2016 Design & Fabrication of ACV International Journal of Emerging Trends in Science and Technology **3** pp 3995 – 4015.

[14] Uma Maheswara Rao S V and Surya Prakash V S 2014 Development of a Integrated Air Cushioned Vehicle (Hovercraft) *International Journal of Modern Engineering Research* **4** pp 21 – 28.

[15] Rajmanovah, Bebeto S R, Jensin J and Muthusamy 2014 Air Assisted Directional Control of a Hovercraft *International Journal of Mechanical And Production Engineering* **2** pp 88 – 91.

[16] Nair A H, Pai K P, Varghese M, Thomas S B and George T S 2014 Hovercraft Based Farming System *International Journal For Research & Development in Technology* **1**.

[17] Amiruddin A K, Sapuan S M and Jaafar A A 2011 Development of a hovercraft prototype with an aluminium hull base *International Journal of the Physical Sciences* **6**, pp 4185 - 4194.

IOP Conf. Series: Materials Science and Engineering **595** (2019) 012049 doi:10.1088/1757-899X/595/1/012049

[18] Amiruddin A K, Sapuan S M and Jaafar A A 2008 Analysis of glass fibre reinforced epoxy composite hovercraft hull base *Materials and Design* **29** pp 1453 – 1458.

[19] Aiman A F, Tarmizi A M, Bakri B, Hussin M S, Hafiezal M R M and Fairuz M A 2012 A Study on Fiber Glass Composite Hovercraft Hull with Different Wall Thickness Using Generative Structural Analysis *International Review of Mechanical Engineering* **6** pp 1307 – 1311.

[20] Kishan G, Rao B V S, Ravinder Reddy P and Srinivasulu N V 2016 Structural Analysis of Dyneema – Epoxy Based Hovercraft Hull Using Finite Element Method *International Journal on Mechanical Engineering and Robotics* **4** pp 55 - 64.

[21] China Plywood Agency, *Aircraft Plywood* - https://www.plywood.cc/2013/02/01/aircraft-plywood/ (accessed on 26.02.2019).

[22] Plywood Association of Australia Ltd, *Marine Plywood and Boats* - http://australply.com.au/images/pdf/Marine%20Plywood%20And%20Boats.pdf (accessed on 26.02.2019).

[23] Bss Timbers Ltd, *Bendy Plywood* - http://www.bbstimbers.co.nz/wp-content/uploads/2014/08/Bendy-Plywood.pdf (accessed on 26.02.2019).

[24] Glemco Inc, *6061-T6 Aluminum. Material Notes* - http://www2.glemco.com/pdf/NEW_MARTERIAL_LIST/Alumina%206061-T6.pdf (accessed on 26.02.2019).

Acknowledgments

The authors would like to acknowledge the technical resources offered by the Laboratory of Computer-Aided Fluid Engineering (equipped with technical resources with the financial support of the grant ENERED, POSCCE-A2-O2.2.1-2009-4, ID 911), from the Department of Fluid Mechanics, Fluid Machinery and Fluid Power Systems, "Gheorghe Asachi" Technical University of Iasi, Romania.

IOP Conf. Series: Materials Science and Engineering **595** (2019) 012050 doi:10.1088/1757-899X/595/1/012050

Trifold PV-T-TEG (photovoltaic-thermal-thermoelectric generators) panel characterization overview

A G Lupu[1], C E Panaite[1], V M Homutescu[1], D T Balanescu[1] and A Popescu[1]

[1]Automotive and Mechanical Engineering Department, "Gheorghe Asachi" Technical University of Iasi, Iasi, Romania

E-mail: aristotel.popescu@tuiasi.ro

Abstract. The anticipated depletion of conventional fuels (crude oil, coal) reserves as energy sources triggered extended researches for renewable energy sources. In the last few decades, technologies for energy conversion from solar to electric and thermal energy evolved rapidly. Some recent photovoltaic (PV) panel cooling methods involve thermoelectric (TE) elements. The authors proposed a trifold panel that employs both thermal (T) and TE cooling to lower the temperature of PV panel operation and extract energy from waste heat. Recently, an important number of publications tackled various aspects of PV-TE and PV-T-TE hybrid systems, especially for building integration applications. This paper attempts to cover important topics, with emphasis on aspects related to design and characterization of mobile applications.

1. Introduction

Energy crisis from the 1970s and the increase of pollution awareness in the 1990s triggered an intensive search for energy from sustainable and renewable sources. Many important technological development steps performed lately toward energy conversion from solar, wind, biomass, geothermal and hydropower sources increased the efficiency of such processes. Yet, the most available and, at the same time, the most underused resource remains the solar energy. The share of solar energy in global electricity production is about 2%, [1, 2].

Solar energy harvesting technologies involve conversion to thermal or/and electrical energy.

Solar thermal technologies for small-to-medium applications use various types of collectors to heat up the fluid agent to prepare domestic hot water (DHW) or for space heating purposes: flat panel (FPC), evacuated tube (ETC), or heat pipes (HP-ETC). Industrial-size systems operate at much higher temperatures and use the thermal agent either in heating applications or in conversion to electrical energy. The concentrated solar power (CSP) plants include either Linear Fresnel Reflectors (LFR), parabolic trough collectors (PTC) or dish reflector (PDR), or heliostat field concentrator (HFC).

Solar electric technologies rely largely on photovoltaic cells for direct conversion to electricity. Although based on a physical phenomenon discovered almost two centuries ago (Becquerel, 1839), the research in this area increased in the last few decades, when the production costs decreased based on new technologies and materials. Yet, in recent years, a new conversion method emerged, i.e. use of thermoelectric (TE) elements in solar applications. Thermoelectric effect discovered around beginning of the 19th century, encompasses physical phenomena discovered by Volta (1794), Seebeck (1821), Peltier (1834) and Thomson (Lord Kelvin, 1851) and include the Joule heating effect, as well. Dealing in the past with very small voltages, in the order of hundreds of millivolts, the effect of converting a certain temperature gradient into electricity became more interesting recently, due to new materials.

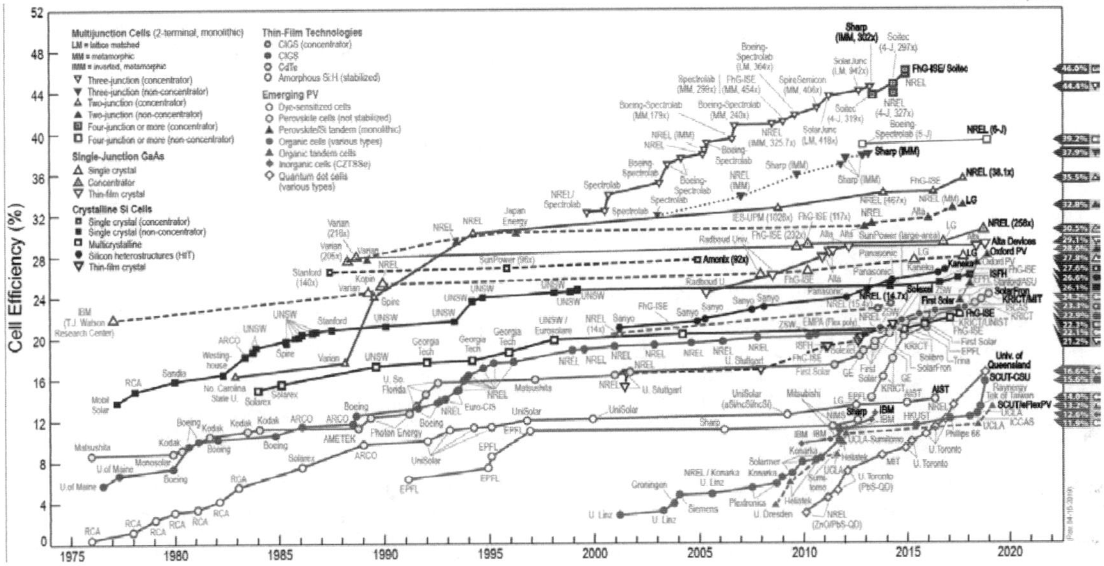

Figure 1. The NREL best efficiency PV cells, in April 2019 [3].

The large number of publications in photovoltaics area, on new materials, production methods and technologies, conversion efficiency, demonstrate the increased scientific research interest to enhance the conversion efficiency of solar energy into electricity. Annually, NREL compiles available data on PV cell efficiency and publishes the most recent findings [3]. Figure 1 demonstrates that, for the last four decades, there was a clear trend in research to increase cell efficiency, which peaked in 2015 [4] with a conversion efficiency of 46% for a special four-junction configuration with optical concentrator and in laboratory conditions. The commercially available units range from 10% to about 25%, with a trend towards lowering costs for both materials and production.

From the first theoretical studies on PV cells and arrays published in the early 1980s [5, 6], it was obvious that a PV system should operate at the lowest possible temperature. The simplest PV model configuration is a single diode non-linear system, with the characteristic (*I-V*) curve mathematically described as [7]:

$$I = I_L - I_o \left[\exp\left(\frac{V + IR_s}{a} \right) - 1 \right] - \frac{V + IR_s}{R_{sh}} \tag{1}$$

where a, the diode quality factor, is expressed as:

$$a = \frac{nkT}{q} \tag{2}$$

where n is the ideality factor, k is the Boltzmann constant, T is the absolute temperature, and q is the elementary charge, [8]. Therefore, the operational temperature is an important factor that influences the PV cell performance.

Numerous published studies, both analytical and experimental, reviewed in [1], revealed that the open circuit voltage of a PV cell decreases with increasing cell temperature. Even if current increases slightly, the decrease in voltage is more significant and the power decreases accordingly, figure 2.

The research intensified in the area of controlling the operational temperature and optimizing the PV panel performance, as an estimated 80-85% of the incident solar energy represents waste heat, i.e.

thermal energy that had to be removed. By tapping into this waste energy, the overall efficiency of the solar panel (electrical and thermal) improves greatly.

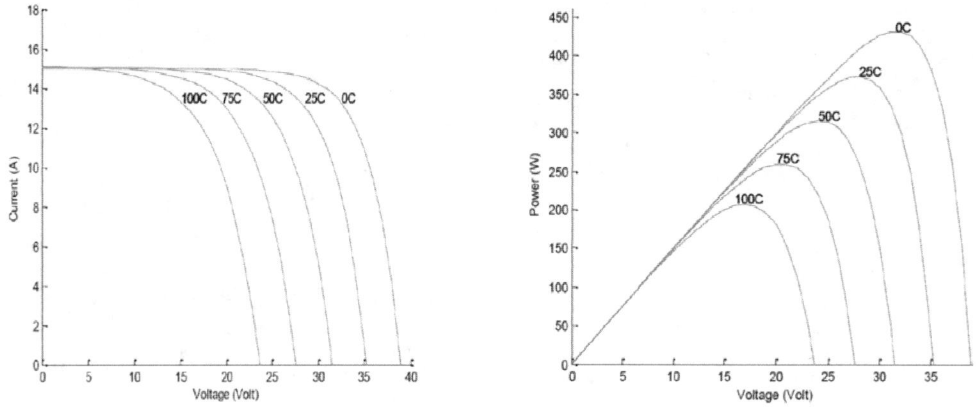

Figure 2. Temperature dependence of (*I-V*) and (*P-V*) curves, at constant irradiance of 1000 W/m^2 [9].

The newly designed hybrid systems may alleviate the disadvantages of a stand-alone PV system (low electrical output, no thermal output from waste heat, high operational temperature, and low overall system stability), despite manufacturing complexity and higher costs. An extensive literature research [1] that included recent papers and conclusions from other reviews demonstrated the benefits and limitations of active cooling methods and technologies known to date, figure 3.

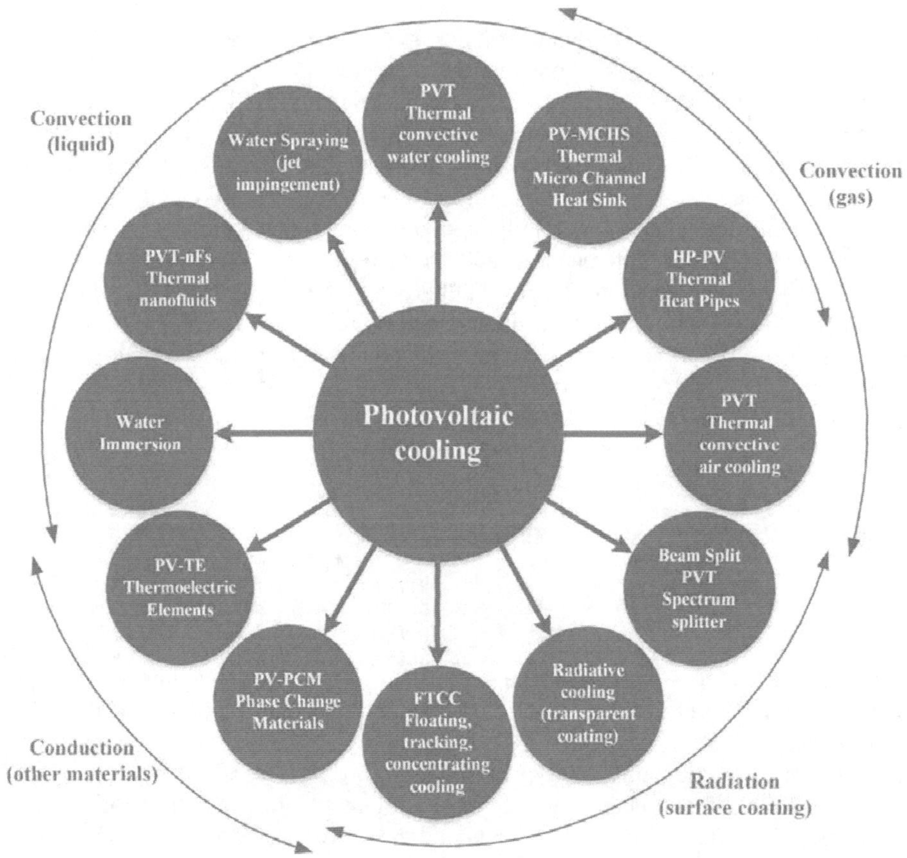

Figure 3. PV cooling methods and technologies known to date.

IOP Conf. Series: Materials Science and Engineering **595** (2019) 012050 doi:10.1088/1757-899X/595/1/012050

On the convective methods side, natural fluid circulation (as passive methods) is more economical, but less efficient. Similarly, the use of liquids is more effective, but incurs technological complications (and higher costs), higher pumping power, as well as corrosion, leak proofing and freezing problems. In addition, special liquids, or even water, may not be readily available in some areas. Newer methods (HP, MCHS, nFs, jet impingement, water spraying) may improve thermal output and overall system performance, but still have high production costs, material characteristics not completely controlled, and localized cooling, in some cases.

Characterized as passive cooling, the radiative methods are partially efficient, as the heat is rejected from the PV panel frontal surface. The spectrum splitting methods incurs some energy losses due to either liquids, thin film depositions or successive reflections, depending on specific technology.

The PCM methods are more successful in medium-to-large size applications, as energy is stored in the molten materials and used continuously day and night. However, the systems include supplemental equipment for conversion of solar to mechanical and then electrical energy.

During the last decade, the studies on TE-based solar systems grew exponentially, as the materials and technologies used for TE manufacturing became less expensive and with better properties. The TE modules are capable to convert the energy from the low-temperature waste heat at the backside of PV panel producing extra electricity at a minimum system cost and controlling the system temperature.

Taking the extra step towards hybridization, the authors proposed and work towards developing a complex structure of a trifold collector panel that encompass a PV panel, TE elements and a thermal recovery unit, to maximize the conversion efficiency of solar in both electrical and thermal energy.

2. Thermoelectric elements

A thermoelectric (TE) device relies on TE materials featuring the specific physical transport properties of direct conversion between thermal and electrical energy.

Independent of Alessandro Volta's experiment in the late 1700s, Thomas Seebeck demonstrated in 1824 that a temperature gradient causes simultaneously a heat flux and an electrical current. In certain materials, mainly metals and semiconductors, the charge carriers (electrons and holes) may move freely within the lattice, carrying thermal energy (heat) as well. In the presence of a thermal gradient, energy is carried from the high-energy (hot) side towards the low-energy (cold) side. As a result, the charge-carrier-based heat transfer generates also a charge gradient (electrostatic potential). The two phenomena are directly proportional, connected by the Seebeck coefficient, α [V/K], or thermo-power.

A decade later, in 1834, Jean Charles Peltier associates his name with the reverse phenomenon, i.e. a voltage (potential) gradient generates an electrical current and a heat flux, the proportionality constant being the Peltier coefficient, π [V].

Later, Lord Kelvin connected the two coefficients by:

$$\pi = \alpha T \tag{3}$$

The main issue with these thermoelectric effects is that they are about two orders of magnitude smaller than the Joule effect (heating) for current flow through a conductor, [10]. Therefore, the use of a single element is impractical, as well as multiple elements connected in parallel. The more efficient and practical solution is the series electrical connection and parallel thermal connection of alternating n-type and p-type materials (electron or hole doped, respectively), figure 4. The substrates should be electrical insulators, but good thermal conductors.

The heat flow within a thermoelectric material is characterized by three components: the material's thermal conductance (heat conduction), the Peltier heat absorbed at the hot side, and Joule heating of the material. The last component is equally divided among the hot and cold sides of the material. To characterize the thermoelectric capabilities of a material, the figure of merit, Z [1/K] is introduced as:

$$Z = \alpha^2 \cdot \frac{\sigma}{\kappa} \tag{4}$$

being proportional to the ratio of conductivities, electrical σ [S/m] vs. thermal κ [W/mK].

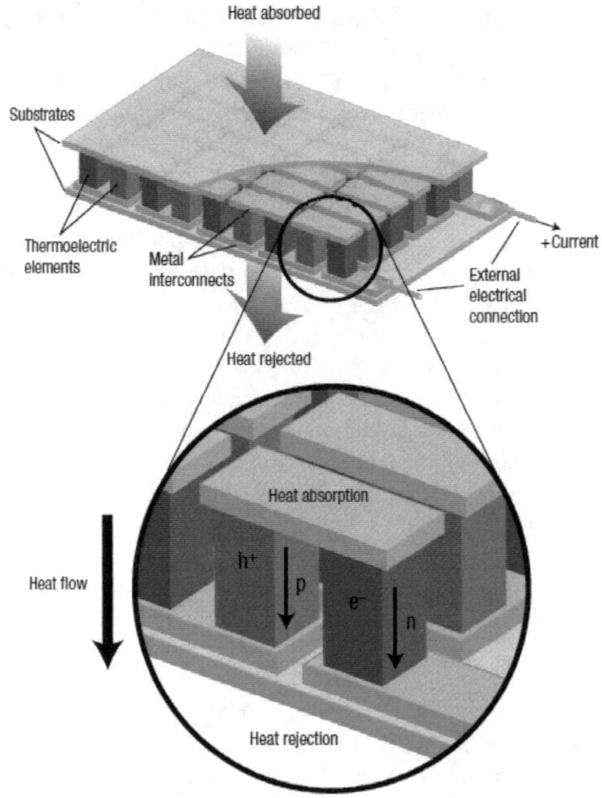

Figure 4. Thermoelectric module components [11].

Electrical conductivity depends on carrier concentration, and is the inverse of electrical resistivity, ρ [Ωm], i.e. $\sigma = 1/\rho$. Total thermal conductivity represents the sum of two contributions, from electrons κ_E and from phonons or lattice vibrations, κ_L. The contribution from lattice vibrations is approximately constant, while electrons contribution varies proportionally with carrier concentration. Good TE materials should exhibit high α and low ρ and κ, but high α implies high ρ, whereas low ρ implies high κ. Therefore, to maximize the figure of merit for TE materials the research should focus on finding the optimum carrier concentration, for high electrical and low thermal conductivities. Best compromise in this area seems to be highly doped semiconductors.

In order to compare the material characteristics, the dimensionless figure of merit ($Z{\cdot}T$) is used:

$$ZT = \frac{\alpha^2 \cdot \sigma \cdot T}{\left(\kappa_E + \kappa_L\right)} = \frac{\alpha^2 \cdot T}{\rho \cdot \kappa} \quad (5)$$

For the same material, the dimensionless figure of merit is dependent of the charge carrier type, i.e. *n*-type (electrons doped) or *p*-type (holes doped). Scattered values for different materials are presented in literature, and Tim Hogan from Michigan State University compiled fifty years (1957 – 2007) of publishes data into two separate graphs, figure 5. Even though the research in TE materials increased in the last decade, especially in the area of nanomaterials and ceramics, the most used and lowest cost material remains Bi_2Te_3. Both graphs show that at normal operational temperatures in solar application of TE devices, i.e. 300 K – 400 K, Bi_2Te_3 has the highest value for the material figure of merit.

At this point, it is important to observe that some authors [11] distinguish between TE efficiencies, material figure of merit (zT) and device figure of merit (ZT). For the ideal case when the TE properties of both *n*-type and *p*-type materials are similar and temperature independent, then $z = Z$.

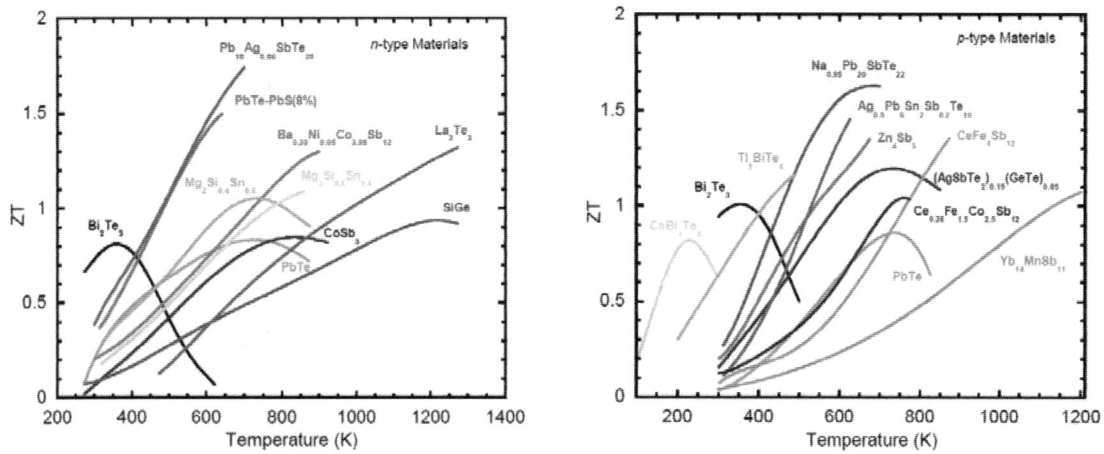

Figure 5. Figure of merit for various thermoelectric materials, *n*-type and *p*-type [12].

The above observation is important, because material's efficiency value is useless by itself, since a TE device is composed of an array of TE *n*-type and *p*-type couples. If contact resistance and radiation effects are neglected, the dimensionless figure of merit for a TE couple, $Z_{np}T_M$, is defined as, [13]:

$$Z_{np}T_M = \frac{\left(\alpha_p - \alpha_n\right)^2 \cdot T_M}{\left[\left(\rho_n \cdot \kappa_n\right)^{1/2} + \left(\rho_p \cdot \kappa_p\right)^{1/2}\right]^2} \tag{6}$$

If the transport properties are similar for both *n*- and *p*-type materials, then the dimensionless figure of merit becomes:

$$Z_{np}T_M \approx \frac{Z_nT_M + Z_pT_M}{2} \tag{7}$$

where T_M is the average temperature if the TE couple operates between T_H the hot-side temperature and T_C the cold-side temperature.

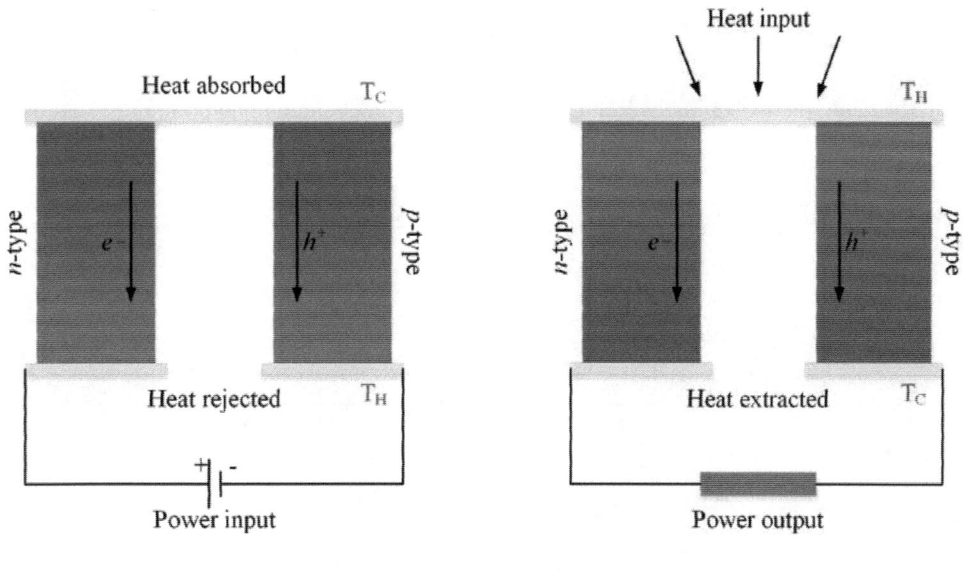

Figure 6. Thermo-Electric Cooling (TEC – Peltier) vs. generation (TEG – Seebeck).

A TE couple, or module, may function two ways, figure 6.

As a refrigerator or cooler (TEC), under the influence of a voltage gradient, an electrical current is supplied to the circuit and a heat flux occurs within the module (Peltier effect). Heat Q_C is absorbed from the top side at T_C producing a cooling effect, and Q_H is rejected at T_H producing a heating effect, figure 6 left. For refrigeration devices, the coefficient to performance, *COP*, is defined as the ratio of cooling power (useful result, Q_C) vs. input power (consumed energy, *P*):

$$COP = \frac{Q_C}{P_{TEC}} = \frac{Q_C}{Q_H - Q_C}. \tag{8}$$

The optimization process of *COP* yields the maximum *COP* value as:

$$COP_{\max} = \frac{T_C}{T_H - T_C} \cdot \frac{\sqrt{1 + Z_{np}T_M} - T_H/T_C}{\sqrt{1 + Z_{np}T_M} + 1} = COP_{Carnot} \cdot \frac{\sqrt{1 + Z_{np}T_M} - T_H/T_C}{\sqrt{1 + Z_{np}T_M} + 1} \tag{9}$$

As a power generator (TEG), under the influence of a temperature gradient, an electrical current is generated within the circuit (Seebeck effect). The heat Q_H is supplied on the module top at T_H and the bottom is cooled by extracting Q_C at temperature T_C, figure 6 right. For power generation devices, the energy conversion efficiency, η, is defined as the ratio of power output (useful result, P) vs. net heat input (consumed energy, Q_H):

$$\eta_{TEG} = \frac{P_{TEG}}{Q_H} = \frac{Q_H - Q_C}{Q_H} \tag{10}$$

The optimization of conversion efficiency yields the maximum *COP* value as:

$$\eta_{\max} = \frac{T_H - T_C}{T_H} \cdot \frac{\sqrt{1 + Z_{np}T_M} - 1}{\sqrt{1 + Z_{np}T_M} + T_C/T_H} = \eta_{Carnot} \cdot \frac{\sqrt{1 + Z_{np}T_M} - 1}{\sqrt{1 + Z_{np}T_M} + T_C/T_H} \tag{11}$$

The output from a single TE module is not practically useful. For real-life applications, multiple TE modules connected electrically in series and thermally in parallel, electrically isolated on top and bottom by two ceramic substrates, form a TE device (figure 4). The equivalent electrical circuit is very simple [14] and is modeled in similar manner for both cases, TEC and TEG.

For solar PV cooling applications, multiple TE devices connected on the backside of the panel absorb the waste heat and thus control the operational temperature of the PV, figure 7.

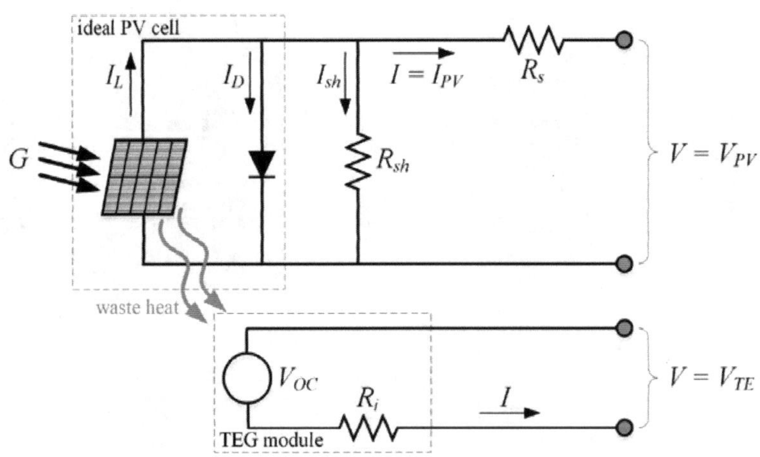

Figure 7. Equivalent electrical circuit for a PV – TE device.

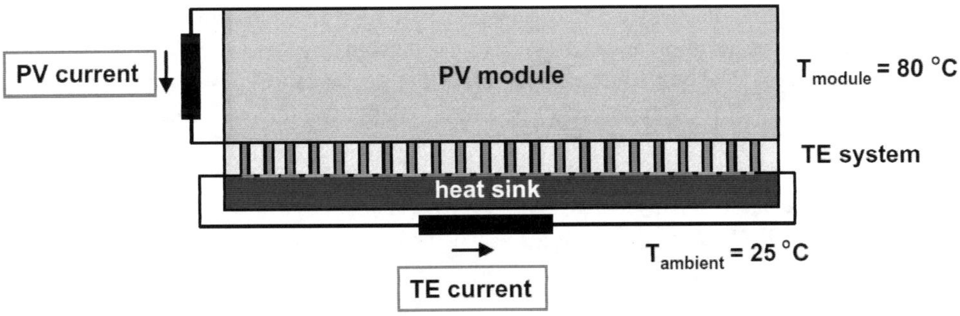

Figure 8. Early PV-TE devices setup concentrated only on PV cooling [18].

3. PV-TE or PVT-TE?

Reviews of advancements on TE devices published in recent years [15 - 17] are mainly concerned with materials properties and manufacturing technologies, but mention future trends and applications, such as waste heat energy recovery, cooling, power generation, thermal energy sensors or aerospace.

Although solar applications with TE were considered since late 1990s, the coupling of TE elements with PV cells represents a relatively new research area.

The early PV-TE applications were concerned with PV cooling solutions rather than increasing the conversion efficiency from solar to electric energy. The goal was to control the PV panel operational temperature, and thus, increase the conversion efficiency.

Lowering the PV temperature, the PV-TE efficiency increased to 11-14% in outdoor applications and up to 23% in laboratory tests [18]. However, measurements demonstrated that only 10% of these values (i.e. 1-1.5%) represent the TE modules contribution. Other research [19] published similar conclusions, as PV-TE and PVT-TE cooling solutions revealed that overall PV efficiency improved, but the TE contribution to electrical production was a mere 1%. Also, [20, 21] accounted only for PV efficiency improvement (1-18%) due to TE cooling, with a 6-26% panel temperature drop.

During the second part of the current decade, the advancements in materials and technologies allowed a new perspective over TE module usage in solar PV applications.

Materials for PV cells and TE modules have better conversion efficiency and lower manufacturing costs. The research in the PV cells area conducted to newer materials, which, even if exhibit lower conversion efficiencies than Si-based materials, are much cheaper to obtain and to embed in PV modules. Also, for TE materials, reviews [15, 17] identified the parameters that improve the thermal to electrical conversion: Seebeck coefficient, electrical and thermal conductivities, and thus figure of merit of material and device, energy gap and band structure, charge carrier concentration and mobility, material composition (metal based, semiconductors, ceramics, polymers), diffusion, thermal expansion coefficient, brittleness, compressibility.

The analysis and characterization of a PV-TE system performance [22, 23] quantify the power output of PV and TEG systems combined and separated. The most important parameters identified were open circuit voltage V_{OC} , short circuit current I_{SC} , PV power output, TE power output, operational PV module temperature, environmental temperature, etc.

The idea of solar cogeneration devices, i.e. simultaneous production of both thermal and electrical energy from a single source (solar energy), implied the partial recovery of energy contained in the waste heat from the backside of the PV panel. The photovoltaic-thermal (PVT) devices developed witnessed a sharp increase in total conversion efficiency, since solar-to-thermal conversion has an efficiency of minimum 30%. Several well-known reviews [24 30] identified PV cooling methods with emphasis on the thermal conversion as well.

Early ideas of using PVT and TE for cooling [31] simply compared the methods and concluded that fluid cooling are far superior as process efficiency, in direct parallel comparison. However, at the same time, authors note that PVT behavior is not optimum and consider the idea that both cooling methods are useful for specific applications.

IOP Conf. Series: Materials Science and Engineering **595** (2019) 012050 doi:10.1088/1757-899X/595/1/012050

Experiments proved that PV water-cooling is very effective in increasing conversion efficiency, [32], and it increases even more by adding TE devices, but only at lower irradiation values. As solar irradiation increases, the TE output cannot compensate the decrease in PV electrical output, the main benefit being the recovery of thermal energy from the PV panel waste heat. Similar research indicated that a higher concentration ratio results in higher power production from TEG module due to increased absorbed heat flux, [33]. Additionally, the PV cells temperature is maintained at optimum operational value by decreasing thermal resistance between the PV cells and TEG module. Simulations indicate an increased efficiency of TEG modules, the total power production of integrated PV-TE system being augmented.

4. Trifold panel characterization

The conclusions from solar cogeneration analyses mentioned above, triggered in-depth research, both theoretical and experimental, on solar cogeneration and even tri-generation (combined cooling, heat and power – CCHP).

Last couple of years several researchers concentrated their efforts in this hot-topic area. Theoretical concepts proposed various combinations depending on specific applications and availability of parts and technologies: compact PV-T-TE, compact PV-TE-T, and spectral beam splitting for incident solar radiation (PV-TE and T).

The PV-T-TE configuration is useful in applications that require preparation of domestic hot water along with electricity production, [34]. A fluid directly cools the backside of the PV panel, absorbing most of the energy from the waste heat rejected. On the other side of the fluid channel, the TE tap into the remaining thermal energy. This design is also useful for the high solar irradiation case [35], when PV panel may attain higher temperatures and requires an intensive temperature control.

The most common configuration is the PV-TE-T, where TE modules glued directly on the backside of the PV panel absorb the heat, cooling it down, figure 9. The working fluid may be air, water, glycol or nanofluids. The setup design using air does not raise many technical problems, but has lower efficiency. The literature reports experimental values of 4.83% and 46.16% for average electrical and thermal efficiencies, respectively, for a triangular setup with concentrator, [36]. The contribution of the TE modules represented a mere 3.3% of the total electrical power generated, i.e. 0.16% of total energy conversion. On the other hand, theoretical models [37] demonstrate maximum attainable values of up to 84% and 12% for thermal and electrical efficiencies, depending on working fluid mass flow rate and density of TE modules attached. If the goal is concentrated on controlling and optimizing the PV panel operational temperature [35], then PV-TE-T solution achieves lower conversion efficiencies of 43.06%, 27.68% and 15.38% for total, electrical and thermal efficiency, respectively. However, the system eliminates the fluctuations in operational temperature and maintains it in the 343-355 K range.

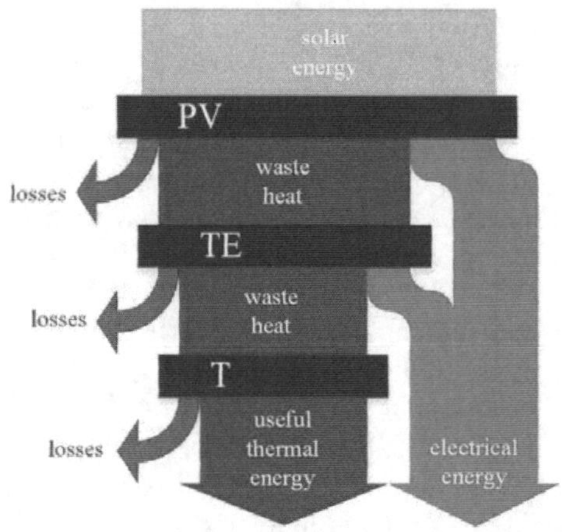

Figure 9. Energy flow in a trifold PV-TE-T solar panel.

Use of nanofluids, [38], further improves the total conversion efficiency, even though authors noted a decrease of about 3% in thermal conversion efficiency. The system control represents the main advantage, by easily adjusting the mass flow rate and volume fraction of the nanofluid. A comparison of all fluids demonstrated [39] the better performance of water-based system when compared to air-based ones with 2-3% and 21-22% for overall electrical and thermal efficiencies, respectively, and mentioned the benefits of using nanofluids.

The solution of spectral beam splitting for incident solar radiation relies on separating the useful 600-1100 nm bandwidth for the PV panel, while the rest of the spectrum is converted into high-grade thermal energy within solar thermal collectors, [34]. This requires a much larger surface area for both panels, electrical and thermal, as well as some technological complications, [40].

The complexity of a trifold solar system involves a large number of parameters that influence the design and operation of all system components (PV, TE, T). Some of these parameters are geometry structure and material composition, heat transfer characteristics (material properties and temperatures), environmental characteristics (location, climate, ambient temperature, and wind speed), solar concentrator system, nanofluid parameters (mass flow rate, volume fraction), exergy performance etc.

5. Conclusions

This paper performed an overview of the PV cell operations and cooling requirements, of TE module design and function and of the current solutions for solar cogeneration and tri-generation systems. The main conclusion is that trifold PV-TE-T hybrid systems exhibit a much better conversion performance when compared with stand-alone PV systems or hybrid PV-T or PV-TE systems.

Even though results of theoretical and experimental analyses demonstrate that the presence of TE modules in the proposed systems does not greatly influence the electrical conversion efficiency (power generation), it improves on controlling and maintaining the operational PV temperature within safety range. Therefore, the system will exhibit higher electrical and thermal efficiency, as well as higher system stability and security. Other ideas may further improve on system performance, such as use of nanofluids or nano-structured materials.

The manufacturing costs of entire system are higher than for individual components, but given the evolution of scientific research and the capability of simultaneous production of electrical, thermal and possibly satisfying cooling requirements, may prove to be an economical solution for specific mobile application in remote locations or for emergency disaster relief forces.

Future research should be concerned with developing materials for both PV cells and TE modules with better characteristics (conversion efficiency, figure of merit) and with lower manufacturing costs, performing theoretical analyses on energy efficiency, exergy efficiency, economic and environmental, as well as performing theoretical simulations and experimental validation of models.

6. References

[1] Lupu A G, Homutescu V M, Balanescu D T and Popescu A 2018 A review of solar photovoltaic systems cooling technologies *IOP Conf Ser: Mater Sci Eng.* **444** 082016

[2] REN21 2018 *Renewables 2018 Global Status Report* (Paris: REN21 Secretariat)

[3] NREL https://www.nrel.gov/pv/ (accessed on 22.04.2019)

[4] Fraunhofer ISE 2016 *Photovoltaics Report* (Freiburg: Fraunhofer Institute for Solar Energy Systems ISE)

[5] O'Leary M J and Clements L D 1980 Thermal-electric performance analysis for actively cooled concentrating photovoltaic systems *Sol Energy.* **25** 401-6

[6] Evans D L 1981 Simplified method for predicting photovoltaic array output *Sol Energy.* **27** 555-60

[7] Tian H, Mancilla-David F, Ellis K, Muljadi E and Jenkins P 2012 A cell-to-module-to-array detailed model for photovoltaic panels *Sol Energy.* **86** 2695-706

[8] Meyer E L 2017 Extraction of saturation current and ideality factor from measuring V_{oc} and I_{sc} of photovoltaic modules *Int J Photoenergy* doi.org/10.1155/2017/8479487

[9] Mboumboue E and Njomo D 2013 Mathematical modeling and digital simulation of PV solar panel using MATLAB software *Int J Emerging Technol Adv Eng.* **3** 24-32

[10] Enescu D 2019 Thermoelectric Energy Harvesting: Basic Principles and Applications *Green Energy Advances* doi:10.5772/intechopen.83495

[11] Snyder G J and Toberer E S 2008 Complex thermoelectric materials *Nat Mater.* **7** 105-14 doi:10.1038/nmat2090

[12] Hogan T 2019 Introduction to Thermoelectrics 1 (*accessed online on April 13, 2019)* at: https://www.egr.msu.edu/~hogant/Introduction%20to%20Thermoelectrics%201.pdf

[13] Tritt T M 2011 Thermoelectric Phenomena, Materials, and Applications *Annu Rev Mater Res.* **41** 433–48

[14] Lineykin S and Ben-Yaakov S 2007 Modeling and analysis of thermoelectric modules *IEEE Trans Ind Appl* **43** 505–12

[15] Elsheikh M H, Shnawah D A, Sabri M S M, Said S B M, Hassan M H, Bashir M B A and Mohamad M 2014 A review on thermoelectric renewable energy: Principle parameters that affect their performance *Renew Sustain Energy Rev.* **30** 337–55

[16] Enescu D and Virjoghe E O 2014 A review on thermoelectric cooling parameters and performance *Renew Sustain Energy Rev.* **38** 903–16

[17] Twaha S, Zhu J, Yan Y and Li B 2016 A comprehensive review of thermoelectric technology: materials, applications, modelling and performance improvement *Renew Sustain Energy Rev.* **65** 698–726

[18] van Sark W G J H M 2011 Feasibility of photovoltaic-thermoelectric hybrid modules *Appl Energy.* **88** 2785-90

[19] Najafi H and Woodbury K 2013 Modeling and analysis of a combined photovoltaic-thermoelectric power generation system *J Sol Energy Eng.* **135** 031013

[20] Verma V, Kane A and Singh B 2016 Complementary performance enhancement of PV energy system through thermoelectric generation *Renew Sustain Energy Rev.* **58** 1017-26

[21] Kane A, Verma V and Singh B 2017 Optimization of thermoelectric cooling technology for an active cooling of photovoltaic panel *Renew Sustain Energy Rev.* **75** 1295-305

[22] Kossyvakis D, Voutsinas G and Hristoforou E 2016 Experimental analysis and performance evaluation of a tandem photovoltaic-thermoelectric hybrid system *Energy Convers Manag.* **117** pp 490-500

[23] Ruzaimi Ariffin M, Shafie S, Hassan W Z W, Azis N and Ya'acob M E 2017 Conceptual design of hybrid photovoltaic-thermoelectric generator (PV/TEG) for automated greenhouse system *2017 IEEE 15th Student Conference on Research and Development (SCOReD)* 309-14

[24] Kalogirou S A and Tripanagnostopoulos Y 2006 Hybrid PV/T solar systems for domestic hot water and electricity production *Energy Convers Manag.* **47** 3368-82

[25] Chow T T 2010 A review on photovoltaic/thermal hybrid solar technology *Appl Energy.* **87** 365-79

[26] Avezov R R, Akhatov J S and Avezova N R 2011 A review on photovoltaic–thermal (PV-T) air and water collectors *Appl Sol Energy.* **47** 169-83

[27] Zhang X, Zhao X, Smith S, Xu J and Yu X 2012 Review of R&D progress and practical application of the solar photovoltaic/thermal (PV/T) technologies *Renew Sustain Energy Rev.* **16** 599-617

[28] Michael J J, Iniyan S and Goic R 2015 Flat plate solar photovoltaic–thermal (PV/T) systems: a reference guide *Renew Sustain Energy Rev.* **51** 62-88

[29] Siecker J, Kusakana K and Numbi B P 2017 A review of solar photovoltaic systems cooling technologies *Renew Sustain Energy Rev.* **79** pp 192-203

[30] Joshi S S and Dhoble A S 2018 Photovoltaic -Thermal systems (PVT): Technology review and future trends *Renew Sustain Energy Rev.* **92** pp 848-82

[31] Rockendorf G, Sillmann R, Podlowski L and Litzenburger B 1999 PV-hybrid and thermoelectric collectors *Sol Energy.* **67** 227–37

IOP Conf. Series: Materials Science and Engineering **595** (2019) 012050 doi:10.1088/1757-899X/595/1/012050

[32] Yang D and Yin H 2011 Energy conversion efficiency of a novel hybrid solar system for photovoltaic, thermoelectric, and heat utilization *IEEE Trans Energy Convers.* **26** 662-70

[33] Kiflemariam R, Fekrmandi H and Lin C 2014 Embedded microfluidic/thermoelectric generation system for self-coooling of electronic devices *Proc 2014 COMSOL Conference* Boston US

[34] Popescu A, Panaite C E, Lupu A G, Atanasiu M V, Vlachokostas C and Michailidou A 2014 Self-sustained, independent trifold solar energy conversion system for isolated locations in hot climate areas *Applied Mechanics and Materials.* **659** 421–4

[35] Yin E, Li Q and Xuan Y 2019 Feasibility analysis of a concentrating photovoltaic-thermoelectric-thermal cogeneration *Appl Energy* **236** 560–73

[36] Mohsenzadeh M, Shafii M B and Jafari M H 2017 A novel concentrating photovoltaic/thermal solar system combined with thermoelectric module in an integrated design *Renew Energy.* **113** 822–834

[37] Nazri N S, Fudholi A, Bakhtyar B, Yen C H, Ibrahim A, Ruslan M H, Mat S and Sopian K 2018 Energy economic analysis of photovoltaic–thermal-thermoelectric (PVT-TE) air collectors *Renew Sustain Energy Rev.* **92** 187–97

[38] Lekbir A, Hassani S, Ab Ghani M R, Gan C K, Mekhilef S and Saidur R 2018 Improved energy conversion performance of a novel design of concentrated photovoltaic system combined with thermoelectric generator with advance cooling system *Energy Convers Manag.* **177** 19–29

[39] Dimri N, Tiwari A and Tiwari G N 2019 Comparative study of photovoltaic thermal (PVT) integrated thermoelectric cooler (TEC) fluid collectors *Renew Energy.* **134** 343–356

[40] Li Y, Witharana S, Cao H, Lasfargues M, Huang Y and Ding Y 2014 Wide spectrum solar energy harvesting through an integrated photovoltaic and thermoelectric system *Particuology.* **15** 39–44

Acknowledgments
This work was partly supported by the project POSCCE-A2-O2.2.1-2009-4-ENERED, ID nr. 911, co-financed by the European Social Fund within the Sectoral Operational Program "*Increase of Economic Competitiveness*".

IOP Conf. Series: Materials Science and Engineering **595** (2019) 012051 doi:10.1088/1757-899X/595/1/012051

Energy characteristics of the char obtained by two steps pyrolysis of chicken manure

E Pop[1], L Mihăescu[1], G Lăzăroiu[2], I Pîșă[1], G Negreanu[1] and V Berbece[1]

[1] Department of Thermodynamics, Engines, Thermal and Refrigeration Equipment, Faculty of Mechanical Engineering and Mechatronics, Politehnica University of Bucharest, Romania.
[2] Department of Power Systems, Power Engineering Faculty, Politehnica University of Bucharest, Romania.

Email: elena.pop@upb.ro

Abstract. The thermal treatment of chicken manure is an ecological alternative of neutralizing this kind of waste with energy recovery. In this regard, a two-stage pyrolysis process was chosen, the first stage having the purpose of preheating and moisture elimination. The drying process up to the starting point of volatile matter release, about 200-250°C was taken into consideration. As a result of the moisture elimination in the first step of the pyrolysis process, a gas fuel riched in CO, CH_4, H_2 will be obtained. The experimental determinations consisted in the proximate and ultimate analysis of the initial residue and tests of the pyrolysis process in a pilot plant that allows the pyrolysis of a quantity of waste in range of 40-200g. A sample of a 140 g was subjected to the pyrolysis process at 500°C and the char (an organic, solid, consistent and odourless product) was obtained. The carbon (C), nitrogen (N), hydrogen (H) and sulphur (S) contents of the final product of pyrolysis were determined by a Costech 4120 Elemental Analyser that allows us to calculate the Lower Heating Value (LHV) of the char. For the evaluation of the energy potential, a theoretical balance of the energy consumption in the process of pyrolysis and the energy developed by combustion of the char, has been made. This theoretical balance was verified by determining the energy developed by combustion of char based on the composition determined in the laboratory. The laboratory results confirmed the existence of an energy potential of the pyrolysis product, which offers a technological alternative.

1. General considerations

As the waste from poultry is significant, two main questions arise - first regarding ecology and the second - the possibility of energy generation.

In the past, these wastes have been used as agricultural fertilizers, especially in the field of vegetable crops, but their storage creates problems related to groundwater and odours in the atmosphere, especially odours resulting from the high ammonia content.

Experience has shown that the most difficult process is to handle waste when it is intended to be incinerated due to its very high moisture content (on average 50%).

In the agricultural production sector, if a hay bed or wood chips placed above the floor are used, the handling capacity of poultry manure is increased (generally, this bed represents about 10-12% of the manure weight) as well as the general energy characteristics. Three technologies can be considered for energy production: direct combustion in steam or hot water boilers, pyrolysis and gasification.

The authors of the paper conducted theoretical studies regarding the direct combustion conditions in a boiler furnace and experimental studies on pyrolysis. Taking into account the physical characteristics of poultry manure, gasification appears to be the least applicable.

2. Physical-chemical and energy characteristics of poultry manure

These characteristics have been determined in the laboratory of the Department of Thermodynamics, Engines, Thermal and Refrigeration Equipment, Faculty of Mechanical Engineering and Mechatronics of Politehnica University of Bucharest, Romania. There have been made determinations of both, pure and mixed poultry manure with solid biomass bed, in a mass ratio of 10-14%. Experimental determinations have been performed for poultry manure from the central region of Moldova and from the central region of Muntenia.

Figure 1 and 2 show the general aspects of the poultry manure obtained from a local intermediate storage area compared to storage with cereal straw bed.

Figure 1. The aspect of pure poultry manure.

Figure 2. The aspect of poultry manure with cereal straw bed.

A small difference between the average density of pure poultry manure compared to the poultry manure with biomass bed have been noticed. The values are presented below:

- Bulk density of pure poultry manure: 88-142 kg/m^3;
- Bulk density of poultry manure with straw bed: 79-127 kg/m^3;

Water content is very high, averaging 35-50%.

The technical analysis of poultry manure is presented in Tables 1 and 2.

Table 1. The results of the technical analysis for poultry manure with biomass bed.

Characteristics	UM	Test I	Test II	Test III	Test VI	Test V
Moisture content, W^i_t	%	35.2	36.2	37.4	39.1	40.2
Volatile matter, V^i	%	50.3	48.2	47.2	45.7	46.1
Ash content, A^i	%	6.8	6.2	6.7	6.1	6.1
Fixed carbon, C^i_f	%	7.1	8.8	8.1	9.1	7.6

Table 2. The results of the technical analysis for pure poultry manure.

Characteristics	UM	Test I	Test II	Test III	Test VI	Test V
Moisture content, W^i_t	%	37.1	37.2	36.9	37.2	36.9
Volatile matter, V^i	%	50.2	50.2	49.9	50.1	50.4
Ash content, A^i	%	6.2	6.1	6.3	6.2	6.2
Fixed carbon, C^i_f	%	6.5	6.5	6.9	6.5	6.5

Index “i” refers to the initial state considered in the intermediate storage warehouse.

The tests indicate a small amount of combustible solid mass in the form of fixed carbon and a high ballast content (sum of water content and ash).

The high content of ballast, in range of (43-46.3)%, shows a difficult application of direct combustion technology indicating that two-stage pyrolysis technology is applicable (the first stage representing only the elimination of moisture).

Table 3. Ballast content of poultry manure.

Characteristics	UM	Test I	Test II	Test III	Test VI	Test V
Ballast content of pure poultry manure	%	43.3	43.2	43.2	43.4	43.1
Ballast content of poultry manure with biomass bed	%	42.6	43.0	44.1	45.2	46.3

Table 4. Elemental analysis.

Type of sample	*C* %	*H* %	*N* %	*S* %	*O* %	*A* %	*Wt* %	*LHV* kJ/kg
pure poultry manure	12.1-16.1	4.3-4.9	1.6-1.8	1.6-1.8	34.5-36.2	6.1-6.8	36.1-40.1	5910-3810
poultry manure with biomass bed	12.3-22.5	4.3-5.2	1.4-2.3	1.7-2.0	35.1-37.4	6.1-12.9	34-40	5940-3840

LHV is the main indicator of the efficiency of energy recovery for this type of waste and is calculated basis on the elemental analysis with the following relation:

$$LHV = 339C^i + 1029.1H^i - 109\left(O^i - S_c^i\right) - 25.1W_t^i, \quad kJ/kg \tag{1}$$

The results presented in table 4 show low values of LHV for both samples, which implies the need of a fuel support (gas or liquid) for direct combustion.

3. Experimental pyrolysis test

The tests have been performed using a chamotte furnace equipped with a thermometer and thermostat, according to the solution presented in Figure 3.

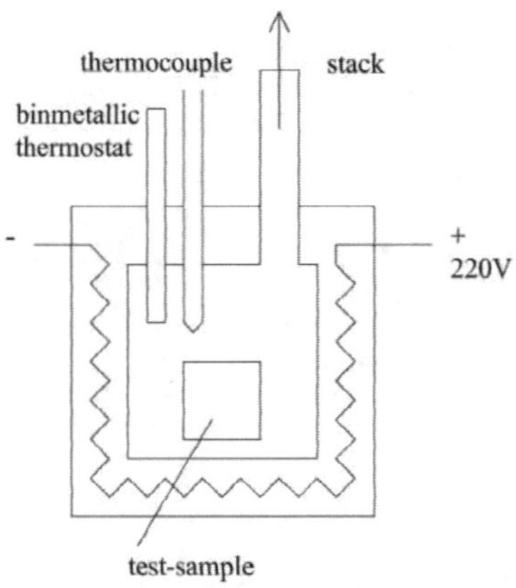

Figure 3. The scheme of the pyrolysis installation.

The poultry manure samples had a weight of (120-160) g and were placed on a 60 x 120 mm tray type device. The char was obtained by heating the samples for 2 hours at (500-580)°C. The char was obtained by heating the samples for 2 hours at 900°C.

Figure 4. The aspect of char obtained from poultry manure.

The char obtained is environmentally friendly, does not smell and does not adhere to the surfaces it comes into contact with. The drying process was carried out at temperatures of (200-250)°C, in order to reduce the time. Furthermore, burning experiments were carried out with the produced char.

4. Determinations of pyrolysis char composition

For the characterization of the biochar obtained from poultry manure, the following instrumentation has been used: a COSTECH ECS 4010 analyzer to perform the ultimate analysis of the samples, for determination of the content of elements such as Carbon, Hydrogen, Nitrogen and Sulphur (Standard deviation: 0.1% abs, CHNS simultaneously for 2-5 mg sample).

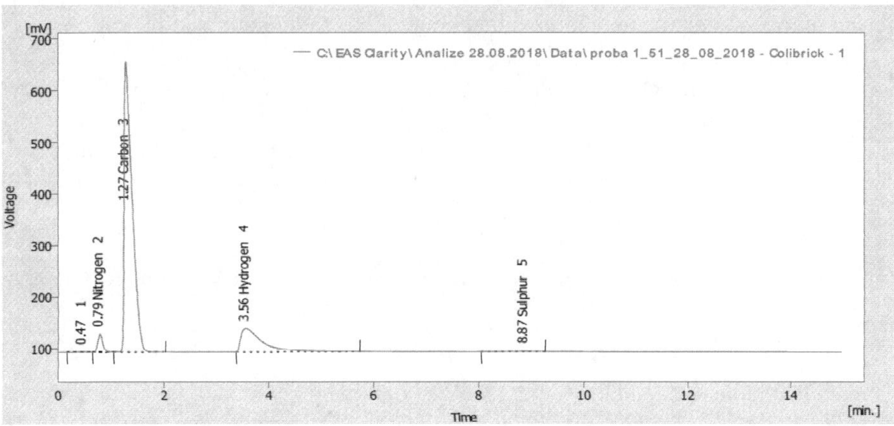

IOP Conf. Series: Materials Science and Engineering **595** (2019) 012051 doi:10.1088/1757-899X/595/1/012051

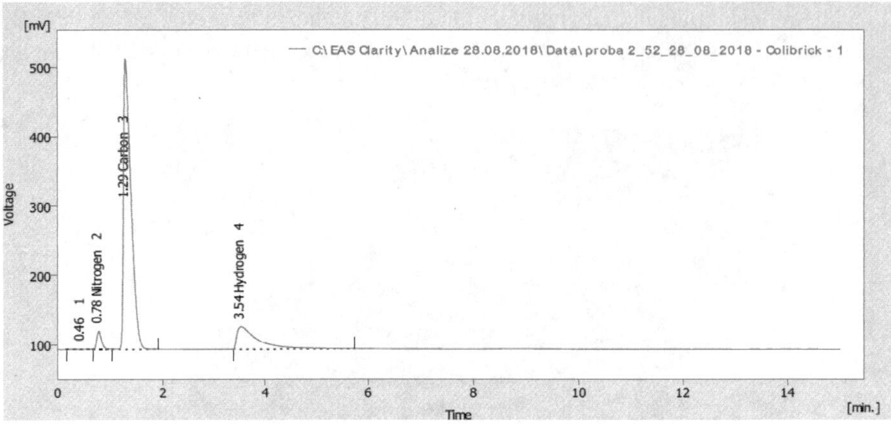

Figure 5. CHNS chromatograms of poultry manure char.

Table 5. CHNS elemental analysis for poultry manure char.

Type of sample	*C* %	*H* %	*N* %	*S* %
Poultry manure char	23.81-24.71	1.37-1.45	2.11-2.16	0-0.1

A very small amount of sulphur was detected. It is noted that in the pyrolysis process, an increase of *C* content occurred concurrently with the *H* loss. Also *N* content in the char increased in pyrolysis process, related to the raw feedstock. Ash content of the biochar from poultry manure has been determined using a Nabertherm furnace.

5. The energy balance of the semi-pyrolysis process

The heat balance was performed for a two-stage semi-pyrolysis fuel unit, the first stage comprising of only the heating and the removal of water content.

As the gas exhausted in the pyrolysis process has a high amount of moisture, the placement of a dryer before the pyrolysis reactor is recommended, while the temperature in drying process does not exceed the 250°C value. In this case, the temperature in the reactor is maintained in the range (450-580)°C. Figure 6 represents the scheme of the low temperature pyrolysis process with the dryer.

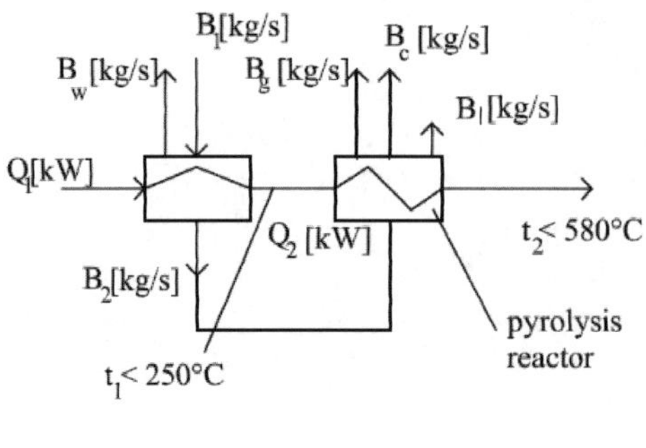

Figure 6. The pyrolysis process with preceding drying.

The flow rate of the material (fuel) subjected to pyrolysis is *B*, kg/s. The heat for the drying process will be determined by the equation:

$$Q_1 = B_1 \cdot \left(c_1 \cdot \Delta t_1 + 2510 \cdot \frac{W_t^i}{100} + c_V \cdot \Delta t_1 \cdot \frac{W_t^i}{100} \right), kW \quad (2)$$

where $\Delta t_1 = t_1 - t_i$ and t_i is the initial temperature. The last term of the equation is the heat of water vapour released to the temperature t_1.

$$B_2 = B_1 \cdot \frac{100 - W_t^i}{100}, \text{ kg/s} \quad (3)$$

$$B_W = B_1 \cdot \frac{W_t^i}{100}, \text{ kg/s} \quad (4)$$

The resulting char quantity respects the following relation:

$$B_c = B_1 \cdot \frac{100 - W_t^i - V^i}{100}, \text{ kg/s} \quad (5)$$

The quantity of fuel gas will be:

$$B_g = B_1 \cdot \phi \cdot \frac{W_t^i}{100}, \text{ kg/s} \quad (6)$$

It was considered that in the pyrolysis phase the devolatilization has been realized in parts marked with φ. For a temperature above 550°C, the devolatilization value is (0.65-0.8).

It is stated that the volatiles contained in the initial matter were not released during the drying phase.

The heat consumed for the pyrolysis:

$$Q_2 = \cdot B_2 \cdot c_2 \cdot \Delta t_2 \quad (7)$$

with $\Delta t_2 = t_2 - t_1$. A calculation regarding the energy efficiency for processing of mass unit of fuel indicates that the specific heat consumed in the drying process was assimilated with that of the peat, a fuel that is similar to the poultry manure:

$$c_1 = 0.01 \cdot W_t^i \cdot c_W + \left(1 - 0.01 \cdot W_t^i\right) \cdot c_C, \text{ kJ/kg} \quad (8)$$

where $c_c = 1.2$ for temperatures above 250°C. For poultry manure with W_t^i=40% and V^i=27%, the result is:

$$c_1 = 0.01 \cdot 40 \cdot 4.18 + (1 - 0.01 \cdot 40) \cdot 1.2 = 2.4 \text{ kJ/kg·K}$$

For $\Delta t_2 = 250 - 20 = 230°C$, the result for the heat consumption for the dryer, with the considered mass flow rate B_1= 1 kg/s, was the following:

$$Q_1 = 2.4 \cdot 230 + 2510 \cdot \frac{40}{100} + 1.3 \cdot 230 \cdot \frac{40}{100} = 1670 \text{ kW}$$

The mass flow rate of waste entering the pyrolysis reactor will become:

$$B_2 = 1 \cdot \frac{100 - 40}{100} = 0.6 \text{ kg/s}$$

The mass flow rate of combustible gas exhausted from the pyrolysis reactor:

$$B_g = 0.6 \cdot 0.75 \cdot \frac{27}{100} = 0.12 \text{ kg/s}$$

The solid mass flow rate of pyrolysis waste (the char rate) will be:

$$B_c = B_1 \cdot \frac{100 - W_t^i - V^i}{100} = 0.33 \text{ kg/s}$$

The heat consumed for the pyrolysis, based on the data:

$$\Delta t_2 = t_2 - t_1 = 540 - 250 = 290°C$$

$$c_2 = 1.4 \text{ kJ/kg·K}$$

obtained from the case of the anhydrous peat at $t_m = 400°C$:

$$Q_2 = 0.33 \cdot 1.4 \cdot 290 \cdot = 134 \text{ kW}$$

For the entire pyrolysis process, the consumed heat will be:

$$Q = Q_1 + Q_2 = 1670 + 134 = 1804 \text{ kW}$$

The heat produced by burning the char containing ash in the percentage A^i=7%, will be:

$$Q_{char} = B_c \cdot H_c = 0.33 \cdot 30000 \cdot \frac{100-7}{100} = 9200 \text{ kW}$$

The theoretical value is confirmed by the calorific value determined based on the composition of the char: $C^l = 23.81\%$, $H^l = 1.37\%$, $LHV_i^i = 9481.3$ kJ/kg

The amount of biochar will be:

$$B_{char} = B_c + \frac{A^i}{100} = 0.33 + \frac{7}{100} = 0.4 \text{ kg/s}$$

For this quantity of char obtained from poultry manure, the thermal power becomes:

$$P_t = 0.4 \cdot 9481.3 = 3800 \text{ kW}$$

Even if the energy value of the resulting combustible gas is not considered. It should be noted that the energy value of the heat produced by the burning of the char obtained from poultry manure is greater than that consumed for the heating. This results in technical and economic feasibility in using the pyrolysis technology of the poultry manure.

6. Conclusions

The paper is a continuation of a series of experiments and analyses regarding the possibility to use the pyrolysis as a technology for energy capture from the poultry manure.

The previous experimental data were the foundation for the numerical application regarding the energy efficiency of the pyrolysis process.

The numerical application has evidently demonstrated the existence of an active thermal energy with respect to the consumptions imposed by the pyrolysis process.

Measurements on the composition of char obtained from poultry manure confirmed the existence of a positive thermal potential in energy consumption for the internal pyrolysis process even if it is lower than the theoretical predictions.

IOP Conf. Series: Materials Science and Engineering **595** (2019) 012051 doi:10.1088/1757-899X/595/1/012051

References

[1] Dalolio F S et al. 2017 Poultry as biomass energy: A review and future perspectives *Renewable and Sustainable Energy Reviews* **76** p 041-949

[2] Petric I, Šestan A and Šestan I 2009 Influence of wheat straw addition on composting of poultry manure.*Process Safety and Environmental Protection* **87(3)** p 206-212

[3] Cantrell K B, Hunt P, Uchimiya M, Novak J M and Ro K S 2012 Impact of Pyrolysis Temperature and Manure Source on Physicochemical Characteristics of Biochar *Bioresource Technology* **107** p 419-428

[4] Edwards D R and Daniel T C 1992 Environmental impacts of-farm poultry waste disposal – a review.*Biores Technol* **41** p 9-33

[5] Mihăescu L, Prisecaru T and Oprea I 2002 *Cazane şi turbine* (in Romanian) (Ed Perfect, Bucureşti) p 190

[6] Pîşă I 2013 *Generatoare de abur* (In Romanian) (ed. Politehnica Press, Bucureşti) p 430

[7] Neaga C and Epure Al 1988 *Calculul termic al generatoarelor de abur. Îndrumar* (In Romanian) (Ed. Tehnică Bucureşti) p 314

IOP Conf. Series: Materials Science and Engineering **595** (2019) 012052 doi:10.1088/1757-899X/595/1/012052

Equilibrium moisture content of pumpkin seeds

I Kiryakov, M Georgieva and D Atanasov

University of Food Technologies – Plovdiv, Bulgaria

E-mail: kirjakoff@abv.bg

Abstract. The equilibrium moisture content of Pumpkin Seeds is studied and the sorption and desorption curves at temperature 20°C are obtained. The strain measurement method for taking down of sorption curves is used for the aim. Analytic dependences describing the sorption and desorption curves are derived. Values of equilibrium moisture contents for temperatures higher than 20°C (40°C, 60°C, 80°C) are obtained by using the method of Pass-Slepchenko. The results are presented in graphical and table form.

1. Introduction

Prolonged contact between a food product and the wet gas at constant external conditions leads to the establishment of thermal and diffusive equilibrium, which was maintained throughout the period of storage. The temperature of the product becomes equal to the ambient temperature, as well as the partial vapour pressure of the liquid is equal in the material and the environment. At this point the moisture of the material is constant and it is called equilibrium moisture content.

The equilibrium moisture content of a product W_P can be measured for series of values of the relative humidity of the air at a constant temperature. The functional dependence of $W_P = f(\varphi)$ at t=*const.* has been plotted according the data obtained in the coordinate system φ, W_P, and it represents an isotherm of sorption or desorption, depending upon whether it is obtained, by moistening or drying a wet material to an equilibrium state [2, 3].

The equilibrium dependencies are a valuable source of information about the type and the format of the connection of the water with the material, for determining the acceptable level of saturation of the drying agent at the exit of the drying chamber etc.

1.1. Aim of the work

The aim of this study is to determine experimentally the equilibrium moisture content of pumpkin seeds and to obtain dependences describing the sorption-desorption isotherms at different temperatures. They will be used in determining the conditions for storage and sizing of drying plants.

2. Materials and methods

The object of this study was pumpkin seeds (*Cucurbita maxima*) of the "Sampson F1" variety were cultivated in a field situated close to Plovdiv (Bulgaria) – harvest 2017. Pumpkin seeds have been rich in protein, glyceride oil, phytosterols and vitamins. The seeds have been a source of carotenoids, tocopherols and trace elements (K, Mg, Mn, Zn, Se, Co, Cr and Mo). Seeds have been a major product for the oil-producing industry in the production of vegetable oil after cold pressing. The resulting oil has been a high content of unsaturated fatty acids. All these qualities have determined the application of pumpkin seeds to the food industry and medicine [5, 6].

The equilibrium moisture content of pumpkin seeds has been experimentally determined by the strain measurement method. This method is long and labour-consuming but it provides high accuracy measurements.

The method comprises the following: a certain amount of material has been placed in an exicator at a constant temperature (20°C) and relative humidity. The sample has been periodically weighed. Upon reaching a constant weight the equilibrium moisture content of the material has been determined. An analytical balance with accuracy of 0.0001 g has been used for determining the mass of the material. The equilibrium moisture content of the product has been determined by the weight method by drying the sample to constant mass in an atmospheric oven at 100 ± 2 °C.

The constant humidity of the air into each sealed vessel (exicator) has been maintained by means of saturated aqueous solutions of some salts, which have the ability to maintain a constant pressure of water vapour in the atmosphere above them. The relative humidity and the salts used in the experiment at a temperature of 20°C are shown in Table 1.

Table 1 Salts used in the experiment and the corresponding relative humidity maintained at 20°C.

Salts	NaOH	LiCl	CH_3COOK	$MgCl_2$	K_2CO_3	NaBr	NaCl	Na_2SO_4
φ, %	6.98	11.14	23.1	32.1	43.9	58.7	75.4	86.9

Dependences of the $W_P = f(\varphi)$ type (equation (1)) have been obtained from the experimental results, taken at 20°C. A standard statistical program Table Curve 2D v.5.01 has been used for determining the abovementioned dependences. Values of the equilibrium moisture content of pumpkin seeds for accurate values of relative humidity have been calculated from the resulting equations for sorption and desorption.

However, for dimensioning of the drying processes and systems, it is necessary to have a sorption (desorption respectively) isotherms for a sufficiently wide temperature range. For determination of the equilibrium moisture content at other temperatures (40, 60 and 80°C) the Pass and Slepchenko's methods have been used [2]. These methods give good results for cereals and wood in the temperature range (-20°C ÷ 100°C). On the basis of the data received, analytical dependences have been deduced. The equilibrium moisture content of the product has been calculated using these equations for the same values of relative humidity.

3. Results and discussion

The results for the equilibrium moisture content of pumpkin seeds are presented in Table 2 and Table 3. The values at 20°C have been obtained experimentally, and those at 40, 60 and 80°C have been defined theoretically by the Pass and Slepchenko's method.

Except in tabular form, analytical consideration of sorption and desorption isotherms is also interesting. Due to the large number of factors that affect the sorption process, the describing of the equilibrium isotherm with a single equation is not possible. Many empirical and semi-empirical equations valid in a certain range of relative humidity for specified products have been derived in practice. The equations of the BET (Brunauer, Emmett and Teller) or GAB (Guggenheim, Anderson and de Boer) type are the most commonly used [4]. These equations have a common disadvantage that except their theoretical basis, they contain constants which must be experimentally determined [1, 4].

The resulting analytical relationships describing the sorption-desorption curves have the following form:

$$W_P = a + b\varphi + c\varphi^2 + d\varphi^3 + fe^{-\varphi} \quad (1)$$

where: φ - relative humidity of the air, %; W_P - equilibrium moisture of pumpkin seeds, %; a, b, c, d and f are coefficients whose values are given in Table 4 and Table 5.

Correlation coefficients - r^2 and the average statistical error are presented in the same tables. The resulting equations are valid in the range $0 < \varphi < 0.9$.

Table 2. Equilibrium moisture content of pumpkin seeds W_P, %, determined by sorption at different temperatures.

φ, %	Temperature, °C			
	20	40	60	80
10	2.76	2.10	1.73	1.53
20	4.46	3.82	3.19	2.87
30	5.51	4.90	4.19	3.78
40	6.25	5.66	4.98	4.51
50	7.00	6.41	5.83	5.30
60	8.08	7.47	6.97	6.40
70	9.82	9.17	8.65	8.07
80	12.54	11.82	11.13	10.56
90	16.57	15.75	14.66	14.12

Table 3. Equilibrium moisture content of pumpkin seeds W_P, %, determined by desorption at different temperatures.

φ, %	Temperature, °C			
	20	40	60	80
10	3.73	2.74	2.37	2.11
20	5.67	4.91	4.36	3.78
30	6.80	6.17	5.57	4.85
40	7.52	6.95	6.37	5.63
50	8.24	7.66	7.12	6.44
60	9.35	8.71	8.18	7.58
70	11.25	10.52	9.90	9.37
80	14.34	13.50	12.66	12.13
90	19.02	18.07	16.81	16.17

Table 4. Coefficients for the equation of sorption.

Coefficients	Temperature, °C			
	20	40	60	80
a	0.090934154	-0.59139928	-0.42551847	-0.50496383
b	0.32581379	0.32799478	0.25914591	0.2460491
c	-0.0064648998	-0.0064289349	-0.0047628187	-0.0046895179
d	5.421925×10^{-5}	5.3361224×10^{-5}	4.1620998×10^{-5}	4.1785683×10^{-5}
f	-0.090763689	0.59396246	0.43397981	0.51179484
r^2	0.9921	0.9993	0.9978	0.9964
Fitt Std Err	0.46595	0.15747	0.27083	0.33416

Table 5. Coefficients for the equation of desorption.

Coefficients	Temperature, °C			
	20	40	60	80
a	0.5819142	-0.73260932	-0.74871477	-0.47772246
b	0.38858797	0.42719159	0.38115402	0.31523576
c	-0.0080424875	-0.0086413011	-0.007489932	-0.0061530981
d	6.6682343×10^{-5}	6.9072976×10^{-5}	6.0195917×10^{-5}	5.228519×10^{-5}
f	-0.58189573	0.73631271	0.75153411	0.48464056
r^2	0.9941	0.9993	0.9986	0.9962
Fitt Std Err	0.44784	0.181184	0.26511	0.40997

The data from Tables 2 and 3 are presented graphically in Figures 1 and 2. As can be seen the increase in temperature leads to a shift of the equilibrium isotherms to the left in the entire temperature range. There is also a pronounced sorption hysteresis.

The isotherms have clearly depicted *S* - shaped character, which indicates the existence of three types of connection of moisture to the material - monomolecular adsorption, polymolecular adsorption and capillary condensation. The analysis of the character of the sorption and desorption curves shows that the adsorption bound moisture in the material is a minor portion (up to $W_P = 2\%$ at 20°C) and decreases with increasing the temperature, which is testified by convex section of the curve in the direction of the X-axis [1]. This convex section decreases at high temperatures, which means that the major part of the moisture is related to the capillary and polymolecular adsorption.

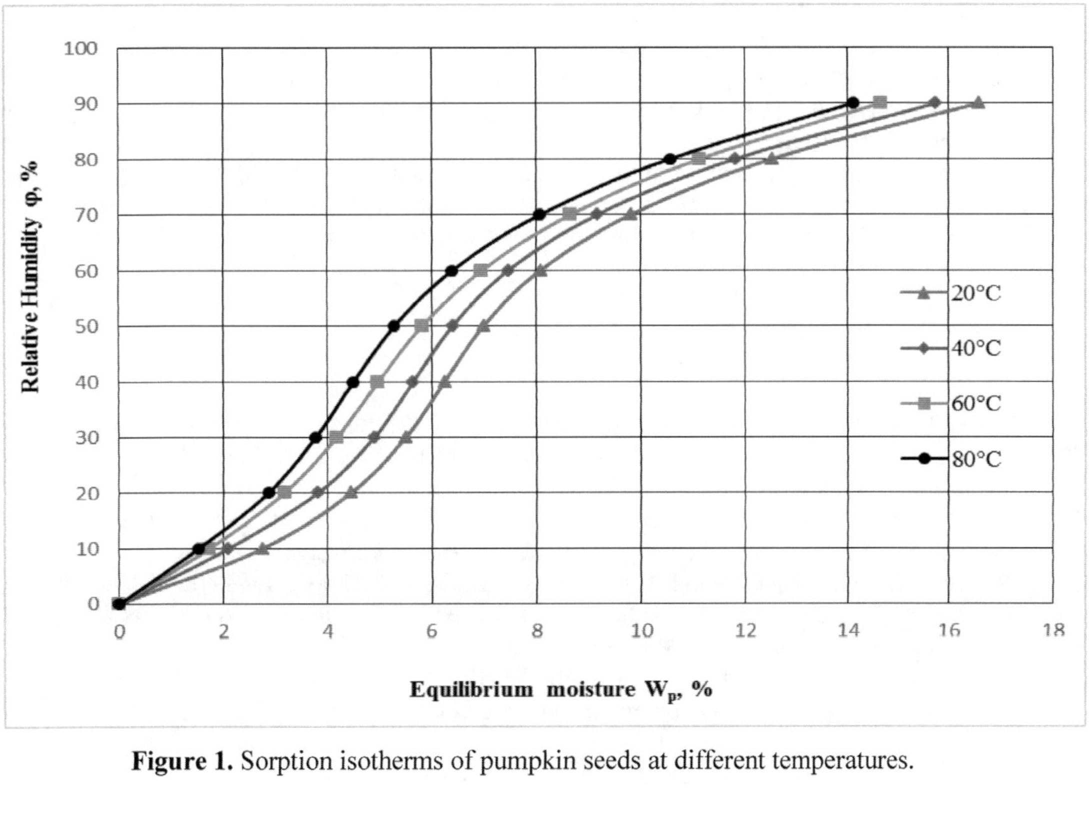

Figure 1. Sorption isotherms of pumpkin seeds at different temperatures.

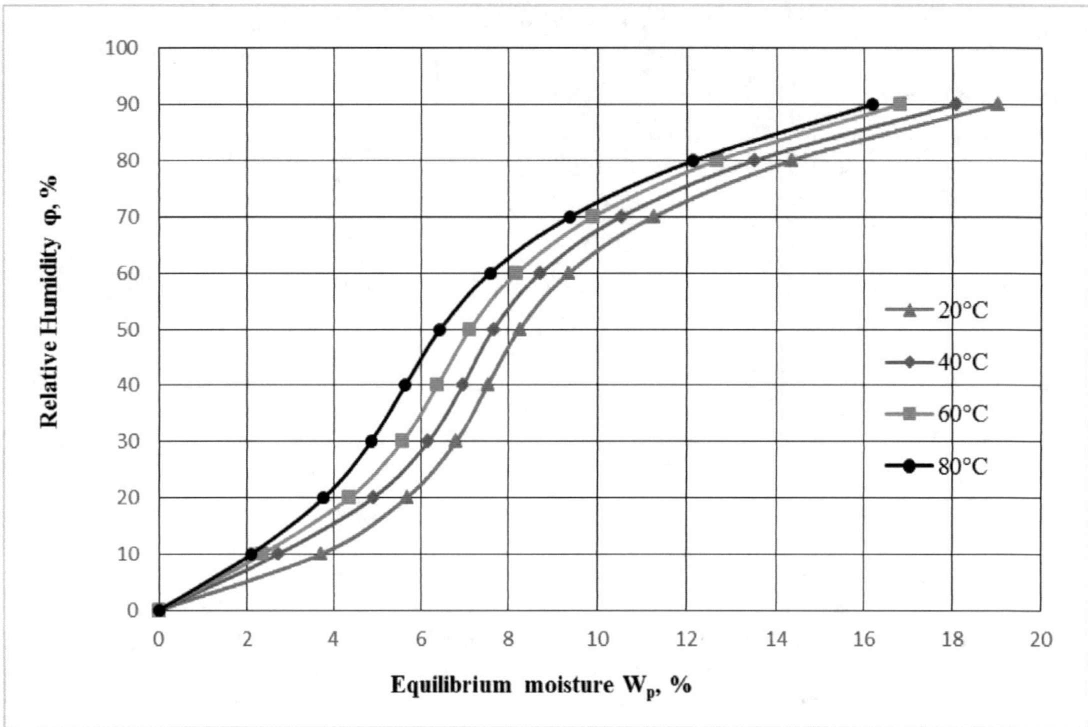

Figure 2. Desorption isotherms of pumpkin seeds at different temperatures.

4. Conclusions

1. Experiments for determination of the equilibrium moisture content of pumpkin seeds at a temperature of 20°C have been conducted. The values of the equilibrium moisture content at higher temperatures - 40, 60 and 80°C have been recalculated by Pass and Slepchenko's methods. The results are presented in tabular form - Table 2 and Table 3.

2. An analytical relationship (equation 1) for calculating the equilibrium humidity of pumpkin seeds for a range of relative humidity $0 < \varphi < 0.9$ and range of temperatures $t = 20 \div 80$°C has been obtained.

3. Sorption and desorption isotherms of pumpkin seeds at various temperatures have been presented in graphic form.

4. The results may be useful in determining the storage conditions and design of drying equipment and installations for the production of pumpkin seeds.

References

[1] Filonenko G et al. 1971 Drying of food plant materials (in Russ.) *Food industry* Moscow
[2] Ginzburg A 1973 Fundamentals of the theory and technology of drying foods (in Russ.) *Food Industry* Moscow
[3] Minchev M 2012 Drying equipment (in Bulg.), Plovdiv, *Academic Publishing House of UFT*
[4] Nevenkin S 1993 Drying and drying equipment (in Bulg.) *Engineering* Sofia
[5] Petkova Zh Y, Antova G A and Angelova-Romova M Y 2018 Development of lipid damage of pumpkin seed oil stabilized with different antioxidants during long-term storage *Bulgarian Chemical Communications* **50** Special Issue C 112
[6] Petkova Zh Y and Antova G A 2014 Phospholipid composition of Cucurbitaceae seed oils, *Bulgarian Chemical Communications* **46** Special Issue A 100

IOP Conf. Series: Materials Science and Engineering **595** (2019) 012053 doi:10.1088/1757-899X/595/1/012053

Wastewater sludge to energy production. A review

N Vatachi

Thermal Systems and Environmental Engineering Department, "Dunarea de Jos" University of Galati, Romania

E-mail: nicusor.vatachi@ugal.ro

Abstract. Global economic and social development involves increasing demands for water use for various purposes: domestic, public, industrial. Wastewater treatment in these processes results in increasing volumes of sewage sludge. The elimination of this sludge is a permanent concern at global level because of the dangers to the environment, consisting of the high content of organic pollutants, toxic and heavy metals. However, a significant amount of renewable energy can be recovered from sludge from sewage treatment, with several paths being possible: anaerobic digestion, combustion and co-combustion, pyrolysis, gasification. Each of these pathways is characterized by advantages and disadvantages, benefits and technological limitations. This paper proposes a review of current conversion methods and technologies for energy recovery from sewage sludge, anaerobic digestion, incineration and co-combustion especially. Processes present in current literature and practice are described, resulting in the influence of the properties of sludge fuels and their impact on the energy recovery process when these fuels are used in combustion plants. There are also scientific researchers conducted to improve operational and environmental performance and cost competitiveness.

1. Introduction

The recovery of energy through current waste management methods is of particular importance due to its impact on the minimization of their storage space and the optimization of alternative energy generation. Energy conversion technologies can be highlighted in Figure 1 [1, 2, 4-6], resulting in the possibility of sludge conversion with natural gas, liquid fuel, chemical, heat and/or electricity generation.

Anaerobic digestion is a widely used method of biological conversion because of its low cost and the ability to use high-humidity organic waste without reducing high calorific value of the produced biogas (methane and carbon dioxide). Sewage sludge is a solid, semi-solid or liquid waste generated by wastewater treatment plants. The physical properties (low ratio of solid matter in the liquid) of the slurry require thickening and dehydration to facilitate transport and logistics during the treatment processes. These processes help to increase the concentration of solid particles in the sludge to about (10-25)% by weight from the predominantly liquid original state (<3% by weight solid) [1-4]. Pre-treatment of raw waste water removes large particles such as gravel and stone but also the finest sands in specialized installations. Continued processing with the use of sedimentation tanks where the primary sludge is generated under the gravitational force [2, 5].

Sludge from sedimentation and decantation is sent to digesters where biogas is generated by oxygen-free fermentation. Biogas from the digester can be further cleaned and upgraded to produce bio-methane that can be a direct substitute for natural gas or can be converted to thermal and electrical energy by cogeneration using thermal reactors.

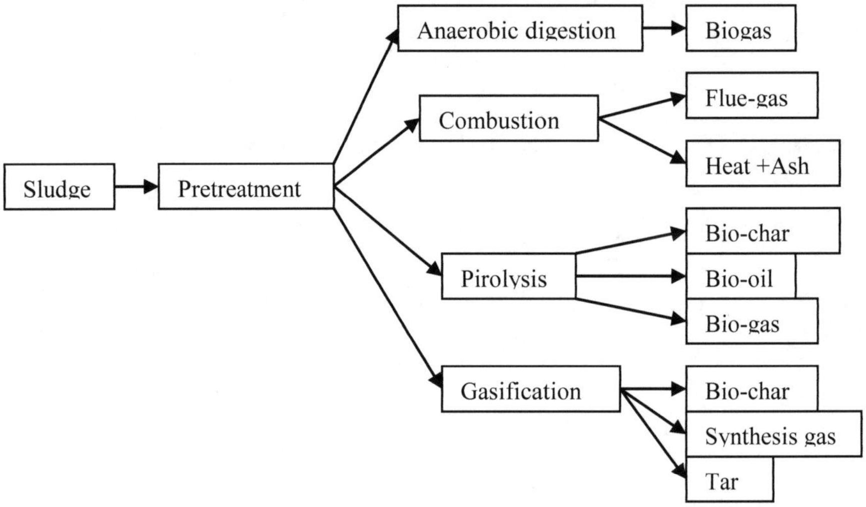

Figure 1. Potential Sludge-to-Energy Recovery Routes.

On the other hand, thermos-chemical conversion paths, such as burning or incinerating sludge, are characterized by shorter reaction times, ranging from seconds to min. These processes require low-moisture sludge and drying of the sludge can be done with high energy consumption. Fast and controlled decomposition of 80% of the inert, partially or total oxidized organic matter causes these thermal processes to be considered more beneficial compared to anaerobic digestion [7]. However, the cost of thermal technologies is still significantly higher.

Incineration is a process currently used for sewage sludge, but was not intended for energy recovery but is used to reduce the volume of waste and destroy the harmful elements in its composition [8, 9]. The resulting heat recovery from the combustion gases however, converts traditional incinerators into typical combustion systems that use heat from complete oxidation of organic matter at high temperatures (800-1150)°C. Heat extracted is used to heat working fluids (usually water) that can be used directly or are used to generate electricity through a steam turbine.

Unlike combustion, pyrolysis takes place under completely inert (no oxygen) conditions at moderate to high temperatures (300-900)°C to produce pyrolytic oil, carbon dioxide and non-condensable gases (CO, H_2, CO_2, CH_4 and light hydrocarbons) [7, 9].

The operating temperature, heating speed and residence time greatly influence the quality of the product and the energy content of the pyrolytic products. Bio-oil can be processed and used as liquid fuel or converted by synthesis to gas (CO and H_2) for chemical production, while biofuel, non-condensable gases and bio-oil can be used as solid, gaseous and liquid fuels for the production electric and thermal combustion.

Alternatively, the biofuel can be used in adsorption or catalyst applications. Finally, we mention the gasification, the thermo-chemical conversion of the organic compounds by partial oxidation (lower oxidant than stoichiometric requirements) at high temperatures (650-1000)°C to maximize the production of gaseous products, especially synthesis gas (CO and H_2) [10].

The energy contained in the product gas varies from 4-28 MJ/Nm^3 depending on the gas and temperature agent. The derived gas can be directed to a variety of end uses such as direct heat combustion and electricity generation using a combined cycle gas turbine. By further modernizing the synthesis process or using chemical processes the gas can be converted to liquid fuels by processes such as Fischer-Tropsch synthesis.

2. **Anaerobic digestion**

By anaerobic digestion we understand a process during which the organic compounds in the mud to be treated are converted into methane and carbon dioxide in microbiological processes. This takes place in four stages: hydrolysis, acidogenesis, acetogenesis and metanogenesis. During hydrolysis insoluble organic compounds and high molecular weight compounds are converted to soluble organic, such as amino acids and fatty acids. The compounds produced by hydrolysis are taken up in the acidogenesis phase where acidogenic bacteria turn into volatile acids, ammonia, carbon dioxide and other byproducts. Acetogenesis further digests the acidogenesis product with acetogens, the final products being acetic acid, carbon dioxide and hydrogen. Hydrogen is the one that, through its partial pressure in the mix, controls this conversion. Finally, in the methanogenesis stage, methane is generated by means of two groups of obligatory anaerobic bacteria.

Therefore, anaerobic digestion is performed by three different groups of bacteria, namely hydrolytic and acetogenic bacteria, the group of hydrogen producing bacteria, acetic hydrogen and carbon dioxide, and the methanogenic methane group producing methane and carbon dioxide. These groups use in their metabolic process the product generated by the previous group.

The whole process of microbial digestion is slow and requires a long retention time (Fig. 2). Throughout it, the whole organic material is digested, less stable woody materials since the anaerobic micro-organisms are unable to degrade lignin.

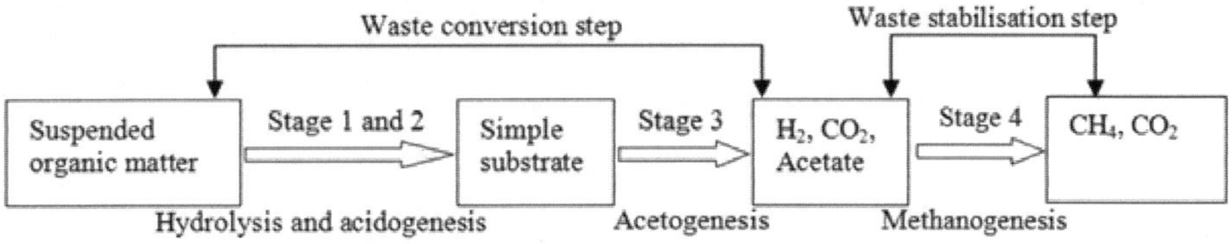

Figure 2. Stages of anaerobic digestion.

Therefore, microbiology anaerobic digestion is complex and delicate, each of the groups of bacteria with their own optimal working conditions. These groups are sensitive and may optionally be inhibited by a number of waste and environmental factors.

The quantity of methane that can be produced depends on **waste composition** and in particular the degradable organic content. The theoretical maximum methane yield can be estimated from Eq. (1):

$$C_cH_hO_oN_nS_s + y\,H_2O \rightarrow x\,CH_4 + (c-x)CO_2 + n\,NH_3 + s\,H_2S \tag{1}$$

where:

$$x = (4c + h - 20 - 3n - 2s)/8$$
$$y = (4c - h - 20 + 3n + 3s)/4$$

However, all of the organic content may not actually be able to be degraded by the bacteria. The anaerobic fermentation process can be improved by generating a more intimate contact between biodegradable particles and bacteria. This can be accomplished by finely grinding the particles and making a more homogeneous mixture between finely obtained particles and bacteria. Crushing produces an intimate particle-bacterial contact surface and in this way the organic material is more exposed to the action of the bacteria. This method leads to an increase in the fermentation yield of about 20%.

Another way to increase the efficiency of anaerobic fermentation is to ensure an optimal temperature of the mixture of the material that is degradable with the bacteria involved in the process. This is done by thermal sludge pretreatment, ie providing a temperature in the range (150-200) °C and some pressures in the range of (600-2500) kPa. Heat introduced during thermal pretreatment helps solubilize cellular components, accelerating extra cellular enzyme activity acting as a catalyst for

IOP Conf. Series: Materials Science and Engineering **595** (2019) 012053 doi:10.1088/1757-899X/595/1/012053

fermentative reactions. The optimum temperature differs depending on the nature and composition of the sludge [51].

Table 1 shows the results obtained by different researchers in the use of thermal pretreatment to improve anaerobic digestion.

Table 1. Overview of thermal pre-treatment studies.

Reference	Treatment conditions	Comments
Hiraoka et al. [11]	(60–100)°C	-Maximum increase in gas production at 60°C -Maximum VS reduction at 100°C (only 5–10%)
Pinnekamp [12]	(120–220)°C	-ODS reduction of 10–55% for WAS -ODS reduction from 7% to 34% for primary sludge -Maximum gas yield for treatment temperature of 170 °C -Positive correlation between gas yield and treatment temperature
Li and Noike [13]	(62–175)°C (30–60) min	-Increase of sludge solubilisation ratio by 25–45% (optimum at 90 °C) for WAS -Increase of 30% VSS degradation and of 100% methane production (optimum at 170 °C and 60 min) -No further improvement for longer treatment times -Reduction of retention time in digester by 5 days
Tanaka et al. [14]	180°C 60 min	-90% increase of methane production -VSS solubilisation of 30%
Zheng et al. [15]	220°C 30 s	-55% VS reduction during digestion -Increase in gas production of 200% during first 2 days -Total increase in gas production of 80%
Kim et al. [16]	121°C 30 min	-Increase of VS reduction by 30%
Valo et al. [17]	170°C	-59% increase of TS reduction -92% higher gas production
Ferrer et al. [18]	70°C 9–72 h	-Positive effect on gas production -Higher temperature (110–134 °C) did not have any effect
Climent et al. [19]	(70–134)°C 90 min–9 h	-Studied thermophilic digestion -50% increase of biogas production at 70 °C (9 h)

The formed biogas has a high calorific value and is considered as a renewable energy source. It is beneficial to produce as much biogas as possible [20].

Despite these advantages of AD, some limitations are inevitable:

- the organic fraction suffers a partial, not total decomposition;
- slow reaction speeds;
- high cost of digesters;
- the process is sensitive to the influence of various reaction inhibitors;
- the presence of other gases along with biogas, CO_2, H_2S, water vapor;
- the possibility of degradation of the biogas combustion boiler pipes due to the presence in the biogas of volatile siloxanes, which generate monocrystalline silica which is deposited on the pipes;
- present in important quantities of heavy metals in the residual sludge;

Therefore, microbiology anaerobic digestion is complex and delicate, each of the groups of bacteria with their own optimal working conditions. These groups are sensitive and may optionally be inhibited by a number of waste and environmental factors.

In terms of energy, the anaerobic digestion process is, on the one hand, an important consumer (heating and pumping mixing sludge) and on the other hand it can generate methane-rich biogas as a source of renewable energy for other processes. For this reason, there are various researches that have integrated energy recovery systems in wastewater treatment plants, WWTP-CHP-AD.

Figure 3 shows an example of a WWTP-CHP system integrated with an AD fermentation process (WWTP-CHP-AD). The plant treats 21000 m^3 of wastewater per day with the following efficiency figures: 96% for total solids (TSS), 02% for chemical oxygen demand (COD) and 64% for total nitrogen (TN).

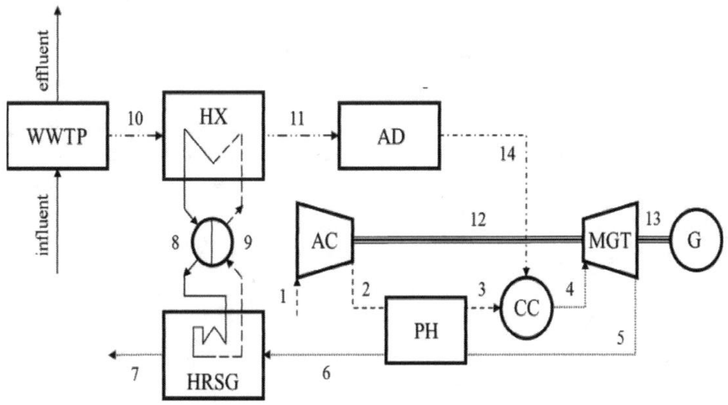

Figure 3. Schematic diagram of the integrated WWTP-CHP-AD system [10].

The process generates excess sludge at primary and secondary decanters. The volume of excess sludge is reduced in the integrated anaerobic fermentation process, which generates significant amounts of methane-rich fermentation gas as a by-product. The anaerobic fermentation and fermentation model is based on the BSM2 simulation model, Benchmark Simulation Model No. 2 [21].

The CHP integrated system covers the heat demand for the AD anaerobic fermentation process and partially the power demand of the plant. In the CHP subsystem, the air is compressed by the AC compressor, then preheated in the PH preheater based on heat recovery from the hot gas stream delivered by the micro gas turbine (MGT). Methane-rich biogas (provided by AD anaerobic fermentation) is injected into a combustion chamber (CC) where it is burnt with compressed air. The high-temperature and high-pressure combustion gases in the CC combustion chamber relax in the MGT gas turbine generating electricity on the one hand with the G generator and on the other hand covering the power consumption of the AC compressor. The flue gas from the MGT passes through the PH to pre-heat compressed air and then passes through the heat recovery steam generator (HRSG) to generate hot steam, which is used to maintain the operational temperature of the heat exchanger, HX, (used as the heat source for the AD process).

3. Combustion

Burning of sewage sludge is similar to combustion of solid fuels. It is about getting heat from the high temperature of fuel oxidation, with the generation of carbon dioxide, water vapour and other gases.

Therefore, combustion technology for residual materials, such as sludge, can be used, mainly for obtaining heat (conventional combustion) or for reducing the amount of waste (incineration). The conventional use of heat generated by combustion technology is heating or generating electricity through thermal engines, while incineration systems may or may not use combustion heat, as their primary purpose is the combustion destruction of harmful elements waste before final disposal or reuse of residual ash in the construction industry. It has been noticed that by incineration a reduction of up to 90% of the volume of mud can be obtained [22].

The use of incinerators is recommended for the disposal of municipal and municipal solid waste and has been imposed lately due to the need to reduce the use of cultivated land for waste disposal. Combustion of solid fuels involves drying, pyrolysis, volatile combustion, carbon burning, ash melting and agglomeration. These stages occur sequentially or simultaneously, depending on the configuration, the conditions of the reactor and the properties of the fuel. For example, some sludge and biomass can start low-temperature pyrolysis (~ 150°C) typical for fuel drying [25, 26].

Combustion of volatile matter dominates heat production from sludge burning, with carbon burning being relatively negligible. Increasing the temperature at the surface of the fuel due to the burning of volatiles may result in the melting of the ash covering the surface of the particles at temperatures of 900 ° C. Carbon combustion occurs with increasing temperature and increases the proportion of gaseous compounds until the combustion takes place rapidly (in an equivalent time or less with that of the devolatilization stage) due to the low carbon content of the sludge [27, 28].

Finally, agglomeration of ash particles is facilitated by the heat from the combustion of carbon, and after reaching the maximum surface temperature the gradual cooling begins, indicating complete firing [29]. This suggests that the burning of all organic matter and the ash melting are necessary to reduce the volume of sludge during incineration while drying, devolatilization and burning of volatile substances are more important for the conventional combustion system for heat generation. Hence, the necessary considerations regarding the efficiency of combustion and incineration can be derived. These systems must be equipped with flue gas cleaning installations to minimize fly ash or hazardous gas emissions.

Different combustion installations such as multiple hearth furnaces, rotary furnaces and cyclones, and fluidized bed furnace, which have a different fuel supply regime, other mode of operation, with its advantages and disadvantages are used. However, the fluidized bed furnace is quite popular for burning wet sludge (35-59% humidity) and dry, due to its simplicity, of low cost, uniform heating, low-pollutant combustion gases (50% CO and NO_x, 40% CO_2), lower residence time and high combustion efficiency [30-32]. Such technologies are also known to reduce the costs associated with extra fuel (about $ 0.2-1 million annually at operational facilities) [33]). Recent studies focus mainly on three areas:
- improving combustion and optimizing reaction parameters by using catalysts and co-mixing with other fuels such as coal, crude oil and biomass [34-39];
- the use of ash and fly ash generated by the process as a fertilizer or in the cement industry [40-42];
- reduction of toxic emissions and CO, NOx, SOx and PAH pollutants, prevention of heavy metal emissions [45].

3.1. Sludge incineration

Different methods are used for the thermal treatment of sludge:

- combustion or monoincineration using multiple-hearth furnace, fluidized bed, combined MHF-FBC, cyclone furnace, smelting furnace, rotary furnace;
- co-combustion with coal in power plants, with coal in FBC power plants, with other fuels, with MSW;
- alternative processes such as wet oxidation, pyrolysis, oil from sludge, fuel from sludge, gasification;

Combustion and co-firing are recommended because the processes can be well controlled during the course of the reactions, and various combustion products purification techniques are possible.
If only incineration is used as a method of destroying sludge, it is also not possible to obtain energy that can be utilized because of their low calorific value.

Figure 4 shows a simple sludge incineration scheme. This, with a low solid content (about 1-4%) is introduced into the mixing and homogenization tank. Thereafter, the stage of thickening and removal of the supernatant follows, at which end the solids content reaches volumes of 3-8%

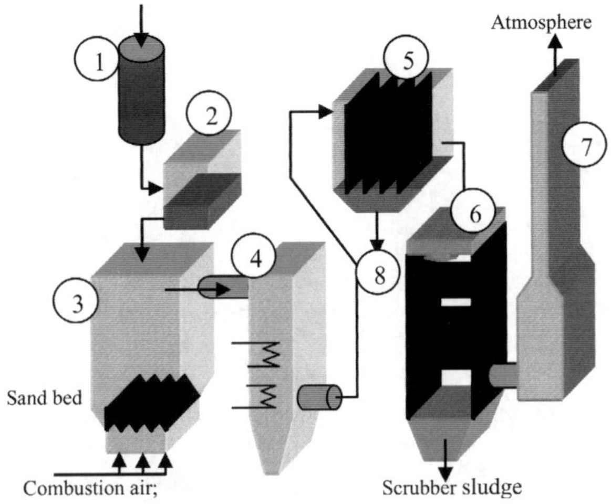

Figure 4. A simple sludge incineration scheme [51].
1- sludge mixing tank; 2- belt press, plate press, centrifuge; 3- fluidised bed incinerator; 4- waste heat boiler; 5- electrostatic precipitator; 6- wet scrubber; 7 - stack; 8 - inorganic incinerated sewage sludge ash.

The dried sludge is then dehydrated using plate or band filters. In this process organic and inorganic additives are added which increase the calorific value of the sludge while reducing the ash content generated. At the end of this stage the solids content increases to 18-35%. The combustion is carried out in a fluidized bed with the introduction of combustion air at temperatures (500-600)°C. Thus, an area with temperatures (800-900)°C is generated in the furnace where the water vaporizes and the volatile metals and the compounds organic burns and turns into combustion gases. The resulting inorganic ash is discharged from the furnace with the hot combustion gases.
After yielding a significant amount of heat to the steam boiler, the gases are cleaned by passing through an electrostatic filter and a wet scrubber. The remaining particles from the gas generate significant amounts of sewage sludge and then the clean gases are evacuated to the stack [43].

3.2. Sludge co-combustion

The construction of a sludge incineration plant is an important investment, given the high costs for the boiler plant and for the flue gas treatment plant. The idea was to use existing coal-fired power plants for incineration of dry co-combustion sludge. In this way, a special sludge incinerator is no longer required, but existing coal combustion plants and combustion exhaust systems are used. The small amount of dry mud compared to the large amount of coal to be burned in the plant makes the sludge incineration effects on air and ash qualities insignificant.

Co-combustion of coal-purification sludge thus becomes a viable option for destroying sludge but also for obtaining significant amounts of renewable energy. In this way, a higher efficiency of the combustion plant is obtained from an economic point of view under approximately similar environmental protection conditions.

Currently co-combustion of coal with sewage sludge is applied, but there are co-incineration plants for combustion of a classic fuel in combination with municipal waste [44].
Co-combustion is agreed to the disadvantage of incineration due to advantages, of which:
- reduction of carbon dioxide emissions;
- a more efficient use of fuels in the location of the installation;
- the use of biomass as a high efficiency residual fuel under controlled environmental conditions;

As potential disadvantages can be enumerated:
- additional investments in combustion gas treatment equipment;
- higher costs in sludge drying installations;
- the possibility of the occurrence of toxic combustion compounds due to the composition and the properties of some sludges, [45].

Combined combustion of fossil fuels with sewage sludge and with different types of biomass is the most efficient way of using waste for energy purposes.

Like any method, the possibilities of application and use can be hampered by different aspects:
- Fueling problems of combustion plants (difficulty in homogenising the fuel mixture before combustion);
- the combined use of several types of fuel in the same combustion plant is difficult to control;
- correlation of combustion quality control with variation of composition of fuel-waste mixtures;

To overcome these difficulties, combustion plants have to be equipped with advanced combustion control systems and advanced exploitation technologies, especially in the handling and fuelling area [46].

Of the most known types of co-combustion plants that are used, some can be highlighted:
- adaptation of an existing coal-fired or suspended coal combustion plant to combined coal-sewage sludge combustion by additional direct feeding of the pulverized sludge [47, 48];
- additional supply by introducing sludge onto the combustion grill of a fluid bed combustion plant [49];
- separate burning of sludge by steam injection coupled to the main charcoal burner;
- using sludge as additional fuel for recharging the combustion plant;
- the use of sludge as a fuel in a post-combustion installation to reduce nitrogen oxide emissions [50].

Although recommended as a method of destroying sewage sludge and other types of waste, co-incineration can generate a series of unintended exploits. Some of these are:
- Energy and volatile production: The use of additional low calorific fuels, obtained on the basis of various additives during preliminary thermal treatments, can cause thermal imbalance of the boiler, with consequences for its operation and reliability;
- Effects on air pollution: substances that generate pollutant emissions, nitrogen, sulphur, chlorine can be converted into pollutants, NO_x, SO_2, HCl to a greater or lesser extent depending on the concentrations in the respective substances of the basic fuels and extra. If the concentration is higher in the additional fuel the emissions will be higher and vice versa;
- Ashes generation: The presence of ash-generating elements may favour the formation of deposits which will reduce the flue gas flow and will contribute to the occurrence of corrosion phenomena;
- Heavy metals: pollutes the atmosphere by volatilizing poisons and soil and waters by ashes poisoning.

4. Conclusion

Anaerobic digestion is a proven technical success and is currently being applied to a variety of wastewater and waste streams, but worldwide application is still limited, thus negating an important renewable energy source. Furthermore, some potential sources that are currently being treated are an excellent substrate for anaerobic treatment and could contribute to the production of renewable energy rather than energy consumption during treatment. Applying anaerobic digestion to produce energy from sludge and solid waste can help to obtain low-cost fuels compared to natural gas. During anaerobic treatment, biodegradable compounds are effectively eliminated, leaving a small number of compounds in the effluent, such as ammonium, organic N compounds, sulphides, organic P compounds and pathogens. Depending on subsequent use, a complementary treatment step is required. Anaerobic treatment is an ecologically and economically proven technique. Technology is relatively new, and in many countries application is still at an early stage. It is extremely important to establish demonstration projects in many parts of the world to show the environmental and economic benefits and to provide the necessary confidence in the technique.

Biomass is a viable alternative to energy fuel for a good time. Studies in the use of biomass as renewable energy can be successfully applied in the energy sector and other categories of organic waste, such as sludge from sewage treatment.

The combined combustion technique of sludge with a conventional fuel in energy steam generators is a way to follow, given the simplicity of technology and the fossil fuel economy. However, prior to being introduced into the combustion plant, the sludge undergoes a variety of thermal, mechanical, ultrasound, filtration, etc. treatments, whereby organic materials in the source water, including suspended and colloidal are concentrated in a small volume, processed mud. It contains with more than 100% more energy that can be recovered compared to sludge from a classic installation without pretreatment.

Although combined combustion has a number of advantages, so the process is efficient, the sludge must be autothermic. For this reason, before choosing the sludge treatment solution, its energy characteristics, such as calorific power and ignition temperature, must be determined. For this we must first establish the energy balance of the mud burning based on the values of the burning heat present in the literature, from which it is to be determined how much it should be dehydrated to become self-thermally and at what costs.

References

[1] Seiple T E, Coleman A M and Skaggs R L 2017 Municipal wastewater sludge as a sustainable bioresource in the United States. *J. Environ. Manag.* **197** pp 673–680

[2] Syed-Hassan S S A, Wang Y, Hu S, Su S and Xiang J 2017 Thermochemical processing of sewage sludge to energy and fuel: Fundamentals, challenges and considerations. Renew. *Sustain. Energy Rev.* **80** pp 888–913

[3] Cieslik B M, Namiesnik J and Konieczka P 2015 Review of sewage sludge management: Standards, regulations and analytical methods *J. Clean. Prod.* **90** pp 1–15

[4] Magdziarz, A, Dalai A K and Kozinski J A 2016 Chemical composition, character and reactivity of renewable fuel ashes. *Fuel.* **176** pp 135–145

[5] Harrison E Z, Oakes S R, Hysell M and Hay A 2006 Organic chemicals in sewage sludges. *Sci. Total Environ.* **367** pp 481–497

[6] Mulchandani A and Westerhoff P 2016 Recovery opportunities for metals and energy from sewage sludges. *Bioresour. Technol.* **215** pp 215–226

[7] Ruffino B et al. 2017 *Enhancement of waste activated sludge (WAS) anaerobic digestion by means of pre- and intermediate treatments.* In Proceedings of the International Conference on Sustainable Solid Waste Management, Limassol, Cyprus, 21–24 June 2017

[8] Nielsen H B, Thygesen A, Thomsen A B and Schmidt J E 2011 Anaerobic digestion of waste activated sludge—Comparison of thermal pretreatments with thermal inter-stage treatments. *J. Chem. Technol. Biotechnol.* **86**, 238–245

[9] Valo A, Carrère H and Delgenès J P 2004 Thermal, chemical and thermo-chemical pre-treatment of waste activated sludge for anaerobic digestion. *J. Chem. Technol. Biotechnol.* **79**, 1197–1203

[10] Sivagurunathan P, Zhen G, Kim S and Saratale G D 2017 A review on bio-electrochemical systems (BESs) for the syngas and value added biochemicals production *Chemosphere.* **177**, 84–92

[11] Hiraoka M, Takeda N, Sakai S, and Yasuda A 1989 Highly efficient anaerobic digestion with thermal pre-treatment. *Water Sci Technol.* **17:54**

[12] Pinnekamp J 1989 Effects of thermal pre-treatment of sewage sludge on anaerobic digestion. *Water Sci Technol.* **21** 97–108

[13] Li YY and Noike T 1992 Upgrading of anaerobic digestion of waste activated sludge by thermal pre-treatment. *Water Sci Technol.* **3-4** 857–66

[14] Tanaka S, Kobayashi T, Kamiyama K and Bildan M L S 1997 Effects of thermochemical pre-treatment on the anaerobic digestion of water activated sludge. *Water Sci Technol.* **8** 209–15
[15] Zheng J, Graff RA, Fillos J and Rinard I 1998 Incorporation of rapid thermal conditioning into a wastewater treatment plant. *Fuel Process Technol.* **56** 183–200
[16] Kim J, Park C, Kim T, Lee M, Kim S, Kim S et al. 2003 Effects of various pretreatments for enhanced anaerobic digestion with waste activated sludge. *J Biosci Bioeng.* **95** 271–5
[17] Valo A, Carrere H and Delgene´ J 2004 Thermal, chemical and thermo-chemical pretreatment of waste activated sludge for anaerobic digestion. *J Chem Technol Biotechnol.* **79** 1197–203
[18] Ferrer I, Climent M, Baeza MM, Artola A, Vazquez F and Font X 2006 *Effect of sludge pretreatment on thermophilic anaerobic digestion.* In: Proceedings of the IWA specialised conference on sustainable sludge management: state-of-the-art, challenges and perspectives, Moscow, Russia, May 29–31, 2006
[19] Climent M, Ferrer I, Baeza M D, Artola A, Vazquez F and Font X 2007 Effects of thermal and mechanical pretreatments of secondary sludge on biogas production under thermophilic conditions. *Chem Eng J.* **133** 335–342
[20] Sridhar Pilli et al. 2011 Ultrasonic pretreatment of sludge: A review. *Ultrasonics Sonochemistry.* **18** pp 1–18
[21] Nopens I et al. 2010 Benchmark Simulation Model No 2: finalisation of plant layout and default control strategy. *Water Sci Technol.* **62** pp. 1967–74
[22] Wu M H, Lin C L, Huang W C and Chen J W 2016 Characteristics of pervious concrete using incineration bottom ash in place of sandstone graded material. *Constr. Build. Mater.* **111** pp 618–624
[23] Ogada T and Werther J 1996 Combustion characteristics of wet sludge in a fluidized bed: Release and combustion of the volatiles. *Fuel.* **75** pp 617–626
[24] Urciuolo M, Solimene R, Chirone R and Salatino P 2012 Fluidized bed combustion and fragmentation of wet sewage sludge. *Exp. Therm. Fluid Sci.* **43** pp 97–104
[25] Ding H H, Chang S and Liu Y 2017 Biological hydrolysis pretreatment on secondary sludge: Enhancement of anaerobic digestion and mechanism study. *Bioresour. Technol.* **244** pp 989–995
[26] Tsai W T 2012 An Analysis of the Use of Biosludge as an Energy Source and Its Environmental Benefits in Taiwan. *Energies.* **5**, 3064–3073
[27] Werther J and Ogada T 1999 Sewage sludge combustion. *Prog. Energy Combust. Sci.* **25** pp 55–116
[28] Dennis J S, Lambert R J, Milne A J, Scott S A and Hayhurst A N 2005 The kinetics of combustion of chars derived from sewage sludge. *Fuel.* **84**, 117–126
[29] Cui H et al. 2006 Fundamental Behaviors in Combustion of Raw Sewage Sludge. *Energy Fuels.* **20**, 77–83
[30] Fytili D and Zabaniotou A 2008 Utilization of sewage sludge in EU application of old and new methods—A review. *Renew. Sustain. Energy Rev.* **12**, 116–140
[31] Kijo-Kleczkowska A, Sroda K and Otwinowski H 2013 Study into Combustion of Sewage Sludge as Energetic Fuel. *Arch. Min. Sci.* **58**, 1085
[32] Batistella L, Silva V, Suzin R C et al. 2015 *Gaseous emissions from sewage sludge combustion in a moving bed combustor.* Waste Manag. **46**, 430–439
[33] Raheem A, Sikarwar V S, He J et al. 2018 Opportunities and challenges in sustainable treatment and resource reuse of sewage sludge: A review. *Chem. Eng. J.* **337**, 616–641
[34] Kijo-Kleczkowska A, Sroda K, Kosowska-Golachowska M, Musiał T and Wolski K 2016 Combustion of pelleted sewage sludge with reference to coal and biomass. *Fuel.* **170**, 141–160
[35] Lin Y, Liao Y, Yu Z, Fang S and Ma X 2017 The investigation of co-combustion of sewage sludge and oil shale using thermogravimetric analysis. *Thermochim. Acta.* **653**, 71–78

[36] Huang L et al. 2016 Thermodynamics and kinetics parameters of co-combustion between sewage sludge and water hyacinth in CO_2/O_2 atmosphere as biomass to solid biofuel. *Bioresour. Technol.* **218**, 631–642

[37] Wang Z, Hong C, Xing Y, Li Y, Feng L and Jia M 2018 Combustion behaviors and kinetics of sewage sludge blended with pulverized coal: With and without catalysts. *Waste Manag.* **74**, 288–296

[38] Kijo-Kleczkowska A et al. 2016 Experimental research of sewage sludge with coal and biomass co-combustion, in pellet form. Waste Manag. **53**, 165–181

[39] Deng S, Tan H, Wang X, Yang F, Cao R, Wang Z and Ruan R 2017 Investigation on the fast co-pyrolysis of sewage sludge with biomass and the combustion reactivity of residual char. *Bioresour. Technol.* **239**, 302–310

[40] Niu X, Shen L 2018 Release and transformation of phosphorus in chemical looping combustion of sewage sludge. *Chem. Eng. J.* **335**, 621–630

[41] Cheng X, Zhang M, Wang Z, Xu G, Ma C 2018 IR and kinetic study of sewage sludge combustion at different oxygen concentrations. *Waste Manag.* **74**, 279–287

[42] Chen Y. Wang T, Zhou M, Hou H, Xue Y and Wang H 2018 Rice husk and sewage sludge co-combustion ash: Leaching behavior analysis and cementitious property. *Constr. Build. Mater.* **163**, 63–72

[43] Shane D and Christopher R C 2013 Recycling and recovery routes for incinerated sewage sludge ash (ISSA): A review. *Waste Management.* **33** pp. 2328–2340.

[44] Luts D, Devoldere K, Laethem B, Bartholomeeusen W and Ockier P 2000 Co-incineration of dried sewage sludge in coal-fired power plants: A case study. *Water Sci. Technol.* **42(9),** 259–268

[45] Bo Leckner 2007 Co-combustion – A summary of technology, *Thermal science*. **II(4)** pp. 5-40

[46] Järvinen T and Alakangas E 2001 *Co-Firing of Biomass - Evaluation of Fuel Procurement and Handling*, in: Selected Existing Plants and Exchange of Information (COFIRING), Part 2, VTT-Energy Altener, January 2001

[47] Hughes E E and Tillman D A 1998 Biomass Cofiring: Status and Prospects *Fuel Processing Technology*. **54(1-3)**, pp. 127-142

[48] Savolainen K 2003 Co-Firing of Biomass in Coal-Fired Utility Boilers, *Applied Energy*, **74(3-4)**, pp. 369-38

[49] Mory A and Tauschitz J 1999 *Co-Combustion of Biomass in Coal-Fired Power Plants in Austria,* VGB PowerTech. **79(1),** pp. 50-55;

[50] Fernando R 2002 *Experience of Indirect Cofiring of Biomass and Coal,* IEA Clean Coal Centre, CCC 64;

[51] Vatachi N 2015 *Sludge from wastewater plants treatment and the power generation*, Annals of "Dunarea De Jos" University of Galati. Fascicle IV. Refrigerating technique, internal combustion engines, boilers and turbines, year XXIX, 2015, (http://www.stim.ugal.ro/Items/Annals/2015/2015-Annals-05.pdf)

IOP Conf. Series: Materials Science and Engineering **595** (2019) 012054 doi:10.1088/1757-899X/595/1/012054

Examination of natural noise signals: preliminary results for aquatic environment

H Zhivomirov[1,2] and I Nedelchev[1,2]

[1] Department of Theory of Electrical Engineering and Measurements, Technical University of Varna, Varna, Bulgaria
[2] SigNautic Lab – Examination of Underwater Noises, Signals and Vibrations of Marine Vessels and Structures, Technical University of Varna, Bulgaria

E-mail: hristo_car@abv.bg

Abstract. The present article treats the signal analysis and the revealing of the properties of noise signals which are caused by cavitation, boiling, flow and motion of water (aquatic environment) in natural conditions. All examined signals are not of androgenic and/or technogenic origin and corresponds to a few natural processes *e.g.*, boiling water, rain, sea surf. All analyses are conducted in the Matlab® environment and include: computation of the mean and RMS values, dynamic range, crest factor, skewness and kurtosis, autocorrelation time; test for stationarity; plot of the oscillogram, correlogram, histogram, periodogram and spectrogram. Finally, some common conclusions are drawn, based on the analyses results.

1. Introduction

The present paper treats the signal analysis and the revealing of the properties of the natural stochastic noise signals and contains some preliminary results for the corresponding processes in the aquatic environment. All examined processes, which are an object of consideration here, have natural essence, that is, they are not of androgenic and/or technogenic origin. By means of the signal analysis one draws a conclusion about the properties of the underlying process.

In the present paper, the term "noise signal" is referred to a signal produced by a stochastic process. The term "coloured noise" used bellow, refers to noise signal whose power spectral density (PSD) is proportional as $\mathrm{PSD} \propto f^{\alpha}$ for $\alpha \in \mathbb{R}$. The definition of the colored noise signal (*e.g.*, white noise) implies an infinity bandwidth, and hence infinity energy and so it is a purely theoretical and non-casual. In the real-world applications the noise signal is considered over a frequency range relevant to the context [1].

Often, it is incorrectly assumed the terms "Gaussian noise" and "white noise" to be used interchangeably. It must be reminded, that Gaussianity refers to the amplitude probability distribution in the time domain, while the term "white noise" refers to the independently distributed power in the frequency domain and neither property implies the other [2].

2. Signal analysis techniques

A specified set of signal analysis techniques are used in order to reveal the signals behaviour and to classify the signals properly. The signals are presented and analysed in three different domains – the classic time and frequency domains (along with time-frequency one) and so-called statistical domain – a special notional (semantic) domain, where one is interested of specific signal parameters as mean level, peak level, dynamic range, probability density function, stationarity *etc.*

The signals of interest are produced by audio recording of real-world analog signals (*i.e.*, mathematically real and time limited), and digitalized with sampling frequency f_s, according to the Nyquist-Shannon-Kotelnikov sampling theorem.

2.1. Revealing of the signal behaviour into the time domain

For revealing of the signal behaviour into the time domain two types of signal representation are used – classic oscillogram and correlogram (a result diagram from autocorrelation analysis). The autocorrelation is calculated as [3]

$$R_{xx}[m] = \frac{1}{N}\sum_{n} x[n]\cdot x[n+m] \tag{1}$$

By means of correlogram one examines the given signal for randomness or independence into the time domain.

2.2. Revealing of the signal behaviour into the frequency domain

A lot more information could be drawn about the signal, using a representation into the frequency domain, especially concerning the signal "colour" *i.e.*, the spectral slope. In order to classify a signal as, for instance, pink or red noise, one must observe the PSD of the signal (since the signal is not deterministic). In this paper, the one-sided PSD is computed using the Welch's modified periodogram [4, 5]

$$\begin{gathered} x_l[m] = w[m]x[m+lH] \\ \dot{X}[k,l] = \frac{2}{f_s\sum_{m} w[m]^2}\left|\sum_{m=0}^{K-1} x_l[m]e^{-i2\pi\frac{mk}{K}}\right|^2 \\ \hat{\dot{X}}[k] = \frac{1}{L}\sum_{l}\dot{X}[k,l] \end{gathered} \tag{2}$$

where $m \in \{1,2,...,M\}$ is the local time index (*i.e.*, an index relative to the start of the signal frame), $M \in \mathbf{N}$ is the window length, $l \in \{0,1,...,L-1\}$ is the frame index, $L \in \mathbf{N}$ is the total number of frames; $H \in \mathbf{N}$ is the hop size (*i.e.*, the advance from one signal frame to the next), $k \in \{1,2,...,K\}$ is the frequency bin index and $K \in \mathbf{N}$ is the DFT size.

Different types of coloured noise signals are distinguished by the roll-off of their PSD. A few of them are listed in Tab. 1.

Another useful instrument for analysis and visualization take place into the time-frequency domain – the one-sided amplitude spectrogram, defined as [6]:

$$Spectrogram = \frac{2}{\sum_{m} w[m] \big/ M}\left|\dot{X}[k,l]\right| \tag{3}$$

where

$$\dot{X}[k,l] = \frac{1}{M}\sum_{m=0}^{K-1} x_l[m]e^{-i2\pi\frac{mk}{K}} \tag{4}$$

is the time-localized two-sided spectrum (*i.e.*, Short-Time Fourier Transform, STFT) of the analysed signal frame l. More about the applied STFT analysis could be found in [6].

Table 1. Information about some colored noise signals [2].

Noise type	PSD slope dependence	Spectral slope rate	Examples
Violet noise	$\propto f^2$	+6 dB/oct, +20 dB/dec	Signal from an acoustic thermal noise of water | Differentiation of a white noise signal
Blue noise	$\propto f$	+3 dB/oct, +10 dB/dec	Signal from a Cherenkov radiation process | High-frequency filtering of a white noise signal
White noise	flat	0 dB/oct, 0 dB/dec	Signal from a white noise process (*e.g.*, Johnson–Nyquist thermal noise)
Pink noise	$\propto 1/f$	-3 dB/oct, -10 dB/dec	Signal from statistical fluctuations of a number of natural processes | Low-frequency filtering of a white noise signal
Red noise	$\propto 1/f^2$	-6 dB/oct, -20 dB/dec	Signal from a Brownian motion (Winner process) | Integration of a white noise signal
Black noise	-	> -6 dB/oct, > -20 dB/dec	Model of the frequency of the natural disasters

2.3. Signal statistics

As far as the analysed signals are stochastic, their statistical properties are of great significance. One is interested in the magnitude statistics and histogram representation (including skewness and kurtosis values) of a given stochastic signal, as well as if it is stationary or not. The properties of interest are defined as in [4] and are shown in Table 2.

Table 2. Signal statistical parameters notations and definitions.

Notation	Name	Definition
X_0	Mean value	$\frac{1}{N}\sum_n x[n]$
X	Root-Mean-Square value	$\left(\frac{1}{N}\sum_n x[n]^2\right)^{\frac{1}{2}}$
S	Skewness	$\dfrac{\frac{1}{N}\sum_n (x[n]-X_0)^3}{\left(\frac{1}{N}\sum_n (x[n]-X_0)^2\right)^{3/2}}$
K	Kurtosis	$\dfrac{\frac{1}{N}\sum_n (x[n]-X_0)^4}{\left(\frac{1}{N}\sum_n (x[n]-X_0)^2\right)^{2}}$
CF	Crest-factor	$20\log_{10}\left(\dfrac{\max(\lvert x[n]\rvert)}{X}\right)$
DR	Dynamic range	$20\log_{10}\left(\dfrac{\max(\lvert x[n]\rvert)}{\min(\lvert x[n]\rvert)}\right)$

The histogram displays the number of samples there are in the signal that have each of these possible values, grouping together samples that have one and the same value and is used to reveal the probability density function (PDF) of the analysed signal and eventually to classified it as some of the know PDF-types (*e.g.*, normal distribution or Laplace distribution).

Whether the signal (*i.e.*, the time sequence) of interest is stationary or not (*e.g.*, has a unit root) could be checked by the Kwiatkowski–Phillips–Schmidt–Shin (KPSS) test [7] and the augmented

Dickey–Fuller test (ADF) [8]. The KPSS test is used for testing a null hypothesis that an observable time series is stationary around a deterministic trend (*i.e.*, trend-stationary) against the alternative of a unit root, while ADF tests the null hypothesis that a unit root is present in a time series sample. In the paper both tests are used for better results.

3. Signal examination results

An experimental examination has been made of over twenty natural sounds recorded with a linear PCM digital audio recorder. Nine of them are listed on Tab. 3 by their name and duration. Objects of consideration are primarily sounds caused by water dynamic movement – cavitation, boiling, flow, motion. For sake of comparison, two additional sounds have been analysed – fire of wooden-burning stove and sound caused by wind in the wood. All sounds are in .wav file format, sampled at $f_s = 44100$ Hz and 24-bit coded.

All recorded signals are analysed in Matlab® environment using a special-developed script for noise analyses, which provides signal visualization in the time, frequency and time-frequency domains via oscillogram, periodogram and spectrogram, respectively. Also it provides visualization of the signal correlogram and histogram along with rich signal statistics: positive and negative peak values, mean and RMS values, signal dynamic range, crest factor, skewness and kurtosis computation, and measurement of autocorrelation time.

At the beginning of the analysis every signal is detrended and normalized to unity peak amplitude. The PSD is computed using the Welch's modified periodogram with the Hamming window with length $M = \frac{1}{500}$ of the signal length, hop size $H = \frac{M}{4}$ and DFT size $K = 2M$. The spectrogram is computed using STFT with Hamming window with length $M = 1024$ samples and the same hop and DFT size as the PSD. Some of the statistical parameters of the examined signals are listed in Tab. 4.

Due to the restricted space, only two of the most informative plots are given below on Fig. 1 ÷ Fig. 18 – the periodogram and the histogram of the signals. The periodograms' frequency ranges are restricted to two decades – from 100 Hz to 10 kHz. For sake of comparison, the histograms (in blue colour) are accompanied with the appropriate superimposed fitted distribution line (in red colour).

Table 3. List of the explored signals.

Signal abr.	Signal full name	T, s
SF	sink_flow	10
BC	boiling _water_cavitation	10
BW	boiling_water	10
UF	urban_fountain	5
WF	waterfall	10
SS	sea_surf	5
RN	rain	10
WL	winter_in_the_leafs	10
WS	wood-burning_stove	10

Table 4. Some statistical parameters of the explored signals.

Parameter	**SF**	**BC**	**BW**	**UF**	**WF**	**SS**	**RN**	**WL**	**WS**
CF, dB	13	13	16	19	18	20	23	15	14
DR, dB	107	76	77	112	80	87	125	125	125
S	0.005	0.006	-0.020	0.008	0.004	-0.140	-0.016	0.007	-0.057
K	3.00	3.08	3.56	3.96	3.22	6.65	4.52	3.33	3.33
Autocorrelation time, s	0.0027	0.0127	0.0045	0.0351	0.0014	3.1372	0.0038	3.2208	1.5433

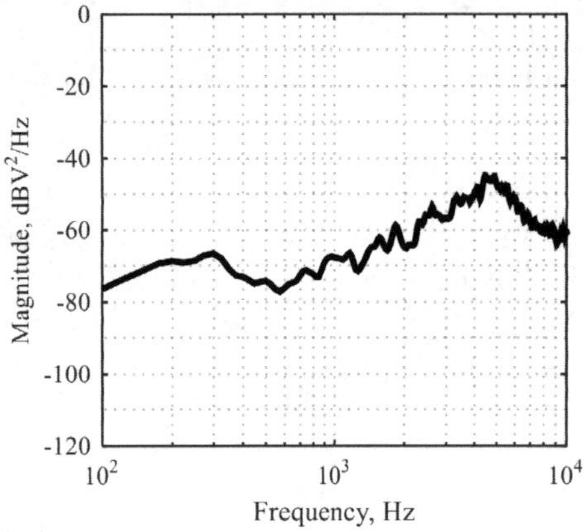

Figure 1. PSD-periodogram of the signal SF.

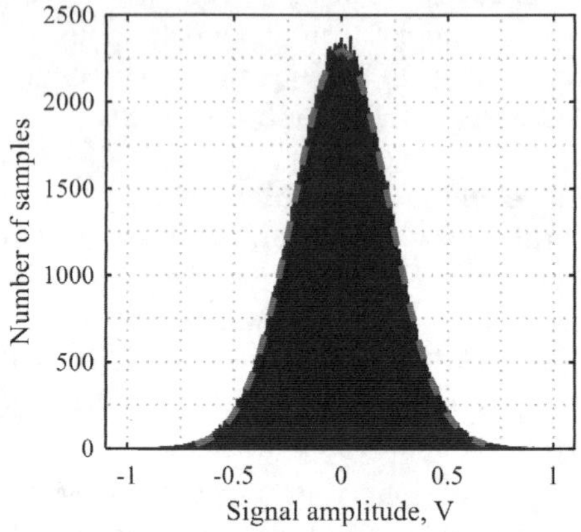

Figure 2. Histogram (blue) and the fitted normal distribution (red) of the signal SF.

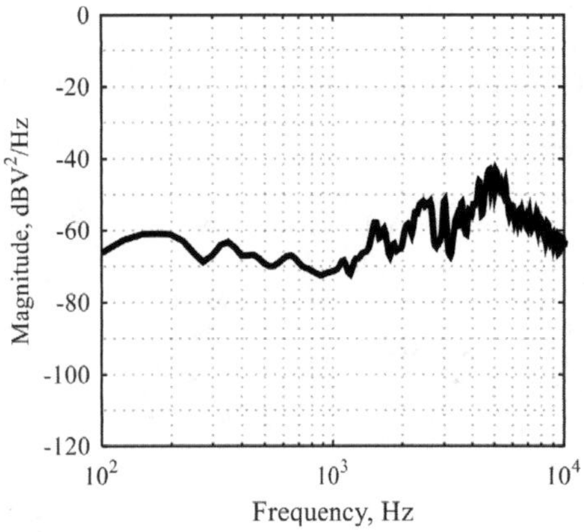

Figure 3. PSD-periodogram of the signal BC.

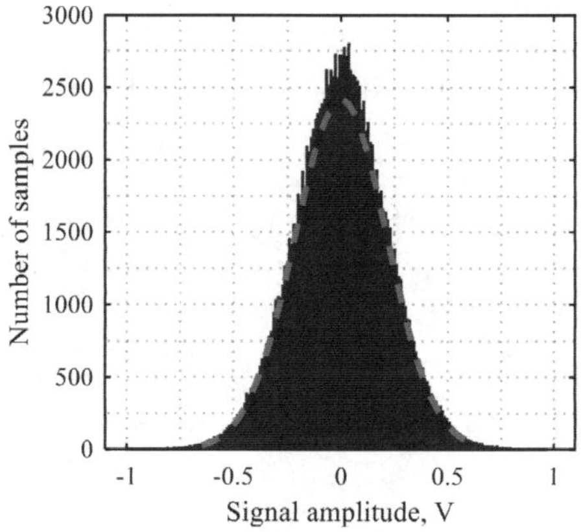

Figure 4. Histogram (blue) and the fitted normal distribution (red) of the signal BC.

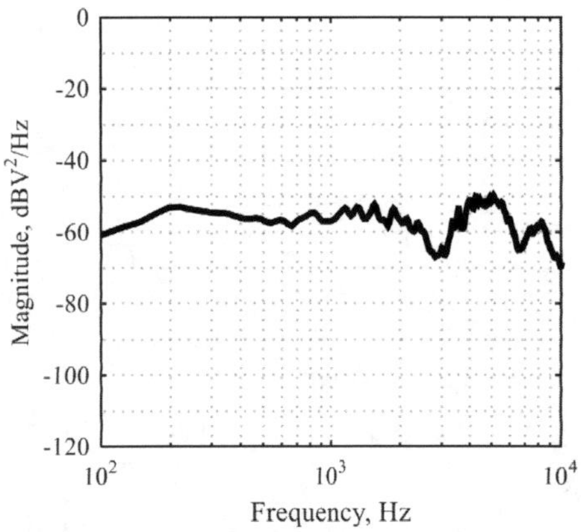

Figure 5. PSD-periodogram of the signal BW.

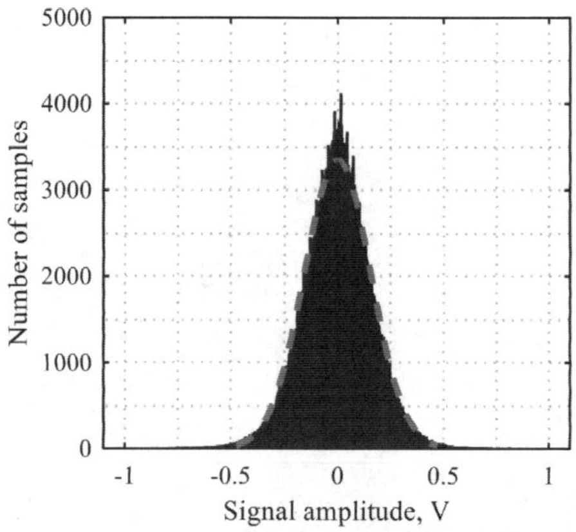

Figure 6. Histogram (blue) and the fitted normal distribution (red) of the signal BW.

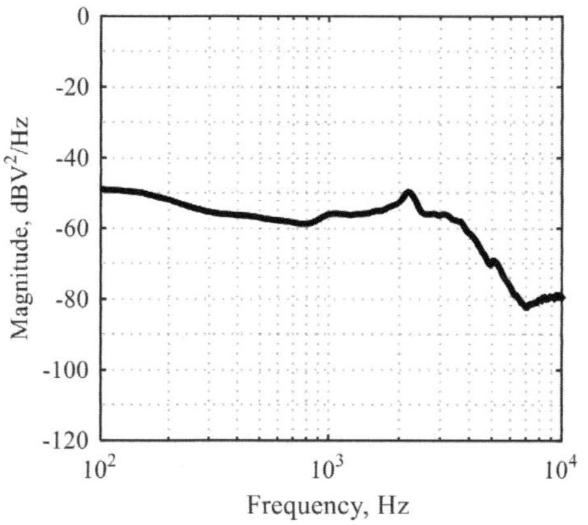

Figure 7. PSD-periodogram of the signal UF.

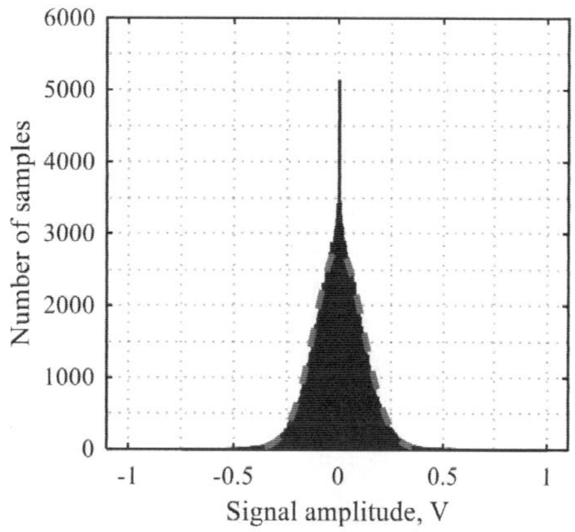

Figure 8. Histogram (blue) and the fitted normal distribution (red) of the signal UF.

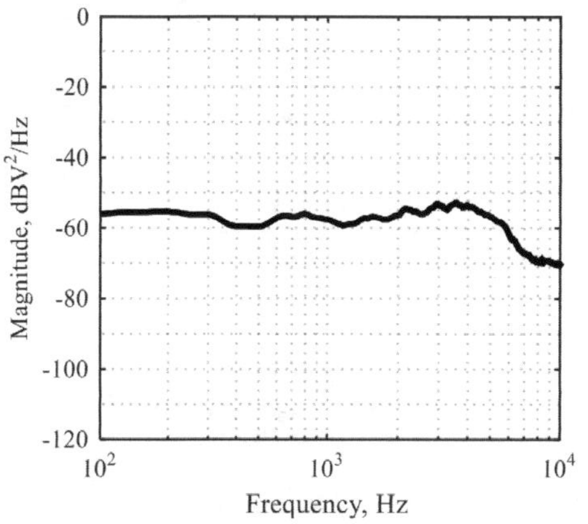

Figure 9. PSD-periodogram of the signal WF.

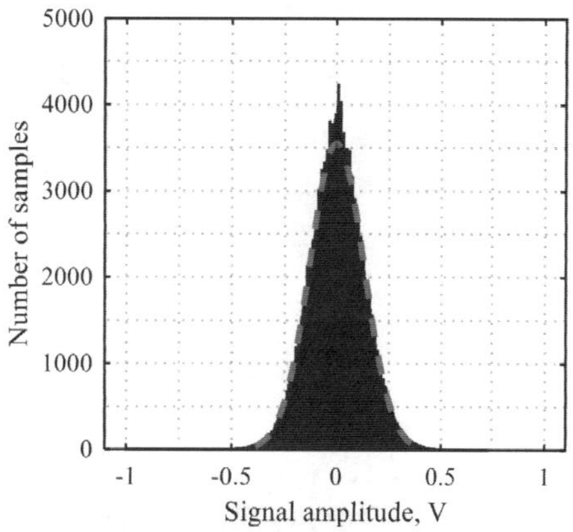

Figure 10. Histogram (blue) and the fitted normal distribution (red) of the signal WF.

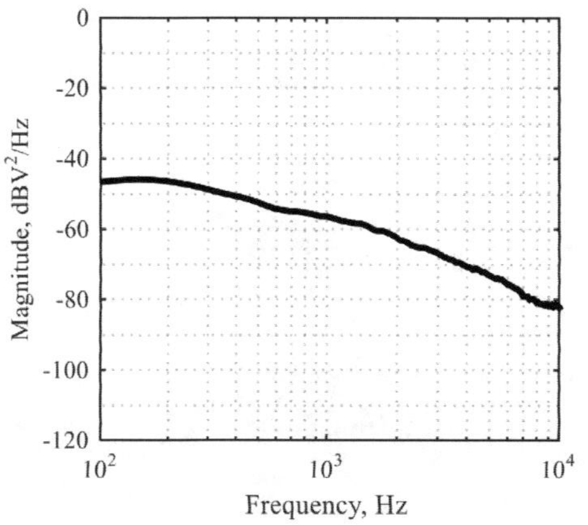

Figure 11. PSD-periodogram of the signal SS.

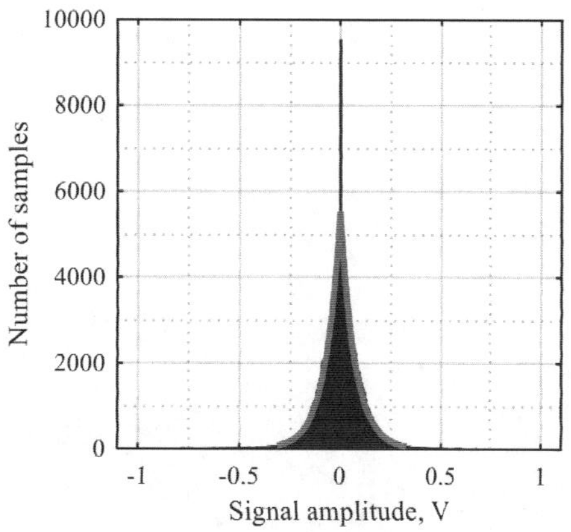

Figure 12. Histogram (blue) and the fitted Laplace distribution (red) of the signal SS.

IOP Conf. Series: Materials Science and Engineering **595** (2019) 012054 doi:10.1088/1757-899X/595/1/012054

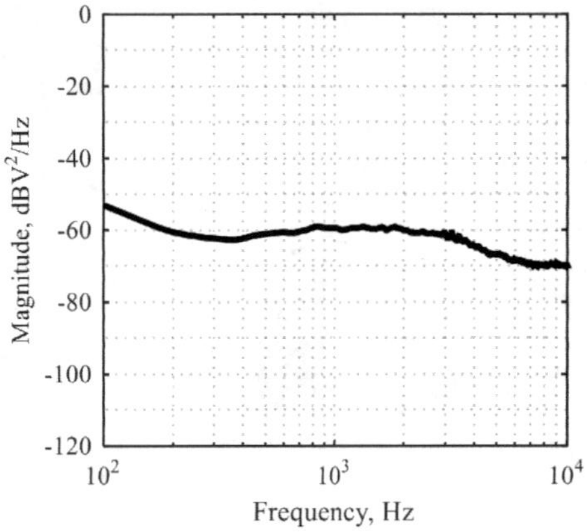

Figure 13. PSD-periodogram of the signal RN.

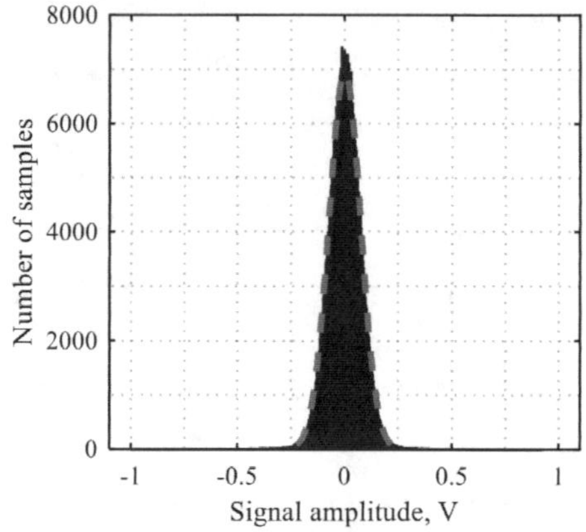

Figure 14. Histogram (blue) and the fitted normal distribution (red) of the signal RN.

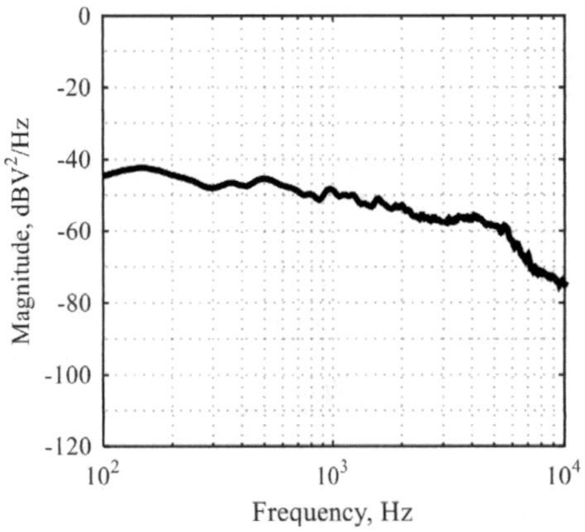

Figure 15. PSD-periodogram of the signal WL.

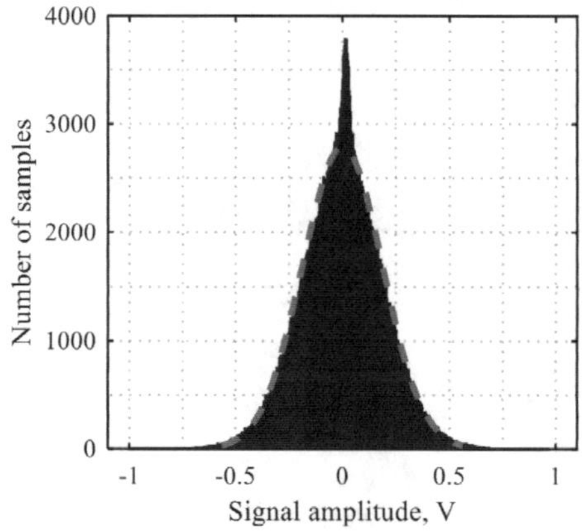

Figure 16. Histogram (blue) and the fitted normal distribution (red) of the signal WL.

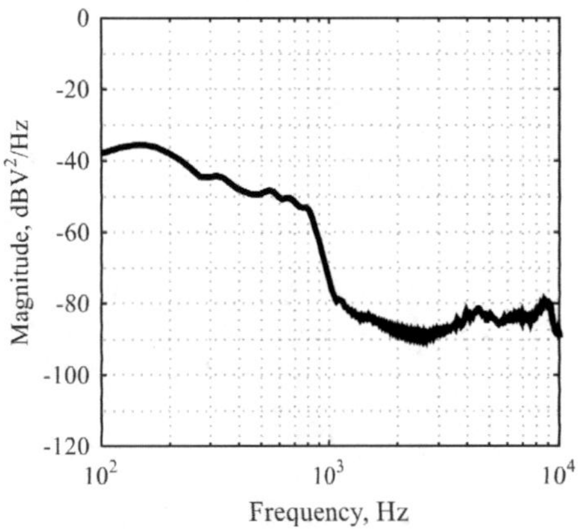

Figure 17. PSD-periodogram of the signal WS.

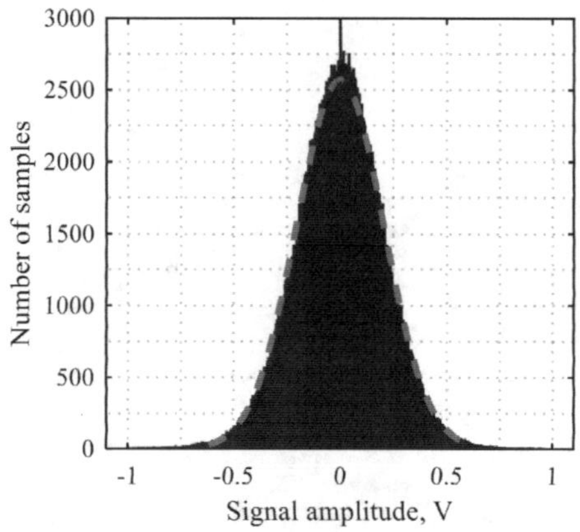

Figure 18. Histogram (blue) and the fitted normal distribution (red) of the signal WS.

The following notes can be taken on the results:

- All signal segments are stationary, according to KPSS and ADF tests, which is also confirmed by the signals' spectrograms;
- The crest factor of the signals varies between 13 dB and 23 dB, while most of them are between 13 dB and 20 dB (approx. 5 to 10 times the RMS value). The signals with the biggest *CF* are the *sea_surf* and the *rain* signals. They also have the biggest kurtosis (*i.e.*, peakedness) of all signals, which could be observed on their histograms, too.
- The dynamic range of the signals varies between 75 dB and 125 dB. Some of the purely nature signals have the biggest *DR – rein, winter_in_the_leafs, and wood-burning_stove*. One of the possible explanations is their "discrete time" pattern – they could be presented as a set of (partially) non-overlapping micro-events.
- The skewness of all signals is close to zero, which means that all signals are symmetric and centred. This could be observed on their histograms, too.
- The kurtosis of the most signals is close to 3 (one of them is perfectly Gaussian – *sinc_flow*), which is an evident for normal (Gaussian) distribution of the most signals, which could be observed on their histograms, too. The bold exceptions are the *rein* and *sea_surf* signals. The last one obtains Laplace distribution of the signal amplitudes. Both of them have well pronounced peakedness.
- All signals obtain no autocorrelation, except the *sea_surf* and *winter_in_the_leafs* signals. These signals are long-term correlated, since in their time-domain pattern a kind of periodicity could be found (so-called "pulsations").
- Regarding the PSDs of the signals:
  - the PSDs of the SF and BC signals are quite similar. Probably this is due to the similar nature of the underlying process – dynamic movement of air-bubbles in water environment;
  - the PSD of the BW signal is somewhat uniform, with resonances at ~3 kHz and ~6 kHz;
  -the PSD of the UF is alike a pink noise (roll-off of –10 dB/dec) at the first frequency decade (100 Hz – 1000 Hz) and then band-limited to about 4 kHz;
  - the PSDs of the WF and RN signals are comparable with the white noise process, which is expected for this type of noise signals;
  - the PSD of the SS signal is like pink noise at the first decade and alike red noise (roll-off of –20 dB/dec) at the second decade (1000 Hz – 10000 Hz), respectively;
  - the PSD of the WL signal is pink up to ~5 kHz, and then band-limited;
  - the PSD of the WS signal approximates red noise up to about 800 Hz and then is band-limited, probably by the volume of the stove, *i.e.* the noise is spatially low-pass filtered.

4. Conclusions

In the presented paper several types of signals, representing the movement of the aquatic environment, are reviewed and some preliminary results from their analysis are showed. Summaries of their nature were made, confirming the preliminary thesis: (i) the majority of signals have a normal (Gaussian) or close to normal probability distribution of the amplitudes; (ii) almost all signals have band-limited PSDs of "coloured" type – mostly like the white, pink and red noises. In addition, some signals obtained very similar PSDs, for instance SF-BC, WF-RN and SS-WL couples, although their different origin.

The obtained *a priori* information assists the further signal processing, for other purposes and applications. This study is a first step from further work on characterization of other noise signals – mainly in the underwater and underground environments. Those results can be used in the processing of signals from measurements in the aquatic environment, as well as in the development of some simulation models.

References

[1] Kasdin N 1995 Discrete simulation of colored noise and stochastic processes and $1/f^{\alpha}$ power law noise generation *Proc. of the IEEE* 5 **83** 802-27
[2] Zhivomirov H 2018 A method for colored noise generation *RJAV* 1 **15** 14-9

[3] Smith J 2008 *Mathematics of the Discrete Fourier Transform (DFT): with Audio Applications* (W3K Publishing) chapter 8 p 189

[4] Brandt A 2011 *Noise and Vibration Analysis: Signal Analysis and Experimental Procedures* (Chichester: John Wiley & Sons) chapter 10 p 211

[5] Smith J 2011 *Spectral Audio Signal Processing* (W3K Publishing) chapter 6 p 248

[6] Zhivomirov H 2019 On the development of STFT-analysis and ISTFT-synthesis routines and their practical implementation *TEM J.* **1 8** 56-64

[7] Kwiatkowski D, Phillips P, Schmidt P and Shin Y 1992 Testing the null hypothesis of stationarity against the alternative of a unit root *Journal of Econometrics* 1-3 **54** pp 159-178

[8] Dickey D and Fuller W 1979 Distribution of the estimators for autoregressive time series with a unit root *J. Am. Stat. Assoc.* **366(74)** pp 427-431

Acknowledgments

ЕВРОПЕЙСКИ СЪЮЗ
ЕВРОПЕЙСКИ ФОНД ЗА РЕГИОНАЛНО РАЗВИТИЕ

This work was supported by the European Regional Development Fund within the OP "Science and Education for Smart Growth 2014 - 2020", Project CoC "Smart Mechatronic, Eco- And Energy Saving Systems And Technologies", № BG05M2OP001-1.002-0023

ОПЕРАТИВНА ПРОГРАМА
НАУКА И ОБРАЗОВАНИЕ ЗА ИНТЕЛИГЕНТЕН РАСТЕЖ

IOP Conf. Series: Materials Science and Engineering **595** (2019) 012055 doi:10.1088/1757-899X/595/1/012055

Generating structural, interactional and processual order through synergetic entanglement is the essence for understanding and using Quantum Biological Thermodynamics with Finite Speed of the Cardio-Pulmonary System

S Petrescu[1], M Costea[1], V Petrescu[1], R Bolohan[2], B Borcila[1] and M Feidt[3]

[1] Department of Engineering Thermodynamics, University Politehnica of Bucharest, Bucharest, Romania
[2] Emergency Center for Cardiovascular Diseases "Dr. Constantin Zamfir", Clinic of Cardiology, Electrophysiology and Arrythmology, Bucharest, Romania
[3] LEMTA, URA CNRS 7563, University of Lorraine, Vandoeuvre-lès-Nancy, France

E-mail: liana5802@yahoo.fr

Abstract. Recently, we invented a new branch of irreversible processes by studying the interaction between Heart and Lungs through thousands of measurements of the heart frequency, lung frequency and oxygen content in the blood for many people of different age. Based on these measurements we discovered a surprisingly simple but powerful formula correlating the two frequencies in *any stationary state* through a *quantum number N*, that has an integer value for healthy people, but non integer for ill persons. Actually, the formula states that the interaction between Heart and Lungs, considered as two macroscopic oscillators, is quantized, which means that it is very much ordered or highly organized. We believe that this discovery has no connection with *Quantum Mechanics,* that describes the behavior of *microscopic systems*. Thus, the aim of the paper is to explain and understand why the *quantization* of the *Cardio-Pulmonary Interaction* occurs in *macroscopic biological system.* We present two diagrams invented by us for describing the quantized stationary states and processes with or without quantum jump between different stationary states depending on person activity. Based on the above mentioned formula and these diagrams, we show that our discovery is in accordance with the new branch of science, Synergetics.

1. Introduction

Recently, we extended the *Thermodynamics with Finite Speed applied to Thermal Machines* [1-13] to "Biological Machines", thereby inventing and building a new branch of irreversible processes [10, 14 - 24] that studies the *interaction between Heart and Lungs*, and called by us *Quantum Biological Thermodynamics with Finite Speed of the Cardio-Pulmonary System* (QBTFSCS).

After doing thousands of measurements of the *Heart frequency* F_H, *Lung Frequency* F_L and *oxygen content* in the blood for many children, young, adult and old persons, we have discovered a very simple correlation of the two frequencies in any *stationary state,* as a function of a *quantum number N*:

IOP Conf. Series: Materials Science and Engineering **595** (2019) 012055 doi:10.1088/1757-899X/595/1/012055

$$\frac{F_H}{F_L} = 2 + \frac{N}{4} \tag{1}$$

where N is an *integer number* for healthy people, but not an integer for ill persons (with heart or lung problems).

The fact that this formula is very simple actually shows that the interaction between Heart and Lungs, viewed *as two macroscopic oscillators is very well organized, and* because of that *is quantized, similarly to the interaction between the electron and proton in the hydrogen atom.* Despite of this similarity, we do believe that this discovery has nothing to do with *Quantum Mechanics,* which describes the behaviour of *microscopic systems*.

Similar research regarding the cardio-pulmonary components functioning and interaction are found in the literature. Some of them correlate the shift in heart rate variability with postures and walking by time–frequency analysis [25]. The reported results based upon eighteen subjects performing four sequential 5-min physical activities (supine, sitting, standing and spontaneous walking) showed heartbeat frequency fluctuations when changing the posture, and that the ratio of low- to high-frequency fluctuation during standing was significantly higher than during supine and sitting.

Another study [26] aimed to detect and compare the changes of heart rate and respiratory frequency during high altitude mountain ascent. Results showed that this activity differently affected the heart rate and the respiration frequency variability, concluding that the respiration frequency had more randomness than the heart rate and altitude increased this difference.

Important results confirming our discovery are provided by the analysis of a set of nonlinear dynamic models for the interactive respiration/cardiovascular mechanism by employing equivalent electric circuits [27]. The authors have found that for a specified activity mode (e.g., sleeping, walking or resting), the heart beat rate can be assumed to be proportional to the breath frequency through a value of constant that is determined by the mode of human body activity.

The present paper will try to explain and argue why the *quantization of the Cardio-Pulmonary Interaction* appears in *macroscopic biological systems*.

The diagrams that we have invented [10, 14 - 24] and used here to describe the *quantized stationary states* and processes with or without *quantum jump between different stationary states* corresponding to different posture of the person emphasize the correlation between the heart and lungs frequencies and the associated quantum number.

The explanation regarding the *quantized macroscopic interaction* between Heart and Lungs is presented in section 3 based on the idea of the *synergetic entanglement of the two oscillators* and also, by using experimental measurements regarding the oscillation periods of these two subsystems.

Conclusions and perspectives are discussed in the last section.

2. Diagrams for describing *stationary states* and *processes* with and without *quantum jump* in the cardio-pulmonary system

Inspired from *Thermodynamics with Finite Speed* [1-13] we invented 5 diagrams for *describing Stationary States* and *Processes with and without quantum jump* between them, in the Cardio-Pulmonary System which are used now in QBTFSCPS, in a similar way like in Thermodynamics of Thermal Machines, in order to describe cyclic diurnal processes of persons of different ages (child, young, adult, old).

In this paper we present only 2 diagrams (Figures 1 and 2), which are the most important, in which we can represent *stationary states* (0, 1, 2, 3,…13, 14, 15, 16) and *processes between stationary states* (0-1, 1-2, 2-3, 3-4,…12-13, 13-14, 14-15, 15-16) *with or without quantum jump*, and the quantum number corresponding to each state.

In these figures starting from state 0, one can read the values of the frequency ratio $R_f = F_H/F_L$ and the quantum integer number N for each stationary state, as function of F_L, in one diagram (figure 1) and as function of F_H in the second diagram (Figure 2), for different postures of a person (SP).

Each of the two diagrams has its importance given by the distribution of the stationary states and processes on the frequency variation range. There is a clear separation of the stationary states on the frequency variation range of each biological machine, and the frequency values F_H and F_L increase progressively with the posture changing. Regarding the processes between stationary state, the diagrams reveal iso - rhythm (1-2, 2-3, ..., 12-13) or iso-pulse processes (6-7).

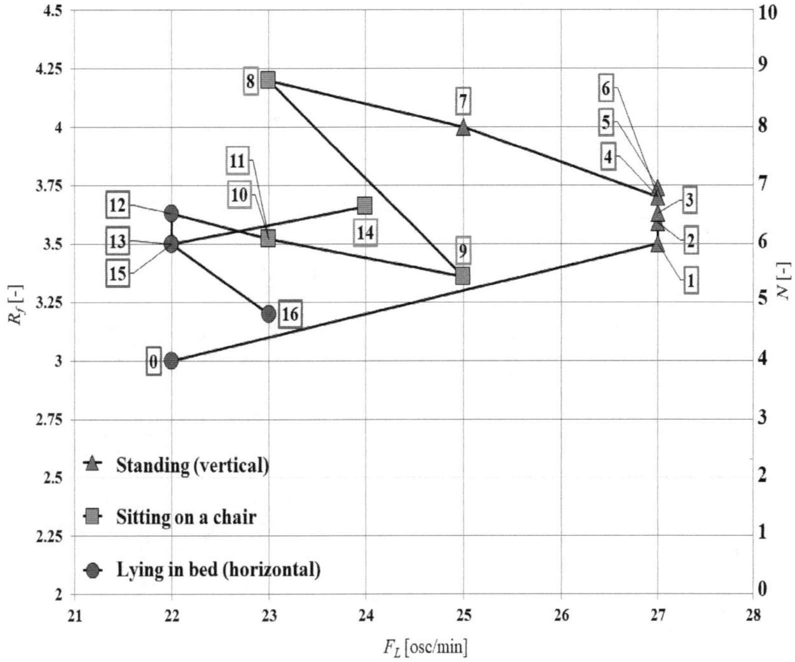

Figure 1. Diagrams R_f, $N = \mathrm{f}(F_L)$ (for SP).

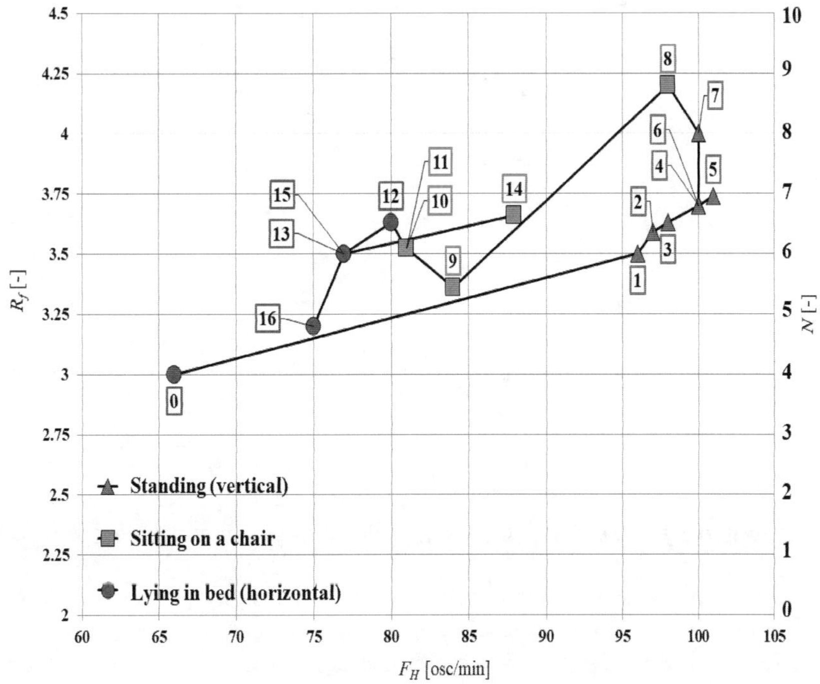

Figure 2. Diagrams R_f, $N = \mathrm{f}(F_H)$ (for SP).

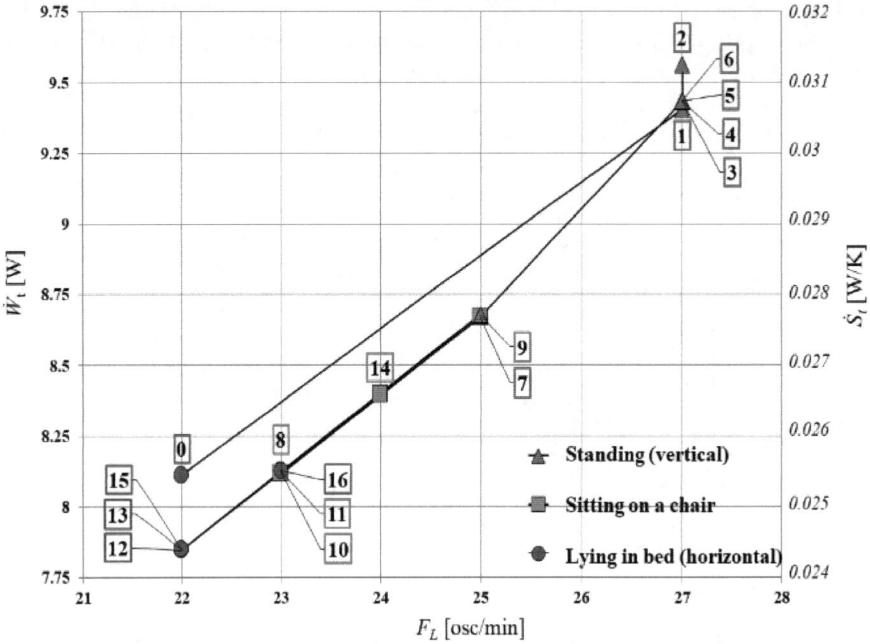

Figure 3. Power $\dot{W}_t = f(F_L)$ and total entropy source $\dot{S}_t$ (for SP).

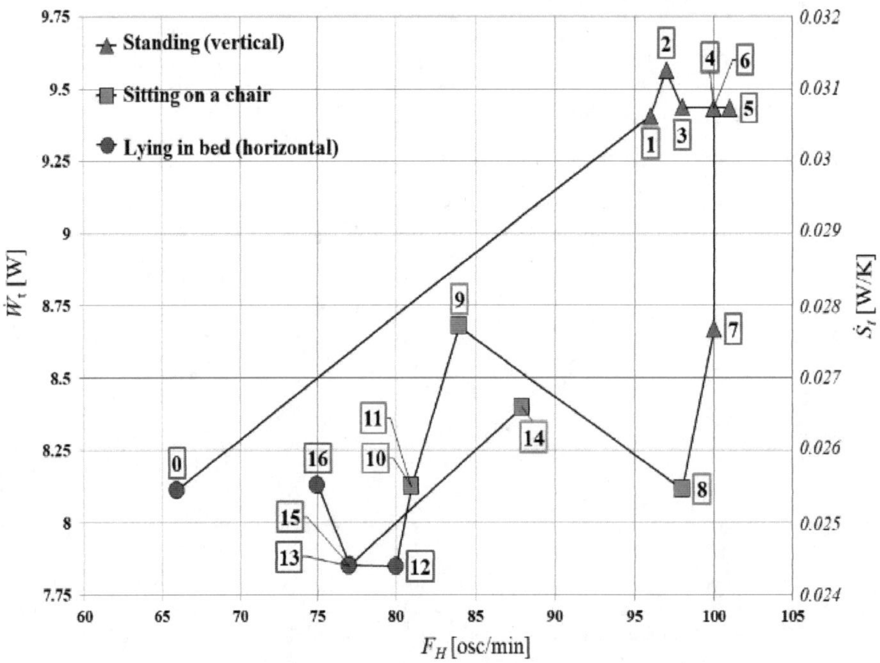

Figure 4. Power $\dot{W}_t = f(F_L)$ and total entropy source $\dot{S}_t$ (for SP).

Based on these diagrams we were capable to compute the variation of the p*ower consumed by heart and lungs,* $\dot{W}_t$, and the t*otal source of entropy* generated in each *stationary state,* $\dot{S}_t$, (Figures 3 and 4).

Note that the same separation previously mentioned appears here in a more obvious way, being question of the increasing demand in energy (power) consumption and consequently, in entropy generated.

IOP Conf. Series: Materials Science and Engineering **595** (2019) 012055 doi:10.1088/1757-899X/595/1/012055

3. The essence for Understanding and using Quantum Biological Thermodynamics with Finite Speed of the Cardio-Pulmonary System Order is through synergetic entanglement

The explanation regarding the *quantized macroscopic interaction* between Heart and Lungs is based on the idea that *the essence of generating structural, interactional and processes order* between these two oscillator subsystems (Heart and Lung) is an effect of *synergetic entanglement* discovered by Hacken [28] when trying to explain how the lasers works.

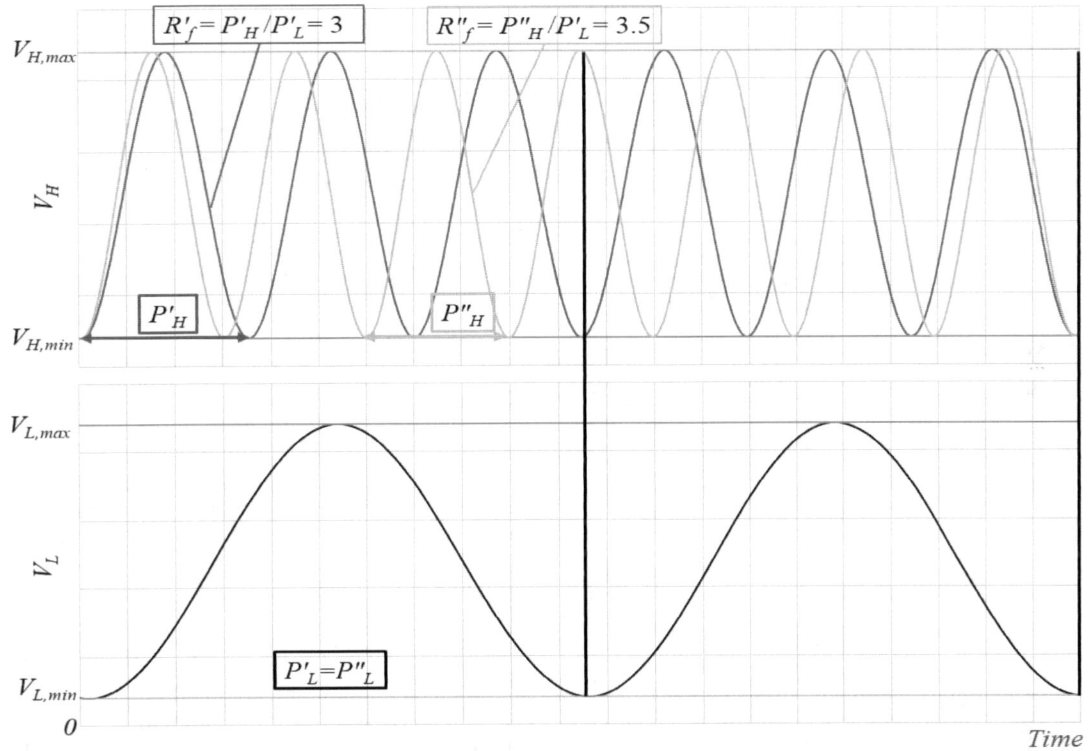

Figure 5. Intuitive explanation of the ordered correlations between oscillations of Heart and Lungs for processes with $R_f = 3$ and $R_f = 3.5$ (for SP).

Based on that discovery he created a new branch of science called Synergetics. In Synergetics he has shown that *between two oscillators either microscopic or macroscopic, an interactional relation can appear which introduce a* ***degree of order***, expressed quantitatively by certain relations between the frequencies of the two oscillators.

An *intuitive explanation of the ordered correlation between the frequency of heart and frequency of lungs oscillation* is presented in figures 5 and 6.

In figure 5 we show the correlation between lungs and heart periods by considering one oscillation of the Lungs for different variation of the volume from V_{min} to V_{max} and back to V_{min} in two stationary states, corresponding to R_f = integer: 3, and to R_f = not integer: 3.5.

Figure 6 illustrates the similar correlation for another person (BB), corresponding to an integer R_f =: 6, and to a not integer $R_f = 5.5$.

One can see that when the number of the *complete oscillations* of the Heart corresponds to only *one oscillation* of the Lungs, *the ratio between the Periods* of these oscillations ($P_L/P_H = F_H/F_L$) will correspond to integer numbers of $R_f = F_H/F_L$.

The explanation of the *not-integer numbers* of R_f, but increasing with 0.5 from an integer number consists in the fact that *a complete oscillation of the Lungs* corresponds (for these cases) to not a complete oscillation, but to *an integer number of oscillations plus a half fraction of the last oscillation* of the Heart.

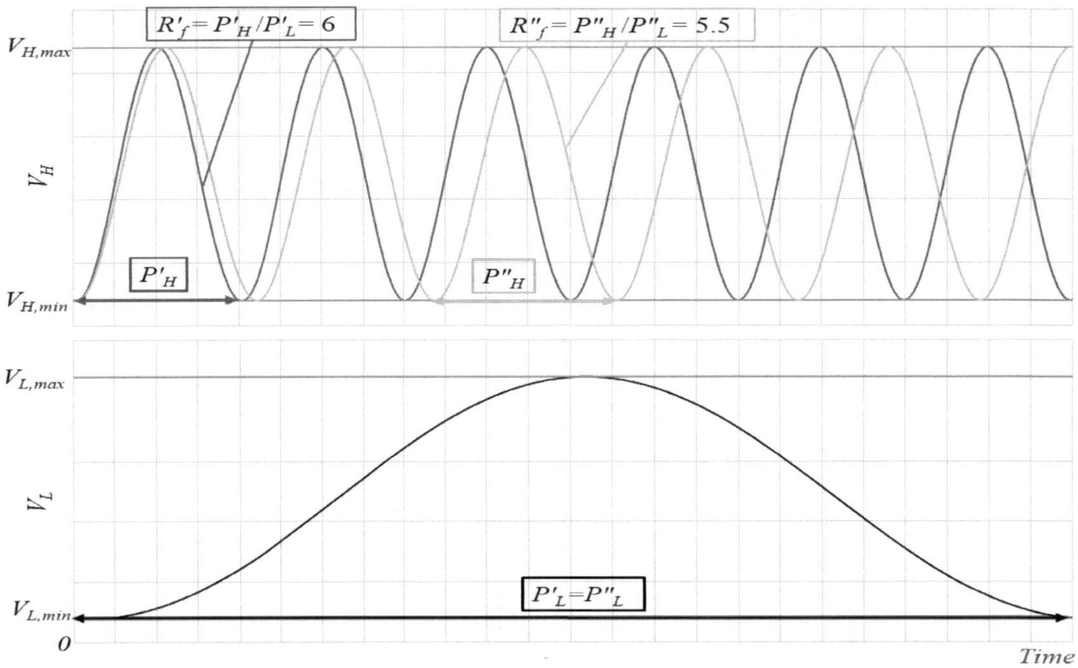

Figure 6. Intuitive explanation of the ordered correlations between oscillations of Heart and Lungs for processes with $R_f = 5.5$ and $R_f = 6$ (for BB).

Thus, in both figures, to one oscillation of the lung corresponds either an integer number (3 for SP and 6 for BB), either a fractional number (3.5 for SP and 5.5 for BB) of heart oscillations.

It is quite obvious that in both cases, despite of the difference in the frequency ratio values R_f and in corresponding values of the quantum number N, that are 4 and 6 respectively (for SP, 79 years old), and 14 and 16 respectively (for BB, 29 years old), the organized interaction of oscillations is similar.

4. Conclusions and perspectives

The discovery of the *quantized formula* that correlates the frequency of the heart - F_H with frequency of the lungs - F_L through a quantum number, which is an integer in *stationary states* of healthy people was the beginning of inventing of a new branch of Irreversible Thermodynamics called by us: *Quantum Biological Thermodynamics with Finite Speed of the Cardio-Pulmonary System* (QBTFSCPS), as an extension of *Thermodynamics with Finite Speed* applied to Thermal Machines.

We have shown that generating structural, interactional and processual order through synergetic entanglement is the essence for understanding and using *Quantum Biological Thermodynamics with Finite Speed of the Cardio-Pulmonary System.*

Two of the five new diagrams that we have invented, where the *stationary states* and *processes between stationary states with* or *without quantum jump* can be studied, are presented. They emphasize the correlation of the heart/lungs frequency ratio with the quantum number in different stationary states that are changing according to the person's posture. Furthermore, a thermodynamic approach of the two biological machines allowed to compute the *power consumed by heart and lungs* and the *total source of entropy* generated in each *stationary state.*

These diagrams drawn by persons in the period of life when they are healthy can help the designers to build quite soon personalized artificial hearts and lungs with a correlated functioning.

Our discovery and the new branch called by us *Quantum Biological Thermodynamics with Finite Speed of the Cardio-Pulmonary System* (QBTFSCS) provide a thermodynamic and synergetic approach of the two components of an important biological system. We intend to extend it in the near future to a triple interaction: heart, lungs and brain.

References

[1] Stoicescu L and Petrescu S 1964 The First Law of Thermodynamics for Processes with Finite Speed in Closed Systems *Sci. Bull. UPB Bucharest* **26(5)** 87-108

[2] Stoicescu L and Petrescu S 1964 Thermodynamic Processes Developing with Constant Finite Speed *Sci. Bull. UPB Bucharest* **26(6)** 79-119

[3] Stoicescu L and Petrescu S 1965 Thermodynamic Processes with Variable Finite Speed *Sci. Bull. UPB Bucharest* **27(1)** 65-96

[4] Stoicescu L and Petrescu S 1965 Thermodynamic Cycles with Finite Speed *Sci. Bull. UPB Bucharest.* **27(2)** 82-95

[5] Stoicescu L and Petrescu S 1965 The Experimental Verification of the New Expression of the First Law for Thermodynamic Processes with Finite Speed *Sci. Bull. UPB Bucharest.* **27(2)** 97-106

[6] Petrescu S 1969 Contribution to the study of thermodynamically non-equilibrium interactions and processes in thermal machines *PhD Thesis* University Politehnica of Bucharest Romania

[7] 1991 *New Sources of Energy* (Helsinki: Helsinki University of Technology)

[8] Petrescu S et al 2006 *Advanced Energy Conversion* – Vol I – II (Lewisburg: Bucknell University

[9] Petrescu S, Costea M, et al 2011 *Development of Thermodynamics with Finite Speed and Direct Method* (Bucharest: AGIR Publishing House)

[10] Petrescu S et al 2015 *Advanced Thermodynamics of Irreversible Processes with Finite Speed and Finite Dimensions*, (Bucharest: AGIR Publishing House)

[11] Petrescu S et al 1993 A Comparison between Optimization of Thermal Machines and Fuel Cells based on New Expression of the First Law of Thermodynamics for Processes with Finite Speed *Proc 1st Int. Thermal Energy Congress ITEC'93* (Marrakech)

[12] Petrescu S et al 2002 Application of the Direct Method to Irreversible Stirling Cycles with Finite Speed *Int. J. Energy Research* **26** 589-609

[13] Petrescu S et al 2013 Effect of irreversibilities estimated with the direct method on the Stirling engine performance, *Proc. Rom. Acad. Tech. Sci Conf. Brașov* (Bucharest: AGIR Publishing House)

[14] Petrescu S et al 2014 Means for Qualitative and Quantitative Description of the Cardio-Pulmonary System Operation within Irreversible Thermodynamics with Finite Speed *Proc. Rom. Acad. Tech. Sci Conf. Brașov* (Bucharest: AGIR Publishing House)

[15] Petrescu S et al 2014 Discovery of Quantum Numbers in the Cardio-Pulmonary Interaction Studied in Thermodynamics with Finite Speed *Proc. Rom. Acad. Tech. Sci Conf. Brașov* (Bucharest: AGIR Publishing House)

[16] Petrescu S et al 2015 From Thermodynamics with Finite Speed towards Biological Quantum Thermodynamics with Finite Speed *Proc. Nat. Conf. on Thermodynamics NACOT'15* Iasi

[17] Petrescu S et al 2015 Stationary Quantum States in Cardio-Pulmonary System, *Proc. Rom. Acad. Tech. Sci Conf.* Galati

[18] Petrescu S et al 2015 Processes with Quantum Jumps in the Cardio-Pulmonary System *Proc. Rom. Acad. Tech. Sci Conf.* Galati

[19] Petrescu S et al 2016 Diagrams Describing Stationary States and Processes in Cardio- Pulmonary System *Proc. Rom. Acad. Tech. Sci Conf.* Targu-Mures

[20] Petrescu S et al 2016 Complete Circadian Cycle of Cardio-Pulmonary System Studied in Quantum Biological Thermodynamics with Finite Speed, *Proc. Rom. Acad. Tech. Sci Conf.* Targu-Mures

[21] Petrescu S et al 2016 Quantum Biological Thermodynamics with Finite Speed of the cardio-respiratory system as a new extension of Thermodynamics with Finite Speed *Proc. Francophone Coll. on Energy and Thermodynamics COFRET 2016* Bucharest

[22] Petrescu S et al 2016 How different we are in terms of Heart-Lung interaction: discovery and description in Quantum Biological Thermodynamics with Finite Speed *Proc. Francophone Coll. on Energy and Thermodynamics COFRET 2016* Bucharest

[23] Petrescu S et al 2018 What is quantum biological thermodynamics with finite speed of the cardio-pulmonary system: a discovery or an invention *Proc. Romanian Acad. Series A – Math. Physics Tech. Sci. Information Science.* **19** 249-254

[24] Borcila B et al 2018 Effect of flue on Stationary States and Processes in the Cardio-Pulmonary System described in Quantum Biological Thermodynamics with Finite Speed *IOP Conference Series: Materials Science and Engineering.* **444(8)** 082023

[25] Chan H L et al 2007 Correlates of the shift in heart rate variability with postures and walking by time–frequency analysis *Computer Methods and Programs in Biomedicine.* **86(2)** 124-130

[26] Wesfreid E and Billat V 2012 Randomness and changes of heart rate and respiratory frequency during high altitude mountain ascent without acclimatization *Physica A: Statistical Mechanics and its Applications.* **391(4)** 1575-1590

[27] Tsai N C and Lee R M 2011 Interaction between cardiovascular system and respiration *Appl. Mathem. Modelling* **35** 5460–5469

[28] Haken H 1983 *Advanced Synergetics, Instability Hierarchies of Self-Organizing Systems and Devices* (Berlin Heidelberg New York Tokyo: Springer Verlag)

Investigation of the effect of drying air temperature on drying time by numerical method for textile bobbins

U Akyol[1], A Karakoca[1], R Shaliyev[2], D Akal[3] and K Uzuneanu[4]

[1] Mechanical Engineering Department, "Tekirdag Namik Kemal University", Çorlu, Tekirdag, Turkey
[2] Civil Engineering Department, "Tekirdag Namik Kemal University", Çorlu, Tekirdag, Turkey
[3] Mechanical Engineering Department, "Trakya University", Edirne, Turkey
[4] Thermal Systems and Environmental Engineering Department, "Dunarea de Jos" University of Galati, Galati, Romania

E-mail: uakyol@nku.edu.tr

Abstract. Drying of the yarn bobbins in the textile sector is a very important stage among the textile finishing processes. Because this process takes a lot of time and causes a large amount of energy consumption. In this study, drying times for different drying air inlet temperatures at a constant drying air pressure were determined with the help of mathematical model for textile bobbin drying process. Finite difference method was used in mathematical model solution and the effect of different space and time steps were taken into consideration during the solution.

1. Introduction

The yarn bobbins must be dried after the dyeing process in the textile industry. It is important to perform this process in a short time with minimum energy consumption in terms of efficiency. For this reason, mathematical models have been emphasized for many years for drying processes. It is significant to develop drying methods for designing the drying process to preserve the quality of the products and shorten the drying time with minimum energy consumption [1]. These models generally focus on simultaneous solutions of heat and mass transfer equations. Ribierio and Ventura (1995) studied on an experimental setup where the drying process was carried out by passing hot air both through the inside and outside of the yarn bobbins [2]. In a study by Lee, et al. (2002), a two-dimensional mathematical model was developed for drying of dyed carpet yarns. Hot air was passed parallel to the axis through porous cylindrical media and heat and mass transfer were calculated considering the energy and mass conservation of the environment [3]. Smith and Farid (2004), in their experimental work, they have obtained correlations that allow the drying times of materials to be determined, taking into account the moving boundary theory for cylindrical geometries [4]. In a study by Hussain and Dinçer (2003), a two-dimensional numerical analysis of heat and moisture transfer during the drying of a cylindrical object was performed using the finite difference approach. Temperature and humidity distributions were obtained in humid bodies for different time periods [5]. Barati and Esfahani (2011) considered heat and mass transfer by conduction and convection in order to model the drying process and developed an analytical solution for the mathematical model they created [6]. In the present work, drying times for different drying air inlet temperatures at a constant drying air pressure were determined with the help of mathematical model for textile bobbin drying process.

2. Materials and methods

The wool yarn bobbins were dried with hot air in the test setup. It passed through the yarn bobbins and the bobbins were dried from only the inside to the outside due to the pressure difference. The volumetric flow rate of the hot air is 500 m^3/h while the temperature and the pressure are 80°C and 1 bar (effective) respectively. The yarn bobbins used in the experiments have an inner diameter of 35 mm and a length of 150 mm.

The wrapped yarn is made up of holes made from polyethylene material that allow air to pass over it. During the tests, 7 thermocouples were placed in the bobbin at even intervals to measure the temperature of the yarn bobbin section (Fig. 1). Akyol et al. (2013) have shown that the temperature change during the length of the bobbin is not important in the experimental study [7]. Therefore, in this study, it is considered that the heat transfer in the bobbin is performed in one dimension in the radial direction.

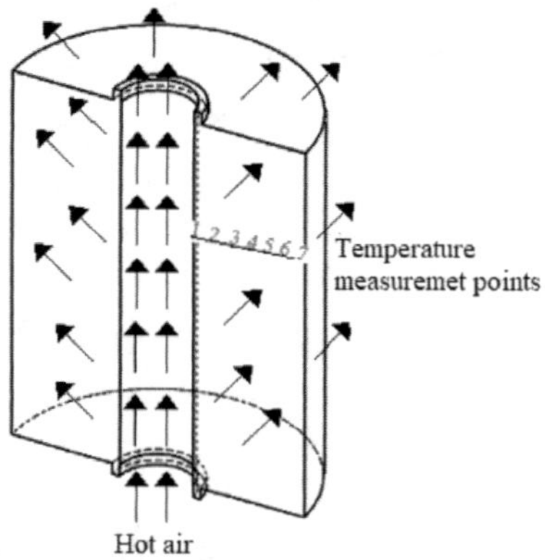

Figure 1. Schematic view of the yarn bobbin used in the experiments and temperature measurement points.

When the mathematical model is written, it is thought that the heat transfer mechanism is realized as follows: Air mass, moving at a specific temperature under a certain pressure has thermal (internal) and kinetic energies. Thermal and kinetic energies are absorbed from the inner surface through the interstices between the yarns by the bulk fluid movement into the bobbin and enter the bobbin volume through spaces between the yarns. Energy can also pass through surfaces by molecular processes. This can happen in two ways: Conduction and mass diffusivity. However, during the passage of the fluid through the material, work is done by pressure and friction forces. Hot air passing through spaces between yarns, collides with yarns and water molecules and transfers some of the energy to them. During the movement, by taking up the evaporating water molecules, the increasing amount of moisture continues to move towards the outer surface while warmed water molecules also undergo forced diffusion [8]. As a result, the mathematical model is expressed as follows:

$$C_{ve}\frac{\partial T}{\partial t} = \frac{1}{r}\frac{\partial}{\partial r}\left(k_e r\frac{\partial T}{\partial r}\right) + P_e(r,T)\frac{\partial T}{\partial r} \quad (1)$$

where: *t, T, r,* C_{ve}, k_e are respectively: time, temperature, radial coordinate, effective volumetric heat capacity and effective thermal conductivity. $P_e(r, T)$ is the function of temperature and coordinate and can be expressed as:

$$P_e(r,T) = (c_{ph}\rho_h + c_{pb}\rho_b)V_a \quad (2)$$

where c_{ph}, c_{pb}, ρ_h, ρ_b, and V_a are the specific heat capacity of the dry air and water vapour, density of the dry air and water vapour, velocity of the fluid in the porous body, i.e., in the yarn bobbin respectively. The mathematical model that expresses the physical process is in the form of Eq. (1). The direct problem is solved by the finite difference method in the initial and boundary conditions as follows:

$$r_{in} < r < r_{out}; \quad 0 < \tau < \tau_m$$

$$T(r,0) = T_i; \quad T(r_{in}) = f_1(\tau); \quad T(r_{out}) = f_2(\tau)$$

3. Results and discussion

Firstly, six different situations (Case 1, 2, 3, 4, 5 and 6) have been considered to examine the effect of time and space step on study. First, the model is solved by determining constant time and constant space steps. The time step was chosen as τ=300 s and the space step h=0.00950 m (Case 1). First, to examine the effect of the space step, the space step is reduced to half value at all points and then the model is solved (Case 2). Afterwards, for the effect of the time step, the time step has been reduced to half its value throughout the entire process (Case 3). Finally, the model is solved in various time and space steps to study the combined effect of these two parameters (Case 4, 5, 6). The test duration is 12900 s for all cases.

Table 1. The time and space step values considered in the study for cases 1, 2 and 3.

Case	Time step τ, (s)	Space step h, (m)
1	300	0.00950
2	300	0.00475
3	150	0.00950

Table 2. The time and space step values considered in this study for different time and space step ranges in Case 4.

Time step τ, (s)	Space step h, (m)
100 (between 0-200 s)	0.00475 (between r=0.033-0.0425 m)
200 (between 200-600 s) 300 (between 600-12900 s)	0.00950 (between r=0.0425-0.090 m)

Table 3. The time and space step values considered in this study for different time and space step ranges in Case 5.

Time step τ, (s)	Space step h, (m)
100 (between 0-200 s)	0.00475 (between r=0.033-0.052 m)
200 (between 200-600 s) 300 (between 600-12900 s)	0.00950 (between r=0.052-0.090 m)

Table 4. The time and space step values considered in this study for different time and space step ranges in Case 6.

Time step τ, (s)	Space step h, (m)
100 (between 0-200 s)	0.00475 (between r=0.033-0.052 m)
200 (between 200-600 s) 300 (between 600-4200 s) 100 (between 4200-4500 s) 300 (between 4500-5700 s)	0.00950 (between r=0.052-0.090 m)

100 (between 5700-7200 s)
300 (between 7200-12900 s)

The examined cases are given in Tables 1-4. The convective term P_e, which includes the convection effect of the air, is the function of the temperature. After determining P_e values, three different temperature regions are considered. These regions are as follows: Before evaporation, evaporation, and after evaporation. The P_e values used in this study were taken as 130 W/m^{2}K before the evaporation region and as 66W/m^{2}K in the evaporation region and as 503 W/m^{2}K after the evaporation region. In addition, the effective volumetric heat capacity (C_{ve}) including phase conversion factor and the effective thermal conductivity (k_e) including both mass transport and thermal conductivity are shown in Figure 23. These two thermophysical properties were taken from [1].

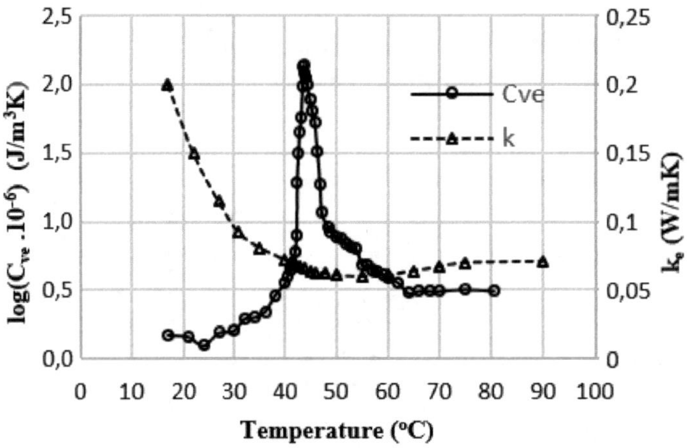

Figure 2. Variation of the effective volumetric heat capacity and the effective thermal conductivity with temperature for wool yarn bobbins (Akyol, 2007).

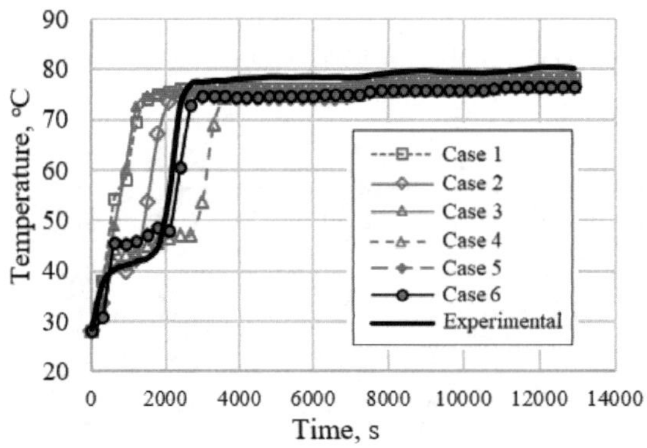

Figure 3. Comparison of experimental results (80°C) and model results (Case 1-6) for temperature measurement point $r_2 = 0.0425$ m

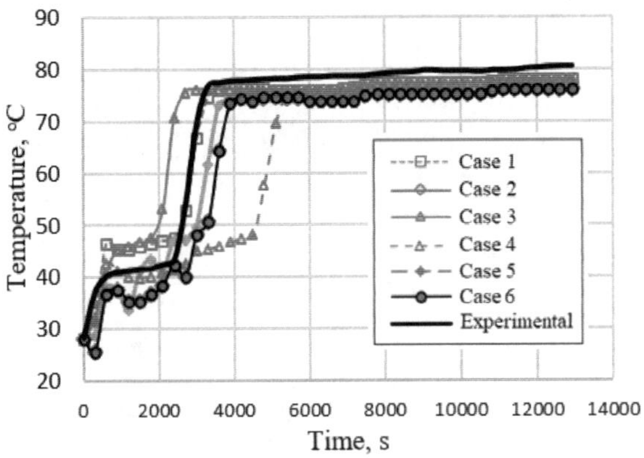

Figure 4. Comparison of experimental results (80°C) and model results (Case 1-6) for temperature measurement point $r_3 = 0.0520$ m

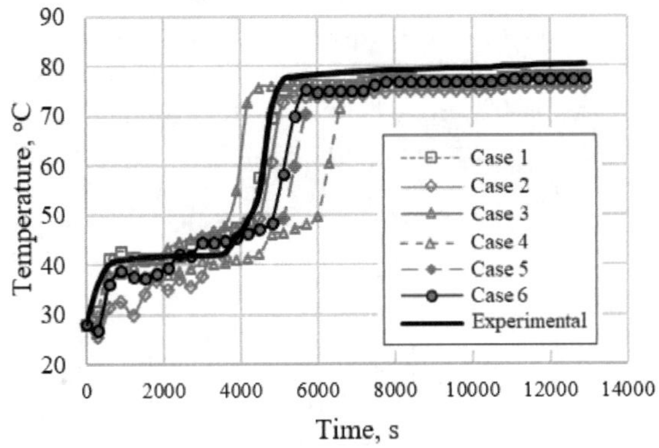

Figure 5. Comparison of experimental results (80°C) and model results (Case 1-6) for temperature measurement point $r_4 = 0.0615$ m

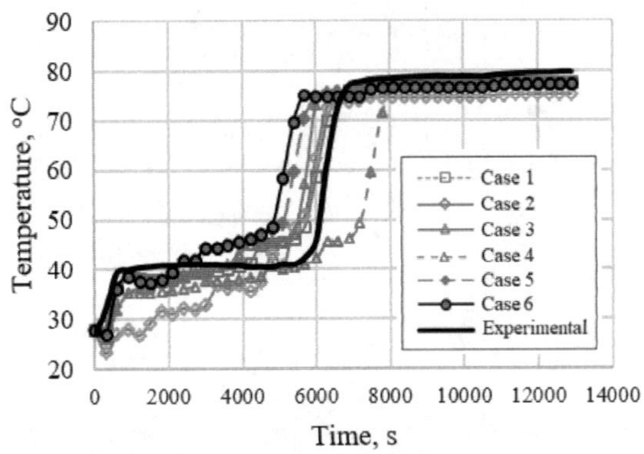

Figure 6. Comparison of experimental results (80°C) and model results (Case 1-6) for temperature measurement point $r_5 = 0.0710$ m

IOP Conf. Series: Materials Science and Engineering **595** (2019) 012056 doi:10.1088/1757-899X/595/1/012056

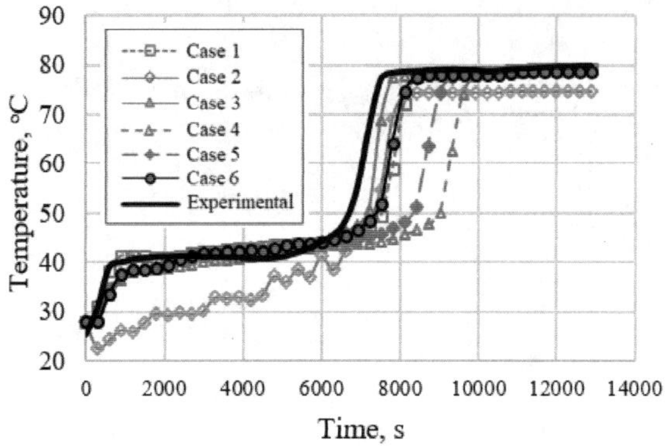

Figure 7. Comparison of experimental results (80°C) and model results (Case 1-6) for temperature measurement point $r_6 = 0.0805$ m

Figure 3-7 shows the comparison of six various cases (Case 1-6) for different temperature measurement points inside the bobbin with the experimental results. For all cases the inlet air temperature (T=80°C) and the pressure of the drying air (P=1 bar) are constant. When these graphs are examined, it is seen that the case, which provides the best fit with the test results is Case 6. The model results show that it is not a good way to select the time steps at small values throughout the entire process. When the temperature field is examined, it is seen that the biggest faults are near the inner surface where the temperature changes sharply at the beginning of the process and at the inlet and outlet regions to the evaporation temperatures. Case 4, Case 5 and Case 6 (different values of h and τ) were solved by the finite difference method and the results were compared with the experimental results to minimize the errors caused by the numerical solution method. The best result was obtained for Case 6. Therefore, considering the effect of space and time steps in the study, the drying time is estimated for the drying air inlet temperatures of 90°C and 100°C by using the conditions in Case 6 by the help of the mathematical model.

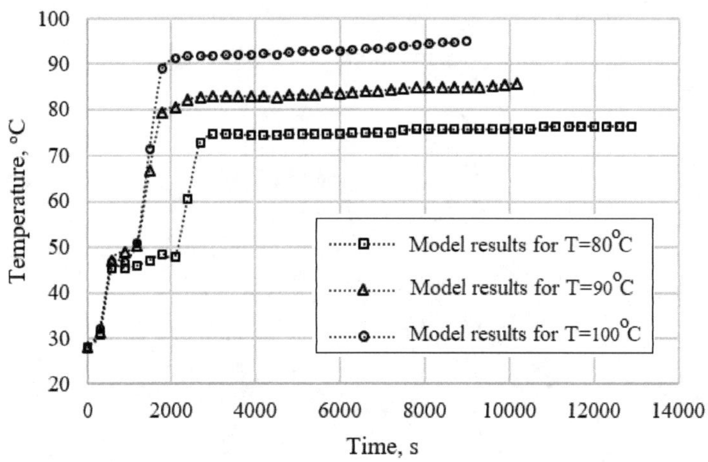

Figure 8. Comparison of model results (T=80°C, 90°C and 100°C) for temperature measurement point at $r_2 = 0.0425$ m

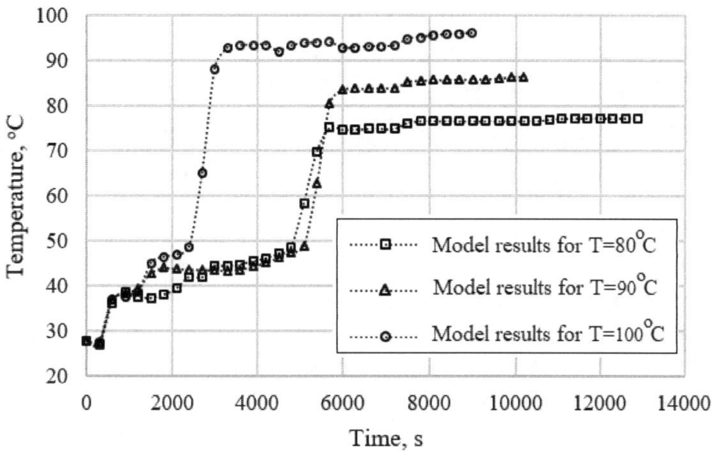

Figure 9. Comparison of model results (T=80°C, 90°C and 100°C) for temperature measurement point at $r_4 = 0.0615$ m.

Figure 8-10 show the variation of the temperature distribution versus time at three different temperature measurement points inside the yarn bobbin for three different inlet air temperatures. As shown in these figures, the yarn bobbin enters a period of phase transformation which takes place at an approximately constant temperature at which the evaporation takes place intensively after a period of warming. After a point in the yarn bobbin has completely dried, the temperature of this point begins to rise rapidly. The drying time of a specific point in the yarn bobbin can be determined from the mathematical model curves by taking this into account.

In Figure 8, when drying at 90°C and 100°C for the 2nd temperature measuring point ($r_2 = 0.0425$ m), these points appear to dry at almost the same time (1200 s). However, this period at 80°C is 2100 s according to the model results. Similarly, when Figure 9 is examined for the 4th temperature measurement point ($r_4 = 0.0615$ m), the drying time of this point for 100 C is 2400 s, while for 80°C and 90°C it is approximately 5100 s. The model results for the temperature measurement point 6 ($r_6 = 0.0805$ m) are given in Figure 10. As seen from the figure, the drying time for 100°C is approximately 3600 s, while drying time is approximately 6900 s for 90°C and 7500 s for 80°C.

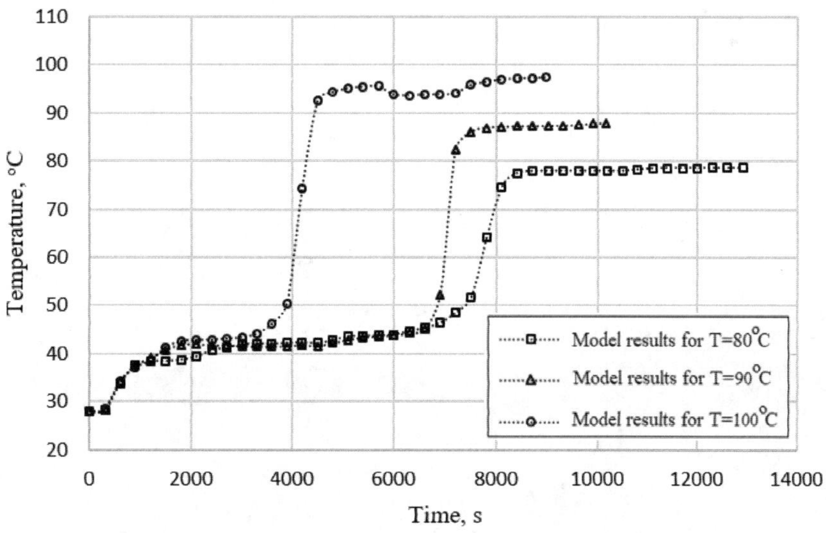

Figure 10. Comparison of model results (T=80°C, 90°C and 100°C) for temperature measurement point at $r_6 = 0.0805$ m.

IOP Conf. Series: Materials Science and Engineering **595** (2019) 012056 doi:10.1088/1757-899X/595/1/012056

4. Conclusions

As can be seen from the results, the drying times at temperatures of 80°C and 90°C are closer to each other at the temperature measurement points away from the internal points. In this case, according to the mathematical model results it can be said that drying the yarn bobbins at 80°C instead of 90°C may be more economical in terms of energy consumption. However, the drying process at 100°C temperature can be preferred by taking into account the energy consumed due to the shortness of the time.

References

[1] Akyol U 2007 İplik bobininin kurutulmasının teorik incelenmesi *PhD* Trakya University Edirne Turkey (in Turkish)

[2] Ribeiro J and Ventura J M P 1995 Evaluation of textile bobbins drying process, *Drying Technology.* **13** pp 239-65

[3] Lee H S Carr W W Beckham H W and Leisen J 2002, A Model of through air drying of tufted textile materials, *International Journal of Heat and Mass Transfer.* **45** pp 357-66

[4] Smith M C and Farid M 2004 A Single correlation for the prediction of dehydration time in drying and frying of samples having different geometry and size *Journal of Food Engineering.* **63** pp 265-71

[5] Hussain M M and Dinçer İ 2003 Two-dimensional heat and moisture transfer analysis of a cylindrical moist object subjected to drying: A finite-difference approach *International Journal of Heat and Mass Transfer.* **46** pp 4033-39

[6] Barati E and Esfahani J A 2011 Mathematical modeling of convective drying: Lumped temperature and spatially distributed moisture in slab. *Energy.* **36** pp 2294-2301

[7] Akyol U Kahveci K and Cihan A 2013 Determination of Optimum Operating Conditions and Simulation of Drying in A Textile Drying Process *The Journal of The Textile Institute* **104** pp 170-77

[8] Akyol U Shaliyev R Cihan A and Kahveci K 2015 Kurutma işleminde iplik bobinleri içerisindeki sıcaklık dağılımının matematiksel model yardımıyla belirlenmesi *20. Ulusal Isı Bilimi ve Tekniği Kongresi, Balıkesir* (in Turkish) pp 114-18

IOP Conf. Series: Materials Science and Engineering **595** (2019) 012057 doi:10.1088/1757-899X/595/1/012057

Kinetics coefficients of solid-liquid extraction of resinoid from tobacco leaves

S Tasheva[1], V Popova[2], T Ivanova[2], A Stoyanova[2] and G Valtchev[1]

[1] Department of Heat Engineering, University of Food Technologies, Plovdiv, Bulgaria
[2] Department of Tobacco, Sugar, Vegetable and Essential Oils, University of Food Technologies, Plovdiv, Bulgaria

E-mail: vpopova2000@abv.bg

Abstract. Cured tobacco leaves (*Nicotiana tabacum* L.) from three commercial types – flue cured Virginia, Burley and Oriental, were processed by extraction to obtain the concentrated aromatic product resinoid, which is used in cosmetics and other areas. The objective of the study was to characterize the kinetics of the process of tobacco resinoid extraction by two basic indices – the coefficient of internal molecular diffusion (D_{int}) and the coefficient of mass transfer (β). For that purpose, extraction was carried out at different temperatures (in the range between 20°C and 70°C), and the process was divided into a series of consecutive 10-minute intervals, for which the respective values of the kinetics coefficients were calculated. Based on the established variations of D_{int} by tobacco type, time and temperature, the respective regression models were derived. The coefficients of mass transfer, as well as the kinetic indices of the extraction process – the Biot diffusion number (Bi) and the Fourier diffusion number (Fo), were calculated for the three tobacco types. Data from the applied calculation procedures are presented in the study.

1. Introduction

The main technological processes in plant material processing – adsorption, absorption, rectification, drying, and extraction – include an interphase mass transition process, which kinetic characteristics need to be specified. In particular, these characteristics comprise process velocity, also called mass transfer velocity, as it determines the transfer velocity of soluble substances from the boundary phases. The velocity of mass transfer processes is often constrained by the molecular diffusion, and in these cases it is referred to as diffusion velocity. For given interphase technological processes related to the convective transfer and molecular diffusion phenomena – including the deriving of extractible substances from plant materials, common kinetics equations can be used. The extractible substances are very diverse by chemical composition or component proportions, depending on their nature (hydrophobic, hydrophilic) and the microstructure of the plant tissues [1-4].

The extraction of aroma-related substances from essential oil bearing plants is a process which provides the obtaining of products that are considered closer in terms of composition and odour to the essential oil in the initial plant material [5-7].

In the solid-liquid extraction, a process of mass transfer from the solid body into the liquid (the solvent) occurs, which unites two separate stages of soluble substances' transition. The first stage involves the extraction of soluble from the solid matrix, and therefore it is described by diffusion that runs within the solid body (e.g. from the centre to the surface of the plant particles). The driving force in this process is the concentration gradient of soluble substances inside the particle and on its surface, as well as the diffusion resistance. The diffusion resistance, in its turn, depends primarily on the structure and the chemical composition of cell walls and protoplasm, the nature of the solvent, and the temperature. A generalized expression of the diffusive properties of an extracted material is the coefficient of internal molecular diffusion. The coefficient takes different values for individual solid bodies (plant particles), and is determined mostly experimentally due to the non-uniformity and complexity of the solid matrices [1-3, 8]. The second stage in the solid-liquid extraction represents a mass transfer process occurring on the solid-liquid boundary, which is characterized by the coefficient of mass transfer. The coefficient is described by the differential equations of motion, and depends on the hydrodynamic conditions for liquids, characterizing the transfer of substances both in the convective and the diffusion fluxes [1, 3, 8].

Generally, the mass transfer process in the extraction of plant materials is described by a model, analogous to Fick's first law of molecular diffusion, in which the proportionality factor is denoted as the coefficient of mass transfer. Essentially, it represents a coefficient of internal diffusion, expressed in the same units as the coefficient of thermal conductivity or the coefficient of molecular diffusion, and is calculated by experimental data [1, 2, 4].

The scientific literature related to the subject provides reference values of the coefficient of internal molecular diffusion of the aromatic product resinoid obtained by extraction from various essential oil bearing plants [3, 9-11].

There are some data about the coefficient of effective diffusion for tobacco and other plant materials [12-14], extracted with ethanol and water, together with calculated values of the constants A, B and H, which are used to determine the concentration of the components being extracted into the liquid phase. In one of these studies, Simeonov and Chilev [13] studied the factors influencing the solid-liquid extraction, and determined directly the diffusion coefficient, and indirectly – the mass transfer coefficient. The mass transfer coefficient did not influence the concentration changes in the solid phase, but depended on the coefficients of internal and external diffusion.

The mechanism and kinetics of the extraction of glyceride oil from tobacco seeds have been investigated by Stanisavljevic et al. [15]. The solvents used in the study were hexane and petroleum ether, the seed-to-solvent ratios were 1:3; 1:5 and 1:10, and the temperature range of extraction included 25°C, 40°C and boiling temperature. The physical and chemical properties of the extracted tobacco seed oils were also characterized.

Tobacco leaves representing various types or commercial grades contain diverse aroma- or biologically active compounds [16], and have been processed to obtain different aromatic products – e.g. essential oil, absolute, concrete, resinoid, and other extracts – used mainly in cosmetics, perfumery or pharmacy. The dominant part of the studies, however, is devoted to the identification of the chemical composition and the activities of the derived products, as well as to the optimization of the extraction procedure (in terms of yield, properties, composition, etc. of the respective phytoproducts). There are few studies that investigated the mechanism of the extraction process and presented data about the coefficients of diffusion and of mass transfer for liquid or concentrated aromatic products, including tobacco resinoids [9-11, 13].

A recent study [17] provides data from calculating the coefficient of internal molecular diffusion and the coefficient of mass transfer, as well as the factors B and μ, for the aromatic product concrete obtained by extraction from flue-cured Virginia, Oriental and Burley tobaccos.

The objective of the study was to characterize the kinetics of the process of tobacco resinoid extraction by calculating two basic indices – the coefficient of internal molecular diffusion and the coefficient of mass transfer.

2. Materials and methods

The modelling of the coefficient of internal molecular diffusion (D_{int}) of the aromatic product resinoid in current study was based on the experimental data published by [9-11]. The resinoids in the reference studies were obtained by laboratory extraction with 95% ethanol of cured leaves of three types of Bulgarian tobacco – flue-cured Virginia, Burley and Oriental. The individual values of D_{int} were calculated for each of the 10-minute intervals of an extraction process with a total duration of 60 min, carried out at temperatures of 20; 30 and 40 °C.

The calculation procedure for the coefficient of mass transfer was adapted from [1, 18] and involved the following steps:

The diffusion coefficient (D, m^2·s^{-1}) was calculated according to the equation:

$$D = 2.83 \cdot 10^{-6} e^{-\frac{2710}{T}} \quad (1)$$

where: T is the temperature, K.

The Fourier diffusion number (Fo_d) was calculated as:

$$Fo_d = \frac{D.\tau}{R^2} \quad (2)$$

where: D is the diffusion coefficient, m^2.s^{-1}; R – solid phase particle size, approximated to an adjusted radius of a cylinder ($R = 5.64.10^{-3}$ m); τ – duration, s.

A series of evenly increasing values of the Biot diffusion number (Bi) was generated, and for each of the values the ratio C_F/C_S was calculated according to the equation:

$$\frac{C_F}{C_S} = B \cdot e^{-\mu^2 Fo} \quad (3)$$

where: C_S and C_F were the initial and the final concentration of the substance (i.e. tobacco resinoid) in the solid, %; μ – a root in the characteristic equation (3), which is a function of the Biot number and was read from reference tables [18]; B – a constant, which is also a function of the Biot number and was calculated according to the equation:

$$B = \frac{2Bi^2}{(\mu^2 + Bi^2 + Bi)\mu^2} \quad (4)$$

The Biot number (Bi) is determined by the mass transfer coefficient, depending on the hydrodynamic conditions related to system porosity, and in particular – to the cell structure of the solid matrix. It is considered a decisive kinetic parameter for the degree of an extraction process including mass transfer through the solid body and the boundary layer into the core of the liquid flux. The value of the Biot number defines the diffusion mode during the extraction [13, 14].

From the plotted function: $\frac{C_F}{C_S} = f(Bi)$, the value of the corresponding Biot number $(Bi)_{exp}$ was taken, for each of the $(C_F/C_S)_{exp}$ ratios obtained from the experiments. Then, the coefficient of mass transfer on the solid-liquid boundary (β) was calculated according to equation (5), in which the respective values of $(Bi)_{exp}$ were substituted:

$$\beta = \frac{Bi.D}{R} \quad (5)$$

Statistical analysis and data processing was done with MicroCal™ Origin 9.1 software.

3. Results and discussion

3.1. Coefficient of internal molecular diffusion of tobacco resinoid

The approximated models expressing the change of the coefficient of internal molecular diffusion (Dint) by temperature and duration of extraction are presented on figures 1÷3, respectively for the Virginia flue-cured, Burley and Oriental types of tobacco.

As it has been stated above, the diffusion coefficient depends both on the structure and the physical properties of the extracted plant material and the solvent used, as well as on the concentration of the extractible substances and the temperature. The increase of extraction temperature creates conditions for accelerated transfer of the extracted substances from the solid into the liquid phase, which in its turn results in increased values of the coefficient of diffusion. The latter varies during the extraction process, due to a situation in which plant tissues undergo physical and chemical transformations altering their permeability [1-3, 8]. As seen from the presented graphical dependencies, the variations of the diffusion coefficient have a different pattern for the three tobacco types, which can be explained by the known differences in cell structure and composition, mostly between the two bright tobaccos (flue-cured Virginia and sun-cured Oriental) and the darker air-cured Burley tobacco [16]. The graphical dependencies had similar profiles in the case of Virginia and Oriental tobaccos, but differed by individual values. The line trends displayed on figure 1 and figure 3 show that the diffusion coefficient increases with temperature. The highest values of the diffusion coefficient were found at 70°C, but with the extension of extraction time there was also a clear trend of increase in the coefficient's values, which suggested that the extraction process was not finalized and its eventual continuation would result in a higher share of extracted substances. The explanation of this trend could be related to the finding that, at higher temperatures, changes occur in the cell structure of the extracted plant material that reduce its hydrodiffusion potential and respectively - increase the coefficient of diffusion. In the case of Burley tobacco extraction, the described variation profile of the diffusion coefficient was not observed. On the contrary, the extension of extraction time at the highest temperature caused a decrease in the coefficient of diffusion, suggesting that the extraction process had been finalized, which in fact was in compliance with literature data [4]. The individual values of the diffusion coefficient were comparable for Burley and Oriental tobaccos, but differed substantially from the values obtained for Virginia flue-cured tobacco.

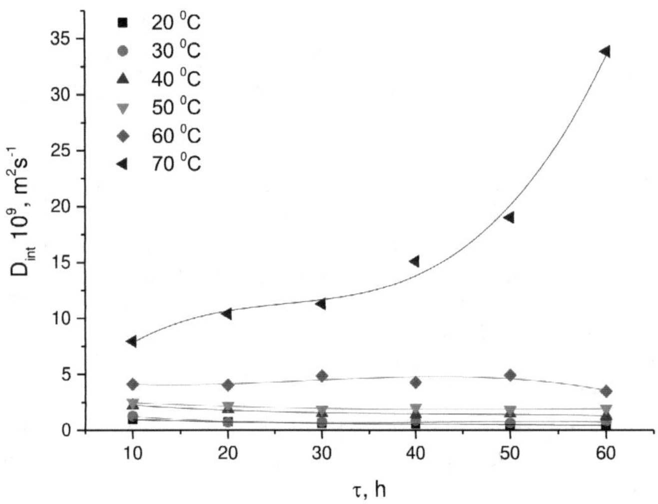

Figure 1. Coefficient of internal diffusion of resinoid vs. extraction time for Virginia flue-cured tobacco.

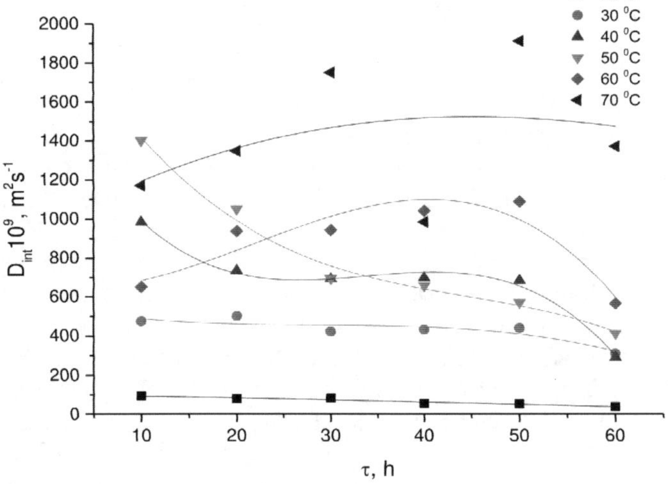

Figure 2. Coefficient of internal diffusion of resinoid vs. extraction time for Burley tobacco.

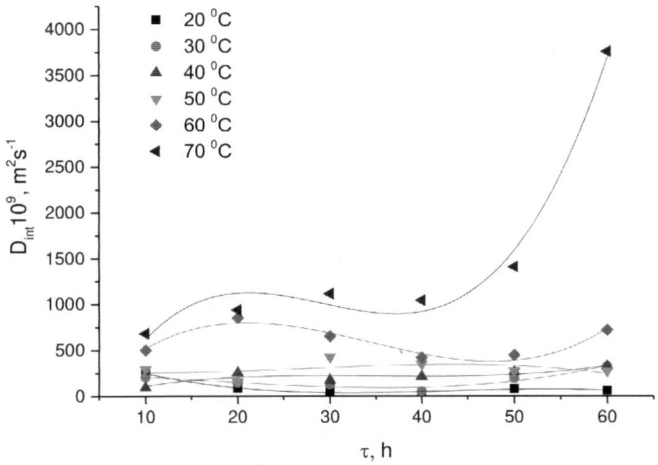

Figure 3. Coefficient of internal diffusion of resinoid vs. extraction time for Oriental tobacco.

A comparison with the results obtained for the aromatic product concrete from the same tobaccos showed very similar trends in the variation of the coefficient of internal molecular diffusion [17]. Obviously, these findings are related to the influence of the internal structure of the extracted plant material, as the diffusion coefficient depends mostly on its properties, as well as on the temperature and duration of the process.

Data from figures 1÷3 were processed to derive the respective regression equations for the three types of tobacco, which allow the calculation of the coefficient of internal molecular diffusion for the applied temperature range. The computed correlation coefficients (R^2) of the models for the three types of tobacco were all within the range 0.75÷0.96. The derived equations with significant coefficients are presented in Table 1. In this table, T means temperature (°C); y means coefficient of internal molecular diffusion (m^2s^{-1}) and x means process duration (s).

Table 1. Equations of the coefficient of internal molecular diffusion of tobacco resinoid.

T [a]	Type of tobacco		
	Virginia flue-cured	Burley	Oriental
20	$y = 1.3221 - 0.0403x$ [bc]	$y = 100.900 - 0.6277x - 0.0098x^2$	$y = 565.9233 - 40.9511x + 1.0144x^2 - 0.0079x^3$
30	$y = 2.1111 - 0.1128x + 0.0029x^2$	$y = 561.800 - 10.3254x + 0.3484x^2 - 0.0041x^3$	$y = 251.4467 - 2.4363x - 0.2325x^2 + 0.0049x^3$
40	$y = 3.0684 - 0.1002x + 0.0022x^2$	$y = 1673.3333 - 94.7269x + 2.9369x^2 - 0.0290x^3$	$y = -116.0167 + 30.7123x - 0.9061x^2 + 0.0087x^3$
50	$y = 2.9847 - 0.0597x + 0.0010x^2$	$y = 2114.4667 - 85.9153x + 1.7566x^2 - 0.0173x^3$	$y = 299.5767 - 8.1085.x + 0.4464x^2 - 0.0054x^3$
60	$y = 4.825 - 0.1195x + 0.0056x^2$	$y = 649.2000 - 6.0933x + 1.1278x^2 - 0.0173x^3$	$y = -410.3733 + 134.6943x - 4.6174x^2 - 0.0049x^3$
70	$y = 0.1457 + 1.1075x - 0.0391x^2$	$y = 956.1333 + 27.3697x - 0.3736x^2 + 0.0010x^3$	$y = -1193.3767 + 270.0574x - 10.0057x^2 + 0.1144x^3$

3.2. Coefficient of mass transfer of tobacco resinoid

Fulfilling the objective of the study, the kinetics of tobacco resinoid extraction was further characterized by the coefficient of mass transfer. The coefficient is known to depend mainly on the velocity and the character of the relative movement of the two phases, as well as on the size and shape of the particles and the contact surface area (e.g. the actual surface involved in the process). Mass transfer is governed by the established hydrodynamic conditions, which in turn depend not only on the nature of the plant material and the solvent, but mostly on the construction of the extraction apparatus and its operation mode. That is why the coefficient of mass transfer takes different values for one and the same plant material when extracted under different conditions [4].

Table 2. Diffusion coefficient (D) and Fourier number (Fo_d) of the extraction of tobacco resinoid.

Time interval (min)	Temperature (°C)	$D\cdot10^9$ ($m^2\cdot s^{-1}$)	$Fo_d\cdot10^3$	$B\cdot10^3$ [a]	μ [b]
0-10	20	27.20	5.12	250.38	1.98
10-20	20			942.51	1.15
20-30	20			897.04	1.02
30-40	20			882.31	1.01
40-50	20			848.95	1.00
50-60	20			826.45	1.00
0-10	30	36.90	6.95	250.38	1.98
10-20	30			942.51	1.15
20-30	30			897.04	1.02
30-40	30			882.31	1.01
40-50	30			848.95	1.00
50-60	30			826.45	1.00
0-10	40	49.10	9.25	250.38	1.98
10-20	40			942.51	1.15
20-30	40			897.04	1.02
30-40	40			882.31	1.01
40-50	40			848.95	1.00
50-60	40			826.45	1.00
0-10	50	64.21	12.09	250.38	1.98
10-20	50			942.51	1.15
20-30	50			897.04	1.02
30-40	50			882.31	1.01
40-50	50			848.95	1.00
50-60	50			826.45	1.00
0-10	60	82.62	15.56	250.38	1.98
10-20	60			942.51	1.15
20-30	60			897.04	1.02
30-40	60			882.31	1.01
40-50	60			848.95	1.00
50-60	60			826.45	1.00
0-10	70	1.05	19.72	250.38	1.98
10-20	70			942.51	1.15
20-30	70			897.04	1.02
30-40	70			882.31	1.01
40-50	70			848.95	1.00
50-60	70			826.45	1.00

[a] calculated according to equation (4).
[b] defined in equation (3) and taken from reference tables [18].

As it has been already stated, the solid-liquid extraction comprises two stages, which define the movement of the extractible substance within the solid body and outside it, into the liquid solvent. The process of mass transfer from the solid body to the solvent is described by the two simultaneously occurring processes – the internal mass conductivity and the external mass exchange. The internal mass transfer is characterized by the coefficient of internal diffusion (coefficient of mass conductivity), and the external mass exchange – by the coefficient of mass transfer [3, 8]. The calculation of the latter kinetic index remains quite a complex task, and its determination typically exploits a similarity theory in the form of the generalized equation of mass transfer including dimensionless complexes – the Biot diffusion number (Bi) and the Fourier number (Fo_d). The coefficient of mass conductivity inside the solid phase is defined by means of a dimensionless complex–– the Fourier number, which stands for a criterion of uniformity of mass transfer over the duration of the process [3].

Table 2 presents the obtained results for the diffusion coefficient (D), the Fourier number (Fo_d), and the factors B and μ, calculated over extraction time and temperature.

These data were used to calculate the Biot number and the coefficient of mass transfer for the resinoids obtained from the three types of tobacco – flue-cured Virginia, Burley and Oriental, and the results are presented in Table 3.

Our results confirmed that the coefficient of mass transfer is proportionate to the coefficient of diffusion in liquids [4].

Data in Table 3 revealed that the values the mass transfer coefficient increase with time and temperature, which is in compliance with the theoretical findings by [2], and with the experimental data for other plant materials or products – e.g. paulownia leaves [19] or tobacco concrete [17]. It is known that Biot number specifies the mode of the extraction process. At high numerical values of the Biot number ($Bi \rightarrow \infty$) the mode of extraction is internal diffusion, i.e. the rate of mass transfer is determined by the velocity of the occurring internal diffusion and the process runs in the intradiffusional area. Conversely, when Biot number takes lower values ($Bi < 1$) process rate is due primarily to the occurring external diffusion (i.e. the mode is external diffusion and the process runs in the external diffusion area) [3, 4, 14].

The values of the coefficient of mass transfer for the resinoid from Burley tobacco were lower compared with the respective values for the other two tobacco types – flue-cured Virginia and Oriental. As it has been stated above, this difference could probably be attributed to the different structure of the processed plant material, which is also a recognized factor in the solid-liquid extraction.

If we draw a parallel to the results about the mass transfer coefficient calculated on the basis of concrete extraction from the same tobaccos [17], it can be found that the diffusion coefficient, the Biot and Fourier numbers, and the factors B and μ take similar values in the extraction of the two aromatic products (concrete and resinoid), as they are determined mostly by the temperature and particle size of the processed plant material. Moreover, the analysis of the results obtained for the aromatic products concrete and resinoid from the three tobacco types studied reveals the complete match of data with regard to the Biot number and the coefficient of mass transfer in the case of Virginia flue-cured tobacco. Regarding the other two types of tobacco (Burley and Oriental) there were some differences in the values of the Biot number and the mass transfer coefficient calculated at one and the same temperature [17]. These findings are explicable by the influence of the solvents used in the extraction of concrete and resinoid (petroleum ether and ethanol, respectively), and not only by the impact of plant material structure, the temperature and the concentration of extractible substances.

Table 3. Biot diffusion number $(Bi_d)_{exp}$ and coefficient of mass transfer (β) of the extraction of tobacco resinoid.

		Virginia flue-cured		Burley		Oriental	
Time interval (min)	Temperature (°C)	$(Bi_d)_{exp}$	$B \cdot 10^9$ (m·s^{-1})	$(Bi_d)_{exp}$	$B \cdot 10^9$ (m·s^{-1})	$(Bi_d)_{exp}$	$B \cdot 10^9$ (m·s^{-1})
0-10	20	0.50	24.09	0.46	22.17	0.05	2.41
10-20	20	0.51	24.57	0.40	19.27	0.06	2.89

20-30	20	0.52	25.06	0.45	21.68	0.07	3.37
30-40	20	0.52	25.06	0.43	20.72	0.06	2.65
40-50	20	0.53	25.55	0.49	23.61	0.06	2.75
0-10	30	0.54	26.02	0.50	24.09	0.06	2.89
10-20	30	0.55	35.96	0.75	49.03	0.17	11.11
20-30	30	0.57	37.27	0.76	49.69	0.18	11.77
30-40	30	0.57	37.27	0.98	63.74	0.19	12.42
40-50	30	0.56	36.61	0.80	52.30	0.26	16.99
50-60	30	0.57	36.94	0.85	55.57	0.25	16.34
0-10	40	0.57	37.27	0.85	55.57	0.30	19.61
10-20	40	0.97	84.38	0.97	84.38	0.42	36.54
20-30	40	0.98	85.25	0.99	86.12	0.42	36.54
30-40	40	0.99	85.69	1.01	87.87	0.43	37.41
40-50	40	0.99	85.86	1.02	88.30	0.42	36.54
50-60	40	0.99	86.12	1.10	95.70	0.45	39.15
0-10	50	1.00	86.56	1.20	104.39	0.50	43.50
10-20	50	1.01	114.87	0.75	85.30	0.98	110.89
20-30	50	1.02	116.00	1.00	113.73	0.98	111.45
30-40	50	1.00	113.73	1.20	136.47	0.99	112.59
40-50	50	1.05	119.42	1.25	142.16	1.00	113.73
50-60	50	1.10	125.10	1.30	147.85	1.20	136.47
0-10	60	1.10	125.10	1.40	159.22	1.20	136.47
10-20	60	3.00	438.92	1.15	168.25	2.80	409.66
20-30	60	3.20	468.18	1.11	162.40	4.00	585.22
30-40	60	3.05	446.23	1.15	168.25	3.98	582.30
40-50	60	3.50	512.07	1.25	182.88	4.10	599.85
50-60	60	3.60	526.70	1.25	182.88	4.15	607.17
0-10	70	4.00	585.22	1.45	212.14	5.00	731.53
10-20	70	4.30	797.53	1.50	278.21	6.00	1112.83
20-30	70	4.50	834.62	1.01	187.33	6.50	1205.57
30-40	70	4.60	853.17	1.02	188.25	6.80	1261.21
40-50	70	4.70	871.72	1.35	250.39	7.00	1298.30
50-60	70	4.80	890.26	1.20	222.57	7.30	1353.94

4. Conclusions

The analysis of the results about the kinetics of the extraction of resinoid from tobacco leaves showed that the coefficient of internal molecular diffusion was influenced mostly by the temperature factor. The regression equations for the three types of tobacco – flue-cured Virginia, Burley and Oriental, were derived, and they could be applied for the calculation of the coefficient of internal molecular diffusion for a wide temperature range (20÷70°C).

The process of mass transfer during the extraction of tobacco resinoid was further characterized by calculating the coefficient of mass transfer, as well as the Biot diffusion number (Bi) and the Fourier diffusion number (*Fo*). The values of the coefficient of mass transfer of the resinoid from Burley tobacco were lower compared with the respective values for the other two tobacco types – flue-cured Virginia and Oriental. It found that the values of the calculated kinetic coefficients (i.e. the coefficient of diffusion, the Fourier number, and the functions B and μ) depended most strongly on the temperature and the structural characteristics of the processed plant material.

IOP Conf. Series: Materials Science and Engineering **595** (2019) 012057 doi:10.1088/1757-899X/595/1/012057

References

[1] Akselrud G and Lisyanskii V 1974 *Extraction in the system "solid – liquid"* (Leningrad: Khimia) (*in Russian*)

[2] Kafarov V 1962 *Basics of mass transfer* (Moscow: Visshaya Shkola) (*in Russian*).

[3] Makarevich N, Bogdanovich N, Tretiakov S, Koptelova E and Kutakova N 2014 Kinetic model for extraction of extractive substances from vegetable raw materials *Proceedings of the National Academy of Sciences of Belarus, Chemical series* **2** 98–102 *(in Russian)*

[4] Romankov P and Kurochkina M 1983 *Extraction from solid materials* (Leningrad: Khimia) (*in Russian*).

[5] Georgiev E and Stoyanova A 2006 *A guide for the specialist in the aromatic and cosmetic industry* (Plovdiv: BNAEOPC) (*in Bulgarian*)

[6] Stoyanova A and Georgiev E 2007 *Technology of essential oils* (Plovdiv: Acad. Publ. House UFT) (*in Bulgarian*)

[7] Nguyen D and Tran T 2005 Extraction and distillation of essential oils *Processing, Analysis and Application of Essential Oils* (Dehradun, India: HKB) pp 1–34

[8] Kasatkin A 1960 *Basic processes and apparatuses in chemical technology* (Sofia: Nauka i Izkustvo) (*in Bulgarian*)

[9] Popova V, Ivanova T, Damyanova S, Tasheva S, Atanasova T and Damyanov D 2009 Coefficients of diffusion in the process of obtaining aromatic extraction products from tobacco. 1. Burley *Sci. Works UFT* **56** 231–5

[10] Popova V, Ivanova T, Damyanova S, Tasheva S, Atanasova T and Damyanov D 2009 Coefficients of diffusion in the process of obtaining aromatic extraction products from tobacco. 2. Virginia flue-cured *Sci. Works Uni. Russe* **48** 165–8

[11] Popova V, Ivanova T, Damyanova S, Tasheva S, Atanasova T and Damyanov D 2010 Coefficients of diffusion in the process of obtaining aromatic extraction products from tobacco. 3. Oriental tobacco *Sci. Works Uni. Russe* **49** 120–3

[12] Simeonov E, Tsibranska I and Minchev A 1999 Solid-liquid extraction from plants – experimental kinetics and modelling *Chemical Engineering Journal* **73** 255–9

[13] Simeonov E and Chilev C 2015 Modelling and kinetics study of solid-liquid extraction from leaves of *Nicotiana tabacum* L. *Journal of Chemical Technology and Metallurgy* **50** 597–600

[14] Simeonov E, Yaneva Z and Chilev C 2017 Investigation of the mechanism and kinetics of extraction from plant materials *Bulgarian Chemical Communications* **49**(2) 399–409

[15] Stanisavljevic I, Lakicevic S, Velickovic D, Lazic M and Veljkovic V 2007 The extraction of oil from tobacco (*Nicotiana tabacum* L.) seeds *CI&CEO* **13**(1) 41–50

[16] Leffingwell F 1999 Basic chemical constituents of tobacco leaf and differences among tobacco types *Tobacco – Production, Chemistry and Technology* ed D Davies and M Nielsen (London: Blackwell Science) pp 265–284

[17] Tasheva S, Popova V, Ivanova T, Stoyanova A and Korolkova N 2018 Kinetics coefficients of solid-liquid extraction of tobacco leaves *Proc. Conf. 7-9 November 2018* part 1 (Voronezh, Russia: Voronezh State Agricultural University) pp 210–8

[18] Menkov N, Dinkov K, Damianov D, Dushkova M, Durakova A and Toshkov N 2009 *A manual for laboratory practice in processes and apparatuses in the food industry* (Plovdiv: Acad. Publ. House UFT) (*in Bulgarian*)

[19] Damyanova S, Tasheva S, Dobreva K, Koleva A, Stoyanova M, Petrova Y and Stoyanova A 2011 Paulownia – a source of biologically active substances. 4. Coefficients of molecular diffusion in the extraction of *Paulownia elongate* leaves *Proc. 7th Int. Conf. 7-9 Sept 2011 Donetsk Ukraine* pp 28–31 *(in Ukrainian)*

Sorption characteristics of subtropical fruit – Lucuma powder

A Durakova[1], A Bogoeva[1], R Krasteva[2], T Gogova[3], A Krasteva[3] and N Georgieva[3]

[1] Department of Process Engineering, University of Food Technologies, Plovdiv, Bulgaria
2 Center of Physical Education and Sport, University of Food Technologies, Plovdiv, Bulgaria
3 Department of Technology of Grain, Fodder, Bread and Confectionery Products, University of Food Technologies, Plovdiv, Bulgaria

E-mail: aldurakova@abv.bg

Abstract. The current scientific research is focused on the Lucuma powder. We were determined the moisture equilibrium data (adsorption and desorption) at three different temperatures: 10°C, 25°C and 40°C and relative humidity from 0.11 to 0.90, using the static gravimetric method of saturated salt solution. The results showed that the sorption capacity decreased when the temperature increases in the conditions of constant water activity. We were used four different mathematical models – modified Chung–Pfost, modified Oswin, modified Halsey and modified Henderson for description of isotherms of Lucuma powder. According to obtained experimental data, we were recommended the modified Halsey model for description satisfactorily the sorption isotherms of powder.

1. Introduction

Recent trends in food production is shown a grow of the plant products, rich in functional components. The aim of the enriching ingredients is to lower the energy value of the product and give it healing, prophylactic and dietary properties [1].

Lucuma known as *Pouteria Lucuma* and *Lucuma Obovata* is a subtropical fruit of the *Sapotaceae* family, most cultivated in Peru, Chile and Ecuador. The exotic fruit has a dark green colour, small and round with a diameter of about 10 cm, the inside of which has a stone and the soft structure is yellow. The fragrance (and even flavors) is exceptionally atypical and easily recognizable. Due to the presence of short chain hydrocarbon alcohols as well as volatile esters, aldehydes that together form the multi-layered odour was discovered/identificated more than 50 flavourings [2]. A major proportion of Lucuma world production (88%) is from Peru's highlands [3]. In the Bulgarian trading network, it is in the form of flour and in the specialized stores for organic and healthy foods is located in the "Superfood" [4]. The term "Superfood" was introduced by manufacturers, nutritionists, and coaches. According to specific personal needs, they offered the use of a similar type of product in individual diets [5]. Nowadays, Lucuma was successfully introduced into ice cream, juice, cakes, biscuits, yogurt, chocolate, baby foods and pies [6, 7].

For last decade, Lucuma impressed the European and world market with its unique nutritional qualities. The exotic Peruvian fruit, often compared to nutritional qualities with the mango, is known as „Gold of the Incas“ [8].

Peruvian fruit is recommended for children consumption because it helps their physical development in adults - to regulate metabolism now it contains fibre, niacin (whose deficiency provokes depression), iron, and a small amount of fat.

Lucuma powder is an excellent substitute for sugar and can help satisfy the desire of sweet foods without affecting the human body (unlike sugar or artificial sweeteners). Despite its sweetness, the Lucuma powder contains only 2 grams of natural fruit sugar per every 11 grams of carbohydrates. This making it a low glycemic index food that can help stabilize your blood sugar levels.

After reference research, we didn`t find information on sorption isotherms (adsorption and desorption) and of Lucuma powder. Thus, the aim of the present study was to determine sorption characteristics of subtropical fruit - Lucuma powder.

2. Materials and methods

Commercial Lucuma powder, produced in Peru, purchased in Bulgaria by "Internet café-BG" ltd, packed by "Zoya bg Organic Shop" was used in this study.

The equilibrium moisture content (EMC) of Lucuma powder was investigated at 10°C, 25°C and 40°C and water activities $a_w = 0.11 \div 0.90$. The static gravimetric method was used [9]. The powder was dried in a desiccator with P_2O_5 at a room temperature for 20 days prior to the beginning of the experiment, for the adsorption process. For determination of the desorption process, the samples were hydrated in a glass jar over distilled water at a room temperature. Samples of 1÷0.02 g were weighed in weighing bottles. Eight saturated salt solutions (LiCl, CH_3COOK, $MgCl_2$, K_2CO_3, $MgNO_3$, NaBr, NaCl, KCl), hold constant water activity environments, were used in the weighing bottles, which were put in hygrostats [10]. All of the used salts were of reagent grade. At high water activities (a_w>0.70) crystalline thymol was placed in the hygrostats to prevent microbial spoilage of the powder. The hygrostats were kept in thermostats at 10°C, 25°C and (40-0.2)°C. Samples were weighed (balance sensitivity = 0.0001g) every three days. Equilibrium was ascertained when three consecutive weight measurements showed a difference less than 0.001g. The moisture content (%) was calculated according to [11].

3. Analysis of data

For verification of the description of sorption isotherms, we were used the followed models:

Modified Chung-Pfost $$a_w = exp\left[\frac{-A}{t+B}\exp(-CM)\right] \quad (1)$$

Modified Halsey $$a_w = exp\left[\frac{-\exp(A+Bt)}{M^C}\right] \quad (2)$$

Modified Oswin $$M = (A+Bt)\left(\frac{a_w}{1-a_w}\right)^C \quad (3)$$

Modified Henderson $$1 - a_w = exp[-A(t+B)M^C] \quad (4)$$

where: M is the average moisture content, % d.b.; a_w is the water activity, decimal; A, B and C are coefficients; t is the temperature, °C.

A nonlinear, least squares regression program was used to fit the four models to the experimental data (all replications). The suitability of the equations was evaluated and compared using the mean relative error P (%); the standard error of moisture (SEM) and the randomness of residuals [12]:

$$P = \frac{100}{N}\sum\left|\frac{M_i - \hat{M}_i}{M_i}\right| \quad (5)$$

$$SEM = \sqrt{\frac{\sum(M_i - \hat{M}_i)^2}{df}} \quad (6)$$

$$e_i = M_i - \hat{M}_i \quad (7)$$

where: M_i and $\widehat{M}_i$ are experimentally observed and predicted by the model value of the equilibrium moisture content; *N* is the number of data points; *A*, *B* and *C* are coefficients; *df* is the number of degree of freedom (number of data points minus number of constants in the model).
The monolayer moisture content is calculated using the Brunauer-Emmett-Teller (BET) equation and the experimental data for water activities up to 0.45, for each temperature [10, 13]:

$$M = \frac{M_e C a_w}{(1-a_w)(1-a_w+Ca_w)} \quad (8)$$

where: *M* is the monolayer moisture content, % d.b.; a_w is the water activity, decimal; *C* is the coefficient.

4. Results and discussion

The obtained mean values of EMC, based on triplicate measurements for the respective water activity and temperature, are presented in Table 1 for adsorption and in Table 2 for desorption.

The reason for this is maybe the decomposition of sugars after which the powder starts absorbing larger quantities of water. The same effect also applies to the processes of adsorption and desorption. We were funded the similar results, which are reported in the scientific literature, for many foods [14, 15]. The results of the EMC increase with an increase in the temperature at constant a_w. The effect on this type of powder is also manifested in other food products high in sugars.

Table 1. Equilibrium moisture content M^a (% d.b.) of Lucuma powder by adsorption at different water activities (a_w) and temperatures t (°C).

Sel	10 °C			25 °C			40 °C		
	a_w	M^*	sd^{**}	a_w	M^{**}	sd^{**}	a_w	M^{**}	sd^{**}
LiCl	**0.113**	3.88	0.01	**0.113**	2.99	0.05	**0.112**	2.33	0.04
CH_3COOK	**0.234**	4.60	0.14	**0.225**	3.96	0.20	**0.201**	3.26	0.17
$MgCl_2$	**0.335**	4.76	0.08	**0.328**	4.49	0.05	**0.316**	4.03	0.05
K_2CO_3	**0.431**	8.16	0.19	**0.432**	5.72	0.01	**0.432**	5.13	0.05
$MgNO_3$	**0.574**	9.27	0.15	**0.529**	7.40	0.16	**0.484**	7.01	0.14
NaBr	**0.622**	9.45	0.06	**0.576**	8.17	0.02	**0.532**	7.54	0.20
NaCl	**0.757**	23.21	0.08	**0.753**	14.10	0.10	**0.747**	13.97	0.40
KCl	**0.868**	39.79	0.20	**0.843**	20.77	0.16	**0.823**	19.57	0.30

Table 2. Equilibrium moisture content M^a (% d.b.) of Lucuma powder by desorption at different water activities (a_w) and temperatures t (°C).

Sel	10 °C			25 °C			40 °C		
	a_w	M^*	sd^{**}	a_w	M^*	sd^{**}	a_w	M^*	sd^{**}
LiCl	**0.113**	4.41	0.04	**0.113**	2.75	0.08	**0.112**	2.13	0.08
CH_3COOK	**0.234**	5.16	0.17	**0.225**	3.83	0.25	**0.201**	3.59	0.08
$MgCl_2$	**0.335**	5.92	0.05	**0.328**	4.40	0.24	**0.316**	3.76	0.06
K_2CO_3	**0.431**	8.47	0.20	**0.432**	4.83	0.17	**0.432**	4.68	0.26
$MgNO_3$	**0.574**	9.25	0.15	**0.529**	6.91	0.21	**0.484**	6.29	0.13
NaBr	**0.622**	10.99	0.21	**0.576**	7.66	0.04	**0.532**	6.89	0.14
NaCl	**0.757**	22.10	0.23	**0.753**	13.02	0.18	**0.747**	12.79	0.11
KCl	**0.868**	38.12	0.06	**0.843**	17.90	0.04	**0.823**	16.50	0.07

Figure 1 gives the experimental data obtained after adsorption and desorption at 10°C. The sorption isotherms have an *S*-shape profile. The hysteresis effect is statistically significant, at a level of significance α=0.05, in the water activity range 0.4÷0.85.

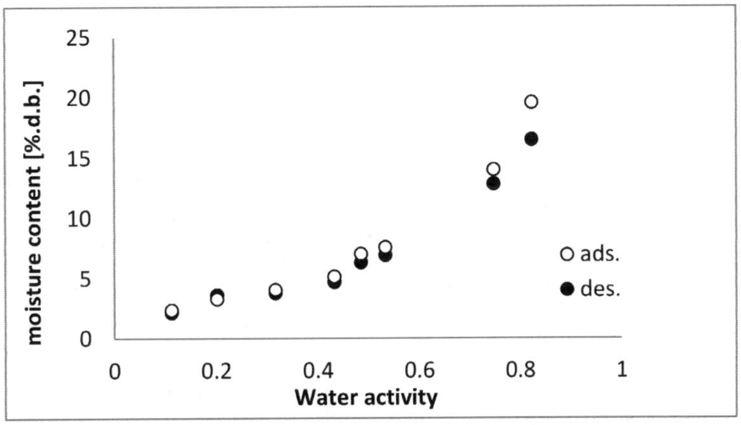

Figure 1. Comparison of isotherms at 10°C, Desorption and Adsorption.

The coefficients for the three-parameter modified models, P and SEM values are presented in Table 3 for adsorption and Table 4 for desorption.

Table 3. Model coefficients (*A, B, C*), mean relative error (P, %), standard error of moisture (SEM) and Correlation coefficient (R) for adsorption.

Model	*A*	*B*	*C*	*P*	*SEM*	*R*
Oswin	9.88225	0.10658	0.75756	23.77	2.18	0.97
Halsey	**2.20304**	**0.00940**	**1.17891**	**10.03**	**1.29**	**0.99**
Henderson	0.000688	78.3072	1.09329	21.59	3.08	0.96
Chung-Pfost	408.4274	0.16189	135.584	24.12	4.69	0.95

Table 4. Model coefficients (*A, B, C*), mean relative error (P, %), standard error of moisture (SEM) and Correlation coefficient (R) for desorption.

Model	*A*	*B*	*C*	*P*	*SEM*	*R*
Oswin	10.52716	0.13661	0.70741	23.95	2.19	0.97
Halsey	**2.492462**	**0.01674**	**1.24790**	**10.07**	**1.31**	**0.99**
Henderson	0.000989	39.6782	1.17033	20.31	2.75	0.96
Chung-Pfost	206.1237	0.17545	51.2206	25.24	3.27	0.95

The results show that the lowest P and SEM values were obtained with the Halsey model. The graphical analysis of the residues demonstrates that the distribution is random for both models, which means that both models are suitable for the description of Lucuma poweder sorption isotherms (Fig. 2 and 3). We recommend the Halsey model, because of its lower values of the coefficients.

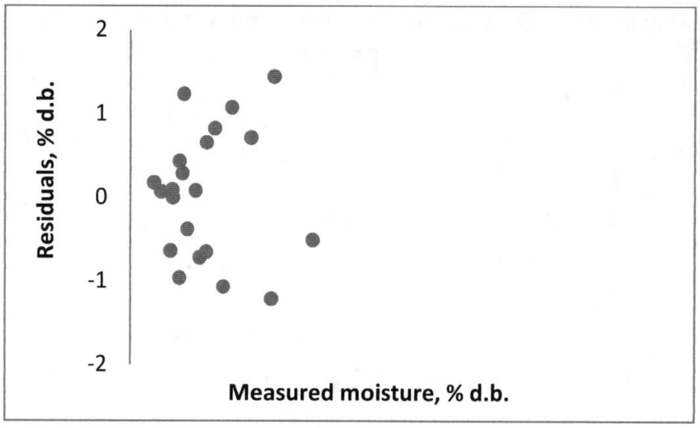

Figure 2. Plot of residuals fit of modified Halsey model to adsorption data.

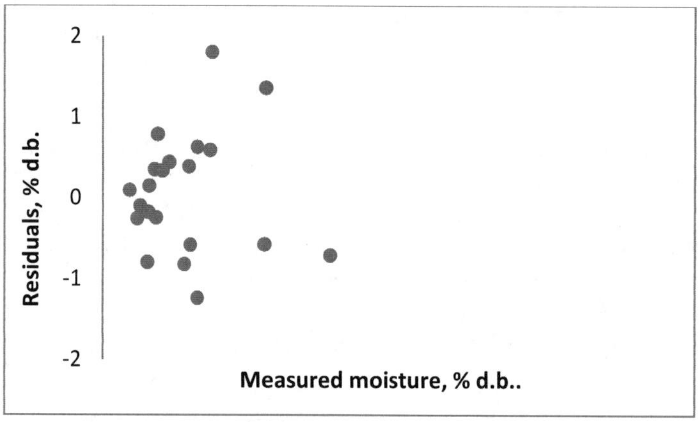

Figure 3. Plot of residuals fit of modified Halsey model to desorption data.

The model (8) is linearly transformed for calculation of the BET monolayer moisture content [16]:

$$\frac{a_w}{(1-a_w)M} = P + Qa_w \tag{9}$$

Based on the coefficients of the linear equation, the monolayer moisture content for the respective temperature is calculated and the results are presented in Table 5.

Table 5. BET monolayer moisture content monolayer moisture content (% d.b.) of Lucuma powder at several temperatures.

t (°C)	Adsorption	Desorption
10	4.18	4.17
25	2.93	2.56
40	3.65	3.21

5. Conclusions

The sorption capacity of Lucuma powder decrease with an increase in temperature at constant water activity. The modified Halsey model is suitable for describing the relationships between the equilibrium moisture content, the water activity and the temperature of the Lucuma powder. According to sorption isotherms obtained for 10°C, 25°C and 40°C, monolayer moisture content is calculated with BET equation.

References

[1] Verbeke W 2006 Functional foods: Consumer willingness to compromise on taste for health? *Food Qual. and Prefer.* **17(1)** 126

[2] Taiti C, Colzi I, Azzarello E and S Mancuso 2017 Discovering a volatile organic compound fingerprinting of Pouteria lucuma fruits *Fruits* **72(3)** 131

[3] Aguilar D S 2015 *Economics Lucuma as an exotic high quality fruit imported into Portugal and the UE 2015* (A Work Project, presented as part of the requirements for the Award of a Master's Degree in Management from the NOVA - School of Business and Economics 1-36)

[4] Mukta N, Sunita M and Aparna S 2017 Different Types of Super Food Product Its Sensory Evaluation Storage and Packaging *International Journal of Advance Research, Ideas and Innovation in Technology* **3(6)** 812

[5] U.S. Department of Health and Human Services and U.S. Department of Agriculture 2015–2020 *Dietary Guidelines for Americans* 8th Edition December 2015.

[6] Dini I 2011 Flavonoid glycosides from Pouteria obovata (R. Br.) fruit flour *Food Chemistry* **124** 884

[7] Fuentealba C, Gálvez L, Cobos A, Olaeta J A, Defilippi B G, Chirinos R, Campos D and Pedreschi R 2016 Characterization of main primary and secondary metabolites and in vitro antioxidant and antihyperglycemic properties in the mesocarp of three biotypes of Pouteria lucuma *Food Chem* **190** 403

[8] Caballero M G A and Del Carmen Ticse Aguilar A M 2017 *Caracterización de macrocomponentes en pulpa congelada de tres biotipos de lúcuma* (Pouteria lucuma) Universidad San Ignacio De Loyola 1-126.

[9] Wolf W, Spiess W E L and Jung G 1985 Standardization of isotherm measurements (COST-Project 90 and 90 bis). In: Stimatos, D. and Multon, J. L. (Eds.) *Properties of Water in Foods in Relation to Quality and Stability* (Martinus Nijhoff, Dordrech 661-679)

[10] Bell L and Labuza T 2000 Determination of moisture sorption isotherms. Moisture Sorption: Practical Aspects of Isotherm Measurement and Use *The American Association of Cereal Chemists*, (Inc. St. Paul MN USA 33-56)

[11] AOAC 1990 Official Methods of Analysis 960.39 15th ed. (Association of Official Analytical Washington DC)

[12] Chen C C and Morey R V 1989 Comparison of four EMC/ERH equations *Transactions of the ASAE* **32(3)** 983

[13] Brunauer S, Emmett P H and Teller E 1938 Adsorption of gases in multimolecular layers *Journal of the American chemical society* **60(2)** 309

[14] Al-Muhtaseb A, McMinn W and Magee T 2002 Moisture sorption isotherm characteristics of food products: a review *Food and bioproducts processing* **80(2)** 118

[15] Durakova A G and Menkov N D 2005 Moisture sorption characteristics of chickpea flour *Journal of food engineering* **68(4)** 535

[16] Brunauer S, Deming L S, Deming W E and Teller E 1940 On a theory of the van der Waals adsorption of gases *Journal of the American chemical society* **62(7)** 1723

Acknowledgments

This study was part of the Fund Scientific Research "FS N07/18-H", director project: Assoc. Prof. PhD Albena Georgieva Durakova, University of Food Technologies, Plovdiv from Bulgaria.

IOP Conf. Series: Materials Science and Engineering **595** (2019) 012059 doi:10.1088/1757-899X/595/1/012059

Effects of wind speed, relative humidity, temperature and air pressure on PM10 concentration for an urban background area

S Paraschiv and L S Paraschiv

Faculty of Engineering, “Dunarea de Jos” University of Galati, Romania

E-mail: sparaschiv@ugal.ro

Abstract. The paper analyses PM10 concentration data collected in Galati at the urban background type station from 1 January 2018 till 31 December 2018. The trend of PM10 concentration during the year was investigated considering the influence of recorded meteorological factors (temperature, wind velocity, pressure and relative humidity). Hourly PM10 concentrations ranged from 0.03 to 223.51 μg/m^3 and exceeded the EU standard limit of 40 μg/m^3. An analysis was carried out to identify the most important meteorological variables that influence the seasonal dynamics of PM10 concentrations. The highest correlation was obtained for wind speed. On average, there are approximately 308 hours annually when the threshold of 40 μg/m^3 is exceeded, 58.44% of those occur in autumn.

1. Introduction
Air pollution with particles can have both anthropic and natural sources [1-3]. Anthropogenic sources include combustion, secondary particles formed in the air by chemical reactions of gaseous pollutants and re-suspension of soil and dust [4-7].

The PM10 includes inhalable particles that are small enough to penetrate deep into the respiratory system and affect human health. Effects of exposure to PM10 pollution, both in the short-term (hours, days) and long-term (months, years) are: worsening of asthma, mortality due to cardiovascular and respiratory diseases and lung cancer. The people most susceptible to illness are those with pre-existing pulmonary or cardiac conditions, as well as the elderly.

2. Data and methods
This paper analyses the evolution of the PM10 concentration for the year 2018 in Galati, Romania. The data were collected at the GL - 2 air quality monitoring station (Figure 1). The site is classified as an urban background site, located near a pedestrian street and a parking zone (Figure 1). The data for PM10 (μg/m^3), pressure (mbar), temperature (°C), wind speed (m / s), wind direction and relative humidity (%) were collected from January 2019 to December 2018. More than 93% of the available data for PM10 and meteorological parameters for the study period were valid.

On average, there are about 308 hours annually when the 40 μg/m^3 threshold is exceeded, being distributed over the year as follows: 14.61 % occur in winter, 5.84 % occur during spring, 21.1 % occur in summer and 58.44 % of these occur during autumn.

The following data were recorded in Galati at the urban background station: the average annual humidity was 79%, the average air pressure was 1013.6 mbar and the average wind speed was 0.4 m/s.

IOP Conf. Series: Materials Science and Engineering **595** (2019) 012059 doi:10.1088/1757-899X/595/1/012059

The average annual temperature in Galati in 2018 was 13.18°C, the lowest temperature has been recorded in March with a temperature of -11.41°C and the maximum temperature of 33.88°C has been recorded in August.

Figure 1. Location of the Galati and the measurement station inside the city.

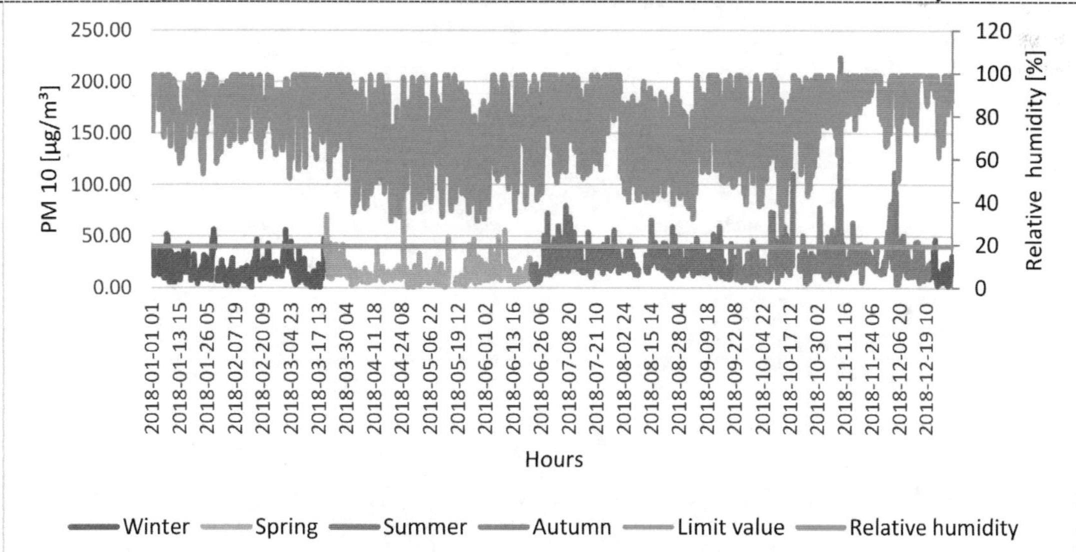

Figure 2. Time variations of PM10 (µg / m³) (lower-line) and relative humidity (%) (upper-line) from January 2018 to December 2018.

Figure 3 shows the time variation of the meteorological parameters: pressure, temperature and wind speed.

IOP Conf. Series: Materials Science and Engineering **595** (2019) 012059 doi:10.1088/1757-899X/595/1/012059

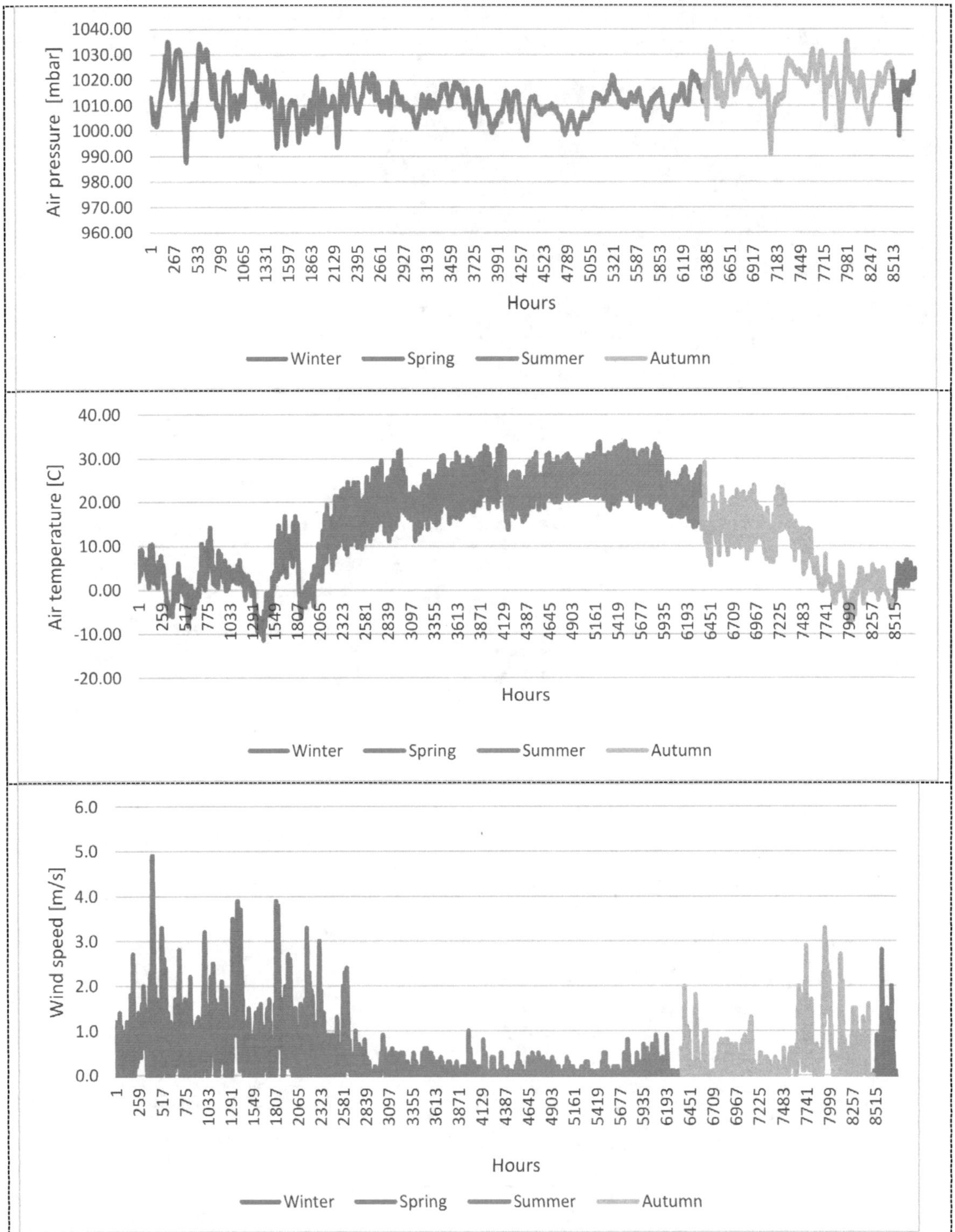

Figure 3. The hourly data of air pressure (mbar), air temperature (°C) and wind speed (m/s) from January 2018 to December 2018 in Galati.

The following graphs illustrate the direct correlation between wind speed and PM10 concentration in the selected urban background monitoring site selected from Galati in 2018. In order to better visualize the impact of wind speed on PM10 concentration variation a shorter period of time was selected, 15 November - 23 December, 2018.

It can be noticed that during this period, the fluctuations in hourly concentrations in figure 4 show a positive correlation with the wind speed, especially with the stronger winds or with windless periods.

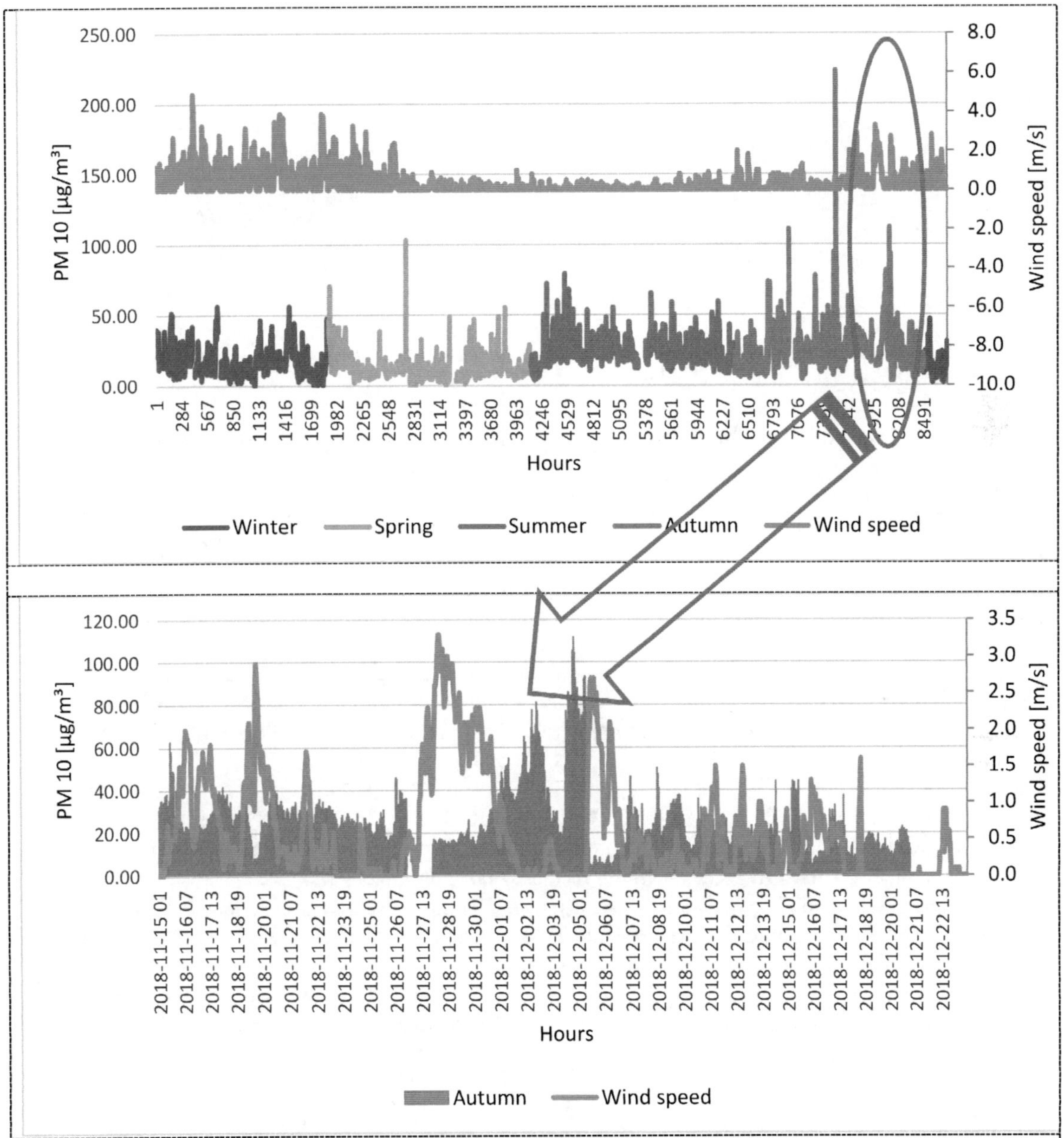

Figure 4. The relationship between PM10 and wind speed.

Hourly weather conditions (wind direction and wind speed at 10 meters above ground level) and hourly PM10 concentrations were obtained from local monitoring networks. Wind direction data were divided into sixteen classes.

In order to analyse the distribution of the PM10 concentration during the year 2018, the values recorded during 8735 hours were used, of which 8074 were valid data, invalid data 544 and 117 missing data, the result is presented in figure 5.

To analyse distribution of the wind direction we have used the data recorded over 8735 hours, out of which 8635 valid data, 94 invalid data and 6 missing data, which were presented in figure 6.

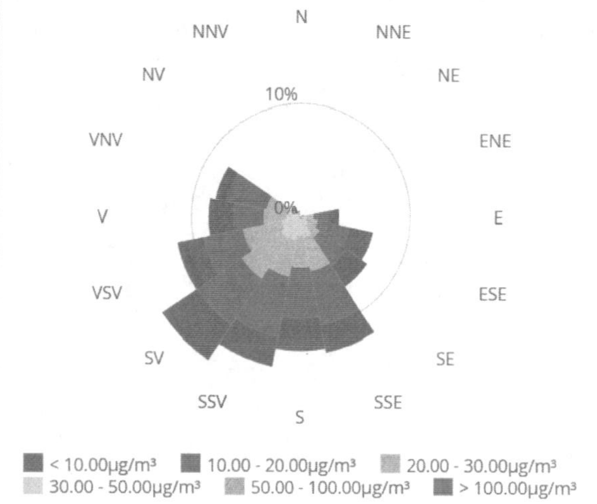

Figure 5. Distribution of PM10 concentration.

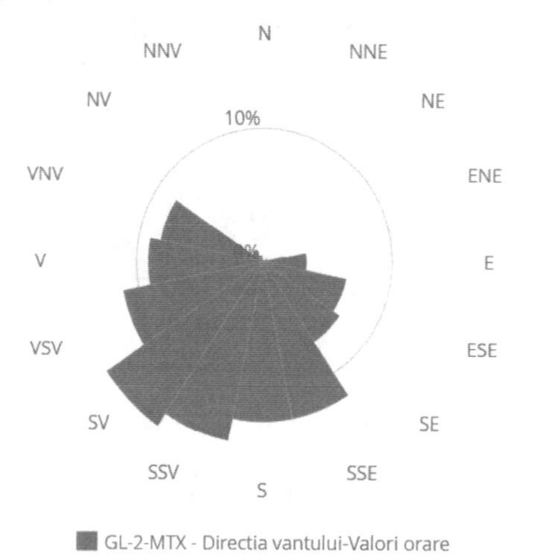

Figure 6. Distribution of wind direction.

3. Conclusion

Research has shown the dependence of daily PM10 concentration on meteorological conditions, mainly on wind speed variation. The highest concentration of PM10 was observed in summer and autumn, and in all seasons of the year, the strongest impact on PM10 pollution was exerted by wind speed.

The PM10 concentration in the air has shown annual seasonal variability. The concentration of PM10 in winter and spring was lower due to higher wind speeds than in summer and autumn. In Galati, increased PM10 levels in the air during the summer and autumn occurred in conditions with high air temperatures, low relative humidity and low wind speed, mainly from the NE direction.

References

[1] Souzana A, John S E, Panayiotis K Y, Savvas K, Joel S and Petros Koutrakis 2014 PM10 Concentration levels at an urban and background site in Cyprus: The impact of urban sources and dust storms, *J Air Waste Manag Assoc.* D **64** 12): 1352–1360

[2] Kalbarczyk R, Kalbarczyk E and Raszka B 2015 Temporal changes in concentration of PM10 dust in Poznań, middle-west poland as dependent on meteorological conditions A*pplied ecology and environmental research* **16** (2):1999-2014.

[3] Lokman H T, Pinar S, Omar A, Ferhat K and Gürdal T 2008 Effect of Meteorological Parameters on Fine and Coarse Particulate Matter Mass Concentration in a Coal-Mining Area in Zonguldak, Turkey *Journal of the Air & Waste Management Association*, **58** 4, 543-552

[4] Jianhua W and Susumu O 2015 Effects of Meteorological Conditions on PM2.5 Concentrations in Nagasaki, Japan *Int. J. Environ. Res. Public Health* **12** 9089-9101

[5] Paraschiv S, Constantin D E, Paraschiv S L and Voiculescu M 2017 OMI and ground-based in-situ tropospheric nitrogen dioxide observations over several important European cities during 2005–2014, *Int. J. Environ. Res. Public Health* **14** 11 1415

[6] Paraschiv S and Paraschiv S L 2017 A review on interactions between energy performance of the buildings, outdoor air pollution and the indoor air quality *Energy Procedia* **128** 179-186

[7] Alahmadi S, Al-Ahmadi K and Almeshari M 2019 Spatial variation in the association between NO2 concentrations and shipping emissions in the Red Sea *Science of The Total Environment* **676** 131